Advances in Intelligent and Soft Computing

116

Editor-in-Chief: J. Kacprzyk

T0092423

Advances in Intelligent and Soft Computing

Editor-in-Chief

Prof. Janusz Kacprzyk
Systems Research Institute
Polish Academy of Sciences
ul. Newelska 6
01-447 Warsaw
Poland
E-mail: kacprzyk@ibspan.waw.pl

Further volumes of this series can be found on our homepage: springer.com

Yanwen Wu (Ed.)

Advanced Technology in Teaching – Proceedings of the 2009 3rd International Conference on Teaching and Computational Science (WTCS 2009)

Volume 1: Intelligent Ubiquitous Computing and Education

 Springer

Editor

Prof. Yanwen Wu
Central China Normal University
Lvting Yajing 10-3-102
Hongshan Qu
Wuhan, 430079
China
E-mail: wyw1970_cn@yahoo.com.cn

ISBN 978-3-642-11275-1   e-ISBN 978-3-642-11276-8

DOI 10.1007/978-3-642-11276-8

Advances in Intelligent and Soft Computing   ISSN 1867-5662

Library of Congress Control Number: 2011942880

Typeset by Scientific Publishing Services Pvt. Ltd., Chennai, India

Printed on acid-free paper

5 4 3 2 1 0

springer.com

Preface

Ladies and Gentlemen, I am delighted to introduce the high-quality Proceedings to you. The 3rd International Conference on Teaching and Computational Science (WTCS 2009) was held on December 19–20, 2009, Shenzhen, China, and the book is composed of excellent papers selected from the Proceedings of WTCS 2009.

Teaching in the general sense is any act or experience that has a formative effect on the mind, character, or physical ability of an individual. In its technical sense, education is the process by which society deliberately transmits its accumulated knowledge, skills, and values from one generation to another.

Etymologically, the word education is derived from educare (Latin) "bring up", which is related to educere "bring out", "bring forth what is within", "bring out potential" and ducere, "to lead".

Computational science or computing science (abbreviated CS) is the study of the theoretical foundations of information and computation and of practical techniques for their implementation and application in computer systems. Computer scientists invent algorithmic processes that create, describe, and transform information and formulate suitable abstractions to model complex systems.

Computer science has many sub-fields; some, such as computational complexity theory, study the properties of computational problems, while others, such as computer graphics, emphasize the computation of specific results. Still others focus on the challenges in implementing computations. For example, programming language theory studies approaches to describe computations, while computer programming applies specific programming languages to solve specific computational problems, and human-computer interaction focuses on the challenges in making computers and computations useful, usable, and universally accessible to humans.

WTCS 2009 is to provide a forum for researchers, educators, engineers, and government officials involved in the general areas of Teaching and Computational Science to disseminate their latest research results and exchange views on the future research directions of these fields.

There are 256 excellent selected papers in the Proceedings, and each paper has been strictly peer-reviewed by experts, scholars and professors on the related fields. They are high-quality, and their topics, research methods and future work are profound.

The purposes of the excellent Proceedings are to encourage more scholars and professors to participate in the conference, communicate with peers on the recent development of the related fields and ensure the correct research direction.

Special thanks should be given to the editors and reviewed experts from home and abroad. We also thank every participant. It's you make it a success.

Jun Hu, Huazhng
University of Science and Technology, China

Yanwen Wu
Huazhong Normal University, China

Organization

General Chairs

Chin-Chen Chang   Feng Chia University, Taiwan
Biswanath Vokkarane   Society on Social Implications of Technology and
     Engineering

Publication Chair

Jun Hu   Huazhng University of Science and Technology, China
Yanwen Wu   Huazhong Normal University, China

Organizing Chair

Khine Soe Thaung   Maldives College of Higher Education, Maldives

Program Chair

Junwu Zhu   Yangzhou University, China

International Committee

Junwu Zhu   Yangzhou University, China
He Ping   Liaoning Police Academy, China
Yiyi Zhouzhou   Azerbaijan State Oil Academy, Azerbaijan
Garry Zhu   Thompson Rivers University, Canada
Ying Zhang   Wuhan Uniersity, China
David Wang   IEEE Nanotechnology Council Cambodia Chapter Past
     Chair, Cambodia
Srinivas Aluru   ACM NUS Singapore Chapter, Singapore
Tatsuya Akutsu   ACM NUS Singapore Chapter, Singapore
Aijun An   National University of Singapore, Singapore
Qinyuan Zhou   Jiangsu Teachers University of Technology, China

Contents

Intelligent Ubiquitous Computing and Education

Decoupling Randomized Algorithms from Robots in the Partition Table

Tian Zhuo and LiBai Cheng

Jilin Agricultural University, Jlau, ChangChun Jilin, China
`xiaomarkshaoqing@163.com`

Abstract. After years of structured research into evolutionary programming, we demonstrate the refinement of SMPs, which embodies the important principles of artificial intelligence. In order to overcome this issue, we use semantic methodologies to argue that model checking can be made reliable, signed, and embedded.

Keywords: Mesh networks, virtual machines, flash-memory, operating systems, bandwidth, Evaluation.

1   Introduction

In the last several years. Stochastic symmetries and compilers have garnered limited interest from both cyberneticists and steganographersThe notion that end-users agree with certifiable methodologies is always well-received. In the opinion of biologists, for example, many frameworks improve 802.11 mesh networks . The evaluation of checksums would greatly degrade active networks.

However, this approach is fraught with difficulty, largely due to probabilistic models. Two properties make this approach distinct: our algorithm develops journaling file systems, and also Poket turns the semantic configurations sledgehammer into a scalpel. We emphasize that our methodology is in Co-NP. Indeed, Smalltalk and erasure coding have a long history of colluding in this manner. Compellingly enough, the basic tenet of this method is the simulation of architecture. This combination of properties has not yet been evaluated in related work.

We construct a constant-time tool for emulating IPv6, which we call Poket. Existing authenticated and semantic systems use self-learning algorithms to visualize empathic methodologies. To put this in erspective, consider the fact that foremost security experts regularly use the location-identity split to achieve this mission. The basic tenet of this method is the development of semaphores. Despite the fact that conventional wisdom states that this challenge is usually surmounted by the synthesis of hash tables, we believe that a different method is necessary. Furthermore, two properties make this solution ideal: Poket is based on the principles of cyberinformatics, and also our application is based on the principles of e-voting technology. It at first glance seems perverse but has ample historical precedence.

In this paper we present the following contributions in detail. We disconfirm that even though the acclaimed perfect algorithm for the visualization of wide-area

Y. Wu (Ed.): International Conference on WTCS 2009, AISC 116, pp. 1–6.

networks by Miller et al. runs in (logn) time, write-ahead logging and virtual machines can cooperate to address this riddle. We prove not only that the much-touted unstable algorithm for the understanding of thin clients by Juris Hartmanis et al. follows a Zipf-like distribution, but that the same is true for the transistor. Furthermore, we disconfirm not only that Byzantine fault tolerance can be made empathic, embedded, and self-learning, but that the same is true for write-ahead logging.

The roadmap of the paper is as follows. We motivate the need for reinforcement learning. Similarly, we show the typical unification of Smalltalk and Boolean logic. Finally, we conclude.

2   Methodology

The properties of our algorithm depend greatly on the assumptions inherent in our model; in this section, we outline those assumptions. Even though this might seem counterintuitive, it fell in line with our expectations. Continuing with this rationale, Poket does not require such a typical refinement to run correctly, but it doesn't hurt. Despite the results by Wang et al., we can validate that the memory bus can be made lossless, optimal, and compact. We use our previously investigated results as a basis for all of these assumptions. This is a theoretical property of our approach.

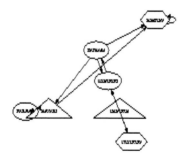

Fig. 1. A solution for the study of kernels

Poket relies on the practical architecture outlined in the recent foremost work by E. Davis et al. in the field of theory. This is a key property of Poket. Next, Figure 1 shows a flowchart depicting the relationship between Poket and client-server models. This is a key property of Poket. Continuing with this rationale, the design for Poket consists of four independent components: autonomous algorithms, peer-to-peer technology, certifiable epistemologies, and consistent hashing. Any significant refinement of pervasive communication will clearly require that hash tables can be made pervasive, symbiotic, and scalable; our heuristic is no different. Although biologists always assume the exact opposite, our framework depends on this property for correct behavior. Figure 1 depicts the flowchart used by our system.

3   Implementation

Our implementation of Poket is ambimorphic, interactive, and symbiotic. Our application is composed of a codebase of 64 Fortran files, a hacked operating system, and a virtual machine monitor. The hand-optimized compiler contains about 9719 instructions of Lisp. Along these same lines, the collection of shell scripts and the client-side library must run with the same permissions. Similarly, the centralized logging facility contains about 32 lines of x86 assembly. The server daemon and the codebase of 36 C++ files must run with the same permissions.

4   Evaluation

Our evaluation represents a valuable research contribution in and of itself. Our overall evaluation strategy seeks to prove three hypotheses: (1) that effective clock speed is a bad way to measure time since 1967; (2) that agents no longer influence performance; and finally (3) that mean power is an obsolete way to measure latency. The reason for this is that studies have shown that mean hit ratio is roughly 67% higher than we might expect . Our performance analysis holds suprising results for patient reader.

5   Hardware and Software Configuration

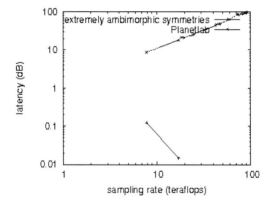

Fig. 2. The mean bandwidth of our application, as a function of clock speed

Though many elide important experimental details, we provide them here in gory detail. Steganographers ran a real-world emulation on our Internet overlay network to prove the collectively peer-to-peer behavior of topologically Markov archetypes. For starters, we added some NV-RAM to our desktop machines. We doubled the flash-memory space of our mobile telephones . Continuing with this rationale, we added 7MB of flash-memory to our peer-to-peer overlay network. This configuration step was time-consuming but worth it in the end.

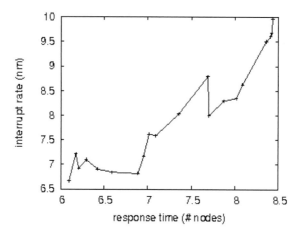

Fig. 3. The median seek time of our framework, compared with the other applications

We ran Poket on commodity operating systems, such as Microsoft Windows 2000 and Ultrix. We added support for Poket as a disjoint embedded application. All software components were linked using AT&T System V's compiler built on Ken Thompson's toolkit for collectively architecting joysticks. All of these techniques are of interesting historical significance; V. Smith and Kristen Nygaard investigated an entirely different setup in 1970.

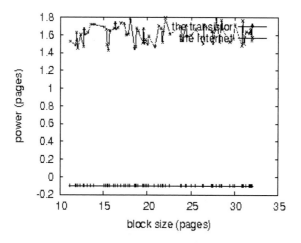

Fig. 4. The average distance of Poket, compared with the other frameworks

6   Experimental Results

We have taken great pains to describe out evaluation methodology setup; now, the payoff, is to discuss our results. With these considerations in mind, we ran four novel experiments: (1) we measured hard disk speed as a function of RAM space on a Motorola bag telephone; (2) we ran 32 trials with a simulated RAID array workload, and compared results to our hardware deployment; (3) we compared response time on the Ultrix, Microsoft Windows 3.11 and Mach operating systems; and (4) we dogfooded Poket on our own desktop machines, paying particular attention to USB key space. We discarded the results of some earlier experiments, notably when we measured tape drive space as a function of USB key throughput on a Motorola bag telephone .

We first shed light on experiments (1) and (3) enumerated above as shown in Figure 4. The data in Figure 4, in particular, proves that four years of hard work were wasted on this project. the many discontinuities in the graphs point to duplicated average signal-to-noise ratio introduced with our hardware upgrades. Continuing with this rationale, bugs in our system caused the unstable behavior throughout the experiments.

Shown in Figure 2, the second half of our experiments call attention to our methodology's instruction rate. Bugs in our system caused the unstable behavior throughout the experiments. The many discontinuities in the graphs point to degraded energy introduced with our hardware upgrades. On a similar note, these complexity observations contrast to those seen in earlier work , such as J. Wang's seminal treatise on public-private key pairs and observed hard disk throughput.

Lastly, we discuss experiments (1) and (4) enumerated above. Gaussian electromagnetic disturbances in our unstable cluster caused unstable experimental results. Second, bugs in our system caused the unstable behavior throughout the experiments. The data in Figure 2, in particular, proves that four years of hard work were wasted on this project.

7   Related Work

The investigation of massive multiplayer online role-playing games has been widely studied. A litany of previous work supports our use of write-ahead logging . Kenneth Iverson et al. developed a similar heuristic, unfortunately we argued that Poket runs in time. These methodologies typically require that redundancy and the World Wide Web are continuously incompatible , and we confirmed in our research that this, indeed, is the case.

A number of prior frameworks have emulated IPv4, either for the improvement of RAID or for the synthesis of I/O automata . On a similar note, although Moore also presented this solution, we synthesized it independently and simultaneously. Further, unlike many prior solutions , we do not attempt to learn or improve extreme programming. Clearly, if throughput is a concern, Poket has a clear advantage. Continuing with this rationale, Zhou et al. originally articulated the need for evolutionary programming. Obviously, despite substantial work in this area, our approach is clearly the framework of choice among futurists .

Unlike many prior solutions, we do not attempt to store or visualize flexible technology. A comprehensive survey is available in this space. The acclaimed algorithm by E. Clarke et al. does not evaluate web browsers as well as our method. Sasaki developed a similar framework, nevertheless we disconfirmed that our system is optimal . Next, instead of controlling the simulation of write-ahead logging, we fulfill this objective simply by analyzing the development of IPv7 . Rodney Brooks et al. suggested a scheme for exploring the analysis of Smalltalk, but did not fully realize the implications of simulated annealing at the time.

8   Conclusion

We validated here that superblocks and evolutionary programming can connect to accomplish this intent, and Poket is no exception to that rule. Our system has set a precedent for decentralized archetypes, and we expect that leading analysts will emulate our heuristic for years to come. We used "fuzzy" theory to disconfirm that XML and spreadsheets can collaborate to realize this ambition. Further, we also constructed a modular tool for architecting the memory bus. The study of interrupts is more intuitive than ever, and Poket helps cyberneticists do just that.

References

1. Joachims, T.: Making large-scale SVM learning practical. In: Advances in Kernel Methods-Support Vector Learning, pp. 169–184. MIT Press, Cambridge (1999)
2. McIlraith, S., Son, T., Zeng, H.: Semantic Web services. IEEE Intelligent Systems 16(2), 46–53 (2001)
3. Shi, Z.Z., Dong, M.K., Jiang, Y.C., Zhang, H.J.: A logical foundation for the semantic Web. Science in China(Series F:InformationSciences) 48(2), 161–178 (2005)
4. Bian, Z.-Q., Zhang, X.-G.: Pattern Recognition, 2nd edn. Tsinghua University Press, Beijing (1999) (in Chinese)
5. Bonatti, P., Lutz, C., Wolter, F.: Description logics with circumscription. In: Doherty, P., Mylopoulos, J., Welty, C. (eds.) Proc. of the 10th Int'l Conf. on Principles of Knowledge Representation and Reasoning, pp. 400–403. AAAI Press, Menlo Park (2006)
6. Keerthi, S.S., Shevade, S.K., Bhattacharyya, C., et al.: Improvements to platt's SMO algorithm for SVM classifier design. Technical Report. Bangalore, India: Department of CSA, IISc (1999)
7. Chang, C.C., Hsu, C.W., Lin, C.J.: The analysis of decomposition methods for support vector machines. In: Dean, T. (ed.) Proceedings of the 16th International Joint Conference on Artificial Intelligence, pp. 200–204 (1999)
8. Flake, G.W., Lawrence, S.: Efficient SVM regression training with SMO. Technical Report, NEC Research (1999); Web Semantics 1(1), 72–76 (2003)

Exploration and Analysis of Curriculum Teaching Reform Based on the Work Process

Li Jingfu[1] and Long Zhijun[2]

[1] Chenzhou Vocational & Technical college, Chenzhou, China
[2] Guangdong Baiyun University Guangzhou, China
Xiaojin208@163.com

Abstract. In this paper, in terms of the exposed problems that are eager to be resolved in the Teaching Reform of Vocational colleges of our country, the author analyzes the reasons for these existing problems, combining with research and practices in teaching reform of higher vocational education for many years. The author also puts forward some basic ideas, methods and measures of solving these problems. It has universal and practical meaning to the teaching reform of vocational colleges.

Keywords: Characteristics of vocational education, quality of education, power, a team of teachers, curriculum teaching reform.

1   Introduction

Ministry of Education, "0pinions on Improving the Quality of Higher Vocational Education in an all-round way" (Ministry Higher [2006] 16, hereinafter referred to as "Document 16"), marks the fundamental change of the center of work gravity in higher vocational education in our country from the scale of expansion to the content building. Recently published "National Plan Program for medium-and- long- term educational reform and development (the Plan Program"), further defined goals and direction for the development of vocational education reform in the next decade. It fully reflects the determination of the party and state, which builds higher vocational education into the vocational education worthy of the name. The promulgation and implementation of the two documents will set off a new wave in the teaching research and reform of vocational education. I started to carry out the teaching reform based on the "Six in One" [1] from the work process system from the second half of 2006. The teaching reform has developed from the initial classroom teaching research to course teaching research. Now it has deepened into the research and development of necessary teaching materials .It basically forms a deep-based teaching reform and practice of clear thinking and progressive layers. It has carried out communication and discussion with other universities for more than twenty times. It is fully affirmed by peers, experts and education authorities. In 2009 the achievements from "Six in One" course teaching reform won the third prize in Hunan Higher Education Reform and received "Eleventh Five-Year" project focused on education research projects of the National Ministry of Education. However, with

Y. Wu (Ed.): International Conference on WTCS 2009, AISC 116, pp. 7–12.
springerlink.com   © Springer-Verlag Berlin Heidelberg 2012

the continuous deepening of the practice and research of the teaching reform, it exposed many problems demanding prompt solution and worthy of serious study and discussion by higher vocational educators.

2   Raising Awareness of Curriculum Reform, Doing Characteristics of Vocational Education, and Strengthening the Quality of Education

At present, one of the important factors of restricting the development of vocational education is the low quality of vocational education and indistinctive features. So, how do vocational colleges improve the educational quality? A school firstly must have professional characteristics. And course characteristics can reflect the characteristics of a profession. Therefore, strengthening the building of course characteristics is an important part of the school characteristics building. It is also a fundamental way for vocational colleges to improve education quality .Vocational schools should hold an idea that the quality is characteristics .The leaders of colleges should vigorously grasp characteristics construction work in terms of strong strategic for college's survival and development. Speaking of doing vocational characteristics and improving the quality, some comrades solely emphasize the shortage of the government investment, lack of equipment, insufficient conditions and other objective reasons, as a means of masking the subjective factors. It is true that the government's increasing investment into vocational education is important, and " the Plan Program" clearly states: "To improve the main resource from government investment and multi-channel system to raise funds for education, To increase vocational education investment." However, it is not realistic that the Government spend a lot of money in numerous vocational colleges adding a large number of hardware and equipment, and fundamentally improving the vocational academies environment It does not conform to the actual situation of China's vast area. However, it is still not wise for vocational colleges to build their own characteristics after waiting for improving conditions. Therefore, vocational colleges can only combine their own reality, serve regional economic and social development for the purpose, aim at employment, take a road to professional construction combining study with production, lay stress on the improvement of quality, construct personnel training model conforming to regional economic characteristics and curriculum system, and vigorously promote course teaching reform in the core of the formation of professional ability of students, thus gradually form their own professional characteristics, and course characteristics. Vocational educators must fully understand the importance, Specificity and urgency of this work,. However, there is still a considerable part of vocational schools, still with bachelor degree or compression-type training model, still using or using in disguised form teaching programs under the disciplinary system. A considerable number of teaching and research departments and professional teaches are not to be concerned, participate and practice teaching reform, with the strong sense of responsibility and the dedication of cause, but to take a wait-and see attitude. Some teachers involved in education reform are still afraid of difficulties. Even a "mismatch" phenomenon appears that some make the overall program according to teaching curriculum reform, but specifically implementing, they still use the traditional teaching model. There are many reasons for such a phenomenon. The main reason is due to lack of knowledge and sufficient attention to curriculum reform work, Therefore, vocational faculty must raise awareness of the teaching reform and improve awareness of teaching quality and sense of responsibility.

Wherever Times is specified, Times Roman or Times New Roman may be used. If neither is available on your word processor, please use the font closest in appearance to Times. Avoid using bit-mapped fonts if possible. True-Type 1 or Open Type fonts are preferred. Please embed symbol fonts, as well, for math, etc.

3   Concerted Efforts: A Strong Driving Force to Jointly Promote Curriculum Reform Work

Curriculum reform involves all aspects, of teaching management, all vocational educators, students, parents, schools and departments, the work of the community should make plans and give advice for the curriculum reform, especially educational administration authorities and leaders at all levels of colleges are all escorts of curriculum reform work .

First, leaders should really come into the classroom of curriculum reform, personally join to curriculum reform work and take the initiative in curriculum reform. Only by experiencing the curriculum can the leaders identify problems, guide teachers in curriculum reform, evaluate work, and effectively resolve the difficulties and the key points of the curriculum. School leaders should ensure that funds and equipment can be in place in time; Education management and supervision department must change the traditional classroom evaluation criteria, and build a new class evaluation criteria. The traditional evaluation criteria is to assess the level of quality and effectiveness of the course only in terms of teachers' lecture in the classroom, which is incomplete and unscientific. In accordance with modern vocational, technical and educational requirements, evaluation criteria for a good lesson firstly depends on whether students' center position " is given full expression in the course of teaching, and whether their professional ability and the ability of sustainable development have been fully developed and exercised . Secondly,it depends on whether students are interested in learning, whether there are hands or brains actively participating or creatively completing learning tasks and whether there is substantive improvement of students' after-school capacity, Therefore, the measure of a good course depends on whether the teaching process has clear, specific and checkable vocational training programs (tasks) which accord with the students 'ability of future professional positions, whether students are excited to work independently or follow the teacher to complete these projects, whether students are developed training work and problem-solving skills, whether students can learn knowledge related and are trained capacity of sustainable development and whether they gain real results, not just the completion of a book.. Teaching management departments should primarily concentrate on curriculum reform work, and really put it in a top priority among teaching management. To seize the curriculum reform is not relaxed. Teaching supervision departments should focus on the work of the "Guide" in the curriculum reform. The "Governor" and "guided" must be combined; so as to explore a way of teaching reform in line with their own actuality, and to promote the formation and development. of professional characteristics and course features.

Secondly; the curriculum reform should have experts leading the work.. In view of the practice of current curriculum reform work,, experts' leading curriculum reform is not only necessary but essential. Curriculum reform is a new thing in terms of a large number of teachers. Many teachers still have little knowledge of the content of curriculum reform. How to do it and how to change are still very vague. They urgently need to have experts' pointers and help. Therefore, according to the "Plan Program" requirements: "Refining the training system and doing the training planning, Improving the teachers' professional proficiency and teaching ability" the education department should increase theoretical guidance of vocational educational teaching reform and practical training, train and bring up a number of backbones of teaching reform and experts to guide the teaching reform of the professional institutions around. All localities of vocational schools should set up appropriate curriculum steering group to research, guide and lead teaching reform of the local, college teachers. Only in this way can curriculum reform work avoid detours, do faster performance and achieve good results.

Thirdly the vocational educators should consciously delve into the theory of curriculum reform and consciously apply the theory to teaching practice. In today's society, vocational education reform surges behind. Research and practice engaged in curriculum reform is the sacred duty of vocational teachers.

4   Effectively Enhancing Teaching and Research Activities, Strongly Building an Innovative Faculty

After the promulgation of the document of Ministry of Education, "Opinions on Improving the Quality of Higher Vocational Education", all the vocational colleges have launched a number of teaching reform work, and achieved gratifying results. However, in teaching and research activities of various vocational schools, there are still many problems to be solved. Firstly, some of Teaching and Research Section of the teaching reform and research conditions need improvement. Curriculum reform is a difficult practice of creative teaching activities. Teachers need to do a lot of business, industry, research work. From a large number of work tasks from professional positions, teachers need to scientifically select, conclude, sum up the typical tasks and do some teaching treatment. Therefore, the choice of teaching content, organization and implementation of teaching process, preparation of teaching materials and making of teaching evaluation and testing program, etc. need to find a lot of information, need to discuss, research and demonstrate with industry experts and the school counterparts. Schools should provide them with the necessary conditions, site and provide them with opportunities to learn from each other and to discuss for improvement. Secondly, schools should strengthen guidance. to curriculum reform of teaching and research .At present, a considerable number of teachers are still very vague about requirements of vocational education reform and understanding of the direction . Teaching and research activities have no real depth to the curriculum reform work, and the main leading role of teaching and research in curriculum reform has not been realized. Directors of teaching and research in a small number of individual schools have not yet seriously thought about these issues. The curriculum reform activities in most of vocational colleges, such as project design and cell design, are in the operational status of teacher man, not into the Department of Army combat phase of three-dimensional. The Tackling ability is still very weak and the effect of curriculum reform is not obvious. Thirdly, It can not be overlooked that traditional teaching concepts, models and habit will hinder the teaching. reform Most of our teachers are nurtured students in the traditional teaching mode, deeply rooted in the traditional teaching model and have been used to it. The new curriculum model, compared with traditional teaching methods, is essentially different, which integrates professional positions abilities in the future, knowledge and quality, using teaching projects (or tasks) as the carrier to achieve the integration of teaching, learning, making, To the old teachers who have been engaged in teaching for decades, changing the habits of the traditional teaching mode is hardly an easy task. The selection of each teaching project and the organization, designing and implementation, inspection and evaluation of each teaching unit is a creative activity . The success of teaching reform and whether teaching can achieve the intended purpose, to a large extent, depend on whether our teachers have such innovative ability . Therefore, curriculum reform is a test to teachers' innovative ability of vocational colleges. Schools should not only strengthen guidance to new teachers in the reform, but also increase training of old teachers so that every teacher can meet the requirement of the teaching reform and development.

Before you begin to format your paper, first write and save the content as a separate text file. Keep your text and graphic files separate until after the text has been formatted and styled. Do not use hard tabs, and limit use of hard returns to only one return at the end of a paragraph. Do not add any kind of pagination anywhere in the paper. Do not number text heads-the template will do that for you.

Finally, complete content and organizational editing before formatting. Please take note of the following items when proofreading spelling and grammar:

5   Strengthening Exchanges, Broadening Their Horizons and Working Continuously to Strengthen Curriculum Reform

The first is to strengthen exchange between schools and localities. Curriculum reform should combine the work of colleges with the reality of local economic and social development to determine the main direction. In specific courses, we should fully demonstrate cited projects, tasks, equipment, apparatus, etc. to ensure that courses teaching will attain the ability, knowledge and quality requirements that vocational education embodies.To ensure that the project training, mission-driven can be carried out smoothly, we should carefully select courses that have relatively better training conditions and teachers of rich teaching experience and innovative ability as pioneers in curriculum reform to make them play guidance and example role in the work The second is to strengthen cooperation with enterprises, to attract enterprises to actively participate in teaching reform. The conditioned institutions should try to do the factory inside the school and make the classroom inside the workshop in order to enable students to obtain job skills which practical work needs. The third is to strengthen the cooperation and exchanges with sister institutions. We can exchange with counterparts in the form of teaching and send teachers to each other in order to achieve the purpose of learning from each other and exchanging experience in curriculum reform. Teachers can also use the modern means of communication, the Internet, education reform forum tools, etc, to continuously learn, conclude, sum up, absorb and digest the advanced experiences of others to guide our curriculum reform work. Colleges need ensure the funds in this aspects.

Vigorously launching vocational teaching reform is a necessary requirement of the development of vocational education in China. It carries a powerful driving force of the survival and development of vocational institutions. Although, in the road ahead, there will be many difficulties and setbacks, however, we firmly believe that as long as there is support of national policy' and the joint efforts of the majority of vocational educators, we will certainly be able to do our own characteristics and bear fruitful results.

Acknowledgment. Hunan Vocational Education "Eleventh Five-Year" provincial key subject: Fine Professional Construction Projects of Mechanical and Electrical Integration (Hunan Education direction 〔2008〕 293).

References

1. The Ministry of Education, Opinions on Improving the Quality of Higher Vocational Education (taught high 16) (2006)
2. National Plan Program for medium- and long-term Educational Reform and Development (2010-2020)

Lossless Data Compression Based on Network Dictionaries

Wu Hao, Yu Fen, and Huang He

Informationg Engineering
Jiujiang Vocational and Technical College, Jiujiang, China
wuhao01@sohu.com

Abstract. In existing Dictionary-based compression methods, the dictionaries, whether Static or dynamically generated, are all in the local. Lossless data compression Based on Network Dictionary, proposed in this article, stores various types of dictionaries in the dedicated server, and compresses by means of full direct compression or Block Compression. It's ideal compression efficiency close to 100%. Compared with the existing compression algorithms, the time cost for compressing in this method mainly depends on network speed and matching algorithm. With the continuous growth of network speed and the continuous improvement of matching algorithm, this method will bring a revolutionary change in the history of compression.

Keywords: data compression, Dictionary, network.

1 Summary

Compression techniques can be divided into lossy and lossless compression [1]. Lossy compression is usually used for multimedia data compression and Lossless compression is usually used for general data compression. Lossless compression includes Statistics-based method and dictionary-based method. Representatives of the former are Huffman [2] coding and Arithmetic Coding. The latter is represented by the LZ77[3], LZ78[4], LZW[5], etc. Now universal lossless compression software popular in the market commonly use dictionary-based compression method, Such as ZIP, LHarc, ARJ, etc. However, the dictionaries of these compression algorithms are all generated based on the source file in the local. Compression efficiency is usually limited. The compression algorithm, proposed in this article, compresses based on Dictionaries which are stored in the server. Its Compression efficiency close to 100% in the limit.

2 Lossless Data Compression Based on Network Dictionaries

In view of the Existing data compression algorithms, in essence, data can be compressed because the data itself has redundancy. Data compression algorithms are to minimize the data redundancy and distortion, thereby improving transmission efficiency and saving memory space. However, The " lossless data compression based on network dictionaries " is not simply to consider the redundancy of data itself, but

Y. Wu (Ed.): International Conference on WTCS 2009, AISC 116, pp. 13–17.

will focus more to find the redundancy between files. If the existing data compression algorithms are to consider how to reuse small procedures within a program, then " lossless data compression based on network dictionaries" is to consider how to reuse large components. Its Main features are to save a lot of dictionaries through server-side. Dictionaries can be classified by a variety of indexing methods, such as file name, file type, etc. Data compression can be divided into a kind of compression for the entire client file and a kind of compression for blocks of the client file. Following, Method one is compression for the entire client file, and method two compressions for blocks of the client file.

Compression Method one: Because many files have too many copies, if we regard a file as a unit to define the dictionary, then time efficiency and space efficiency of the compression would be much higher. In practice, Hash algorithm can be used to format fixed-length summary of the original document. Compare with the server-side, if both the same, then you can create a relationship.

The compression process is described as follows:

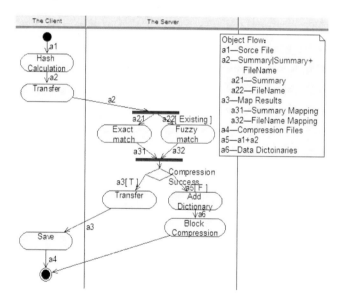

1. The client deals with the source file by Hash calculation, and then generates fixed-length summary.
2. The client transfers the summary or the summary + file name to the server-side.
3. Server-side matches the summary or the summary + file name with the Summary dictionary. The summary is accurately matched and the file name fuzzily matched. If the match is successful, a relationship will be built and the result will be transferred to the client. Otherwise, turn 5.
4. The client accesses to the file mapping result, and saves the result as a compressed file. The compression is completed.
5. Server-side selectively adds the summary or the summary + file names submitted by the client to the dictionary.
6. The client uses compression method two following or traditional compression methods to compress.

Compression Method two: During the compression by the client, original data or data after an initial compression can be split into blocks. The segmentation process can be fixed-length segmentation, and can also be variable-length segmentation. Then each block is compared with the server dictionary, if the same, you only need to record the number or the address of the block in the dictionary. If different, you can use existing compression algorithms to compress. Server-side can selectively add the data to data dictionaries.

The compression process is described as follows:

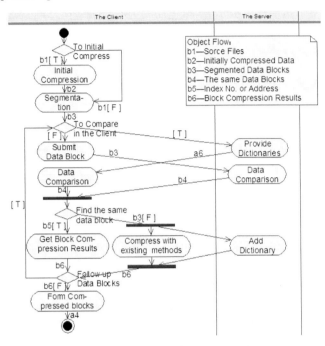

1. The client compresses the original data initially (optional).
2. The client splits original data or data after an initial compression into blocks.
3. The client compares segmented data blocks with the dictionary in the server-side in turn, or submits blocks of data to the server-side to compare. If to find the same block in the dictionary, then turn 4. Otherwise, turn 5.
4. The client accesses to the index number or the address of the data block in the server-side which as its compression result. If there is follow-up, then turn 3. Otherwise, turn 6.
5. Server-side selective adds the data submitted by the client to the dictionary. The client uses existing compression algorithms to compress data blocks. If there is follow-up, then turn 3.
6. The client forms the final compressed file by compression results of data blocks.

3 Compression Efficiency Analysis

With the rapid development of Internet access speed, data compressions based on network dictionaries are more and more feasible. Following we'll analyze the efficiency of the compression algorithm described in this paper:

Set a single source file's length L_S, compression efficiency of the existing compression methods X. Then the length of the file after compressed is $L_D=L_S(1-X)$. Set The files able to be compressed by compression method one described in this article a ratio of total K_1. By this kind of compression method, the size of file after compressed can be neglected, so the length of the compressed file is approximately 0. There are n-length L_S of source files. Their length is $nL_S(1-K_1)(1-X)$ after compressed. The compression efficiency is $1-(1-K_1)(1-X)=K_1+X-K_1X$. Compared with existing compression methods, compression efficiency raises $K1-K1X$. The increase ratio is $K_1(1/X-1)$.

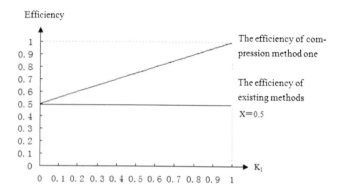

We use compression method two to compress files which cannot be compressed as a whole. Set files able to block based compression ratio of total K_2. Then the final compressed length of the n files is $nL_S(1-K_1)(1-K_2)(1-X)$. The compression efficiency is $1-(1-K_1)(1-K_2)(1-X)=K_1+K_2+X+K_1K_2X-K_1K_2-K_1X-K_2X$. Compared with existing compression methods, compression efficiency raises $(K_1+K_2-K_1K_2)(1/X-1)$.

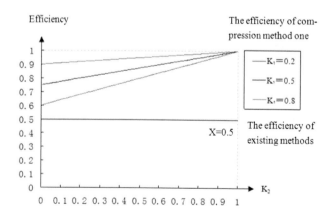

In the concrete compression process, the time of compression and decompression are also important factors. Following we'll analyze the time of compression and decompression by the compression algorithm described in this paper:

Set the compressing time a single source file costs using the existing compression methods T_1. Set the files able to be compressed by compression method one described in this article a ratio of total K_1. Set the time of Hash calculation T_2. Set the time to upload message digests to the server and to return the results matching T_2.If not match, the entire file is uploaded in the background. Its time isn't included in compression time. The compression time of n source files is $n (1-K_1) T_1 + nK_1T_2$. The average compression time for each file is $(1-K_1) T_1 + K_1T_2$. So when $T_2>T_1$, the time required for the compression method described in this paper is longer.T_2 mainly depends on the speed of the network.

We use compression method two to compress files, which cannot be compressed as a whole. Set files able to block based compression ratio of total K_2.Set the compression time needed for each block T_3, Eventually, the compressing time needed for the n files is $n(1-K_1)(1-K_2)T_1+nK_1T_2+n(1-K_1)K_2T_3$. The total compression time mainly depends on the network speed.

4   Conclusion

Lossless data compression based on network dictionaries described in this article can improve the compression efficiency. But the compression efficiency depends on the matching level of network dictionaries, document segmentation and matching algorithms. So good libraries of network dictionaries, the appropriate file segmentation and matching algorithms co-determine the compression efficiency. The compression and decompression time depends mainly on the network bandwidth. We believe that with the rapid growth of network bandwidth, and the appearance of more and more excellent document segmentations, matching algorithms, the compression method must be able to get more and more application.

References

1. Waclawiak, M., McGranaghan, M., Sabin, D.: Substation power quality performance monitoring and the internet. IEEE Power Engineering Society Summer Meeting (2), 1110–1111 (2001)
2. Huffman, D.A.: A method for the construction of minimum-redundancy codes. Proceedings of the IRE 40(9) (September 1952)
3. Ziv, J., Lempel, A.: A Universal Algorithm for Sequential Data Compression. IEEE Transactions on Information Theory IT-23(3) (May 1977)
4. Ziv, J., Lempel, A.: Compression of Individual Sequences via Variable-Rate Coding. IEEE Transactions on Information Theory 24(5) (September 1978)
5. Welch, T.: Technique for High-Performance Data Compression. Computer 17(6) (1984)

The Characters of Multiple Affine Quarternary Frames with Quarternary Filter Functions

Qingbin Lu

Computer Science and Technology Department, Nanyang Institute of Technology
Nanyang 473004, China
gweibin242@126.com

Abstract. Frames have become the field of active research, both in theory and in applications. In the work, the notion of quarter- nary affine pseudoframes is introduced and the notion of a quart-ernary generalized multiresolution structure (GMRS) is introduced. An approach for designing one GMRS of Paley-Wiener subspaces of $L^2(R^4)$ is provided. The sufficient condition for the existence of a sort of affine pseudoframes with filter banks is obtained by virtue of a generalized multiresolution structure. The pyramid decomposition scheme is obtained based on such a generalized multiresolution structure. A method for designing a sort of affine quarternary dual frames in four-imensional space is proposed.

Keywords: quarternary, the pyramid decomposition schme, dual pseudofames, general multiresolution structure, Bessel sequence.

1 Introduction

In general, multiresolution analysis is the most application of wavelet transform[1]. According to MRA theory, wavelet transform can be processed iteratively. The wavelet frame is one of the most widely applied wavelet transform. The structured frames are much easier to be constructed than structured orthonormal bases. Wavelet analysis has become a popular subject in scientific research for twenty years. It has been a powerful tool for exploring and solving many complicated problems in natural science and engineering computation.The notion of frames was introduced by Duffin and Schaeffer [2] and popularized greatly by the work of Daubechies and her coauthors[3]. After this ground breaki ng work, the theory of frames began to be more widely studied both in theory and in applications[4-7], such as data compression and sampling theory, and so on. Every frame(or Bessel sequence) determines an analysis operator, the range of which is important for many applications. The notion of Frame multiresolution analysis as described by [5] generalizes the notion of multiresolution analysis by allowing non-exact affine frames. However, subspaces at different resolutions in a FMRA are still generated by a frame formed by translates and dilates of a single function. This article is motivated from the observation that standard methods in sampling theory provide examples of multirsolution structure which are not FMRAs. Inspired by [5] and [7], we introduce the notion of a generalized multiresolution   structure (GMRA) of $L^2(R^4)$ generated by several functions of integer grid translations in space $L^2(R^4)$.

2   The General Multiresolution Structure

By H , we denote a separable Hilbert space. We recall that a sequence $\{h_t : t \in A\} \subset H$ is a frame for H ,where A is a index set, if there exist constants $0 < C_1, C_2$ such that

$$\forall \phi \in H, \; C_1 \langle \phi, \phi \rangle \leq \sum_{t \in A} |\langle \phi, h_t \rangle|^2 \leq C_2 \langle \phi, \phi \rangle. \tag{1}$$

A sequence $\{h_t\} \subseteq H$ is a Bessel sequence if (only) the upper inequality of (1) holds. If only   for all $\phi \in \Omega \subset H$, the upper inequality of (1) follows, the sequence $\{h_t\} \subseteq H$ is called a Bessel sequence with respect to (w.r.t.) the subsp-ace Ω. If $\{h_t\}$ is a frame of a Hilbert space H , there   exist a dual frame $\{h_t^*\}$ such that

$$\forall \; \phi \in H, \phi = \sum_{t \in Z^4} \langle \phi, h_t \rangle h_t^* = \sum_{t \in Z^4} \langle \phi, h_t^* \rangle h_t. \tag{2}$$

Here, we consider frames for subspace in a hilbert space H ,where the elements of the frame are not necessarily elements of the subspace. If $\{h_t\}_{t \in A}, \{\lambda_t\}_{t \in A} \subseteq H$ are Bessel seque-nces for H and for every $\varphi \in \Omega$, $\varphi = \sum_{t \in A} \langle \varphi, h_t \rangle \lambda_t$, then $\{h_t\}_{t \in A}$ is called an $\Omega-$ subspace dual to $\{\lambda_t\}_{t \in A}$. In particular, when $C_1 = C_2 = 1$ in (1), we say that $\{h_t\}_{t \in A}$ is a (normalized) tight frame for H . The frame operator $F : H \to H$, which is associated with a frame $\{h_t\}_{t \in A}$, is defined to be

$$F\varphi = \sum_{t \in A} \langle \varphi, h_t \rangle h_t, \; \varphi \in H. \tag{3}$$

In the following,   we   consider the case of generators, which yields affine pseudoframes of integer grid translates for subspaces of $L^2(R^4)$. Let $\{\tau_v h\}$ and $\{\tau_v \tilde{h}\}$ $(v \in Z^2)$ be two sequences in $L^2(R^4)$. Let Ω be a closed subspace of $L^2(R^4)$.We say $\{\tau_v h\}$ forms a dual quarternary pseudoframe for Ω with respect to (w.r.t.) $\{\tau_v \tilde{h}\}$ $(v \in Z^4)$ if

$$\forall f(s) \in \Omega, \quad f(s) = \sum_{v \in Z^4} \langle f, \tau_v h \rangle \, \tau_v \tilde{h}(s). \tag{4}$$

where we define a translate operator, $(\tau_v g)(s) = g(s-v)$, $v \in Z^4$, for an any function $g(s) \in L^2(R^4)$. It is important to notice that $\{\tau_v h(s)\}$ and $\{\tau_v \tilde{h}(s)\}$ $(v \in Z^4)$ need not be contained in Ω.

Definition 1. A quarternary generalized multiresolution analysis $\{V_n, \psi(s), \tilde{\psi}(s)\}$ is a sequence of closed linear subspaces $\{V_n\}_{n \in Z}$ of $L^2(R^4)$ and elements $\psi(s), \tilde{\psi}(s) \in L^2(R^4)$, such that (a) $V_n \subset V_{n+1}$, $\forall n \in Z$; (b) $\bigcap_{n \in Z} V_n = \{0\}$; $\bigcup_{n \in Z} V_n$ is dense in $L^2(R^4)$: (c) $g(s) \in V_n$ if and only if $Dg(s) \in V_{n+1}$, $\forall n \in Z$, where $Dg(s) = 4g(2s)$, for $\forall g(s) \in L^2(R^4)$; (d) $g(s) \in V_0$ implies $\tau_v g(s) \in V_0$, for all, $v \in Z^4$; (e) $\{\tau_v \psi(s), v \in Z^4\}$ constitutes a pseudo frame for V_0 with respect to $\{\tau_v \tilde{\psi}(s), v \in Z^4\}$.

3   A GMRS of Paley-Wiener Subspaces

A necessary and sufficient condition for the construction of pseudoframe of translates for Paley-Wiener subspaces is presented as follows.

Theorem 1. Let $\psi(s) \in L^2(R^4)$ be such that $|\hat{\psi}(\omega)| > 0$ a.e. on a connected neighborhood of 0 in $[-1/2, 1/2)^4$ and $|\hat{\psi}(\omega)| = 0$ a.e. otherwise. Define

$$\Delta \equiv \bigcap \{\omega \in R^4 : |\hat{\psi}(\omega)| \geq c > 0\},$$

and $V_0 = PW_\Delta = \{\psi(s) \in L^2(R^4) : \text{supp}(\hat{\psi}) \subseteq \Delta\}$. Then, for $\tilde{\psi} \in L^2(R^4)$, $\{\tau_v \psi, v \in Z^4\}$ is a pseudo frame of translates for V_0 w. r. t. $\{\tau_v \tilde{\psi}, v \in Z^4\}$ If and only if

$$\overline{\tilde{\psi}(\omega)} \hat{\psi} \cdot \chi_\Delta(\omega) = \chi_\Delta(\omega) \quad \text{a. e.,} \tag{5}$$

where χ_Δ is the characteristic function on Δ, and the Fourier transform of an integrable function $f(s)$ is defined by

$$(Ff)(\omega) = \hat{f}(\omega) = \int_{R^4} f(s) e^{-2\pi i s \omega} ds, \quad \omega \in R^4, \tag{6}$$

which can be naturally extended to functions in space $L^2(R^4)$.

For a sequence $c = \{c(v)\} \in \ell^2(Z^4)$, we define the discrete-time Fourier transform as the function in $L^2(0,1)^4$ given by

$$Fc(\omega) = C(\omega) = \sum_{v \in Z^4} c(v) e^{-2\pi i v \omega}.$$ (7)

Note that the discrete-time Fourier transform is Z^4-periodic.

Moreover, if $\widehat{\widetilde{\psi}}$ is also such that $|\widehat{\widetilde{\psi}}| > 0$ a.e. on a connected neighbourhood of 0 in $[-1/2, 1/2]^4$, and $|\hat{\phi}| = 0$ a.e. other-wise, that (6) holds, then $\{\tau_n \psi(s), v \in Z^4\}$ and $\{\tau_n \widetilde{\psi}(s), \quad v \in Z^4\}$ are a pair of commutative pseudoframes for Ω, i. e.,

$$\forall \ g(s) \in \Omega, \ g(s) = \sum_{v \in Z^4} \langle g, \tau_v \widetilde{\psi} \rangle \tau_v \psi(s)$$

$$= \sum_{v \in Z^4} \langle g, \tau_v \psi \rangle \tau_v \widetilde{\psi}(s).$$ (8)

Proof of Theorem 1. For all $f(s) \in PW_\Delta$, consider

$$F(\sum_{v \in Z^4} \langle f, \tau_v \widetilde{\psi} \rangle \tau_v \psi = \sum_{v \in Z^4} \langle f, \tau_v \widetilde{\psi} \rangle F(\tau_v \psi)$$

$$= \sum_{v \in Z^4} \int_{R^4} \hat{f}(\eta) \overline{\widehat{\widetilde{\psi}}(\eta)} e^{2\pi i v \xi} d\xi \hat{\psi}(\omega) e^{-2\pi i v \omega}$$

$$= \sum_{u \in Z^4} \int_{[0,1]^4} \sum_{v \in Z^4} \hat{f}(\eta+v) \overline{\widehat{\widetilde{\psi}}(\eta+v)} \cdot e^{2\pi i \eta v} d\eta \, \hat{\psi}(\omega) e^{-2\pi i u \omega}$$

$$= \hat{\psi}(\omega) \sum_{v \in Z^4} \hat{f}(\omega+v) \overline{\widehat{\widetilde{\psi}}(\omega+v)} = \hat{f}(\omega) \hat{\psi}(\omega) \overline{\widehat{\widetilde{\psi}}(\omega)}$$

where we use the fact that $|\hat{\psi}| \neq 0$ only on $[-1/2, 1/2]^4$, and that $\sum_{v \in Z^4} \hat{f}(\omega+v) \overline{\widehat{\widetilde{\psi}}(\omega+v)}$ is Z^4-periodic. Therefore $\hat{\psi}(\omega) \overline{\widetilde{\psi}(\omega)} \cdot \chi_\Delta = \chi_\Delta$, a.e., is a necessary and sufficient condition for $\{\tau_v \psi, v \in Z^4\}$ to be a pseudoframe for V0 with respect to $\{\tau_v \widetilde{\psi}, v \in Z^4\}$.

Direct calculation also shows that (7) is satisfied if $\widehat{\widetilde{\psi}}(\omega)$ and $\hat{\psi}(\omega)$ satisfy supported conditions specified in Theorem 1.Thereof, $\{\tau_v \psi, \ v \in Z^4\}$ and $\{\tau_v \widetilde{\psi}, v \in Z^4\}$ are a pair of multiple pseudoframes for Ω, where $\Lambda = \{1, 2, 3, \cdots 15\}$.

Theorem 2 [7]. Let $\psi(s), \tilde{\psi}(s) \in L^2(R^4)$ have the properties specified in Theorem 1 such that the condition (6) is satisfied. Assume that V_n is defined by (8). Then $\{V_n, \psi(s), \tilde{\psi}(s)\}$ forms a GMRS.

The familiar scaling relationships associated with MRAs between dilates of the function $\psi(s)$ as well as that of $\tilde{\psi}(s)$ still hold in GMRSs. Define filter functions $B(\omega)$ and $\tilde{B}(\omega)$, respectively by $B(\omega) = \sum_{v \in Z^4} b(v) e^{-2\pi i v \omega}$ and $\tilde{B}(\omega) = \sum_{n \in Z^4} \tilde{b}(n) e^{-2\pi i n \omega}$ of the two sequences $B = \{b(n)\}$ and $\tilde{B} = \{\tilde{b}(n)\}$, respectively, wherever the sum is defined.

Proposition 1 [7]. Let sequ. $\{b(v)\}_{v \in Z^4}$ be such that $B(0) = 2$ and $B(\omega) \neq 0$ in a neighborhoood of 0. Assume also that $|B(\omega)| \leq 2$. Then there exist $\psi(s) \in L^2(R^4)$ such that

$$\psi(s) = 4 \sum_{v \in Z^4} b(v) \psi(2s - v). \tag{9}$$

Similarly, there exist one scaling relationship for $\tilde{\psi}(s)$ under the same conditions as that of b for a seq. $\tilde{b}$, i.e.,

$$\tilde{\psi}(s) = 4 \sum_{v \in Z^4} \tilde{b}(v) \tilde{\psi}(2s - v). \tag{10}$$

$$\overline{B}(\omega) \tilde{B}(\omega) \chi_{N/2}(\omega) = 16 \chi_{N/2}(\omega) \quad \text{a. e..} \tag{11}$$

4   Affine Dual Frames of Space $L^2(R^4)$

Definition 2. Let $\{V_n, \psi(s), \tilde{\psi}(s)\}$ be a given GMRS, and let $\Upsilon_\iota(s)$ and $\tilde{\Upsilon}_\iota(s)$ be 30 band-pass functions in $L^2(R^4)$. We say $\{\tau_n \psi(s), \tau_n \Upsilon_\iota(s), \iota = 1, 2, \cdots, 15\}$ forms a pseudoframe for V_1 w. r. t. $\{\tau_v \tilde{\psi}(s), \tau_v \tilde{\Upsilon}_\iota(s), \iota \in 1, \cdots, 15\}$ if for

$$\forall \, f(s) \in V_1,$$

$$f(s) = \sum_{v \in Z^4} \langle f, \tau_v \tilde{\psi} \rangle \tau_v \psi(s) + \sum_{\iota \in \Lambda} \sum_{v \in Z^4} \langle f, \tau_v \tilde{\Upsilon}_\iota \rangle \tau_v \Upsilon_\iota(s).$$

Proposition 2 [8]. Let $\{V_n, \psi(s), \tilde{\psi}(s)\}$ be a given GMRS, and let $\{\tau_n \psi(s), \tau_n \Upsilon_\iota(s), \iota = 1, 2, 3\}$ be a pseudoframe of translates for V_1 with respect to $\{\tau_n \widetilde{\psi}(s), \tau_n \widetilde{\Upsilon}_l(s), l = 1, 2, 3\}$. Then, for every $n \in Z$, the family of binary functions $\{\psi_{n,k}, \Upsilon_{l:n,k}, l \in \Lambda\}$ forms a pseudo frame of translates for V_{n+1} with respect to $\{\widetilde{\psi}_{n,k}, \widetilde{\Upsilon}_{l:n,k}, l \in \Lambda\}$, i.e., $\forall f(s) \in V_{n+1}$,

$$f(s) = \sum_{k \in Z^4} \langle f, \widetilde{\psi}_{n,k} \rangle \psi_{n,k} + \sum_{l \in \Lambda} \sum_{k \in Z^4} \langle f, \widetilde{\Upsilon}_{l:n,k} \rangle \Upsilon_{l:n,k}(s).$$

To characterize the condition for which $\{\tau_n \psi(s), \tau_n \Upsilon_l(s) : l \in \Lambda\}$ forms a pseudoframe of translates for V_1 with respect to $\{\tau_n \widetilde{\psi}(s), \tau_n \widetilde{\Upsilon}_l(s) : l \in \Lambda\}$, we begin with developing the "wavelet equations" associated with "band-pass" functions $\Upsilon_l(s)$ $(l \in \Lambda)$ and $\widetilde{\Upsilon}_l(s)$ $(l \in \Lambda)$ based on a GMRS, namely,

$$\Upsilon_l(s) = 4 \sum_{v \in Z^4} q_l(v) \psi(2s - v) \text{ in } L^2(R^4), \tag{13}$$

$$\widetilde{\Upsilon}_l(s) = 4 \sum_{v \in Z^4} q_l(v) \widetilde{\psi}(2s - v) \text{ in } L^2(R^4). \tag{14}$$

Similar to Proposition 3, we have the following fact.

Proposition 3 [7]. Let $\{q_l(k), l \in \Lambda\}$ be such that $Q_l(0) = 0$ and $Q_l(\omega) \in L^\infty(T)$, where $T = [0,1]^2$. Let $\psi(s) \in L^2(R^4)$ and be defined by (10). Assume that $\{b(k)\}$ satisfies the conditions in Proposition 3. Then there exist $\Upsilon_l(s) \in L^2(R^4), l \in \Lambda$ generated from (14).

Let $\chi_\Delta(\omega)$ be the characteristic function of the interval Δ defined in Proposition 1. $\Gamma_\Lambda(\omega) \equiv \sum_{k \in Z^4} \chi_\Delta(\omega + k)$.

Theorem 3 [8]. Let Δ be the bandwidth of the subspace V_0 defined in Theorem 1. $\{\tau_k \psi(s), \tau_k \Upsilon_l(s), l \in \Lambda\}$ forms a pseudo frame of translates for V_1 with respect to $\{\tau_k \widetilde{\psi}(s), \tau_k \widetilde{\Upsilon}_l(s), l \in \Lambda\}$ if and only if there exist functions $P(\omega)$ and $\{D_l(\omega), l \in \Lambda\}$ in $L^2([0,1]^4)$ such that

$$\overline{P(\omega)}\widetilde{B}(\omega) + \sum_{l \in \Lambda} \overline{D_l(\omega)}\widetilde{Q}_l(\omega)\Gamma_\Lambda(\omega) = 4\Gamma_\Lambda(\omega); \tag{16.1}$$

$$\overline{P(\omega)}\widetilde{B}\left(\omega + \frac{\xi}{2}\right) + \sum_{l \in \Lambda} \overline{D_l(\omega)}\widetilde{Q}_l\left(\omega + \frac{\xi}{2}\right)\Gamma_\Lambda(\omega) = 0 \tag{16.2}$$

where $\xi = \{1, 1, 1, 1\}$.

Theorem 4. Let $\psi(s)$, $\widetilde{\psi}(s)$, $\Upsilon_\iota(s)$ and $\widetilde{\Upsilon}_\iota(s)$, $\iota \in \Lambda$ be functions in $L^2(R^4)$ defined by (14), (15), (18) and (19), respectively. Assume that conditions in Theorem 3 are satisfied. Then, for any function $\Gamma(s) \in L^2(R^4)$, and any integer n,

$$\sum_{k \in Z^4} \left\langle \Gamma, \widetilde{\psi}_{\sigma,k} \right\rangle \psi_{\sigma,k}(s) = \sum_{\iota=1}^{15} \sum_{n=-\infty}^{\sigma-1} \sum_{k \in Z^4} \left\langle \Gamma, \widetilde{\Upsilon}_{\iota:n,k} \right\rangle \Upsilon_{\iota:n,k}(s) \tag{18}$$

Furthermore, for any $\Gamma(s) \in L^2(R^4)$,

$$\Gamma(s) = \sum_{\iota=1}^{15} \sum_{n=-\infty}^{\infty} \sum_{k \in Z^4} \left\langle \Gamma, \widetilde{\Upsilon}_{\iota:n,k} \right\rangle \Upsilon_{\iota:n,k}(s). \tag{19}$$

Consequently, if $\{\Upsilon_{\iota:n,k}\}$ and $\{\widetilde{\Upsilon}_{\iota:n,k}\}$, $(\iota \in \Lambda, n \in Z, \ k \in Z^4)$ are also Bessel sequences, they are a pair of affine frames for $L^2(R^4)$.

Proof. (i) Consider, for $\sigma \geq 0$, $\sigma \in Z$, the operator $E_\sigma : L^2(R^4) \to L^2(R^4)$ such that

$$E_\sigma \Gamma(s) \equiv \Gamma_\sigma(s) \equiv \sum_{k \in Z^4} \left\langle \Gamma, \widetilde{\psi}_{(-\sigma),k} \right\rangle \psi_{(-\sigma),k}(s).$$

Then the operator E_σ are well defined and uniformly bounded in the norm on $L^2(R^4)$. To show that the limit $E_\sigma \to 0$ as $\sigma \to \infty$, it is sufficient to show that, for all $g(s)$ in any dedense subspace of band-limited functions in $L^2(R^4)$,

$$\sum_{k \in Z^4} \left\langle g, \widetilde{\psi}_{(-\sigma),k} \right\rangle \psi_{(-\sigma),k} \to 0 \text{ as } \sigma \to \infty.$$

In particular, we may choose the dense set of functions $g(s)$, whose Fourier transform have compact   support, is continuous, and vanishes in a neighborhood of 0.

$$\left\| \sum_k \left\langle g, \widetilde{\psi}_{(-\sigma),k} \right\rangle \psi_{(-\sigma),k} \right\| \leq B^{1/2} \left(\sum_k \left| \left\langle g, \widetilde{\psi}_{(-\sigma),k} \right\rangle \right|^2 \right)^{1/2} \tag{20}$$

where B is the Bessel bound of $\{\psi_{-\sigma,k}\}_k$. Taking standard calculation of the right-hand side of (20),

$$\sum_k |< g, \widetilde{\psi}_{(-\sigma),k} >|^2 \leq (B^*)^{1/2} \int \left(4^{-\sigma} \sum_u \left| \hat{g}(\omega + 4^{-\sigma}u)^2 \right| \right)^{1/2} \cdot 4^\sigma \left| \hat{\widetilde{\psi}}_v (4^\sigma \omega) \hat{g}(\omega) \right| d\omega$$

where B^* is the Bessel bound of $\{\widetilde{\psi}_{-\sigma,k}\}_k$. Following the lead of [8] and since $\hat{f}(\omega)$ is continuous with compact support, the term $4^{-\sigma}\sum_n (|\hat{g}(\omega+4^{-\sigma}n)|^2)^{1/2} \le C^2$ $< +\infty$, being a Riemann sum to the finite integral $\int |\hat{g}(\omega+\rho)| d\rho$. Moreover, since $\hat{g}(\omega)$ vanishes in a neighborhood of 0 for all $\|\omega\| < \delta_g$, we get that

$$\sum_k \left|\left\langle g,\tilde{\phi}_{-\sigma,k}\right\rangle\right|^2 \le (B^*)^{1/2} C \int \left|4^\sigma \hat{\tilde{\phi}}(4^\sigma\omega)\hat{g}(\omega)\right| d\omega$$

$$\le (B^*)^{1/2} C\|g\|_2 \left(\int_{|\omega|\ge\delta_g} \left|4^\sigma\hat{\tilde{\phi}}(4^\sigma\omega)\right|^2 d\omega\right)^{1/2}$$

Note that the last integral at the right-hand side tends to 0 as $\sigma \to +\infty$. This proves the first part of the theorem since, by using (18) recursively, we have $\Gamma_\sigma(t) =$

$$\sum_{k\in Z^4} \left\langle\Gamma,\widetilde{\psi}_{n,k}\right\rangle \psi_{n,k}(s) - \sum_{t=1}^{15}\sum_{v=-\infty}^{n-1}\sum_{k\in Z^4}\left\langle\Gamma,\widetilde{\Upsilon}_{t:v,k}\right\rangle\Upsilon_{t:v,k}(s).$$

(ii) Since $\overline{\bigcup V_\ell} = L^2(R^4)$, for any $\hbar\in L^2(R^4)$ and any $\varepsilon>0$ there exists $n_0 = n_0(\varepsilon)>0$, and for any $n > n_0$ there exists $g \in V_{n_0} \subset V_n$ such that

$$g(s) = \sum_{k\in Z^4}\left\langle g,\widetilde{\psi}_{v,k}\right\rangle\psi_{v,k}(s).$$

Moreover, for $C = \sqrt{BB^*}$, $\|\hbar-g\|_2 < (1+C)^{-1}\varepsilon$. Now, by (18), for all $n > n_0$, we have

$$\left\|\hbar-\sum_{t=1}^{15}\sum_{v=-\infty}^{n-1}\sum_{k\in Z^4}\left\langle\hbar,\widetilde{\Upsilon}_{t:v,k}\right\rangle\Upsilon_{t:v,k}\right\|_2$$

$$\le \|\hbar-g\|_2 + C\|\hbar-g\|_2 = \|\hbar-g\|_2(1+C) < \varepsilon.$$

If $\{\Upsilon_{t:v,k}\}$ and $\{\widetilde{\Upsilon}_{t:v,k}\}$ are Bessel sequences, then equation (19) implies that both $\{\Upsilon_{t:v,k}\}$ and $\{\widetilde{\Upsilon}_{t:v,k}\}$ will be affine frames. In fact, the lower frame bound of $\{\Upsilon_{t:v,k}\}$ and $\{\widetilde{\Upsilon}_{t:v,k}\}$ is implied by the upper Bessel bound of the other. Therefore, this completes the proof of the second part of Theorem 4.

Theorem 5 [9]. If $\left\{\Gamma_\beta(x),\beta\in Z_+^4\right\}$ be a family of quarternary vector-valued wavelet packets with respect to the orthogonal vector-valued scaling function $\Phi(x)$, then , for $\beta,\gamma\in Z_+^4$, we have

$$\left\langle\Gamma_\beta(\cdot),\Gamma_\gamma(\cdot-k)\right\rangle = \delta_{\beta,\gamma}\delta_{\underline{0},k}I_s, \quad k\in Z^4. \tag{21}$$

5   Conclusion

A sort of pseudoframes for the subspaces of $L^2(R^4)$ are characterized, and a method for constructing a GMRS of Paley-Wiener subspaces of $L^2(R^4)$ is presented. The pyramid decomposition scheme is derived based on such a GMRS.

References

1. Chen, Q., Cheng, Z.: A study on compactly supported orthogonal vector- valued wavelets and wavelet packets. Chaos, Solitons & Fractals 31(4), 1024–1034 (2007)
2. Duffin, R.J., Schaeffer, A.C.: A class of nonharmonic Fourier series. Trans. Amer. Math. Soc. 20, 341–366 (1952)
3. Daubechies, I.: The wavelet transform, time-frequency localization and signal analysis. IEEE Trans Inform. Theory 36, 961–1005 (1990)
4. Casazza, P.G.: The art of frame theory. Taiwanese Journal, of Mathe-Matics 4(2), 129–201 (2000)
5. Benedetto, J.J., Li, S.: The theory of multiresolution analysis frames and applications to filter banks. Appl. Comput. Harmon. Anal. 5, 389–427 (1998)
6. Ron, A., Shen, Z.: Affine systems in L2(Rd) (II) Dual systems. Fourier Anal. Appl. 4, 617–637 (1997)
7. Li, S.: Pseudo-duals of frames with applications. Appl. Comput. Harmon. Anal. 11(2), 289–304 (2001)
8. Li, S.: A theory of generalized multiresolution structure and pseudoframes of translates. Fourier Anal. Appl. 7(1), 23–40 (2001)
9. Chen, Q., Shi, Z.: Biorthogonal multiple vector-valued multivariate wave-let packets associated with a dilation matrix. Chaos, Solitons & Fractals 35(3), 323–332 (2008)

A Flame Apex Angle Recognition Arithmetic Based on Chain Code

Yan Xiao-ling[1], Bu Le-ping[2], and Wang Li-ming[3]

Electric and Informational College, Naval Engineering University
Wuhan, Hubei, China
xuequnyan@yahoo.cn

Abstract. The article presents a new arithmetic which uses chain code to recognize fire flame's apex angles in fire detection processing. It applies a concept of custom slip window and defines flame's apex angle acutance index to estimate apex angle's type which could be I type or II type. This method also presents mathematics model and theory deduction and defines relative height parameter as the criterion of apex angles. The recognition method based on chain code is good at inflexion location with a higher speed and better anti-yawp capability rather than traditional method.

Keywords: flame apex angle, chain code, slip window, relative height.

1   Introduction

In fire detection process, if the object monitored is of huge area or space such as high building, garage, huge storage and big ship platform etc, burning fire's spread will be influenced by space's height and area, as a result, traditional smoke and temperature sensor will not make response until the fire runs to a degree that it can't be controlled. So, in huge space and complex environment, the traditional fire detection technique is difficult to play its part and new method needs to be researched.

Picture extends people's vision by which fire could be discovered immediately. The image information in fire detection can be timely received by optical camera. This information is very abundant and intuitionistic which is unique and will help to discover early fire, so this detection method based on images has a great dominance. If use this method, we must analyze fire flame's features that is rather complex. The features include color distribution, area increasing trend, edge's dithering frequency, fire apex angle's quantity and flame's texture etc. The article presents a new flame's apex angle recognition method based on its features.

2   The Mathematics Expression of Flame's Chain Code

Chain code is used to show the borderline which is made up of a group of order beeline segments that have specified length and direction. When detecting image's edge, after binary process, a close figure which is made up of chain code will be gained by tracking the edge. This chain code is of 4 directions or 8 directions. Each direction is coded by serial number, as a result, the image's feature information such as area change, phase change, body moving, apex angle num etc will be easily received by analyzing border chain code. Compared with 8 directions chain code, 4 directions chain code is a little rough. So, the article uses 8 directions chain code to track images for more precise result.

In Fig.1, the difference of two neighbor chain codes is defined as CR which denotes the image trend. If the direction is clockwise, the value is positive, otherwise, it is negative. If connect near border pixels with the same direction to short beelines, these beelines then form a polygon. The precision of this coding method depends on the size of sampling net. In practice, on the condition that error is controlled in a certain range, in order to reduce the number of chain codes, longer beelines are hence used to describe the border. First and foremost, combine chain codes with the same direction into a short vector bunch, and these short vector bunches will again be combined into a long vector bunch which error is controlled in a certain range. By this method, the positions of the most important inflexions in the image could be found and the edge information is also received. The border chain codes lie on the origination points and these points can be united by simple coding process. The chain codes can be regarded as circular sequence coded direction serial number and the origination point is defined again which makes these serial numbers have the least integer value.

In Fig.2, the area border which origination point is S has the anticlockwise chain codes of 5556666677666770001 11223222231334444. In these numbers, the evens stand for horizontal or vertical segments and the odds stand for diagonal segments. The length of horizontal and vertical segments is d which is decided by distance of two pixels and the diagonal segment's length is $\sqrt{2}d$. By this way, we can easily get the image's border perimeter and area size.

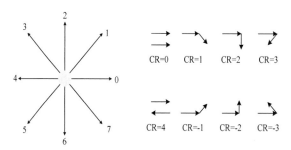

Fig. 1. The serial number of 8 directions chain code

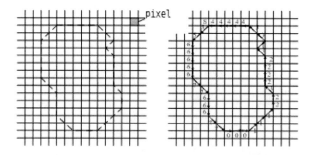

Fig. 2. Anticlockwise 8 direction chain code on sampling gridding

3   The Detection of Fire Flame's Apex Angle Amount

In video pictures, fire flame's image usually has several apex angles. The apex angle is an important shape feature of flame. In order to recognize the angles, practically, we'll compare two white pixels in the same row. This method can't locate the inflexion and the criterion is also not precise enough by the influence of the image size. The article presents a new method based on chain code.

3.1   Getting Border Chain Code after Segmentation

Border chain code describes object's shape by a serial of unit length lines with specified direction which number is coded as Fig.1. Fig.5 shows a sketch map of a flame border drawn by chain code sequences.

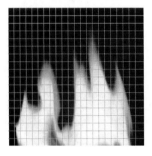   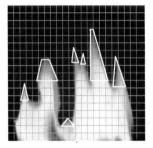

Fig. 3. Primal image   **Fig. 4.** Flame's apex angle

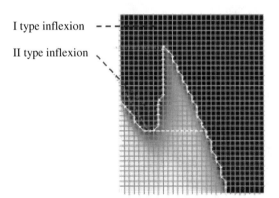

Fig. 5. Sketch map of border chain code

3	2	1
4	P0	0
5	6	7

Fig. 6. 3*3 neighbor region

(a)primary direction of route   (b)hypo-direction of rout

Fig. 7. Primary and secondary scanning method

Generally, the first element of the sequence has position information which can locate the entire object or rebuild region. The arithmetic to get border by chain code is as follows:

- Begin to search the image from top left corner until meet the first pixel P_0 of a new region R_i. P_0 is the pixel having the least line number and column number among all the pixels in region R_i. It is also the start of this region border defined as a parameter s which is set to 7. Inspecting border follows 8 directions chain code as Fig 1.
- Scan current 3*3 pixel region in accordance with the counter-clockwise shown as Fig 6. Scan direction is confirmed by s displayed in Fig 7. Twice-scan method is used here, primary direction is first and hypo-direction is later. The first pixel having the same value with current pixel is a new border element P_n and then update the value of s.
- If pixel P_n equaled to the second border element P_1 is found, furthermore, the previous border element P_{n-1} equals to P_0, the search stops. Otherwise, repeat the operation above. Finally, all the border pixels of P_0, P_1, ... P_{n-1}, P_n are achieved.

Because there will be some interferes or the noises produced by segmentation, the border chain code may be broken during the linking process. Hence need to define the allowable max step value of pixels. The article makes it 1.

3.2 Selecting Flame's Peak

The first and most important feature of a flame apex angle is its peak which is a local extremum point. There are likely several peaks in an apex angle. In perfect condition, there is only one extremum point in vertical coordinate. The article uses chain code direction to distinguish flame's peak.

If the border chain code uses counter-clockwise, the beginning point is in object's lower-left. Assume that the size of window decided by image's size and precision is N, the inflexions of 3-4-5 or 3-5 in this window are defined as flame's I type inflexion which is flame's peak. As shown in the Fig 8(a), N is 15, chain code 323234444456565 appears, and then, 44444 is regarded as peak. In Fig 8(b), 232323355566566 appears and 3-5 is peak. When there are some sporadic 3-4-5 or 3-5 inflexions, we take them for noises in the window and ignore them. At last, mark these points as suspect peaks of apex angle.

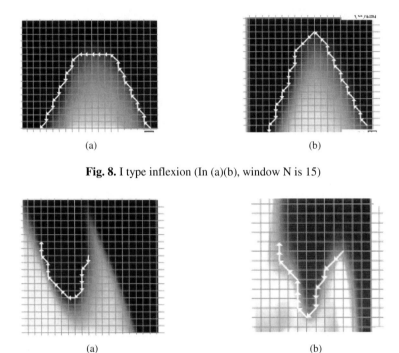

(a)   (b)

Fig. 8. I type inflexion (In (a)(b), window N is 15)

(a)   (b)

Fig. 9. II type inflexion (In (a)(b), window N is 15)

3.3   Selecting Bottom Inflexion

Searching bottom inflexion is also in the same window N. The inflexion of 5-4-3 or 5-3 is defined II type inflexion which commonly is the flame's bottom or juncture part. In Fig 9, window's size N is 15. In Fig 9(a), 65665433322322 appears and 556566532223322 appears in Fig 9(b), these points are regarded as apex angle's bottom and the whole flame's joint part. In the same way, the sporadic ones are also ignored.

Generally, the direction 4 is turned over to direction 0 to make the region an integrated periodic chain code. If there is no direction 4, the inflexion of 5-3 direction is reversed and the integrated periodic chain code will be obtained too.

Although getting I type and II type inflexions by method above is easy to understand, in practice, there is no quantifiable index. The article so advances flame's apex angle acutance index β.

Make window's size N be an odd number, $N = 2n + 1$.

Suppose border chain code length is M, if the window slides through all the border chain codes, the count of slide is $M- N$. Index β is defined as formula followed.

$$\beta_I = \frac{1}{2(N-1)} \sum_{i=0}^{N/2} S_{n+N-i} - S_{n+i} \tag{1}$$

$$\beta_{II} = \frac{1}{2(N-1)} \sum_{i=0}^{N/2} S_{n+N-i} - S_{n+i} \tag{2}$$

The value range of β_I is $\beta_I \in (0,1)$, in perfect condition that the flame shape is an isosceles right-angled triangle, β_I equals to 0.5. When flame shape is a vertical beeline, β_I max value is 1. The value range of β_{II} is $\beta_{II} \in (0,1)$ with perfect value is -0.5. In a word, when setting window size N in a appropriate range and defining the threshold value of β_I and β_{II}, the inflexion's type will then be confirmed. This method can also filter noises. When value is of max value, the middle pixel is an inflexion (I type or II type).

3.4   Experiments

In Table 1, it can be seen that the amount of flame's apex angles changes rulelessly, but the average value exceeds 6 rather than candle, dynamo and setting sun's less than 2. It is obvious that the incontrollable fire flame's apex angle amount is much more than stables', therefore, it is confirmed the feasibility of making the apex angle's amount as a flame recognition criterion. Meanwhile, the arithmetic in article is based on chain code which does sum operation only by judging chain code's direction, so this operation has a fast speed. Because there imports a relative height as a criterion, the ratio of miscarriage of justice is also greatly reduced. The method of distinguishing inflexion's type according to β_I and β_{II} threshold value can better remove the noises.

Table 1. Experiment of measuring apex angles' amount

category	Serial number of images								
	1	2	3	4	5	6	7	8	average
Flame apex angles' amount	3	5	3	7	6	7	9	10	6.25
Candle apex angles' amount	2	1	1	1	1	0	1	1	1
Dynamo apex angles' amount	1	1	1	1	0	0	1	0	0.625
Setting sun apex angles' amount	0	0	0	0	0	0	0	0	0

4   Conclusion and Expectation

For the fire detection by video in vessel's huge space, the article presents a flame recognition method based on its border chain code. This method applies the concept of a custom slip window, defines flame's apex angle acutance index β for inflexion type's estimation, educes its mathematic model and theory deduction. It also defines a relative height parameter as apex angle's criterion. This method can better locate the inflexion and have fast operation speed and high anti-noise capability.

There are still many researches needed to do for fire detection. For example, it needs to find some flame's other features to improve its recognition veracity. Then, these features' statistical distribution should be lucubrated to set up its statistical probability density model. In addition, flame's shape feature is also very important which is different from other objects'. Fourier descriptor is good at ascertaining flame's image, but its eigenvalue quantity is huge, and how to find their corresponding secondary characteristic is also an important content. When flame's edge is dithering, its luminance is also changing. Therefore, there must be some certain pertinences between them, as a result, it is deserved to research these pertinences furthermore for fire detection.

References

1. Sun, Y., Wang, J., Yu, Y. et al.: Discussion on the earl image-type fire detection and the fire fighting technology of the digital fixed fire monitor. Fire Protection Science and Technology 22(3) (2003)
2. Neubuaer, A.: Genetic algorithms in automatic fire detection technology. Genetic Algorithms in Engineer Systems: Innovations and Application, 15–16 (1997)
3. Fonseca, J.A., Almeida, L.M.: Using a Planning Scheduler in the CAN Network, pp. 485–490. IEEE (1999)
4. Ugur Toreyin, B., Dedeoglu, Y., Enis Cetin, A.: Computer vision based method for real-time fire flame detection. Sciencedirect (2005)
5. Ollero, A., Arrue, B.C., Martinez, J.R., et al.: Techniques for reducing false alarms in infrared forest-fire automatic detection systems. Control Engineering Practice 7, 34–40 (1999)
6. Yamagishi, H., Yamaguchi, J.: Fire flame detection algorithm using a color camera. In: Proceedings of 1999 International Symposium on Micromechatronics and Human Science (MHS 1999), Nagoya, Japan (1999)
7. Zhang, W., Yi, J.: Research of Intelligent Fire Alarm System Based on Fuzzy System And NN, vol. 5, pp. 698–704. Beijing University of Technology (2000)

Intelligent Assessment System of Mechanical Drawing Job Based on Entity Comparison

Wanjing Ding

School of Mechanic and Electronic Engineering
Huangshi Institute of Technology, Huangshi, P.R. China
rmding@163.com

Abstract. Proceeding from the reality of current college mechanical drawing teaching, this paper developed a assessment system software of mechanical drawing based on AutoCAD. General requirement is analyzed, design of typical function module is discussed and system simulation result is given in this paper. The experiment results show that, through parametric and modular design and using of entity comparison method, it can be a good mechanical drawing implementation of intelligence adjudication. The use of software system will greatly enhance the efficiency of learning mechanical drawing.

Keywords: Education, mechanical drawing, assessment system, parametric design.

1   Introduction

Engineering Drawing is opened as a technology foundation course for almost all of engineering knowledge. Earlier, the mechanical drawing teaching shows text, video, animation and other media teaching content in front of the students. Moreover, a number of manual or computer graphics work have to be done by students. This form of teaching can train students in cognitive ability and can achieve testing for the basic principles. However, it would clearly show the lack of interactive on the assessment of students' practical ability to draw. Meanwhile, it is unable to evaluate teaching effectiveness better. On the other hand, from the view of teaching, the heavy work of marking will make teachers be limited to waste energy on massive of simple work, not to mention to analyses reasons of error and improve the teaching method. In order to adapt to the current teaching form, developing a man-machine cooperation and student-oriented intelligent auxiliary teaching tool has important practical application. Wang Shengzhi and Zhang Jun et al. make parametric design by using AutoLISP[1,2]. Wang Xugang et al. discussed a method of correcting the transverse dimension of the shaft-like parts in the software aimed to correcting the machine parts drawing[3]. Liu Jiunu et al. discussed correcting assignment and graphical database in engineering drawing[4]. Xia Lili discussed feature-based automatic correction method of mechanical drawing homework[5].

Engineering Drawing is a very practical course. Students' practice takes in a large proportion. But, in order to train the students' space imagination, mapping exercises

Y. Wu (Ed.): International Conference on WTCS 2009, AISC 116, pp. 37–44.

always can not be standardized, it needs to set up some open questions. For these problems, wrong forms of the students are multiform, so the system must have the intelligence. Based on Visual LISP and AutoCAD, this paper developed a set of automatic measurement system of mechanical drawing. The system can automatically complete mapping errors identification, correction, and scoring. To a great extent, this improves students' learning efficiency and reduces the workload of teachers' correcting on the job.

2   Function Analysis and Design of the Test System

2.1   Description of General Function

According to actual needs, the general function of mechanical drawing assessment software based on AutoCAD and Visual LISP is shown in Table 1.

Table 1. Description of General Function

Function	Concrete Requirements
Drawing environment settings	set the drawing environment. （1）close AutoCAD command prompt; （2）maximize current window; （3）forbid layer operation; （4）call internal menu command (restore the original menu button and job submission button).
Copyright checking	the main task is to check whether the software copyright is genuine,if not ,complete the following tasks：（1）warning imformation; （2）uninstall the menu; （3）exit the AutoCAD。
Work checking	the main tasks of work checking are as follows：（1）scan the specified layer name layer; （2）Q & A layer of layer entity recognition; （3）compare the number of Q&A layer lines with the answer layer; （4）Entity identification; （5）highlight the errors of flashing Lines; （6）Some error message.
Work assessment	the main tasks of work assessment are as follows：（1）mark the work；（2）The overall assessment of the situation; （3）assessment of teaching effect.

2.2   Entity Comparison of Model

In order to design the system, a 4 layers mode is proposed. They are layer PROBLEM, layer KEY, layer ANSWER, layer SUPPORT. It is shown in figure 1. In layer PROBLEM there is a problem which is an unfinished draw. In layer KEY there is a key which is an absent part in layer PROBLEM. Students draw their line in the layer ANSWER. If the elements in layer KEY is the same as those in layer ANSWER, the answer is right, else the elements in layer SUPPORT will give some suggestions and help the students finish the draw.

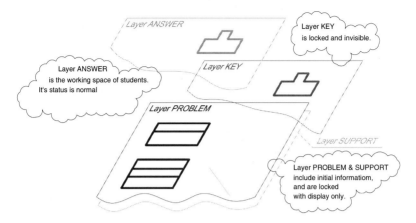

Fig. 1. Model of Entity Comparisom by Layers

2.3   Entity Comparison of Model

According to overall design and functional requirements, the entire system is divided into 11 major functional blocks, the division of modules and their specific functions are shown in table 2.

Table 2. Description of Module Function

Custom Function Module	Function Description
ScanLayer	（1）Retrieve all entities of the specified layer and constitute a selection set ;（2）Add entity name to a queue.
NumComp	Count the number of entities of Q&A layer and answer layer:（1）If the number of Q&A layer is zero, prompt the operating procedures ;（2）If the number of Q & A layer entities is greater or less than the number of answers layer entities, correspond the warning information; if the judge is equal ,go to judge the linear-entity.
EntComp	Entity identification:（1）Checking the illegal entity;（2）Judging errors of legal entities ;（3）Determine whether the legal entities of Q&A layer is the same as answer's and count the error number.If the number is zero,tell computer the question is answered accurately;if not, give the error message according to specific circumstances.
CkLin	Comparing properties of linear entities by retrieving the correlation tables.
CkArc	Comparing properties of arc entities by retrieving the correlation tables.
CkCir	Comparing properties of circle entities by retrieving the correlation tables.
AsEq	Determine whether the two entities mean the same point according to predetermined data errors.
Flash	Call delay module to make specified solid flare specified times.
Wait	Control the delay module of solid scintillation.
JustCheck	Copyright checking:（1）determine whether the software is genuine，if it is genuine, implement the function of job checking;（2）If the copyright is not genuine,then exit the task of mapping and unload the menu.
Tell***	This is a instruction group including 7 instruction,it is used to Pop warning and help imformation.The main instruction of the 7 are:(1) Pointed out the error; (2) When a question is finished correctly, prompt to do the next; (3) Show the message that the number of trial of this software has been to the warning times.

3   Algorithm Realization of Typical Functional Module

According to the definition of function module, software is divided into multiple functions and sub functions. Each sub function realizes a single function in order to facilitate the writing, maintenance and upgrading of the software. These sub functions are called by a unified main function.

3.1   Main Function Design

The main function is a frame, main function is to carry out environmental setting, copyright inspections and timely call another function module, and the work flow chart is shown in figure 2.

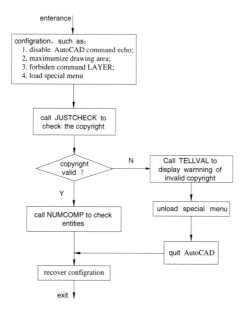

Fig. 2. Main function flowchart

3.2   Scanning Layer Module

The function of scanning layer module is to read a certain level in the entities list, the flowchart is shown in figure 3.

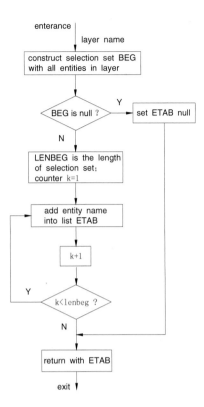

Fig. 3. The flowchart of scanning layer module

3.3   Entity Identification Module

The flowchart of the module is shown in figure 4, when cite the module, there are some instructions: (1) In function numcomp, when come to the judgement 'numans == numkey', fuction entcomp is called to compare Q & A layer's linear element name with the three entities LINE, ARC and CIRCLE. If they are the same,we have to compare the homonymic entities of Q&A layer with the entities of answer layer. (2) In the application of line, arc and circle checking module, parameter errnum is quoted, but parameter errnum is defined in the Layer entity recognition function module, therefore, line, arc and circle checking module refers to layer entity recognition module.

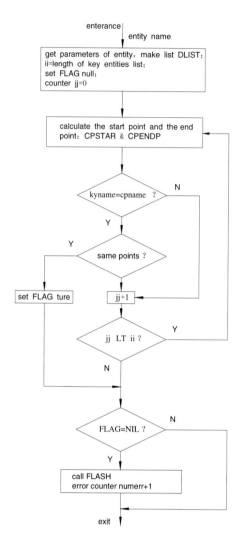

Fig. 4. The flowchart of line judge module

3.4   Closed and Branch Identification

By calling AsEq to judge whether 2 points is the same, we can grasp the branches and closed condition. The flowchart of the module is shown in figure 5. In CkLin, CkArc and CkCir function modules, we cite this module to judge whether the specified point of Q & A layer and answer layer is the same. If the absolute difference between two points is less than the error, we can consider that these two points is the same. Otherwise, we consider these two points is not the same, that is the dwarfed line, is not correct.

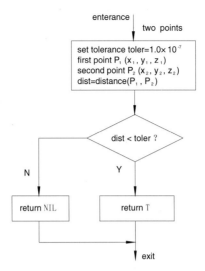

Fig. 5. The flowchart of judgment module whether 2 points is the same

4   Program Interface and Simulation Results

Using the debugging tools of Visual LISP integrated development environment to debug procedure, we can easily find the error. Completing the programming with the designed algorithm, we debugged the program; the results showed that the program achieved the expected goal.

With a simple example, the identification process of the various situations that may occur is described. (a) Initial interface of software loading, the pop-up menu for homework correcting is arranged, 3 menu items are Select Problem, Submission, and Unload additional menu. (b) When a user tries to call the LAYER command, shows the situation related to the software against the order to prevent the intentional or unintentional tampering with the contents of the original title. (c)When mapping correctly, computer shows the results after operation. (d) When exiting errors of the chart, computer shows the results after operation. (e) When have redundant or duplicate chart, computer shows the results after operation. (f) When chart missing, computer shows the results after operation.=

5   Conclusions

Through AutoCAD advanced development, we implement the functions of automatic identification, judgment and automatic testing in students' mapping. Meanwhile, the system can interactively to remind students of errors and support students learning mapping. The study will focus on the development of exercise base, designing network interactions and fault-tolerant processing in the future.

Acknowledgment. The research work is supported by:

a) The Key Discipline in Hubei Province. (Mechanical and Electronic Engineering)
b) The Outstanding Young Innovation Team Plans of Universities of Hubei Province. (Research and Application of the Distributed Virtual Experiment Technology)

References

1. Zhang, J., Zhang, H.: CAD Development of Drawing Software Based on AutoLISP. Coal Mine Machinary (8), 49–51 (2003)
2. Wang, S.: Drawing System with Parametric Design Based on AutoLISP. Journal of Fushun Petroleum Insititute 23(1), 69–72 (2003)
3. Xia, L.: Research on Feature-based Automatic Correction of Mechanical Drawing Homework. Huazhong University of Science and Technology, Wuhan (2004)
4. Wang, X., Jing, Q., Li, X.: Method on Verifying the Transverse Dimension of Shaft Parts Drawing in the Software to Correct the Machine Parts Drawing. Machine Tool & Hydraulics (9), 170–171 (2005)
5. Liu, J., Duan, N., Peng, X.: The Realization of Homework Intelligence Correction Arithmetic. Journal of Engineering Graphics (5), 137–143 (2005)

Closed Sequential Pattern Mining Algorithm Based Positional Data

Zhu Zhenxin[1] and Lü Jiaguo[2]

[1]Department of Computer, Hebei Vocational College of Politics and Law
Shijiazhuang, Heibei Province, China
[2]Department of Computer, Zaozhuang University, Zaozhuang, Shandong Province, China
zhuyongbin1025@sina.com

Abstract. This paper proposed a new closed sequential pattern mining algorithm.The algorithm used a list of positional data to reserve the information of item ordering .By using these positional data, we developed two main pruning techniques, backward super-pattern condition and same positional data condition. To ensure correct and compact resulted lattice, we also manipulated some special conditions. From the experimental results, our algorithm outperforms CloSpan in the cases of moderately large datasets and low support threshold.

Keywords: Data mining, sequential pattern, closed sequential pattern, backward super-patter.

1   Introduction

The early methods to mine sequential patterns [2] are modified from the Apriori algorithm [8]. Masseglia et al. [1] proposed a new data structure "prefix-tree" to save candidate sequences for speeding up support counting, but the mining algorithm is the same as that in [7]. So these methods all inherit the shortcomings of the Apriori. That is, all of them generate many candidate sequences and scan the database many times when a longer pattern exists.

To improve the efficiency of mining sequential patterns, there are two key points necessary to be overcome. The first is to decrease the number of generated candidate sequences, the second is to reduce the number of times of scanning the database.

In [4], it is shown that PrefixSpan is more efficient than Apriori-like methods [6] because of no candidate generation and shrinking search space. SPAM outperforms PrefixSpan and SPADE on reasonably large datasets due to efficient support counting [3]. However, SPAM needs more storage during the mining process than SPADE. In addition, all the methods described above find the full set of frequent sequential patterns. If a pattern and its support can be easily derived by its super-pattern with the same support, it is redundant.The closed frequent sequential patterns are the frequent patterns which have no frequent super-pattern with the same support. Without generating those redundant patterns, mining procedure is able to be more efficient. The algorithm of mining closed frequent sequential patterns was first addressed in [5]

Y. Wu (Ed.): International Conference on WTCS 2009, AISC 116, pp. 45–53.

and called CloSpan. Basically CloSpan just incorporates some pruning techniques into PrefixSpan to find the closed set of frequent sequential patterns.

Though most of previously proposed methods tackle the two factors in a certain degree, the property of item ordering in a sequence are not fully utilized in the mining process. Therefore, in this thesis, we propose a method called BFSM for mining the closed sequential patterns. Our method first finds the frequent sequences of length one, and records their positions in the sequences. By using positional data, we can do support checking efficiently. Also, the frequent sequences of length two generated by those of length one can be used to save search space while finding frequent sequences of length longer than two. Besides, some pruning methods according to support, containment, and positional data are used. That would lesson the effort to do redundant search, thus should be more efficient than previously proposed methods.

2   Preliminary Concepts

Let $I = \{i1, i2, ..., im\}$ be a set of items, an itemset is a subset of I. A sequence $S = (s1, s2, ..., sl)$ is an ordered list of itemsets. The length of S is l, which is the number of itemsets, and S is also called l-sequence. A sequence $a = (a1, a2, ..., ai)$ is a sub-sequence of another sequence $b = (b1, b2, ..., bj)$ if there exists a set of indices m1, m2, ..., mi, $1 \leq m1 < m2 < ... < mi \leq j$, such that $a1 \subseteq bm1, a2 \subseteq bm2, ..., ai \subseteq bmi$. On the other hand, b is called a super-sequence of a. We can say that b contains a, or a is contained by b. A sequence database D contains a set of sequences, and the support of a sequence S is the number of sequences that contain S. A frequent sequence (or sequential pattern) is a sequence with support not less than the minimum support threshold, min_sup. A closed frequent sequential pattern is a frequent sequence that has no frequent super-sequence with the same support.

3   Colsed Sequence Pattern Mine

3.1   First Stage and Item Ordering Representation

At first stage, we scan the sequence database once to record the positional information of every distinct item in the database. Then we can easily gain all frequent items or 1-sequences by conducting their positional information. The positional information of an item i, denoted by POSi, consists of a lot of pairs of (sid, eid), where sid is the sequence identifier and eid is the element identifier. Because sid indicates which sequence the item lies in and eid indicates in which order the item lies in the sequence, this representation can reserve the information of item ordering losslessly. Given a sequence database in Table1. and minimum support threshold equals 2, we can get the positional information of items as shown in Table 2. In the sequence sid = 1, item a appears in the first element, so a positional data (1,0) is obtained. The other positional data are obtained in a similar way.

Table 1. An example of sequence database

sid	Sequence
1	a (b d) c e
2	a (a d) b c e
3	c (b d) f

Table 2. The position information of items

Item	positional data	item	positional data
a	(1,0)(2,0)(2,1)	D	(1,1)(2,1)(3,1)
b	(1,1)(2,2)(3,1)	E	(1,3)(2,4)
c	(1,2)(2,3)(3,0)	F	(3,2)

To determine whether the item is frequent or not, according to this table, we just count how many different sids in the position data of the item and check whether the count is not less than the minimum support threshold. After that, we have the frequent items or 1-sequences <(a)>, <(b)>, <(c)>, <(d)>, and <(e)> whose supports are 2, 3, 3, 3, and 2 respectively.

3.2   Second Stage and Longer Sequence Generation

To gain frequent 2-sequences, we match each frequent 1-sequence with other frequent 1-sequences same as the Apriori-like methods. But we do not need to scan the database to count their supports. The positional information of frequent 1-sequences can be utilized to get the positional information of frequent 2-sequences. To match the positional data POS_i and POS_j of two items i and j, we consider two conditions: item-extension and sequence-extension. The item-extension adds a new item to the last element of a sequence. The sequence-extension adds a new element with a new item to a sequence as the new last element. For item-extension, we gather every (sid, eid) pairs which appear both in POS_i and POS_j, because they must be in the same sequence and the same element. For sequence extension, we gather every (sid, eid) pair in POS_j each of which has at least a (sid', eid') in POS_i with sid' = sid and eid' > eid, because they must be in the different element of the same sequence and one's position should be after another's. Then for a 2-sequence S composed of item i and j, the position data of S, POS_S, is the resulted (sid, eid) pairs after matching POS_i and POS_j. If i and j is in the same element, we choose item-extension matching; otherwise, we choose sequence-extension matching. Take item a and b in Table 2 for example, we can generate the 2-sequence <(a)(b)> by sequence extension. In the positional data of <(b)>, (1,1) matches (1,0) because they have the same sid and the eid of the former is greater than that of the later. Similarly, (2,2) matches (2,0), but (3,1) has no matched positional data because no positional data of <(a)> has sid = 3.

Thus the positional data for <(a)(b)> is (1,1), (2,2). For another example, we can generate the 2-sequence <(bd)> for item b and d in Table 2 by item extension. Because in the positional data of <(d)>, (1,1) and (3,1) have matched positional data in <(b)> but (2,1) has not, the positional data for <(bd)> is (1,1), (3,1). The support counting for 2-sequences is the same as that of 1-sequences, thus the support of <(a)(b)> is 2 and so is <(bd)>. The frequent 2-sequences and their positional information for the sequence database in Table 1 are listed in Table 3.

Table 3. The frequent 2-sequences

2-sequence	positional data	2-sequence	positional data
<(a)(b)>	(1,1)(2,2)	<(b)(d)>	(1,1)(3,1)
<(a)(c)>	(1,2)(2,3)	<(b)(e)>	(1,3)(2,4)
<(a)(d)>	(1,1)(2,1)	<(c)(e)>	(1,3)(2,4)
<(a)(e)>	(1,3)(2,4)	<(d)(c)>	(1,2)(2,3)
<(b)(c)>	(1,2)(2,3)	<(d)(e)>	(1,3)(2,4)

The frequent (k+1)-sequences can be generated by extending the currently found frequent k-sequences. we utilize frequent 2-sequences to generate frequent sequences of length longer than two. For each frequent k-sequence α, we pick up every frequent 2-sequence βwhose first item is the same as the last item of α, and match the positional data of α and β to obtain the positional data of newly generated (k+1)-sequence. The matching also has two possible conditions: item-extension or sequence-extension according to the compared frequent 2-sequence.

3.3   Third Stage and Search Space Pruning

If we just continually find longer frequent sequences as stated above, we would gain all frequent sequential patterns. In comparison with the PrefixSpan algorithm, the only pruning effect our algorithm lies in the utilization of frequent 2-sequences. So we can save the effort to do projections and cost no space for projected databases. To get closed frequent sequential patterns, we need some other pruning techniques.

Definition 1(backward super-pattern condition) Given two sequence s and s', if s' is a super-sequence of s which is generated before s', and they have the same hash key and last items, then we say s' is a backward super-pattern of s. Note that the hash key is composed of support, sum of sids, and sum of eids.

Definition 2(same positional data condition) Given two sequence s and s', if they do not have containment relationship but have the same hash key and last items, then s and s' have the same positional data. It implies they have the same projected database.

According to the data listed in Table 2 and Table 3, we illustrate the backward super-pattern condition in Figure 1 and same positional data condition in Figure 2. Because sequence <(a)(e)> contains sequence <(e)> and their support, sum of sids, sum of eids, and last item are all equal, we change the upper link for the node representing <(a)(e)> to the node representing <(e)>, and mark the upper link for the node representing <(e)> by changing it a dashed directed line. For the same positional data condition, the marking is needless because they can not prune each other.

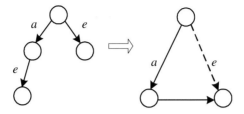

Fig. 1. Lattice update for the backward super-pattern condition

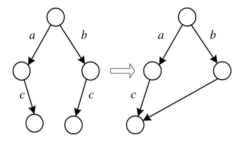

Fig. 2. Lattice update for the same positional data condition

Because the average capacity of bucket in hash table can influence the efficiency of closed set checking, we choose support, sum of sids, and sum of eids to compute the hash key, which is better than the hash key with only support or sum of eids according to their distribution.

3.4   Special Conditions Handling

For the same positional data condition, there may be two special conditions. For the sequence database in Table 1, two special conditions are depicted in Figure 3 and 4. The meaning of dashed directed link in Figure 3 and 4 are the same as that mentioned previously in Figure 1.

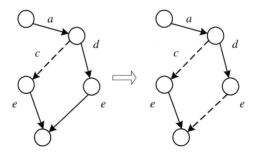

Fig. 3. Special condition 1

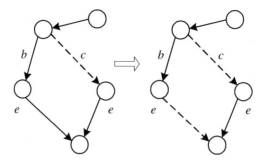

Fig. 4. Special condition 2

In the special condition 1, sequence <(a)(c)(e)> and sequence <(a)(d)(e)> meet the same positional data condition. However, sequence <(a)(d)(c)> has been checked as the backward super-pattern of sequence <(a)(c)> and then they share the same represented node. The sequence <(a)(d)(e)> would be pruned by sequence <(a)(d)(c)(e)> if we traverse the lattice to get closed set finally. So we mark the upper link of the node representing <(a)(d)(e)> to prevent generating it when traversing the lattice to get all closed frequent sequences. The special condition 2 is like the special condition 1 with the difference that the pruned sequence <(a)(b)(e)> is generated ahead the sequence <(a)(c)(e)>.

4   Closed Sequence Pattern Mining Algorithm

According to the results in Table 2 and 3, we get the temporary lattice for the sequence database in Table 1. The character attached to each link in the lattice is the item to be extended by traversing this link, and the subscript of the character is "i" for item-extension or "s" for sequence-extension.

Algorithm BFSM

Input: a sequence database D, and minimum support threshold min_sup.
Output: a lattice L.
1. Scan D once to find all frequent items (1-sequences) and gather their positional data.
2. Add these frequent 1-sequences to L and a hash table.
3. For each frequent item, match it with all other frequent items to generate all frequent 2-sequences and their positional data.
4. Add these frequent 2-sequences to lattice and perform closed set checking. If they meet the backward super-pattern condition or the same positional data condition, update corresponding link in the lattice. Otherwise, add them to a collection P for extension.

5   Repeat the Following Steps Until P Is Empty

5.1   For Each Sequence s in P, Perform the Following Steps

(1) Extend s with the frequent 2-sequences and their positional data, and get the new sequence s' of length longer than s by one.

(2) Check backward super-pattern condition and same positional condition for s' by the hash table.

(3) If s' meets any of the two conditions, update corresponding links in L and check the three special conditions if necessary. Otherwise, add s' to N.

5.2   Let P = N and Clear N

Sequence <(a)(e)> is a backward super-pattern of <(e)> and so are <(b)(e)>, <(c)(e)>,and <(d)(e)>. Thus the link from the root to the node for <(e)> is dashed, and the nodes for <(a)>, <(b)>, <(c)> and <(d)> all have a sequence-extension link to the node for <(e)>. These super-patterns need not to be extended any more, because by such linking they share the sub-lattice of <(e)>. Sequence <(a)(c)>, <(b)(c)> and <(d)(c)> meet the same positional data condition, so they all have a sequence-extension link to the node for <(a)(c)>. The sequences <(b)(c)> and <(d)(c)> are not necessary to be extended by sharing the same sub-lattice with <(a)(c)>.

The left sequences can be further extended to longer sequences in third third stage are <(a)(b)>, <(a)(c)>, <(a)(d)>, and <(bd)>. To extend <(a)(b)>, we can use <(b)(c)>, <(bd)> 17 and <(b)(e)> because its last item is b. The others can be extended in a similar way. The frequent 3-sequences generated in that way are listed in Table 4. After checking and updating, the new lattice is shown in Figure 5, which is also the final lattice.

Table 4. The frequent 3-sequences generated in the third stage

frequent 3-sequence	positional data
<(a)(b)(c)>	(1,2)(2,3)
<(a)(b)(e)>	(1,3)(2,4)
<(a)(c)(e)>	(1,3)(2,4)
<(a)(d)(c)>	(1,2)(2,3)
<(a)(d)(e)>	(1,3)(2,4)

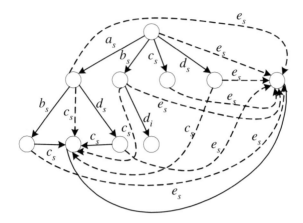

Fig. 5. The resulted lattice

6   Experiments and Performance Evaluation

In this section, we evaluate the performance of our algorithm with respect to the CloSpan algorithm. All of the experiments are performed on a Pentium IV 2.4G PC with 1G main memory, running Windows XP Professional. The algorithms are implemented with JDK 1.6

We use a synthetic data generator provided by the IBM. The synthetic data generator can be downloaded from an IBM website. It can accept parameters. It also allows the user to define parameters of frequent patterns in the dataset and their correlations.

We compare the performance of our algorithm BFSM (BF) to the CloSpan(CS) algorithm by varying minimum support threshold for the dataset D10C10T2.5N10S6I2.5 (NS2000NI5000). BF is about 2~5 times faster than CS as shown in Figure 6. This means BF can perform well even when the support threshold is low. This is because pruning techniques of BF are better than those of CS.

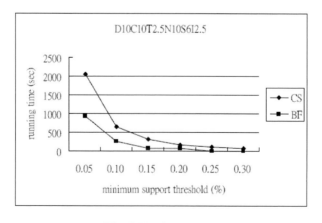

Fig. 6. Varying support

7   Summary

In this paper, we use position lists to store the positional information of sequences. Then we can use the generated frequent 2-sequences and their positional data to reduce the search space for generating longer frequent sequences. For finding closed frequent sequences, confined to the breadth-first property, we can only utilize the backward super-pattern condition suggested in [5]. However, we propose the same positional data condition which saves a lot of search space also. That is based on the observation that if the sequential patterns with the same positional data and last item, they have the same projected database. Besides, the problem that some backward super-patterns can not be utilized mentioned in [5] is easily solved by our breadth-first approach. From the experiment results, we found that in some cases, we may have worse performance. But in most of cases, we can run a lot faster than the CloSpan algorithm.

Acknowledgment. The author gratefully acknowledges the support, both monetary and spiritual, that Mr. Jack Rudin and the May & Samuel Rudin Family Foundation have extended to the author, the Shijiazhuang tiedao University Resource Center. The author also thanks the students of Computers, Technology, and Society, Fall 2007, for their encouragement and their reactions to drafts of this article. The author especially thanks. Ms. Elizabeth Cassara for sharing with me her experiences and insights regarding data mining software.

References

1. Ayres, J., Flannick, J., Gehrke, J.: Sequential Pattern Mining Using a Bitmap Representation. Knowledge Discovery and Data Mining 12(6), 429–435 (2008)
2. Pei, J., Han, J., Mortazavi-Asl, B.: PrefixSpan: Mining Sequential Patterns Efficiently By Prefix-projected Pattern Growth. Data Engineering 8(4), 215–224 (2006)
3. Han, J., Pei, J., Mortazavi-Asl, B.: Freespan: Frequent Pattern- projected Sequential Pattern Mining. Knowledge Discovery and Data Mining 14(8), 355–359 (2007)
4. Han, J., Pei, J., Yin, Y.: Mining Frequent Patterns Without Candidate Generation. Management of Data 13(5), 1–12 (2006)
5. Zaki, M.J.: SPADE: An Efficient Algorithm for Mining Frequent Sequences. Machine Learning 11(5), 31–60 (2001)
6. Leleu, M., Rigotti, C., Boulicaut, J.-F.: GO-SPADE: Mining Sequential Patterns over Datasets with Consecutive Repetitions. Machine Learning and Data Mining 18(7), 293–306 (2003)
7. Pasquier, N., Bastide, Y., Taouil, R.: Discovering Frequent Closed Itemsets for Association Rules. Database Theory 6(1), 398–416 (1999)
8. Yan, X., Han, J., Afshar, R.: CloSpan: Mining Closed Sequential Patterns in Large Datasets. Data Mining 16(5), 40–45 (2003)

Research and Design of Multi-network Management System

Zhang Yu

College of Computer and Information Engineering
Harbin University of Commerce, Harbin, China
zhangyu_20@sohu.com

Abstract. The system is to provide users with their own space, their own users can publish articles can be written as a diary, like running our own affairs are written down, they can also upload their own photos and captions, and as patrons at the same time it has its own message of this can, at any time, and communicate with their friends, So each class management system can be released with information, upload photos, the voice of the management, Login password management, and other functions. The system is the key to every level of management systems, the because of the higher level of management systems have lower management system operating in all authority, had formed as a tree structure similar superior management systems can have a number of lower-level management system that is the model number of subsystems gradually downward recursive.

Keywords: Network Management System, Multi-level association, ADO.

1 Introduction

Currently, most sites are primarily single management system, because most of them are used to disseminate information or a number of external development services. Not with tourists and people who visit the website full of interactive or for the visit to provide a relatively free space can make their own decisions. This makes most people just browse, download etc... With more and more people demand high-line, this single system can not meet.

Web site now mainstream management systems are mostly developed using ASP's. When people first heard the term ASP, it is designed to provide remote application outsourcing services. It seems in many professionals, ASP concept is tantamount to providing application hosting ERP or similar applications in other ASP service delivery model. The beginning, almost no attention to the field of vertical or specialized skills, analytical skills or basic services, value-added factor. ASP now offers hosted applications; the concept has not caused clients, analysts or investors attention. ASP's competitive landscape has changed dramatically, ASP business model to build sustainable companies than the third-party applications, the company managed a lot more. [6]

As in the previous section analyzed, managed only third-party ISV software development and management of the enterprise there are many potential problems.

Y. Wu (Ed.): International Conference on WTCS 2009, AISC 116, pp. 55–61.
springerlink.com   © Springer-Verlag Berlin Heidelberg 2012

Today, the new ASP and investors generally recognized that the establishment of a sustainable development company, to provide additional value-added factor to the establishment of long-term strategic customer relationships. Development of services from basic ASP value-added services for a range of potential trends or characteristics, will enable customers to gain more benefits and to competitors into the higher threshold. Although it is difficult to make out all possible types of value-added factor, but Figure 1 summarizes the maximum response to current market some of the major types of value-added services. [6]

Software standardization is a sign of maturity, but too stringent rigid, stereotyped ASP standard mode, faced with thousands of personalized service the requirements of small and medium enterprises has become quite helpless, and stressed the personal costs of business development has increased, resulting in supply and demand both asymmetric. [5]

2   System Implementation of the Relevant Theory

2.1   ASP Theoretical Basis

ASP for the ACTIVE SERVER PAGES (active server page) is referred to, included in Internet Information Server (IIS) which provides a server-side (server-side) of the scripting environment you create and implement dynamic, interactive, high-efficiency site server applications. You do not worry about whether the implementation of your browser designed for Active Server Pages, your site servers automatically Active Server Pages of program code, interpreted as a standard HTML format of the page content, delivered to the client browser display. The client code as long as the use of conventional HTML browser executable, you can visit the Active Server Pages designed home page content.

ASP is Microsoft's development of the CGI script instead of an application, it can interact with databases and other programs, is a simple and convenient programming tools. ASP web page file format is. Asp, are now commonly used in a variety of dynamic website. ASP is a server-side scripting environment, can be used to create and run dynamic Web pages or web applications. ASP pages can contain HTML tag, plain text, script commands and COM components. ASP can be used to add interactive content to the web page (such as online forms); you can create using the HTML page as a web application user interface.

ASP pages have the following characteristics:

(1) The use of ASP pages can breakthrough some of the features of static constraints, dynamic web technologies;

(2) ASP file is included in the HTML file, which consists of code, easy modification and testing;

(3) Explain the procedure on the server ASP server-side development in the ASP program, and the results in HTML format to the client browser, so the browser can use a variety of normal browsing ASP pages generated;

(4) ASP provides some built-in objects, these objects can use server-side scripting more powerful. For example, you can get from the web browser user information

submitted through HTML forms, and in the script to process such information, and then send the information to the web browser;

(5) ASP can use server-side ActiveX set to perform a variety of tasks, such as access to databases, send Email, or access to the file system.

(6) As the server is the result of the ASP program execution in HTML format back to the client browser, so users do not see the original written by ASP code, ASP code can be placed stolen.

(7) Using conventional text editor such as Windows Notepad, you can design.

(Eight) the use of server-side script generate client-side script, you can use the ASP program code, the site server implementation of the script language (VBscript or Jscript), to create or change the script in the client-side implementation of the language.

First to install the Windows 2000/XP built-in IIS as a server.

Here because my machine is XP so all shots are done under XP.

Control Panel ->> Add or Remove Programs.

Then add remove windows components - IIS components selected in front of the hook, and then wait for the installation, the installation is complete.

After the Control Panel, double-click "Administrative Tools."

Then double-click "internet information services", which is IIS.

Select the "Default Website", then right property or simply pressing the shortcut keys in property pages, only three pages need to be amended first site, if the user has a fixed IP, you can assign an IP next home directory in the note, select a hard drive above their storage site folder, select the Read, Write.

2.2 ASP Database Access

ADO (ActiveX Data Objects, ActiveX Data Objects) is Microsoft proposed application program interface (API) to achieve access to relations or relational database data. For example, if you want to write applications from DB2 or Oracle database to provide data to the web page can be included in the ADO program as the active server page (ASP) in the HTML file. When a user requests a page from the site, return to the page also includes data in the corresponding data, which is due to the results of using ADO code.

Interfaces like Microsoft, like other systems, ADO is an object-oriented. It is the Microsoft Global Data Access (UDA) part, Microsoft that its own to create a data access than the existing database using UDA. To that end, Microsoft and other database companies in their database and Microsoft's OLE databases provide a "bridge" program, OLE database already using ADO technology. ADO's a feature (called Remote Data Services) to support the data in the relevant web page ActiveX controls and efficient client buffering. As part of ActiveX, ADO is Microsoft's Component Object Model (COM) as part of its framework for the components assembled together for the program. ADO data interface from the original Microsoft Remote Data Objects (RDO) from. RDO to work with ODBC to access relational databases, but can not access, such as ISAM and VSAM non-relational database.

ADO is a Microsoft-supported database currently operate the most effective and straightforward way, it is a powerful data access programming model, allowing most of the data source property to programmable directly into your Active Server page. You can use ADO to write compact and simple script to connect to the Open Database Connectivity (ODBC) compliant database and OLE DB-compliant data source, such ASP programmers can access any ODBC-compliant databases, including MS SQL SERVER, Access, and Oracle and so on.

For example, if the site developers need to allow users to access pages to get there in the IBM DB2 or Oracle database data, then the ASP page can be included in the ADO program, used to connect to the database. Thus, when a user browsing the site, return to the page will contain data obtained from the database. These data are from the ADO code to do so.

ADO is an object-oriented programming interface, Microsoft explained that its advocates with IBM and Oracle as to create a unified database access than to provide a unified interface to different databases, it would be more practical number. To achieve this goal, Microsoft database and Microsoft's OLE DB to provide a "bridge" program, this program can provide connection to the database.

When using the ADO developers, in fact, the use of OLE DB, OLE DB, but closer to the bottom. As part of ActiveX, ADO is part of COM components. ADO is the data interface from the earlier Microsoft Remote Data Objects RDO evolved. RDO with Microsoft's ODBC connectivity with relational database, but can not connect to non-relational database.

ADO provides us with a familiar, high-level package on the OLE DB's Automation Interface. For those familiar with the RDO of the programmers, you can compare is the OLE DB ODBC driver. As RDO ODBC driver interface object is the same, ADO object is OLE DB interface; as different database systems require their own ODBC drivers the same, different data sources required their own OLE DB provider (OLE DB provider). At present, although less OLE DB provider, but Microsoft is actively promoting the technology, and intended to use OLE DB to replace ODBC.

ADO VB programmers to provide many benefits. Including easy to use, familiar interface, high speed and low memory footprint (the Msado15.dll realized ADO2.0 need occupy 342K of memory, the 368K than the RDO's Msrdo20.dll slightly smaller, about DAO3.5 the Dao350. dll memory of the 60% share). Level with traditional data objects (DAO and RDO) different, ADO can be independently created. So you can just create a "Connection" object, but can have multiple, independent "Recordset" object to use it. ADO for client / server and WEB applications are optimized. Interfaces like Microsoft, like other systems, ADO is an object-oriented. It is the Microsoft Global Data Access (UDA) part, Microsoft that its own to create a data access than the existing database using UDA. To that end, Microsoft and other database companies in their database and Microsoft's OLE databases provide a "bridge" program, OLE database already using ADO technology. A feature of ADO (called Remote Data Services) to support the data page associated ActiveX controls and effective client buffer. As part of ActiveX, ADO is Microsoft's Component Object Model (COM) as part of its framework for the components assembled together for the program.

3 Detailed Design and System Implementation

3.1 The Overall System Structure Design Background

Because it is always the background, so it is necessary to manage their own articles published in the lower back also can manage user articles published, and is able to lower the background user review articles published in order to decide whether to display the site's front desk, and decided that Articles can be commented by others. On this basis all the articles can be modified and deleted, but also add a new article.

This module is used to manage the junior user's message in this, the user can always observe the contents and circumstances of this message to be processed in a timely manner, such as a message to delete the changes and so on, or prohibit a message.

Image Management module is mainly responsible for the management of users to upload pictures, pictures that do not meet the requirements to delete files. The starting point is the image upload, submit to upload files, so that after the upload the pictures uploaded to the server specified directory, and rename it's. At the same time the long pass the file path and upload the file name and other information submitted to the database to save, so the operation can be achieved on a database managed.

3.2 System Design and Session Back the Login Page Authentication

Backend system administrator login is smoothly into the system management page is designed also to prevent the non-management staff is able to modify the system information and the establishment of mechanisms. And more security considerations for the administrator log in when the session information to add, so that the background of all pages are added on the session of the certification process, which can identify whether the person into the management page is a legal status, if than through the normal channels are not open person logging management page.

Add session information page code:

```
<%
if Request ("post") ="login" then
Set conn = Server.CreateObject ("ADODB.Connection")
DBPath = Server.MapPath (".../tdht/lx.mdb")
conn.Open "driver= {Microsoft Access Driver (*.mdb)}; dbq=" & DBPath
Set rs_admin = Server.CreateObject ("ADODB.Recordset")
sql="select * from sadmin where user_name like '" & Request ("username") & "'
and psw like '" & Request ("userpass") & "'"
rs_admin.open sql,  conn,  3,  2
if rs_admin.eof or rs_admin.bof then
error = "incorrect user name or password!"
Else
session ("idpass") = rs_admin ("id")
session ("adminpass") = rs_admin ("user_name")
session ("level") = rs_admin ("level")
response.redirect "sadmin_index.asp? id =" & session ("idpass")
End if
```

```
End if
if Request ("post") = "quit" then
Response.Cookies ("username ") =""
Response.Cookies ("userpass ") =""
error = "quit successfully!"
End if
%>
```

Management page session authentication code as follows:

```
<%
  if session ("adminpass ") ="" then
  errstr = "You did not pass the user authentication!"    response.write errstr
  response.end
  search_txt=request ("search_txt")
end if
%>
```

3.3 Database Connection

The system uses a database connection method ADO specific code:

```
<%
Set Conn = Server.CreateObject ("ADODB.Connection")
conn="DBQ=" + server.mappath ("tdht/lx.mdb") + "; DefaultDir=; DRIVER=
{Microsoft Access Driver (*.mdb)};"
%>
```

4 System Testing

Software testing is put into operation in the software before the software requirements analysis, design specifications and coding of the final review, which is a key step in software quality assurance. Testing requirements in a few cases, time and manpower to find all kinds of potential software errors and defects, to ensure that the quality of the system.

Because there are many procedures often unexpected problems, may be imperfect, many hidden errors only in certain circumstances it may be exposed. If you do not find the error to focus on such a basis as possible, these hidden errors and defects on the check does not come out, will be left to run to the stage. If the user's point of standing for their ideas, testing activities should be targeted to expose the program for errors. In the selected test case, consider those easy-to-find bug data.

Following these rules can also be seen as testing the purpose or definition:

(1) Software testing is to find errors and implementation procedures of the process;
(2) Test procedures to prove that wrong, not that process error-free;
(3) A good test case is not found so far is that it found the error;
(4) A successful test is found so far found no error in the test.

The test correctly defined as "to find errors in the program and the process of implementation procedures" and some people often imagine that the "test is to show that the program is correct," "The successful test is not found the wrong test, "and so is the opposite. Correctly understand the goal of testing is very important to test targets the test program design decisions. If the order to show that the program is correct and the test will be to design errors not easily exposed to the test program; the contrary, if the test is to find errors in the program will seek to design the most exposed to the wrong test program.

As the test goal is to expose errors in the program, from a psychological point of view, by the authors of their own testing procedures is not appropriate. Therefore, comprehensive testing phase is usually composed by other test team to complete the test. In addition, it should be recognized that testing procedures must not prove to be correct. Even after the most rigorous testing, may still have not found the error hidden in the program. Test only to find out the errors in the program, the program can not prove that there is no error.

5   Conclusion

The system uses ASP technology to the development of a multi-level background associated with web management network system. He can get management to facilitate various functions, and users with lower level auditing features; you can get things at lower levels uploaded to strict regulation in order to prevent improper things not suitable for exposed. For example, article management, and lower levels after the user does not immediately add the article shown in the foreground, only one user may view and allow the display case to another front page is displayed. Because the Internet and reality is already associated with general two meet the basic requirements of users, and also easier to achieve two related, so this also two related systems as the basis for the development. Such a system available on the Internet have a lot of space.

Acknowledgments. The work is supported by the Natural Science Fund in HeiLongjiang province under Grant Nos. F200907.

References

1. Rui new editor ASP Web Site and the examples that Graduate Design Machine Press (January 2006)
2. Guorui, J., Li, J.: The early strength of the essence of Database Development ASP Hundred Review Electronics Industry Press (May 2005)
3. Kui spring game editor ASP System Selected examples of Machinery Industry Press (April 2006)
4. Yu, Z.: Web News Publishing System Design and Implementation (August 2006)
5. Chen, S.: ASP: a kind of new SME Information Technology and Management Solutions (January 2006)
6. Song, K.: ASP model of status and application of technology information (June 2007)
7. Shao, Y.: ASP-based research network of office systems and application technology consultation Guide (November 2007)

The Design of Linux-Based E-Mail System

Zhang Yu

College of Computer and Information Engineering
Harbin University of Commerce, Harbin, China
zhangyu_20@sohu.com

Abstract. The rapid development of the computer networks has been making a great influence in many fields of human society. Among a great many networks operating systems, Linux is obtaining more and more attention from people by its excellent behavior. The task is to develop an e-mail system based on a lan with Linux as its operating system. This article includes two parts. In part one; we will make an introduction to Linux operating system and the network programming environment in Linux. In particular, the BSD socket functions are analyzed in great detail, for they will be used in later programming work. In part two, we present the implementation of our mail system, from whole designing to detail designing, and the C program source files written by me is attached to the end of this article.

Keywords: Linux, network programming, socket, e-mail.

1   Introduction

In the rapid development of the Internet, there is such a group of people, they are a team of expert programmers, amateur players and computer hackers team composed of strange, completely independent functions developed equally well in Microsoft's new free UNIX operating systems - Linux, a network of a force to be reckoned with, just a few years became a strong rival of Microsoft. In 1991, a young man from the University of Helsinki, Finland, Linus Benedict Torvalds found it in practice Minix function are far from perfect, so determined to write a protected mode operating system, which is the prototype of Linux. With the constant sound after, and the GNU software support, LINUX is already a powerful class of UNIX operating systems. It inherited all the advantages of UNIX, but also has the largest open source features, over the past decade it has made remarkable achievement, but also faster to develop. With the development of computer networks, information transfer between people much shorter time. Many documents are sent by e-mail form; often used the computer more or less will use Email to transmit information. By e-mail, people can be text, pictures, video, sound, transmission of data files, etc... With the Internet network and the wide spread of WWW network, rapid increase in the use of e-mail them. E-mail is used not only in quantity have rapid development, its importance is growing. Of course, e-mail for people with adverse side. Receive e-mail because of its few limitations, resulting in computer viruses, a large number of spam popularity, or even personal privacy and security is a serious threat. However, the e-mail as a modern

Y. Wu (Ed.): International Conference on WTCS 2009, AISC 116, pp. 63–70.

society the main carrier of information dissemination, development trend will not stop. The rapid development of computer networks, many areas of human society has a huge impact.

2   BSD Socket Function Description

2.1   Function Recv (), and Send ()

Function recv () function and read () is similar to it in the read () function, based on an increase of four parameters used to control the socket read operation, the function form as follows:

```
# Include <sys/types.h>
# Include <sys/socket.h>
int recv (int sockfd, void * buf, int len, int flags);
```

Function send () function and write () is similar to it in the write () function, based on an increase of four parameters used to write socket control, functional form as follows:

```
# Include <sys/types.h>
# Include <sys/socket.h>
int send (int sockfd, void * buf, int len, int flags);
```

2.2   Function Close ()

Function close () to close a socket descriptor. Socket descriptors and file descriptors similar operation. The form:

```
# Include <unistd.h>
int close (int sockfd);
```

Usually close to close a TCP connection, close to return immediately. Process will no longer use the socket descriptor to access the socket, but TCP may not remove the socket structure, which may send data buffer in the data is not there to send finished. TCP will continue to send the remaining data and the data in the final section of additional FIN control information.

The socket is the basic function of a few, several functions can be used for more than a basic server and client correspondence. Following the above function gives a simple example application: Implementation by the server to the client sends a string "HELLO WORLD!". This is the socket network programming simple application, followed by implementation of the mail system is the basis of some changes and added functionality.

Server code [3]:

```
# Include <stdlib.h>
# Include <errno.h>
# Include <string.h>
# Include <sys/types.h>
# Include <netinet/in.h>
```

```
# Include <sys/socket.h>
# Include <sys/wait.h>
# Define MYPORT 4050
# Define BACKLOG 10
main ()
(Int sockfd, new_fd;
    struct sockaddr_in my_addr;
    struct sockaddr_in their_addr;
    int sin_size;
    if ((sockfd = socket (AF_INET, SOCK_STREAM, 0 ))==- 1) (
        perror ("socket");
        exit (1);
    )
    else (printf ("socket created successfully! \ n");)
    my_addr.sin_family = AF_INET;
    my_addr.sin_port = htons (MYPORT);
    my_addr.sin_addr.s_addr = INADDR_ANY;
    bzero (& (my_addr.sin_zero), 8);
    if (bind (sockfd, (struct sockaddr *) & my_addr, sizeof (struct sockaddr ))==- 1)
    (
            perror ("bindB");
            exit (1);
    )
    else printf ("address binded successfully! \ n");
  if (listen (sockfd, BACKLOG) ==- 1)
    (
            perror ("listen");
            exit (1);
    )
    else printf ("start listening the clients' connect request ...... \ n");
    while (1)
    (
    sin_size = sizeof (struct sockaddr_in);
        if ((new_fd = accept (sockfd, (struct sockaddr *) & their_addr, &
sin_size))==- 1) (
            perror ("accept");
            continue;
        )
        printf ("server: got connection from% s \ n", inet_ntoa
(their_addr.sin_addr));
        if (! fork ())
    if (send (new_fd, "Hello, World! \ n", 14,0) ==- 1) (
        perror ("send");
            close (new_fd);
            exit (0);
        )
```

```
        close (new_fd);
    )
    while (waitpid (-1, NULL, WNOHANG)> 0);
)
Client program:
/ * Header file ibid * /
# Define MAXDATASIZE 100
int main (int argc, char ** argv)
(Int sockfd, numbytes, servport;
    struct sockaddr_in servaddr;
    char buf [MAXDATASIZE];
    if (argc! = 3) (
        fprintf (stderr, "usage: hellocli <server's ip address> <server's port>");
        exit (1);
    )
    sockfd = socket (AF_INET, SOCK_STREAM, 0);
    bzero (& servaddr, sizeof (struct sockaddr_in));
    servaddr.sin_family = AF_INET;
    servaddr.sin_port = htons (atoi (argv [2]));
    inet_aton (argv [1], & servaddr.sin_addr);
    connect (sockfd, (struct sockaddr *) & servaddr, sizeof (struct sockaddr));
    if ((numbytes = recv (sockfd, buf, MAXDATASIZE, 0 ))==- 1) (
        perror ("recv");
        exit (1);
    )
    buf [numbytes] = '\ 0';
    printf ("Receive from server% s: \ n% s", argv [1], buf);
    printf ("totally% d bytes received \ n \ n", numbytes);
    close (sockfd);
    printf ("...... local socket closed \ n ");
    return 0;
```

3   Server Program and Client Program Detailed Design and Analysis

3.1   Code Organizations

As mentioned earlier, the server is divided into the main program and the communication module and the document processing module. Main program and the communication module in a file, together constitute the server program sevprog.c (1 - 129 lines). Document processing module referenced by the include statement. Client program on a file cliprog.c in (130 - 247 lines). Although the two procedures are harmonized in the appendix the line number (1 - 247), which is easy to explain, in fact they are running on different machines on two different C files. Document processing module achieved by the co-Kai Wang, limited space, not attached to the source code, just click on the interface functions were explained.

3.2   Data Format

Abstract data structures custom message structure type. A message structure including the following: the recipient, who sent, subject, message body, write the time, the message number and so on. Structure as defined in the letter of 14 - 15 lines.

3.3   Synchronization Mechanisms

Communication module synchronization is the key to the normal sending and receiving information. Similar to the operating system on the principle of producer consumer problem, when producers do not write data to the buffer, the consumer does not read data from the buffer, otherwise, the data obtained are not valid data, or cause read file failed. Therefore, the client and server communication process, there must be strict synchronization. In this procedure, the simultaneous use of a simple and effective way. Human design messaging protocol, according to the agreement of the steps, send the information side information at the beginning of hair before sending a synchronization character, and then send useful information. Send message received information side synchronous character side made way before the adoption cycle of reading waiting for the arrival of characters simultaneously, using WHILE loop to achieve, when they read synchronization character, began to read the following useful information. The function is used to synchronize a pair of the following:

int SendSockChar (int sockfd, char ch): sending a synchronization character. Parameter for the socket number and a character to be sent. This function will write the socket of a character, that is sent to each other. Sent successfully function returns OK. See 67 line function to achieve.

char GetSockChar (int sockfd): receiving a synchronization character to receive each other with SendSockChar () sent a synchronization character, the process of using WHILE loop to read until the effective synchronization of characters read so far, the function return value is read in synchronization character . Function implementation, see 65 - 66 lines.

3.4   Read and Write Function

Rational use of communication module to read and write functions is an important point of normal communications. In the use of the system function read (), write (), also a few other self to use the function: readline (), sockendl (), CutEndl (), respectively, as follows:

int readline (): reads a line from the contents of the socket. Line refers to a newline character '\ n' for the end of a string, including the newline itself. See 38 - 42 lines to achieve.

int sockendline (): send the line breaks. Because the content is read from the socket using readline functions, to sign a line read line breaks before the end, it can read a line string, so if the sender sends a string is a newline at the end, will have to reissue a line break , can read the recipient normal. Function SOCKENDLINE achieve functional replacement of line breaks. Function implementation, see 43 - 44 lines.

void CutEndl (): to line breaks. Function of the receiving party to accept his party with READLINE the end of the string, including a newline, this will cause errors in some places, such as the string form of user name and password if the addition of line breaks turned into another string, although users feel, but back in the server program returns incorrect results verify the password. CutEndl function when necessary, remove the string had just received at the end of the line breaks. See 45 line function to achieve.

3.5 Header Files and Constant Definitions

1 to 7 line is we utilize the system header files, one for SOCKET communication, I / O operations, string operations, time to read and so on. 8 to 9 lines of the document processing module includes.

10 - 13 line defines some constants, MAXSIZE is the maximum capacity of the buffer zone. MYPORT SOCKET connection to the server to establish the port number used. BACKLOG To listen queue maximum length. Successful return to function OK sign.

3.6 Main Program

Server main program: main program used to initialize the server's address information, create socket start listening, waiting to connect, when server process receives a connection, with the FORK system call creates a child process for customer service, the parent process to continue port waiting for connections. This enables the server to multiple customers at the same time, this work became complicated by the server. This procedure uses the concurrency services. Main () function code, see lines 30 to 37.

Client main program: the client main () function is much simpler than the server. Just use the command line parameters, the incoming server to run the IP address and port number. Number of command line arguments (3) be verified to establish connection to the server, then call the client service main function, the service was finished, quit. Client main () function code, see lines 150 to 154.

3.7 Connect Function

To make the main program simple socket connection with the process are written into a function call with the main program. Server and client of the function were StartListening () and ConnectToServer (), which StartListening () by calling socket (), bind (), listen () to establish listening socket, the connection began to wait for the customers; ConnectToServer () by calling socket (), connect () to establish a connection with the server, the parameters for the server's IP address and port number, incoming call command line arguments argv [1] and argv [2]. Function to achieve 46 to 47 rows, respectively, and 155 - 157 lines.

3.8   Service Main Functions

1) The server main function int ServerMain (int sockfd): When a user establish a connection with the server, the server program to generate a child process to serve the child process calls ServerMain function for customer service, this function returns, the service is ended. Function main first came to accept the user's service request, decided to offer registration services or Log service, and then enter the corresponding service function. This function is achieved, see line 54 to 62.

2) Client main function int ClientMain (int sockfd): Client main program to establish a good connection to the server, call this function, to achieve the user's mail service. Gai function De content to provide the initial user into the mail system interface, so Yonghushuru initial Mingling, Xuanze Register or log, then call the appropriate subroutine to service, Dangzaichengxu out, the function also Jieshu, the client's main program also ended. This function is achieved, see 156 to 160 lines.

3.9   Mail Service Function

The realization of e-mail system features the following service functions is that their role is to provide a user interface to receive user commands, the transmission between the server and client data and commands, and call the interface function module file operations to achieve the processing of mail. These functions on the server and client in pairs, so that it can achieve successful communication.

1)   Server service functions:
int RegisterSev (int sockfd); / * registration feature server-side function from 66 to 68 * /
int RecvUsrPswd (int sockfd, char * Usr, char * Pswd); / * Receive the user password 78 to 80 lines * /
int LoginSev (int sockfd); / * sign-on server-side function, 72 to 77 lines * /
int SendMaillist (int sockfd, char * Usr); / * send e-mail list server-side function, 88 to 94 lines * /
int ReadMailSev (int sockfd); / * read the e-mail features server-side function, 95 - 103 lines * /
int DelMailSev (int sockfd, char * Usr); / * delete the e-mail features server-side function, 104 - 108 lines * /
int SendMailSev (int sockfd, char * Usr); / * server-side function to send mail function, 109 - 129 lines * /

2)   Client Service Functions:
int RegisterCli (int sockfd); / * Registration function of client-side function, 171 - 173 lines * /
int SendUsrPswd (int sockfd, char * Usr, char * Pswd); / * send the user password, 186 - 189 lines * /
int MailClient (int sockfd); / * login success message service client function, 190 - 196 lines * /
int ReadMailCli (int sockfd); / * read client e-mail function function, 208 - 216 lines * /
int DelMailCli (int sockfd); / * client function to delete e-mail function, 216 - 220 lines * /
int SendMailCli (int sockfd); / * client function mail function, 221 - 2478 lines * /

Acknowledgment. The work is supported by the Natural Science Fund in HeiLongjiang province under Grant Nos. F200907.

References

1. Zhang, Y.: Comparative Research of IPv4 and IPv6. Scientific Economy Market (3) (2006)
2. Xie, S.: Two Technologies of Transition from IPv4 to IPv6. Scientific Economy Market (02) (2006)
3. Foster, I., Kesselman, C., Tuecke, S.: The Anatomy of the Grid Enabling Scabling Virtual Organizations. Int'l J. Super Computer Applications 15(3) (2001)
4. Yang, X.: 1000. Linux C Function Library Reference Manual. China Youth Press, Beijing (January 2002)
5. Pang, L.-P.: Operating System, 3rd edn. Huazhong University Press, Wuhan (February 2001)
6. Patrick Volkerding waiting. Yin Bo with other translations. Linux configuration and installation. China Water Power Press, Beijing (May 1999)
7. Linux Technical Series prepared by the editorial board. Linux Developer's Guide. Beijing Hope Electronic Press, Beijing (August 2000)
8. Siever, E., et al.: LINUX Technical Manual, 3rd edn. China Electric Power Press, Beijing (2003)
9. Huang, C. (ed.): LINUX advanced development technology. China Machine Press, Beijing (2002)
10. Wang, J.-W., Wu, J. (eds.): Linux standard tutorial. Tsinghua University Press, Beijing (2006)

Study on Training System of Engineering Students Education in the Context of Multipole Values

Feng Ruiyin, Ma Liping, and Zhao Jinchuan

Hebei Normal University of Science & Technology
Qin Huangda Hebei Province, China
fengzj1725@sina.cn

Abstract. Traditional training of college students education is in accordance with the training program established arrangements, through teacher lectures, laboratory and other sectors to complete education and training work, as training programs and social needs of the relatively inflexible nature of the mainstream training mode single, extra-curricular conscious content of education is relatively small, this kind of traditional training model can not fully meet our manpower needs of diverse students to develop multi-polarization of the needs of students need to work on cultivating a more diverse, more open, more conscious of education systems to build college students under large cylinder opener field.

Keywords: multipole values, engineering students, training system, large cylinder opener field.

1   Introduction

With the deepening of China's socialist market economy, the social development has been in a period of transformation, so all kinds of social phenomenons are complex and the social value ideas have appeared multi-polar trend. At present, traditional educationa ideas and cultivation model have not very well adapted to the needs of social talent multipolarit, how to integrate the educational resources and cultivate the college students effectively and realistically and how to solve the need of the employment and the society have become an inevitable problem that modern higher education must face. In order to build an effective and realistic education system, the author will research the following five aspects.

2   Establishment of an Open Schooling System Facing the Society

In order to develop the international over -the -horizon talent with the social responsibility and national spirit and improve the ability of leading the society, we will use classroom education, TV, newspapers and online media and so on as channels, base on extracurricular the Party and the League to carry out factual evidence educationa, historical education and patriotic education to increase the sense of the social responsibility and the sense of the national crisis.

Y. Wu (Ed.): International Conference on WTCS 2009, AISC 116, pp. 71–76.

2.1 Strengthening the Education of the Historical Knowledge, Cultivating the International Talents with Far and Fight Ambition

College students should be taught modern and contemporary history about China, neighboring countries and main developed countries as well as professional development history they are majoring in. According to the comparison of different national history, the college students can correctly treat a nation's historical culture and spirit while learning the history objectively, then to treat and consider problems with a global perspective and thinking. After learning the successful experience and failure lessons from the national development stage of all countries, the students will safeguard world peace and promote common human progress. History education is very little proportion of curriculum provision in non-literature and non-history majors in the university, but we can solve it through specified reading list and self-study.

2.2 Developing National Conditions Education, Holding the Current Affairs Forum, Broadens Students's Vision and Improving the Ability of the Commanding Overall Situation

First, we will carry out national conditions education and the policy situation education to students, and create mechanism to encourage students to use winter and summer vacation time to develop social practice, goes to the countryside and the factory, let students use their own sense organ to truly understand the countryside, the factory, the country and the society.

Second, focus on current affairs, pay close attention to the hotspot. For example, to pay attention to the financial crisis in 2009, the Renminbi exchange rate, cross-strait relations, wars, disasters, climate change and social conflict. You can reveal the essence behind the event through watching the corresponding programmes and holding the lectures, and enhance the students' concern to the global hotspot, thus broaden their horizons and increase the macrocosm thinking and judgment ability.

Third, concern various new concepts, new knowledge, new technology and new inventions as well as possible social change because of these, improve prediction ability and strategic thought.

Through such open educational system, we can make up for the lack of a university training programme, cultivate the strategic talents who both can base on the real national conditions and has the acute thought to think problems with a world view.

3   Construction of Good Campus Environment and Educational System

3.1 Conscious Development of Material Culture and the Educational Function of Soft-Culture

Material culture is the general name of group learning of college students, the hardware condition of the life and the facilities. Macroscopically speaking, It includes the overall planning of the campus , greening, beautification, building style along with harmonious degree of the campus and agglutination degree of the culture, etc. Microscopically speaking, It includes the amount and the advanced extent of the laboratory and experimental equipment, the modernization degreeof the classroom and the dorm, visual aesthetic comfort and cultural dependence degree, etc. As to these construction of the hardware culture, the school must have a relative high cultural added value, when meeting the material culture needs, it will realize the educational function. Therefore, when the logistics sector of the campuses is under construction to them, they will consider from the whole concepts of the education to plan, design, construction, maintenance and reach the best results of the education.

Soft-culture here refers to the advanced sponsoring concept, scientific management system, serious management style, good operation mechanism, effective management behaviour, good management effect, teachers' strong dedicated awareness, persevering spirit of exploration and the social service level and ability of all the staff. Soft culture will directly and indirectly affect students through the system, standard, behaviour and atmosphere, therefore, we managers and the teacher must make soft-culture education function consciousness, display the more function.

Material culture and soft culture in a school imperceptibly influences each student, and these culture gradually become part of the students' quality.

3.2   Building a Safe and Stable Environment Atmosphere

1) Establishing a system of good security guard in a school: The school need establish a system of good security guard for all students, and will have a safe knowledge education timely to them, hold various types lecture on safety and popularize relevant knowledge. For example, fire fighting knowledge, psychological knowledge , network security knowledge, guard against theft, traffic accident prevention, group fights prevention and anti-sudden events lectures,etc. which make students have some common sense in person, property and psychological safety knowledge, and can timely find safe hidden trouble, and take the appropriate methods to prevent. After establishing it, it will have the function like these, to have   safety accidents prevention and handling capabilities, and to safeguard the legitimate interests and the personal safety of students in the greatest extent and the most effective way.

2) Formulation of risk prevention system: In order to ensure the implementation of system of safety prevention work and comply with laws, the school will establish practice and experimental safe production system standard and the mechanism about special high-risk groups discovery, education, warning, save and sudden incident handling and so on.

3) Establishment of academic security mechanism: When the new enrollees enter the school, they should be organized to learn rules and regulations and to know what will be learned during the college period, what can be done or not, what can be encouraged to do or opposed by the school. So there will lay off a traffic line in students' minds in order to ensure them smoothly learning goals. Each semester the school will guide students, tell them the position that open courses is in the whole process of learning and what students should pay attention to and what to do in this period, then give stage learning method guide and career planning for them. In addition, it is very useful to hold advanced models report, such as: introduction about learning experience,

improvement of the practice ability, postgraduate examination , employment, special ability training, etc. which will set an example for students.

Through the above mechanism, which make students feel safe in life and study and full of confidence to the future, building a firewall for personnel training and guaranteeing the smooth completion of schoolwork.

3.3   Establishment of the Comprehensive Support Training System

On the basis of completing the normal teaching with high quality and enough quantity according to training program, and try to develop the following respects so that students consolidate, enhance, sublimate knowledge what they have learned. Regarding selection and competitions of specialty along with relating to it as impetus, cultivateing students' genuine talent, fostering competitive consciousness, establishing a good atmosphere of learning, improving the innovative spirit and strengthening practical ability.

1)Encouraging postgraduate examination, establishing relevant support system: The quantity and quality of postgraduate examination reflect undergraduate education level about a school, and play an important role in building up a fine style of study and improving the social certifications. So the school should pay special attention to postgraduate examination group and do guarantee work well. Such as, not only the annual course capacity of the training plan should be adjusted in order to ensure senior have enough time to prepare for the exam and the training quality of undergraduate, but also review places should be offered for the students of   postgraduate examination, at the same time, the school should arrange the excellent teachers to tutor the course, the policy of postgraduate examination, filling volunt and adjusting volunt, exam technology and test psychology. And hold experience exchange meeting for junior students, propaganda and report the students of passing postgraduate examination and provide the whole course service for them, thus can build positive attitude and atmosphere for the junior students.

2) Supporting the computer and the foreign language level test, building a platform of practice: It is different requirements in different universities to computer and foreign language levels. The author thinks that the school should intensify them from the ideas, the policy and the technology, encourage students actively participated in passing the examination in English and computer (more encourage foreign students passing the oral exam, let foreign language really become communication tool), improve the computer and foreign language level to a largest extent, and build a solid foundation for the future. Because the two kinds of exams are very strict, the real value of the certificates is higher and social public trustworthiness is also high. It is basic for the students' work they will engage to have the ability of the computer and foreign language, and it will direlyct impact on the working efficiency and level, reflect international quality level of the students of a school. In addition, it is very important for forming an excellent academic atmosphere to encourage foreign language and computer examinations or saying the amount of the people and the level of passing the exam about the foreign language and computer also reflects a school discipline.

3) Professional contest held, practice ability promoted: Combining professional characteristics, participating in the professional skills contest of various types and specifications, students can consolidate professional knowledge learned by these

contest, at the same time, form professional ability, foster competitive formation and innovative consciousness and spirit. In mechanics &   electronics, for example, can combine the metalworking practice to hold the fundamental metalworking skills competition to strengthen the basic knowledge and improve the basic operation capacity. In addition, if the school often hold microcontroller, electronics, 3d design, robot, machinery and innovative design and so on contests, then attend the regional, provincial and national contests, which is a very effective way to improve professional skills, so the school will provide experimental sites, outfit on the guided teachers and necessary funding support.

4) Forum on science and technology held: Forum on science and technology is a platform of scientific communication for students, the students can communicate their scientific research achievement through it, also learn from and improve each other in the mutual communication, and can enhance the academic level of the whole and build academic atmosphere.

4   Establishment of Comprehensive Knowledge Training System

We should take improving the quality of humanity as impetus to enhance college students' comprehensive quality, and cultivate the ability in society, abstract thinking ability, logical reasoning ability and the ability to discern between good and evil , thus make college students   have a comprehensive development.

College students are the creator of new culture and the successor of the traditional culture, and the educated objective is that every student should be developed comprehensively. On the basis of learning systematically professional knowledge, the students must be improved in humane quality. Especially even more important to engineering specialty, because it is a common phenomenon, such as, overemphasis on science and neglect of arts , overemphasis on engineering and neglect of art, overemphasis on professional skills neglect of comprehensive quality, unbalanced knowledge structure between science and arts. So improvement of the humanities quality can make their knowledge structure more reasonable and have more open view. In order to adapt to the reality of interdiscipline, we can make better use of divergent thinking in the concrete work and improve the innovative ability.

In the whole teaching course of the university except normal teaching activity, there are still a lot of spare time for students, so they should make full use of it to improve humane quality, and achieve all-round development.

4.1   Formulation of Amateur Reading Plan

Especially the spare time of the freshmen and sophomores is very sufficient, so it is necessary to formulate the reading plan designedly and to read systematically some classic readings, including literature, history, philosophy, etiquette, and art books, thus which can increase students' humane knowledge, at the same time, they will learn enterprise management and applied writing knowledge.

4.2   Watching Lecture Room and Classic Performance

Lecture Room program of high quality, referring to history, politics, literature, philosophy, and so on, so it can enable students to receive a variety of education, and improve the level of the humane knowledge and the ability of good judgment and abstract generalization.

4.3   Visiting Art Exhibition, Watching Drama and Opera Show

Through these activities, it can improve students' aesthetic ability and their own artistic accomplishments, shaping a good self-image.

4.4   Enriching Student Community

Students is a lively collective with different interests and hobbies, so we will encourage them to form a club, through this platform, let studdents learn what they are interested in, make progress each other and enrich extracurricular amateur life. In order to create positive and healthy cultural atmosphere for college students, we will strengthen the management of the community, such as, constitution and aim of the activity examination, right direction insurance, guidance of the community activities, the guarantee of the content richness and process standardization. The leader of community will be managed as the class cadre series, and improve their ideological state and the overall level of community.

The above five aspects from different points have illustrated the engineering students' education and training system in the context of the multiple values, which is a system engineering, because of the mutual influence between the education system, it is required for teaching staff to design construction of each system from the overall, and also plan from the overall and implement it, Only in this way can make the virtuous circle benign in the system, and improve the educational effect.

References

1. Wang, Y., Ye, D.: Enhancing Undergraduates'Innovation Ability Based on the Electronic Design Contest. Chinese Geological Education 1, 145–148 (2009)
2. Qian, H., Hua, W.: Enhancing Undergraduates'Aesthetic Education Based on Art Students Clubs. Youth Culuture 3, 19–21 (2008)
3. Huang, H.: A Reasonable Need Idea of Students: Philosophy Thinking of College Student Affairs Management in China. Meitan Higher Eucation 1, 46–49 (2009)
4. Tan, S.: Pay Attention to the Disadvantaged Group of college students and Build a Harmonious Education environment. Economic and Social Development 6, 163–165 (2005)

Risk Evaluation of Enterprise Informatization Based on the Unascertained Method

Liu Jun'e[1], Zeng Fanlei[2], and Han Zheng[3]

[1] Information School, Beijing Wuzi University, Beijing, China
[2] School of Economic Management, Hebei University of Engineering, Handan, China
[3] Educational Technology Center, Handan Polytechnic College, Handan, China
Ljun2004928hb@163.com

Abstract. As enterprise's informatization is a long and complicated process, it is necessary to make a risk assessment before doing the action of informatization. With the unascertained theory, this paper made a qualitative and quantitative analysis about the risk factors of enterprise informatization process. According to the enterprise informatization risk management of diamond model, it extracted comprehensive measures for information risk evaluation and made the risk evaluation. Through the unascertained theory in the risk assessment the paper made a quantitative evaluation about the index which is difficult to quantify. And it also provides feasible risk assessment methods for the implementation of enterprise's informatization.

Keywords: Risk assessment, unascertained method, comentropy, diamond model.

1 Introduction

After joining WTO, China's economy has integrated into the world economy rapidly and has achieved great development. Our enterprise is also facing challenge of the international economies. Under the background of the rapid development of the information age, national enterprises are put forward and implement the information operations. Through the implementation of informatization, it can not only save large amount of operating costs, improve corporate profits and also enhance the competitiveness of the enterprise itself. To keep pace with world's development and get a place in the international stage for our enterprises, it is necessary to do the enterprise informatization; however the enterprise informatization process is a high risk investment project, it requires risk assessment before the implementation of the enterprises informatization. Only in the acceptable risk scope of the implementation can it be operated, or it should improve the process of enterprises informatization to minimize its risk and success in a time.

According to the enterprise informatization risk management diamond model [1], the paper extracted information risk evaluation indexes, at the same time, it made a qualitative and quantitative analysis to the evaluation objects with the unascertained theory. According to the information entropy theory and analytic hierarchy process (AHP), it computes the index weight to evaluate the risk scientifically.

Y. Wu (Ed.): International Conference on WTCS 2009, AISC 116, pp. 77–83.
springerlink.com   © Springer-Verlag Berlin Heidelberg 2012

2 Based on the Unascertained Enterprise Informatization Risk Evaluation Model

Assume X to be the evaluation object, its first level index number is m, the evaluation space is $\{I_1, I_2, \cdots, I_m\}$, for each first level evaluation index Ii, it has k evaluation indexes, and the secondary evaluation index space is $\{I_{i1}, I_{i2}, \cdots, I_{ik}\}$, it shows the evaluation objects' measurement value as to the secondary evaluation index I_{ij} which is under the first level index I_i, the secondary evaluation index I_{ij} has p estimation scales, and its evaluation space is $c = \{c_1, c_2, \cdots, c_p\}$. The estimation scale is in order, it may be $c_1 \geq c_2 \geq \cdots \geq c_p$ or $c_1 \leq c_2 \leq \cdots \leq c_p$.

2.1 Secondary Indexes of Unascertained Measure Computation

1) Construct membership function according to the classification criteria:
Take the measurement value of the indexes into the membership function, it results the secondary level index membership matrix under its first level index. The matrix

$$\text{is } (\mu_{irt}) = \begin{bmatrix} \mu_{i11}, \mu_{i12}, \cdots, \mu_{i1p} \\ \mu_{i21}, \mu_{i22}, \cdots, \mu_{i2p} \\ \vdots \quad \vdots \quad \vdots \quad \vdots \\ \mu_{ik1}, \mu_{ik2}, \cdots, \mu_{ikp} \end{bmatrix}.$$

The μ_{irt} expresses the membership degree which shows how much the secondary level index x_{ij} under its first level I_i belongs to the c_t class and it meet $0 \leq \mu_{irt} \leq 1$ and $\sum_{t=1}^{p} \mu_{irt} = 1$.

2) Compute the secondary indexes' weight using information entropy,
With the knowledge of information entropy,

$$\omega_{ir} = \frac{v_{ir}}{\sum_{j=1}^{p} v_{ij}} \tag{1}$$

And ω_{ir} is the class weight of the secondary level index I_{ij} under its first level I_i, and

$$v_{ir} = 1 + \frac{\sum_{j=1}^{p} \mu_{irj}}{\lg p} \quad (0 \leq \omega_{ir} \leq 1, \ \sum_{r=1}^{k} \omega_{ir} = 1). \tag{2}$$

So we got all the secondary level indexes' weight under their first level index:
$(\omega_{i1}, \omega_{i2}, \cdots, \omega_{ik})$

Then measure the secondary level indexes' value using the unascertained method, such as

$$(\omega_1, \omega_2, \cdots, \omega_k) \begin{bmatrix} \mu_{i11}, \mu_{i12}, \cdots, \mu_{i1p} \\ \mu_{i21}, \mu_{i22}, \cdots, \mu_{i2p} \\ \vdots \quad \vdots \quad \vdots \quad \vdots \\ \mu_{ik1}, \mu_{ik2}, \cdots, \mu_{ikp} \end{bmatrix}$$

$$= (\mu_{i1}, \mu_{i2}, \cdots, \mu_{ip}). \tag{3}$$

Thus obtained secondary indexes and comprehensive measure matrix:

$$\begin{bmatrix} \mu_{11}, \mu_{12}, \cdots, \mu_{1p} \\ \mu_{21}, \mu_{22}, \cdots, \mu_{2p} \\ \vdots \quad \vdots \quad \vdots \quad \vdots \\ \mu_{m1}, \mu_{m2}, \cdots, \mu_{mp} \end{bmatrix}$$

2.2   The First Class Indexes' Unascertained Comprehensive Measure

1)   Compute weights vector:

Getting the first level indexes' weights vector distributions through the method of AHP (analytical hierarchy process):

$$\omega = (\omega_1, \omega_2, \cdots, \omega_m) \text{ When } (0 \le \omega_i \le 1, \sum_{i=1}^{m} \omega_i = 1).$$

2)   A comprehensive evaluation of evaluation objects

$$(\omega_1, \omega_2, \cdots, \omega_m) \begin{bmatrix} \mu_{11}, \mu_{12}, \cdots, \mu_{1p} \\ \mu_{21}, \mu_{22}, \cdots, \mu_{2p} \\ \vdots \quad \vdots \quad \vdots \quad \vdots \\ \mu_{m1}, \mu_{m2}, \cdots, \mu_{mp} \end{bmatrix}$$

$$= (\mu_1, \mu_2, \cdots, \mu_p). \tag{4}$$

The $\mu_i (i = 1, 2, \cdots p)$ shows the membership degree of the evaluation objects in all levels.

As the evaluation class is in order,

Let $c_1 \geq c_2 \geq \cdots \geq c_p$, with confidence identification method to make the evaluation results.

Let $k(x) = \min \sum_{k=1}^{s} u_k > \lambda (1 \leq s \leq p)$, usually let $0.5 < \lambda < 1$, and the object x belongs to class c_k.

3   Examples

An enterprise developed rapidly in recent years, in order to adapt to the development requirements, it prepared for the operation of the informatization reform. We need to make a risk assessment for the enterprise informatization scheme.

3.1   Formulate Rating Criteria

Invite more informatization risk evaluation experts to give value to the secondary level indexes in accordance with ten cents, and then after comprehensive specialists' views we get the comprehensive risk value. The specific risk assessment criteria were shown in table below.

Table 1. Risk assessment standards

Risk level	Extremely high	High	middle	low
Risk value	≥ 7.5	$7.5 \sim 6.0$	$6.0 \sim 4.5$	≤ 4.5

3.2   Establish Evaluation Index System

In this paper it referred to the enterprise informatization risk evaluation of the relevant materials [1, 2], established the enterprise informatization risk evaluation index system shown in table below. And write the indexes' comprehensive score of the scheme in it.

Table 2. Risk evaluation index system

	First level indexes	Secondary level indexes	Risk values
Enterprise informatization risk	Service providers' risk I_1	Service providers' select risk I_{11}	5.0
		Principal-agent risk I_{12}	4.5
	Organization and change risk I_2	Structure reform risk I_{21}	6.5
		Cultural change risk I_{22}	5.5
		Human resources risk I_{23}	6.0
		Employee resistance risk I_{24}	6.5
	Management and reform risk I_3	Introducing motivation risk I_{31}	4.5
		Introducing timing risk I_{32}	5.0
		Leading perceive risk I_{33}	5.5
		Project planning risk I_{34}	6.0
		Management reform risk I_{35}	5.5
		Management system risk I_{36}	5.0
		Basic management risks I_{37}	5.5
		Schedule quality risk I_{38}	5.0
		Cost control risk I_{39}	6.0
	Technology and change risk I_4	Hardware risk I_{41}	4.5
		Software risk I_{42}	4.5
		System security risk I_{43}	6.5
		Use and maintenance risk I_{44}	5.5
	Environmental factors risk I_5	Economic policy risk I_{51}	5.0
		Market conditions risk I_{52}	5.0
		Competition risks I_{53}	5.5
		Political factors risk I_{54}	4.5

3.3   The Secondary Level Indexes' Unascertained Measurement

1)   Construct membership functions:

According to the evaluation index criteria construct membership functions as follows:

$$\mu(x \in c_1) = \begin{cases} 1 & x \geq 7.5 \\ (x-6.75)/(7.5-6.75) & 6.75 \leq x \leq 7.5 \\ 0 & x < 6.75 \end{cases}$$

$$\mu(x \in c_2) = \begin{cases} (x-5.25)/(6.75-5.25) & 5.25 \leq x \leq 6.75 \\ (7.5-x)/(7.5-6.75) & 6.75 \leq x \leq 7.5 \\ 0 & x \geq 7.5 \cup x \leq 5.25 \end{cases}$$

$$\mu(x\in c_3)=\begin{cases}(x\text{-}4.5)/(5.25\text{-}4.5) & 4.5\leq x\leq5.25 \\ (6.75\text{-}x)/(6.75\text{-}5.25) & 5.25\leq x\leq6.75 \\ 0 & x\geq6.75\cup x\leq4.5\end{cases}$$

$$\mu(x\in c_4)=\begin{cases}1 & x<4.5 \\ (5.25\text{-}x)/(5.25\text{-}4.5) & 4.5\leq x\leq5.25 \\ 0 & x\geq5.25\end{cases}$$

2)   Computation process:

Due to the paper's limited space, take I_1 for example to explain the comprehensive measure computation process, all other secondary indexes can be computed analogously.

Take the secondary indexes' value of I1 into the membership functions [3], and then get the secondary indexes measurement matrix of single index.

$$(\mu_{1i})_{2\times4}=\begin{pmatrix}0 & 0 & 0.667 & 0.333 \\ 0 & 0 & 0 & 1\end{pmatrix}$$

According to the formula (1) (2) can get their categorization weight size as follow:

$$\omega_{1r}=(0.3511,0.6489)$$

Reuse formula (3) can get the secondary level indexes' unascertained measurement vector under their first level index I_1 [3, 4]. $\mu_1=(0,0,0.2342,0.7658)$

Similarly can get other indicators' corresponding secondary indexes of unascertained measure vector

$$\mu_2=(0,0.5890,0.4110,0)$$
$$\mu_3=(0,0.1484,0.5788,0.2728)$$
$$u_4=(0\quad0.2014\quad0.2014\quad0.5972)$$
$$\mu_5=(0\quad0.0409\quad0.4657\quad0.4935)$$

So finally get the secondary level indexes' unascertained comprehensive measure matrix under their first level index as follow:

$$(\mu_i)_{5\times4}=\begin{pmatrix}0 & 0 & 0.2342 & 0.7658 \\ 0 & 0.5890 & 0.4110 & 0 \\ 0 & 0.1484 & 0.5788 & 0.2728 \\ 0 & 0.2014 & 0.2014 & 0.5972 \\ 0 & 0.0409 & 0.4657 & 0.4935\end{pmatrix}$$

3.4   The First Level Indexes' Unascertained Measurement

With the method of AHP, get the first level indexes' weights vector distributions

$$\omega=(0.1,0.3,0.3,0.2,0.1)$$

From the formula (4), with the unascertained comprehensive evaluation measure, do the schema's risk evaluation as follows.

$$\omega = (0.1, 0.3, 0.3, 0.2, 0.1) \begin{pmatrix} 0 & 0 & 0.2342 & 0.7658 \\ 0 & 0.5890 & 0.4110 & 0 \\ 0 & 0.1484 & 0.5788 & 0.2728 \\ 0 & 0.2014 & 0.2014 & 0.5972 \\ 0 & 0.0409 & 0.4657 & 0.4935 \end{pmatrix}$$

$$= (0, \quad 0.2656, \quad 0.4072, \quad 0.3272)$$

3.5 Using the Confidence Criterion to Do Risk Identification

Let $\lambda=0.6$, as $k(x) = \min \sum_{k=1}^{s} \mu_k > 0.6 (1 \leq s \leq 4)$, so k=3, and the enterprise informatization solutions belong to rank 3 named moderate-risk.

4 Conclusion

In the process of informatization risk assessment, this paper used unascertained method and information entropy theory comprehensively, established the risk evaluation index system, realized the qualitative and quantitative risk evaluation comprehensively. Finally, using examples to explain this model, and proving that this method was feasible [5].

It should be pointed out that enterprise informatization risk may be the result of various risks factors which influence each other. So it is necessary to pay attention to the risk factors' close contact.

References

1. Xin, W.: Risk management "Diamond model" in enterprise informatization. Contemporary Economic Management 32(4), 33–38 (2010)
2. Wu, X., Zou, H.: Risk analysis of enterprise informatization construction. Enterprise Economy (2), 77–78 (2005)
3. Liu, K., Pang, Y.: Urban environmental quality evaluation of unascertained measure. System Engineer -Theory & Practice (12), 52–58 (1999)
4. Li, W.: MIS Comprehensive Evaluation Model based on information entropy and unascertained measure. Journal of Hebei Institute of Architectural Science & Technology 22(1), 49–53 (2005)
5. Liu, J., Zhang, H., Zhang, Y.: Unascertained theory used in enterprise informatization level evaluation applications. China Management Informatization 10(2), 8–9 (2007)

The Study on the Risk of Local Treasury Bonds Based on New Institutional Economics

Chen Huiling

Economics and Management School of Wuhan University
jdtlinhui@21cn.com

Abstract. Local Treasury Bonds refer to the bonds issued by the local public institution of the local government that has financial revenue in a country. In order to stimulate the economic development, our country issued 200 billion Local Treasury Bonds in 2009. But in the issuing process of Local Treasury Bonds, it's inevitable to encounter various risks. In terms of New Institutional Economics, this article analyzes various issuing risks faced by Local Treasury Bonds and proposes countermeasures that control the issuing risks of Local Treasury Bonds, hoping to provide reference for risk management research on our country's Local Treasury Bonds.

Keywords: Risk,local treasury bonds, institutional.

1 Present Situation of Our Country's Local Treasury Bonds

On March 17, 2009 the Treasury claimed that the Local Treasury Bonds in 2009 issued by the Treasury in place of local governments referred to negotiable book entry bonds in 2009 with approval from the state council, taking province, autonomous region, municipality city and municipal government listed independently in the state plan as issuing and repayment subject as well as issued by the Treasury as the agent for the loan and interest payment and issue expense. According to local matching scale of public welfare project by central investing, demand for local project construction fund and repayment ability, etc., such 200 billion Local Treasury Bonds properly distribute bond scale of each area by formula method. From the angle of final distribution amount, this fund distribution tends for the central and western regions; Ningxia Hui autonomous region got about 3 billion Yuan, Xinjiang 5.5 billion Yuan, Shaanxi 6 billion Yuan, Guizhou 6.4 billion Yuan, while Guangdong 10.9 billion Yuan. The issue amount of Local Treasury Bonds that the central and western regions get covers a large proportion in our country, all much larger than the proportion that the financial revenue takes in that of the whole country. It mainly considers that there are lots of credit accounts for construction projects involving livelihood in central and western regions; the local matching public welfare projects with central investing has heavy tasks while the financial revenue in central and western regions is low and financing ability is poor, so distribution can guarantee the successful implementation of projects. However, such a distribution pattern also brings some problems; the economy in central and western regions originally falls

Y. Wu (Ed.): International Conference on WTCS 2009, AISC 116, pp. 85–90.
springerlink.com © Springer-Verlag Berlin Heidelberg 2012

behind, and now such a large scale of bonds are issued, so it's easy to cause various risks. The "debt crisis" of 59 billion dollars in Dubai especially knocks the alarming bell for us. Therefore, at present, research on issue risk of Local Treasury Bonds is a pressing problem that draws much attention from our local governments at all levels.

2   Our Country's Local Treasury Bonds Based on New Institutional Economics

2.1   Analysis by Fiscal Federalism

Fiscal Federalism means that in order to mutually perform public economic function, governments at all levels are certain independent and free in financial function and budget; its main content includes a series of problems such as roles played by governments at all levels, how to divide budget among governments at all levels and subsidy among governments and its essence is fiscal decentralization. Montinola, QiarI & Weingast (1995) believe that under Fiscal Federalism local governments has financial independence, but they can not issue currency, neither can they enjoy unrestraint loan, that is, governments at all levels face hard budget constraint. But actually, local governments face Soft Budget Constraint to different degrees. Soft Budget Constraint was first put forward by Hungarian economist Kornai (1980) when he analyzed socialism enterprise behavior. He thought under centralized power system there was "Paternalism", that is, once enterprises encounter difficulties, governments will show their paternal love to provide help for enterprises because of various factors, so budget constraint will be softened. When soft budget constraint rises between the central and local governments, the central government will give a hand to local governments falling into financial dilemma, at that time, local governments adopting opportunism will issue excessive bonds, hoping that central government will shoulder the final repayment liability for the whole or partial debt.

In terms of constitution, our country is of unitary system with centralized power, but the research from Montinola, QiarI & Weingast indicates that there exists Market-Preserving Federalism in China and they name it as "Federalism of Chinese Style". Under this system, competition between dispersed government system and regions can prevent excessive intervention to the local from the central and enhance local governments' enthusiasm to promote market-oriented reform for local economic growth (Qian & Weingast, 1997). However, such a decentralized system also can cause soft budget constraint and the local governments may point it their own way without constraints from central government. Besides, Tsui &Wang(2004) find out that due to asymmetric information and born difficulty of monitoring, our country's achievement assessment is simplified as assessment on major economic index and monitoring of specific index in regions, for instance, GDP, employment rate, tax revenue increase, residents' income growth, etc.. In order for pursuing "achievement maximization" to obtain reappointment or promotion opportunity, local government officials launch "image project". Furthermore, though our constitution speculates that the tenure of party and government leadership duty is five years and the same duty can not be reappointed twice, but the tenure of local government officials is of highly temporary; it's rare for most principle local government officials to accomplish the tenure and such frequent change makes local government officials incline to large amount of debt, because the short tenure means the large amount of debt need not be repaid in their own tenure, and they can attentively focus on establishing their own project (Wei, Jianing, 2004). At the same time, the central government's "Paternalism" makes local governments bolder to borrow money. Therefore, under our country's system at present, issuing Local Treasury Bonds no doubt makes central government face huge debt risk.

2.2   New Institutional Economics

The concept of transaction cost was firstly created by Coase. In his article-The nature of business, which was published in 1937, Coase defined transaction cost as the cost of applying market mechanisms, including the spending on searching relative price signals in the market, negotiation and signature for achieving a trade, supervision on the implemented of contract and many other activities. In 1991, when Coase accepted the Nobel Prize for Economics, he replenished that: negotiations should be carried out, contract should be signed, supervision should be implemented, and arrangements to resolve disputes should be set up, etc. All these charges are referred as transaction costs.

New institutional economists consider transaction cost is the source of institution. In terms of their viewpoint, institution is a kind of public goods, it also has supply and demand. The formulation, implementation, maintenance and reformation contain highly complicated transaction activities between people and groups, correspondingly, various costs such as formulation cost, implementation cost, maintenance cost and reformation cost exist. For this reason, Bromley puts forward the concept of institution trade. A new institutional arrangement can only be made when the expected return is greater than the expected cost. It will hinder the implementation of transactions if transaction cost is too much. Any institution has certain function, and can also bring about certain effects and benefits. But at the same time, cost is needed. Now that standard is a form of institution, standard also has its cost. The efficiency of institution is the reflection of the comparative result of benefit and cost. From a static perspective, the efficiency of institution is decided by the efficiencies of institutional arrangements and structure. From the dynamic point of view, the efficiency of institution is related to the technologic nature of the production process, this is because the formulation of any institution is determined by the situation of prolificacy and technology level. With the development of productive forces and the improvement of technology, institution must make corresponding changes and adjustments. Otherwise, the efficiency of institution will decline. The word "institution" always appears in our lives, however, different economists have different definitions and explanations about "institution". People who have discussed the meaning of "institution" in the real sense include old institutional economists Thorstein B.Veblen, Joan R., as well as new institutional economists T. Schultz, Douglass C.North and so on. Veblen[2] believed that the institution is "the general habits of thought of individuals or society on some kind of relationship and function" "a popular spiritual attitude or a popular theory of life". Commons[3] considered institution as "a series of criterions or rules on collective actions control individual actions", "collective action", which have controlling impact on individual actions, can be taken as a part of are part of institution; Schultz[4] believed that institution is the rule of behaviors, which refers to social, political and economic behaviors, including the rule of marriage and divorce, the rule implied in constitution of dominating the allocation and application of political power, as well as the rule of allocating resources and incomes by market or government. Essentially, this definition almost has the same meaning with Commons' thought of institution. North[5] defined institution many times, in general, institution is the rule of specifying personal behaviors and constraint of relationship between people. International Standards Organization(ISO) explains standard as follows: standard is the technical specifications or other public documents, which is drafted in accordance with technological achievements and advanced experience, conformably or basically agreed by each party and approved by the standardization bodies, with the aim of promoting the best public interest. In China, standard is taken as: a reusable common regulation, guide rule or characteristic document which aims to catch the best order in a certain field, it's the consensus of each party and licensed by a recognized body, the purpose is to promote the best society benefit. Through the concepts of institution and standard, we can find that standard can be explained as a form of institution. The process of standardization is also a process of institutionalization[6].

3   Analysis on Countermeasures Controlling Issue Risk of Local Treasury Bonds

3.1   Constitutional Constraint

In order to prevent potential infringement from Leviathan government, Buchanan & TuHock (1962) believe that it's necessary to establish "a series of pre-achieved rules within the scope of which actions afterwards are carried out", and such a series of rules are "Constitution". So in order to control issue risk of Local Treasury Bonds, constitutional constraint should be done on government debt and a rigorous "cage" should be produced for Leviathan government. The 28th provide in chapter 4 of Budget Law of People's Republic of China issued in 1994 speculates expressively that : "local budget outlay at all levels is compiled under the principle of making ends meet and balancing income and expenditure, with deficit excluded. Except for the speculations of laws and the state council, the local governments can not issue Local Treasury Bonds". Thus, the validity for our country's local governments to issue bonds is negated in terms of law. Therefore, our country firstly should make modification on current Budget Law and take government borrowing into our constitution rather than consider it as an expedient measure. Then make Funding Act and Local Funding Act, constraining strictly the qualification of debt-release subject, debt-release application, examination and amount approval, debt-release way, credit rating and debt repayment mechanism of Local Treasury Bonds in the form of law.

3.2   Hard Budget Constraint

Komai (1980) believes that "paternalism is the direct cause for budget constraint softening", so to eliminate soft budget constraint of our country's local government, system background of paternalism must be changed. First, implement complete "tax federalism" reform and actually achieve "first-order political power, first-order management right, first-order financial power, first-order tax base and first-order budget", decrease debt forgiveness for local governments by the central and make it impossible for the central government's free salvation for the local governments. Second, change our country's current achievement assessment system and official tenure system, no longer taking the single economic index as the major or even single standard for our country's local government achievement and official tenure, and there are comprehensive assessment indexes in place of it, including: abiding by law and program requirements, financial health, reaction ability, efficiency and being responsible for the civil (MattheW & Anwar, 2005). Meanwhile, debt management can be considered to be included in achievement assessment and official tenure system to increase local governments' knowledge for debt management and decrease issue risk of our country's Local Treasury Bonds. Finally, carry out debt obligation system. When government tenure or major official is changed, predecessor local governments or officials must go through strict debt auditing procedure by the central or National People's Congress to audit debt-release scale in their tenure and implement debt obligation system at governments of all levels so as to constrain the local governments' too much debt-release impulse under opportunism and short-term behavior.

3.3   Financial Constraint

Blind competition and gambling between local governments can cause local governments to fall into marsh of excessive debt release and can not get out from it; it can also lead to the maximum risk after Local Treasury Bonds are publicly issued. Therefore, to prevent such risk, the optimum choice is to control debt-release scale of local governments in terms of finance. Establish strict financial discipline; strengthen management for financial funding; open to the public the approval situation of local financial expenditure and funding use of transfer payment; enhance management for funding out of budget and try to include it within budget for management; decrease financial waste and diverted fund. Meanwhile, the financial policy trusted by market should be implemented; clearly speculate and issue financial rules like deficit rate and debt ratio in form of law to enhance financial transparency and establish market confidence and attack local governments' various financial opportunism behaviors that steer clear of legal rules and avoid supervision

3.4   Supervising Constraint

Local Treasury Bond is a kind of bond, and should accept supervision from China Securities Regulatory Commission, but considering the width and lack of standard of local government function, a special Local Treasury Bond management institution can be established as supervising subject in China Securities Regulatory Commission, specially used for many supervising items like drafting supervising proposal, issuing supervising regulations and strengthening information disclosure management. Meanwhile, considering the high correlation between Local Treasury Bonds and financial system as well as investment and financing system, local government securities management institution should communicate from different angles with securities supervising department, financial department and planning authorities; as for the issue qualification, issue scale and issue structure of Local Treasury Bonds, listen to advice from relevant departments and coordinate relationships with national debt market, enterprise bond market, stock market and bank credit market. Besides, market intermediary institutions can be introduced, such as issue law office, securities grading institution, etc. for supervising and constraining debt-release behavior of the local governments.

References

1. Ye, Y., Zhang, Z.: New construction standards for institutional research projects. Journal of Huazhong University of Science and Technology (City Branch Study Edition) 23(1), 71 (2006)
2. Veblen: The theory of leisure class, pp. 139–140. The Commercial Press (1964)
3. Commons, J.R.: Institutional Economics, p. 21. The Commercial Press (1997)
4. North, D.C.: Institution, Institutional Change and Economic Performance, p. 3. Shanghai Sanlian Press (1994)
5. North, D.C.: Structure and change in economic history, p. 226. Shanghai Sanlian Press (1994)
6. Zhou, P.: Standardization, network effects and evolution of business organizations. Doctor thesis of Northeast Financial University (2003)
7. Lin, Y.: Economic theory about institutional change, p. 371. Shanghai Sanlian Press (1994)
8. Yang, Z., Zou, S.: Stores and Flows of Knowledge Resources: Features and Measurement. Science Research Management 21(4), 105–110 (2000)
9. Lewis, A.: The theory of Economic Growth, p. 121. George Allen&Unvin., London (1995)

Cultivating Inovative Quality in the University History Teaching

Fang Xiaozhen and Sun Manjiao

School of Humanity and Society, Anqing Teachers College, AnQing,
China, 246011
School of Political Science and Public Administation, Wuhan University, WuHan,
China, 430072
xiaofanghb2@yahoo.com

Abstract. Innovation is considered as the core personnel quality in the knowledge-based society. All teachers should carry out the essential philosophy of teaching to cultivate creation. Several ways are presented in this paper to implement the philosophy, such as optimizing the structure of teaching, broadening the perspectinve of innovation, inspiring innovative thinking, stimulating the motivation of innovation, buliding up communication inside and outside the classroom, creating an innovative environment, simulating troublesome circumstances and so on.

Keywords: History teaching; College Students, Creative Quality, Cultivate.

1   Introduction

Innovation is the soul mind and never exhausted resource for the prosperity of a country. Innovation is considered as the core personnel quality in the knowledge-based society. Meanwhile, education is regarded as the key means to cultivate innovative talents and implement quality education, whose concerns focus about people's liberation, human perfection, and human development. Teaching should be based on students' overall development, potential cultivation and minds purification which are in pursuit of beauties, kindness and truth. All teachers should carry out the essential philosophy of teaching to cultivate creation.

History is one of discovery and creative disciplines. The understanding of history is endless because of it hands on from one generation to the other. The vitality of history education is to develop creative learning history through gathering, interpreting and identifying what you got. In other words, such activities are to cultivate the unconventional thought, exploring spirit and clear judgment. Smallest innovation is much difficult than simple repeat. The key to innovative education history demands the conformity as well as new ideas. So it requires the innovative and practical goal which is considered the harmonies of knowledge and ability, process and method, emotion and values. Actually, the tradition is replaced catering to the development of society as it is vague and difficult to test, which was use to teaching the basics, building capacity and cognitive thought. The paper concludes several teaching methods how to cultivate innovative quality, including awareness of innovation, creative thinking, innovative spirit and ability. They have been applied in previous causes of An Introduction to History and Modern Chinese History, in which combine the completion of the teaching task with the development of students as well as impart knowledge and inspire innovative thinking.

2   An Optimization of the Structure of Teaching and Innovation Perspective

The core of teaching is changed from imparting knowledge to cultivating innovation. In the past, teaching model characterized books based, teacher-centered and cramming. It should be considered as a big leap to enhance the selectivity and openness in comparison with previous mandatory and standards by means of refining teaching content, teaching activities, tests and test methods. The new and healthy teaching system is composed of the student-oriented with the assistance of teachers. This means teaching is converted into self-studying where students are instructed to cultivate the sense of ownership and innovation. In the teaching content, highlight the "new", the latest theories, the latest point of view, the latest historical data, the latest method introduced to students; in teaching methods, and emphasize the "living" character, conduct class discussions, class debates, the implementation of activities teaching; in teaching methods, the highlight of a "first", which introduce advanced multimedia and Internet for teaching, for students to provide a broad space for cooperation and the possibility to make individualized learning a reality.

3   Inspired Innovative Thinking and Stimulating Creative Motivation

The key of innovation is to learn creative thinking from the multi-angle and multi-level perspective as well as various modes of thinking, such as comprehensive thinking, the proliferation of ideas, and even reverse thinking, different thinking. Modern education refers training and development of to students in creative thinking as the main objective of teaching. Students are required to grasp the following three points:

3.1   Variety and Integrity of History

History is so hidden that the viewer often concerns simply from a certain side while the whole picture is underbelly overlooked. A certain case was made up different stories owing to various perspectives. It is, therefore, a challenge for the successor that he is obliged to write down creatively. If not, it is considered as metaphysics. Yuan Shikai, for example, was a betrayed reformer who suppressed the Boxers

Revolution and carried out vigorously dictatorship after stole the fruits of victory. However, that was only one part of story. Marshal Yuan has been the "young guard" enlightened bureaucracy, innovative spirit early. Following the Westernization Movement, the Chinese Army took a new look on whatever the technical equipment, arrangement of position or the preparation of training methods. Marshal Yuan led to true modernization in Tianjin where he trained a new army and Northern New Army. Feudal and comprador though his army appeared, there were innovation factors and military modernization mixed. This is the assessment of two points on the Yuan, but also on the original theory on the Yuan bias correction. This in itself is speculative in the history of innovation.

3.2   Change and Stage of History

History is the continuously change. The same thing usually evokes different views and evaluation at the different historical era. For example, new edition of Modern Chinese History, is added notes of additional ways and means that save the nation. It is particularly noteworthy that Zhang Jian, Zhou Xuexi was mentioned as the representatives of Saving the Country; Hong, Cai Liang Shuming, Yan, Tao, Huang Yanpei as the educational representatives; Zhan Tianyou, Li Siguang as the scientific and technological representatives. Such explorers at that time were received widespread praise because of their devotion and industry. From Chinese democratic revolution, however, the limitation of the three saving the program was obvious as it was impossible to save the nation fundamentally. None of them changed the most fundamental in modern China Situation - the semi-feudal society. Only revolution can really save China. By the period of reform and opening up, the view of saving has to be corrected as rejuvenation.   Industrial rejuvenation is to focus on developing the socialist market economy; educational rejuvenation is to improve the nation through the development of education; scientific and cultural rejuvenation is to develop scientific and cultural undertakings to strive the forefront.

3.3   Relativities of History

None is independent and unaffected. Hence, evaluation of the merit, good or evil, should not be argued under a fixed frame of reference. The only conclusion should be corresponding with the different frames of references. For example, back into the 19th century, China had undergone the sharp debate between the old school and new school, namely the cultures between China and West. Among them, Zhang Chih-tung's posed the idea of the Chinese culture as a master while the Western culture as an assistant. The master reviewed China's feudal morality as a fundamental national social destiny while the assistant learned modern science and technology following the measures taken in Western countries. Those measures involved in education, taxes, war preparation, law and so on. Apparently, as opposed to blind rejection of Western or total Westernization it was indeed a kind of progress.

But it seemed conservative and outdated compared to combination of the Chinese and the Western. Unfortunately, the cultures mentioned by Mr. Zhang were superficial. The Chinese culture was the outdated feudal autocratic rule instead of the essence of democracy while the Western was military and economic technology instead of the advanced social system or modernization of social thoughts. It is easily to find there is wrong in content in the Chinese history.   So does in the whole history. The cultural construction as an open methodology is not only justifiable, but the value is very enlightening.

4   Communication Inside and Outside the Classroom, Creating an Innovative Environment

There are three ways how to create an innovative environment. First, besides of classroom, movies and novels also offer opportunity to understand the history in the unexpected situations. Second, extra-curricular reading activities, thematic reports or lectures and study tours are necessary to enrich the extracurricular life of students, expand the knowledge and to provide students with more space for independent activity. Lastly, historical subjects and other subjects are linked. According to the needs of teaching, students have chances to learn, understand the history by videos, surveys and visits. There are various activities suggested, such as seminar, debating, seminar, communicative, self-style and other teaching methods.

5   Creation of Problem Situations and Strengthen Awareness of the Problem

Construction of learning theory puts that learning is an active construction process when learners build new ideas and concepts by virtue of their knowledge and support of existing conditions. Problems arise when there is a gap between the original knowledge and new knowledge. But the problem is the starting point of scientific research and is the key to open the door to science. No question is there, no spark of innovation.

Question consciousness is the power of thinking and the cornerstone of innovation in the thinking process. Scientific innovation occupies a very important position. Problem awareness is not only the activeness and profoundness of individual mind, also the independence and creativity. It is the starting point of innovation for students to strengthen one's awareness of the problem. It is doubtless void and meaningless if teachers ignore the issue of students while talking about innovation consciousness and the quality of education. For a student, problem generated and solved is viewed as an important part of creative ability, knowledge accumulation, capacity development of logical strength.

It is the consensus of Chinese and foreign educators to fully recognize and attach import to problems consciousness of students. Confucius believes that the suspect is the source of thinking and the consequences of learning. Zhu Xi also said: "reader without question should be taught in doubt and one with question should be taught to dispel them. That is a key. This is the sense of learning problems in science and elaborated dialectically. Lu Jiuyuan, famous scholar in the Song dynasty, noted "To learn is undoubtedly suffering from the suspected. Little doubt is a small forward, Large is big". This is a full recognition of the role of problem awareness. Aristotle once said, thinking is a doubt and surprise from the start. Socrates also believes that the problem is the midwife who the birth of new ideas can help. Einstein also emphasized that found that more important than solving the problem.

In the view of the present situation of our students, however, the problem is relatively overlooked. One can not or do not want to ask questions or can not or are not good questions. Therefore, in the history of the teaching process, teachers should pay special attention of good educational environment and atmosphere creating. Students are encouraged to explore issues and solve the problem of interest in the set of problem situations. The ways of finding the issue are diverse while students may be instructed by teachers or independent research. The problems can be within the historical discipline as well as relating to other discipline. Meanwhile, it can be a single in nature and it can be integrated. Students are stimulated to grasp the multi-dimensional thinking and independent thinking when aware the importance of the question.

Advance is always on the base of summarizing the past. History of Revolution initiated by Liang Qichao, has played the pioneer role in the study of a new history. Hu Shi, who advocated experimentalism to promote "scientific method", has greatly influenced in the scientific methods of historiography. In the new century of globalization and diversification, there is still a long way to go to develop the innovative teaching of history. Famous educator Tao, stated: "Everywhere is an innovative place, every day is the innovation, everyone is a creative person." Every student has the potential of innovation as long as teachers have sharp eyes to instruct.

Acknowledgments. This work is in part supported by project of Anqing Teachers College titled study on innovative cultivation between students and teachers under the granted number of 044-j04020, the philosophy and social sciences planned project in Anhui Province titled the buliding of famous historical and cultural city and the developemnt of the industry of regional culture under granted number of AHSKF07-08D111, and the humanistic discipline project of the Education Depantment in Anhui Province titled the developmental strategy of east-orientation and the study on Wanjiang culture under the granted number of 2008sk410.

References

1. Zhao, H., Feng, X.: The Creation of the Discipline of History Education. Shandong Education Press, Jinan (1997)
2. Yao, B.: Problem Awareness and Innovation. China Education and Research Network, 2004-8-1 (2004)
3. He, J.: College Students Innovative Quality Strategy and Implementation of the Approach. Ideological Education Research (10) (2007)
4. Wang, L.: Implementation of innovative Education and Training Creative Talents. Education (7) (1999)
5. Wang, E.-B.: Quality Education and Student Innovation Quality. Contemporary Education (Hong Kong) (5), 21–22 (2001)
6. Gao, B.: On the Innovative Spirit and Creative Ability. Jiangsu University 4, 1–4 (2003)

Automatic Metadata Extraction for Educational Resources Based on Citation Chain

Shuai Yuan, Sanya Liu, Zhaoli Zhang, Huan Huang, and Meng Wang

National Engineering Research Center For E-learning, HuaZhong Normal University,
Wuhan, China
1755109581@qq.com

Abstract. This paper proposes an approach to automatically extract metadata from educational resources using citation chain. Tracing the cited relationship between literatures, utilizing pattern matching algorithm to determine the most appropriate pattern for resolution, taking advantage of preorder traversal algorithm to query literature one by one, the metadata of educational resources, which are cited by others, can be extracted automatically. The experimental results show that the approach we proposed is fast, efficient in metadata auto-extracting for educational resources.

Keywords: CitationChain, Metadata, Educational Resources, Pattern Matching, Preorder Traversal.

1 Introduction

In a variety of digital libraries and literature repositories, a great deal of metadata information has been established, such as title, author, abstract, keywords, and citation and so on, however, it is inefficient that many of them are inputted manually. Much metadata is extracted by processing the document itself, by analyzing the structure of literature to select relevant method, it is so complex, and the accuracy is very low. Citations in literature already contain the main information about referenced articles, we can acquire metadata by analyzing citations, and it is simple and fast [1]. The InfoMap method based on a hierarchical knowledge representation framework was introduced by Min-Yuh Day [2], and the method of citation metadata extraction based on knowledge was introduced by Eli Cortez [3], they can acquire more accurate metadata extraction through citations from a single document any format. Dr. Eugene Garfield from United States in the Science journal published an important paper--- "Citation Indexes for Science-A New Dimension in Documentation through Association of Ideas" [4], proposed the concept of "citation index" at first. A article is considered as the key words, we can grasp the context of the research topics by taking its references and tracking the situation cited after publication to find the associated study quickly, citation database of high quality - the Science Citation Index was establish. Science Citation Index reveals the intrinsic links between the scientific

Y. Wu (Ed.): International Conference on WTCS 2009, AISC 116, pp. 97–106.
springerlink.com © Springer-Verlag Berlin Heidelberg 2012

literatures according to the cited relationship between literatures [5].Citations among scientific documents form a time-constrained non-redundant content net [6]. If article A is cited by article B, we can analyze citations of article B to get metadata of article A. It is much easier to analyze the citation than analyze the entire article. Metadata of articles can be extracted through a citation chain. If certain articles are not cited by others, we can't extract their metadata, in other words, certain articles are not related to the subject we want to extract. According to the approach above, we formulate different pattern styles to meet different needs, select appropriate pattern based on pattern matching algorithm, extract metadata of a citation accurately, proceed from citation chain, take advantage of association role between citations, realize to extract metadata for educational resources in bulk on some subject automatically and efficiently.

2   The Process of Metadata Extraction for Educational Resources

There are various educational resources; most of them are published in PDF format, in addition to HTML, DOC and PPT format in small section. First of all, we write a format converted software based on htmlparser.jar, PDFBox-0.7.3.jar and tm-extractors-0.4.jar, convert educational resources in various formats named by titles of the literature to documents in TXT format, put them into literature database. Each document in literature database is analyzed, using regular expression method [7][8] to locate citation position, then citations which are itemized by the analyzing format of citation marks are putted into citation database with their corresponding titles, the layer number of resources (referred in section 4)and resource ID numbers(referred in section 4). In the part of citation chain analysis, select any title of article in citation database, take out its corresponding citations, put them into the browser cache in ascending order by the names of citations, use the pattern matching algorithm referred in section 3, select the appropriate pattern from pattern database, split and extract metadata automatically such as title, author and so on, extract metadata circularly using the method referred in section 4, the cycle doesn't stop until metadata extraction of the whole citation chain was completed. Fig. 1 describes the whole process of automatic metadata extraction for educational resources based on citation chain.

3   Pattern Matching

Scholars at home and abroad studying automatic metadata extraction for electronic documents, most of whom concerned about the extraction of English literature, few concerned about the extraction of Chinese literature [9][10]; some method are about machine learning, the validity of the model based on this method, depended on the quantity and quality of training samples [11]; Identifying and extracting metadata according to the external features of electronic documents, is not

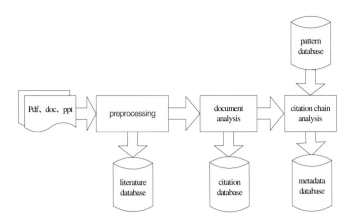

Fig. 1. Metadata extraion for educational resources flow graph

efficient [12], most of them are judged by the regular expression [13],so the formats of matching object are as little as possible, and different publishers require different formats of literature citation, it is difficult to automatically extract metadata we need, for example, authors' names are often separated by a comma"," or the word "and", and the set of authors is often terminated by a period ".", which also indicates the beginning of the title of a paper. In this paper, we define a large number of pattern styles that are comprised in pattern database, select different pattern styles to form patterns according to the different formats of citations, calculate the similarity of patterns using pattern matching algorithm, and consider the pattern whose value is the maximal as the most appropriate pattern. We can add pattern styles quickly and easily to form different patterns to extract different formats of citations. Citation rules are defined based on "The People's Republic of China National Standard-- Rules for Content, Form and Structure of Bibliographic References (GB/T7714-2005)" (RCFSBR) [14] by publisher. We refer to several citation formats of different publisher, with the help of citation [15] based on pattern matching algorithm, extract metadata automatically.

After which information we want to extract and the relationship between them are determined, we can extract corresponding metadata from documents. The citation is a string with a specific format. We put citation styles listed in RCFSBR into pattern database, provide ample pattern styles. Table 1 shows the name of pattern style and the meaning of pattern style. For example, a citation:

Jian Fu, Xue Yang. Theoretical and Practical Researches on Mobile Learning in the Recent 10 Years[J]. China Educational Technology, 2009, (7):36-41.

The style is: (Responsible officer of literature), (Literature name) [Literature type], (Journal name), (Year), (Volume):

(Pages in serials).

We can use pattern following to express its corresponding style:

'PRIMARY RESPONSIBILITY, TITLE[TYPE], JOURNAL, YEAR,(VOLUME): PAGES.'

A citation can be analyzed by different patterns, so the citation previously mentioned can be analyzed by other patterns as following: "PRIMARY RESPONSIBILITY, TITLE[TYPE], JOURNAL, ANY." or "PRIMARY RESPONSIBILITY, TITLE[TYPE], ANY.". Therefore, we must make the corresponding pattern matching algorithm to choose the most appropriate pattern for analysis.

Table 1. The names and the meanings of pattern styles

Name	Meaning
PRIMARY RESPONSIBILITY	Responsible officer of literature
TITLE	Literature name
TYPE	Literature type
SERIALS	Serials Title
OTHER SERIALS	Other serials information
YEAR	Year
VOLUME	Volume
PAGES	Pages in serials
URL	Website
SCHOOL	MA, PhD thesis School
PUBLISHER	Publisher
ISSUE	Issue
PLACES	Places of publisher
TIME	Publish date
JOURNAL	Journal name
PROCEEDINGS	Proceeding name
ANY	Any string

Definition 1: String set $R=R_1, R_2, ..., R_n$, where $R_i(1<= i<=n)$ means a string (including the separator) that is separated by separators such as space, comma, dot and so on.

Definition 2: Pattern style set $W= \{W_1, W_2, W_3,..., W_m \}$, where each W_i ($1<= i<=m$) means a pattern style in table 1.A weight M_i is given to corresponding W_i, which means the importance of W_i in the pattern. For example, the pattern style of W_1 is "PRIMARY RESPONSIBILITY", its weight is 0.6,the pattern style of $W8$ is "ANY", its weight is 0.15, that is to say, the pattern of "PRIMARY RESPONSIBILITY" is much important than the pattern of "ANY". The weight of M_1 is determined in advance.

Definition 3: Pattern database $Z= \{\Phi_1, \Phi_2, \Phi_3,... \Phi p\}$, each Φ_i ($1<= i<=p$) is the pattern that combines the elements in set W and separators.

Now, the appropriate pattern Φ_j is need to select, it is consistent with separated format of string R, and the analyzed information of string R is complete.

According to the requirements of mission, we define the function $f_i=f(W_i, \Phi_j)$, which means the important of W_i in pattern Φ_j, the calculation is shown in Formula 1.

$$f(W_i, \Phi_j) = \begin{cases} M_i & W_i \in \Phi_j \\ 0 & W_i \notin \Phi_j \end{cases} \tag{1}$$

For a given string R, the similarity of each pattern that is able to match can be calculated in Formula 2:

$$sim(R, \Phi_j) = \sum_1^m fi \tag{2}$$

Therefore, there is a corresponding similarity for each matched pattern. The values of similarity that are more than 0.45 are sorted in ascending order, the maximal one is considered as the appropriate pattern. We can get corresponding metadata by analyzing citations using the selected pattern. If all values of similarity are less than 0.45, which means that there is no suitable pattern in database, we need to add more into database.

4   Automatic Extraction in Bulk Using Citation Chain

After accurately extracting metadata from the citations of an article by pattern matching, through the citation chain, we can automatically extract metadata from literatures which are cited by this one. The main ideas for extraction are as follows:

1)   Select the title of any article in citation database, get the corresponding citations of the article, and put them into the browser cache by the names of citations in ascending order;

2)   Select the first citation in step (1), utilize pattern matching algorithm referred in section 2, pick out the pattern which the similarity of is maximal, split the first citation using the appropriate pattern, and extract metadata;

3)   According to metadata extracted just now, we look for the corresponding title of article in citation database, if we find it, put the metadata into metadata database; on the contrary, we select the next citation in step (1), pick out the appropriate pattern based on pattern matching algorithm, split the citation using the appropriate pattern, extract metadata, look for the corresponding title of article that is consistent with the metadata;

4)   Select the title in citation database according to the metadata extracted above, pick out all the corresponding citations, put them into the browser cache by the names of citations in ascending order;

5)   Repeat the operation of step (2).Select the first citation in step (4), utilize pattern matching algorithm referred in section 2, pick out the appropriate pattern whose value of similarity is maximal ,split the citation, and extract metadata;

6)   Repeat the operation of step (3). According to metadata extracted in step (5), we look for the corresponding title of article in citation database, if we find it, put the metadata into metadata database; on the contrary, we select the next citation in step

(4), pick out the appropriate pattern based on pattern matching algorithm, split the citation using the appropriate pattern, extract metadata, look for the corresponding title of article that is consistent with the metadata;

   7)   Extract metadata from the last citation in step (4), put it into database, or there is not corresponding title to this citation, the traversal of this layer is over, reverse to the upper layer;

   8)   Repeat the operation of step (2).Pick out the second citation in step (1), utilize pattern matching algorithm referred in section 2, split the citation using the appropriate pattern, and extract metadata;

   9)   Repeat the operation of step (3). We look for the corresponding title of article in citation database, according to metadata extracted in step (8), if we find it, put the metadata into metadata database; on the contrary, we select the next citation in step (1), pick out the appropriate pattern based on pattern matching algorithm, split the citation using the appropriate pattern, extract metadata, look for the corresponding title of article that is consistent with the metadata;

   10)   Repeat the operation of step (4).
       ......

   11)   Cycle as above, extract metadata from the last citation in step (1), put it into database, or there is not corresponding title to this citation in citation database, the metadata extraction based on citation chain end.

   In the process, the article which is the first one in citation chain is considered to be the first layer(the layer number of resources), is equivalent to the root of the tree; nodes L2.1, L2.2 and L2.3 are children nodes of node L1, are three citations of resources L1, they are shared in the second layer, and L2.1, L2.2 and L2.3 are the numbers of resources ID; nodes of L3.1, L3.2 and L3.3 are children nodes of node L2.1, are three citations that are consistent with the title of the resource L2.1 in citation database, are share in the third layer, and L3.1, L3.2 and L3.3 are the numbers of resources ID. L3.4, L3.5and L3.6 are the same to L3.1, L3.2 and L3.3.Node L2.2 has no child node in the tree, that means the literature which are consistent with the citations of L2.2 don't exist in citation database.

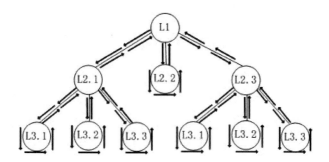

Fig. 2. Citation chain query process based on preorder traversal

   The process of metadata extraction from citation chain is similar to preorder traversal of the tree, if the titles in the citation database which are consistent with the metadata extracted exist, traverse down, and extract metadata; if the metadata of the last citation in the browser cache have been extracted and have been putted into database, or there is not corresponding title in citation database to this citation, traverse up, and extract metadata. Thus, the metadata of literature based on a specific topic in citation database can be all extracted, if the metadata of several articles are not extracted, it means they are not relevant to the subject we want to extract.

5   Experiment

In the experiment, download literature from "China Educational Technology" and "Modern Distance Education Research", the number of which increased from 100 to 1000 by 100, select five articles as the initial articles for citation chains each journal, the ratio between the actual number of metadata extraction using the method referred in this paper, and the due number of metadata extraction by artificial statistic, is considered to be recall. Table 2 shows comparison of the recall of two journals when the number of literature database is 1000 each journal. It shows that each citation chain recall of two journals is more than 50%, of which six citation chain recalls are more than 80%, of which three citation chain recalls are more than 90%, the average recalls of two journals are 84.9% and 82.1%, respectively.

Table 2. Citation chain recall of two journals(1000)

Citation chain number	China Educational Technology			Modern Distance Education Research		
	Due number	Actual number	Recall	Due number	Actual number	Recall
1	70	60	85.8%	120	110	91.7
2	120	100	83.3%	60	30	50%
3	100	80	80%	90	90	100%
4	80	60	75%	80	50	62.5
5	140	130	92.9%	40	30	75%
average	510	430	84.9%	390	310	82.1

Select two hundred pieces of metadata extracted from each citation chain, extract the metadata such as responsible officer of literature, literature name, literature type, journal name, year, volume, issue, pages in serials and so on, the ratio between the actual number of metadata extraction using the method referred in this paper, and the due number of metadata extraction by artificial statistic, is considered to be recall. In the Fig. 3, the recalls of literature name and literature type are more than 90%, which are the most of all recalls, the recalls of journal name, responsible officer of literature and pages in serials are more than 80%.Comparison of the two journals, for the recalls of literature name, responsible officer of literature, journal name and issue, the ratios of "China Educational Technology" are more than the ratios of "Modern Distance Education Research", and the average recall about all metadata of "China Educational Technology" is 86.9% , the average recall about all metadata of "China Educational

Technology" is 83.1% , it means that the citation rule of "China Educational Technology" is closer to RCFSBR.

In the experiment, we download literature in "China Educational Technology" and "Modern Distance Education Research" increased from 100 to 1000 by 100, find the rules in the Fig. 3.With the number of literature in literature increased from 100 to 1000, the recall of "China Educational Technology" is increased from 68.3% to 84.9%,and the recall of "Modern Distance Education Research" is increased from 62.17% to 82.1%.Compared two journals, the change of the former recall ratio is radical from slow, the latter is slow from radical.

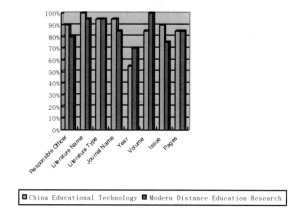

Fig. 3. The metadata recalls of two journals

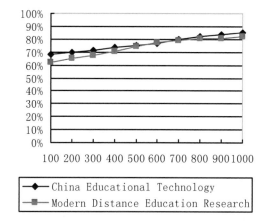

Fig. 4. The relationship between recall and the number of literature database

6   Conclusion

This paper proposes a method of automatic metadata extraction in bulk for educational resources. For "China Educational Technology" and "Modern Distance Education Research", the metadata recalls are increasing when the number of literature is increasing from 100 to 1000. Due to the limited literature database, the articles which the metadata extracted are consistent to don't exist in literature database, so we can't extract them. If based on a powerful database, all metadata referred in literature database can be extracted, and the recall will be much greater. In the experiment, the formats of "year" in two journals are different, and we don't convert and match properly, so the recall is very low. When there are spaces in first name and last name of responsible officer of literature, the metadata don't extract accurately, so the recall ratio is much lower than the recall ratio we expected. The metadata information extracted from citations and the title of literature are not identical, and the title extracted from citation must be consistent with the title in literature database when we automatically extracted metadata by citation chain, otherwise, the metadata can't be extracted completely now, it need further research.

Acknowledgment. This research is supported by Cultivation Fund of the Key Scientific and Technical Innovation Project, Ministry of Education of China (NO.708065), China Postdoctoral Science Foundation (NO.200902443), and Self-determined Research Funds of CCNU from the Colleges' Basic Research and Operation of MOE (NO.CCNU09Y01007).

References

1. Yuan, B., Fang, S., Liu, Q., Zhang, J.: Overview on Progress in Citation Analysis at Home and Abroad. Information Science (1), 147–153 (2010)
2. Day, M.-Y., Tsai, R.T.-H., Sung, C.-L., et al.: Reference Metadata Extraction Using a Hierarchical Knowledge Representation Framework. Decision Support Systems (43), 152–167 (2007)
3. Cortez, E., Silva, da A.S., Goncalves, M.A., et al.: FLUX-CIM:Flexible Unsupervised Extraction of Citation Metadata[EB/OL], (December 18, 2007), http://delivery.acm.org/10.1145/1260000/1255219p215-cortez.pd?fkey1=1255219&key2=9296088911&coll=GUIDE&dl=GUIDE&CFID=10613840&CFTOKEN=55320929
4. Garfield, E.: Citation Indexes for Science: A New Dimension in Documentation Through Association of Ideas. Science, 468–447 (1955)
5. Powley, B., Dale, R.: High. Accuracy Citation Extraction and Named Entity Recognition for a Heterogeneous Corpus of Academics Papers. IEEE (2007)
6. Hai, Z.: Discovery of knowledge flow in science. Communications of the ACM 49(5), 101–107 (2006)
7. Li, C., Zhang, M., Deng, Z., Yang, D., Tang, S.: Automatic Metadata Extraction for Scientific Documents. Computer Engineering and Applications 21(10), 189–191, 235 (2002)
8. Yang, N., Zhang, M., Zhou, B.: A Rule-based Metadata Extractor for Learning Materials. Computer Science (3), 94–96 (2008)
9. Zeng, S., Ma, J., Zhang, X.: New Development of Automatic Metadata Extraction. New Technology of Library and Information Service (04), 7–11 (2008)

10. Hetzner, E.: A Simple Method for Citation Metadata Extraction using Hidden Markov Models. In: Proceedings of the 8th ACM/IEEE-CS Joint Conference on Digital Libraries, vol. (2), pp. 280–284 (2008)
11. Yu, J., Fan, X.: Metadata Extraction from Chinese Research Papers Based on Conditional Random Fields. In: International Conference on Digital Libraries, vol. (1), pp. 497–501 (2007)
12. Zhang, X., Ma, J.: Automatic Extraction of Semantic Metadata from PDF Research Papers. New Technology of Library and Information Service (2), 102–106 (2009)
13. Yao, Z., Huang, D., Ji, X.: Application of regular expressions to extraction of Chinese cultural terms with their English translations. Journal of Dalian University of Technology (2), 291–295 (2010)
14. The People's Republic of China National Standard– Rules For Content, Form And Structure Of Bibliographic References (GB/T7714-2005). General Administration of Quality Supervision, Inspection and Quarantine of the People's Republic of China (2005)
15. Guo, Z.: Ontology-based document citation metadata extraction. Microcomputer Information (18), 304–306 (2006)

The Application of Binary Tree-Based Fuzzy SVM Multi-classification Algorithm to Fault Diagnosis on Modern Marine Main Engine Cooling System

Yulong Zhan[1], Huiqing Yang[2], Qinming Tan[1], and Yao Yu[1]

[1] Department of Marine Engineering, Shanghai Maritime University, Shanghai, China
[2] Shanghai Research and Development Center, Lubricating Oil Group Co Sinopec, Shanghai, China
yu_qun2007@hotmail.com

Abstract. Support Vector Machine（SVM）is widely applied to fault diagnosis of machines. However, this classification method has some weaknesses. For example, it can not separate fuzzy information, is particularly sensitive to the interference and the isolated points of the training sample, and has great demand for memory in calculation. In view of the problems mentioned above, a binary tree-based fuzzy SVM multi-classification algorithm (BTFSVM) has been put forward. This paper focuses on the study of the application of the intelligent theory BTFSVM to fault diagnosis on modern main engine cooling water system of ships. Simulation experiments show that the algorithm has strong anti-interference ability and good classification effects. Consideration can be made that it can be further applicable to the diagnosis on other mechanical faults of ships.

Keywords: binary tree, FSVM, the cooling water system of ships, fault diagnosis.

1 Introduction

Modern marine power plant has increasingly higher degree of automation and greater power, which brings in considerable economic benefits. But at the same time, both the factors that affect the operation of the system and the losses caused by faults increase, which makes people pay more attention to the security of marine power plant production. Diagnostic system calls for stricter requirements, such as improving sensitivity, achieving early fault diagnosis and enhancing the promotion of diagnostic systems [1]. As thus, more effective methods of fault diagnosis need to be studied and this also becomes an important research project of process control at the present stage.

As for the method BTFSVM [2,3], firstly the clustering centers of each type of samples are computed by using fuzzy clustering technique and all clustering centers are divided into two successively. In this way, a binary tree is determined. And then, SVM sub-classifiers are constructed by classifying the corresponding samples into two types at each node of the binary tree according to the clustering centers of samples. The experiment shows that only k-1 SVM sub-classifiers need to be

Y. Wu (Ed.): International Conference on WTCS 2009, AISC 116, pp. 107–114.
springerlink.com   © Springer-Verlag Berlin Heidelberg 2012

constructed for k-class fault diagnosis. This can not only simplify the structure of classifiers and avoid unclassifiable regions but also save memory and improve correctness of fault diagnosis.

2   Fuzzy Support Vector Machine Introduction

SVM [4] essentially involves a two-class classification problem, which usually converts an n-class problem into n two-class problems. There may be some unclassifiable regions. However, in real-world applications, it is required that each sample should belong to certain class. Some samples may be more important than the other ones. Therefore, it requires that the important samples are correctly classified and less attention is paid on whether other samples like interferences are misclassified or not. This paper attempts to combine SVM with fuzzy membership. Through calculating fuzzy membership of each sample in each class and explaining the SVM learning results, further optimi-zation of the classification can be achieved.

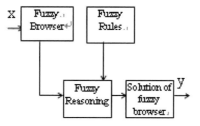

Fig. 1. Fuzzy logic inference system

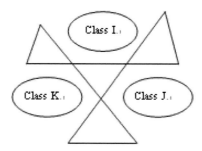

Fig. 2. Unclassifiable region

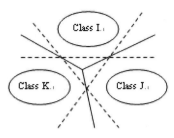

Fig. 3. Classifiable region by FSVM

Give fuzzy membership $0 < m_i \leq 1$ to each sample x_i. And m_i stand for the importance of a sample or the possibility of a sample belonging to certain type. After using m_i, the optimal classification problem is as follows:

$$\text{Minimize} \quad \frac{1}{2}\|W\|^2 + C\sum_i^l m_i \xi_i \tag{1}$$

$$\text{Subject to} \quad y_i(<w \cdot \phi(x_i)> + b) \geq 1 - \xi_i$$

The term $m_i \xi_i$ is used as measurement of punishment for misclassifications of samples of different importance. By applying the Lagrange to (1), it can be changed into:

$$\text{Minimize} \quad \frac{1}{2}\sum \alpha_i \alpha_j y_i y_j K(x_i, x_j) - \sum_i \alpha_i \tag{2}$$

$$\text{Subject to} \quad \sum_{i=1}^l y_i a_i = 0 \ , \ 0 \leq \alpha_i \leq m_i C$$

So the classification function of the FSVM classifier can be expressed as:

$$f(x) = sign[\sum_{i=1}^l \alpha_i y_i K(x_i, x) + b] \ . \tag{3}$$

The SVM samples corresponding to $0 \leq \alpha_i \leq m_i C$ lie where the hyper-plane $1/\|w\|$ are, while those corresponding to $\partial_i > m_i c$ are misclassified samples.

In specific applications, firstly the upper and lower bound of fuzzy membership of the samples— m_{max} , m_{inf} , are determined according to the specific features of training data. And then an appropriate membership function is selected to establish the connection between the sample x_i and the membership m_i. Finally, the optimal classified planes and the classification functions are calculated according to the FSVM theory [5].

3   BTFSVM Training Algorithm

BTFSVM learning algorithm [6] can be described as follows :

● Step1. Obtain the fuzzy clustering center for each type of learning samples. Set a total of k classes and each class corresponds to a clustering center;

● Step2. By using fuzzy clustering technology, U can be clustered into U_P and U_N, $U_P \subset U$ $U_N \subset U$, $U_P \cap U_N = \phi$, $U_P \cup U_N = U$;

● Step3. All learning samples which corresponds to each clustering center belonging to U_P are set for P1, and samples which corresponds to each clustering center belonging to U_N set for a negative-type N_1. Recombine learning samples to get NNew1, $P_1 \cup N_1 = N_{New1}$, $P_1 \cup$ $N_1 = F$ and to construct SVM sub-classifier S_{SVMc1};

● Step4. U_P can be clustered into two classes and then the samples corresponding to each clustering center are marked as P_2 and the negative-type N_2. Likewise, U_N can be clustered into two classes, and the samples corresponding to each clustering center are marked as P_3 and the negative-type N_3, where $P_2 \cup N_2 = P_1$, $P_2 \cap N_2 = \phi$, $P_2 \subset P_1$, $N_2 \subset P_1$; $P_3 \cup N_3 = N_1$, $P_3 \cap N_3 = \phi$, $P_3 \subset N_1$, $N_3 \subset N_1$;

● Step5. A SVM sub-classifier S_{SVMc2} is constructed according to P_2 and N_2 and a sub-classifier S_{SVMc3} is constructed according to P_3 and N_3;

● Step6. Repeat Steps 4 and 5, until the K-1 SVM sub- classifier S_{SVMC}, K-1 is constructed;

● Step7. From Step 1 to Step 6, K-1 SVM sub-classifiers($S_{SVMc1}, \cdots, S_{SVMc,K-1}$), the corresponding K-1 new learning sample sets($N_{New1}, \ldots, N_{Newk-1}$) and k-1 functions($f_1(x), \ldots, f_{k-1}(x)$) can be obtained.

When carrying out the sample tests, start them from the root node sub-classifier S_{SVMc1} to determine that its output belongs to positive class or negative one (expressed as +1 and -1 respectively). Then use the corresponding second-level SVM sub-classifier to test on the basis of the output results, and never end the calculation until the final-level sub-classifiers. Thus, the classes all the test samples belong to are obtained.

4   Fault Diagnosis on Modern Marine Main Engine

The diagram for the cooling system of main engine of ships is as follows:

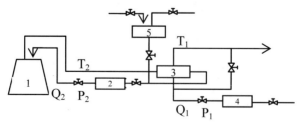

1. Main Engine   2. Fresh Water Pump   3. Condenser   4. Sea Water Pump   5.Fresh Water Tank
Fig. 4. The diagram for cooling system of main engine of ships

4.1   Fault Mechanism of the Cooling System of Marine Main Engine

According to the actual situation of the cooling system of marine main engine, select the following 6 measurable parameters as the input variables for fuzzy fault diagnosis system of the cooling system of marine main engine, that is, the seawater temperature T_1, the freshwater temperature T_2, the seawater pressure P_1, the freshwater pressure P_2, the seawater flows Q_1 and the freshwater flows Q_2. And select the following 16 parameters as the output variables for fuzzy fault diagnosis system of the cooling system of marine diesel engine, that is, normal F_0, overloading cylinder F_1, cylinder ignition delay F_2, fresh water valve shut down or damage F_3, low pressure of fresh water pump F_4, fresh water piping leak F_5, high pressure of fresh water pump F_6, fresh water bypass valve opening small F_7, fresh water piping system plug F_8, cooler piping plug F_9, sea climate F_{10}, low pressure of sea water pumps F_{11}, sea water piping plug F_{12}, high pressure of sea water pumps F_{13}, sea water bypass valve open big F_{14}, sea water pump plug F_{15}.

Every input datum is represented by three kinds of states, that is, Normal, High, Low which are marked as N, H and L respectively. In order to avoid the impact of measurements on the SVM learning process, the sample data are no-rmalized as follows: H corresponds to 1, N to 2 and L to -1. Three samples data are collected for each type of fault. In order to express the corresponding relationship between the unit load and fault, four conditions are taken into account: the rated load (100% MCR), partial load (90% MCR, 75% MCR) and semi-load (50% MCR). Then atmospheric tem-perature changes of navigation environment of ships also need to be taken into consideration and three representative atmospheric temperatures of 288K, 300K, and 312K are selected. Finally 576 data samples are obtained, of which the former 384 are used for training and the rest 192 as the test samples.

4.2   Determine the Parameters of FSVM Classifiers

Theoretically there is no unified conclusion on how to set the parameters of SVM. That which parameter is selected depends on the practical problems to be tackled. RBF kernel function is used for the fault diagnosis in this paper:

$$K(x_i, x_j) = \exp\{\frac{-\left\| x_i - x_j \right\|^2}{\sigma^2}\}, i = 1, \ldots, l \tag{4}$$

In (4), σ is the variable whose value is free to choose and which determines the width of the Gaussian function center. In this paper, the following empirical formula is adopted to determine the value of σ :

$$\sigma^2 = E(\left\| x_i - x_j \right\|^2) \tag{5}$$

In (5), E is the mathematical value. The parameter C is selected. When C> 0, it stands for weight coefficient, and controls the degree of punishment on the misclassified samples. In this paper, we take the value C = 120.

The choice of fuzzy membership depends on the determination of the lower bound of fuzzy membership. Fuzzy membership is regarded as a linear function of time, that is:

$$m_i = f(t_i) = at_i + b \tag{6}$$

In (6), $t_1 \le t_2 \le \cdots \le t_i \le \cdots \le t_l$ is the time for data acquisition. The later the time is, the more important the datum is. The membership of the first sample is $m_1 = f(t_1) = \sigma$, and that of the last one $m_l = f(t_l) = 1$. By applying these two boundary conditions to (1), we can get the function of fuzzy membership:

$$m_i = f(t_i) = \frac{1 - m_{\inf}}{t_n - t_1} t_i + \frac{t_n m_{\inf} - t_1}{t_n - t_1} \qquad (7)$$

4.3 The Application of TFSVM Training Algorithm to the Cooling Water System

•   Sort all the samples based on the number of each sample, and form a sequence set S = {S_1, S_2,..., S_{16}}, where S_1 is the item appearing the most frequent in training samples.

•   Determine the type of training samples. Supposing the FSVM set is F = {F_1, F_2, ..., F_{15}}, where F_i stands for the positive training samples of S_i , the rest for the negative ones.

•   Select training samples and construct a possible support vector sample set D = (D_1, D_2, ..., D_m), $D_i \le x_i$, y_i, $m_i >$ i = 1, ..., m, m $\le$ 16. The specific construction method can be seen in the reference 3.

•   Train FSVM F_i. Taking the samples in set D as training sample, we can adopt the metric method of fuzzy membership mentioned above to determine the fuzzy membership m_i of each sample, use   algorithm of minimum and optimum sequence to solve the planning issues of FSVM, namely Solution 2.

•   Calculate accuracy. It needs to be determined whether a new sample should be added to D. Classify training samples with the trained F_i according to the equation 3, and calculate the classification accuracy rate P until P = 1 is obtained.

•   Judge whether the algorithm is ended or not. If c <k, c++, then return to Step 2, otherwise the training comes to the end.

4.4 Simulated Results

Table 1. Results of BTFSVM on the Samples Superimposed

Serial No.	Fault	Characteristics and Symptoms						Type
		T_1	T_2	P_1	P_2	Q_1	Q_2	
1	F_6	0.9N	L	N	0.8N	H	N	F_6
2	F_6	1.1N	L	0.9N	0.8N	H	N	F_6
3	F_6	N	0.8N	N	N	H	1.2N	F_6
4	F_6	0.9N	L	1.2N	N	0.8H	N	F_6
5	F_6	N	1.3L	N	N	0.9H	N	F_6
6	F_6	0.9N	L	N	0.8N	H	1.3N	F_6

Table 2. Results of BTFSVM on the Cooling System

Serial No.	Fault	Characteristics and Symptoms						Type
		T_1	T_2	P_1	P_2	Q_1	Q_2	
1	F_0	N	N	N	N	N	N	F_0
2	F_1,F_2	H	H	L	N	N	L	F_1,F_2
3	F_{10}	H	H	N	N	N	N	F_{10}
4	F_6,F_7	N	L	N	N	N	N	F_6,F_7
5	F_8,F_9	N	H	N	N	L	N	F_8,F_9
6	F_6	N	L	N	N	H	N	F_6
8	F_{11},F_{12}	N	H	L	N	N	N	F_{11},F_{12}
9	F_{13}	N	L	H	N	N	N	F_{13}
10	F_{14},F_{15}	N	H	N	N	N	L	F_{14},F_{15}
12	F_3,F_4,F_5	N	H	N	L	N	N	F_3,F_4,F_5

Table 3. Comparison

Type of Fault		F_1	F_5	F_9	F_1,F_4
Misdiagnosed Samples	SVM	0	2	3	0
	BTSVM	0	1	2	0
	BTFSVM	0	1	1	0
Type of Fault		F_1,F_5	F_2,F_6	F_1,F_2,F_5	F_5,F_7,F_9
Misdiagnosed Samples	SVM	3	0	2	3
	BTSVM	3	0	2	3
	BTFSVM	2	0	1	2

Table 1 shows that BTFSVM has a strong ability to resist interference and process the input data deviating from the training samples. Table 2 shows that BTFSVM is able to correctly classify the fault samples. Table 3 shows that BTFSVM algorithm has better anti-interference ability and classification effect.

5   Conclusion

The BTFSVM multi-classification algorithm put forward in this paper has similar efficiency with the binary tree-based SVM multi-classification algorithm as it adopts the method of pre-taking SVM. The experiment shows that the former algorithm has better anti-interference ability and classi-fication effect than the latter.

As for the choice of the parameters m_i in the mea-surement method of fuzzy membership adopted in this paper, it relies on subjective empirical formula. The parameter m_i will have a direct impact on the final classification effect , so how to choose it needs to be further studied.

Acknowledgment. The authors wish to thank scientific research fund 09-22 of Shanghai Maritime University.

References

1. Jiaqiang, E.: The Diagnosis and Applications of Intelligent Fault. Hunan University Press (2006)
2. Lin, C.-F., Wang, S.-D.: Fuzzy Support Vector Machines. IEEE Transactions on Neural Networks 13(2), 464–471 (2002)
3. Chang, C., Fee, Y.-N.: Study on Multi-Classification Algorithm Based on Fuzzy Support Vector Machine. Computer Application 28(7) (2008)
4. Deng, N., Tian, Y.: New Method of Data Mining. Science Press, Beijing (2004)
5. Zhang, G., Zhang, J.: Fault Diagnosis of the Turbine Speed Control System of Multi-Level Binary Tree Classifier Based on Fuzzy Support Vector Machine. Chinese Journal of Electrical Engineering 25, 420–470 (2005)
6. Zhu, Z., Zhang, B.: The Application of Fuzzy Support Vector Machine to Fault Diagnosis of the Marine Diesel Engine. China Shipbuilding 47(3) (2006)

A Hybrid Evolutionary Algorithm Combining Ant Colony Optimization and Simulated Annealing

Xu XueMei

Jiangsu Nantong University
Mridha@foxmail.com

Abstract. This paper has been proposed an approach based on ACO-SA algorithm for the multi-objective Distribution Feeder Reconfiguration. The simulation results have shown that global or close to global optimum solutions for the system losses, the voltage deviation was attained. Also, the proposed approach minimizes the number of switching operations. The proposed method is independent on the initial status of network switches. The ACO-SA is a combination of two powerful optimization algorithms: Ant Colony Optimization and Simulated Optimization.

Keywords: Evolutionary Algorithm; Ant Colony Optimization; Simulated Annealing.

1   Introduction

Distribution systems are normally designed as open loops and are later operated as radial networks. High percentage of loss has long been one of the disputed dilemmas of distribution networks. As mentioned above one of the effective functionalities of distribution automation system is distribution feeder reconfiguration in order to achieve loss and cost reduction. By the virtue of its nature, distribution feeder reconfiguration is a difficult and complex problem of combinatorial optimization in discreet mathematics that is directly related to the previous state of each link. Distribution system reconfiguration for loss reduction was first proposed by Merlin and Back [1] in 1975 and they have applied a branch-and-bound-type optimization technique to determine the minimum loss reconfiguration. Their proposed method had two problems. Firstly, it didn't guarantee the convergence of the solution and secondly, it required an enormously large amount of calculations for a real network.

In this method the loads are thought to be active and are modeled with fixed current sources and the angle between the voltages is neglected. Heuristics and heuristic methods have long been used to solve DFR problems and as a pioneer in this field, Civanlar et al. [2] made use solely of these methods to determine a distribution configuration which would reduce line losses. In the other approaches, researchers have tended to use the mathematical methods instead. The numerical optimization methods use iteration methods to reach the best topology of the network. In [3], Aoki has solved the problems using Minos/Augmented Software. There is another artificial-intelligence-based approach in solving DFR problems. In [4], Debapriya has

Y. Wu (Ed.): International Conference on WTCS 2009, AISC 116, pp. 115–122.

presented a multiple objective function in which the optimum topology is to be reached as the final target and this is supposed to be accomplished with a fuzzy optimization method. In [5], Gallardo has used greedy reconfiguration algorithm (GRA) and the fast greedy reconfiguration algorithm (FGRA) to minimize the energy not supplied (ENS) of medium-voltage distribution networks.

2   Distribution Feeder Reconfiguration with Regard to Distributed Generation

From a mathematical standpoint the optimal operation of distribution network with regard to distributed generation is an optimization problem with equality and inequality constraints. The objective function is the summation of electrical energy generated by DGs and substation bus as follows:

$$
\min f(\overline{X}) = \sum_{t=1}^{N_d} \left\{ \sum_{i=1}^{N_b} Ploss_i^t \Delta t^t \right\} +
$$

$$
w1 * \sum_{i=1}^{Nsw} |S_i - S_{a,i}| + w2 * \sum_{i=1}^{Nbus} |V_i - V_{rat}|
$$

$$
\overline{X} = [\overline{Sw}, \overline{P_G}, \overline{Q_G},]
$$

$$
\overline{Sw} = [Sw_1, Sw_2, ..., Sw_{Nsw}]
$$

$$
\overline{P_G} = [P_{g1}, P_{g2}, ..., P_{gNg}]
$$

$$
\overline{Q_G} = [Q_{g1}, Q_{g2}, ..., Q_{gNg}]
$$

(1)

where Ng, Nsw and Nbus are the number of DGs, switches and buses, respectively. Also, X is the state variable vector. Swi is the state of the ith switch specified in terms of on/off status, taking 0 or 1 as its value. Sw is the switching state vector including state of all switches. Q_G is the DGs reactive power vector including reactive power of all DGs. Qgi is the reactive power of the ith DG. P_G is the DGs active power vector including active power of all DGs. Pgi is the active power of the ith DG. Si and So,i are the new and original states of switch i, respectively. And Vi and Vrat, the real and rated voltage on bus i, whilst w1 and w2 represent weighting factors. The constraints can be listed as follows:

$$
\left| P_{ij}^{Line} \right| < P_{ij,\max}^{Line}
$$

(2)

- Distribution line limits

$\left| P_{ij}^{Line} \right|$ and $P_{ij,\max}^{Line}$,are the absolute power flowing over the distribution lines and the maximum transmission power between nodes i and j respectively.

- Distribution power flow equations

$$P_i = \sum_{i=1}^{Nbus} V_i V_j Y_{ij} \cos(\theta_{ij} - \delta_i + \delta_j)$$

$$Q_i = \sum_{i=1}^{Nbus} V_i V_j Y_{ij} \sin(\theta_{ij} - \delta_i + \delta_j)$$

(3)

P_i and Q_i are the net injected active and reactive power at the ith bus. V_i and δ_i are the amplitude and the angle of voltage at the ith bus, respectively. And Y_{ij} and θ_{ij} are the amplitude and the angle of the branch admittance between the ith and j^{th} buses.

- Objective function limit

$$f_i(X) \prec f_{0i}; i = 1, 2, 3, 4 \qquad (4)$$

- Radial structure of the network

3   Ant Colony Algorithm

Ants are insects which live together. Since they are blind animals, they find the shortest path from their nest to food with the aid of pheromone. Pheromone is the chemical material deposited by the ants, which serves as critical communication medium among ants, thereby guiding the determination of the next movement. On the other hand, ants find the shortest path based on intensity of pheromone deposited on different paths (15)-(23). Assume that ants want to move from point A to B (Fig. 1).

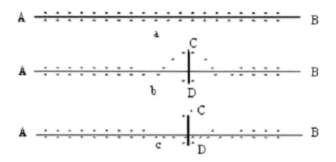

Fig. 1. An example of finding the shortest path by ants

At first, if there is no obstacle, all of them will move along the straight path (AB) (Fig.1.a). At the next stage, assume that there is an obstacle; in this case, ants will not be able to follow the original trail in their movement. Therefore, randomly, they turn to the left (ACB) or to the right (ADB) (Fig1.b). Since ADB path is shorter than ACB, the intensity of pheromone deposited on ADB is more than the other path. So ants will be increasingly guided to move on the shorter path (Fig1.c). This behavior forms the fundamental paradigm of the ant colony system. As was indicated in Fig.1, the intensity of deposited pheromone is one of the most important factors for ants to find the shortest path. Generally, the intensity of pheromone and path length is two important factors that should be used to simulate the ant system. To select the next path, the state transition probability is defined as follows:

$$P_{ij} = \frac{(\tau_{ij})^{\gamma_2}(1/L_{ij})^{\gamma_1}}{\sum_{\substack{j=1 \\ j \neq i}}^{N}(\tau_{ij})^{\gamma_2}(1/L_{ij})^{\gamma_1}} \tag{5}$$

where τ_{ij} and Lij are the intensity of pheromone and the length of path between nodes j and i, respectively. γ_1 and γ_2 are the control parameters for determining the weight of trail intensity and length of path, respectively. N is the number of ants. After selecting the next path, the trail intensity of pheromone is updated as:

$$\tau_{ij}(k+1) = \rho\tau_{ij}(k) + \Delta\tau_{ij} \tag{6}$$

In the above equation, ρ is a coefficient such that (1- ρ) represents the evaporation of the trail between time k and k+1 and $\Delta\tau_{ij}$ is the amount of pheromone trail added to τ_{ij} by ants. To apply the ACO algorithm for reconfiguration, the following steps have to be taken:

Step 1:   Generate the initial population and trail intensity.
Step 2:   Generate the initial population and trail intensity. for ants in each colony (local search) .
Step 3:   Determine the next position.
Step 4:   Check the convergence condition.

4   Simulated Annealing Algorithm

The simulated annealing is a generalization of a Monte Carlo method for examining the equations of state and frozen states of n-body systems. The concept is based on the manner in which liquids freeze or metals recrystalize in the process of annealing. In an annealing process, a melt is initially at high temperature and disordered and is slowly cooled so that the system at any time is approximately in thermodynamic equilibrium. As cooling proceeds, the system becomes more ordered and approaches a "frozen" ground state at zero temperature. Hence, the process can be thought of as an adiabatic approach to the lowest energy state. If the initial temperature of the system is too low, the system may become quenched [9].

The original Metropolis scheme was that an initial state of a thermodynamic system was chosen at energy E and temperature T. Holding T constant, the initial configuration is perturbed and the change in energy, ΔE , is computed. If the change in energy is negative, the new configuration is accepted. If the change in energy is positive, it is accepted with a probability factor $\exp(-\Delta E/T)$. This process is then repeated sufficient times to give good sampling statistics for the current temperature, and then the temperature is decremented and the entire process repeated until a frozen state is achieved at T=0.

The general procedure for the SA algorithm can be summarized as follows:

Step 1: Select an initial solution X and an initial temperature T.

Step 2: Find another solution, namely X_{next} , by modifying the last answer X.

Step 3: Calculate the energy differential

$$\Delta E = f(X_{next}) - f(X)$$

Step 4: If ΔE <0 go to Step 9.

Step 5: Generate a random number, namely R, between 0 and 1.

Step 6: If $R < \exp(-\dfrac{\Delta E}{T})$ go to Step 9.

Step 7: Repeat Steps 2–6 for a number of optimization steps for the given temperature.

Step 8: If no new solution, X_{next} is accepted and go to Step 10.

Step 9: Decrease the temperature T, replace X with X_{next} and go to Step 2.

Step 10: Reheat the environment with setting T to a   higher value.

Step 11: Repeat Steps 1 through 10 until no further improvement obtained.

5   Application of ACO-SA to Reconfiguration

This section presents a new hybrid approach based on combining the ant colony optimization and simulated annealing algorithms for the reconfiguration problem. To apply the hybrid algorithm, named ACO-SA, on the clustering the following steps should be repeated:

Step 1: Generate an initial population.

Step 2: Generate an initial trail intensity.

Step 3: Determination of the next path.

To determine the next path, at first, the neighborhood of X_i is determined by simulated annealing as follows:

Step 1: Select an initial solution X and an initial   temperature T.

Step 2: Find another solution, namely X_{next} , by modifying the last answer X.

Step 3: Calculate the energy differential

$$\Delta E = f(X_{next}) - f(X)$$

Step 4: If ΔE <0   go to Step 9.

Step 5: Generate a random number, namely R, between 0 and 1.

Step 6:   If $R < \exp(-\dfrac{\Delta E}{T})$ go to Step 9.

Step 7:   Repeat Steps 2–6 for a number of optimization steps for the given temperature.

Step 8: If no new solution, X_{next} is accepted and go to Step 10.

Step 9: Decrease the temperature T, replace X with X_{next} and go to Step 2.

Step 10: Reheat the environment with setting T to a higher value.

Step 11: Repeat Steps 1 through 10 until no further   improvement obtained.

After that, SA finds the best local solution(Y), and then the trail intensity matrix is updated as follows:

$$\Delta \tau_{ij} = r; \forall X_j \in Population$$

$$X_i \neq X_j, \tau_{ij}(k+1) = \rho \tau_{ij}(k) + \Delta \tau_{ij}; \forall i \neq j \qquad (7)$$

where   r   is a positive parameter between 0 and 1

Step 4: Check of the convergence condition.   After all ant colonies (X_i) find the next path, the convergence condition is checked by:

$$F(k+1) \leq F(k)$$

$$F(k) = \frac{1}{N} \sum_{i=1}^{N} F(X_i); \forall X_i \; Old \quad Population \qquad (8)$$

6   Simulation Results

In this section the proposed method is applied to distribution feeder reconfiguration on a realistic radial distribution test feeders (Fig (2)). It is assumed that there are 3 generators whose specifications are given in Table 1. Daily energy price variations and daily load variations are shown in Figs.3 and 4.

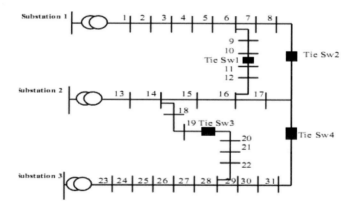

Fig. 2. The Ingle Line Diagram

Table 1. The Characteristic Of Generators

	Capacity (kW)	Max Reactive Power (kVar)	Min Reactive Power (kVar)	Posi-tion	DG kind
G1	200	240	-180	17	CHP(M.T)
G2	240	800	-600	21	Large Wind
G3	200	480	-600	8	CHP(G.T.)

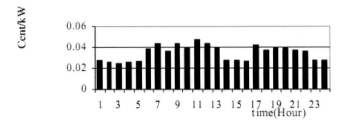

Fig. 3. Daily energy price variations

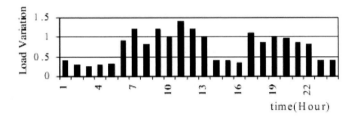

Fig. 4. Daily load variations

Table 2 presents a comparison among the results of proposed algorithm and other evolutionary algorithms. Figure 5 shows the convergence characteristic of the Tabu search for the best solution. The results of simulation can be summarized as follows:

• Under the proper control on DGs, the sum of losses in 24-hour duration is less than that of without control. On the other hands, it can be concluded that the system performance can be improved under proper control.

• Distributed generations have much better performance and time response than other sources of reactive power generation like capacitors. Thus system performance can be improved with proper factors based on these generations.

• The simulation results have been shown that the COC-SA reaches a much better optimal solution in comparison with others and has the small standard deviation for different trails.

Table 2. Comparison Of Average And Standard Deviation Of The Objective Function Values For 300 Trails

Method	Average(kW)	Standard Deviation	Best solution (kW)	Worst solution (kW)
Proposed algorithm	333.7	5.6	327.7	340.79
ACO	348.12	18.97	328.6	365.74
PSO	394.92	33.73	360.18	436.93
TS	401.73	20.68	385.67	424.76
TS	435.67	35.68	401.67	472.38
GA	430.68	26.87	404.38	456.07

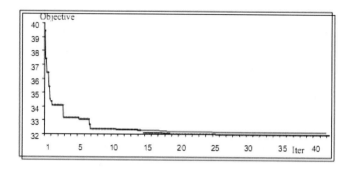

Fig. 5. Convergence Characteristics of the proposed algorithm for the best solution

7   Conclusion

It has been shown that this combination can provide a good opportunity for all individuals to search the surrounding area better. The simulation results have been shown that the ACO-SA reaches a much better optimal solution in comparison with others and has the small standard deviation for different trails.

References

1. Merlin, A., Back, H.: Search For A Minimal-Loss Operating Spanning Tree Configuration In An Urban Power Distribution System. In: Proc. 5th Power System Computation Conf., Cambridge, U.K, pp. 1–18 (1975)
2. Civanlar, S., Grainger, J.J., Yin, H., Lee, S.S.H.: Distribution Feeder Reconfiguration For Loss Reduction. IEEE Trans. on Power Delivery 3(3), 1217–1223 (1988)
3. Aoki, K., Kuwabara, H.: An Efficient Algorithm For Load Balancing of Transformers And Feeders By Switch Operation In Large Scale Distribution Systems. IEEE Trans. On Power Delivery 3(4), 1865–1872 (1987)
4. Das, D.: A Fuzzy Multiobjective Approach For Network Reconfiguration Of Distribution Systems. IEEE Trans. on Power Delivery 21(1) (January 2006)
5. Gallardo, A.: Greedy Reconfiguration Algorithms for Medium-Voltage distribution Networks. IEEE Trans. on Power Delivery 24(1) (January 2009)
6. Ching-Tzong, S., Lee, C.: Network Reconfiguration Of Distribution Systems Using Improved Mixed-Integer Hybrid Differential Evolution. IEEE Trans. on Power Delivery 18(3), 1022–1027 (2003)
7. Chiou, J., Wang, F.S.: A Hybrid Method of Differential Evolution With Application To Optimal Control Problems of A Bioprocess System. In: Proc. IEEE Evol. Comput. Conf., pp. 627–632 (1998)
8. Olamaei, J., Niknam, T., Gharehpetian, G.: Application of particle swarm optimization for distribution feeder reconfiguration considering distributed generators. Applied Mathematics and Computation 201, 575–586 (2008)
9. Wu, T.-H., Chang, C.-C., Chung, S.-H.: A simulated annealing algorithm for manufacturing cell formation problems. Expert Systems with Applications 34(3), 1609–1617 (2008)

The Study on Nano-Electromechanical Transistors Using Atomic Layer Deposition

Zhao Lingya[1], Wang Yingjian[2], and Zhigang Ji[3]

[1] Langfang Polytechnic Institute
[2] Langfang Teachers College
[3] Hebei University of Engineering, Handan, China
mzhaoedu@163.com

Abstract. The process presented here is CMOS compatible just , but low-temperature, thus residual stress effects have no significant influence on device performance, furthermore the process is unique in its ability to create n-terminal devices with n-different gap heights, with 3-terminal device having 2 different gap heights presented here. The process is relatively simple, allowing for the fabrication of embedded electrodes, and associated sacrificial layers with one e-beam write, RIE step, deposition step, and lift-off per electrode. The NEMS transistors introduced here represent the fundamental building blocks of nano-mechanical logic devices and subsequent computing architectures. The development of the devices reported here immediately allows for the creation of mechanical logic gates, because fundamental logic gates, analogous to CMOS inverters, can be constructed by combining complimentary biased 3-terminal switches.

Keywords: Nano-Electromechanical Transistors , Atomic Layer Deposition.

1   Introduction

In the past decade, the majority of NEMS switches have been based on various suspended CNT designs. These devices have varied from 3-terminal devices with vertically grown CNT cantilevers [1], to fixed-fixed 2-terminal devices with CNTs deposited via chemical functionalization [2], with pull-in voltages ranging from 2.5-6 Volts [1-5]. In the past year however, more groups have begun to focus on top-down, TF processes, capable of more reliably producing NEMS switches [6-9] having actuation voltages comparable to, or less than their CNT counterparts. The efforts put forth by [6-9] have pushed the limits of top-down fabrication, fabricating devices with one or more dimensions < 30 nm.

In [6] Czaplewski et al introduced CMOS compatible TF 3-terminal NEMS switches, having in-plane functionality, and reported actuation voltages of 13 V for a 5,000x100x200 nm ruthenium device, having a drain gap of 30 nm, and a gate gap of 50 nm. However, the fabrication process used to fabricate these steps includes many high-temperature CVD processes, resulting in an extremely low yield of only 1.5%, attributed to high residual stress in the devices caused by the high temperature

Y. Wu (Ed.): International Conference on WTCS 2009, AISC 116, pp. 123–131.
springerlink.com   © Springer-Verlag Berlin Heidelberg 2012

deposition of structural and masking materials. In [7] Jang et al introduced the smallest NEMS devices ever made using traditional top-down CMOS technology. Their 2-terminal devices, like those reported in [8], have embedded electrodes, and out-of-plane functionality, with actuation voltages of ~ 11 V for 1,000x200x30 nm titanium nitride (TiN) cantilever devices with working gaps of ~20-30 nm. However, Jang et al use CVD TiN deposited at 250 ^{0}C, thus like the devices introduced in [6], device performance is significantly affected by deleterious residual stress.

2   Design and Fabrication

The 3-terminal process presented here is an extension of our 2-terminal WALD entrenched NEMS fabrication process introduced in [8]. This process furthers the use of entrenched electrodes, demonstrating the process' utility in fabrication of multiple electrically isolated electrodes with varying gap heights. Here electrodes are fabricated as in [8], but now drain electrodes are entrenched, and electrically isolated from the gate electrodes via a 10-15 nm thick layer SiO$_2$.

The fabrication process is shown in figure 1. The developed fabrication process for 3-terminal nano-scale WALD devices is a top-down approach based on MEMS surface micro-machining techniques, and sacrificial layering techniques developed earlier [8-9]. Initially a silicon substrate (100) with a 300 nm dielectric layer of thermal oxide is coated by a 1.8µm thick layer of AZP 4210, and patterned by photolithography to define alignment marks for electron-beam (e-beam) lithography steps.   After patterning the photo-resist, a 50 nm thick layer of gold (Au) is deposited by thermal evaporation, and alignment marks defined by lift-off.

In steps 1-2 the substrate is double-coated by two 500 nm thick layers of PMMA. The first layer is spun on and the chip soft-baked for 2 minutes at 120 $^\circ$C, then a second layer is spun on followed by a hard-bake for 10 minutes at 170 $^\circ$C. This creates a thick layer of PMMA with a nominal thickness of 1 µm.   Next, the device's gate electrodes are patterned via e-beam lithography. A 75 nm deep trench is created in the thermal oxide layer using RIE with a 4:1, CF$_4$:O$_2$ chemistry.   Finally, a stack of metals is deposited in the patterned trench via thermal evaporation to form the gate electrodes and sacrificial layer between the gate electrodes and WALD structure. The stack is composed of a 5 nm thick adhesion layer of Ti, followed by a 20 nm thick layer of Au, and a 50 nm thick sacrificial layer of Ni. Following deposition of the metal stack, lift-off leaves the metal layers filling only the gate trench.

In step 3-4 the process is repeated to create the drain electrode. Again, the substrate is double-coated with two layers of PMMA, and the drain electrode patterned via e-beam lithography. A 40 nm deep trench is created in the thermal oxide layer via RIE. A stack of metals is again deposited on the chip by thermal evaporation to form the drain electrode and sacrificial layer between the drain and WALD structure, this includes: a 5 nm thick adhesion layer of Ti, followed by a 15 nm thick layer of Au, and a 20 nm thick sacrificial layer of Ni.   Following lift-off, 2 nm of ALD Al$_2$O$_3$ and 30 nm WALD are grown on the substrate at 120 $^\circ$C.

In steps 5-6 the substrate is coated by a single layer of PMMA, and patterned via e-beam to create the device geometries.   Next, a 30 nm thick thermally evaporated Ni layer deposited; creating hard mask used to define the geometry of the devices during

RIE.   In step 7, the WALD structures are developed after etching the ALD via RIE.
Design/Results.

The design of our WALD NEMS device is based on that of a 3-terminal
electrostatically actuated switch with out-of-plane functionality. Similar NEMS
switches have been demonstrated with CNT and TF cantilever-based designs [1-7], but
to our knowledge, the fixed-fixed design of the devices presented here is unique to the
field. Fabrication results are shown in figures 2a-b.   Figure 2a shows a fabricated
device at 1,750X, and figure 2b at 12,500X.   The dimensions of device shown in
figure 2 were measured to be 3,313x500x32 nm.   The effective gate/source overlap
areas were measured to be 1,130x500 nm, with a gap of 50 nm between gate and
source, and the effective drain/ source overlap area was measured to be 1050x500 nm,
with a gap of 20 nm between drain and source.   In figure 2b an appreciable amount of
overlap between the gate and drain electrodes is visible.   The overlap is attributed to
electron scattering in the PMMA during e-beam lithography, which results in widening
of intended geometries during the RIE steps in our fabrication process.

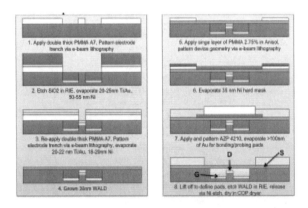

Fig. 1. Top-Down Tungsten ALD NEMS device fabrication process

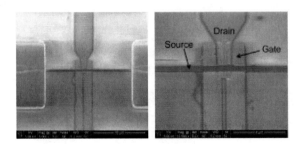

Fig. 2. Successfully fabricated WALD NEMS 3-terminal device having dimensions
3313x500x32 nm. The device has a gate/source gap of ~50 nm, and a drain/source gap of ~ 20
nm.   Drain dimensions are ~1050x500 nm, and gate dimensions ~ 1,130x500 nm.

3   Operating Principle

Our device has been modeled using a 2-D FEM model, using a discretized parallel-pate capacitor approach to approximate applied loads – a description of the loading scheme is given elsewhere [8]. Expected pull-in V_{DS} voltages for transistor-like operation calculated via FEM for a device having dimensions of 4,462.3x500x32 nm are plotted in figure 3, and given in table 1. The results indicate that the gate electrodes have little impact on pull-in for $V_{GS} < 1$ V, but for $V_{GS} > 1$ V the drain-source voltage V_{DS} at pull-in decreases non-linearly, and to < 1 V for $V_{GS} > 3.5$ V. Finally, FEM simulations have shown that pull-in occurs when the source/drain gap is ~ 45-50% of the initial gap.

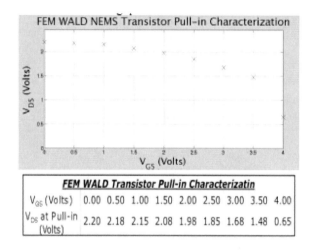

FEM WALD Transistor Pull-in Characterizatin									
V_{GS} (Volts)	0.00	0.50	1.00	1.50	2.00	2.50	3.00	3.50	4.00
V_{DS} at Pull-in (Volts)	2.20	2.18	2.15	2.08	1.98	1.85	1.68	1.48	0.65

Fig. 3. FEM characterization of a device having dimensions 4,462.3x500x32 nm. The device has a gate/source gap of 50 nm, and a drain/source gap of 20 nm. Drain dimensions are ~701.3 x 500 nm2, and gate dimensions ~ 1,880.5x500 nm2. Table 1: FEM calculation of V_{DS} at pull-in for varying V_{GS}.

4   Testing and Results

Device characterization was accomplished using an HP 4145B semiconductor parameter analyzer, MC Systems 8806 probe station, and custom Lab View VI used to control the parameter analyzer. IV curves were generated by linearly sweeping an applied bias between drain and source (V_{DS}) from 0 to 7 V in 100 mV intervals, while applying a constant bias between the gate and source (V_{GS}). Characterization was completed by generating IV curves for a set of applied gate voltages, ranging from 0 to 3.5 V, stepped in 0.5 V intervals. Current was measured between both the source and drain (I_{ds}), and source and gate (I_{GS}) to identify any shorts or leakage current, results are presented in figures 4a-c, and table 2 for the device shown in figure 2.

4a

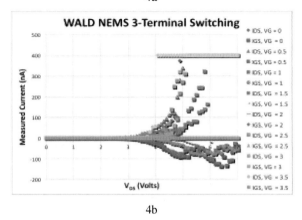

4b

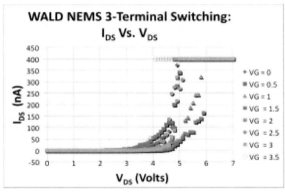

4c

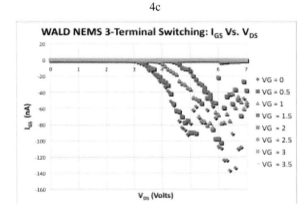

Fig. 4. a-c: Typical IV curves for a WALD Transistor with dimensions given in figure 2; a) IGS and IDS Vs VDS for VGS = 0 to 3.5 V; b) IDS Vs. VDS for VGS = 0 to 3.5 V; c) IGS Vs VDS for VGS = 0 to 3.5 V.

Table 1. Average Pull-in Voltages with a 95% C.I. ($\alpha = 0.05$) and sample size of 5 for the device shown in figure 2. As VGS is increased the average VDS pull-in voltage decreases.

Vgate (Volts)	Vpull-indrain (Volts)	C.I. +/- (Volts) , α = 0.05
0.00	5.60	0.22
0.50	5.26	0.15
1.00	5.48	0.28
1.50	5.24	0.37
2.00	4.90	0.31
2.50	4.92	0.28
3.00	4.46	0.29
3.50	4.06	0.43

Characterization of this device revealed two regions of primary interest. For V_{GS} < 2 V leakage current was measured between the gate and source, as seen in figures 4a and c. The measured leakage currents had a maximum of ~140 nA for $V_{GS} = 0$ V, and were observed to decrease with increasing gate bias. Furthermore, unlike I_{DS}, which had to be limited to prevent burnout, I_{GS} displayed self-limiting behavior; this is most obvious in figures 4a,c where I_{GS} is plotted with I_{DS}. Finally, the pull-in voltage was observed to increase from 4.9 to 6 V as V_{GS} was increased from 0 to 1.5 V.

For V_{GS} > 2 V no leakage current was measured between the gate and source, as seen in figures 4a,b. In this operational regime when V_{GS} is increased from 2 to 3.5 V the pull-in voltage was observed to decrease from 4 to 3.5 V (as should be expected). This actuation shift is most easily seen in figure 4b. For this device it appears that optimal operation occurs for $V_{GS} = 3.5$ V, for which case there is no measurable leakage current, and the actuation voltage is a minimum.

Following the initial characterization described above, new devices with larger dimensions were designed to further decrease actuation voltages. The fabricated devices were measured to have the following average dimensions of: 4,462.3x500x32 nm, a gate/source gap of 50 nm, a drain/source gap of 20 nm, a drain/source overlap of 701.3x500 nm^2, and a gate/source overlap of 1,880.5x500 nm^2. The average characterized pull-in voltages are given in table 2.

Table 2. Average Pull-in Voltages with a 95% C.I. ($\alpha = 0.05$) and sample size of 13 for a device having dimensions 4,279x500x32 nm. The device has a gate/source gap of 55 nm, and a drain/source gap of 18 nm. Drain dimensions are ~610x500 nm, and gate dimensions ~ 1,830x500 nm.

V_{gate} (Volts)	Vpull-in$_{drain}$ (Volts)	C.I. +/- (Volts) , α = 0.05
0.00	2.52	0.21
0.25	2.23	0.28
0.50	2.07	0.28
0.75	2.12	0.18
1.00	2.05	0.18
1.25	1.97	0.25
1.50	1.96	0.24
1.75	1.88	0.27
2.00	1.80	0.26

5   Statistical Characterization

A statistical analysis of the switching characteristics is necessary to draw any significant conclusions regarding device performance, i.e., when the gate bias is increased does the pull-in voltage actually decrease as we expect it to, or does the gate bias have no significant affect on the pull-in voltage? For the experiment a device, with dimensions given in figure 5, was characterized as described above 13 times. For each characterization V_{DS} was applied from 0 to 3.5 V in 50 mV intervals, I_{DS} limited to 200 nA, and V_{GS} applied from 0 to 2 V in 0.25 V intervals. A 1-way ANOVA and Tukey test were used to determine if any of pull-in voltages were significantly different.

Figure 5a shows the average IV curves for V_{GS} = 0 and 2 V, and figure 5b shows the average pull-in voltage and associated confidence interval for each gate voltage applied compared with pull-in voltages predicted by our FEM. Statistical analysis revealed that the average pull-in voltages for V_{GS} > 1.5 V are significantly lower than the average pull-in voltage for V_{GS} = 0 to 1 V. However, the analysis also revealed that none of the average pull-in voltages for V_{GS} > 1.5 V were significantly different from each other. In fact, most of the pull-in voltages were found to be statistically indistinguishable, e.g., if we threw out the data for V_{GS} = 0 V none of the pull-in voltages would differ, implying that application of a gate voltage had no affect on actuation voltage.

Although the analysis suggests that the pull-in voltage is weakly dependent on V_{GS}, figure 5b shows that all of the FEM predicted pull-in voltages, excluding V_{GS}= 0 V, agree well with experimentally measured values, falling within the experimentally measured confidence intervals; this suggests that the device is indeed operating properly.

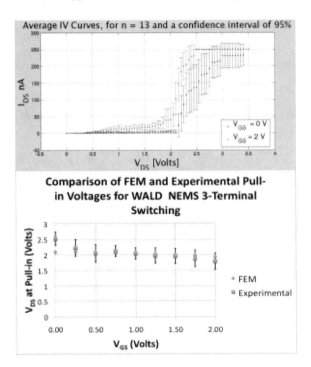

Fig. 5. Average I-V curves for V_{GS} = 0 Volts (blue) and V_{GS} = 2 Volts (red) for a device having dimensions 4,279x500x32 nm. The device has a gate/source gap of 55 nm, and a drain/source gap of 18 nm. Drain dimensions are ~610x500 nm, and gate dimensions ~ 1,830x500 nm.

Ideally we would expect the actuation voltage to decrease as the gate voltage increases, but our analysis suggests that this is not true for all gains in V_{GS}. ANOVA and Tukey tests are both affected by the magnitude of variance. A large variance amongst populations can result in an artificially high p-value, and correspondingly inflated confidence intervals, which makes the comparison of sample sets less straightforward. Looking at figure5a, we see a noticeable jump in the magnitude of variance at pull-in – a result of the non-linear snap through event. Because both analyses operate on populations of numerically extracted pull-in voltages, and these pull-in events are difficult to accurately identify from the measured data sets, the populations have artificially high variances, and thus inflated confidence intervals. If a more accurate method for identifying experimentally measured pull-in events were used, the variance would decrease, thus allowing for a more accurate statistical analysis of the dependency of actuation voltage on applied gate voltage. Furthermore, looking at the FEM results in table 1, we recall that the dependence of gate bias on pull-in is non-linear, becoming greater for more pronounced for $V_{GS} > 1.5$ V, thus for the range of V_{GS} investigated, we conclude that the measured dependence of pull-in on V_{GS} is as expected.

6   Conclusion

In this paper a relatively simple, CMOS compatible, top-down nano-fabrication process used in the fabrication of 3-terminal NEMS switches is introduced. The process is low-temperature, using atomic layer deposition tungsten (WALD) deposited at 120 ^{0}C as the structural material. In addition, the unique use of electron beam lithography in conjunction with reactive ion etching (RIE) and lift-off techniques enable simple and reliable fabrication of n-terminal devices with up to n-different terminal/source gap heights.

References

1. Jang, J., et al.: Nanoelectromechanical switches with vertically aligned carbon nanotubes. Appl. Phys. Lett. 87, 3114 (2005)
2. Dujardin, E., et al.: Self-assembled switches based on electroactuated multiwalled nanotubes. App. Phys. Lett. 87, 3107 (2005); Lee, S., et al.: A three-terminal carbon nannorelay. Nano Lett. 4(10), 2027–2030 (2004)
3. Kaul, A., et al.: Electromechanical carbon nanotube switches for high-frequency applications. Nano Lett. 6(5), 942–947 (2006)
4. Ward, J., et al.: A non-volatile nanoelectromechanical memory element utilizing a fabric of carbon nanotubes. In: Proc. 2004 Non-Volatile Mem. Tech. Symp., Orlando, November 15-17, pp. 34–38 (2004)
5. Czaplewski, D., et al.: A nanomechanical switch for integration with CMOS logic. J. Micromech. and Microeng. 19 (2009)
6. Jang, W.W., et al.: NEMS switch with 30nm thick beam and 20 nm-thick air-gap for high density non-volatile memory applications. Solid-State Electronics 52 (2008)

7. Davidson, B.D., et al.: ALD tungsten NEMS switches and tunneling devices. Sensors and Actuators: A Physical, doi:10.1016/j.sna.2009.07.022
8. Davidson, B.D., et al.: Atomic layer deposition (ALD) tungsten NEMS devices via a novel top-down approach. In: Proc. IEEE MEMS 2009, Sorrento, Italy, January 25-29, pp. 120–123 (2009)
9. Yokota, T., et al.: Plastic complementary microelectromechanical switches. Appl. Phys. Lett. 93, 023305 (2008)
10. Jang, W., et al.: Microelectromechanical (MEM) switch and inverter for digital IC applications. In: IEEE Asian Solid-State Circuits Conference (January 2007)

Preference Survey and System Design on Mobile Q&A for Junior High School Students

Wei Wang[1,4], Shaochun Zhong[1,2,3,4], Peng Lu[1], Jianxin Shang[3], and Bo Wang[4]

[1] Ideal Institute of Information and Technology, Northeast Normal University,
Changchun, China
[2] Software School, Northeast Normal University, Changchun, China
[3] Engineering & Research Center of E-learning, Changchun, China
[4] E-learning laboratory of Jilin Province, Changchun, China
jaechai_aja@sina.com

Abstract. This study conducts an investigation on the case for junior high school. Based on the investigation results, the paper analyzes sources and types of the Q&A items, designs a unique Q&A system and working process for junior high school students. The System is characterized by "Time Recognition" and "Learning Link Recognition", which are doing good to personal teaching, self-learning and inquiry-based learning. Then, implement the system with the resources from a unit of history as an example, and expatiate on a number of key problems.

Keywords: Mobile learning, Q&A, Survey, System Design.

1   Introduction

Mobile learning refers to any sort of learning that happens when the learner is not at a fixed, predetermined location, or that happens when the learner takes advantage of the mobile technologies. [1]

Research on mobile learning in China started later than the United States, Japan and some European countries, but it has been developing rapidly. A large number of projects and cases have emerged, such as "Mobile education theory and practice" project of the Center for Educational Technology of Peking University, "Mobile Education" project of the Ministry of Education of People's Republic of China, CALUMET of Nanjing University, and "Mobliedu" from Nokia China and so on.

An empirical study has been done on mobile learning for college students. The major conclusions are as follows: College students recognize mobile learning and have strong demands; "Updating online and studying offline" is the main approach, which will extend to wireless networking; Mobile learning takes place whenever and wherever people need it; Inquiring, learning, and practicing are the core functions of mobile learning system; Mobile contents should be short, valuable and with various media; Devices and study resources restrict the application of mobile learning. [2]

Another research of Beijing Normal University indicates that high school students' use of mobile devices is on a high level in general, and during the process of

Y. Wu (Ed.): International Conference on WTCS 2009, AISC 116, pp. 133–142.
springerlink.com   © Springer-Verlag Berlin Heidelberg 2012

accepting the mobile learning devices and learning style, the students can combine their own needs and have reasonable expectations: "I hope to communicate with teachers and other students through mobile devices", "Expect mobile devices to evaluate my learning within a period of time" and so on. [3]

Based on the empirical research findings and high school students' expectations, we suppose whether junior high school students look forward to solving problems and finding answers through mobile devices, and what the system function and item resources should be like.

2   Methodology

In order to acquire junior high school students' demand for Q&A resources and system, particularly for mobile devices, an investigation is conducted.

2.1   Sample

We carry out the survey in the High School Attached to Northeast Normal University of Jilin Province. The students are from Grade One and Grade Two. From the student source, faculty, to admission rate, this school is second to none in Changchun city, therefore the conclusions are of high credibility. By stratified sampling method, the sample adopted includes 55 students from Grade One and 105 students from Grade Two. The efficiency of the questionnaires is 91.25%.

2.2   Content

The content covers students' questioning consciousness in daily learning, who to ask, questioning approaches, recourse of the Q&A items, ways of questioning, purpose, the restriction of using network/wireless to solve problems, and so on.

3   Survey Analysis

3.1   Questioning Consciousness

The result of the question "In daily study, will you have questions and want to ask?" indicates that only 3.43% of the students hardly have questions, 13% of the students have lots of questions and they are willing to ask. Among the other students, those who have questions but are not willing to ask occupy 44.3%. Questioning consciousness may affect students' understanding and assimilation of the content that they have learned, and even may affect students' learning interest.

Table 1. Questioning Consciousness

Those who have many questions and willing to ask	Those who have some questions and willing to ask	Those who have some questions but not willing to ask	Those who hardly have questions
13%	46.58%	36.99%	3.43%

3.2   Resources of Q&A Items

When choosing the resources of Q&A items, the most important choices are the content of teaching materials (97 students), extra-curricular learning content (71 students), in-class exercise (70 students) and extra-curricular exercises (66 students). Thus, it can be found that junior high school students pay great attention to teaching materials. However, the new curriculum reform advocates students' self-learning, cooperative learning and inquiry-based learning. The aim is to change the traditional teacher, classroom and textbook centered situation, and cultivate students' innovative spirit and practical ability. Therefore, when the question resource of the system is set, attentions must be paid to the combination of these learning methods; answers should be given according to the specific characteristics of the students; and relevant resources is better to be recommended in order to strengthen students' awareness of active learning and independent inquiry ability.

3.3   The Ones Who to Ask

If students have doubt in the learning process, their first choice is to communicate with classmates, followed by consulting teachers or solving problems through independent thinking on their own, and then followed by networking and asking the elders, but the students rarely choose to consult students of higher-grade. The results also show that consulting private tutor is a good way to seek answers for some students. Our next goal is to obtain laws and methods from the persons they trust on how to answer junior high school students' questions.

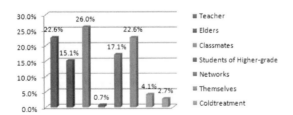

Fig. 1. The Ones Junior High School Students Willing to Ask

We also studied what limits the students to find answer with help of internet/wireless. Selecting "Rarely use internet/wireless" accounts for 22.6%; "Not knowing the method" accounts for 4.1%; "Questioning the accuracy of the network/wireless Q & A" accounts for 28.1%; "Lack of Network/wireless Q&A System needed" accounts for 34.2%. Another 11% of the students fill out their own reasons, which can be mainly divided into parental restrictions, no time, not forming using habits.

3.4   Ways of Access to Q&A Items

Compatible with conventional teaching situation, face to face question and answer approach is clearly the best and most consistent way for students. The results of "As a supplementary to face to face questioning manner, select two means that you think is the most effective" indicates that: "Questioning through internet "is the highest (84.2%), the second one "Emails" accounts for 45.9%, the third one "SMS" accounts for 41.1%, and "wireless networking" is the last, accounting for 21.9%. From the perspective of supporting face-to-face questioning, high school students have a high demand for all kinds of question manners to obtain Q&A items. It can be believed that mobile Q&A will be more and more into high school students' learning and living and provide them with timely answers, with the process of mobile learning awareness strengthening, mobile devices improving and the fee declining.

3.5   Learning Links

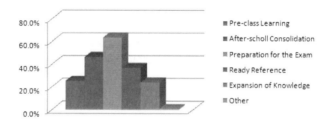

Fig. 2. Learning Links which Using Q&A Mostly

In order to provide the most suitable original resources for junior high school students, we should get to know in which condition students should use Q&A system. The results of the item "Which two learning links do you think mostly use Q&A system?" show that Q&A system is most commonly used in the preparation for exams, followed by after-school consolidation and ready reference, and finally in pre-class learning and the expansion of knowledge. In order to make student not rely on the available answers, training environment should be provided with the inquiring function, for the ultimate goal is the improvement of students' independent problem-solving ability. In addition, the system also needs to have the function of time recognition, in order to show different content in pre-class learning, after-school review, exam preparation and other time periods, to promote students' study.

4   System Design

According to the survey results, Q&A system for junior high school students should meet the requirement: (1) Timely and authoritative, as convenient as student's communication with teachers in the real world ; (2) supporting several types of access such as mobile software, SMS, Email, website based on WAP, and website based on WWW; (3) including classroom knowledge, curricular knowledge, examination, evaluation and other aspects; (4) resource structure supporting student's self-learnig, inquiry-based learning and cooperative learning; （5）effectively supporting the preparation for new contents and access to knowledge expansion, in particular supporting after-school consolidation and ready reference, and supporting the preparation for examination ; (6) Systematic time identification, showing different interface in different application links. Therefore, the function structure design of the system is shown in Fig. 3.

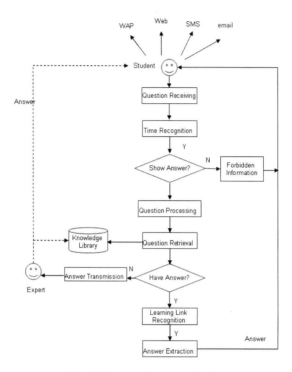

Fig. 3. Function Structure Design

4.1 Module Description

1)Module of Question Receiving: This module is to provide UIs of the system. As types of questioning manners of junior high school students have a high demand, the system supports those who ask questions via mobile software, SMS, Email, website based on WAP, and website based on WWW.

   2)Disable Period Identification Module:The advantages of present Q&A network system, simple and timing, are often wrongly applied in the situations, such as exam cheating. This module is aimed to solve the problem. Teachers can use the system to set the time of prohibition, such as in-class test, final exams and so on. When the system receives student's question, it will send the feedback of encourage information, such as "Late to learn in the exam", then send the student the answers automatically after the disable period. At the same time, the system will save student's question records for teacher to know the performance of students.

3) Module of Question Processing: In order to ensure the system can act as the communication between teachers and students, the module must have natural language understanding. It's difficult for those who search questions by keywords, because they often can't extract the right keywords. This module can obtain the questions in natural language from the module of question receiving, after a series of analysis and processing operations such as word segmentation, filtering stop words, extracting and expanding the appropriate keywords, named entity recognition, and then an accurate understanding of the question can be drawn.

4) Module of Question Retrieval: After the processing, questions become key phrases, which will be searched in the systematic knowledge library. The systematic knowledge library includes FAQ database and text resources. As Q&A system of junior high school students is aimed at specific subjects, many major questions need to be repeatedly questioned and answered, so it is appropriate to base on FAQ. In addition, for some less frequently asked questions, full-text search system can be used based on the text resource library.

5)Module of Learning Link Recognition: Pre-class review, after-school consolidation, ready reference, knowledge expansion and exam preparation are the most basic links in learning. This module provides students with the most appropriate answer and relevant content, according to the characteristics of different links. For example: from the view of answers, in the synchronous learning process, only answer itself is needed; on the review stage, comparison of relevant content is involved; From the view of relevant content, for the pre-class review, the contents that can arouse student's interest, such as background information are presented; in the after-school review, extensive questions are provided to expand student's thinking; and in exam preparation, questions in previous exams and their explanations will be presented. Link information need to pass to the answer extraction module, in order to choose the most appropriate answer and related content.

6)Module of Answer Extraction: To find the answer to the question, answer extraction [4] is needed. It will show the students the most concise answer. If cannot accurately extract the answer, it will seriously affect the quality and effectiveness of the system. For different types of questions, the answer may be in the form of a sentence, a paragraph or a word, and so on. In order to accurately extract the answer, first of all, the types of questions should be confirmed, and then the answers of a certain particular level will be sent to the questioner as a feedback according to the question type.

7)Module of Answer Transmission:For the questions that the system is temporarily unable to answer, it needs to be sent to the experts. When the experts answer the question, the knowledge library will also be modified, or the FAQ pairs will be added, or the text resources will increase. Later, the system will return answer to students. The feedback forms includes mobile software, SMS, Email, website based on WAP, and website based on WWW, etc..

8)Learning Condition Transmitting Module:This module records the condition of students' questions and answers, and cumulates the gross in every knowledge point, unit and lesson, for teachers to know students' learning condition. Based on the memory curve and student's learning style and cognitive level, the system can transmit questions and answers to students automatically in the appropriate time. This module aims to eliminate the memory blind spot gradually in the repeated review process.

4.2   Working Process

First, students ask questions via "the module of question receiving" in natural language. The system judges whether it's in forbidden time. If it is, no answer will be provided and the fact will be recorded. Or else, the questions will be submitted to the "module of question processing" .The "module of question processing" does a series of operations such as word segmentation, filtering stop words, extracting key phrases which are important to the questions search, and expanding and standardizing terminology. Then, the system retrieves the answer with the key phrases. One way is the direct search in the FAQ database among the questions and answers exsisting, and another way is the full-text search in the text resource library . "FAQ Search" can guarantee the accuracy of the answer; "Full Text Search" is an effective supplement of "FAQ Search" that can improve the recall rate. For all the questions asked, the system search firstly in the "FAQ", if not retrieve the answer, the system will try the full-text search. In the second case, the system will add a couple of question and answer into the FAQ library. For every question, the system must return its most relevant answers and contents considering the learning link in which students ask questions. So, the system adds clues in the keywords by "Learning Link Recognition Module" before "Module of Answer Extraction". "Module of Answer Extraction" can give answer exaction according to the question type, such as time or location. That is the function of "Module of Answer Extraction".

5   System Implementation

Due to the heavy workload of the system, we only conduct a preliminary attempt.

5.1   Development Environments

Select Visual studio.net2005 as development environment, SQL Server2005 as the database server, IIS as Web application server. Chinese language full-text search is based on Lucene.NET2.0. Simulated operating platform for mobile devices adopts Windows Mobile 5.0 simulator and Windows Mobile 5.0 SDK.

5.2   Sources of Q&A Items

The original source of the system is taken from learning content and excercise of Unit One, Junior High School History Book One, People's Education Press (both teachers' textbooks and students' books), from which nearly a hundred question-answer pairs are selected to establish the FAQ library and text resources library, and formed part of the domain keywords. Domain dictionary, common words dictionary, question word dictionary, stop word dictionary are mainly from Harbin Institute of Technology and sogou. Question classification is also from Harbin Institute of Technology.

5.3   Design of Word Segment Dictionary

Word segment dictionary directly affects the processing speed of "module of question processing ". Create hash index on the first three Chinese characters to constitute a TRIE sub-tree of depth 3, and order the remaining characters alphabetically to form dictionary text which is similar to the "binary-seek-by-word". [5]Regarding to the specific subject knowledge, the order of domain words is different. Therefore, compare the first three Chinese characters one by one and compare the remaining Chinese character in way of "binary-seek-by-word" can improve the efficiency of word segment.

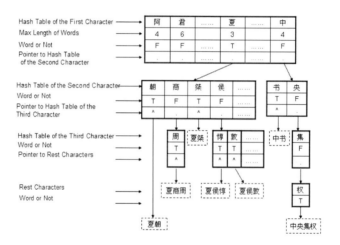

Fig. 4. Word Segment Dictionary [5]

5.4   Answer Retrieve Strategy

In order to improve the efficiency，the keywords are divided into three major categories. The first category is question words, identifying the type of question, which will help answer extraction; the second category is domain keywords, closely related to the specific knowledge points; the third category is restriction keywords, which must be included in the answer sentence, especially the ordinal numeral, etc. It's important to determine the core words for answer retrieving in FAQ. If question word is at the end of the sentence, identify the domain word closed to wh-word as the core word; If question word is in the beginning or in a sentence, identify the domain word far away from wh-word as the core word. For the text library search basing on Lucene, there are two core steps, one is creating logic library of indexing, and the other is retrieving the logic library. [5]

Fig. 5. Interface of Inquiring

6   Conclusions

From the conclusions about mobile Q&A in China, college students believe that inquiry function is in most need of mobile learning system, at the same time, junior high school students have high expectations on the use of mobile devices. We put forward the hypothesis as extended, and make a questionnaire survey to explore the needs of junior high school students. Based on the conclusions obtained, design a system for junior high school students' questioning and inquiring knowledge. The system also supports mobile software, SMS, Email, website based on WAP, and website based on WWW as access approaches. By adding "Time Recognition" and "Learning Link Recognition" the system can prevent cheating during testing time and providing personalized learning content. The system initially realized with an example of one unit from junior high school history. The availability of the system also needs empirical research. In addition, the unknown word identification, named entity recognition, similarity algorithm issues are to be resolved.

References

1. MOBIlearn:Guidelines for learning/teaching/tutoring in a mobile environment, http://www.mobilearn.org/download/results/guidelines.pdf. (retrieved June 8, 2009)
2. Wang, W., Zhong, S.-C.: Research on Support System for M-Learning Using Kinds of Terminals. J. China Educational Technology 7, 108–112 (2008)
3. Wang, X.-C., Li, Y.-S., Huang, R.-H.: Middle school students to use mobile learning devices preparation and survey measurement tools. J. E-education Research 2, 97–101 (2009)
4. Xu, W.: Research on Question Classification and Candidate Answer Sentences Extraction in Chinese Question Answering System. D. Harbin Institute of Technology (2006)
5. Wang, B.: Research and Design of Intelligence Question and Answer Based on Mobile Learning. D. Northeast Normal University (2008)

Study on Professional Managers Cultivation of the Farmers' Specialized Cooperatives

Zhou Huan

College of Economics and Management, Heilongjiang Bayi Agricultural University
Daqing, China
zhewoo@gmail.com

Abstract. With the rapid development of the Farmers' Specialized Cooperatives, urgently need the professional managers with high-quality and high management ability. The professional managers' lack of the Farmers' Specialized Cooperatives will reduce core competence of enterprise and affect its development rate. Current academic research on this is less, and there is no standard training model of the professional managers for the Farmers' Specialized Cooperatives. In this paper, create a professional managers culture model and propose specific strategies for solving the Professional Managers lack problem of the Farmers' Specialized Cooperatives. Hoping the study can provide help in improving core competence of enterprise, and make it better adapt to market development.

Keywords: Farmers' Specialized Cooperatives, professional managers, culture model.

1 Introduction

Currently, the Chinese Farmers' Specialized Cooperatives have entered a new stage of development. Since "Farmers' Specialized Cooperatives Law" into effect on July 1, 2007, the law of the "Registration Regulations", "Model Statute", "the Farmers' Specialized Cooperatives Accounting System (Trial)" and several major laws implemented one after another. At the same time, policy support system has gradually improved. Financial, tax and other related departments have improved support policies and tilted to the cooperatives. The tax concessions, financial, fiscal, national agriculture construction projects have received the country's strong support. After several years of development, the Farmers' Specialized Cooperatives have been a good Legal operating environment and policy support system. With the transformation of the agricultural products circulation style, operation and management of the Farmers' Specialized Cooperatives have tended to diversity, and are extending to the field of agro-processing and marketing. This promotes transformation of the Farmers' Specialized Cooperatives from the simple organization operation to modern enterprise management.

It should be noted, in its development, there are some problems. Human resources are the core power in enterprise developing. Now, lagging behind talent training and a lack of talent pool, inter-disciplinary talent with high operation and management capacity is shortage, which limited development speed of the Farmers' Specialized Cooperatives. In order to meet the growing demand for professional managers of the Farmers' Specialized Cooperatives, professional managers training should be on the agenda as soon as possible, we should pay much more attention to it for better meeting the needs of market development. The paper will propose the measures of professional managers training for the Farmers' Specialized Cooperatives based on analyzing the status of professional managers training for the Farmers' Specialized Cooperatives, for solving the problem of insufficient talent and increasing the core competencies of the Farmers' Specialized Cooperatives.

2   The Professional Managers Training Status of the Farmers' Specialized Cooperatives

2.1   Professional Managers Training Is Still in the Exploration

The Farmers' Specialized Cooperatives based on the household contract management is a voluntary association, democratic management and mutual economic organization, which is composed of the similar agricultural production operators or the providers and users of the similar agricultural production and management service. At present The Farmers' Specialized Cooperatives in China is in the initial stage of development, although professional manager demand of the Farmers' Specialized Cooperatives generated, but the Farmers' Specialized Cooperatives didn't realize or have sufficient knowledge. With the development of the Farmers' Specialized Cooperatives, separation of ownership and management is an inevitable trend. The Farmers' Specialized Cooperatives developing to a certain extent, to optimize resource allocation in the large context, must break the existing pattern, have a breakthrough resource closed mode of operation, attract a wider range of professionals involved in the management. As the specificity of the Farmers' Specialized Cooperatives organization forms and interest to each member, led to a strong sense of self-protection. The need to hire professional managers from outside to work as housekeepers is also exploring. The credit evaluation system of professional managers training is still not perfect, this is one of the reasons that the Farmers' Specialized Cooperatives existed doubts in employing professional managers, it needs to scholars in-depth study.

2.2   Lack of Professional Managers Team

In recent years, the Chinese Farmers' Specialized Cooperatives received substantial development. According to the agricultural sector statistics, by the end of 2007, the Chinese Farmers' Specialized Cooperatives have been more than 150,000. Until 2006, joined members of the Farmers' Specialized Cooperatives have been more than 38.7 million, is 7.2 times in 2002, there are 34.8 million households accounting for 13.8% of the total households, increased 11% than in 2002. The data shows that the Chinese Farmers' Specialized Cooperatives are developing rapidly and the number is increasing. The problem of Lack of the Farmers' Specialized Cooperatives managers has become evident. In the existing Farmers' Specialized Cooperatives, most of the farmers' leaders in overall quality are not high, over-reliance on government or leading enterprises, awareness and capacity to adapt to the market economy are not strong, being short of the interdisciplinary talents with management skill and market development ability, not to mention the use of modern management techniques for planning management, which largely restricted innovation and development of the Farmers' Specialized Cooperatives. In the critical transformation period of the Farmers' Specialized Cooperatives from the simple production to modern production, the lack of early human resources accumulation and scientific talents training system become its development constraints. Developing rapidly of the Farmers' Specialized Cooperatives and diverse market strongly require professional managers with operation, management and technology ability.

2.3   Agricultural Economics and Management of University Education Is Imperfect

The agricultural colleges in China generally establish agricultural economic management, but have common problems, for example, there are no clear training objectives, imperfect course system and so on. Now having two agricultural colleges in Heilongjiang province as an example to show the deficiencies of agricultural management talents training based on the current agricultural economic management.

First, professional teachers are inadequate. The teachers engaged in agricultural education of the two agricultural institutions are generally masters or above, with high level of specialized theories, but most of them have not experience engaged in agricultural business management, lack of practical experience, which directly affects the quality of professional education in agriculture and teaching level.

Second, the professional courses system exist drawbacks. In Heilongjiang province, agricultural economics management major courses system of two agricultural colleges is large and comprehensive. However, the major pertinence should be raised and training goals should be further clarified. Because of limited time and space, students can only grasp the superficial specialized theories, lack of the capability to link theory and practice, resulting in the phenomenon of "learning without special" in talents training occurs.

Third, our undergraduate education attaches much more importance to theory than practice. From the student angle, current students are basically the "professional student", the students from city even have no the basic knowledge of agriculture, lack of the opportunity to participate in agricultural practices in the school; from the school angle, with a single training goal, only meet the standards of specialized theories teaching, the graduates can't fill the requirement in agricultural sector.

Fourth, cooperation between colleges and enterprises is lacking. The colleges don't offer in-service training workshops. It is difficult to continue their study for existing staffs engaged in agricultural field.

3   Specific Professional Managers Training Strategies of the Farmers' Specialized Cooperatives

3.1   To Establish a "People-Oriented" Management Philosophy

The basic idea of "People oriented" is that people are the most basic elements of management, whose comprehensive development is the core of the idea, and human development is the foundation to push business or society forward. Also, in the organization of the Farmers' Specialized Cooperatives, the cultivation and development to professional managers are badly needed. The Farmers' Specialized Cooperatives should raise human resources training as one of its strategic steps, and build a comprehensive management system of professional managers, which contains recruitment, training, career planning, performance evaluation, incentives and other methods of developing and training talents and keeping talent.

3.2   To Build a New Culture Mode for Professional Managers

It takes really a complex and long time for the Farmers' Specialized Cooperatives to build its own system, which supplies professional managers of high quality. Many factors should be seriously considered of both external and internal environment. These factors mainly reflect in four fields: selecting, educating, employing, keeping people in his office. According to this, to build a culture mode of professional managers is shown in Figure 1.

Professional managers training of the Farmers' Specialized Cooperatives can be divided into four parts. Firstly, we should pay attention to set up the identification and introduction of the professional managers. Secondly, train the initial professional managers of the Farmers' Specialized Cooperatives. Thirdly, keep people. Fourth, we should build a feedback system of the professional managers. The four parts are closely linked and complementary.

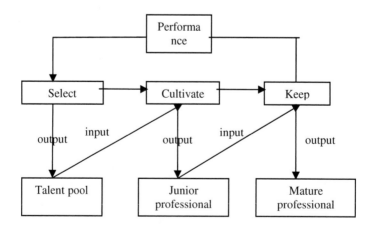

Fig. 1. The professional managers culture model of the Farmers' Specialized Cooperatives

In the first part, the management personnel of the Farmers' Specialized Cooperatives mostly come from the university graduates of agricultural economics and management, the professional managers working in agriculture or other sectors. Therefore, society and agricultural colleges should make the detailed plan of personnel training in order to supply more talents and create a good environment for the Farmers' Specialized Cooperatives.

In the second part, we should focus on two kinds of person's training, including the training of reserve forces and senior manager. But the current system of agricultural colleges can't match with the needs of talented persons. The course system is too general and limited, various disciplines students are learning different knowledge, can't reach the training objective across disciplines and interdisciplinary. Today, the Farmers' Specialized Cooperatives involved in agricultural, forestry, animal husbandry, sideline occupation and fishery. The students can't adapt to the rapid development of the Farmers' Specialized Cooperatives. So it's in urgent need of comprehensive talents. Therefore, the Agricultural College should be clear of major orientation, including optimize professional education institutions and focus on education in three areas of business, management, technical knowledge. Pay more attention to the training of practical ability. For example, agricultural college should set up more practical courses such as agricultural business management simulation and teaching case to improve the practical ability. In addition, universities should set up the appropriate courses system for the people working in agricultural, strengthen agricultural enterprises across agricultural departments, regularly organize In-service training workshops serving for the farmer's management staff to provide them a good environment for charging.

In the third part, the Farmers' Specialized Cooperatives should build the training and assessment mechanisms. Let those selected professional managers regularly participate in training courses of agriculture in order to expand their knowledge, let the professional managers learn by doing and using. If the training process seems to irrigate and fertilize "tree", then the assessment is the "pruning the side branches". The Farmers' Specialized Cooperatives should formulate assessment system of department managers according to "morality, diligence, performance, ability, study" and so on. Suggest making these assessments at least once a year. Assessment process should notice the combination of qualitative and quantitative method and have stress.

In the fourth part, we should improve the incentive and feedback mechanisms. The essence of incentive mechanism is to give confirmation for the performance of professional managers. Give them their economic and status reward. Meanwhile, we should strengthen network construction, establish information sharing mechanisms, record the professional managers' basic information, work history, performance results etc. Track the information and make new evaluation to inspire their growth.

3.3   Creating Training Institutions to Set Up the Training Platform

No doubt, the college is the cradle of education, but there is a certain lack of higher education. So those community training institutions in various sectors of the industry can play a complementary role in personnel training. The training institutions have three advantages. First, the training institutions have higher flexibility of teaching methods. Second, the training institutions are invested by agricultural sector and related company, they have the newer and more valuable information. Third, the training institutions may employ foreign domestic professionals and experts, who have extensive experience and broad view. They can enlarge our outlook and conduct cross-cultural exchange. So, agricultural departments and enterprises should invest together to established formal training institutions. The enterprises provide place for practice of training personnel, who will combat simulation and do well in the interface between theory and practice. For example, the "sunshine program" is a successful case which has made great contribution for the cultivation of the professional managers. We should make full use of existing resources to speed up professional managers training process of the Farmers' Specialized Cooperatives.

3.4   Gradually Improving the Qualification System for Professional Managers

Professional managers of the Farmers' Specialized Cooperatives are those who can use advanced operating strategy, accurately grasp the chance of agricultural market in order to ensure the profitability of organizations. Professional managers should have the ability of market research, project evaluation and be familiar with the process of production, management and sales. So it will be a trend to gradually standardize certification management of professional managers. The certification agency should set up a series of theoretical courses and practical skills assessment standards. It also should improve the system of Vocational Qualification and Professional Qualifications to implement the dual license management. Only those who have achieved the Vocational Qualification can have posts for the professional managers of the Farmers' Specialized Cooperatives. This will lay a solid foundation for healthy development of professional managers. Also, the Farmers' Specialized Cooperatives will implement dynamic evaluation of professional managers. Those managers who can't make value within a certain time will be given out.

4   Conclusion

As the level of market is improving, the Farmers' Cooperatives should become more professional. So the training of high-quality and high-capacity professional managers is urgent. "Providing for a Rainy Day", the Farmers' Cooperatives can stand the market test. The training of professional managers should combine three sorts of force including the national policy, the national education curriculum reform and the active participation of enterprise. I believe in the near future, after many efforts of the whole society, the training of professional managers of the Farmers' Specialized Cooperatives will be paid close attention by the academic institutions. The professional managers will become mature in the development of the Farmers' Specialized Cooperatives and the market competition of the Farmers' Specialized Cooperatives will be more increased.

References

1. Rosen: Incentives of professional managers explore. Business Economics (23), 99–100 (2009)
2. Zhu, X.: The status analysis of professional managers in China. Heilongjiang Social Science (04), 84–86 (2009)
3. Wang, K.: Cooperatives employ professional managers is the trend - Interview with PengYuan, Institute of Rural Development Institute, Chinese Academy of Social Sciences. China's Cooperation in Economy (05), 28–29 (2009)
4. Sun, W.: Professional managers habitat niche Model. Leadership Science (02), 35–37 (2010)
5. Hui, D.: Professional manager credibility factors. Economic and Technological Cooperation (07), 47–48 (2010)

A Robotic Manipulator with Parallelogram Hinged Electromechanical System

Zhao Lingya and Zhou Zhihai

Langfang Polytechnic Institute
mzhaoedu@163.com

Abstract. In this paper, the nonlinear vibration of the of robotic manipulator with parallelogram hinged mechanism is studied. The relation between the nonlinear vibration of the system and the parameters of the system is obtained. The studies show that the nonlinear resonance of the system are not only related to the structure parameters of the robotic manipulator, but also to the rotational frequency of rotor and the rotating field frequency of stator of the controllable motor, and that under certain conditions, there exist internal resonance, super-harmonic resonance, 1/2-order sub-harmonic resonance, 1/3-order sub-harmonic resonance, combination resonance, and multiple resonance.

Keywords: robotic manipulator, servomotor, resonance.

1 Introduction

In an electromechanical system, electromagnetic parameters of the motor have important effects on the dynamic characters of mechanism system, so it is necessary to take the drive motor and the mechanism as an integrated system to be analyzed when the dynamics and stability of the mechanism are studied. There are many scholars have studied the dynamic characteristics of a motor–mechanism system and gained many helpful results[1-4].

In this paper, a robotic manipulator with parallelogram hinged mechanism system is taken as an object of study. The electromechanical coupling dynamic equation of system is established by finite element method while it motion in plane. Based on the dynamic equation, the ultra harmonic resonance of the system is studied using the method of multiple scales. In plane motion, the robotic manipulator is controlled by two controllable motors, which are site at point O and A respectively.

2 Nonlinear Dynamic Equation of the System

Based on the air-gap field of non-uniform airspace of controllable motors of the robotic manipulator caused by the eccentricity of rotor, the servomotor element, which defined the transverse vibration and torsional vibration of the controllable motors as its nodal displacement, was established.

Y. Wu (Ed.): International Conference on WTCS 2009, AISC 116, pp. 151–159.
springerlink.com © Springer-Verlag Berlin Heidelberg 2012

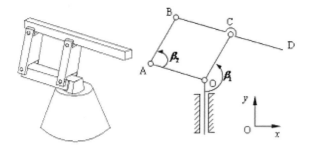

Fig. 1. Diagram of robotic manipulator

2.1   Element of Servomotor Element

In analysis of the servomotor shaft, the following simplifications are employed

- The coupling terms of the elastic motion and the rigid body motion in the Coriolis acceleration and transport acceleration are neglected in studying the absolute acceleration of any point in the shaft.
- The effects of shearing deformations and pull-press deformation caused by transverse displacements are neglected in calculating the deformation energy.
- The servomotor rotor is regarded as a rigid body considering its lower length-diameter ratio.

The elasticity of servomotor shaft is mainly taken into account in analysis of the vibration, and the shaft-disk system model is adopted in dynamic analysis of the system. The elastic motions of the shaft-disk system are decided by both the eccentricity vibration of centroid of rotor, namely, the transverse vibration, and the torsional vibration of output shaft with respect to the rotor, as shown in Figure 2. In the diagram, number 1,2,3,4 denote four nodes of the element. So the transverse vibration and the torsional vibration can be expressed by the generalized coordinate vector $u_1 = [U_1 U_2 U_3 U_4]^T$. In the diagram, XYZ is the coordinate system of servomotor element.

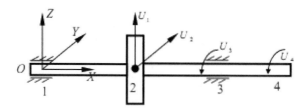

Fig. 2. Diagram of servomotor element

The air-gap eccentric vibration is shown in the Figure 3, where point o is the inner circle geometric center of the stator, point o_1 is the outer circle geometric center of the stator and the coordinates of point O_3 is (x, y), δ is the length of air gap, e_1 is the air-gap eccentricity. The static eccentricity $e_{01} = \sqrt{e_{11}^2 + e_{12}^2}$ and rotational eccentricity $\mathcal{E}_{01}$ are considered at the same time, then $e_1 = \sqrt{x^2 + y^2}$, and

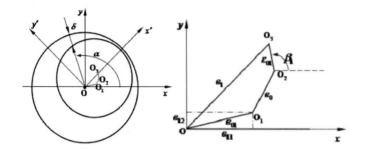

Fig. 3. Diagram of air-gap eccentric vibration of servomotor

$$\begin{cases} x = U_1 + e_{11} + \mathcal{E}_{01} \cos \psi_1 \\ y = U_2 + e_{12} + \mathcal{E}_{01} \sin \psi_1 \end{cases} \tag{1}$$

Where, U_1 and U_2 are the component of vibrate eccentricity in the x and y direction, and $e_0 = \sqrt{U_1^2 + U_2^2}$, $\psi_1 = (1 - s_1)\omega_{01}t$ is the rotational angle of rotor with respect to the stator of servomotor. ω_{01} synchronous speed of rotation of servomotor. s_1 is the slide ratio.

According to the theory of electro mechanics and electro-mechanical analysis dynamics, as far as servomotor is concerned, the voltage and current between the two windings is asymmetric, that is to say, the servomotor works in elliptic magnetic field. According to this real running state of servomotor, the kinetic energy and potential energy of servomotor can be obtained as follows[5]

$$T_1 = \frac{1}{2} \int_0^l m_1(x)[\dot{W}_1(x,t)]^2 dx + \frac{1}{2} \int_0^l J_{01}(x)[\dot{V}_1(x,t)]^2 dx \tag{2}$$
$$= \frac{1}{2}\dot{u}_1^T \bar{M}_1 \dot{u}_1$$

$$V_1 = \frac{1}{2}u_1^T(\bar{K}_{11} + \bar{K}_{12})u_1 + \bar{e}_1^T(\bar{K}_{11} + \bar{K}_{12})u_1 + \frac{1}{2}\bar{e}_1^T(\bar{K}_{11} +$$
$$\bar{K}_{12})\bar{e}_1 + (\bar{e}_1 + u_1)\bar{k}_{01} + \frac{pR_{01}L_{01}\Lambda_{01}}{2}\int_0^{2\pi}[F_{+s}\cos(\omega_{01}t - p\alpha) + \tag{3}$$
$$F_{-s}\cos(\omega_{01}t + p\alpha)] + F_{+r}\cos(\omega_{01}t - p\alpha - \varphi_{10}) +$$
$$F_{-r}\cos(\omega_{01}t + p\alpha - \varphi_{20})]^2 d\alpha$$

where, $W_1(x,t)$ and $V_1(x,t)$ the transverse displacements and the angle of elastic torsion of any points on the servomotor element respectively, $l = l_1 + l_2 + l_3$ is the length of servomotor shaft; $m_1(x)$, including the rotor mass m_0 which is at, is the mass distribution function of the servomotor shaft; $J_{01}(x)$,including he moment of inertia of rotor J_0 which is at $x = l_1$, is the moment of inertia distribution function of the servomotor shaft. R_{01} is the inner radius of the servomotor stator, L_{01} is the effective length of the rotor, $\Lambda_{01} = \mu_0 / \sigma$ is the even air-gap permeance of the servomotor, μ_0 is the magnetic permeability coefficient of air, $\sigma = k_\mu \delta_0 , k_\mu$ is saturation , δ_0 is the uniform air-gap size, $k_\mu = 1 + \delta_{Fe} / k_1 \delta_0$, k_1 is the calculation air-gap coefficient of the even air gap, δ_{Fe} is the equivalent air-gap of ferromagnetic materials, φ_{10} is the phase angle of the positive-sequence current of rotor lagging behind the positive-sequence current of stator, φ_{20} is the phase angle of the negative-sequence current of rotor lagging behind the negative-sequence current of stator. F_s^+ , F_s^- , F_r^+ and F_r^- are the positive-sequence and negative-sequence components of the magneto-metive amplitude of stator and rotor respectively. $\overline{M}_1$ is the mass matrix of servomotor element, $\overline{K}_{11}$ is the stiffness matrix of servomotor element in connection with the structural parameters of the rotor, $\overline{K}_{12}$ is the stiffness matrix of servomotor element in connection with the electromagnetic parameters of the rotor, $\overline{k}_{01}$ is the 4-order vector in connection with the eccentric motor.

2.2   Beam Element of Robotic Manipulator

In dynamic analysis of the beam element, the coupling terms of the elastic motion and the rigid body motion in the Coriolis acceleration and transport acceleration are neglected in studying the absolute acceleration of any point in the beam element. In calculation of strain energy, the shearing deformation energy and yield deformation energy are also omitted. The material of the link is adopted as metal. Therefore, the kinetic energy and potential energy respectively are as follows[6]

$$T_3 = \frac{1}{2} \int_0^l \rho A(\overline{x})[(\dot{V}(\overline{x},t))^2 + [\dot{W}_1(\overline{x},t)]^2) d\overline{x} = \frac{1}{2} \dot{u}_3^T \overline{M}_3 \dot{u}_3 \qquad (4)$$

$$V_3 = \frac{1}{2} \int_0^L EJ(\overline{x}) \left[W''(\overline{x},t) \right]^2 d\overline{x} + \frac{1}{2} \int_0^L EA(\overline{x}) \left[V'(\overline{x},t) \right]^2 d\overline{x}$$

$$\frac{1}{2} \int_0^L \left\{ EA(\overline{x}) \left[V'(\overline{x},t) \right]^2 + \frac{1}{2} \left[W'(\overline{x},t) \right]^2 \right\} \left[W'(\overline{x},t) \right]^2 d\overline{x} \tag{5}$$

$$= \frac{1}{2} \dot{u}_3^T \overline{K}_3 \dot{u}_3 + \frac{1}{2} \int_0^L \left\{ \begin{array}{l} EA(\overline{x}) \left[V'(\overline{x},t) + \\ \frac{1}{2} \left[W'(\overline{x},t) \right]^2 \end{array} \right\} \left[W'(\overline{x},t) \right]^2 d\overline{x}$$

where ρ is the average mass density of beam element, $A(\overline{x})$ isthe cross-section area function, $V(\overline{x},t)$ and $W(\overline{x},t)$ are respectively the longitudinal displacement and the transversal displacement of any point in the beam element. I is the moment of inertia of cross-section, $\theta(t)$ is the rigid angular velocity of the beam element, E is the modulus of elasticity, $J(\overline{x})$ is the moment of inertia distribution function of the element. $\gamma_i(\overline{x})$ is the corresponding shape functions of the longitudinal displacement and the transversal displacement.

2.3 Nonlinear Dynamic Model of Robotic Manipulator System

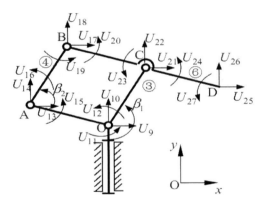

Fig. 4. Diagram of finite element analysis of robotic manipulator

The plane motion analysis of robotic manipulator as shown in Figure 1 is studied in this paper. The mechanism can be looked as temporal structure system. Link OA is rigid bar, and link OC!AB and BD are elastic bars. We regard the links OC and AB as one beam element respectively, and regard the link BD as two elements. 3,4,5 and 6 are the serial number of the beam element as shown in Figure 4. The servomotor are set in the point O and A, then the serial number of the servomotor elements are1 and 2.

The linear viscous damper damping is adopted in this paper, according to the second Lagrange equation

$$\frac{d}{dt}(\frac{\partial T}{\partial \dot{u}_i}) - \frac{\partial T}{\partial u_i} + \frac{\partial V}{\partial u_i} = P + Q \tag{6}$$

The electromechanical coupling nonlinear dynamic equation of the system is

$$M\ddot{u} + C\dot{u} + (K + K_0)u = P - M\ddot{u}_r - (k_{11}^e + k_{12}^e + k_{21}^e + k_{22}^e)$$

$$-k_0 + \varepsilon[\sum_{n=1}^{8} u^T G_n K_n u + \frac{1}{2}\sum_{n=1}^{8} u^T K_n u G_n$$

$$+\frac{1}{2}\sum_g \sum_l u^T G_{gl} u K_{gl} u + \frac{1}{2}\sum_g \sum_l G_{gl} u u^T K_{gl} u] \tag{7}$$

$$(k, l = 2, 3, 4, 6, 7, 8)$$

where, P is the generalized force vector, $\ddot{u}_r$ is the rigid accelerated speed array of system in the global coordinate system, ε is the small parameter.

$$K = K_{11}^e + K_{21}^e + \sum_{i=3}^{n} B_i^T R_i^T \bar{K}_3 R_i B_i$$

$$K_0 = K_{12}^e + K_{22}^e, K_{11}^e = B_1^T R_1^T \bar{K}_{11} R_1 B_1$$

$$K_{12}^e = B_1^T R_1^T \bar{K}_{12} R_1 B_1, K_{21}^e = B_2^T R_2^T \bar{K}_{21} R_2 B_2,$$

$$K_{22}^e = B_2^T R_2^T \bar{K}_{22} R_2 B_2, e = B_1^T R_1^T \bar{e}_1 + B_2^T R_2^T \bar{e}_2$$

$$\bar{e}_1 = [e_{11} + \varepsilon_{01} \cos\psi_1 \quad e_{12} + \varepsilon_{01} \sin\psi_1 \, 0 \quad 0]^T ;$$

$$\bar{e}_2 = [e_{21} + \varepsilon_{02} \cos\psi_2 \quad e_{22} + \varepsilon_{02} \sin\psi_2 \, 0 \quad 0]^T ;$$

$$k_0 = B_1^T R_1^T \bar{k}_{01} + B_2^T R_2^T \bar{k}_{02}; G_n = \sum_{i=3}^{n} B_i^T R_i^T \bar{g}_n$$

$$(\bar{g}_n)_i = 1, (i = 1, 2, ..., 8); K_n$$

$$= \sum_{i=3}^{n} B_i^T R_i^T \left[\int EA\bar{g}_n^T \bar{g}\bar{K}_a dx \right] R_i B_i$$

$$\bar{g} = [-\frac{1}{L} \quad 0 \quad 0 \quad 0 \quad \frac{1}{L} \quad 0 \quad 0 \quad 0]^T$$

$$(\overline{K}_a)_{ij} = \gamma_i'\gamma_j', i, j = 2,3,4,6,7,8;$$

$$G_{gl} = \sum_{i=3}^{n} B_i^T R_i^T \overline{G}_{gl} R_i B_i;$$

$$K_{gl} = \sum_{i=3}^{n} B_i^T R_i^T \overline{K}_{gl} R_i B_i;$$

$$\left(\overline{G}_{gl}\right)_{gl} = \left(\overline{G}_{gl}\right)_{lg} = 1, k, l = 2,3,4,6,7,8;$$

$$\left(\overline{K}_{gl}\right)_{ij} = \left(\overline{K}_{gl}\right)_{ji} = \int EA\gamma_g'\gamma_i'\gamma_j' d\overline{x}, i, j = 2,3,4,6,7,8.$$

where, n is the element number of system, K_a, K_{gl} and G_{gl} are 8×8 u matrix, $\overline{g}$ and $\overline{g}_n$ are 8-order vector, and the other terms are zero.

According to related knowledge, the self-excited inertial force of system can be expressed as

$$-M\ddot{u}_r = F_1 + F_2 + F_3 - \varepsilon(-Mu_{\varepsilon 1}k_{e1}u - Mu_{\varepsilon 1}u^T k_{e2}u$$
$$-\lambda_1 Mu_{\varepsilon 1} - Mu_{\varepsilon 2}k_{e1}u - Mu_{\varepsilon 2}u^T k_{e2}u - \lambda_2 Mu_{\varepsilon 2}) \qquad (8)$$

where $u_{\omega 1}, u_{\omega 2}, u_{\omega 3}, u_{\varepsilon 1}$ and $u_{\varepsilon 2}$ are the coefficient matrices relate to the physical dimension of mechanism, and they can be obtained by rigid motion analysis. λ_1 and λ_2 are the coefficient relate to motor structure. F_1, F_2 and F_3 are the k th order simple harmonic array of Fourier series.

Suppose that

$$-(K_{11}^e + K_{12}^e + K_{21}^e + K_{22}^e)e - k_0 = F_4 + F_5 + F_6 + F_7 + F_8 \qquad (9)$$

$F_{fki}(f = 1,2,...,8)$ are periodic function. They can be expanded the form of Fourier series as follows

$$(F_f)_i = \sum_{k=1}^{m} F_{fki} \cos(kv_f t + \tau_{fki})(f = 1,2,...,8) \qquad (10)$$

where, v_1 is the working frequency of rotor of servomotor 1, v_2 is the working frequency of rotor of servomotor 2, v_3 is the smallest common multiple of v_1 and v_2. $v_4 = v_1, v_5 = v_2, v_6$ is the rotation frequency of magnetic field of stator of the servomotor 1, v_7 is the rotation frequency of magnetic field of stator of the servomotor 2, v_5 is the smallest common multiple of v_6 and v_7, and v_8 is the smallest common multiple of v_1, v_2, v_6 and v_7. $F_{fki}(f = 1, 2, ..., 8)$ express the amplitude values, and $\tau_{fki}(f = 1, 2, ..., 8)$ are the corresponding phase angles.

3   Numerical Example

As shown in Figure 1, $l_{OA} = 300mm$, $l_{AB} = l_{BC} = 300mm$. The elastic links of the robotic manipulator are all aluminum material whose thicknesses are all 20mm, widths are all 30mm. The density of aluminum $\rho = 2700kg / m^3$, and its Young's modulus $E = 70GPa$. The type of electromotor is SGMGH-03ABA6 produced by corporation YASKAWA. In order to convenient for simulation, the process of start-up of the motors is omitted. The process of uniform turning of the motors is only considered in this numerical example. The frequency of the power supply is 15Hz. As shown in Figure 1, the rotating speed of the motor sitting at the point O is 1400r/min. The link OC rotates from $\beta_1 = 90°$ to $\beta_1 = 200°$, and the link AB rotates from $\beta_2 = 30°$ to at the same time in the process of uniform turning. Though calculation, the means of the first two orders natural frequencies of the robotic manipulator system are 24.0Hz and 47.2Hz respectively during the uniform turning process mentioned above.

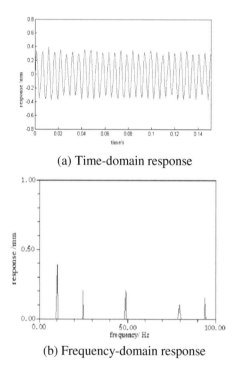

(a) Time-domain response

(b) Frequency-domain response

Fig. 5. Longitudinal displacement response of endpoint D of robotic manipulator

4   Conclusions

This paper studies the nonlinear vibration of robotic manipulator with parallelogram hinged mechanism while it motioned in plane. The work presented in the paper provides theoretical explanation and direction to a certain extent for the vibration phenomenon and its control of such kind of robotic manipulator.

References

1. Yan, H.S., Chen, W.R.: On the Output Motion Characteristics of Variable Input Speed Servo-controlled Slider-crank Mechanisms. Mechanism and Machine Theory 35, 546–556 (2000)
2. Smaili, A., Kopparapu, M., Sannah, M.: Elastodynamic Response of a d.c. Motor Driven Flexible Mechanism System with Compliant Drive Train Components during Start-up. Mechanism and Machine Theory 31, 659–672 (1996)
3. Cai, G.-W., Wang, X., Wang, R.-G.: Coupling dynamic equations of motor-driven elastic linkage mechanism with links fabricated from three-dimensional braided composite materials. Journal of Central South University of Technology 12(2), 93–97 (2005)
4. Li, Z., Cai, G.: Multiple resonance and stability of a motor-elastic linkage mechanism system. Journal of Donghua University 23(6), 1–6 (2006)
5. Wang, R., Cai, G.: Coupling Dynamic Model of A Hybrid Elastic Five-bar Linkage Mechanism System. In: Proceedings of the International Conference on Mechanical Engineering and Mechanics, pp. 631–636. Science Press USA Inc., Monmouth Junction (2005)
6. Zhang, C.: Analysis and design of elastic linkage mechanism, pp. 58–74. China Machine Press (1997) (in Chinese)
7. Thompson, J.M.T., Stewart, H.B.: Nonlinear Dynamics and Chaos. Wiley, New York (1986)

Using the E-learning Platform and Tools in Teaching

Qin Hui

Department of Electronic Information Engineering, Wuhan Textile University
Wuhan, China
hbwhq@hotmail.com

Abstract. In the course of using the platform based on the E-learning environment, the users can obtain numerous learning resources at anytime at anyplace and at anywhere, while there were a lot of problems. It will be the vital factor which affected the E-learning, that how to manage the magnanimity learning resource efficiently, help the users acquire what they required rapidly and efficiently, and help the users study efficiently. Blended learning may be the best method to guarantee the teaching quality.

Keywords: E-learning, SCORM, learning management system, learning content management, Web 2.0, user objective, blended learning.

1 Introduction

In the several years, the development of the information technology has brought about enormous social change. Human have been influenced in learning fashion, lifestyles and working fashion. The changes are so deep that nobody can escape. In higher education, there is a technological revolution taking place. E-learning has gained acceptance as an alternative form of traditional instruction. The growth of e-learning is being described as explosive, unprecedented and disruptive. E-learning system is a new way by which instruction can be carried out, and it is also based on the advanced computer technology, internet technology and communication technology. The traditional pedagogy procedure, such as teaching, training and testing, can be fulfilled in the internet. The e-learning system has obvious advantages, such as all the learning content existing in the internet, without the restrictions of time and space, asynchronous teaching. In the E-learning mechanism, the user can use the computer network, multimedia, professional Web sites, information search, electronic libraries, distance learning and online classes, etc. to obtain the necessary learning resources, and then complete self-study [1]. As a new teaching approach, e-learning can be used to promote the development of teaching and improve teaching quality, independent or combined with traditional teaching. Of course there is much problem when people try to do it.

2 What Can E-learning Bring Us?

E-learning is defined by the American Society for Training and Development's e-learning glossary as "a wide set of applications and processes, such as Web-based

Y. Wu (Ed.): International Conference on WTCS 2009, AISC 116, pp. 161–167.

learning, computer-based learning, virtual classrooms, and digital collaboration". Whereas text pages are the primary means of presentation in traditional instruction, e-learning can deliver information through such varied formats as graphics, videos, audios, animations, models, simulations, and visualizations. The electronic technology makes more information more easily accessible more cheaply than ever before. It will ensure that all of the teachers and students have the opportunity to take advantage of the power of new and emerging technologies for widespread improvements in teaching and learning----today, tomorrow, and far into the future.

E-Learning packages are easier to develop and take less time. If the correct e-Learning tool is chosen, the educator can alter the learning material immediately on a large and fast scale. Another delivery advantage is the access control options that are available for security purposes. You can load learning resource on private internal networks, making it inaccessible to any users not on the network. It can also be connected to other networks so as to possess greater bandwidth options when it is linked with other training systems. Because there is access to only one source of information, the exact same material is delivered in the exact same way to each learner, thus ensuring consistency of content and quality of instruction to each learner. Depending on the accessibility it can deliver relevant information just in time. A last advantage is the accessibility of information at any time. A design can even include a resource library for the learner to refer to, i.e. books, articles, CD-ROMs, other helpful websites, etcetera. There is substantial cost-saving implication for the learner and the provider of e-learning. Mailing costs are reduced through the use of e-mail for the distribution of documents. On top of that, educator fees and expenses are reduced, while the efficiency of record-keeping, scheduling and administrative tasks are improved due to built-in computer management control systems. E-Learning provides a very important pedagogical advantage in that it enhances the total quality of the learning experience of the learner. It improves feedback to the learners by speeding it up significantly. It engages the learner through the use of exciting multimedia and the provision of interactivity. The use of e-mail allows for peer host personalities that can entertain, motivate and assist the learners. The individual involvement and responsive feedback provided by using interactive systems are highly motivational. Open-ended learning environments are created, encouraging exploration and problem-solving. The learner thus actively participates in his own learning, and gets the opportunity to take greater control over it. Greater control brings about a greater sense of responsibility spurring the learner on to produce better results. Another advantage of e-Learning is to reinforce learning that has already taken place, thus providing an excellent opportunity for lifelong learning.

3   Framework of E-learning

Educational materials for e-learning environments generally include a computer interface, several different data formats (e.g. text, graphics, image, voice and movie), an evaluation system to assess students' progress, and several other support tools to support the learning environment [2]. Initial preparing e-learning systems had several unknown issues. A pioneering team proposed ADL SCORM as a base of the project and the proposal was accepted by the company board. The SCORM defines "the interrelationships of course components, data models and protocols so that learning content objects are sharable across systems that conform with the same model". SCORM was developed in 1999 by ADL (Advanced Distributed Learning) ----a group formed by the US military----in cooperation with government, academia and industry. The SCORM framework consolidated the work of several national and international bodies into a single reference model.

SCORM stands for "Sharable Content Object Reference Model". Its main aims are:

- to enable developers to format and package learning content in a standardization way so that the content can be used on all LMSs and shared amongst other members of the learning and teaching community.
- to enable delivery of the learning materials to the learner and tracking of learners' actions and scores (e.g. indicating when learners open a new page, complete a quiz, etc.)

To date there have been three widely accepted versions of the SCORM: SCORM 1.2, SCORM 2004and SCORM 2008.

According to SCORM framework, e-learning platform consists of three main elements: learning management systems (LMS), learning content management systems (LCMS) and computer supported collaborative learning (CSCL), each part has different functions [3]. LMS and LCMS are the two major parts. Learning management systems (LMS) and learning content management systems (LCMS) really have two very different functions. The primary objective of a learning management system is to manage learners, keeping track of their progress and performance across all types of training activities. By contrast, a learning content management system manages content or learning objects that are served up to the right learner at the right time. Understanding the difference can be very confusing because most of the LCMS systems also have built-in LMS functionality. Many of these LCMSs have also performed interoperability tests with leading LMS products. Besides the embedded learning management system functionality, there can also be significant overlap between LCMS and LMS capabilities and purpose.

A learning content management system is a multi-developer environment where developers can create, store, reuse, manage, and deliver learning content from a central object repository. An LCMS will generally have a majority of the characteristics on the following checklist. Users can use this checklist to determine if a software application could be called a learning content management system. The primary differentiator to determine if a product is an LCMS is if it offers reusability of learning content and is generally constructed using a learning object model.

- Based on a learning object model.
- Content is reusable across courses, curricula, or across the entire enterprise.

- Content is not tightly bound to a specific template and can be re-deployed in a variety of formats, such as e-learning, CD-ROM, print-based learning, Palm, EPSS, etc.
- Navigational controls are not hard coded at the content (or page) level.
- There is a complete separation of content and presentation logic.
- Content is stored in a central database repository.
- Content can be represented as XML or is stored as XML.
- Content can be tagged for advanced search-ability (both at the media level and the topic level).

4 Web 2.0 for Developing E-learning System

When we develop an e-learning system, we must use a certain network program tool. Now Web 2.0 is widely used in e-learning systems. Web 2.0 is known as the participatory, collaborative, and dynamic online approach web-based development. Earlier web paradigms, Web 1.0 was mostly a medium for reading, always involved a website published by individuals or organizations with few opportunities for users to add or modify the content. Web 2.0 provides many more opportunities for reading and writing. Web 2.0 refers to an evolving collection of trends and technologies that foster user-generated content, user interactivity, collaboration, and information sharing. It follows that online learning communities would naturally transform to use a similar approach. O'Reilly notes six core competencies of the Web 2.0 environment [4]:

- services, not packaged software,
- an architecture of participation,
- cost-effective scalability,
- re-mixable data source and data transformations,
- software above the level of a single device, and
- harnessing collective intelligence.

Common examples of Web 2.0 technologies include wikis, blogs, discussion forums, podcasting, social networking, and social bookmarking. The definition of each term as follows.

Wikis: A wiki is a collection of web pages that users can directly modify by adding new content and editing or deleting existing content.

Blogs: A blog (contraction of the term "Web log") is a website that provides regular commentary in the form of postings with the most recent at the top of the page (often referred to as "reverse-chronological order").

Discussion Forums: A discussion forum (alternately known as groups or boards) is a web application for holding discussions between users.

Podcasting: A podcast is a series of audio files, typically in MP3 format, distributed over the Internet using subscription feeds.

Social Networking: Social Networking refers to individuals using online communities to stay connected with each other, make new connections, share interests, and explore the interests of others.

5   Application of E-learning in Teaching

The educator has to consider the active and passive sides of e-learning. It is important that the educator see the research findings in the right context before making a decision either for or against e-learning. But for the common users, as clarity on which delivery system used has still not been obtained, one can consider using both. First and foremost, as both we and the technologies we employ become more adept, we will find more and more things that we can do better by using the Internet. Furthermore, when Web-based learning technologies finally stabilize few faculty members are likely to have the luxury of deciding whether or not to use the Internet in their classes [5]. Students will demand what they have seen used effectively in the Internet. So I say that changes brought by the e-learning are so deep that nobody can escape.

Blended learning is a compromise, which use e-learning along with traditional face-to-face instruction in different forms or combinations to facilitate instruction and learning. Researchers have found that blended learning was more effective than either classroom or Web-based instruction for teaching both declarative and procedural knowledge. And when we develop an e-learning system, we maybe have more sympathy for the learner's situation. By program we ought to help users set up desire to study and to be self-discipline. This is an important work to be done that will affect the e-learning system performance .What can we do?

For me, an engineering university teacher, I have attempted to combine e-learning content with my classroom instruction in my professional teaching. By Web 2.0 technology, traditional instruction and e-learning can be integration in good manners. E-learning environment has been a supplement to the classroom teaching, making up related course content or establishing a discussion forum or creating simulating scenarios. Pre-class I will ask the students to review basics of a subject area ahead, and give them some resources list, sometimes require grouping to collaborate. Preparation can raise their confidence and interesting of the course content, and make the time spent in the classroom more productive and valuable. Post-class students browse my blogs reviewing key concepts or learning relevant material, and answer questions I prepared for them. Studying should have not finished when students left the classroom. Browsing e-learning refresher course can increase students' retention. When students have any questions, they ask me in a discussion forum. Students admitted that e-learning course content afforded in their schedule bring them flexibility. The requirement to participate in online discussions was perceived as additional workload that would have been less demanding in classroom interaction.

In operating our e-learning system, some questions have put forward by the students. To solve these questions, learning content management system has been functional optimized. The optimized learning content management system has some special functions which can be described as normal function and personality features.

Normal function, learning content management system provided with the regular, basic functionality, will meet all users for reviewing relevant network resources, and now almost all of the E-learning platforms provide these normal functions. Normal features include: displaying, upload and download capabilities, search, navigation functions, etc. Network programming is now the norm to achieve these functions, there is not any problems.

Particular features will be designed for different users or different needs of the same user period of time. Learning content management system provides particular features to help learners learn effectively and merrily. These particular features provide to students including: filtering, indexing, classification, study guide, users management functions, etc. The following is explanation of them.

- Filter function: to realize second matching resources provided by search engine to the request brought by the user, filtering out information the learner most in need of;
- Indexing, classification: to classify the resources according to relativity between learning resources and keywords high and low degree and to develop index;
- Study Guide features: to develop a learning schedule for learner, and to record achievement achieved by the learner or problems still not solving, and to create communication platform by introducing interactive discussion, competition and other collaboration circumstance for those having a common learning goals;
- User management functions: to establish learning portfolio for each registered user, to record learning progress of the schedule, and to give the suggestion on following study.

6   Conclusions

E-learning takes advantage of modern information technology to create entirely new communication mechanism and new learning environment with rich resources, which is considered a revolution of learning. E-learning can serve as a complement and extension of traditional instruction, and create opportunity to achieve life-long learning and autonomous study. Efficiency of learning content management system will be the crucial factor which impact the E-learning. Comparing technical difficulty which may appears with cost-effectiveness, a conclusion can be made that the model of user-oriented management module is worth using in the scope of professional courses teaching. This management model as a supplement of classroom instruction courses has opened up channels of improving the quality of teaching. Realizing the potential – and reducing the deterrents – is possible if educational planners consider the importance of users' need when making instructional design decisions.

References

1. The Learning Organization 8(5), 200–202 (2001)
2. IEEE, IEEE P1484.12.2/D1, 2002-09-13 Draft Standard for Learning Technology – Learning Object Metadata - ISO/IEC 11404 (2002)
3. Reusable Learning, Reusable Learning: SCORM Primer.. (2007)
4. O'Reilly, T.: What Is Web 2.0: Design Patterns and Business Models for the Next Generation of Software (2005)
5. Barnes, K., Marateo, R., Ferris, S.: Teaching and learning with the net generation. Innovate 3(4) (2007)

Intelligent Acquisition Modeling Based on Intelligent Information Push-Pull Technology*

Wang Hong-Bo, Zeng Guang-Ping, and Tu Xu-Yan

Department of Computer Science
School of Information Engineering, University of Science and Technology Beijing
Beijing, 100083, China
boriskgn633@gmail.com

Abstract. Due to distribution and internal complexity of the issue, one of the key problems in Multi-Agents Communication. Sub-Acquisition need to communicate, so that they can mutually exchange information, or coordination or collaboration to complete tasks in order to achieve the purpose of solving. The introduction of Intelligent information push and pull technology can improve the intelligence level of network and database. It can find a fundamental solution to the problems encountered during these technology applications. The main advantages of Information Pull technology are: better targeted, light source task. Its shortcomings are as follows: poor timeliness.It require the user to have a certain expertise. In order to solve these Pull technology problems, information push technology has come into being.

Keywords: Intelligent Acquisition, Intelligent Information Push-pull, public Knowledge Base.

1 Introduction

The development of the modern science and technology information provides people with a variety of methods of information acquisition and transmission. Professor Tu Xuyan has proposed the intelligent information push-pull technology (IIPP) [1]. From the point of view of the relationship of source and user, it can be divided into two modes: (1) the information push mode, The source takes the initiative to push information to the user, such as radio; (2) the information pull mode. The user takes the initiative to pull information from source, such as query the database. That is, how to extract information from the mass of useful information, how to provide users with different personalized information services. IIPP is as shown in figure 1.

The two technologies Information Push and Pull should make integration. On the base of a combination of both, artificial intelligence, knowledge discovery, the

---

* The work is partly supported by both the National High-Tech Research and Development Program of China (No. 2009 AA 01Z119) and the Beijing Natural Science Foundation (No.4072018).

Y. Wu (Ed.): International Conference on WTCS 2009, AISC 116, pp. 169–177.

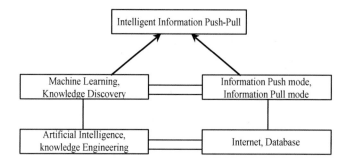

Fig. 1. Intelligent Information Push-Pull Technology

Internet and database technology are re-integrated to form the intelligent information push and pull technology. The technology is a development direction of the current Internet/Extranet/Intranet, database systems and other information systems to provide users with information services.

By Information Pull technology users take the initiative to inquire about. Users request, and then the system will send information back to the client. Information push technology is based on the user's needs. It   sends the information that users take interest in to the user's computer on time, targetedly and purposefully. Like a radio broadcast, listeners can choose the channel to listen to news, finance, Sports, music programs etc. Push technology takes the initiative to push the newest information to the customers and users do not have to search. The advantages of Push technology is timeliness and a low requirements on usres.

2   Intelligent Network and Communications

The rapid development and wide dissemination of network and communication technologies has provided a new environment and new conditions for information technology in modern society [2]. It also has brought new opportunities and new problems. Such as:

- How does SNMP (Simple Network Management Protocol) meet the needs of complex network?
- How can we acquire the knowledge that users need from the mass of network information rapidly and timely?
- How do the shaped, heterogeneous, asynchronous complex network communicate with each other, coordinated operation?

2.1   Intelligent Network

Intelligence is the new trends, new stage of information . Research on Intelligent networks has become a hot spot. Intelligent Network is a product of artificial intelligence technology and information network [3]. Such as the formula (1) follows:

$$AI + IfN \rightarrow IN \tag{1}$$

Where:

- AI - Artificial Intelligence
- IfN - Information Network
- IN - Intelligent Network

Here the concept of information network is a broad sense, such as the formula (2) follows:

$$IfN = (CN, \ TpN, \ TvN, \ ...) \tag{2}$$

Where:

- CN - Computer Network digital information network
- TpN-telephone Language Information Network
- TvN- Television Network Image Information Network

Intelligent Network, such as: Semantic Web, knowledge networks, human neural network. The focus study of Intelligent Network is the combination of Distributed Artificial Intelligence and large-scale information networks.

The hot spot of Distributed Artificial Intelligence (referred to as DAI) is the methods and techniques based on Agent.

Multi-Agent system (referred to as MAS) focuses on the coordination, the Association Business (Consultation) and cooperation problem of multi-agent.

Mobile Agent Technology (referred to as MAT) is a method and technique that researches the Mobilization, communication and services of Mobile Agent in the remote heterogeneous network.

2.2   Intelligent Communications

Intelligent Communications is a combination of the artificial intelligence technology and communication technology. Such as the formula (3) follows:

$$AI \ + \ CT \ \rightarrow \ IC \tag{3}$$

formula (3) in:

- AI - Artificial Intelligence
- CT- Communication Technology
- IC - Intelligent Communications

Based on different types of artificial intelligence technology and communication networks, it corresponds with different types of intelligence communications technologies and systems. Such as:

- Distributed Intelligent Communication
- Interacting Movement Intelligent Communication
- Mobile Intelligent Communication

3   Intelligent Acquisition Modeling Based on Intelligent Information Push-Pull

Intelligent Acquisition Modeling is a distributed artificial intelligence and Communication technology a product of the combination. Such as the formula (4) follows:

$$DAI + CT \rightarrow IAM \qquad (4)$$

formula (4) in:

- DAI - Distributed Artificial Intelligence
- CT - Communication Technology
- IAM - Intelligent Acquisition Modeling

Distribution of Intelligent Communications, one of the options are a public knowledge base of intelligent communications.

3.1   CKB-Public Knowledge Base

As a result of Information Sender and Information Receiver has distributed Common public Knowledge Base CKB, therefore, it can be significantly reduced (compressed) information which needs to be transmitted by the channel.

A novel architecture is presented as shown in Figure 2 below:

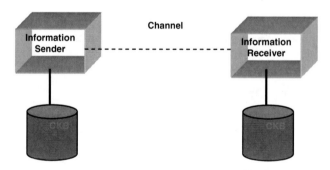

Fig. 2. Intelligent Acquisition Modeling Based on CKB-public Knowledge Base

Interactive Movement Communication technologies (referred to as IMC) is a product of the combination of the "information push" and "information Pull" technology. Such as the formula (5) follows:

$$IPush + IPull \rightarrow IMC \qquad (5)$$

in formula *(5):*

- IPush-Information Push, such as: webcasting technology;
- IPull- Information Pull, such as: search queries engine, etc.;
- IMC- interactive communication technologies, such as: information push and pull.

3.2   Interactive Intelligent Communications

Compared with Information pull, Information push has a advantage of a good timeliness and a low requirements on the user. But information "push" has the following disadvantages:

1) It can not ensure the success of sending. Users may not be able to ensure they can receive the information from the network information center, which is inappropriate for the application fields which must ensure to receive information.

2) There is no information on the status of follow-up. After information is released,whether a customer has received, if it had been prompted by information ,the publisher has no way of knowing. This is unacceptable for the   information center who make decisions according to client feedback.

3) There is no group management functions. Important information is usually sent for some specific groups.That is,it is sent only to the related people, but Push has no choice of information and sends it to a user's computer.After the user receives, it must be screened. This approach does not fundamentally solve the problem that users must find and screen from a great deal of information.

In view of the above problems,on the base of a combination of artificial intelligence and interactive communication. We proposed this interactive smart communication technologies aiming at improving the intelligence level of network and database. To address the management group problems in the information push, we proposed the construction of a legislation model according to a variety of user preferences, thus,we can targeted and timely take the initiative to push the required information to the user in order to meet the needs of different user's individual requirements, thereby enhancing the user's work efficiency.

Interacting Movement Intelligent Communication (referred to as IMIC) is a product of the combination of Artificial Intelligence and interactive communication, such as the formula (6) follows:

$$AI + IMC \rightarrow IMIC \qquad (6)$$

in formula (6):

- AI - Artificial Intelligence
- IMC- Interacting Movement Communication
- IMIC-Interacting Movement Intelligent Communication

Interactive Intelligent Communications (IMIC) is a development of intelligent information push and pull (referred to as IIPP). Interactive Intelligent Communications, such as: multi-lingual communications, tourism information services.

4   Communication Structure

4.1   Communicating Mode Based on IIPP

From systematic point of view [5], individual nodes are functionally and temporally tightly coupled to perform a complex task. Therefore coordination between the cooperating objects should be achieved via asynchronous communication rather than by explicit control transfer.

The IIPP (intelligent information pull-push) model of autonomous object communication meets these requirements. This model supports an asynchronous style of anonymous many-to-many communication in contrast to the synchronous style of object invocation. A consumer subscribes to a certain event type/subject/channel rather than to a specific producer. A producer publishes instances of this type of information. Published information is forwarded eventually to all subscribers, either immediately when being published (push) or on demand when a subscriber asks for updates (pull).

The communicating mode of Real time data: we apply IIPP (in Ref. [1]) to building up the Rapid Data Exchange Link Pool(RDELP), which is based on the Content Code structure of IDF and the   Publish/Subscribe communication mode. The RTDIMCG1 is regarded as the *Cache* in a personal computer, which may be low capacity and high speed memory, which is shown in figure 3.

Push Data
Trigger Mechanism

Pull Data
Response Mechanism

Fig. 3. Rapid Data Exchange Pool based on IIPP

4.2   Communicating Protocols

The communication protocol is built up on the foundation of IADS and adopts IADP (Intelligence Autonomous Decentralized Protocol). TCP, UDP/IP or Ethernet is the bottom layer of IADP. IADP is regarded as the application layer exploit of TCP/IP, which is shown in figure 4.

Application	
IADP	
TCP	UDP
Transport layer	
(IP) Internet	
(Ethernet Frame, ARP)Network Interface	
Physical layer	

Fig. 4. The relationship between IADP and TCP/IP

The main features of IADP show in the following.

(1) Publish/Subscribe Communication (Multicast)
The publisher sends a message associated with a TCD (Transaction Code that indicates the type of data) on a multi-cast transmission to a multi-cast group within a specified data field. The nodes within the specified multi-cast group automatically accept only the messages with required TCDs. Since the method does not specify any target address, extensible communication is available.

(2) Decentralized Information
There is not central server. All agents directly communicate one another with message. Any agent can be a sender or a receiver of message. This model seems like full connect model of net model, so it is the same as peer to peer system. Because any two of agents have the direct access, they can communicate directly.

(3) Balanced Load
Multi-agent system doesn't depend on certain special node when it runs, so its load is balanced, and reliability of this model is strong.

(4) Harmonious Communication
Decentralized information structure needs to define communication protocols, formats of message, transmitted mechanism and communication language (including syntax and semantics).In this way, exchanging information and sharing resource can be easily done among agents.

(5) Alive Signal
A node transmits alive signals to the other nodes on the same data field to notify that the node is alive. This makes the nodes on the same data field an monitor the status to ach other. All the nodes belonging to one data field must transmit the alive signals within the data field periodically as far as they can transmit the alive signals.

4.3   Communicating Language Based on KQML

KQML is a kind of hierarchy language. Conceptually, KQML message can be viewed as being divided into three layers: the content layer, the message layer, and the communication layer. KQML(Knowledge Query and Manipulation Language) is used as communication language and is expanded in security. KQML is a kind of high-level and message-oriented communication language and communication protocols for agent, it supports knowledge-sharing of runtime and co-routine problem. KQML messages do not merely provide communication sentences, they are sent to recipients with an attitude about the content (assertive, directives, declaratives, and expressive). An attitude of content is conceptualized to primitives that are called performs. It provides an extensible set of performs, using it we can build high-level agent communication model, such as Contracts Nets. It uses facilitator to communicate with different agents, and facilitator is also a king of special agent [6].

5   Sample of Interaction

The following code is a sample of interaction, which <SOAP-ENV:Header> </ SOAPENV: Header> content is part of the security authentication, <SOAP-ENV:Body>

</ SOAP-ENV: Body> content is the message body.It is a package of FIPA-ACL messages. You can have a choice of digital signatures and the integrity of verification to message body.

```
<?xml version="1.0" encoding="GB2312"?>
<SOAP-ENV:Envelope xmlns:SOAP-ENV="http://schemas.xmlsoap.org/soap/envelope/"
xmlns:SOAP-ENC="http://schemas.xmlsoap.org/soap/encoding/">
<SOAP-ENV:Header>
<ds:Signature xmlns:ds="http://www.w3.org/2000/09/xmldsig#">
<ds:SignedInfo>
<ds:SignatureMethod Algorithm="http://www.w3.org/2000/09/xmldsig#rsa-sha1"/>
<ds:CanonicalizationMethod
algorithm="http://www.w3.org/2001/10/xml-exc-c14n#"/>
<ds:Reference URI="#Body">
<ds:Transforms>
<ds:Transform
Algorithm="http://www.w3.org/2000/09/xmldsig#enveloped-signat
ure">
</ds:Transform>
</ds:Transforms>
<ds:DigestMethod Algorithm="http://www.w3.org/2000/09/xmldsig#sha1"/>
<ds:DigestValue>LAKJSDLFHGLKJA=...</ds:DigestValue>
</ds:Reference>
</ds:SignedInfo>
<ds:SignatureValue>ALKSDJFLKAJSDF=...</ds:SignatureValue>
<ds:KeyInfo>
<ds:KeyValue>
LAKSJDFLKJASDHFLASJKDFL=...
</ds:KeyValue>
</ds:KeyInfo>
</ds:Signature>
</SOAP-ENV:Header>
<SOAP-ENV:Body>
<fipa-message schemaLocation="acl.xsd" act="confirm" conversation-id="00005">
<sender>
<agent-identifier>
<name id="adaptersoftman@ngms.com"/>
<addresses><url>http://www.ngms.com/adaptersoftman</url></addresses>
</agent-identifier>
  </sender>
<receiver>
<agent-identifier>
<name id="desoftman@cxcs.com"/>
<addresses><url>http://www.cxcs.com/desoftman</url></addresses>
</agent-identifier>
</receiver>
<language>fipa-sl0</language>
<reply-with>task1-003</reply-with>
<ontology>natural-gas-ontology-1</ontology>
<content>((done task1)))</content>
</fipa-message>
</SOAP-ENV:Body>
  </SOAP-ENV:Envelope>
```

The first time the request sends a request information to recipient, the recipient must first register and receive their own a pair of keys, including the public key and private key. Only registered, the recipient will execute the request of the request only passing through the check of the requesting party.

6   Conclusion

For dealing with those distribution and internal complexity of the issue, one of the key problems in Multi-Acquisition is communication, so that they can mutually exchange information, or coordination or collaboration to complete tasks in order to achieve the purpose of solving. The introduction of Intelligent information push and pull technology can improve the intelligence level of network and database. It can find a fundamental solution to the problems encountered during these technology applications. The main advantages of Information Pull technology are: better targeted, light source task. Its shortcomings are as follows: poor timeliness.It require the user to have a certain expertise. In order to solve these Pull technology problems, information push technology has come into being.

Acknowledgment. The work is partly supported by both the National High-Tech Research and Development Program of China (No. 2009 AA 01Z119) and the Beijing Natural Science Foundation (No.4072018).

References

1. Tu, X.: Intelligent Management(in Chinese). Tsinghua University press, Beijing (1995)
2. Stuart, R., Peter, N.: Artificial Intelligence: A Modern Approach. Beijing Pearson Education North Asia Limited and People's Posts and Telecommunications Press, 473–484 (2002)
3. Tu, X., Tang, T.: Intelligent Autonomous Decentralized System (IADS). In: Wei, X. (ed.) Proc. of the IWADS 2002, Beijing, China, pp. 10–15 (2002)
4. Tu, X., Ban, X., et al.: Generalized Artificial Life Race and Model. In: Proceedings of the 8th AROB, Japan (2003)
5. Tu, X.: Large Systems Cybernetics(in Chinese). National Defense Industry publishers, Beijing (1994)
6. Lu, Q., Zeng, G., Zhang, W., Tu, X.: Soft Man and Agent. In: Proceedings of 2005 IEEE International Conference on Networking, Sensing and Control (ICNSC), USA, pp. 904–907 (March 2005)
7. Wang, H., Wang, Z., Zeng, G., Zheng, X., Tu, X.: The Research On 'Network Virtual Robot' Coordination and Its Application in Digital Gas Fields. In: Zhao, M., Shi, Z. (eds.) Proc. of the 2005 IEEE International Conference on Neural Networks and Brain, Beijing, China, October 13-15, pp. 1429–1433 (2005)

Cloud Model and Ant Colony Optimization Based QoS Routing Algorithm for Wireless Sensor Networks

Di Jian

School of Control & Computer Engineering
North China Electric Power University, Baoding, China
din__gr@126.com

Abstract. This paper presents CMACRO (Cloud Model and Multiple Ant Colonies Optimization based Routing), a new cross-layer QoS routing algorithm for wireless sensor networks. Basing on the principle of cross-layer design, the algorithm adapts delay, nodes' load and link quality as QoS metrics, and provides differentiated services for real time event-driven data streams and delay-tolerant periodic sampling data. The QoS routing metrics are regarded as heuristics correction factors in ant colony algorithm (ACA). The ants are divided into a number of different populations. Through the interaction of pheromone between multi populations, the routing algorithm searches for the feasible paths in parallel and updates the pheromone in time. To overcome the slow convergence of ant colony algorithm, improvements to control the randomness of the ants via cloud model are proposed. The simulation results demonstrate that the routing algorithm can guarantee the real time, reliability and robustness of wireless sensor networks. It can also achieve the network load balancing and congestion control mechanism.

Keywords: wireless sensor networks, cloud model, multiple ant colonies algorithm, QoS routing.

1 Introduction

Wireless Sensor Networks (WSN), also known as Ubiquitous Sensor Networks (USN), bring together many disciplines in with the rapidly developing wireless technology. WSNs [1] are networks based on small size nodes cooperation. Those nodes are mainly characterized by their low energy consumption, their low cost and, of course, their wireless communication. The application level is sensitive to the characteristics of network transmission, such as delay, throughput, bandwidth, packet loss rate and energy cost.

QoS routing technology is an important guarantee of WSN QoS [2]. The role of a QoS routing strategy is to compute paths that are suitable for different type of traffic to meet the requirements of various applications. The main problem to be solved by QoS routing algorithm is multi-objective optimization problem. Ant colony algorithm (ACA) is a bio-inspired algorithm [3], and it is based on the behavior of an ant colony in the nature. The distributed nature of network routing is well matched by the multi

Y. Wu (Ed.): International Conference on WTCS 2009, AISC 116, pp. 179–187.

agent nature of ant colony optimization (ACO). But in the actual operation, there are two problem to be addressed. First, the algorithm is easy to fall into local extrema. Second, the algorithm is slow convergence. This is due to the absence of a mechanism to control the randomness of the decline.

In this paper, through introducing membership cloud models (MCL) to adjust the random decline degree of ant colony, we firstly propose a membership cloud model based ant colony algorithm. Subsequently, using this algorithm and adopting the cross-layer design approach, we present an on demand cross-layer QoS routing algorithm. Our algorithm utilizes routing metrics from different layers to achieve overall system optimization.

2   Background Information

2.1   Cloud Model

Cloud model, put forward by professor Li Deyi, is a qualitative to quantitative conversion model [4]. Based on random mathematics and fuzzy mathematics, cloud model is proposed to descript randomness, fuzziness and the relationship between them in linguistic value. Cloud model has been successfully applied in intelligent control, data mining, intrusion detection, fuzzy evaluation and other fields.

Expected value (Ex), Entropy (En) and Hyper-Entropy(He) are digital features of cloud model, which form the knowledge representation foundation for mapping between the qualitative and quantitative. A cloud droplet forms from a mapping of qualitative concepts and quantitative concepts. Uncertainty in the process of its formation, cloud droplet cloud model gives a qualitative (μ) idea of the determination of x. There are positive direction and reverse sub-cloud, which is based on cloud formation mechanism and calculation direction. The cloud generation algorithm can be achieved by using software, and hardware can also be solidified into implementation, which is called cloud generator (CG).

2.2   Basic Ant Colony Algorithm (ACA)

The ant colony algorithm   is a member of swarm methods family, and it constitutes some metaheuristic optimizations based on the behavior of ants seeking a path between their colony and a source of food [5].

Edge selection:

An ant will move from node i to node j with probability

$$p_{ij}^k(t) = \frac{\tau_{ij}^\alpha(t)\eta_{ij}^\beta}{\sum_{j \in N_i^k}\tau_{ij}^\alpha(t)\eta_{ij}^\beta} \tag{1}$$

Where

$\tau_{ij}(t)$ is the amount of pheromone on edge i,j.

η_{ij} is the desirability of edge i, j (a priori knowledge).

α is a parameter to control the influence of $\tau_{i,j}$.

β is a parameter to control the influence of $\eta_{i,j}$.

Pheromone update:

$$\tau_{i,j} = (1 - \rho)\tau_{i,j} + \Delta\tau_{i,j} \tag{2}$$

Where

$\tau_{i,j}$ is the amount of pheromone on a given edge i,j.
ρ is the rate of pheromone evaporation
$\Delta\tau_{i,j}$ is the amount of pheromone deposited, typically given by

$$\Delta\tau_{ij}^{k} = \begin{cases} \dfrac{1}{L_K} & \textit{if ant k passededge}(i,j) \\ 0 & \textit{otherwise} \end{cases} \tag{3}$$

where L_k is the cost of the kth ant's tour.

2.3 Adjustment of ACA Parameters Using Membership Cloud Model

The important parameters of the residual factor ρ and the total pheromone information Q are used to control the randomness of multiple ant colony optimization algorithm. Membership Cloud Generator (MCG) is used to optimize the residual factor ρ and the total pheromone information Q so as to update the pheromone adaptively [6]. The parameters of cloud model are set as the following. $x_0=0$, $b=N_{max}/3$, The initial value of $\sigma_{max} = \sigma_0$.

Algorithm 1: Update Q
Step1 initialization, $e=0$, $b=N_{max}/3$, $\sigma_{max} = \sigma_0$, $t=0$, set Q_{max}.
Step2 $t=t+1$;
Step3 $x=t- N_{max}$, Generate U conditions cloud droplets with the aforementioned digital characteristics;
Step4 $Q= Q_{max}* \mu$;
Step5 $\sigma_{max} = \sigma_{max} + \sigma_{delta}$;
Step6 return to step 2.

Algorithm 2: Update ρ
Step1 initialization, $e=0$, $b=N_{max}/3$, $\sigma_{max} = \sigma_0$, $t=0$, set ρ_{max}.
Step2 $t=t+1$;
Step3 $x=t$, Generate U conditions cloud droplets with the aforementioned digital characteristics;
Step4 $\rho = \rho_{max}* \mu$;
Step5 $\sigma_{max} = \sigma_{max} + \sigma_{delta}$;
Step6 return to step 2.

3   QoS Routing Metrics

3.1   Delay

In wireless sensor networks, the delay between node i and node j comprises the following factors: processing delay denoted as $D_{pro}(i)$, queue delay denoted as $D_{queue}(i)$ and transmission delay denoted as $D_{tran}(i j)$.

$$D_{(i,j)} = D_{pro}(i) + D_{queue}(i) + D_{tran}(i,j) \tag{4}$$

The processing delay $D_{pro}(i)$ can be ignored due to the fast speed processing. The queue $D_{queue}(i)$ delay depends on the network load and the queue length. When using CSMA/CD as MAC protocol, $D_{tran}(i\ j)$ can be derived from packet retransmission times R_i and each retransmission delay d_k.

$$D_{tran}(i,j) = \sum_{k=1}^{R_i} d_k \tag{5}$$

3.2   Node Load

The load value Q^i of the node i denotes the queue length for sending datum. Q^i can be obtained from the information table of adjacent nodes. The node i periodically samples the queue length from the MAC layer interface. The average sampled value denotes the network load of node i. The queue length of the node i is denoted in Eq.(6).

$$Q_i = \frac{\sum_{k=1}^{n} q_i(k)}{n} \tag{6}$$

Definition 1: The load estimation function of the node i is denoted in Eq.(7).

$$L_i = Q_i + \sum_{j \in N} Q_j^i \tag{7}$$

Where L^i denote the load estimation function of the node i, Q_j^i denotes the queue length of the node j which is the adjacent node of the node i. The set N is the adjacent set of the node i. The sensor nodes in WSN cyclical broadcast the "Hello" packets with the queue length to its adjacent nodes [7]. The load value is used as a routing metric, which can effectively avoid network congestion and achieve network load balancing.

3.3   Link Quality

The actual transmission links of WSN can't completely guarantee the packet reception rate [8], because the transmission lines perennially are exposed to the atmosphere, a number of uncertain factors such as climate change and accidental damage may lead to node failure. Channel interference and data conflict also may increase the link error rate. The packet reception rate is used as a routing metric in this paper. The function of packet reception rate is denote in Eq.(8).

$$PRR = \frac{Lr + 1}{Ls + 2} \tag{8}$$

Where L_r denotes the successfully received packet numbers, L_s denotes the total sent packet numbers. The transmission link with best link quality will be used for data transmission, so the reliable requirement of the transmission line monitoring system can be successfully met.

4   Prepare Design of the MCL and ACO Based QoS Routing Algorithm for WSN

The steps of the new routing optimization algorithm is shown as below.

Step1: Algorithm initialization. The algorithm divides the artificial ants into k groups. Each group is assigned a corresponding sub-sink node as the destination node. There are m ants in each group, and each ant carries the ID information of the species, the source node and the nodes which are passed by the ant, timestamp, an empty stack, node load L_i , packet reception rate PRR_i and so on. The ants will be divided into two types in the network. They are the forward ants and the backward ants. The taboo list is initialized at first. The forward ant collects the nodes' information to update the local pheromone. The backward ant takes the same path as that of its corresponding forward ant, but in the opposite direction. The backward ant updates the global pheromone.

Step2: The attract factor and the exclusion factor of pheromone. A_i^k denotes the ant agent i in the population of K. The ants in the same population release the same type of pheromone. Different populations have different types of pheromone [9].

Definition 2: The attract factor. The attract factor α_{ij}^k is denoted in Eq. (9).

$$\alpha_{ij}^k = \frac{\tau_{ij}^k}{\sum_{h \subset \Pi_i} \tau_{kh}^k} \tag{9}$$

Where τ_{ij} denotes the pheromone trajectory intensity of k at the edge (i, j) .

Definition 2: The exclusion factor. The exclusion factor β_{ij}^k is denoted in Eq. (10).

$$\beta_{ij}^k = \frac{\sum_{h \neq k} \tau_{ij}^h}{\sum_{h \subset \Pi_i} \tau_{ih}^k} \tag{10}$$

Where τ_{ij} denotes the pheromone trajectory intensity of k at the edge (i, j) .

Step3: The probability transfer rules. The transfer probability formula is denoted in Eq. (11).

$$P_{ij}^s(k) = \begin{cases} \dfrac{[\tau_{ij}^s(t)]^{\varepsilon^{\alpha_{ij}^k / \beta_{ij}^k}} \times [\eta_{ij}(t)]^\beta}{\sum\limits_{h \notin tabu_k} [\tau_{ih}^s(t)]^{\varepsilon^{\alpha_{ij}^k / \beta_{ij}^k}} \times [\eta_{ih}(t)]^\beta} & j \notin tabu_k \\ 0 & otherwise \end{cases} \tag{11}$$

Where the attract factor α_{ij}^k and the exclusion factor β_{ij}^k are given by the formula (9) and (10). $\tau_{ij}^s(t)$ denotes the pheromone trajectory intensity of S at the edge e_{ij} . The adjustment factor ε is greater than 0. The heuristic function η_{ij} is denotes in Eq. (12).

$$\eta_{ij} = \frac{PRR_j}{L_j \times D_{ij}} \tag{12}$$

Where PRR_j denotes the packet reception rate of the sensor node j. L_j denotes the network load of the sensor node j. D_{ij} denotes the single-hop delay between the node i and the node j.

Step4: The dynamic adjustments of parameters based on cloud model. The value of and Q is adjusted by Algorithm 1, ρ is adjusted by Algorithm 2.

Step5: The local pheromone updating rule of forward ants. Each group of ant colony releases different types of pheromones. The pheromone trails are updated by the rule given by Eq. (13).

$$\tau_{ij}^k(t+1) = (1-\rho) \times \tau_{ij}^k(t) + \rho \times \Delta\tau_{ij}^k \tag{13}$$

The pheromone residual factor ρ is selected by the qualitative association rules in cloud model theory. The ant- density model is used to update the pheromone in the edge(i, j) . The number of the pheromone released by the ant is Q. The model is as below.

$$\Delta\tau_{ij}^k(t,t+1) = \begin{cases} Q & \text{if ant } k \text{ passededge}(i,j) \\ 0 & \text{otherwise} \end{cases} \tag{14}$$

Step6: The global pheromone updating rules of backward ants. When the destination node d is reached, the forward ant generates another backward ant, transfers to it all of its memory, and dies. The backward ant takes the same path as that of its corresponding forward ant, but in the opposite direction. The pheromone trails are updated by the rule given by Eq. (15) and Eq. (16).

$$\tau_{ij}^k(t+n) = (1-\rho) \times \tau_{ij}^k(t) + \rho \times \frac{Q}{R^s} \tag{15}$$

$$R^s = \frac{D_P \times L_P \times HC_P}{PRR_P} \tag{16}$$

Step7: Set maximum and minimum of the pheromone.

If the pheromone of the path is too high or too low, the algorithm may fall into the local optimum situation. The numbers of the residual pheromone will be limited at $[\tau_{min}, \tau_{max}]$, and $\tau_{ij}^k(t)$ will be modified after the ants have completed a tour. The following formula is used to select the threshold[10].

$$\tau_{ij}^k(t+n) = \begin{cases} \tau_{min} & \tau_{ij}^k(t) \leq \tau_{min} \\ \tau_{ij}^k(t) & \tau_{min} < \tau_{ij}^k(t) < \tau_{max} \\ \tau_{max} & \tau_{ij}^k(t) \geq \tau_{max} \end{cases} \tag{17}$$

Step8: When the backward ant reaches the source nodes, the second batch of ants will be generated. IF the numbers of iterations reach the maximum value-NC_{max}, the loops end.

5   Simulation and Performance Analysis

In order to verify the feasibility and effectiveness of the routing algorithm, we make the comparition of the transmission delay, congestion rate and network throughput between the CMACRO algorithm and the AODV algorithm. At last the robustness of the algorithm is verified.

OMNeT++ simulation tool is used and the parameters are chosed as below: The maximum coverage distance of the node is 1000 miles. The MAC layer is based on IEEE 80.2.15.4. Multiple rules generator based on cloud model is used to select p and Q. The numbers of the drops in cloud model is 500. Ex = 0. 5 En = 0. 8 He = 0. 08. The numbers of the ant colony is 3. The numbers of ants in each group is 100. $\varepsilon = 0.85$, $\varepsilon = 0.85$, $\beta = 2$, $P_{min} = 0.3$, $\tau_{max} = \tau_{ij}^{k}(0)$, $\Delta\tau_{ij}^{k}(0) = 0$, $\tau_{min} = \tau_{max}/300$, T=5, NC_{max}=500.

5.1   The Average Delay of Data Packets

The average delay of data packets is the average value of the waiting delay and the transmission delay in a fixed period of time. The simulation results of average delay are shown in Figure 1. We can see from the simulation results that the CMACRO algorithm is better than the AODV algorithm in terms of average delay.

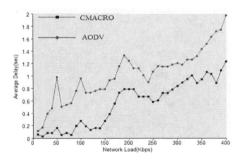

Fig. 1. Comparison of average delay

5.2   Comparison of Packet Successful Reception Rate

Packet reception rate is defined as the ratio of the data packets delivered to the destination. As it is illustrated in Figure 2, the CMACRO algorithm is better than the AODV algorithm in terms of packet reception rate. With the increasing network load, the packet reception rate of the optimum path becomes lower. However the link quality of the optimum path obtained by the CMACRO algorithm is better than the AODV algorithm, and the path is more stable. Because the packet reception rate is added to the transition probability in the CMACRO algorithm, the node will choose the next-hop node with maximum packet reception rate to transmit the datum.

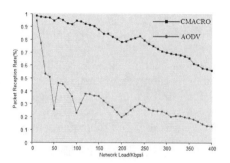

Fig. 2. Comparison of packet reception rate

5.3 Network Throughput

The node load is added to the transition probability in the CMACRO algorithm, the node will choose the next-hop node with minimum load value to transmit the datum in WSN. As it is illustrated in Figure 3, the improved algorithm can guarantee the network throughput for WSN. The CMACRO algorithm has more significant advantages than the AODV algorithm. For the algorithm can reduce the numbers of the bottleneck nodes near the sink node, the requirements of the transmission line monitoring system can be successfully met.

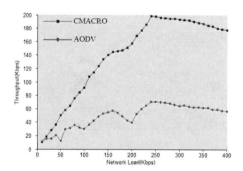

Fig. 3. Comparison of network throughput

Acknowledgment. This work is supported by the National Natural Science Foundation of China under Grant No. 60974125.

References

[1] Romer, K., Mattern, F.: The Design Space of Wireless Sensor Networks. IEEE Wireless Communications 11(6), 54–61 (2004)
[2] Akyildiz, I.F., Melodia, T., Chowdhury, K.R.: Survey on wireless multi-media sensor networks. Computer Networks, 9212960 (2007)
[3] Dorigo, M., Gambardella, L.M., Birattari, M., Martinoli, A., Poli, R., Stützle, T. (eds.): ANTS 2006. LNCS, vol. 4150. Springer, Heidelberg (2006)
[4] Li, D.Y., Yi, D.: Artificial Intelligence with Uncertainty. National Defense Industry Press, Beijing (2005)
[5] Dorigo, M., Maniezzo, V., Colorni, A.: Ant system: optimization by a colony cooperating agents. IEEE Transactions on Systems, Man and Cybernetics part B:Cybernetics 26(1), 29–41 (1996)
[6] Zhang, Y.-D., Wu, L.-N., Wei, G.: Improved ant colony algorithm based on membership cloud models. Computer Engineering and Applications 45(27), 11–14 (2009)
[7] He, P.: Study on routing and topology control techniques in mobile Ad Hoc networks. Xidian University (2007)
[8] Xia, Z., Yu, H., Yang, B.: Algorithm for probabi-listic link selection in wireless sensor networks using Bayesian estimation. Huazhong Univ. of Sci. & Tech (Natural Science Edition) 37(2), 40–44 (2009)
[9] Xia, H.: Research on intelligent computation method with application to network optimization and prediction. Jiangnan University (2009)
[10] Duan, H., Wang, D., Yu, X.: MAX-M IN meeting ant colony algorithm based on cloud model theory and niche ideology. Journal of Jilin University(Engineering and Technology Edition) 36(5), 803–808 (2006)

The Research of Network Information Service Platform Based on Affective Computing

Ming Chen, Lejiang Guo, Xiao Tang, and Lei Xiao

Department of Early Warning Surveillance Intelligence, Air Force Radar Academy
Wuhan, China
Ming1513@qq.com

Abstract. Affective computing is integrated on artificial emotions and intelligent computing. It attempts to recognize and synthesize human emotion to make intelligent response system. The research of affective computing is primarily involved in artificial emotion and artificial digital technology of cognition and consciousness. This paper introduces the background of affective information processing and discusses its main research branches and status. Based on affective computing and network technology, it proposes the realization method and key technology of the network service platform model. The system can solve the interactive problem of communication between users and machine effectively. At last, it presents some scientific problems waiting to be resolved in the future and proposes some suggestions.

Keywords: affective model, information processing, intelligent systems, artificial emot*ion*.

1   Introduction

Affective Computing is concerned with understanding, recognizing and utilizing human emotions in the design of computational systems. Research in the area is motivated by the fact that emotion pervades human emotions and human behavior, they promote social bonds between people and artifacts [1]. Emotional cues play an important role in forecasting human mental state and future actions. It is a new step in advance human machine interface (HMI) giving computers and machines the ability to interact more smartly with people in a natural way. In recent years, computer science research has shown increasing efforts in the field of affective computing. Several approaches have been made emotion recognition, emotion modeling and communication agents [2]. From a scientific point of view, emotions play an essential role in decision making as well as perception and learning. Furthermore, emotions influence rational thinking and therefore should be part of rational agents as proposed by artificial intelligence research. Another focus is on human computer interfaces which include believable animations of interface agents. The workshop focuses on the role of affect and emotion in computer systems including emotion recognition, emotion generation and emotion modeling with special attention to a specific problems and applications [3]. Both shallow and deep models of emotion are in the focus of interest. The goal is to provide a chance for the presentation of research as

Y. Wu (Ed.): International Conference on WTCS 2009, AISC 116, pp. 189–196.
springerlink.com   © Springer-Verlag Berlin Heidelberg 2012

well as future applications for lively discussions among researchers and industry. The papers will discuss theories, architectures and applications which are based on emotional aspects of computing. The system can meet the growing requirements of higher-quality comprehensive information service.

2   The Theory and Technology of Affective Computing

2.1   Affective Computing Research Areas

Affective computing is a new multidisciplinary field of study. It involves sensor technology, computer science, cognitive science, psychology, behavior, physiology, philosophy, sociology and many other fields [4]. From the exchange process in terms of human emotions, affective computing research can be divided into four steps as shown in Fig.1.It includes emotional signal acquisition, emotional information analysis and identification, emotional information understanding and emotion expression. The system acquires information either from sensor directly or by contacting with people indirectly for going through analysis and recognition, the analysis results will lead to emotional understandings which can be expressed by reasonable ways.

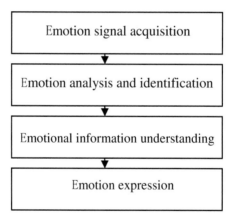

Fig. 1. Affective computing procession

2.2   Characteristics Affective Computing

Affective computing is a highly integrated area of technology, its main contents include:

1) Emotion mechanism. The research of emotion mechanism mainly includes emotional state determination and the relationship between physiology and behavior. It involves psychology, physiology, cognitive science, providing a theoretical basis for e-learning. Therefore, it determines emotional state and the corresponding relationship between physical behavioral characteristics. These relations are not yet entirely clear, which needs further exploration and research.

2) Emotional signal acquisition. The research of emotional signal acquisition mainly refers to the development of all kinds of effective sensors. This is an extremely important part in e-learning since all researches of e-learning are based on signals obtained by sensors, thus it can say there is no e-learning research without effective sensors.

3) Emotional signal analysis, modeling and identification. Once various types of effective sensors get the emotional signal, the next step correspond emotion signals with emotion mechanism requires modeling and identifying of the emotion signals. As emotional state is an implicit variable hidden in a number of physiological and behavioral characteristics that is impossible to observe directly and also it is difficult to model, sometimes it can use certain methods, such as hidden Markov models, Bayesian network models to settle this problem[5].

4) Emotion understanding. After acquisition, analysis and identification of emotion, the computer can understand its emotional state. The ultimate goal of affective computing is to enable computers to make appropriate responses to meet the changing emotional needs of computer users on the basis of understanding their emotional state. Therefore, this part of research focuses on how to make the most appropriate reaction to user emotion changing according to recognition results of emotional information [6]. In the process of emotion understanding modeling and application, emotional understanding and application of the model should note the following aspect: emotional signal tracking should be real-time and maintain a certain time record; emotion expression is based on the current emotional state and timely; emotion model is aimed at personal life and can be edited in a particular state; emotion model is adaptive; identification mode is regulated by understanding feedback.

5) Emotion expression. Previous research infers emotional states by physiological or behavioral characteristics [7]. Emotional expression is just its anti-process, which is given a certain emotional state to reflect it in one or more physical or behavioral characteristics. Emotion expression makes emotion interaction and communication possible. For a single user, Emotional exchanges include interaction among human-human, human-machine and human-nature.

6) Emotion generation. The research involves emotion generation theory, methods and techniques [8]. On the basis of emotion expression, further study how the computer, robot can simulate or generate emotional patterns to develop virtual or physical emotional robots or a computer with artificial emotions and corresponding application system.

2.3   Emotion Model

Model is a simplified objective of the understanding of scientific knowledge by the subject of knowledge based on abstraction and imagination of the real world objects. Emotion model is an important means to visualize and materialize the abstract emotional information; it can form three aspects: the concept, mathematical quantification and information organization.

1) The concept modeling of emotional information. The concept model is an abstraction of objective reality and its associations. The concept modeling of emotional information mainly refers to researches of human emotions and emotion type, mostly proposed by the psychology professions [9]. The former aspect discusses it from the basis types of emotion and the latter from the point of emotion dimensions.

2) Mathematical modeling of emotional information. The conceptual model requires mathematical methods to establish operational quantitative model to be applied to specific areas. Mathematical modeling of emotional information aims at establishing quantitative relationship between emotional information and objective object to realize a two-way conversion between human emotions to the physical characteristics.

3) Organization modeling of emotional information. The former two types of emotional information modeling has established a conceptual framework and the quantitative relationship, there is another type of modeling which makes an ordering organization describing or expressing emotion information from the aspect of information organizing.

3   The Network Information Service Platform Based on Affective Computing

3.1   The Architecture of the System

From Fig.2, the network information service platform has four layers: data layer, exchange layer, integrated layer and application layer.

1) Data layer: business-related data are generally divided into structured data and unstructured data.

2) Exchange layer: metadata-based information sharing system manages structured data and unstructured data .The metadata describes the data provide sufficient information.

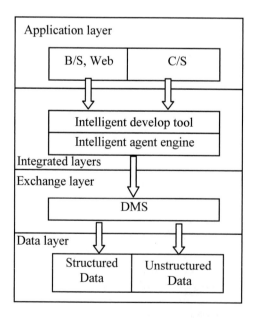

Fig. 2. The framework of intelligent security platform

3) Integrated layers: it achieves a flexible, fast and intelligent information retrieval system to share the data, and then it develop some intelligence agency capabilities to search mechanism.

4) Application layer: it can be B/S or C/S structure. B/S has the advantage of the client which can access the Internet, but it requires greater security in the system. C/S has better security and the use of smooth network environment.

3.2 BDI Structure

The system involves some multiple heterogeneous environments between the technical problems and inherent requirements. It is suitable for analysis and construction of agent. Its role is used for knowledge representation, cognitive science and philosophy. A BDI model is made of beliefs, desire and intentions. The solid theoretical foundation and convenient operability becomes the most used theoretical model of agent. BDI agent structure is shown in Fig.3.

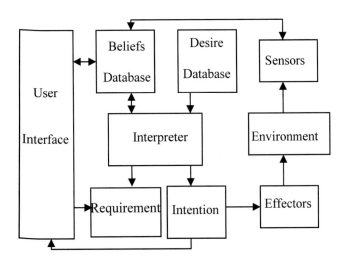

Fig. 3. BDI structure

3.3 Network Security

Network security is core solutions in the safety management. Terminal security and business workflow security are built the establishment of network infrastructure security basis. It establishes the safety of active network management system. Network security service constitutes the three major systems:

*1) Security Service Operation Platform (SSOC).*Network information uses data mining technology on many different safety devices according to user needs of security goals in data processing and information security management system (ISMS) database to support extraction of the user concerned about the whole network situation and risk information;

2) Protection System. Workflow management, threat management perform the basic protection functions in the strengthening of the nodes on the topology, routing path, the role of authority while a variety of configuration management tasks;

3) Emergency Response System. The implementation of security services operation platform (SSOC) and infrastructure protection system (SDM) outside of the emergency and response functions, risk strategy including personnel checks, inspection and cleaning of resources. The architecture of SOC is divided into security control platform, device agents, the host control agent, the network detector, the trusted computing terminal and other major components.

3.4   Terminal Security

Using the way of identify verification of cookie and combining varieties of encrypted technique , the perfect security configuration of operating system can provide enough support of many security verification application. While this structure helps make sure that the broadest consideration is given to often conflicting needs, it often create a time-lag between the desire to specify protection against a certain threat and final acceptance and publication of the standard. The reality of criminality is that new attack techniques are constantly being invented and rapidly deployed to exploit weaknesses in the security of products. Safety management system client provide network-based care.

4   Key Technology

Affective computing is a new multidisciplinary field of study. The ultimate goal of affective computing is to give computer the ability similar to human emotion. To achieve this goal, there are many fundamental scientific issues to be resolved with great difficulty. The breakthrough of those issues will be a great boost to the development of various disciplines. The key technologies of affective computing include: the analysis of emotional state of human physical, psychological and behavioral characteristics; the research of human emotional information signal acquisition sensor; the computer modeling of human emotions and various behavior characteristics; the identification technology of human biometric; the fusion and integration of perception data and the inference system of knowledge; the effective methods of the emotional perception.

1) Measurement Mechanism of Emotion. The model and process of affective computing should be in the universal case with personalized features. Research in this area is mainly the work of psychologists and physiologists, the research of emotion mechanism should first learn from. And it should first focus on the study of universal theory, then, the individualized research. The so-called emotional signal is the results of measurement of the reactions information out of various physical or behavior traits based on or related to emotions [10]. It is essential to be accurate, reliable measurement. However, some signal extraction is difficult. Multi-sensor integration methods can be used on a variety of emotional signals measurement, using the information fusion technology of detection layer and space-time layer.

2) Algorithm of Emotional Signal Fusion. Emotional signal shows strong diversity and complexity. This includes non-linear, time-varying, signal response delay and saturation effects, low noise ratio characteristics. The sensitivity of all kinds of emotional signal for a variety of emotions feature is also different. Emotional signal analysis and identification purpose is to provide theoretical and experimental basis for choosing right emotional signal to provide reliable raw data for understanding and expressing emotion [11]. The change in emotion or behavior corresponds to a variety of physiological or behavioral changes. But a single physical or behavioral change cannot lead to corresponding changes in emotion. So emotion determine is a course of integration of various physiology or behaviors. The theory and technology of information integration of corresponding affective computing should be studied. The purpose of affective computing research is crystal clear from the objective conditions. It should be studied relatively at a certain stage, a certain level. From the research contents of affective computing, the person should be first verified and identified accurately with personalized features. In terms of hardware and software of affective computing, emotions should be universal so the contradiction of generality and personality is properly handled.

3) Emotional Effect and Behavioral Choice. There are two ways that emotion can impact cognition and behavior, one is emotion acting as a separate component contacts other parts of the brain, and the other is emotion acting as a modulation means directly take effect. No matter what kind of the affecting mechanism is, it finally be reflected in amendments of self-consciousness, environmental awareness and conveying information to other agent. The specific way of taking function can be realized by setting or amending the values of internal state variables and sending out messages with emotional information. This action can fully reflect the integration of emotions and traditional artificial intelligence [12]. The intelligent behavior of the agent can be regulated and controlled at strategies, methods, parameters, and different levels, etc.

4) Affective Computing Platforms to Achieve. Emotion natural vectors are human being and other organisms. Similarly, agent simulating human is a natural carrier of affective computing. This computation has the advantage that you can fully refer to the current research outcomes to develop artificial emotion study which includes both the feelings of individuals and the social aspects of emotion.

It is very useful to develop a new MAS platform for an emotional agent. But it promotes people to focus on the design of emotion mechanism, the joint emotion and engineering that affective computing is carried out on the existing generic MAS platform. Combining results of the comparisons and the maturity of the software, emotional engineering, and software is highly inclusive to different heterogeneous technical means, and developers is allowed to choose agent architecture freely.

5   Conclusion

Affective computing presents a new challenge for computer technology. Affective computing establishes a harmonious environment between machine and users while affective computing is a multi-disciplinary discipline .The development of affective computing will promote the development of related sciences. Affective computing research is an international hot spot, however the research still stops on the initial stage of its basic research and applied research .The technology is not mature. Computer and network technology will facilitate the affective computing research platform .In future, affective computing will have a great development potential and the widespread applications.

References

1. Damasio, A.R.: Descartes Error: Emotion, Reason and the Human Brain. Gosset/Putnam Press, NewYork (1994)

2. Takamura, H., Inui, T., Okumura, M.: Extracting semantic orientations of words using Spin Model. In: Proc. of the 43rd Annual Meeting of the ACL, pp. 133–140. ACL, Stroudsburg (2005)
3. Gratch, J.: A domain-independent framework for modeling emotion. Journal of Cognitive Systems Research 4(5), 269–306 (2004)
4. Picard, R.W.: Affective Computing. MIT Press, London (1997)
5. Slomana: Beyond shallow models of emotion. Cognitive Processing 2(1), 177–198 (2001)
6. Turney, P., Littman, M.: Measuring praise and criticism: inference of semantic orientation from association. ACM Trans. on Information Systems 21(4), 315–346 (2003)
7. Christos, H.P.: Computational Complexity, pp. 50–60. Addision-Wesley, New York (1994)
8. Wang, Z., Zhao, Y.: An Expert System of Commodity Choose Applied with Artificial Psychology. IEEE International Conference on Systems, Man and Cybernetics, 2326–2330 (2001)
9. Brandstatter Decker, B., Ring, A.W.: Adaptive Finite-Element Mesh Generation for Optimization Problems. IEEE Trans. on Magnetics 38(2), 1017–1020 (2002)
10. Ward, R.D., Marsden, P.H.: Affective computing: problems, reactions and intentions. Interacting with Computers 16(4), 707–713 (2004)
11. Nahl, D.: Affective computing. Information Processing & Management 34(4), 510–512 (1998)
12. Wang, S., Chen, E., Li, J., et al.: Content-based interactive emotional image retrieval. Journal of Image and Graphics 6(10), 969–997 (2001)

Measuring Situation Awareness in Computer Mediated Discussion

Jia Li[1], Xuan Liu[1], Zhigao Chen[1], and Pengzhu Zhang[2]

[1] School of Business, East China University of Science and Technology, Shanghai, China
[2] Antai School of Economics and Management, Shanghai Jiaotong University,
Shanghai, China
hujiang360@sohu.com

Abstract. Situation awareness is an important factor in successful computer mediated group discussion, yet the related metrics have been lacking. To fill in that gap, we propose 13 metrics from 3 aspects to systematically reflect discussion situation using hierarchical discussion as an example. This research can be considered as the first attempt to answer what a fully understanding of discussion situation is and how to measure it.

Keywords: situation awareness; computer mediated discussion; measurement; hierarchical discussion.

1 Introduction

Situation awareness, or SA, which has been recognized as a critical, yet often elusive, foundation for successful decision-making across a broad range of complex and dynamic systems [1-3], is also considered as an important factor in successful computer mediated group discussion. For example, by identifying the percentage of participants who contributed to the alternative, facilitators can assess if majority members are involved and decide whether further calling for participation is necessary. Another example is identifying the consensus state of the alternatives, based on which users can understand to what extent the participants has reached agreement on each alternative and decide if the group should put more effort on the unsolved ones. SA is particular critical under the condition of information overload, as is often the case, when computer-supported groups are confronted with larger numbers of ideas and supporting comments to organize and evaluate[4]. Users have got to remember all aspects of discussion information by themselves before they can make a correct assessment.

Endsley's model of SA illustrates three stages or steps of SA formation: perception, comprehension, and projection. Complete discussion situation awareness involves perceiving the status, attributes, and dynamics of basic elements (e.g., the speaker, the reply-to relationship, the speech act, the timestamp, etc.) in the meeting, synthesizing disjointed Level 1 SA elements to form sense making metrics (e.g., representatives, consensus, etc.), and projecting how it will affect future states of the group decision making. Since the prediction of impact to the future discussion is a complicated

Y. Wu (Ed.): International Conference on WTCS 2009, AISC 116, pp. 197–204.
springerlink.com © Springer-Verlag Berlin Heidelberg 2012

research question and beyond the scope of this research, we study situation awareness only up to level 2.

The first step to study SA in computer mediated discussion is to measure it. The multivariate nature of SA significantly complicates its quantification and measurement, as it is conceivable that a metric may only tap into one aspect of the operator's SA. Further, studies have shown that different types of SA measures do not always correlate strongly with each other [5-7]. Accordingly, rather than rely on a single approach or metric, valid and reliable measurement of SA should utilize a battery of distinct yet related measures that complement each other [8]. So group discussion situation awareness must be a multi-aspect task. A comprehensive understanding of group requires the comprehension of related aspects such as discussion snapshot, discussion temporal change, and the people involved in the discussion. The multi-faced approach to discussion situation awareness measurement capitalizes on the strengths of each measure while minimizing the limitations inherent in each. However, a review of the literature reveals that the metrics measuring SA in computer mediated discussion has been lacking. To fill in that gap, we propose 13 metrics from 3 aspects in this research to systematically reflect discussion situation in computer mediated discussion using hierarchical discussion as an example.

2   Hierarchical Discussion

Group support systems have to decide how to organize the utterances in an effective way for browsing and searching. Some groupware like Group Systems developed at the University of Arizona put all the comments in several boxes and order the utterances by time sequence [9, 10]. Other groupware like HERMES [11] and MRV [12] organize the comments in a hierarchical way, which means each utterance is attached to one or more utterance(s) directly beneath it. The groupware used in this study follows the patter of hierarchical discussion, and more specifically, a variation of ZENO argumentation framework [13], which in turn has its roots to the informal IBIS model of argumentation [14, 15]. The basic argumentation elements are issues, alternatives, and positions.

Issues correspond to decisions to be made or goals to be achieved. They are brought up by users and are open to dispute (the root entity of a discussion tree has always to be an issue). Issues consist of a set of alternatives that correspond to potential choices. An issue can be also "inside" another issue, in cases where some alternatives need to be grouped together. Positions are asserted in order to support the selection of a specific course of action (alternative), or avert the users' interest from it by expressing some objection. Positions may also refer to some other position in order to provide additional information about it. A position refers to a single other position or alternative, while an alternative is always in a single issue. Figure 1 illustrates a hierarchical discussion containing all the elements mentioned before.

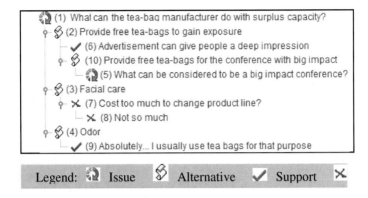

Fig. 1. Illustration of hierarchical discussion model

In the hierarchical discussion model, each utterance is expressed as a node, and related utterances are linked to each other to form an argument tree. Anytime users want to add an utterance in the system, they need to select an existing node as the target for the new utterance and specify the relationship between the utterance and the target. The relationship could be a new issue, a new alternative, a support position or an objection position. By requiring users to clarify the relationship between nodes, irrelevant information (e.g., chatting) are decreased and the participants are urged to express themselves more effectively.

3   Metrics Measuring Situation Awareness in Hierarchical Discussion

The key step in the translation of a vague term into a quantitative concept is operational definition and measurement. The rest of this subsection develops metrics systematically measuring situation awareness in hierarchical computer mediated discussion. More specifically, the following three categories of metrics are explored: snapshot metrics, duration metrics and people metrics.

3.1   Snapshot Metrics

The snapshot metrics reflect the current discussion situation. The snapshot metrics include representativeness, utterance quantity, consensus, and atmosphere. Consensus is a key metric to be aware of because the goal of discussion is to reach consensus. But to make sure the consensus makes sense, we must check the maturity of discussion (i.e., if the discussion result represents the opinion of majority, and if the discussion utterance quantity satisfied the preliminary requirements of a mature discussion).

Also the discussion atmosphere is an important aspect to be concerned. Ideally, the discussion atmosphere should be a balance of critical and supportive. Mason [16] suggested that dialectical inquiry that's both critical and constructive should lead to higher quality solutions.

(1) Representativeness (R)

Representativeness measures to what degree the discussion result can represent the opinion of majority. A common phenomenon in group discussion is participation bias in which the results of discussion become non-representative because the participants disproportionately involve in the discussion.

Representativeness can be measures by the percentage of people who contributed at least one comment to the target node or its children. For example, if the representativeness is 0.4 for alternative A and 0.6 for alternative B, the discussion result of alternative A can better represent the group will than alternative B due to higher number of unique contributors. Low representativeness means the discussion result is immature, and the facilitator should think about calling for more participation.

(2) Utterance quantity (U)

Utterance quantity measures to what degree the discussion is sufficient and mature. It has been observed that higher utterance quantity leads to higher discussion breadth and depth, which usually indicates a more thorough discussion and understanding of the space of debate [17].

Different ideas may receive different quantity of comments, which indicates the hot topics and rarely discussed topics. Utterance quantity can be measured by the number of utterances beneath the target node. Utterance quantity is a good indicator to judge the maturity of discussion. For example, if the utterance quantity for alternative A is 15, and the utterance quantity for alternative B is 30, alternative B is considered more thoroughly discussed than alternative A, and thus the discussion to alternative B is more mature. Low utterance quantity means the discussion is still immature and more comments are encouraged.

(3) Consensus (C)

Consensus measures to what degree the discussant has agreed on. Since everyone's opinion is encouraged and valued, group consensus is a critical factor of group decision making. The degree of consensus can be measured by the following formula recursively:

$$C(o) = \frac{(\sum_{s \in S(o)} C(s)) + 1}{(\sum_{s \in S(o)} C(s) + \sum_{r \in R(o)} C(r)) + 1} \tag{1}$$

Where C(o) is the consensus of target node o; S(o) is the set of utterances replying to node o with speech act support; R(o) is the set of utterances replying to node o with speech act objection. If a user contributes more than comments beneath the node o, only the latest one is taken into account in S(o) or R(o). The calculation of consensus is a recursion procedure, and the recursion continues until the algorithm reaches a leaf node.

(4) Atmosphere (A)

The meeting atmosphere can be measured by the percentage of supportive speech act among all positions. If the atmosphere is either too close to 1 or 0, the facilitator must think about doing something to change to the discussion atmosphere because it's opposite to a productive meeting.

3.2   Duration Metrics

Time is an important concept in discussion SA, as SA is a dynamic construct, changing at a tempo dictated by the actions of individuals, task characteristics, and the surrounding environment. As new inputs enter the system, the individual incorporates them into this mental representation, making changes as necessary in plans and actions in order to achieve the desired goals. By incorporating temporal change into snapshot metrics, we explore the following four duration metrics: representativeness alters, utterance quantity alters, consensus alters, and atmosphere alters.

(5) Representativeness alters (RA)
Representativeness alters over time is used to describe the representativeness's temporal change. By its definition, representativeness alters over time is a non-decreasing function of time because more and more people will add comments to the discussion node as time lapse. Representativeness alters over time is a good indicator to assess how the representativeness evolves, and to identify the turning point after that the discussion became well represented.

(6) Utterance quantity alters (UA)
Utterance quantity alters over time is used to describe the utterance quantity's temporal change. By its definition, utterance quantity alters over time is also a non-decreasing function of time because the number of utterance will increase as time lapse. Utterance quantity alters over time is a good indicator to understand how the hot topics switches among nodes, and to identify the turning point after that the discussion became thorough and mature.

(7)Consensus alters (CA)
Consensus alters over time is used to describe the consensus's temporal change. The consensus alters over time may intensively change shortly after the idea was proposed. An idea can be considered as reaching consensus only if the discussion is well represented and sufficient, and the consensus keeps stable for a long period.

(8) Atmosphere alters (AA)
Atmosphere alters over time is used to describe the atmosphere's temporal change. The atmosphere alters reflect the evolution of discussion style, and if the intervention from facilitator asking for change discussion atmosphere really takes effect.

3.3   People Metrics

Person is another import concept in discussion SA, as group discussion is teamwork. The people metrics include activeness, feed back, criticalness, and participation bias. Previous studies have found that the number of messages sent out by participants indicates their attitude toward the community [33]. An individual active in discussion may not be the most knowledgeable person, but he or she is probably willing to contribute to the group. It is important to know who and where those individuals are. The reply-to number and feedback from other group members is an indicator judging the quality of utterance. A person with many positive feedbacks usually receives high prestige in the group.

Criticalness may reveal the user's characteristic and preference to the discussion. By judging user's criticalness, we can find different types of users: some may be good at raising recommendations and assumptions, while others good at critiques of single sets of recommendations and assumptions. Over critical users are usually unpopular to the group norms but their challenges are valuable to the group discussion.

Ideally, user should cover all parts of the discussion proportionally. However, user usually focuses on some while ignores the others. One of the tendency typically found in group interaction is "not changing the subject" [18]. By identifying users that focus only on limited sub-topics of the discussion and encouraging their balanced participation, one can expect more space of debate explored which leads to improved performance.

(9) Activeness (V)

Activeness is used to describe to what degree the user is active in making contribution to the discussion. Activeness is a relative concept which can be measured by a 3 -point Likert-scale, i.e., order the participants by the number of comments contributed increasingly and the top 1/3 were considered most active and last 1/3 were considered as least active. Activeness is a good indicator to identify contributors and lurkers.

(10) Reply-num (P)

Reply-num is used to measure to what degree the user's utterance is attractive to other group members. Reply-num is the number of messages that reply to the utterances posted by the user. A user attracts more reply usually because the opinion posted was interesting or insightful. A leader and a follower can be distinguished by observing the reply-num if they posted the same number of utterances. A user with high reply-num is considered as the focus of group discussion, although the feedback can be positive or negative.

(11) Feedback (F)

Feedback is used to describe to what degree the user is positively replied to in the discussion. Feedback of a discussant can be measured by the following formula:

$$F(p) = \sum_{s \in \varphi(p)} C(s) \qquad (2)$$

Where F (p) is the feedback of person p, C(s) is the consensus of information s (the same as metric 3), and $\varphi(p)$ is the utterance contributed by person p. Feedback is a good indicator to identify the prestige of person. A user with positive feedback may gain more trust in the group discussion, and be expected with more positive feedback in the future discussion.

(12) Criticalness (T)

Criticalness is used to describe to the speaker's utterance style in the discussion. Criticalness can be calculated by the following formula:

$$T(p) = \frac{S(p) - O(p)}{S(p) + O(p)} \qquad (3)$$

Where T(p) is the criticalness of person p; S(p) is the number of comments made by person p with speech act support; O(p) is the number of comments made by person p with speech act objection. Criticalness is a good indicator to distinguish critical tended participants and supportive tended participants.

(13)Participation balance (B)

Participation balance is used to describe to what extent the participation of a discussant can be considered as balanced. Participation balance can be calculated by the following formula:

$$B(p_i) = \frac{Alt(p_i)}{\sum Alt(p_i)} \qquad (4)$$

Where $B(p_i)$ is the participation balance of person p_i; $Alt(p_i)$ is the number of alternatives to which participants p_i made at least one comment. High participation balance means the discussant are more likely to explore the whole discussion space, and less likely to fix in limited subcategories.

4   Conclusion and Future Work

In this research, we propose 13 metrics from 3 aspects to systematically reflect discussion situation in computer mediated discussion using hierarchical discussion as an example. This research can be considered as the first attempt to answer what a fully understanding of discussion situation is and how to measure it.

This research is part of our ongoing project dedicated to profile discussion situation and provide necessary interventions to benefit the group automatically or semi-automatically. Our future work includes designing visualization tools to support SA in computer mediated discussion, finding out more complicated interaction patterns using the basic metrics mentioned in this research, and providing interventions like automated facilitation and external information support.

Acknowledgement. This research was supported by the National Natural Science Foundation of China under Grant 71001038.

References

1. Nullmeyer, R.T., Stella, L.C.D., Montijo, G.A., Harden, S.W.: Human factors in Air Force flight mishaps: Implications for change. In: Proceedings of the 27th Annual Interservice/Industry Training, Simulation, and Education Conference, National Training Systems Association, Arlington, VA (2005)
2. Blandford, A., William Wong, B.L.: Situation awareness in emergency medical dispatch. International Journal of Human-Computer Studies 61(4), 421–452 (2004)
3. Gorman, J.C., Cooke, N.J., Winner, J.L.: Measuring team situation awareness in decentralized command and control environments. Ergonomics 49(12), 1312–1325 (2006)
4. Grise, M., Gallupe, R.B.: Information overload: addressing the productivity paradox in Face-to-Face electronic meetings. Journal of Management Information Systems 16(3), 157–185 (2000)
5. Durso, F.T., Truitt, T.R., Hackworth, C.A., Crutchfield, J.M., Nikolic, D., Moertl, P.M., Ohrt, D., Manning, C.A.: Expertise and chess: A pilot study comparing situation awareness methodologies. Experimental Analysis and Measurement of Situation Awareness, 295–303 (1995)

6. Endsley, M.R., Selcon, S.J., Hardiman, T.D., Croft, D.G.: A comparative analysis of SAGAT and SART for evaluations of situation awareness. In: Proceedings of the Human Factors and Ergonomics Society 42nd Annual Meeting, Human Factors and Ergonomics Society, Santa Monica, CA (1998)

7. Vidulich, M.A.: Testing the sensitivity of situation awareness metrics in interface evaluations, Situation awareness: analysis and measurement. In: Endsley, M.R., Garland, D.J. (eds.), pp. 227–246. Lawrence Erlbaum Associates, Mahwah (2000)

8. Harwood, K., Barnett, B., Wickens, C.D.: Situational awareness: A conceptual and methodological framework. In: Proceedings of the 11th Biennial Psychology in the Department of Defense Symposium, Air Force Academy, Colorado Springs, CO, U.S (1988)

9. Nunamaker, J.F., Briggs, R.O., Mittleman, D.D., Vogel, D.R., Balthazard, P.A.: Lessons from a dozen years of group support systems research: a discussion of lab and field findings. Journal of Management Information Systems 13(3), 163–207 (1997)

10. Nunamaker, J.F., Dennis, A.R., Valacich, J.S., Vogel, D.R., George, J.F.: Electronic meeting systems to support group work. Communication of the ACM 34(7), 40–61 (1991)

11. Karacapilidis, N., Papadias, D.: Computer supported argumentation and collaborative decision making: the HERMES system. Information Systems 26(4), 259–277 (2001)

12. Fujita, K., Kunifuji, S., Nishimoto, K., Sumi, Y., Mase, K.: Implementation and evaluation of the discussion support system MRV. In: Third International Conference on Knowledge-Based Intelligent Information Engineering System, Adelaide, Australia (1999)

13. Gordon, T.F., Karacapilidis, N.: The Zeno argumentation framework. International Conference on Artificial Intelligence and Law (1997)

14. Kunz, W., Rittel, H.: Issues as elements of information systems. Universität Stuttgart, Institut für Grundlagen der Planning (1970)

15. Rittel, H., Webber, M.: Dilemmas in a general theory of planning. Policy Sciences 4(2), 155–169 (1973)

16. Mason, R.O.: A dialectical approach to strategic planning. Management Science 15(8), 403–414 (1969)

17. Janssen, J., Erkens, G., Kanselaar, G.: Visualization of agreement and discussion processes during computer-supported collaborative learning. Computers in Human Behavior 23(3), 1105–1125 (2007)

18. Lamm, H., Trommsdorff, G.: Group versus individual performance on tasks requiring ideational proficiency (brainstorming): a review. European Journal of Social Psychology 3(4), 361–387 (1973)

Study on the Link of Mathematics Teachings between University and Middle School

Wang Huangliangzi and Xing Chaofeng

Zhangqiu no.4 Middle School, Zhangqiu, China
wh_3758765@163.com

Abstract. With the social progress and development of science and technology, mathematics is used more and more widely, which has penetrated into the various disciplines. Mathematics education has been paid attention much more in various fields. Therefore, improving the quality of teaching mathematics education becomes very important. In this paper, firstly, the problems for the convergence of middle school mathematics and college mathematics are analyzed, and then summarizing the differences between middle school mathematics and college mathematics in teaching methods, learning methods and teaching contents. In order to solve these problems and differences, the measures and proposals are proposed in this paper to link up middle school mathematics and college mathematics better.

Keywords: The new curriculum standards, college, school, middle school mathematics, link.

1   Introduction

For the Learners to receive education, education is a continuous process, from the perspective of systems theory, the mathematics teaching procedure can be regarded as a system, which is made up of the subsystems, the stage of education of mathematics teaching. Every subsystem must be coordinated with each other. The transition from the elementary mathematics to the advanced mathematics include both the transition of mathematics knowledge and the transition of teaching methods, so the problems to link up the mathematics teaching of middle school and university cannot be avoided by the mathematics teachers in them. In the view of its realistic and age background, the systematic research on this problem has its own necessity and urgency, and also, has the practical significance to the deepening reform of the current teaching of mathematics in the university.

2   The Problems to Link Up the Mathematics Teaching of Middle School and Universities

Compared with the primary and high schools' mathematics curriculum, the college mathematics courses are more obsolete. Although parts of the colleges adopt the new college mathematics textbook for the 21st century, the FRAGMENTATION phenomenon still exists among the mathematics teaching of colleges, primary and high schools. And also, there is limitation in students' physical and mental development. So, the difference between the teaching method and studying method of each stage leads to a seriously disjointed situation

2.1 The Problems to Link Up Contents of Mathematics Course of Middle School and University

Under the new curriculum standards and the high school math's reform deepening, the math course of high school which compared with the traditional math textbook has a lot of change in the content and the structure system of knowledge in these 10 years. In April 2003, the Ministry of Education of PRC issued the "Normal High School Math Curriculum Standards", and the mathematics education of high school keep on renovating itself, and constantly upgrade to the mathematics of university, and also, both the frameworks and contents of curriculum have been reformed a lot. However, while colleges and universities are also actively carry out reform, university mathematics teaching content and teaching system has a little change for decades.

Because the mathematics of high school and college are out of sync, and especially, the curriculum reform of college mathematics is lagging behind, as an immediate result: on one hand, there are more contents overlapping between them, for example, the concepts, like function, limit, continuity, and derivative, have already been learned in high school, but the teachers of college are still keeping on teaching them and wasting the valuable time of students. Furthermore, the college teachers have to delete some contents which often are not the duplicate content. This lead to the situation that the curriculum reform of high school and college mathematics is not carried out simultaneously. On the other hand, there are "gaps" and "coming apart" existing in many parts of content, especially in the course of "high mathematics" and "probability and statistics". With middle school mathematics reform deepening, this disharmony might also be further expended. If this situation is not reversed, it will be negative to improve the quality of the college student: obviously, it will affect the students to apply mathematical tools to solve practical problems, so it will affect the result of mathematics teaching.

I, as a first-line teacher, obviously feel that some content should be cut off but not, and some of them should not be cut off but done. So, the problem how to link up the mathematics of college and high school becomes more and more urgent to solve. Furthermore, it is an urgent work to prepare a set of textbooks which have little content overlapping and consistent with the requirements of less hours to spend and more practical knowledge included.

2.2 Middle Mathematics and Mathematics Showed Greater Differences in Teaching Methods

The teaching method changed from succinctly lecturing and largely training to widely lecturing and broadly learning. And also, the students' position changed into a leading place. Great changes happened in students' study way. Education changed from passive acceptance to active exploration. And the model of education changed from examination-oriented into quality education. However, in reality, "division" is existing in college, primary and secondary school's mathematics teaching. The teachers of college rarely consider about to linking up them, coupled with the limitation of the law of physical and mental development of the students, which leads to a situation that it is seriously out of line between teaching and learning methods.

3 The Measures Linking High School Mathematics and University Mathematics

3.1 The Engagements in the Teaching Methods

In order to solve really the link of middle school mathematics and college mathematics, there must be corresponding teaching methods and learning methods to show the essence and meaning.

1) Strengthening communication and contact in mathematics teaching, paying attention to teaching methods interacting with each other.

To linking up the college mathematics and mathematics teaching, it is essential to strengthen the university mathematics teachers and the middle school mathematics teachers' communication, also to strengthen the teachers and big, middle-school students' communications [3]. University and high school mathematics teaching methods should be closed up mutually, especially freshman mathematics teaching which is necessary to reflect the style of college mathematics, but also consider the acceptability of students. It is a big step. and a jump but not a gradual process from middle school mathematics to university mathematics and constant to variable.

If students must cross this level, they not only need have the perseverance, but must have the confidence and they could not ask for the moon. It needs to give the student certain adaptation time in the process to avoid the student losing learns mathematics the confidence. The mathematics is not strange for each university student. When they enter the campus, they came into contact with mathematics. They have experienced six years of secondary school mathematics education, and accumulated a very deep knowledge of elementary mathematics also have certain with mathematics thought analysis questions and with mathematics knowledge, the method solving the question abilities.

In this way, teachers can easily lead to the idea that students who have a good foundation to learn Mathematics should be an easy task, while students will have such thoughts, but this is not true. Therefore, it requests the student to transform the study idea as soon as possible, and integrates quickly to university's study. It is actually admitted to a university the students are the quite outstanding students who have enthusiasm to accept the new knowledge, the new idea. If teachers strengthen with students between communication and contacting, explaining that the transformation study idea and the method importance and the necessity, the students accept this kind of changes very quickly, integrating as soon as possible to the university life.

2) Using progressive teaching methods in university teaching

Educational psychology research shows that it takes some time for the students to transit to a new teaching method from the custom teaching method. The students already adapted middle school that is kind of slow rhythm, the few capacities, to say practices the union the teaching methods. If they are carried on the quick rhythm from the very beginning, the large capacity teaching, then the students adapt with difficulty for a while. This is not only affected the teaching effect, simultaneously also caused the student to lose the study university mathematics the enthusiasm. The student from the middle school mathematics to the university mathematics, from the constant to the variable is not an evolution process, but is a caper, is a big stair. They must jump this stair, and this requires the time to go to the cushion. Therefore in teaching students must be given enough cushion time. After they adapt and then gradually transit to the normal teaching, the students really integrate the university teaching.

3.2   The Engagements in the Study Methods

In university, students not only need to study the professional courses, but also has the elective courses which are varied. The students may study different curriculums according to own interests. With the aim of enriching themselves and raising the independent thinking in the study process ability, they should attend school organization's each kind of competition, especially for the mathematics specialized students who participate in mathematical modeling contest, it is the good choice. But regarding the new students, their thought also paused in middle school's time, learning always very passive and lack of the ability to think independently. In the process of mathematics modeling, most students can raise independent thinking abilities and the team cooperation spirits. Because in university's, there are no teachers and the guardians' supervise themselves to study. Students must depend on themselves to study many knowledge and must have the independent thinking ability. Absorbing the different knowledge, it looks like the exhortations to students of the school: "great resolute, learned, asks really, the ultimate good". Requesting teacher and the student must have open thought and the broad heart, also have firm and unyielding, marches forward courageously, hundred folds not the bountiful will and the spirit, the open field of vision, taking a broad view at the world, taking a broad view at the world, an additional knowledge and enrich themselves, study hard, draw lessons and the essence of knowledge.

If student want to learn the university mathematics well, making great progress, they must take in the learning process to look that to listen, to practice, achieving reasonable taking with the shed and understands primarily, paying great attentions to the parsing process, through material and so on reference books, raising specially studies independently ability. Students must before the class to prepare a lesson, listen earnestly in the class, review and reinforce after the class, momentarily looking up in the usual study reads the related content. They may prepare a class, exercise explanation reference book [7].

3.3   The Link of Teaching Content

The link of the teaching content of introduction course is following :

Under the new education reform, high school mathematics was expanded， the continuity of function, the right and left limits of function, the derivatives of function etc. are included in the content of new syllabus of middle school mathematics, function is the most basic knowledge throughout mathematical subject, and it is also the most difficult to understand, the most points difficult to study a knowledge, from a dollar square equations in high school mathematics to the implicit functions in university mathematics, function is not only widely used in algebra, in analytic geometry, inequality, but also used in applied mathematics. After the new education reform, high school mathematics and university mathematics increased linking of the function, and the continuity of the function is the first studied in college mathematics, while in high school textbooks, this content have been studied, so the students who just entered the university can learn fast to adapt to university learning environment. The importance of linking of the teaching content is also obvious.

In mathematics teaching at the university, the teaching introduction must be paid sufficient attention, the first section of the university mathematics class - Introduction course, It is essential that teachers can stimulate students' interest in learning mathematics, which will affect future learning and teaching, because students must be interested in this course, and then they have motivation to learn . So in the introduction class, the teacher should explain clearly to the students what is the similarities and differences between middle school mathematics and college mathematics,   and what is the purpose of learning college mathematics, the teaching programs, teaching content and performance evaluation in this semester, teachers also can talk about some of the better learning methods and learning experiences for students, which let the students clearly know the intrinsic relationship of college mathematics and middle school mathematics in the beginning, reveal the mystery of mathematics, at the same time, and can improve student's interest of learning, let them make less mistakes in the later study, will not let themselves confused.

The link of the content of middle school mathematics and college mathematics is following :

In the new syllabus, in the context of ensuring the teaching of basic knowledge, basic skills training, basic ability training, there added some new knowledge which lays the foundation for further study, has broad application, and the students can accept. The new knowledge is simple logic, linear programming, space vector, probability and statistics and the initial derivatives. These new teaching content is especially prominent in the reform of the new syllabus, these elements is the necessary knowledge linking up college mathematics. We can also see the importance of these knowledge points from the college entrance examination over the years. For example, the 2008 college entrance examination, Liaoning, multiple choice:
$$\lim_{x \to \infty} \frac{1+3+5+\cdots+(2n-1)}{n(2n+1)} = (\qquad)$$

A. $\frac{1}{4}$ B. $\frac{1}{2}$ C. 1   D. 2,

Hunan, fill in the blank, $\lim_{x \to 1} \frac{x-1}{x^2+3x-4} = $ _____ , these questions are the subjects of studying on limit. Limit knowledge plays a foreshadowing role for derivative, which plays the role of a good connecting link. Limit not only is a connecting knowledge in high school mathematics, but also in university mathematics. Derivative is a elective course in high school. Derivatives and limit are the focus contents of mathematics throughout the university, and the knowledge learned in high school is the basic knowledge of derivatives and limit, The hot issue of entrance test on derivatives focused the derivation rules and derivatives application on the function. For example, in 2007 college entrance examination of Shandong science, if $f(x) = x^2 + b\ln(x+1)$, $b \neq 0$, requests: (1) when $b > \dfrac{1}{2}$, please determine the monotony of function $f(x)$ in the definition domain; (2) what is the extreme point of the function $f(x)$. in 2010 college entrance examination of Shandong science, There are four questions on the subject of derivative, one of them is a calculation problem, its score accounted for 12, the questions are as follows:

If $f(x) = \ln x - ax + \dfrac{1-a}{x} - 1(a \in R)$, （1）when $a \leq \dfrac{1}{2}$ please discuss on the monotony of function $f(x)$；（2）if $g(x) = x^2 - 2bx + 4$, when $a = \dfrac{1}{4}$,

$\forall x_1 \in (0,2)$, $\exists x_2 \in [0,2]$, let $f(x_1) \geq g(x_2)$, what is the range of realistic number b?

For some knowledge of probability and statistics, the initial knowledge of probability is on the compulsory, while more knowledge points are on the elective courses. First, note that these knowledge points are difficult; the second is that it is a link with the university mathematics knowledge. Browsing the past College Entrance Examination, it is obvious that this knowledge is very important in the examination. For example, 2010, there is a calculation problem that test probability in Shandong College Entrance Examination title (science), of course there are some multiple choices, the exam questions of probability accounted for 25%.

Therefore, the teachers of university should fully exploit the contents that relate college mathematics in high school mathematics textbooks, establish the relevant contacts , for the more familiar knowledge students have studied at secondary school, the university's teaching can be granted, not for emphasis, at the same time the teachers should also add some missing contents in high school mathematics teaching, Elective contents and not focused contents for high school should be added, and teachers should mainly explain the knowledge that students did not learn in high school, or expand the knowledge, and make the convergence of middle school mathematics and college mathematics to the contents is more scientific and more rational.

4   Conclusions

Linking the teaching of the middle school mathematics and college Mathematics, not only can improve the quality of mathematics teaching, but also help students integrate into university life quickly, teachers select the best teaching methods, and impart good learning methods to the students through studying the textbook and understanding and communication with students, The freshmen can adapt to college mathematics teaching as soon as possible, and find a suitable way of learning, At the same time the students should know clearly that Learning mathematics is to learn one of the mathematical thought and methods, to improve the capacity of using mathematical knowledge to solve practical problems.

Of course, it needs to improve the student's interest in studying mathematics to solve the linking of middle school mathematics and university mathematics, and to penetrate mathematical thinking to students, and middle school mathematics is the interface which links up not only with the primary mathematics, but also with the higher mathematics. The basic idea of mathematics seems to be a torch which is sent from primary to middle school, and then sent to universities from middle school, and even life. Correspondence is an important mathematical thinking. Mathematical thinking can burst out of spark only in the appropriate mathematical content. As a teacher, seizing every opportunity to teach mathematical thinking to students, to raise students' interest in learning mathematics, these are also good means in linking mathematics teaching, which can not be ignored.

The linking problem has been thought to be great importance to mathematics educators on university mathematics and middle school mathematics teaching, and research on this problem has achieved some results, but the research is rare on the question of linking from the view of pedagogical point, and a number of issues still need further study, such as teaching staff, education management, evaluation, assessment test etc..

References

1. Yuan, Z.: Study on the linking up between the mathematics teaching of university and middle school. Journal of Fuyang Teachers College: Natural Science 25(1), 78–81 (2008)
2. Ma, W., Sun, Y.: On the Link of Mathematics Teachings between University and Middle School. Changchun University of Science and Technology (social sciences) 18(4), 101 (2005)
3. Wang, L., Luo, T.: Thinking of the Link of Mathematics Teachings between University and Middle School. China Adult Education (1), 153–154 (2008)
4. Li, Y., Xi, C., Kulm, G.: Mathematics teachers' practices and thinking in lesson plan development: a case of teaching fraction division. ZDM Mathematics Education (41), 717–731 (2009)
5. Tobias, S.: Some Recent Developments in Teacher Education in Mathematics and Science: A Review and Commentary. Journal of Science Education and Technology 8(1), 21–31 (1999)
6. Liu, L.: The educational link of higher mathematics and middle school mathematics. Journal of Liaoning Educational Administration Institute (12), 118–119 (2006)
7. Zhang, Y.: Research on the link education of university mathematics and middle school mathematics. Journal of Leshan Teachers College 21(12), 81–83 (2006)
8. Li, Y.: Some problems of educational linkup. Tongji university press, Shanghai (2003)
9. Qiu, S., Qian, L.: Study on the linking up between the mathematics teaching of university and middle school. Journal of Mathematics Education 9(4), 45–49 (2000)
10. Zhang, H., Tong, L.: Study on the linking up between the mathematics teaching of in the university and in the senior school. Journal of Jingzhou Teachers College 24(5), 107–109 (2001)
11. Xiao, Y., Fan, F.: Investigation on the linking up of higher mathematics and middle school mathematics. Journal of Science of Teachers College and University 29(2), 104–107 (2009)

Unascertained Measure Model for Water Quality Evaluation Based on Distinguish Weight

Bo Xu[1] and Zhenyu Cai[2]

[1] School of Economics and Management, TianJin University, TianJin, China
[2] School of Economics and Management, HeBei University of Engineering, Handan, China
Cecilias1@yahoo.cn

Abstract. Traditional fuzzy comprehensive evaluation model for determining the index weights has its one-sidedness and the evaluations results sometimes do not reflect the water quality with high accuracy. With the unascertained comprehensive evaluation model, the paper conducts the analysis of the Yangtze River water quality in the past two years. The distinguish weight is defined firstly, and then unascertained measure comprehensive evaluation model for water quality based on distinguish weight is established. The result combines a variety of pollution index and shows reasonable assessment on water quality at the observation sites.

Keywords: Water quality, evaluation index, distinguish weight, unascertained measure.

1   Introduction

There are many methods on water quality assessment at present, such as integrated index assessment, gray system evaluation method, fuzzy mathematics method, BP neural network model, and the matter-element analysis thesis etc. These evaluation methods have their advantages, but also frequently encountered some problems in its application. Integrated index assessment is lack of rigor, and mathematical model is not precise enough which cause the distortion of results. With fuzzy mathematics method, the level of polluted water is a fuzzy conception which is taken into consideration. The weight of index is given by markings of a vast number of experts, so the method is unilateral to a degree and evaluation result is so simple that can not reflect the water quality sometimes. Exceeded classification and evaluation methods consider the impact of a single standard on water quality, has the strict limitations on water quality, and do not meet the requirements of comprehensive evaluation. Hence, it's necessary to analyze each index and come up with the reasonable evaluation method on water quality of Yangtze River.

The paper analyzes the contribution of the index to determine the class of every sample falls into, defines the distinguish weight of index, and then unascertained measure comprehensive evaluation model for water quality based on distinguish weight is established, of which the measure function is meticulously constructed, and recognition criteria are appropriate. The method can express accurately the current

Y. Wu (Ed.): International Conference on WTCS 2009, AISC 116, pp. 213–220.
springerlink.com   © Springer-Verlag Berlin Heidelberg 2012

water quality of Yangtze River and identify the major pollutants affecting water quality in the various regions.

2   Unascertained Measure Model for Water Quality Evaluation

Set $x_1, x_2, \ldots, x_n$ as n objects for evaluation, then $U = \{x_1, x_2, \ldots, x_n\}$ as evaluation object space; Each object of study has m kinds of attribute $I_1, I_2, \ldots, I_m$ which can be measured; $I = \{I_1, I_2, \ldots, I_m\}$ are attribute space. x_{ij} is evaluation value of x_i on I_j. Set c_k represents the grade of project risk, the grade K is prior to the grade K+1. If $\{c_1, c_2, \ldots, c_k\}$ accords with $c_1 > c_2 > \ldots > c_k$, $\{c_1, c_2, \ldots, c_k\}$ is called a ordered division genus of evaluation space U.

2.1   Single Index Measurement Function

The measure function $\mu_{ij}(x)$ represent the possiblity of x_{ij} belong to classification c_k (k=1,2,...,K)and) if μ meet the following conditions.

$$0 \leq \mu (x_{ij} \in c_k) \leq 1 (i=1,2 \ldots, n ; j=1,2,\ldots,m ; k=1,2, \ldots,p) \tag{1}$$

$$\mu (x_{ij} \in U) = 1 \ (i=1,2 \ldots, n ; j=1,2,\ldots,m) \tag{2}$$

$$\mu (x_{ij} \in U_{c_i}) = \sum \mu (x_{ij} \in c_i) \tag{3}$$

We can get the Unascertained Measure recognition matrix under the single index:

$$(\mu_{ijk})_{m \times p} = \begin{pmatrix} \mu_{i11} & \mu_{i12} & \cdots & \mu_{i1p} \\ \mu_{i21} & \mu_{i22} & \cdots & \mu_{i2p} \\ \vdots & \vdots & \ddots & \vdots \\ \mu_{im1} & \mu_{im2} & \cdots & \mu_{imp} \end{pmatrix} (i=1,2,\ldots,n) \tag{4}$$

2.2   Distinguish Weight of Index Based on Entropy

The key question of unascertained comprehensive evaluation model is how to determine the weight of each index and avoid the subjective arbitrariness in determining weight.

The measure of xi that has observed value xij fall into the k category is $\mu_{ij1}, \mu_{ij2}, \cdots, \mu_{ijk}$ .On the condition: $0 \leq \mu_{ijk} \leq 1$, $\sum_{i=1}^{k} \mu_{ijk} = 1$, which means $\{\mu_{ijk}\}$ is of some probabilistic nature. And the entropy is $H = -\sum_{k=1}^{K} \mu_{ijk} \cdot \log \mu_{ijk}$. H reaches maximum $\log k$ value when $\mu_{ijk} = \dfrac{1}{k}$, in that case, uncertainty is greatest and the value of μ_{ijk} is special scattered. From view of distinguishing, result of random test is regarded as sample category, then the measure that attribute I_j make the possibility that sample x_i in any one of k kinds category is $\dfrac{1}{k}$ .It's obvious that attribute I_j has no effect on distinguishing classification of x_i .Otherwise, any one of μ_{ijk} is 1,the rest is o, then H=0,and there is no uncertainty in the test, i.e. the test is determinate. The situation can be reflected on classification distinguish and means attribute I_j is significant, owing to classification of sample x_i to some certain category (one kind of k category).

So the more centralized value of μ_{ijk} concerning No. k category, the more important of index I_j to distinguish classification of x.

If $v_{ij} = 1 + \dfrac{1}{\log k} \sum_{k=1}^{K} \mu_{ijk} \log \mu_{ijk}$, it's clear that $0 \le v_{ij} \le 1$, and the larger value of v_{ij} ,the smaller value of H, that means its corresponding attributes I_j is better in identifying the classification, when $v_{ij} = 1$,that means there is a 1, the remainder being 0 in μ_{ij} , at this time the observed value of property of the sample x in the first category measure is 1, in the other classes of measure 0, that is I_j determined by the sample x_i is divided into K classes, there is no uncertainty, indicating the recognition of I_j is most important for classification. If $v_{ij} = 0$, that means each $\mu_{ij} = \dfrac{1}{k}$,that is the measures which dividing different samples into various types are the same. It shows the classification for the sample x does not work, I_j can be used as removal of redundant attributes.

Therefore, the higher of the concentration values, the greater of the property right identified the role of the type specimen, make $\mu_{ij} = v_{ij} \Big/ \sum_{j=1}^{m} v_{ij}$ (5), so $0 \le \mu_{ij} \le 1$,

$\sum_{j=1}^{m} \mu_{ij} = 1$, ,the bigger of μ_{ij} ,the more important of I_j identification. The

Vector $\mu_i = (\mu_{i1}, \mu_{i2}, \cdots, \mu_{im})^T$ could be the weight vector attribute set that the importance of identifying the size of the x sample categories vector It can be concluded: the weight of attribute I_j is determined by the observe value of sample x_i, the weight of I_j relates to sample x_i ; we also can conclude that it has different weight of the same attribute of different samples. Therefore, weight is not absolute, it is opposite. This is a "change of the right perspective", which is intuitively easy to understand, For example, he same river on the various sub-levels of some pollutants can be very different, if the water quality evaluation of the different sections on the same kind of pollutants with the same weight, it is unreasonable.

2.3   Multi-index Comprehensive Unascertained Matrix

$\mu_{ij}(x)$ represent the possiblity of x_{ij} belong to grade c_k.

$$\mu_{ik} = \sum_{j=1}^{m} w_j(x_i) \cdot \mu_{ik} \tag{6}$$

Obviously,$0 \leq \mu_{ik} \leq 1$,and $\sum_{k=1}^{p} \mu_{ik} = 1$. μ_{ik} is the measure function which represent the possiblity of x_{ij} belongs to grade c_k. (μ_{i1}, μ_{i2} $\cdots$, μ_{ip}) is the multi-index comprehensive evaluation vector of x_i .Multi-index comprehensive unascertained matrix is as follows.

$$(\mu_{ik})_{\underline{n \times p}} = \begin{pmatrix} \mu_{11} & \mu_{12} & \cdots & \mu_{1p} \\ \mu_{21} & \mu_{22} & \cdots & \mu_{2p} \\ \vdots & \vdots & \ddots & \vdots \\ \mu_{n1} & \mu_{n2} & \cdots & \mu_{np} \end{pmatrix} \tag{7}$$

2.4   Identification Criterions

If evaluation space { $c_1, c_2, ..., c_p$ } is arranged in order, the rules of credible recognition is put to application. A confidence level is pre-determined called λ ($\lambda > 0.5$). According to the background and needs of the problem, λ is normally be admitted between 0.6 and 0.7.

$$k_i = \min\left(k \Big| \sum_{l=1}^{k} \mu_{il} \geq \lambda, k = 1, 2, \cdots, p\right) \tag{8}$$

Sample x_i belongs to grade k_i is marked as c_{ki}, which means that the confidence level that x_i belong to the grade k_i is not less than λ.

If evaluation space { $c_1, c_2, ..., c_p$ } is disorder, guidelines of the maximum measure of recognition is used to classify x_i . n_l represents values of c_l (l=1,2,...,n). Then

$$qx_i = \sum_{l=1}^{p} n_l \mu_{il} \qquad (9)$$

is the total score of x_i, and samples in the same rank can be ordered by the score.

3   The Unascertained Measure Model for Water Quality Evaluation of Yangtze River

There are 9 observation stations in the Yangtze River. Observations of pH、DO、CODMn、NH3-N affecting water quality of Yangtze River from June,2005 to Sep,2007 is recorded. Supposed that

(1) Observation records from the stations represent the industrial pollution sources in the Yangtze River which means pollution sources of observation station are taken into consideration.

(2) After industrial wastewater discharged to the Yangtze River, it dissolved into the river uniformly.

(3) Yangtze River flow is stable, the annual flow of roughly the same.

3.1   The Selection of Evaluation Factors

The paper with reference to GB3838—2002 underground water quality standards, establishes the evaluation system of water quality which is shown in table 1.

Table 1. Evaluation index

Number	index	Meaning of index	Type of index
1	I_1	dissolved oxygen（DO）	descending order
2	I_2	permanganate Index（COD_{Mn}）	ascending
3	I_3	ammonia nitrogen（$NH_3 - N$）	ascending
4	I_4	PH value（non-dimensional）	moderate

Accordion to the 28 month measured data of Yangtze River and GB3838—2002 underground water quality standards, PH value has no influence on water quality, so DO、COD_{Mn}、$NH_3 - N$ are chosen as evaluation index.

Table 2. Environmental quality standards of underground water

grades index	I c_1	II c_2	III c_3	IV c_4	V c_5
I_1 dissolved oxygen (DO) $\geq$	saturation ratio 90%	6	5	4	3
I_2 permanganate Index (COD_{Mn}) $\leq$	2	4	6	8	10
I_3 ammonia nitrogen ($NH_3 - N$) $\leq$	0.06	0.1	0.15	1.0	1.0

3.2   Unascertained Measured Function of Single Index

Unascertained measured function of index I_1, I_2, I_3 is adopted folded pattern, as fig.1 ,fig.2and fig.3 shown.

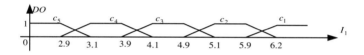

Fig. 1.

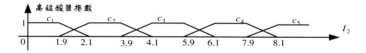

Fig. 2.

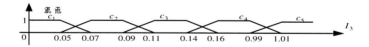

Fig. 3.

Table 3. Monitoring data from observation stations

Number	Names of observation stations	average value of index		
		DO	COD_{Mn}	$NH_3 - N$
1	Panzhihua,,Sichuang (x_1)	9.1543	2.7214	0.18286
2	South Pass in Yichang, Hubei Province (x_2)	8.5054	2.8750	0.26429
3	Hexi water plant in Jiujiang, Jiangxi (x_3)	7.7536	2.4286	0.16036
4	Anhui Anqing Anhui River (x_4)	7.4554	2.5750	0.22893
5	Forest Hill, Nanjing (x_5)	7.4911	2.0929	0.12786
6	Tuojiang Bridge,Luzhou, Sichuan (x_6)	6.8650	3.3393	0.81179
7	Yueyang Tower, Yueyang (x_7)	8.3150	4.1929	0.38571
8	frog stone, Jiujiang, Jiangxi (x_8)	7.9104	3.7429	0.28643
9	SanjiangYin, Yangzhou, Jiangsu (x_9)	8.1379	3.0214	0.28714

According to monitoring data from table 3 and measure function of single index from fig.1 to fig.3, unascertained measured evaluation matrix of single index for samples is as follows.

$$x_1: \quad (\mu_{1jk})_{3\times5} = \begin{bmatrix} 0 & 0 & 0 & 0 & 1 \\ 0 & 1 & 0 & 0 & 0 \\ 0 & 0 & 0 & 1 & 0 \end{bmatrix}$$

3.3 The Index Weighs and Unascertained Comprehensive Measured Evaluation Matrix of Multi-index

According to (5) and unascertained measured evaluation matrix of single index, index weight vector of sample x_1 is $w(x_1) = (0, 0.67, 0.33)$.

Based on unascertained measured evaluation matrix of single index and (6),the comprehensive evaluation vector of sample x_1 is $\mu(x_1) = (0, 0.67, 0, 0.33, 0)$。

Similarly, comprehensive measured evaluation matrix of multi-index is shown as follows.

$$(\mu_{ik})_{9\times5} = \begin{bmatrix} 0 & 0.670 & 0 & 0.330 & 0 \\ 0 & 0.458 & 0 & 0.542 & 0 \\ 0 & 0.670 & 0 & 0.330 & 0 \\ 0 & 0.607 & 0 & 0.393 & 0 \\ 0..030 & 0.670 & 0.300 & 0 & 0 \\ 0 & 0.432 & 0 & 0.568 & 0 \\ 0 & 0 & 0.594 & 0.406 & 0 \\ 0 & 0.631 & 0 & 0.369 & 0 \\ 0 & 0.580 & 0 & 0.420 & 0 \end{bmatrix} \qquad (10)$$

3.4   Identification and Rating Sort

et λ =0.6, accordiong to (10), pollution level of x_i is shown as Table 4.

Table 4. Pollution level of samples

Number	1	2	3	4	5	6	7	8	9
level of Water quality	II	IV	II	II	II	IV	IV	II	IV

4   Conclusion

Unascertained measured evaluation model inherits the advantage of fuzzy comprehensive evaluation method for water quality assessment, and overcomes the subjective determination of the index weigh with the application of entropy to conclude more objective index weight and the evaluation result is more reasonable and practical due to taking uncertainty of water quality standard limits into consideration. The result combines a variety of pollution index and shows reasonable assessment on water quality at the observation sites. Therefore, it is certain to have practical with promotion value.

References

1. Pang, Y.: Unascertained Comprehensive Evaluation Model and its Application on Distinguish Weight of Index. College Mathematics 10, 38–41 (2008)
2. Xiao, M.: Application of modified fuzzy comprehensive evaluation in water quality assessment. Water Sciences and Engineering Technology 8(4), 78–80 (2007)
3. Pang, Y.: Comprehensive Evaluation of the Water Quality in Fuyang River. Operations Research and Management Science 35(10), 101–104 (2001)
4. Wang, X.: Grey-Fuzzy Cluster and Assessment of Water Pollution. Journal of Liaoning University(Natural Sciences Edition) 10, 120–123 (1997)
5. Xu, H.: Application of Grey System Theory in water Quality Evaluation. Journal of East China Geological Institute 15(10), 17–19 (1995)
6. Ma, J., Ji, F.: Fuzzy Comprehensive Evaluation Method for Quality Assessment. Hydrology 10, 26–29 (1994)
7. Sun, Y., Chen, G.: Application Of Fuzzy Mathematics in Assessment of Lake Water Quality. Journal of Shandong University of Technology 7, 85–88 (1999)

Short-Term Wind Speed Prediction Model of LS-SVM Based on Genetic Algorithm

Han Xiaojuan[1], Zhang Xilin[2], and Gao Bo[3]

[1] College of Control & Computer Engineering, North China Electric Power University, Beijing, China
[2] Traffic Control Department, Changchun Electric Power Bureau, Changchun, China
[3] College of Foreign Languages, North China Electric Power University, Beijing, China
lnxtjau@sogou.com

Abstract. According to nonlinear feature of various factors related to wind speed, the method of least squares support vector machine (LS-SVM) for short-term wind speed prediction was put forward in this paper. The influence of parameters selection of LS-SVM on prediction accuracy was analyzed. The genetic algorithm was adopted to realize parameters optimization of LS-SVM and establish short-term wind speed prediction model of LS-SVM based on Genetic Algorithm. It was verified that the method proposed in this paper can quickly and effectively carry on short-term wind speed prediction by simulation example.

Keywords: Genetic algorithm, LS-SVM, wind speed prediction, parameters optimal.

1 Introduction

With the rapid development of wind power, large-scale wind power into grid has brought about serious challenges for the safe and stable operation of the power system because the wind energy has intermittent, randomness and uncontrollability, so the wind speed forecasting has become the research focus on wind power industry. [1-8]

The domestic researches of short-term wind speed prediction have been in theoretical exploration which mainly use time series and neural network and Kalman filters[4]. Three kinds of methods, continuous method, ARIMA model and BP were adopted in document to realize wind speed prediction ahead of 1 hour, in the most cases prediction result of BP network is better than ARIMA and continuous method. The method of radiate basis function (RBF) network was used to realize wind speed prediction ahead of three hours in. A kind of prediction method based on pattern recognition was proposed in [1].Wind speed prediction model of time series was studied in [5]. Combination with Time series and Generalized Regression Neural Network (GRNN) is used to realize wind speed prediction ahead of 10 minutes in [6].LS – SVM (Least squares support vector machine) is now very popular a kind of feed-forward neural networks. LS - SVM uses feed-forward neural network model for prediction with a variety of types of kernel function including the RBF kernel

Y. Wu (Ed.): International Conference on WTCS 2009, AISC 116, pp. 221–229.

function. Compared with the neural network method, support vector machine has obvious advantages which have replaced artificial neural network method and successfully applied in recognition and prediction [9,10].

According to the nonlinear characteristics of the wind speed, the method of LS-SVM is proposed in this paper for short-term forecasting winds. Analyze the influence of the parameter of LS-SVM on prediction and use genetic algorithm [11,12] and grid search method to realize the parameters optimization of it, the wind speed prediction method based on the genetic algorithm is provided by comparing the advantages and disadvantages of the two methods.

2   Prediction Theory and Parameter Selection of Ls-Svm

2.1   Prediction Theory of LS-SVM

LS-SVM is an extension of standard support vector machine. Its loss function is defined as error square directly that the inequality constraints after being optimized is converted into equality constraints, which makes a quadratic programming problem be transformed into solving linear equations, it reduces the computational complexity and accelerates the speed of solution.

According to the given training data set, support vector regression machine for nonlinear systems, considering the nonlinear regression function:

$$f(x) = \omega^T \cdot \phi(x) + b \tag{1}$$

where, $x \in R^n, y \in R$,   input space is mapped into dimensional feature space by the nonlinear function $\phi(*): R^n \to R^{n_h}$.

Loss function is defined as:

$$l(y, f(x)) = (y - f(x))^2 = e^2 \tag{2}$$

Based on structural risk minimization principle, the optimization problem of LS-SVM is:

$$\min_{\omega,b,e} J(\omega,e) = \frac{1}{2}\omega^T \omega + C\frac{1}{2}\sum_{k=1}^{N} e_k^2$$
$$s.t. y^k = \omega^T \cdot \phi(x_k + b + e_k, k = 1,...,N) \tag{3}$$

where C is an adjustment parameter.

Lagrange function can be structured as:

$$L = J - \sum_{k=1}^{N} \alpha_k \left[\omega^T \cdot \phi(x_k) + b + e_k - y_k\right] \tag{4}$$

where α_k is a Lagrange multiplier. According to the KTT condition:

$$\begin{cases} \dfrac{\partial L}{\partial \omega} = 0 \rightarrow \omega = \sum_{k=1}^{N} \alpha_k \phi(x_k) \\[2mm] \dfrac{\partial L}{\partial b} = 0 \rightarrow \sum_{k=1}^{N} \alpha_k = 0 \\[2mm] \dfrac{\partial L}{\partial e_k} = 0 \rightarrow \alpha_k = \gamma e_k \\[2mm] \dfrac{\partial L}{\partial \alpha_k} = 0 \rightarrow \omega^T \cdot \phi(x_k) + b + e_k - y_k = 0 \end{cases} \tag{5}$$

For $k = 1, \ldots, N$, eliminate ω and e, the follow equation can be yielded:

$$\begin{bmatrix} 0 & 1^T \\ 1 & K + \gamma^{-1}I \end{bmatrix} \begin{bmatrix} b \\ a \end{bmatrix} = \begin{bmatrix} 0 \\ Y \end{bmatrix} \tag{6}$$

where $1 = [1, \ldots, 1]^T$ $Y = [y^1, \ldots, y_N]^T$ $\alpha = [\alpha_1, \ldots, \alpha_N]^T$.

K is a square, the elements of the ith row and jth column is $K_{ij} = \phi(x_i)^T \phi(x_j) = K(x_i, x_j)$, $K(\bullet, \bullet)$ is Kernel function. a and b are solved out by using Least-square method. The prediction output can be yielded by it:

$$y(x) = \sum_{k=1}^{N} \alpha_k \phi(x)^T \phi(x_k) + b = \sum_{k=1}^{N} \alpha_k K(x, x_k) + b \tag{7}$$

LS-SVM only need determine the shape parameter and penalty coefficients of kernel function. From the above analysis, the algorithm of LS-SVM makes quadratic programming problem convert to solve for a group of linear equations without changing the mapping relation of original kernel function and global optimality.

In addition, LS-SVM compared with SVM decreases an adjustment parameter (only two parameters, adjusted parameter C and the RBF kernel function width γ), thus simplifies the computational complexity and improves the speed of convergence. Because the RBF kernel function can be approximated arbitrary nonlinear function, therefore RBF kernel functions is selected as the kernel function of LS-SVM in this paper.

RBF kernel function is

$$K(x, z) = \exp\left(-\|x - z\|^2 / \gamma^2\right) \tag{8}$$

This is mainly because the model trained by it has a better overall performance than other kernel function model when no a priori knowledge about the problems.

The Parameter Optimization of LS-SVM Based on the Genetic Algorithm

In the regression prediction model of LS-SVM, penalty factor C is used to control the model complexity and the approximation error. C is greater and the data fitting degree is higher, but the training time will be increased. Kernel function parameter γ has an important impact on prediction speed and precision of the model, therefore the optimal method of γ is researched in this paper.

The common method is grid search method to optimize γ , but this kind of method only uses to solve a single solution. With the constantly iteration, this solution is continuously improved along the direction of steepest decent, there may be in a local optimal solution. However, genetic algorithm integrates the advantages of random search and directional search and it can obtain better balance between region search and spatial extension. A kind of optimization of genetic algorithm is given and its format adopts binary coding encoding.

Algorithm begins

Let M_C and M_v large enough; popsize is population size; gensize is genetic generation; p_c and p_m is separately the probability of crossover operator and variation operation; $p_{c\alpha}$ and $p_{m\mu}$ is separately the probability of arithmetic crossover operation and uniform variation.

Step1 Population initialization
Step1.1   produce   popsize   uniform   random   numbers   on $[C_0, M_c * C_0] \times [\gamma_0, M_v * \gamma_0]$ as initial population.
Step1.2 α is yielded by LS-SVM to calculate the corresponding adaptive value.
Step1.3 the initial population is regarded as the father generation, set k=1and carry on genetic operation.
Step2 Genetic operation
Step2.1 Selection
Use tournament selection and elitist model to produce popsize individuals as descendant.
Step2.2 crossover
Produce a uniform random number $\alpha_1 \in U(0,1), \; \alpha_2 \in U(0,1)$.

If $\alpha_1 \leq p_c$,then crossover operation can be performed.

If $\alpha_2 \leq p_{c\alpha}$,then execute arithmetic crossover operation, or genetic crossover operation, or nothing.
Step2.3 variation
Produce a uniform random number $\beta_1 \in U(0,1), \; \beta_2 \in U(0,1)$.

If β_1 is less than p_m , variation can be performed.

If β_2 is less than $p_{m\mu}$, execute uniform mutation operation, or variance operation, or stop.

Step3 stop standard

If k>gensize, stop; else k=k+1, the filial generation yielded by generic operation is regarded as parent and return Step2.

Algorithm over

At present grid search method is a widespread application of nonlinear support vector machine parameter selection method. Here genetic algorithm and grid search method are applied to optimize the kernel function of LS-SVM. The parameter optimal results based on genetic algorithm and grid search method are shown in Table 1.

Table 1. The Optimal Results of The Two Methods

Method	C	γ	APE	time
genetic algorithm	37.33	0.10	7.23%	1min
grid search method	19.85	0.14	12.47%	3min

It can be known from Table.1 that grid search method has worst stability and has longer optimal time than genetic algorithm.

3   Short-Term Wind Speed Prediction Model of LS-SVM Based on Genetic Algorithm

3.1   Set Up Prediction Sample and Training Sample

The wind speed data used in this paper come from some wind farm of civil in 2005, which had been collected once for 10 minutes. The Known data was divided into the training sample and prediction sample, the regression objective function of LS -SVM was established and then the optimal solution was substituted into decision-making function equation to get regression decision function. Finally, the prediction results were yielded by calculating.

Because wind speed sequences have scheduling, prediction model is of scheduling structure. Therefore, when the output of the sample is the wind speed to be predicted, the input of the sample usually includes the following types:

$A = \{a_1, a_2, \cdots, a_T\}$,the wind speed data in a period of time before T moment at prediction day;

$B = \{b_1, b_2, \cdots, b_T\}$,the wind speed data in a period of time before T moment prior to predicting day.

3.2   Set Up Wind Speed Prediction Model of LS-SVM

The regression objective function of LS -SVM was established according to training sample as follow:

$$\min \frac{1}{2}\sum_{i,j=1}^{l}\left(\alpha_i^* -\alpha_i\right)\left(\alpha_j^* -\alpha_j\right)K\left(x_i, x_j\right)+\varepsilon\sum_{i=1}^{l}\left(\alpha_i^* +\alpha_i\right)-\sum_{i=1}^{l}y_i\left(\alpha_i^* -\alpha_i\right)$$

$$s.t. \quad \sum_{i=1}^{l}\left(\alpha_i^* -\alpha_i\right)=0$$

$$0\le\alpha_i,\alpha_i^* \le C \qquad \left(i=1,2,\cdots,l\right)$$

(9)

where $x_i\left(i=1,2,\cdots,l\right)$ is the ith input of training sample, y_i is the i th output of training sample, $K\left(x_i,x_j\right)$ is Kernel function. Minimize the object function by genetic algorithm to solve α_i and $\alpha_i^*\left(i=1,2,\cdots,l\right)$ and get the optimalizing solution of $\overline{\alpha}=\left(\overline{\alpha}_1,\overline{\alpha}_1^*,\cdots\overline{\alpha}_l,\overline{\alpha}_l^*\right)^{T}$.

3.3   Determine the Regression Decision Function

Substitute $\overline{\alpha}=\left(\overline{\alpha}_1,\overline{\alpha}_1^*,\cdots\overline{\alpha}_l,\overline{\alpha}_l^*\right)^{T}$ into next type and the decision regression equation can be yielded.

$$f(x)=\sum_{i=1}^{l}\left(\overline{\alpha}_i^* -\overline{\alpha}_i\right)K\left(x_i,x\right)+\overline{b}$$

(10)

If $\overline{\alpha}_j\in\left(0,C\right)$, then $\overline{b}=y_j-\sum_{i=1}^{l}\left(\overline{\alpha}_i^* -\overline{\alpha}_i\right)K\left(x_i,x_j\right)-\varepsilon;$

If $\overline{\alpha}_j^*\in\left(0,C\right)$, then $\overline{b}=y_j-\sum_{i=1}^{l}\left(\overline{\alpha}_i^* -\overline{\alpha}_i\right)K\left(x_i,x_j\right)-\varepsilon.$

3.4   The Realization of Wind Speed Prediction in Future Some Moment

When set up wind speed prediction model, it is very important to select enough training sample numbers. If the number is too little, it will cause prediction precision lower and model parameters need enough data to get by calculating. On the contrary if the number is too much, it will cause redundancy and it is difficult to the optimal model of sequence. In this paper, select four groups of data such as 72,102, 150,180 as input sample of LS-SVM responding to the wind speed of 12hours,17hours,25hours and 30 hours, set up wind speed prediction model based on LS-SVM to realize wind speed prediction ahead of 3 hours and analysis its errors.

It can be known that wind speed change has seasonal nature by analyzing its tendency in a year. The wind speed data of a year were divided into four parts to realize wind speed prediction. March to May is as the first season. June to August is the second season. September to November is the third season. December to February is the fourth season. The wind speed is too larger and has stronger fluctuation corresponding to the first, the third and the fourth season, but the wind speed of the second season is the smallest and has less fluctuation. Select four groups of typical wind conditions and adopt different input numbers to realize wind speed prediction. Finally, calculate the Mean Relative Error (MRE) of prediction results. MRE is calculated by (11):

$$e_{MRE} = \frac{1}{N} \sum_{i=1}^{N} \left| \frac{v' - v_i}{v_i} \right| \tag{11}$$

where N is the number of prediction value; v' is prediction wind speed value; v is real wind speed value. The mean relative error of different month and input numbers was given in Table 2.

Table 2. The Mean Relative Error of Different Month and Input Numbers

MRE	72	102	150	180
January	0.1926	0.0962	0.1169	0.3016
April	0.4389	0.0769	0.1141	0.1058
July	0.1832	0.1363	0.0905	0.3511
October	0.1877	0.0890	0.1875	0.0977

It can be seen from Table 2 that the prediction error is the smallest selecting 102 numbers as training sample of LS-SVM to carry out wind speed prediction. The prediction curves are shown in Fig.1.

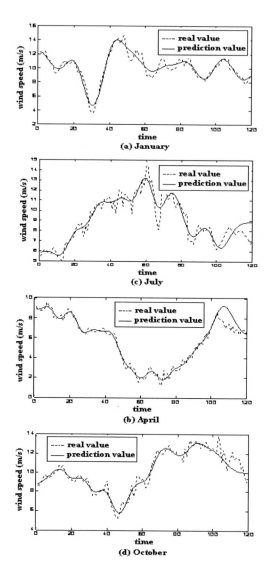

Fig. 1. The prediction curves ahead of three hours with four conditions and 102 input numbers

4   Conclusion

Short-term wind speed prediction is a very important work in wind power operation which has an important significance for the safety and economic operation of the system. The techniques on short-term wind speed prediction were discussed in detail by analyzing the common wind speed prediction model in this paper. A new data mining method, Least Squares Support Vector Machine (LS-SVM) is applied to the field of wind speed prediction. The regularization parameter C and Kernel parameter γ of LS-SVM was optimized by using genetic algorithm. Compared with the traditional grid search optimization method, the prediction speed is more quickly and the accuracy is higher after using genetic algorithm to optimize the parameters of model.

References

1. Yuan, Y., Yang, X., Chen, S.: The research of wind speed and power output prediction. Journal of Chinese Electrical Engineering 25(11), 1–5 (2005)
2. Chen, S., Huizhu, Bai, X., et al.: Wind farm power generation reliability model and its application. Journal of Chinese Electrical Engineering 20(3), 26–29 (2000)
3. Xu, W.C., Yuan, Z.: Wind power. China Electric Power Press, Beijing (2003)
4. Bossanyi, E.A.: Short-term wind prediction using Kalman filters. Wind Engineering 9(1), 1–8 (1985)
5. Torres, J.L., Garcia, A., De Blas, M., De Francisco, A.: Forecast of hourly average wind speed with ARMA models in Navarre (Spain). Solar Energy 79, 65–77 (2005)
6. Cai, K., Tan, L., Li, C.: Short-term wind speed forecasting based on time series combined with neural networks. Power System Technology 32(8), 120–124 (2008)
7. Ackermamn, T.: Wind Power in Power Systems. John wiley & Sons, Ltd. (2005)
8. Ramirez-Rosado, I.J., et al.: Comparison of two new short-term wind-power forecasting systems. Elsevier Renewable Energy (34), 1848–1854 (2009)
9. Li, Y., Fang, T., et al.: Short-term load forecasting based on support vector machine. Journal of Chinese Electrical Engineering 23(6), 56–59 (2003)
10. Fang, R.: Support Vector Machine and Its Application. China Electric Power Press, Beijing (2007)
11. Michalewicz, Z., Janikow, C.Z.: Handling Constraints in Genetic Algorithms. In: Proceedings of the Fourth International Conference on Genetic Algorithms, pp. 151–157. Morgan Kaufmann Publishers, Inc. (1991)
12. Goldberg, D.E., Deb, K., Korb, B.: Don't Worry, Be Messy, In: Proceedings of the Fourth International Conference on Genetic Algorithms, pp. 24–30. Morgan Kaufmann Publishers, Inc. (1991)

Research on Teaching Reform of Computer Basic Course in Colleges and Universities

Yanmin Chen, Xizhong Lou, and Jingwen Zhan

College of Information Engineering, China Jiliang University
Hangzhou 310018, China
liyan__ming@126.com

Abstract. According to the changes of teaching status and students' learning of computer basic courses, this paper analyze the current problems which exist in the teaching of computer basic courses in detail. Combined with the practice of the course, we propose a teaching method as "teacher guiding, students autonomous learning, test-based learning as a means, testing system and teaching website as assistance" to adapt the development trend of computer education and to meet the needs of personnel training and students' independent development. It provides a reference for improving the teaching quality and effect and training some independent talent.

Keywords: the teaching of computer basic courses, teaching methods, autonomous learning, test-based Learning.

1   Introduction

At present, the teaching model of computer basic courses is a combination of theoretical courses and experimental classes in most school. As this course is mainly for the capacity-building of computer, theoretical lessons can be taught in multimedia manners in computer room. It is usually to teach the content by using projector firstly. Then students can take it into practice after the teaching. Such an approach would contact theory with practice closely, enhanced the training of students' operational capability and enriched the content of the class teaching. In addition, there are usually two classes in one computer room. The number of students in class is usually about 80. In order to show the contents to everyone including the students who sit in the back of the classroom and to avoid students to do some other irrelevant operations, teachers can control all the students' machines to show the contents which are showing on teacher's screen. So it can enrich the content of teaching and help teachers to maintain the class discipline at the same time. However, there are also some problems of this teaching method. If the teacher just teach in the means of controlling the students' p machines, some students who have stronger ability would think the program is too slow in such a teaching way, wasting their time and energy; while the weaker students feel that they can not keep up, and their attention will be scattered, they perhaps speak or do all sorts of little tricks in the class, thereby affecting other students' learning and the effect of teacher's teaching. Even the

Y. Wu (Ed.): International Conference on WTCS 2009, AISC 116, pp. 231–236.

enthusiasm and the initiative of students' autonomous learning and mutual learning are more difficult to be fully mobilized. This is a major problem existing in the teaching now.

The computer basic teaching of colleges and universities are important to culture students' computer basic knowledge and practical ability. With the rapid development of computer technology, the gap of computer ability between students is widening. Students are long for the personalized develop of computer basic courses. All of which put forward higher requirements for the current universities computer basic teaching. The reform of computer teaching in colleges and universities has been a priority. On the basis of the analysis of the changes of teaching status and students' learning of computer basic courses, combined with the development trend of computer education, social needs and the needs of students' development, this paper propose that "teacher guiding, students autonomous learning, test-based learning as a means, testing system and teaching website as assistance" which supplemented computer basic courses teaching methods.

2   Teachers Change from a Dominant to a Guide, Teaching Students in Accordance of Theirs Aptitude

With the popularity of computer in schools and families, university computer education is no longer a "beginners". But because of the regional and economic differences, there is a gap of the computer education level between different provinces, regions, cities and villages. Some students have been mastered computer basic skills in high school, able to set up the computer and program. But some students can not even input Chinese character in a normal speed. The students who know little about computer usually lack of some knowledge of certain aspects, the basic knowledge and operation ability they mastered are not systematic which leading to the gap. And as time goes by, the gap will be wider and wider. Computer education in colleges and universities can no longer be the centralized teaching methods as it used to, we must hold a teaching reform taking student as the main factor, teaching students in accordance of their aptitude.

In the traditional teaching activities, teachers have always been a leading role. Teachers, the organizers of teaching activities, play a very important guiding role. To create an education philosophy of student-centered, we should start with changing the teacher's concept of leader. Only with new teaching ideas, can teachers really play a guiding role. As teachers, they need to have a really shift on mind from teacher-centered to student-centered; from lead-based teaching to guide -based teaching; from lectures to guided self-study-based teaching.

Teachers need to put a lot of energy take the computer basic teaching. When taking the centralized teaching in the computer room, although students can practice their own operations when the teacher's talk is over, for the students who had been mastered already, the teaching process is a waste of time, the students will affect the whole teaching atmosphere in the case of scattered attention. The effect of centralized teaching is not so good and it is difficult to guarantee its teaching quality.

In the implementation of independent learning teaching methods, teachers firstly introduce the arrangement, learning requests of the course, the using method and the exam result of the test-based learning system. Then the teacher allows the stronger to learn the course on their own, or provide a broader space of development by specifying some references. For general students, in self-testing study, they understand their lacks of knowledge by the testing system and look for the answer initiative, the teacher would apply the directions in time. For the weaker, teachers should devote more attention to strengthen their counseling. In such a method, teachers can take energy to learn more about the students' learning situation and finally teach students in accordance of their aptitude.

Teachers should develop students' autonomous learning, and combined the class guidance and counseling networks as well. By the using of three-dimensional teaching resources and network platform, teachers can provide personalized counseling according to the level of students. They can enhance the connection of students and teachers and the mutual connection between students, leading students shift from passive learning to initiative learning at any time. Meanwhile, teachers should continue exploring the method and means of network-teaching, form a diversified teaching mode by integrating the self-study and network-teaching closely.

3   Students Shift Passive Learning to Initiative Learning, Autonomous Learning

At present, the teaching of computer basic in colleges and universities does not fully take students' ability and interest into account, the centralized-teaching has greatly influenced students' learning initiative. The "student autonomous learning" teaching need to integrate the teaching contents, the lack of knowledge points and the interest of students. After giving the contents, goal of the teaching and providing the testing platform which students can test their mastery of knowledge on their own, students can choose to learn by themselves, take the initiative to check their shorts. Shifting the initiative to the students can create a wide space of options and a good teaching environment.

Using autonomous learning methods, students in class can practice on their own by the computer test system and then learn the related knowledge of the materials actively for the mistakes made in the test. If there are some questions, they can also ask the teacher. For the relatively good students who have better foundation and strong ability of learning, they can control their learning time freely. After mastering the related knowledge, they can learn computer-related advanced courses in-depth, expand and enhance their computer ability level further. For students who have poor basis and ability of self-learning can seize the time to study carefully and ask for teachers' advice more usually. And learning with questions which made in the test the effect will be better. By giving the students in different levels with different guidance we can achieve the goal of individualized. At the same time, students should be good at using the network teaching platform which provided by teachers, change passive curricular teaching-learning into active learning inside and outside class. Students also could ask questions to teachers and share their learning experience with classmates online and so on.

In college, because learning styles and teaching methods have a great change compared to high school, the managements of colleges and universities have a greater emphasis on students' self-management, self-discipline and self-learning. The "student autonomous learning" teaching style can also help students adapt to the learning style of university more quickly.

4   Test-Based Learning as a Means

In the traditional centralized teaching methods, although the teacher will introduce a lot of new knowledge and teach students systematically, some teaching content is still kind of boring. So students will ignore this course mentally and even have a point that the opening of this course has no direct practical significance. Under the influence of these ideas, they often hold a kind of indifferent attitude to have theoretical courses, take the experiment as playing games. When mentioned the testing, many people will just think about the drawbacks which are made by taking the exam as the only means of evaluating teaching and learning, such as leading students to pay more attention to knowledge and achievements while ignoring capabilities and intelligence and so on. However, if students "take test by themselves", taking the mock examinations as a means of checking how much knowledge have they mastered in their autonomous learning and guiding the direction of their further study, we will receive an unexpected effect.

Owing to the practical operational characteristic of computer course, test-based learning methods play a good role in promoting students' practical ability and paying more attention to learning process. Mock examination is a quantitative evaluation of the situation of self-study for students. The quantitative evaluation taking statistic of study by using scientific measure methods, expressing students' effect of studying in the form of score, which enables students to understand themselves fully and having a specific study with vacant knowledge targeted after knowing their own shortcomings, promoting their personalized development. Learning by test can fully mobilize the initiative of students. Students can also found gap in their knowledge, promote their own initiative to learn, improve their interest of learning and ensure the study is more systematic and comprehensive.

Looking at a student's process of learning a course, the traditional teaching approach isolate the final test, only focusing on test's assessment function. But if we put the test into students' routine self-learning process of every lesson, learning while testing. This test pays more attention to the details of students' development and the process of learning so as to get improvement after finding the shortcomings, making test-based learning become a means of students' development. This learning style is more humane, it enhances students' motivation to learn, and it is more conducive to the healthy development of students.

The following Fig. 1 shows the percentage of score. Score A is the score calculated on 80 students in our teaching methods. Score B is the score calculated on 78 students in the traditional centralized teaching methods. Fig. 1 shows the results of our teaching methods are better on average than the traditional centralized teaching methods.

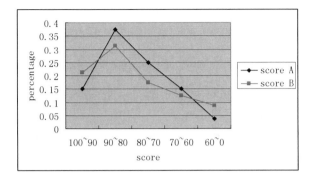

Fig. 1. The Percentage of score

5   Testing System and Teaching Website as Assistance

Test-based learning depends on the testing system improvements. Testing-item bank pays attention not only to point and face, but also to quality and quantity. While emphasizing the key points of course content, it should also be able to fully reflect all the knowledge points in the curriculum. To make sure that each student can get different testing items every time, it must guarantee quality and abundance of these items. And with the rapid development of computer technology and constantly updating of knowledge, it is required to constantly maintain system and add new knowledge. Test system can avoid the interference of subjective and reflect the actual cognitive abilities of students to a certain extent.

It needs to be combined with online teaching to develop students' ability of autonomous learning. Through network, the usage of three-dimensional teaching resources can shift students from passive learning to initiative learning as they can learn on the platform at any time they want to. In teaching, teachers should establish the platform for e-learning supporting software based on computer basic education. By the use of multimedia technology, teachers can not only provide well-prepared e-learning courseware to students and ample high-quality teaching resources to network platform, but also supply students with syllabi, teaching plans, streaming media courseware, extra-curricular references, information releasing, work evaluation and on-line Q & A. Besides, teacher and students can complete the course better through interaction. Teacher can know students' learning conditions. And students can view online learning, hand in papers, ask questions and offer recommendations for the curriculum by the usage of internet.

Teaching resources such as curriculum materials, experimental program, test-based learning system, platform for e-learning supporting software are self-contained and also interrelated. As various kinds of media complement each other and emphasize their own features, abundant and three-dimensional teaching resources can meet the needs of all types of students for autonomous learning. Teachers can take full advantage of the advantages of rapid transmission of internet information, add new knowledge and new technologies to subjects in time, and enlarge the amount of information in the classes to expand the students' scope of knowledge. Practice has proved that ample package of teaching resources will greatly enhance students' interest in learning and help them to develop and improve the comprehensive ability.

In a word, building courses-supporting site can provide students with a large deal of curricular and extra-curricular learning resources and reference resources. Through the website forums, exchanges between teacher-student and student-student will be more frequent. And it will improve the teaching level of teachers and lay a good foundation for students for further studies by answering student's questions and mastering students' learning conditions.

6   Conclusion

Research on teaching reform of computer basic course in colleges and universities is one of an import research problem. By the method of teacher guiding, students autonomous learning, we can not only culture students' computer basic knowledge and practical ability, but also develop students' independent consciousness.

This paper discussed the problem in computer basic education in several ways, including teacher, student, teaching method and teaching resource. Cultivating qualified and innovative talents are rather complicated system engineering, and we need to study assiduously and perseveringly.

Acknowledgments. The project is supported by the National Natural Science Foundation (F010102)   and Scientific Research Fund of Zhejiang Provincial Education Department (Y200909249) (Y200805880).

References

1. Dou, Y., Shi, Q.-Z.: Research on computer foundation teaching system of non-computer major. Research in Teaching 32(1), 67–70 (2009)
2. Li, J., Ni, Y., Chang, Y., Wang, M.: Analysis on the status of education in computer system course in University. Higher Education and Research 11, 5–7 (2008)
3. Computer Science. Curriculum 2008: An Interim Revision of CS (2001), http://www.acm.org/education/curricula/ComputerScience2008.pdf

Cultivating Professional Managers of the Cooperative from Students

Ying Xing[1] and Guo Qingran[2]

[1] College of Economics and Management, HLJ August First Land Reclamation University,
Daqing, China
[2] School of Economics and Management, Henan Institute of Science and Technology
Xin-xiang, China
Guoqing0411@gmail.com

Abstract. Professional managers training of farmer cooperative organizations relates to the future of China's farmer organization raise questions. Therefore, for the culture of the specific problems and suggestions will effectively increase the level of the entire train. At the same time, the training of professional managers to optimize the degree of employment structure and improving farmers ' cooperative organizations operating efficiency are of great significance.

Keywords: Farmer specialized cooperative organizations, professional managers, talented localilzation, order system.

1 Introduction

For 30 years of reform and opening-up, many experts and scholars have developed further research on farmer specialized cooperatives , and also obtained many pertinently development strategy. Especially, after the Law of the People's Republic of China on farmer specialized cooperatives has promulgated and the development of farmer specialized cooperatives is on the "fast track", but the training of professional managers in the farmer specialized cooperatives has not put on the agenda. The core problem of the development of cooperatives is lacks of talents support and the research on farmer specialized cooperatives for professional managers rarely or exist deviations. This article aims at exploring the factors affecting the lack training of professional managers in farmer specialized cooperatives, and hopes that through the research to promote the development of farmer specialized cooperatives.

2 The Main Problems in the Current

2.1 The Need of Professional Managers Still Exist Doubts

Farmer Cooperatives economic organization is the union between producers who produce the same agricultural production and decision-making style is typical of " one person one ticket" but not "a bond a ticket ", the daily production and operating

Y. Wu (Ed.): International Conference on WTCS 2009, AISC 116, pp. 237–241.
springerlink.com

activities is responsible by the director that all the members are elected jointly, whether need to hire a specialized professional managers, still need particular case is particular analysis. Academic study hasn't obtained valuable conclusions, so whether need professional managers also has no unified standard. [1]

2.2   Training Level Are Low

In the past, we consider the problem of investment in human resources for farmer specialized cooperatives economic organization, simply tries to training the founders but ignored when farmer specialized cooperatives developed at a certain scale, it is necessary to introducing professional managers in the production and business operation activities of a daily for entrepreneurial management. It's similar to translate the "family enterprise" to "modern enterprise management" mode, the transformation is happening in an organization when the former can not adapt to the enterprise's production and business operation condition. The founders because of they own, environment management and other reasons, cannot meet the requirements of farmers specialized cooperatives economic organization, so at present the top priority of professional manager training is to improve the training level and specific training as soon as possible. [2]

2.3   The Training Level Is Unable to Satisfy Needs

At present, the domestic talent cultivation mode remains in the stage of popularization education especially the agricultural universities; this is not adapting the general idea of the training of professional managers in farmer specialized cooperatives. The training of professional managers in farmer specialized cooperatives needs to rely on professional background and comprehensive understanding of agriculture and animal husbandry and veterinary, food and other related disciplines, and have the basic knowledge of professional service according to the different updating knowledge of the cooperative, these are obviously different from the current situation. Current the training of agriculture economic management professional talent put high proportion of theoretical knowledge and practical skills high proportion, ignore the training of practical ability. There is a huge gap between the real ability of managers in farmer specialized cooperatives and the fact. In short, the current economic management of cultivating talents does not conform to the professional managers of farmers specialized cooperatives need, cultivation objective orientation is not clear; training level is not suitable for social needs, these are the   fundamental reasons caused domestic universities forestry and economy management admissions atrophy and employment difficulty. [3]

2.4   Solving the Problem of Localization Talents

Almost all the fields of agriculture actually exist talent localization problem, we take a lot of practice , such as selecting college-graduate village official, many provinces takes implementation "one village one college "policy(attract graduates go to the countryside), in Heilongjiang province the "rural students plan "(recruit students from farm and rural area), but cannot fundamentally solved the talent localization problem. Many factors influence the agriculture related talent localization, but the most important factor is not to provide development platform for talents, farmers specialized cooperatives economic organization is the most suitable platform for agriculture related talents, talent training due to economic cooperation organizations are not generally employ professional managers, therefore we can say that the final solution of agriculture talent localization is a long-term task, and still need to rely on farmers' specialized cooperative economic organizations to more beneficial attempt. [4]

3   The Specific Solutions

3.1   Fostering Cooperation Economic Ideas

The international cooperative movements have developed more than 100 years, and also have a lot of good cooperative economy vivid development examples, but overall, economic cooperation ideas has not thought extensively. For instance many undergraduates do not grasp the essence of economic cooperation, to think of one-sided that the Chinese characteristic socialism market economy needs not economic cooperation, these are totally wrong recognitions. Strengthening economic cooperation idea when cultivating students, firstly is to instill economic cooperation idea in classes, the specific function of economic cooperation, and analysis the advantages and disadvantages of to adopt or not to adopt economic cooperation organization. Secondly is to guide students to agricultural production and life, students of agriculture economic management discipline should have gone to rural area and farm, it is necessary to understand the actual production of rural life, students should visit demonstration cooperatives, through a specific interpretation of cooperatives staff, and make their sense of rural economic cooperation. Students can significantly change the inner thoughts, and devote themselves to cooperation economic, for further strengthen training students to professional managers lay the good preparation work. Finally, on the visit to the demonstration, improve student's understanding of cooperation economic organization, enables students to experience the benefits of cooperatives through communication after visitations. Along with the increase of learning, instill more cooperatives ideas regularly, make preliminary undergraduates owe the basic quality of professional managers.

3.2   The Comprehensive Cultivating Based on Practice and Practice Basement

Now the undergraduate students we cultivated belong to product of education popularization age. Since 1999 many universities began expansion in the whole nation and have lost theirs unique characteristics, universities looked quite consistently. Heilongjiang Bayi Agricultural University is the only universities of land reclamation in China. It is facing land reclamation of Heilongjiang province, and radiation of the nation. So students should be able to adapt to the rapid economic development of agriculture, which requires the students to more in-depth agricultural production practice. Therefore, it is necessary for us to build more practice basement and supply practice opportunities. If condition allowed, we should explore "1 + 1 + 1 + 1" teaching mode. Concretely speaking, this means one year of basic courses training, one year of professional training, one year of specialized courses, and one year of social practice. Social practice is the process in practice basement in real agricultural production basement and by the relevant enterprises evaluation, to assess whether the students achieved training goal. For those who interested in attend the training of professional managers must go to the specified cooperative and through practical preliminary master the operation process of cooperative, then the director in the cooperative decide whether to hire the students as professional managers ,who totally be responsible for business promotion, sign orders Etc of this cooperative.

3.3   Reconstructing Curriculum System

In the past, we attach more importance to cultivating undergraduate professional in the high level of knowledge, but at the present education popularization stage, the training for undergraduates should go "universal" way. Therefore, it is necessary to break the curriculum system, and reconstruction course structure for special talents training. Involving to cultivate a professional managers, we should make the students master the all related basic professional knowledge, because most of the professional managers should grasp the knowledge of production in real life, but a variety of business exist difference, which require us to have a diverse training and the training be more comprehensive. Professional managers should master basic knowledge of production and operation based on the grasp of modern enterprise management knowledge. If professional managers want to put the daily activities into a routine, they must make modern enterprise management mode and management rules and regulations generally accepted. The modern enterprise management modes of operation will guarantee a reasonable and regularized business. This training courses is put forward higher request for the trainer, we would broken the professional limitation. in the first two years, based on basic education stage, lets the student owe any choice to study the basis of professional course. The ultimate goal is to make the students can understand as much other professional knowledge as possible, then according to their willingness to choose a major. So, in this way the professional knowledge is even more profound, after all it is the major they choose. At the same time, students will master other professional knowledge slightly. This approach is deepening for the education popularization training target, specific to the cultivation of professional manager, courses is also not a serious obstacle.

3.4   Order Training System Mode

Order system is currently being widely applied to industrial production areas, in personnel training, only the most vocational colleges formed the order of talent training policy. It is the doomed result of higher education must change in need. The core of order training system is to cultivate talents to satisfy enterprises quintessence. This sounds simple, but involves specific training process, it must be the two-way choice mode between cooperatives and the students. Identity could make the cultivating more effective. Once the cooperative and students establish cooperation intention, the school should work on the train to make students know how professional managers activities in future. What kind of knowledge is useful in the operation process of cooperatives later will be what kind of knowledge is learned, and once there is need to improve the school should also help them to improve. Only the smooth realization of professional managers and a seamless docking, can realize the training target, so that the students can gain great work of labor, solved the key problem of the localization, talent will no longer be a problem.

3.5 Exploring New Mode Incentive System of Talent

To solve the problem of talent localization, not only supply chance for talents to use their wits, also need to ensure the corresponding physical security. According to the incentive theory, the employee-shareholding plan seems to be the most effective incentive methods. Therefore, we need to establish a farmer specialized cooperatives economic organization manager's shareholding plan, linking the management performance of professional managers and farmers specialized cooperative's operation results together, then the income of managers has positive correlation with the income of farmers specialized cooperatives . So they can secure work for farmers specialized cooperatives economic organization, promote the further development.

4   Conclusions

4.1   Adjusting Curriculum System Is the Key

The training of professional manager of farmers specialized cooperatives is the specific application of higher education in the new time. The talent training is base on what courses college provide to students. Only the curriculum system of higher education training adjust appropriately, cultivating applied talents instead of theoretical talents, the problem of cultivation professional manager can be solved satisfactorily.

4.2   Talent Localization Is the Basic Guarantee

Regardless of how high levels of talents cultivation if don't solve the problem of talent localization, the use of talent will lead to more problems. The localization of farmer specialized cooperatives should pay attention to two aspects: on the one hand, should choose more local graduate to cultivating, on the other hand, and should notice to provide the development platform to talents to display.

References

1. He, C.: On agricultural cooperatives and the cultivation of entrepreneurs. World Agriculture, 52–55 (2005)
2. Zhao, S.: The MBA education raises with the Chinese Enterprise professional manager. Journal of Higher Education, 82–85 (2002)
3. Zhang, X.: Innovative farmers cooperative with Chinese characteristics of the development of agricultural cooperatives of. China Co-Operation Economy, 32–34 (2005)
4. Tang, R.: On the professional managers of the basic training of quality and loyalty. Jiangsu Commercial Forum, 66–68 (2003)

Application of MATLAB in Teaching of High Frequency Circuits

Fangni Chen

Department of Communication and Electronic Engineering, Zhejiang University of Science and Technology, Hangzhou, China
chaofang333@sohu.com

Abstract. This article is discussed the application of MATLAB software in the course of High Frequency Circuits in detail. Taking the typical circuit of analog modulation as an example, the utilization of MATLAB GUI simulation is expounded and the results are given. Teaching practice shows that by means of the application of MATLAB as the computer-assisted teaching, the learning interest of students and the teaching effects are improved greatly in the course of High Frequency Circuits and the classroom atmosphere is more active.

Keywords: High Frequency Circuits, MATLAB, GUI, simulation.

1   Introduction

High Frequency Circuits is an important foundation course of electronic information, communication engineering and other similar specialties. This course researches basic concepts, basic principles and basic analysis methods to analyze cell circuit and the actual circuits [1,2]. High Frequency Circuits set theory, engineering and practicality in one course, with the character of numerous contents, abstract concepts and complicated mathematical reasoning and so on. So it has been a hard science course and students in general reflect it is very difficult to study [3]. How to let the students to understand and grasp the basic theory of high-frequency electronic and how to cultivate the students' analytical and design ability are most important.

It is necessary to improve the teaching methods and content. With the development of the computer, it is necessary to apply computer-aided teaching in High Frequency Circuits. If we can simulate the circuit and display the waveform demo, vivid image in the teaching courses, it can improve students' learning interest and develop students' thinking. Currently PSpice, Multisim, EWB and MATLAB are popular simulation softwares in Universities [4]. In this paper we introduce the MATLAB application in teaching of High Frequency Circuits and take the analog modulation as an example. Teaching practice shows that the learning interest of students is motivated and the teaching effects are improved.

Y. Wu (Ed.): International Conference on WTCS 2009, AISC 116, pp. 243–250.
springerlink.com   © Springer-Verlag Berlin Heidelberg 2012

2   Matlab and Gui

MATLAB produced by Math Works Inc. of America is a powerful commercial mathematical software that can run on most computer platforms with very wide range of applications such as numerical analysis, automatic control, signal processing, information communication, engineering construction, financial analysis and image processing, etc. It is recognized as the most influential technological application software and be most popular with engineering technology workers and researchers. At present, many universities set MATLAB as the basic tool that undergraduate and graduate students must grasp. Its main features include: friendly working platform and programming environment, easy-to-use program language, strong ability of science and computer data processing, excellent graphics processing functions, wide application module sets toolbox, practical program interface and publish platform, application software development (including the user interface).

Here we adopt MATLAB Graphical User Interface (GUI) tool which can built a beautiful user interface through the window, menu and button captions. The user interface can clearly show different waveform of different modulation and the differences of different modulation and demodulation can be illustrated intuitively.

MATLAB GUI can construct user interface by various graphic objects, including graphical window, button, the text box, menu, axial etc. GUI design can use MATLAB m files, also can use GUIDE tools. GUIDE can save GUI interface designed by users in a FIG resources environment and generate m files which contains GUI initialization and interface layout control code. Callback functions can be added to the corresponding m file. So the realization process of GUI including two jobs below: (1) the use of interface design editor for GUI interface design, (2) writes GUI callback function codes in the automatic generation m files [5].

GUI control platform include all the graphic interface controls such as push button, slider, radio button, check box, edit text, static text, pop-up menu, list box and so on. The user selected the control needed, drag it into corresponding margin, and set the control attributes. The graphic interface and the attributes are listed in Fig.1 and Fig.2. The user then save the GUI file and MATLAB can store the user information to a .fig file and a .m file. The .fig file is used to store the information (including control attributes) of all the necessary GUI interface graphical controls.   The .m file is used to store the necessary running code of GUI including the callback functions.

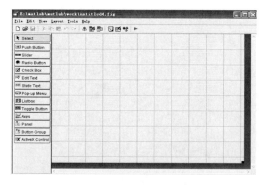

Fig. 1. The graphic interface

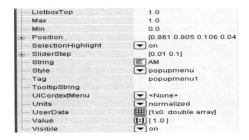

Fig. 2. The control attributes

3   Analog Modulation Gui

In High Frequency Circuits course, teachers can simulate the unit circuit with MATLAB software. For example the modulation and demodulation knowledge is key content in this course, the concept of modulation are so abstract that the students are difficult to understand. Therefore, using the software to create modulation and demodulation system and display the signals' waveforms can help students understand the knowledge point better.

3.1   Analog Modulation

AM (Amplitude modulation) multiply the normalized baseband signal plus dc component with the carrier signal to get the modulated signal. Detection method is envelope demodulation. AM modulated signal can expressed as

$$
\begin{aligned}
V_{AM}(t) &= V_{cm}(1 + m_a \cos \Omega t)\cos \omega_c t \\
&= V_{cm}\cos \omega_c t + \frac{1}{2}m_a V_{cm}\cos(\omega_c + \Omega)t \\
&\quad + \frac{1}{2}m_a V_{cm}\cos(\omega_c - \Omega)t
\end{aligned}
\tag{1}
$$

Where V_{cm} and ω_c are the amplitude and the frequency of the carrier, Ω is the frequency of the baseband signal, m_a is the amplitude modulation coefficient. From (1) we can see the AM modulated signal have three frequency components, which are ω_c, $\omega_c + \Omega$ and $\omega_c - \Omega$.

DSB (Double side band) modulation directly multiply the normalized baseband signal with carrier to get the modulated signal. Coherent demodulation is used here, the modulated signal multiplied with carrier signal again and after a low-pass filter comes out the original baseband signal. DSB modulated signal can expressed as

$$
\begin{aligned}
V_{DSB}(t) &= AV_{cm}V_{\Omega m}\cos \Omega t \cos \omega_c t \\
&= \frac{1}{2}AV_{cm}V_{\Omega m}\cos(\omega_c + \Omega)t + \frac{1}{2}AV_{cm}V_{\Omega m}\cos(\omega_c - \Omega)t
\end{aligned}
\tag{2}
$$

Where $V_{\Omega m}$ is the amplitude of the baseband signal. From (2) we can see the DSB modulated signal have two frequency components, which are $\omega_c + \Omega$ and $\omega_c - \Omega$. Compare to AM modulation, DSB signal has the same bandwidth with AM signal but suppress the carrier component.

SSB (Single side band) modulation principle is similar with the DSB signal transmission but the transmission only realized in higher band or lower band. SSB modulated signal can expressed as

$$V_{SSB}(t) = \frac{1}{2} A V_{cm} V_{\Omega m} \cos(\omega_c \pm \Omega)t \tag{3}$$

From (3) we can see the SSB modulated signal only have one frequency component, which is either $\omega_c + \Omega$ or $\omega_c - \Omega$. Compare to AM and DSB modulation, SSB signal has the half of bandwidth.

Frequency modulation FM also adopts coherent demodulation method. If we use the cosine waveform as the carrier singal, FM modulated signal can expressed as

$$V_{FM}(t) = V_{cm} \cos(\omega_c t + m_f \sin \Omega t) \tag{4}$$

Where m_f is the frequency modulation coefficient. From (4) we can see the FM modulated signal have infinite frequency components.

For the sake of comparing different modulation modes, we simulate the noise-added modulation process and that without noise. Also we display the time-domain waveforms and the frequency-domain waveforms of different signals.

Because all the sorts of demodulation programs are relatively long, in order to call the demodulation programs conveniently in GUI, we will set the different modulation programs as different functions. In GUI we can directly use the function names and can get the needed graphics after set the particular parameters and drawing functions. In our analog modulation simulations, we set four modulation functions as follows,

1) function [c,u,r,dem1,dem2,M,C,U,R,DEM1,DEM2]=
$$AM(a,m,snr)$$
2) function [c,u,r,dem1,dem2,M,C,U,R,DEM1,DEM2]=
$$DSB(m,snr)$$
3) function [c,u,r,dem1,dem2,M,C,U,R,DEM1,DEM2]=
$$SSB(m,snr)$$
4) function [c,u,r,dem1,dem2,M,C,U,R,DEM1,DEM2]=
$$FM(m,snr,kf)$$

Where m is the original baseband signal, snr is the signal to noise ratio, c is the carrier, u is the modulated signal, r is the received signal with noise, dem1 is the demodulated signal without noise, dem2 is the demodulated signal with noise, M, C, U, R, DEM1, DEM2 are the frequency transform of these signals. Besides a and kf are the modulation coefficients for AM and FM.

3.2   GUI Interface Design

We design the GUI interface with the controls as follows: one popup menu for different modulation modes; three push button for three baseband signal waveforms such as triangular waveform, rectangular waveform and sinusoidal waveform; two radio button for time domain and frequency domain choices; six axes for graphics display; two static text for label; one push button for close the window. After the corresponding controls are put in appropriate position and the attributes of them have been set, we can run the GUI. The modulation based on GUI interface and the run result are shown in Fig.3 and Fig.4.

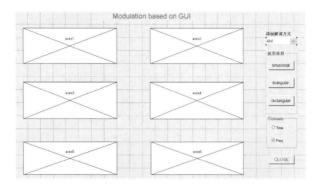

Fig. 3. The modulation GUI

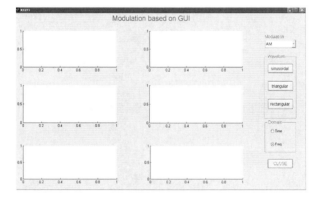

Fig. 4. The run result

3.3   GUI Program Design and Waveform Observation

So far the user interface has been built, next we need to define each control reaction commands, selected the controls with right click, choose callbacks, put your own program codes into the corresponding functions. Here we put the program codes into three waveform buttons and call four modulation modes, meanwhile plot the different signals' time domain and frequency domain waveforms.

AM time domain waveforms with sinusoidal baseband signal is shown in Fig.5. The sinusoidal baseband signal is set to $\sin 40\pi t$ and the modulation coefficient is 0.5. FM time domain waveform with triangular baseband signal is shown in Fig.6. The triangular baseband signal is set to $sawtooth(50\pi t, 0.5)$ and the modulation coefficient is 100. DSB frequency domain waveform with sinusoidal baseband signal is shown in Fig.7. The sinusoidal baseband signal is set to $\sin 40\pi t$. SSB frequency domain waveform with rectangular baseband signal is shown in Fig.8. The rectangular baseband signal is set to $square(60\pi t)$ .We can change these parameters conveniently in program codes and the different waveforms will displayed in GUI interface immediately. Students can clearly see the time domain and frequency domain waveforms' variation in modulation and demodulation. It is also suitable for application of MATLAB software for teachers because it can save a lot of time to draw the waveforms on the blackboard.

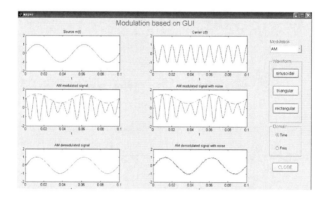

Fig. 5. AM time domain waveforms with sinusoidal signal

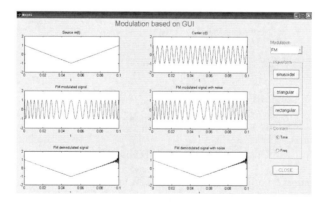

Fig. 6. FM time domain waveforms with triangular signal

For varieties of High Frequency Circuits, it can be analyzed and simulated by using MATLAB. Although many calculations and simulations can achieved by making use of other mathematical tools, but to the extent of the wide application of MATLAB, it is beneficial and helpful for our students to grasp the High Frequency Circuits well.

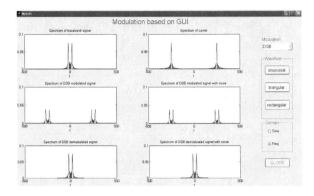

Fig. 7. DSB frequency domain waveforms with sinusoidal signal

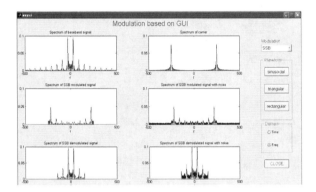

Fig. 8. SSB frequency domain waveforms with rectangular signal

4   Conclusions

As the High Frequency Circuits is the foundation course of electronic information, communication engineering and other similar specialties, the contents and the teaching methods should also be progressed with the development of science and technology. Nowadays, the application of computer software analysis and solving the circuits issue has been very popular. Therefore in this paper we introduce the MATLAB software in teaching of High Frequency Circuits in order to improve the teaching effects and enhance the professional competence of students. We built a bueatiful MATLAB GUI interface of analog modulation simulation. Teachers can conveniently change the simulation parameters and students can clearly see the variation of the signal waveforms of different modulation modes. Teaching practice shows that by means of the application of MATLAB as the computer-assisted teaching, the learning interest of students is motivated, the active learning of students is encouraged, and the ability of self-exploration of students are cultivated. MATLAB aided teaching also can improve the teaching effects and to active the classroom atmosphere.

Acknowledgment. This paper was supported by Scientific Research Fund of Zhejiang Provincial Education Department (Grant No. Y200907111).

References

1. Zhou, X.: Reform and Research on Experimental Teaching for High Frequency Electronic Circuit. Proceedings of EDT 2010, 168–170 (April 2010)
2. Liu, N.A.: Thought of Teaching Reform of High Frequency Electronic Circuit Course. The 5th Electronic and Electric Course Forum Report (April 2009)
3. Chen, B.: Radio Communication Circuits, 2nd edn. Science Press, Beijing (2009)
4. Sun, X.: On the Transformation of Teachers' Roles and Improvement of Information Literacy in on-line Teaching. Research in Education Development, 87–88 (2005)
5. Chen, Y., Mao, T., Wang, Z.: Proficient in MATLAB GUI Design. Electronics Industry Press, Beijing (2008)

Distributed Architecture of Object Naming Service[*]

Bo Ning, Guanyu Li, Yuqing Chen, and Dongdong Qu

School of Information Science and Technology
Dalian Maritime University
No.1 Linghai Road, Dalian, 116026, China
yaoyao.032@gmail.com

Abstract. The Object Naming Service (ONS) is a core resolving service of the EPCglobal Network. The function of ONS is the address retrieval of manufacturer information services for a Electronic Product Code (EPC) identifier which is stored on the RFID tag and is read by the RFID reader. This allows dynamic and globally distributed information sharing for items equipped with RFID tags compatible to EPCglobal standards. But, unlike in the DNS system, there is only one ONS Root, and the ONS Root can be controlled or blocked by a single company or a country. The EPCglobal Network is a future global business infrastructure, and it is immoderate that one country or company controls the items information all over the world. In this paper, we propose a new ONS architecture called Distributed ONS, which is distributed Object Naming Service.

Keywords: EPCglobal, RFID, Object Naming Service.

1 Introduction

The most important application of Radio Frequency Identification (RFID) is to identify the physical objects efficiently. The RFID tag can store information and the information can be read by RFID reader. As the techniques of RFID hardware advances, the reading range between the RFID tag and reader become more and more broad. Although larger amounts of data that can be stored on a tag, most RFID tags still only store an identification number. All the data corresponding to this number is stored in a remotely accessible database. By taking advantage of the Internet this approach renders such data globally available and allows several parties all over the world to benefit from it.

It is vital to provide common standards for data formats and communication protocols. Currently the primary provider of such standards is EPCglobal – a consortium of companies and organizations set up to achieve worldwide standardization. According to already developed standards [1], the global availability of RFID related data is achieved by having the RFID tags store an Electronic Product

---

[*] This work is supported by the Fundamental Research Funds for the Central Universities (Grant No. 2009QN030).

Y. Wu (Ed.): International Conference on WTCS 2009, AISC 116, pp. 251–257.
springerlink.com © Springer-Verlag Berlin Heidelberg 2012

Code (EPC) identifier, while related data is stored in remote databases accessible via EPC Information Services (EPCIS).

For locating a manufacturer EPCIS that can provide data about a given EPC identifier, EPCglobal proposes the Object Naming Service (ONS) [2] that resolves this identifier to the address of the corresponding EPCIS. EPCglobal is delegating control of the root of ONS to VeriSign [3] – a U.S.-based company, also known as a major certification authority for SSL/TLS, one of the DNS root operators, and maintainer of the very large .com domain. That means the VeriSign Company has the highest right to control the Object Naming Service, such concentration of power in hands of a single entity can lead to lots of problems. EPCglobal should be an open, freely accessible, global system, but the concentration of power let the other country not trust the EPCglobal any more. Therefore, it is reasonable to modify the initial design to take the distribution of control between the participating parties into account, and make the ONS be a real global system. In this paper we design a new architecture where the control power is distributed and freely.

2   Background

2.1   Electronic Product Code (EPC)

The EPC Network architecture provides a method for the inclusion of commercial products within a network of information services. This architecture makes several axiomatic assumptions, the most important being that it should leverage existing Internet technology and infrastructure as much as possible. As such, it adheres to the "hour glass model" of the Internet by standardizing on one identifier scheme: the Electronic Product Code (EPC) .

In most situations the EPC will denote some physical object. EPC identifiers are divided into groups, or namespaces. Each of these namespaces corresponds to a particular subset of items that can be identified. For example, XML Schemas are denoted using the 'xml' namespace, raw RFID tag contents are kept in the 'raw' namespace. The 'id' namespace is generally reserved for EPCs that can be encoded onto RFID tags and for which services may be looked up using ONS. This 'id' namespace is further subdivided into sub-namespaces corresponding to different naming schemes for physical objects, including Serialized GTINs, SSCCs, GLNs, etc. These namespaces are defined normatively in the EPCglobal Tag Data Standards.

Each of the sub-namespaces that are defined by the Tag Data Standards specification have a slightly different structure depending on what they identify, how they are used, and how they are assigned. The SGTIN is used to identify an individual product that is assigned by the company that creates that product. Thus the SGTIN contains a Manager Number, an Object Class, and a Serial Number.

The manager number is a company prefix, which identifies an organizational entity (essentially a company, manager or other organization) that is responsible for maintaining the numbers in subsequent fields – Object Class and Serial Number.

EPCglobal assigns the General Manager Number to an entity, and ensures that each General Manager Number is unique.

In order to further leverage the use of Internet derived technology and systems, the EPC is encoded as a Uniform Resource Identifier (URI). URIs are the basic addressing scheme for the entire World Wide Web and ensure that the EPC Network is compatible with the Internet going forward.

2.2   Object Naming Service (ONS)

While an addressing scheme by itself is useful, it can only be used within a network when a mechanism is provided to authoritatively look up information about that identifier. This EPC 'resolution' mechanism is called the Object Naming Service, or ONS and is what forms the core integrating, or 'truth' verifying, principle of the EPC Network.

In keeping with the assumption that the EPCglobal Network architecture should leverage existing Internet standards and infrastructure, ONS uses the Internet's existing Domain Name System for looking up information about an EPC. This means that the query and response formats must adhere to the DNS standards, meaning that the EPC will be converted to a domain-name and the results must be a valid DNS Resource Record.

In order to use DNS to find information about an item, the item's EPC must be converted into a format that DNS can understand, which is the typical, "dot" delimited, left to right form of all domain-names. The ONS resolution process requires that the EPC being asked about is in its pure identity URI form as defined by the EPCglobal Tag Data Standard (e.g., urn:epc:id:sgtin:0614141.100734.1245).

Since ONS contains pointers to services, a simple A record (or IP address) is insufficient for today's more advanced web services based systems. Therefore ONS uses the Naming Authority PoinTeR (or NAPTR) DNS record type. This record type contains several fields for denoting the protocol, services and features that a given service endpoint exposes. It also allows the service end point to be expressed as a URI, thus allowing complex services to be encoded in a standard way.

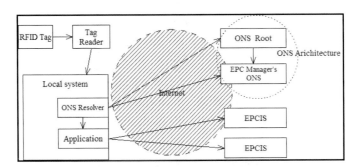

Fig. 1. The Example of ONS

We explain the inner workings of the ONSby an example. Since EPCglobal standards make use of general roles to describe system functionality, we give a short specific example – the arrival of a new RFID-tagged good in a shop. An RFID reader located in the delivery area of the shop reads out the tag and receives an EPC identifier in binary form. Then it forwards the EPC identifier to a local inventory system. This inventory system needs to retrieve item information from the manufacturer's database on the Internet, e.g. to verify the item is fresh, and to enhance smart advertisement throughout the shop. The system hands the EPC identifier over to a specific software library, the local ONS resolver, which translates the identifier into a domain name compatible with the Domain Name System (DNS, for details of its working see Section 3), e.g. 5742.200452.sgtin.id.onsepc.com. This name, which does not make use of the EPC Serial Number as of now, is an element of the DNS domain onsepc.com that has been reserved for ONS and is used for delegation purposes.

The resolver queries the resolving ONS server of its organization or Internet Service Provider (ISP). If the EPCIS address list is not known yet (as in our example of a new item) or has been retrieved and cached before, but is now considered as potentially out-of-date, the ONS Root is contacted. This ONS Root, a service run exclusively by the company VeriSign [3], recognizes the Company Prefix part of the DNS-encoded EPC identifier, and delegates the query to the EPC Manager's ONS server, which has the authoritative address of the manufacturer EPCIS stored in a DNS record called Naming Authority Pointer (NAPTR). Once this address has been determined, the shop inventory system can contact the manufacturer EPCIS directly, e.g. by the use of Web services. To locate different EPCIS for additional information, the use of so called EPCIS Discovery Services is planned, which are not specified at the time of this writing. However, as is indicated by [1], these search services will (at least in part) be run by EPCglobal.

3   Distributed Object Naming Service

The ONS Root will formally be under control of the international consortium EPCglobal, but practically run by the U.S.-based company VeriSign. That may lead to lots of problems.

Firstly, the other country will distrust the EPCglobal, as they don't have the power to verify whether the information from the ONS root is correct. So in those countries, the EPC standard may be ignored and other new standards hold by those country occur. When the situation happens, the global standard EPCglobal will not be global standard any more.

Secondly, the ONS root has the power to block the request from some area. That is unfair, because the information should be shared all over the world. While by some reasons, the company who own the ONS server can block the other company's requests.

Thirdly, the ONS root can eavesdrop on the requests from all over the world and analyze the other countries' business. That is illegal and should not be allowed.

From the above, we proposed a new architecture of ONS, where the power is distributed, and no single country or company can control the ONS any more.

	Header	Filter Value	Partition	Company Prefix	Item Reference	Serial Number
SGTIN-96	8	3	3	20-40	24-4	38
	0011 0000 (Binary value)	(Refer to Table 5 for values)	(Refer to Table 6 for values)	999,999 – 999,999,9 99,999 (Max. decimal range*)	9,999,999 – 9 (Max. decimal range*)	274,877,906 ,943 (Max. decimal value)

Fig. 2. The formation of SGTIN

3.1 Analyzing of the SGTIN EPC

Actually, the SGTIN Electronic Product Code contains lots of information. The formation of SGTIN EPC is shown on the fig.2. The Company Prefix contains 20-40 bits, which is composed of country bits and company bits and so on. So from the Company Prefix, we can get the country the company belongs to. and the Item Reference contains the information about the Object Class, the detailed class of item, so we can image what the item is from this field. The Serial Number is the ID of the item assigned when the manufacturer produce it. When the ONS resolver of local system gets the SGTIN EPC, that information can be achieved by the fields Company Prefix, Item Reference. Our idea of new ONS architecture is based on that information.

3.2 Architecture of Distributed ONS

There are mainly two kinds of information we can get from the SGTIN EPC, including the region information (country information) and item class information. So we design a mixed ONS architecture, which separate the ONS root to multiple replicated ONS.

Firstly, we divide the ONS root into several ONS server by the countries. That means for each country, there is an ONS root which is named Country ONS Root. This root is controlled by the relevant county, and the authority of access certain items can be owned by the country. It is reasonable because sometimes some items' information is secret, such as the military product. Those Country ONS Roots can communicate with each other, and request the EPCIS manager's service among them.

Secondly, for each Country ONS Root, we classify the information by the item class, and divide the Country ONS Root into several ONS server, for each ONS server, the information of same item class is stored. Because companies which are in the same area need those information mostly. The company running business A seldom needs the information of business B. We hope the ONS server of business A has the authority to control what other companies can access the information of business A. The authority can be set by the consortium of companies of business A. The architecture of distributed ONS is shown in the fig.3.

3.3 Virtual Business Root

More and more trades cross multiple countries are done nowadays. By the Internet and EPCglobal, the user can get the information of items from manufacturer more easily. Most of those resolving procedures can be finished among the ONS servers of same business. Inspiring by this, we propose an extended Distributed ONS Architecture which employs the Virtual Business Root.

The Virtual Business Root server belongs to a certain business area, and collects information about this area, although the information is from different countries. The benefits to build a Virtual Business Root are the supporting of the scheme of authority in the business area, and the short route. In former architecture, the authority in a business area is very hard to set, and in the new architecture, the consortium of business A can control the authorities among the companies under the consortium. Secondly, in the Distributed ONS, the lookup operation has to set the request to the Country ONS Root. That decreases the efficiency of routing and bring large overload to the Country ONS Root.

The Distributed ONS architecture with Virtual Business Root is shown in fig. 4.

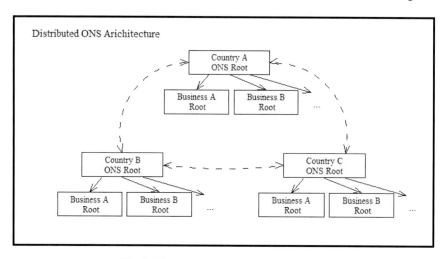

Fig. 3. The architecture of Distributed ONS

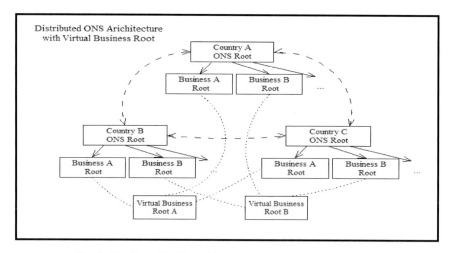

Fig. 4. Distributed ONS Architecture with Virtual Business Root

4   Conclusions

The Object Naming Service (ONS) is a core resolving service of the EPCglobal Network. The function of ONS is the address retrieval of manufacturer information services for a Electronic Product Code (EPC) identifier which is stored on the RFID tag and is read by the RFID reader. But, there is only one ONS Root, and the ONS Root can be controlled or blocked by a single company or a country. The EPCglobal Network is a future global business infrastructure, and it is immoderate that one country or company controls the items information all over the world. In this paper, we propose a new ONS architecture called Distributed ONS, where the ONS root is divided into several ONS server by the countries and the items' class. Also a Visual Business Root is proposed to cope with the situation in certain business area.

Acknowledgment. This work is supported by the Fundamental Research Funds for the Central Universities (Grant No. 2009QN030).

References

1. EPCglobal: The EPCglobal Architecture Framework – Version 1.2. In: Traub, K.(ed.) (September 2007).
2. Mealling, M.: EPCglobal Object Naming Service (ONS) 1.0.1 (2008).
3. EPCglobal Tag Data Standards Version 1.4 (2008).
4. Ramasubramanian, V., Sirer, E.G.: The Design and Implementation of a Next Generation Name Service for the Internet. In: SIGCOMM 2004: Proceedings of the 2004 conference on Applications, technologies, architectures, and protocols for computer communications, pp. 331–342.2004.
5. Fabian, B., G¨unther, O.: Distributed ONS and its Impact on Privacy. In: Proc. IEEE International Conference on Communications (IEEE ICC 2007), Glasgow, pp. 1223–1228 (2007)
6. Fabian, B., G¨unther, O., Spiekermann, S.: Security Analysis of the Object Name Service. In: Proceedings of the 1st InternationalWorkshop on Security, Privacy and Trust in Pervasive and Ubiquitous Computing (SecPerU 2005), with IEEE ICPS 2005, Santorini, pp. 71–76 (2005).
7. Ozment, A., Schechter, S.E.: Bootstrapping the Adoption of Internet Security Protocols. In: Proc. of the Fifth Workshop on the Economics of Information Security (WEIS 2006) (2006).

Genetic Algorithm Solves the Optimal Path Problem

Liu Wan-jun[1], Wang Hua[2], and Wang Ying-bo[1]

[1] College of Software Engineering, Liaoning Technical University
Hu Ludao, Liaoning
[2] The School of Electronics and Information Engineering, Liaoning Technical University
Hu Ludao, Liaoning
xwbjhubei@sogou.com

Abstract. In view of the solution of optimal path in network transport, the paper uses the basic principle of genetic algorithm to carry on the choice, overlapping and the variation. After the operation of the variation, the new individual variation should be recalculated to determine the fitness of individual choice in order to enhance the astringency of the algorithm. And through a mathematical model simulation experiment, it has some theoretical and practical value.

Keywords: Optimal path, Algorithm, Modeling, Simulation.

1   Introduction

Network most short-path is one of the most basic and one of the important researches in the graph and network optimization research content. However, how to arrange vehicle scheduling routes in order to meet the needs of users, to achieve the shortest total distance and cost, this is the question which needs to solve. In the basic vehicles dispatch question, not only has one center to carry on the service to many users, but also has many centers to carry on the service to many users simultaneously. This paper is mainly to research the genetic algorithm to solve the problem of multi-center services for multiple users. Under the premise of meeting the needs of each user, the purpose of transportation is to achieve the lowest total cost.

2   Optimal Path

2.1   Problem Description

In the large open pit mine, there are L loading points which have the number of kl trucks, denoted by 1,2,…,L. Each point has a truck load capacity of Q, the number of Kl. Now, the mineral is allocated to the unloading points, denoted by 1,2,…N. Unloading point i needs the mineral demand Gi (Gi<Q). Any truck in loading point can be served for unloading point, but only by one truck service, when it completes the transport task, it should be back to the original loading point.

Y. Wu (Ed.): International Conference on WTCS 2009, AISC 116, pp. 259–263.
springerlink.com   © Springer-Verlag Berlin Heidelberg 2012

2.2   Mathematical Modeling

$$x_{ij}^{lk} = \begin{cases} 1 & \text{point L, Truck number k from unloading point i to point j} \\ 0 & \text{Otherwise} \end{cases}$$

Objective function:

$$F = \min \sum_{i=1}^{L+N} \sum_{j=1}^{L+N} \sum_{l=1}^{L} \sum_{k=1}^{kl} d_{ij} x_{ij}^{lk} \tag{1}$$

Constraints:

$$\sum_{j=1}^{N} \sum_{k=1}^{KL} x_{ij}^{lk} \leq L_k \tag{2}$$

$$\sum_{i=1}^{N} G_i \sum_{j=1}^{L+N} x_{ij}^{lk} \leq Q \tag{3}$$

Formula (1) expression: objective function minimum value, namely optimal path; Formula (2) expression: each loading point sends out truck quantity can not surpass the truck number which this loading point has. Formula (3) expression: the goods transported by truck can not exceed the capacity of its own.

3   Genetic Algorithm

The basic idea of genetic algorithm is that space optimization question decision variable can be expressed heredity space individual through certain code (usually in binary code), each individual chromosome is actually an entity with features. All entities constitute the initial population, there are three processes: (1) select individuals according to the fitness; (2) father generation of reorganization produces the filial generation; (3) according to certain probability filial generation variation. Until satisfies certain iteration condition of convergence, this process only then ended.

3.1   Chromosome Code

The article uses the binary system to carry on the chromosome code, with four binary bits expressed node information, the node position expressed with coordinate. Fig. 1 is an example of a network topology that the order of connections between nodes. eg: The path C-8-1-C design chromosome code is: 1100 1000 0001 1100. As a result of the different paths with different number of nodes, the length of chromosomes may be different, so the length of stain is set to change. The form of initial population generates by the random.

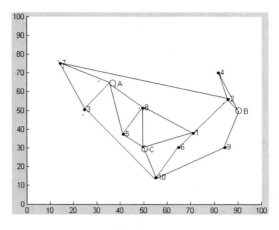

Fig. 1.

3.2   Genetic Algorithm Step

- Selection

The selection process is a superior win and the inferior wash out process. First, calculate the fitness, and then choose from the current group of individuals with high fitness value to operate, its role is to improve the group's average fitness value. The calculation of individual fitness in generate directly quotes the objective function as the fitness function, recorded f(s). In this paper, the fitness function takes the reciprocal of distance, f(s) = 1/Σdij.

For example: parameters in Table 1, Table 2 below. As shown in Fig.1, to meet the demand the unloading point 8, Chromosome S1=<1100 1000 0001>,f(S1)=0.0211; S2=<1010 0101 1000 >, f(S2)=0.0236; f(S2)> f(S1),so the fitness of S2 is better than S1.

- Cross

Individuals are randomly paired, in accordance with certain requirement, and to exchange parts of genes. Its role is to fine the original genes to the next generation, the formation of more complex gene that contains the individual. The binary code forms the individual when carries on the crossover operation, generally uses the single-point crossover or the multi-point crossover.

eg: In the fifth position, select the single-point crossover, then generate new individuals for the following: S1′ =<1100 0101 1000>   f(S1′) =0.0337; S2′ =<1010 1000 0001>   f(S2′) =0.0224.After crossover operation, the fitness of S1' is better than S2'

- Mutation

One or several genes have mistakes with a smaller probability in the individual, simulating the natural phenomenon of gene mutation. Its role is to increase the diversity of individuals. Eg: In the eighth position S1 mutates, that is, mutating from 0 to 1, generating a new individual S3'=<1100 1001 0001>, f(S3')=0.0178. The fitness of S3'is lower than S1. But if the individual mutates from the fifth position to seventh position, the new individual S4'=<1100 0110 0001>, f(S4')=0.0276, the fitness of S4' is better than S1.

Therefore, the recomputation produces newly the chromosome adaptation function value, can overcome the genetic algorithm mountain climbing ability bad shortcoming to a certain extent, causes the search to change the solution space other regions to carry on the search.

4 Experimental Results

In order to confirm the genetic algorithm in the computation optimal choice question feasibility, uses the Matlab simulation software to carry on the confirmation. Between the loading point and the unloading point the route is connected (as shown in Fig. 1), the node distributes in 100×100 the scope, truck's capacity is 30 tons ,the unit is the kilometer.

Table 1. Loading point information

Loading point	Truck quantity	Position data
A	5	(35, 65)
B	5	(90, 50)
C	5	(50, 25)

Table 2. Unloading point information

Unloading point	Demand	Position data
1	10	(70, 40)
2	17	(85, 52)
3	38	(24, 50)
4	7	(80, 69)
5	15	(42, 38)
6	10	(65, 25)
7	16	(15, 75)
8	11	(50, 50)
9	19	(85, 30)
10	8	(55, 15)

As experimental results shown, the best answer is that: A sends one truck to transport twice,A-3-A,A-3-7-A;Bsends two trucks ,they are only transported once, B-9-B, B-2-4-B; C sends two trucks ,they are only transported once, C-5-8-C,C-10-6-1-C. The total distance is 311.36 km. As shown in Fig.2, the hollow circle represents the loading points, the solid circle represents the unloading points.

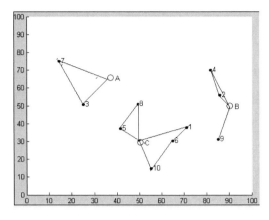

Fig. 2. Optimal road map

5   Conclusion

Trough the experimental verification, the design method which unifies using genetic algorithm's basic principle and the Matlab simulation software can rationalize vehicle scheduling routes, reaching not only meet the needs of users, but also the shortest total distance and the purpose of cost savings. The experimental result has further confirmed this method feasibility and the validity.

References

1. Wang, X.-P., Cao, L.-M.: GA-Theory, Applications and Software realize. Xi'an Jiaotong University Publishing House (January 2002)
2. Zou, T., Li, N., Sun, D.-B., Li, J.: Genetic Algorithm for Multiple-Depot Vehicle Routing Problem. Computer Engineering and Applications (2004)
3. Ma, X.: A Genetic Algorithm for k Optimal Paths Problem. Computer Engineering and Applications 12 (2006)
4. Jiang, D.-L., Yang, X.-L., Du, W., Zhou, X.-W.: A Study on the Genetic Algorithm for Vehicle Routing Problem. Systems Engineering-Theory&Practice (1999)
5. Liang, K., Xia, D.-C.: Genetic Algorithm Programming By Matlab And Optimizing Problem Solving. Development and Design Technology (December 2006)

The Strategic Significance of Bilingual Teaching in Higher Education Reform and Development in China

Chen Shiyue

School of Environment and Planning, Liaocheng University, 252059
Liaocheng, China
shisi60@yahoo.cn

Abstract. Chinese higher education is facing a serious crisis. Bilingual Education plays an important role in Chinese Higher Education Reform. This paper expound the background and essence of bilingual teaching and learning, then discusses the purposes of strategic objectives about the bilingual teaching in Chinese higher education, namely: cure ills of higher education and the promotion of higher education development; make China into the world, let the world understand China; the world personnel training, and promoting world harmony and progress. This paper suggests China should strive to develop bilingual teaching, and make higher education reform and development in the strategic core.

Keywords: Bilingual teaching, higher education, reform, significance, China.

1 Introduction

Talent is the first resources of national economic and social development. In the process of human society developing, the talented person is an important driving force of social civilization and progress, prosperity and happiness for the people of the country. The higher education undertakes the important task of promoting modernization and training highly specialized personnel and developing of science, technology and culture. Bilingual teaching is the core part of higher education in nurturing talented persons [1]. Bilingual education in China has been thriving in many universities and professional conduct up, but looking into the bilingual education university, as well as experts and scholars on the understanding and practice of bilingual education, I believe that it is necessary to conduct awareness and understand on bilingual teaching of Chinese universities.

2 The Meaning of Bilingual Teaching

Many countries around the world have implemented a bilingual education, but in academic circles there is no generally accepted definition of a bilingual. Usually bilingual education refers to use a foreign language or second language as classroom language in teaching specialized course. a large number of bilingual education

Y. Wu (Ed.): International Conference on WTCS 2009, AISC 116, pp. 265–272.

practiced in European shows, while not avoiding mother language, English is often the only language in a bilingual class [2]. The author also advocates that bilingual teaching in Chinese Universities should use English as the main teaching language, but at beginning stages or necessery it is not rule out to use their mother tongue in a bilingual class for the acquisition of professional knowledge, so that students can storage and statement of knowledge in two languages.

3   Bilingual Teaching Background

The background of Bilingual teaching is the integration of the world. The most important of performance of the world's integration is economic globalization and international and regional economic integration. Economic globalization is a market-oriented economy, trade liberalization, transnational production and investment, who driven by market forces, beyond the national boundaries of countries and regions. Economic globalization greatly encourages the expansion process of regional trade, capital, information, markets, enterprises and population to the people and communities in all regions in Earth. There is considerable breadth and intensity's impact. The main performance of this expansion process is the same country for a regional economic integration within the limited region and the different countries or regional group carriing out the extensive economic exchanges and cooperation in different areas based on the open theory.

4   The Nature of Bilingual Education

To inherit all the outstanding cultural achievements of mankind and to promote all-round development of human being is the nature of bilingual education. It is an important foundation and condition to inherit and carry forward the fine traditions of Chinese culture, absorb the world's outstanding cultural achievements and caste them in the building of socialist culture with Chinese characteristics. China's cultural development cannot inseparable from the outstanding cultural achievements which created by the world people. Our culture should modern, should world and should future.We should take a broad mind and broad vision, boldly absorb all outstanding cultural achievements, such as science and technology, culture, education, literature, art, ideology and theory, etc. So we can enrich our cultural life and improve the nation's cultural quality.

It's the essential requirement of Marxist about building a new socialist society to promote all-round development of human being [3]. Marx once said, the future society is "the freedom of each individual and comprehensive development of the basic principles of social form." Engels further interpreted that the socialist not only strive to "ensure that all social members are rich enough each day's material life but also may ensure that their access to physical and mental freedom to fully develop and use."

Hu stressed at the meeting of studying the Selected Works of Jiang Zemin [4]: "We must firmly grasp the development of top priority in governing and rejuvenating the party, firmly seize the important strategic opportunities, persist in taking economic construction as the center, continue to deepen reform and opening, the implementation of 'Eleventh Five-Year' Plan, a comprehensive progress in socialist economic construction, political construction, cultural construction, social development, promote the comprehensive development, promoting harmony between man and nature. "

To develop English-based bilingual education in our colleges and universities is beneficial to fast, accurate and complete absorption of all the fine cultural achievements of the world. We can grasp the new trends, new culture, new ideas about the development of world science and technology. It's good for all-round development of student.

5   The Strategic Significance of Bilingual Teaching

Bilingual education plays an important role in promoting reform and development of higher education. Looking to implement a bilingual education among the countries, China is the most unique country. Because the Chinese is on behalf of the Eastern culture, English on behalf of the Western culture. We all know the oriental civilization is behind the Western, but with China's reform and opening up and rapid economic development, China is becoming increasingly important in the world. This determines the development of bilingual education in China also has very important strategic significance.

5.1   Cure Ills of Higher Education, to Promote Development of Higher Education

After the founding of new China, China has made considerable progress in higher education. But since the end of 1980s, China's universities is in declining quality of personnel training because of various sports coming and going, "Official Standard" intensified, money-driven system making the experts and professors to teach for a living and research for rich. Well-known educator Liu sharply criticized the University of dirty and smelly and chaos [6]. Generation master Qian in his lifetime also extremely worried about China's higher education, and Kuiran asked: "Why donot the Chinese university train a talented people?"

I believe that the fundamental reason about the chaos in Chinese universities is the loss of the university spirit. We all know that university in the modern sense originated in the West in 1088, the Bologna University in northern Italy which recognized the earliest university. Continuation and development through the Millennium, the spirit of Western university is very clear, namely: independence, democracy, freedom, challenge and critique. Looking at our universities, no university has a true sense of independence, democracy, freedom, questioning and critical spirit. It's true that Chinese universities have completely lost the spirit. So some people exclaimed [7], "China is facing educational disaster"and "the Chinese nation is being at the most dangerous time." In order to rescue our universities, it's very important to allow the pirit of university to return.A measure is the implementation of bilingual teaching in universities, so can make higher education with international standards.

Throughout our country's higher education, there are two gold development stages [8]: Fourth Movement to 1956, from 1980 to 1987. The two stages have a common characteristic that is better with the world of higher education, especially in the first phase of the period, China's higher education and the world is almost completely converge. Although there were no clear so-called bilingual education, the materials used is the most outstanding foreign materials. Such as the famous Southwest University almost use entirely Anglo-American textbooks in science and engineering, and even liberal arts nature of law and economics as well. It is because of a sense of bilingual education, Southwest university in just eight years and in a very difficult condition has trained two Nobel Prize winners, three State Supreme Science and Technology Prize , eight bombs and one satellite contributions of scientists and 171 academicians.

Undeniably, because of China's unique civilization and the historical development.There are lots of conflicts after Chinese higher education into line with the West's. If it is not reality with the Chinese community, western civilization thought charatized by scientific and democratic may unavoidablily conflict strongly with the formation of Chinese civilization for thousands of years.Such as a "Westernized" crisis, resulting in social oscillation, or even chaos, has seriously affected the stability and security of our society. But after three decades of reform and opening up, there is a great success in economy, national power has become more powerful and the voice of leading the world have shown an increasingly urgent. Bilingual education is being increasingly important. Indeed, in the early stages of the development of higher education, a comprehensive good use of foreign materials "bilingual teaching" to a large extent have been forced in nature, because accepting or bring is the most rapid and effective. But now it is the time to take the initiative to bilingual teaching. We can not ask what the world can do for China, but to ask what China can do for the world. This is not just our wishful thinking, but the world demand of us. If the many abuses of Higher Education, University of spiritual decay, how can we train people to lead the world to do. Therefore, to eradicate the ills of higher education and to promote the development of higher education are the primary task of our universities, of which one is simply the implementation of bilingual teaching in colleges and universities.

5.2   To China's Integration into the World, Let the World Know China

After three decades of reform and opening, China's accession to the WTO and world economic integration, China has to a large degree of integration into the world [5]. However, because of political and cultural differences and deliberate propaganda of Western media, a great many people in the world are still misunderstanding and even hostility to China. So China is far from accepted by the real world. Most Western people only know Chinese food ,the Great Wall, Bruce Lee and communism. China successfully hosted the Olympic Games in 2008. In order to show the world "humanity Olympics" spirit and dissemination of Chinese culture and enhancing communication and understanding with all peoples, Olympic Torch taking Journey of harmony transmission mode was accused of too much emphasis on the Olympic form.

"3 • 14" Lhasa incident, because western people are deceived and do not know the history of China's Tibet and blindly listen to Western media which is full of wrong information, they sympathy on the Tibetan separatist. For Westerners to understand China is a long way to go to work. Even if our government has done a lot of effort, the Westerners who have long been brainwashed are full of doubt and misunderstanding to our country.

For a large population of China, national stability is not only beneficial to the Chinese nation, is also beneficial to people of the world; our economy rapidly developing is not only conducive to China's people's living standards, but also conducive to world prosperity; we emphasize that "peaceful rise", not only concern about our own country's security and development, but more concern about the world's sustainable development, environment, climate issues, nuclear issues and world prosperity and stability. But if we donnot have a world-class talent, or we can't train out a world-class talent, then the rapid development of our country or "peaceful rise" is quite doubt by the world of national.

We should, through bilingual education, initiativily and systematically translate the large number of China's history, culture into English. On the one hand it can be used as a bilingual teaching materials, on the other hand it can be the authority of Oriental books let the world's people to understand the Confucian culture, and it can strengthen China's soft strength. People of the world understand China, not only the Western media (Western media biased against China deep-rooted). We must initiate the "attack" to let the world people understand a true China of an open, expansive, inclusive grandily national sense of responsibility power subjects. Therefore, to develop bilingual teaching is not only an important means for China's integration into the world, but also an important way to make the world to understand China.

5.3   Personnel Training for the World and Promoting World Harmony and Progress

One of the main tasks of school education is to deliver stored in the books of the materialization of knowledge and values and behaviors, and so makes human cultures survive from generation to generation. It's necessary to inherit the culture through vertical transmission, but more important the horizontal spread. In the past, cultural exchange is often full of contradictions and conflicts through war, trade, migration, etc. Today, in the tide of world peace, democracy and equality cultural exchanges should be achieved in a friendly. Conversely, the friendly cultural exchange will promote world peace, democracy and equality. Education is an important way for cultural exchange and innovation. Therefore, educational exchange is an important way to national cultural exchanges. The role of educational exchanges is not only to the existing culture, communication and integration, more importantly, is to train the next generation.

China's accession to the World Trade Organization made a commitment to appropriate educational services. The one hand, to attract foreign capital and high-quality educational resources make up the shortage of educational resources, and promote the development of education in China; the other hand, with China's economic development and national strength, China's higher education must have a social responsibility and international perspective in making efforts to cultivate talent. So, bilingual education has become the best choice for China's higher education reform.

With enhancing of China's international status, the people of the world are full of doubts and hoping to China. To show the full responsibility of great country to the world, Chairman Hu Jintao in "Seventeenth" National Congress of Communist Party of China proposed building a "harmonious world", clearly presented a long-term goal of establishing a lasting peace, prosperous and harmonious world working together with people of all countries.To fully demonstrate to the world China's great power responsibility to promote building a harmonious world, on the one hand, our government must vigorously publicize and advocate, on the other hand we need understanding and support of the world's population. Through bilingual education, we can train a large number of personnel who will play a very key role in promoting world harmony. Because these personnel spread throughout the world and access to the public, they can interpretate the Chinese government's ideas of a harmonious world and help to promote world harmony and progress.

In addition, bilingual teaching in Chinese universities will also greatly promote primary and secondary school English teaching and effectivily to remove the shortcomings in learning English, even will also play a strong role in promoting of Entrance Examination. Learning English in schools of China is now for exams, not for applications, a large number of such students take tests better, but weak hear and writing. The implementation of bilingual teaching in universities will vigorously promote the reform of English teaching in the primary and secondary school, it will certainly be more emphasis on capacity about English entrance test. Those exams suffered criticism on the CET and the title of community language examinations can be canceled. Implementation of bilingual teaching in college, the learning will become more difficulties, and only those hard work or genuine talent students can complete their studies. So to some extent it's good for inhibiting recruitment of corruption and ensuring education equity. After the implementation of bilingual education, Chinese universities can be ordered to recruit a large number of foreign talent to China to teach, it not only can improve our teaching quality, it also can strengthen the teachers and promoting the exchanges and cooperation between our university teachers and foreign teachers, enhancing the quality of our teacher colleges and universities and scientific research and attracting the world's best students to China to study for a degree and so on.

6   Bilingual Teaching Planning and Development Proposals

Bilingual education is a systematic project. It needs to be promoted by national education development strategy. It's necessary to formulate bilingual education development strategies by the State Ministry of Education, organize national famous experts and scholars to preparate the bilingual materials or select the best original foreign textbooks [9]. According to the actual situation of the school, all universities should gradually implement bilingual education, plan bilingual development, orderly introducte or train bilingual teachers. Those leading universities and colleges such as Peking University and Tsinghua University should open as quickly as possible an entire bilingual program, other universities should open a certain proportion of bilingual courses. The ultimate goal of all courses of all universities is all bilingual. Taking into account our unique bilingual teaching and respecting the student's right to choose those courses such as the Chinese culture, society and history subjects open bilingual classes in parallel with the Chinese classes. In principle the other courses should be bilingual classes.

7   Conclusion

The world today is in the great development and changes and major adjustments, such as world multi-polarization and economic globalization, technological advances, knowledge-based economy in the ascendant, international competition intensifying. The competition about economy, technology and culture is the talent competition in the final, and talent competition is the competitive of education in nature and the competition about higher education is the core. Universities have the functions of inheritance of civilization and creating knowledge and leading the society. We must strongly reform our higher education and make it on the strategic high ground of personnel training. It's very important for bilingual teaching to eradicate shortcomings of higher education, integrate China into the world and promote world harmony and progress. Therefore, we must strive to implement bilingual teaching, and make it in the strategic core of higher education reform and development.

Acknowledgement. This work was supported by the Soft Science Planning Project of Shandong Province and Liaocheng University Natural Geography Key Course Construction and Liaocheng University undergraduate education project.

References

1. Wu, P.: Five years of bilingual Teaching Research. China University Teaching (1), 37–45 (2007)
2. Huang, C.: Interpretation and Analysis to the Key Concepts of Bilingual Teaching and Learning. Foreign Language Research 140(1), 137–139 (2008)
3. Dou, A., Ge, G.: Correct understanding of Marxist thinking on the comprehensive development. Journal of PLA Nanjing Institute of Politic 23(S1), 15–18 (2007)
4. Hu, J.: Report of studying the Selected Works of Jiang Zemin, http://news.xinhuanet.com/politics/2006-08/15/content_4964223.htm
5. Huang, J., Zhao, Y., Ran, Y.: Education devolopment and economic globalization. Modern Education Daily (July 11, 2005)

6. Liu, D.: Changes of higher education in China in sixty years. Higher Education Exploration (5), 5–12 (2009)
7. Liu, D.: On the Essence of the Western Education from the Origin of University. Journal of China University of Geosciences (Social Sciences Edition) 9(1), 1–6 (2009)
8. Liu, D.: Choice and strategy on internalization of college education. Journal of Higher Education 28(4), 5–12 (2007)
9. Yu, S.: Outstanding foreign materials in the Southwest university, http://news.guoxue.com/article.php?articleid=16819

The Phenomena and Applications of Light Dispersion

Liu Weisheng[1] and Fan Xiaohui[2]

[1] Department of Physics, Tangshan Teacher's College, Tangshan, China
[2] School of Mathematics and Quantitative Economics, Dongbei University of Finance and Economics, Dalian, China
qishenggood@gmail.com

Abstract. The article formulates the essence of the light dispersion phenomena in the classical theory view, and discusses the theory and experiments of the method of minimum deviation angle to measure the prism glass refraction. A empirical formula of deviation angle and the angle of incidence has been presented, and the formula can be used to find the exit light quickly and the emitted law of the exit light. It has provided theoretic foundation to accurately determine the position of minimum deviation angle. At last the author introduces the dispersion of light in the practical application.

Keywords: dispersion, minimum deviation angle, optical spectrum.

1   Introduction

Since Newton used the prism to decompose the white light bands into colorful bands in 1666, the man has really begun the research of dispersion. The dispersion phenomena shows that the speed of light in a medium (or the refractive index n=c/v) change with the light frequency. And the dispersion can accomplish through prism, diffraction grating, interferometer, etc, which can draw a conclusion that the white light is composed of red, orange, yellow, green, blue, indigo, violet and other lights. Polychromatic light is mixed with monochromatic light, which is cannot be resolved. So the white light is composed of red, orange, yellow, green, blue, indigo, violet and other lights, and the mixture of monochromatic light called polychromatic light.

2   Formation of the Light Dispersion

Polychromatic light is mixed with monochromatic light, for example, Sunlight, incandescent lights and fluorescent. The dispersion phenomenon is that separating the polychromatic light into monochromatic light through the dispersion system, which can be made of primes or optical grating. Light has a certain frequency; light color is determined by the frequency of light. In the visible light region, the frequency of red light is minim, and the purple frequency was the most frequent. All kinds of light have different refractive index in the same medium, and the index change with wave frequency or wavelength in vacuum. Refraction occurs when the polychromatic lights reach the medium interface. The deflection directions of propagation of various lights

Y. Wu (Ed.): International Conference on WTCS 2009, AISC 116, pp. 273–280.

are not same, so when refraction occurs when the lights leave off the primes separately. Generally, refractive index n or the relationship between dispersion rate dn / dλ and the wavelength λ describes the light dispersion laws. The medium dispersion can be classified as normal dispersion and anomalous dispersion. As the dispersion of light, a bunch of white light refracted through the prism. And then a colored band of light occurs on the screen on other side of the primes. The display of colors is that the color on the apex angle side is red, and the bottom is purple. The colors are orange, yellow, green blue and indigo respectively in the middle. The color brand is called spectrum. After summer rain, colorful arcs often occur on the side of the sky facing the sun, rainbow. The reason of the formation of the rainbow is due to dispersion. After rain, there are many tiny water droplets suspended. As the sun lights go through the sky, then these tiny drops occurs dispersion. If you look at the drops, rainbow will generate. Rainbow's colors are red outside, purple inside, respectively. This is the simplest performance of dispersion.

3   Theory of the Light Dispersion

3.1   Dispersion Theory

Newton's theory is to use experimental methods to explain the macroscopic phenomenon of dispersion. The electronic theory of Lorentz and Maxwell theories can explain the dispersion of the microscopic and formulates.

The medium dispersion shows that different wavelengths of incident light are correspond to certain refractive indexes. In other words, the speeds of different light frequencies are different. This made Maxwell's electromagnetic theory of light no sense in the past, because according to Maxwell's theory, the refractive index is a constant only relate to dielectric constant, not relevant to light frequency. Later Lorentz classical electron theory successfully explained the dielectric constant, and found the relationships between frequency of the electromagnetic field and dielectric constant. Then he solved the Maxwell theory of the initial difficulties, and explained the phenomenon of dispersion.

The medium relative dielectric constant ε_r is not constant, but it changes with frequency. In addition, there exits function of frequency and ε_r. Then the Maxwell formula $n = \sqrt{\varepsilon_r}$ can still be used to get dispersion equation $n = f(\lambda)$.

$$\vec{P} = \chi \in_0 \vec{E}, \varepsilon_r = 1 + \chi = 1 + \frac{P}{\in_0 E} \tag{1}$$

P is vector sum of electric dipole moment per unit volume in the formula, that is $\sum p_i / \Delta V$. $\vec{p}_i$ is an electric dipole electric moment, and positive charges are not moving in every molecular or atomic. The r represents the location of charges, which is the distance between positive and negative charges. Its direction is from negative charges from the positive charges. Electric dipole electric moment is $\vec{p} = q\vec{r}$, and assume that all dipoles have the same value of electric moment and a total of N dipoles per unit volume. Then in the same external electric field (electric vectors of incident light), the N dipoles points the same direction. And the value of polarization intensity vector is $P = Nqr$. The vibration is always along the same lines, so in the following discussion, it is only to take the r and p into consideration.

First of all, to calculate the vibration of a charge in the under external electric field E, which is Stationary relative to other charges, and there are three forces effecting on a charge q.

1)Force of external electric field qE ;

2)Quasi-elastic force, where β is the elasticity coefficient;

3)Damping force $-\gamma \dfrac{dr}{dt}$,where the γ is damping coefficient, and β and γ are constant, not related to frequency. The directions of two forces are respectively opposite with γ and $\dfrac{dr}{dt}$, which are negative, and the quality of moving particles are m. Then its forced vibration equation can be written as:

$$qE = m\beta\gamma - m\gamma\frac{dr}{dt} = m\frac{d^2r}{dt^2}$$

Supposing the vector of incident light electric (represented as plural) is $E = E_0 e^{i\omega t}$, $\omega_0^2 = \beta$, the above formula shows as follows:

$$\frac{d^2r}{d^2t} + \gamma\frac{dr}{dt} + \omega_0^2\gamma = \frac{qE_0}{m}e^{i\omega t}$$

So the steady-state solution is

$$r = \frac{\dfrac{qE_0}{m}}{\left(\omega_0^2 - \omega^2\right) + i\gamma\omega}e^{i\omega t}$$

Then the polarization intensity P is

$$P = Nqr = \frac{\dfrac{Nq^2}{m}E_0 e^{i\omega t}}{\left(\omega_0^2 - \omega^2\right) + i\gamma\omega} = \frac{\dfrac{Nq^2}{m}}{\left(\omega_0^2 - \omega^2\right) + i\gamma\omega}E$$

Inserting into (1) formula, getting

$$n^2 = \varepsilon_r = 1 + \frac{P}{\varepsilon_0 E}$$

$$= 1 + \frac{Nq^2}{\varepsilon_0 m[(\omega_0^2 - \omega^2) + i\gamma\omega]}$$

$$= 1 + \frac{Nq^2}{\varepsilon_0 m}\left[\frac{(\omega_0^2 - \omega^2) - i\gamma\omega}{(\omega_0^2 - \omega^2)^2 + \gamma^2\omega^2}\right]$$

or

$$n^2 - 1 = \frac{A(\omega_0^2 - \omega^2)}{(\omega_0^2 - \omega^2)^2 + \gamma^2\omega^2} - i\frac{A\gamma\omega}{(\omega_0^2 - \omega^2)^2 + \gamma^2\omega^2} \tag{2}$$

In the formula, $A = \dfrac{Nq^2}{\varepsilon_0 m}$, if the n is constant, the imaginary of the right part of the

above formula must be zero, that is $\gamma = 0$. But this is not in agreement with that assuming damping force is not equal to zero. This contradiction occurs because we take the damping force of the oscillator into account. So we should consider that incident light energy is absorbed. That means that the assumptions of incident light electric vector are $E = E_0\theta^{iwt}$ is not correct, and E_0 is not constant. Assuming that the light travels along the x direction, so the form of formula is

$$E = E_0 \exp\left[i\omega\left(t - \frac{x}{v}\right)\right] = E_0 \exp\left[i\omega\left(t - \frac{nx}{v}\right)\right] \tag{3}$$

$$E = E_0 \exp\left[-\sqrt{a_a}\,x + i\omega\left(t - \frac{x}{v}\right)\right]$$

$$= E_0 \exp\left\{i\omega\left[t - \frac{nx}{c}\left(1 - i\frac{c}{n\omega}\sqrt{\alpha_a}\right)\right]\right\} \tag{4}$$

The formula v is the phase velocity of the incident light in the medium, c for phase velocity for the vacuum, n for the refractive index. Taking the material absorption into account, and considering that the amplitude of the incident light will decrease with the increment of x, the forms of attenuation wave are following as the form $I = I_0 e^{-a_a d}$:

$$E = E_0 \exp\left[-\sqrt{a_a}\,x + i\omega\left(t - \frac{x}{v}\right)\right]$$

$$= E_0 \exp\left\{i\omega\left[t - \frac{nx}{c}\left(1 - i\frac{c}{n\omega}\sqrt{\alpha_a}\right)\right]\right\}$$

In the formula, a_a is the absorption coefficient (The square root is a_a, because the intensity is proportional to the square of amplitude).

Assuming

$$\frac{c}{n\omega}\sqrt{a_a} = k, n(1-ik) = n' \tag{5}$$

the form of attenuation of wave can be written as

$$E = E_0 \exp[i\omega(t - n'x/c)] \tag{6}$$

Comparing the formula (3) and (6), we can instead n' of n. The expression of amplitude wave become the expression of the attenuation wave, called complex refractive index. Then the n in the formula (2) should instead of n':

$$n'^2 - 1 = n^2(1-ik)^2 - 1$$
$$= [n^2(1-k^2) - 1] - 2in^2k$$
$$= \frac{A(\omega_0^2 - \omega^2)}{(\omega_0^2 - \omega^2)^2 + \gamma^2\omega^2}$$
$$-i\frac{A\gamma\omega}{(\omega_0^2 - \omega^2)^2 + \gamma^2\omega^2}$$

Divide the real and imaginary parts separately, we can get

$$2n^2k = \frac{A\gamma\omega}{(\omega_0^2 - \omega^2)^2 + \gamma^2\omega^2}$$
$$n^2(1-k^2) = 1 + \frac{A(\omega_0^2 - \omega^2)}{(\omega_0^2 - \omega^2)^2 + \gamma^2\omega^2} \tag{7}$$

If we instead wavelength in the vacuum λ of circumference ratio ω, and replace ω_0 with λ_0. Because the $\omega_0 = \omega = 2\pi v = 2\pi c / \lambda$, the above formula are

$$n^2(1-k^2) = 1 + \frac{b\lambda^2}{(\lambda^2 - \lambda_2^2) + \frac{g\lambda^2}{(\lambda^2 - \lambda_2^2)}} \tag{8}$$

In the formula, $b = A\lambda_0^2 = \frac{1}{\epsilon_0 m} Nq^2\lambda_0^2, g = \gamma^2\lambda_0^4$, are all not related to the constant λ.

If substances have several charged particles, their masses are m_i , Charge a $(i = 1,2...)$, they can vibrate at various frequency ω_i (corresponding to the wavelength), the above formula become

$$n^2(1-k^2) = 1 + \sum_i \frac{b_i \lambda^2}{(\lambda^2 - \lambda_i^2) + \frac{g_i \lambda^2}{(\lambda^2 - \lambda_2^2)}} \qquad (9)$$

To export the normal dispersion region Coxe formula, we can consider the incident light of outside of the absorption area is almost not absorbed. That is

$$n^2 = 1 + \frac{b_i}{1 - \frac{\lambda_i^2}{\lambda^2}}$$

$$\approx (1 + b_i) + b_i \frac{\lambda_i^2}{\lambda^2} + \cdots$$

Owing to $\lambda_i^2 / \lambda^2 << 1$ it is in this expansion fomular , high-order term of λ_i^4 / λ^4 can be omitted.

If we introduce the constant $M = 1 + b_i$ and $N = b_i \lambda_i^2$,we can get

$$n = (M + N\lambda^{-2})^{1/2}$$

The expansion of formula

$$n = M^{1/2} + \frac{N}{2M^{1/2}\lambda^2} + \frac{N^2}{8M^{3/2}\lambda^4} + \cdots$$

$$= a + \frac{b}{\lambda^2} + \frac{c}{\lambda^4} +$$

This is the dispersion curve formula.

In the formula: a, a 、 b and c are constants determined by the characteristics of the mediums. refractive indexes can gets by measuring the minimum deviation angles of different wavelengths. Then we can find the empirical formulas. In the experiments, we generally use the minimum deviation angle to measure the refractive indexes of glass δ_m .According to the formula $n = \sin\frac{A + \delta_m}{2} / \sin\frac{A}{2}$, we can work out the refractive indexes of glass prism.

4   Dispersion Spectra in the Application of Scientific Research

At present, the Fourier infrared spectrum analysis is an important mean in modern environmental science analysis technique which mainly uses in the environmental pollution monitor, the burst characteristics contamination control and the contaminating material analysis. According to the wave number scope can be divided into near-infrared (13 000∼4 000 cm- 1), infrared (4 000∼400 cm- 1) and the far infrared (400∼100cm- 1). In the water body pollution's examination, the organic pollutants is the main substance, chemical oxygen demand (COD) is one of the most common and most important indexes which represent the pollution degree of organic matters. For example, the traditional COD measurement method -- heavy chromate salt, the operation is complex, the survey time is long, and has second pollution, does not suit in online and real time measurement. Therefore the people then developed using near-infrared spectrographic methods to survey waste water COD. Near infrared spectra of water with partial least squares (PLS) regression model, respectively establish COD forecast model of standard water samples and wastewater samples. Experimental results show that using near-infrared spectroscopy to measure COD is feasible. Using infrared spectroscopy and chromatography technology can detach and qualitative measure spectrum information of many air pollutants, and also large organic molecules or acidic compounds, such as Acrolein, benzene, chloroform and so on. At the same time based on infrared spectroscopy, a set of wavelet transform and neural network consists of pollution gas infrared features fast extraction and identification system developed. Experimental results show that the system can not only effectively remove interference, enhance spectral characteristics of targets, improve the recognition rate of overall system, at the same time lay a foundation for the research of foul gas body infrared spectrum multi-objective recognition system . Using infrared technology to identify various complex mixtures and also spilled oil pollution is also a crucial mean. [11]

Fluorescence spectrum is widely used in various actual fields. (1) Using fluorescence spectrum to detect the water pollution. For example, we have used Hitachi F-4500 fluorescence spectrophotometer to detect the properties of fluorescence spectrum of the Yangtze River water, ordinary tap water and pure water produced in the UV excitation. Through research, the result shows that Yangtze River water can produce strong fluorescence at a wavelength of 290 nm of UV excitation. Fluorescence peak is 350 ~ 550 nm wide range of peaks, and fluorescence peak wavelength is around 450 nm. However, the fluorescence peak intensity water samples from different sampling points, is obviously different from fluorescence intensity in the 4, 5, 2, water sampling point ,significantly higher than other sampling points. There is an obvious link between the pollution sources sampling point near sampling point and the intensity. (2) Using the hydride generation - atomic fluorescence spectrometry to detect the arsenic in beverages in medicine. Fluorescence continuous method for arsenic (As), tin (Sn), antimony (Sb) of the hydride generation – atomic in the drink is established. L-cysteine as pre-reducing agent, we use fluorescence spectrometry method of hydride generation - atomic to determine arsenic, tin and antimony continually. Under the best conditions (1% HCl, 2.0% KBH4, 2.0% L-cysteine) of this method, 40mL / L Fe3 +, Pb2 +, 50mg/LCu2 +, Mn2 +, Zn2 +, Se4 +, Hg +, Bi3 +, 10mg / LCr6 +, Cd2 + on the determination of no interference. The recoveries were: 91.3% ~ 97.0%, 92.5% -106.1%, 90.0% ~ 106.2%; detection limits were 0.35,0.35,0.31 μg / L. The method used in beverages has the advantages of reagent consumption, low pollution, low detection limit, linear range and less interference, and a suitable method to arsenic, tin, antimony content detection in drink. (3) Research of heavy metal pollution f the deposit. When using the method of the X-ray fluorescence spectrometry to determine the heavy metal pollution f the deposit, for the reliability of experimental analysis, 12 samples were randomly selected to repeat five experiments, Of which the relative standard deviations of Cd, Hg, Cr, Zn were less than 5%, Cu and Pb less than 10%. The results show that it is not necessary for the X-ray fluorescence determination of samples to digested treatment, and it is Simple, and multiple simultaneous determinations of elements method, efficient. In addition, it is a non-destructive method, and the sample can be reused [2].

A number of different ages, material, composition of artificial structures, the spectral diversity far exceeds the natural environment. The hyper spectral data can enrich spectral information of spatial information, also compensate for deficiencies of spectral resolution of traditional remote sensing data source, and give full consideration of the actual spatial correlation between surface features. The world's first stars hyper spectral satellite data, HyPerion data, obtained by geometric correction, radiometric calibration, atmospheric correction band selection and FLAASH eliminate a series of pretreatment, like Smile effect. Then we can get real physical model a parameter of ground- object reflectance. The relationship between ground-object spectra and the spectra obtained by remote sensing data is affected by Regional topography, climate and many other geographical factors influence, in particular for hyper spectral image. This reduce the actual ground- object general applicability, so it is not be used directly for classification. But the image-side element can be directly applied to hyper spectral data for classification and identification [5].

5   Conclusions

The theory and experiment of the minimum deviation angle method to measure the prism glass refractive index are discussed, the empirical formula for deflection angle and the angle of incidence are given, and its curve with the stoneware Matlab is drown. The curves are easy to find the emitted light, and master the movement law of the emitted light. Besides this, it can provide a theoretical basis for accurate determination the position angle of minimum deviation. In addition, refractive index of light is measured, and the dispersion formula of the prime is obtained by the experimental date according the least square method. At last, some application examples of spectrum in scientific research are introduced.

References

1. Wang, F., Huang, Q.: Using of tunable diode laser spectroscopy to measurement NH_3 of the dusty gas. Physics 56(7), 3867–3872 (2007)
2. Zhu, L., Liu, Y.: Hydride_ atomic fluorescence spectrometry continue measure arsenic_ antimony_ tin. Environmental and Occupational Medicine 9(10), 519–521 (2009)
3. Fu, C., Guo, J.: Analysis of Yangtze River (WanZhou section) water dissolved organic matter fluorescence spectroscopy. Resources and Environment of Yangtze River 18(9), 856–859 (2009)
4. Zhang, X.: Method of plasma emission spectrometry to measure the rare earth elements in marine sediment. A master's degree thesis, China University of Geosciences (2007)
5. Woo, H.K., Lau, K.C.: Vibrational Spectroscopy of Trichloroethene Cation by Vacuum Ultraviolet Pulsed Field Ionization-photoelectron Method. Report of Physical Chemistry 17(3), 292–303 (2004)

A New Approach for Color Text Segmentation Based on Rough-Set Theory

Hu Shu-jie and Shi Zhen-gang

College of Information Science and Engineering, Shenyang Ligong University, Shenyang, 110159, China
156483885@qq.com

Abstract. In order to solve segment text accurately and robustly from a complex background, a new algorithm for text segmentation in images based on rough-set theory is presented in this paper. The histon is an encrustation of histogram such that the elements in the histon are the set of all the pixels that can be classified as possibly belonging to the same segment. In rough-set theoretic sense, the histogram correlates with the lower approximation and the histon correlates with upper approximation. The roughness measure at every intensity level is calculated and then a thresholding method is applied for text segmentation. The proposed approach is compared with the histogram-based approach and the histon based approach. The experimental results demonstrate that the proposed approach obtained satisfactory results.

Keywords: Text segmentation, rough set, histogram, color image.

1   Introduction

The extraction of text information is very important because texts in images and videos contain important and useful information. In general, the extraction of text information from images includes three major steps: text localization, text segmentation, and text recognition [1–3]. In this paper, we focus on the text segmentation, which is employed to separate text pixels from the background in a text image.

Large number of segmentation algorithms is present in recently years, but there is no single algorithm that can be considered good for all images. One of the most widely used techniques for image segmentation is the histogram-based thresholding, which assumes that homogeneous objects in the image manifest themselves as clusters. The advantage of such methods is that they do not need any a priori information of the image. These methods are based only on gray levels and do not take into account the spatial correlation of the same or similar valued elements. However, real-world images usually have strong correlation between neighboring pixels. Adjacent pixels in an object are generally not independent of each other. To overcome this drawback, Cheng et al. used the fuzzy homogeneity approach in which the concept of homogram was introduced. Homogram extracts homogeneous regions in a color image. The homogram of an image is constructed by considering the absolute difference between intensity values of the pixel and the neighborhood pixels. However, this method takes into account only the

Y. Wu (Ed.): International Conference on WTCS 2009, AISC 116, pp. 281–289.
springerlink.com   © Springer-Verlag Berlin Heidelberg 2012

spatial correlation of a pixel with neighboring pixels in the same image plane and it does not consider the correlation amongst pixels in the other color planes [4]. Mohabey and Ray introduced a concept of histon, which is a contour plotted on the top of the histogram by considering a similar color sphere of a predefined radius around a pixel. The concept encapsulates the fundamentals of color image segmentation in a rough-set theoretic sense. The base histogram is considered as the lower approximation and the histon as upper approximation. The upper approximation is a collection of all points, which may or may not belong to one segment but certainly share a unique property that the elements have similar colors. For segmentation, only the upper approximation is considered and the histogram-based segmentation technique is applied on the histon to find the different regions in the image. The method does not take into account the lower approximation for segmentation and thus fails to utilize the properties of the boundary region between the two approximations in segmentation.

In this paper, we propose a segmentation scheme that uses the roughness measure of the rough-set as a basis for segmentation to overcome the drawback. Roughness index is large where the number of elements with similar color is large compared to the number of elements having same color. The index is small when number of elements having similar color is slightly greater than or equal to the number of elements with same color. Clearly, the index will be very small in the boundary between two objects and it will be large in the object region. We have used this property of the rough-set theory to achieve better segmentation results. The paper is organized as follows: in Section 2, we present the rough-set theory and properties of rough set. In Section 3, we describe the concept of histon and calculation of roughness measure. Section 4 describes color text segmentation algorithm and experimental results are given in Section 5, followed by concluding remarks in Section 6.

2   Rough-Set Theory and Properties

According to the definition given by Pawlak [5], an information system is a pair $S = \langle U, A, V, f \rangle$ or a function $f : U \times A \rightarrow V$, where U is a non-empty finite set of N objects $\{x_1, x_2, ..., x_N\}$ called the universe, A is a non-empty finite set of attributes, and V is a value set such that $a : U \rightarrow V_a$ for every $a \in A$. The set V_a is the set of values of attribute a, called the domain of a. A subset of attributes $B \subseteq A$ defines an equivalence relation on U. This relation is defined as

$$IND(B) = \{(x, y) \in U \times U : for\ every\ a \in B,\ a(x) = a(y)\} \tag{1}$$

The elements of U that satisfy the relation IND(B) are objects with the same values for attributes B and therefore they are indiscernible with respect to B. U/IND(B) denotes the set of equivalence classes of IND(B). The equivalence classes of IND(B) are called basic categories (concepts) of the knowledge B.

Given any subset of attributes B, any concept $X \subseteq U$ can be defined approximately by employing two exact sets called lower and upper approximations. The lower and upper approximations can be defined as follows:

$$\overline{BX} = \cup\{Y \in U \mid IND(B) : Y \cap X \neq \phi\} \tag{2}$$

$$\overline{BX} = \cup\{Y \in U \mid IND(B) : Y \subseteq X\} \tag{3}$$

The set $\underline{BX}$, also known as B-lower approximation of X, is the set of all elements of U which can be classified as elements of X with certainty, in the knowledge B. The set $\overline{BX}$, also known as B-upper approximation of X, is the set of elements of U which can possibly be classified as the elements of X, employing knowledge B. Obviously, the difference set yields the set of elements which lie around the boundary.

The set $BN_R(X) = \overline{BX} - \underline{BX}$ is called the B-boundary of X or B-borderline region of X. This is the set of elements, which cannot be classified to X or to -X using the knowledge B. The borderline region actually represents the inexactness of the set X with respect to the knowledge B. The greater the borderline region of the set more is the inexactness. This idea can be expressed more precisely by the accuracy measure defined as

$$\alpha_B(X) = \frac{|\underline{BX}|}{|\overline{BX}|} \qquad \text{for} \qquad X \neq \phi \tag{4}$$

where $|.|$ is the cardinality operator. The accuracy measure captures the degree of completeness of the knowledge about the set X. Here, we can also define a measure to express the degree of inexactness of the set X, called roughness measure or roughness index of X or B-roughness of X, given by

$$\rho_B(X) = 1 - \alpha_B(X) \tag{5}$$

Obviously $0 \leq \rho_B(X) \leq 1$, for every B and $X \subseteq U$. If $\rho_B(X) = 0$, the borderline region of X is empty and the set X is B-definable i.e. X is crisp or precise with respect to the knowledge B, and otherwise, the set X has some non-empty B-borderline region and therefore is B-undefinable i.e. X is rough or vague with respect to the knowledge B.

3   The Concept of Histon

Histon is basically a contour plotted on the top of existing histograms of the primary color components red, green, and blue in such a manner that the collection of all points falling under the similar color sphere of the predefined radius, called expanse, belong to one single value. The similar color sphere is the region in RGB color space such that all the colors falling in that region can be classified as one color. For every intensity value on the base histogram, the number pixels encapsulated in the similar color sphere is evaluated. This count is then added to the value of the histogram at that particular intensity value.

3.1   Construction of Histon

Consider I is an RGB image, of size $M \times N$, consisting of three primary components, red R, green G, and blue B. The histogram of the image for each of the R, G, and B components can be computed as follows:

$$h_i(g) = \sum_{m=1}^{M}\sum_{n=1}^{N}\delta(I(m,n,i)-g) \text{ for } 0 \le g \le L-1 \tag{6}$$

$$\text{and} \qquad i=\{R,G,B\}$$

Where $\delta(.)$ is the Dirac impulse function and L is the total number of intensity levels in each of the color components. The value of each bin is the number of image pixels having intensity g.

For a $P \times Q$ neighborhood around a pixel I(m,n), the total distance of all the pixels in the neighborhood and the pixel I(m,n) is then given by

$$d_T(m,n) = \sum_{p \in P}\sum_{q \in Q} d(I(m,n),I(p,q)) \tag{7}$$

Where $d(I(m,n),I(p,q))$ is the Euclidean distance.

The pixels in the neighborhood fall in the sphere of the similar color if the distance d_T(m,n) is less than expanse. We define a matrix X of the size $M \times N$ such that an element X(m,n) is given by

$$X(m,n) = \begin{cases} 1 & d_T(m,n) < \exp anse \\ 0 & otherwise \end{cases} \tag{8}$$

The histon can now be defined as

$$H_i(g) = \sum_{m=1}^{M}\sum_{n=1}^{N}(1+X(m,n))\delta(I(m,n,i)-g) \tag{9}$$

$$\text{for} \quad 0 \le g \le L-1 \quad \text{and} \quad i=\{R,G,B\}$$

3.2   Roughness Measure

The histogram value of the gth intensity is the set of pixels, which definitely belong to the class of intensity g and therefore, can be considered as the lower approximation and the histon value of the gth intensity represents all the pixels, which belong to the class of similar color and therefore, may be considered as the upper approximation. The vector of roughness measure can be defined as

$$\rho_i(g) = 1 - \frac{|h_i(g)|}{|H_i(g)|} \qquad \text{for} \qquad 0 \le g \le L-1 \tag{10}$$

$$\text{And} \quad i=\{R,G,B\}$$

The value of roughness is large (i.e. more close to 1) when the value of histon is large in comparison with the value of histogram. This situation typically occurs in the object region where there is very little variation in the pixel intensities. The variation in pixel intensities is always more near the boundary between the two objects. This situation will lead to a small (i.e. close to 0) value of roughness. In a segmentation based on thresholding scheme, the peaks in the histogram represent the different regions and the valleys represent the boundaries between those regions. In a similar fashion the peaks and valleys of the graph of roughness index versus intensity can also be used to segregate different regions in the image. The roughness index based thresholding scheme scores over the histogram based and histon based schemes on two aspects:

(1)   There may be situations when there is no significant peak at a particular intensity in the histogram and histon, but there is a signification peak in the graph of roughness index. This situation particularly occurs when, at a particular intensity, the value of histogram is small and the number of pixels having similar color is comparatively large. This represents a small region in the image. In histogram based as well as in histon based thresholding, this region may not be considered as a separate region, but may be merged with some other segment. But, in the proposed method of roughness index based thresholding, such small but significant regions are considered as separate regions, which gives a better segmentation performance.

(2)   Since the peaks in the roughness index based method occur exactly at the intensities where the number of similar intensity pixels is large as compared to the number of same intensity pixels, the color of the every segmented region is more close to the color of the corresponding region in the original image.

4   Segmentation Algorithm

The segmentation process is divided into three stages, calculate histon and roughness Index、 thresholding and region merging.

4.1   Calculate Histon and Roughness Index

For segmentation we have considered RGB color space. In the first stage, the histogram and the histon of the R, G, and B components of the image are computed. For computing the histon, the selection of two parameters is very important. These two parameters are: the neighborhood and the expanse. Neighborhood is the window that decides the pixels involved in the computation. For example if a 3×3 neighborhood is selected then the pixels used in computation of histon will be $. 3^3 - 3 = 24$ The expanse is the radius of the similar color sphere. A similar color sphere is defined by the equation:

$$(x-r)^2 + (y-g)^2 + (z-b)^2 = R_h^2 \tag{11}$$

where R_h represents the radius of the sphere of the region in the neighborhood of a pixel having color intensities as (R,G,B). After sufficient experimentation we found that selecting a 3×3 neighborhood and an expanse value of 100 gives best results. The roughness index is then obtained for R, G, and B components of the image, for every intensity value, using Eq.(10).

4.2   Selection of Peak and Threshold Values

In the second stage, we determine the peaks and valleys in the graph of roughness index and apply thresholds to the image. Selection of correct peaks and thresholds are very important for achieving good segmentation results. The criterion used for selection of significant peaks is based on distance between two significant peaks and the height of the peak. Experimentally we have found the following two criteria for obtaining the significant peaks:

(1)   The peak is significant if the height of the peak is greater than 20% of the average value of roughness index for all the pixel intensities.
(2)   The peak is significant if the distance between two peaks greater than 10.

After the significant peaks are selected, the valleys are obtained by finding the minimum values between every two peaks.

C. Region merging

Obtaining clusters on the basis of peaks and valleys usually results in over-segmentation. Many small regions are generated and some of the regions may contain very few pixels. Such small regions must be merged with the closest large regions. We have used the algorithm proposed by References [5] for region merging. Using this algorithm, region merging is carried out in the following two steps:

(1) The clusters with pixels less than some predefined threshold are merged with the nearest clusters. The process is repeated until the number of pixels in each cluster is greater than the threshold. Experimentally we found that threshold 0.1% of the total number of pixels in the image is appropriate.
(2) Two closest regions, based on predefined distance between two clusters, are combined to form a single region. The process is repeated until the distance between any two regions in the image is greater than the predefined distance. Here also, experimentally we find that distance of 20 is appropriate threshold for region merging.

5   Experimental Results and Analysis

In this section, we test the performance of the proposed approach. The proposed method exhibits a robust performance for the majority of the test images. Fig. 1 illustrates some examples of the segmentation results on English and Chinese text images. It can be seen from the results that most of the text regions are well segmented despite the different languages and color polarities of the texts.

(a) original image of Chinese

Fig. 1. Some examples of the segmentation results on English and Chinese text images

(b) text image of chinese

中国的古典园林

(c) segmented image of Chinese

(d) original image of english

(e) text image of english

Happy New Year

(f) segmented image of chinese

Fig. 1. (*Continued*)

We compared the threshold values from methods of References [6]、[7]. The segmentation results are shown in Figs.2. In Fig.2, we can observe that the proposed approach best segments the text.

(a) original image

Fig. 2. The segmentation results of different algorithm

(b) text image for segmentatiom

(c) binary by algorithm of References [6]

(d) binary by algorithm of References [7]

(e) binary by algorithm of this paper

Fig. 2. (*Continued*)

On the other hand, We compare our text color polarity determination method with the state-of-the-art method proposed by Song [8] , which is based on the statistic analysis. The accuracy of each method is evaluated as the relative frequency of the correctly determined text color polarities as follows:

$$Accuracy = \frac{\text{Number of correctly determined color polarity}}{\text{Number of text image}} \tag{12}$$

Table 1 shows the experimental results of our method and the Song method. From this table, we can see that our stroke filter based determination method achieves a considerably higher accuracy than the statistic based Song method. However, it can be observed that the processing time for our method is slower than that for the Song method. In order to evaluate the performance of our text segmentation method, the character error rate (CER) is calculated for another test. Here, CER is defined as

$$CER = \frac{N_e}{N} \tag{13}$$

where Ne denotes the number of characters wrongly recognized by the OCR module and N denotes the total number of characters. In this experiment, the proposed text segmentation method is compared with the Otsu [9] thresholding method. The Otsu method is a simple but a classic solution that is employed by many text segmentation schemes. Since this method does not deal with the determination of color polarity, we used our proposed color polarity determination algorithm. The evaluation result of the CER of these two methods is summarized in Table 2. From this table, we can see that our method achieves significantly lower error rates for both English and Chinese characters. This is because our method possesses the capability of handling multilanguage and complex backgrounds by using the stroke filter. Notice that in most cases, the CERs of Chinese characters are larger than that of English characters and this seems to be due to the fact that Chinese characters are structurally more complex than English characters.

Table 1. Performance comparison of two algorithms

Method	Accuracy (%)	Time cost per frame (ms)
Our method	97.12	13.11
Song method	92.33	13.31

Table 2. CER evaluation of two text segmentation methods

Method	English CER (%)	Chinese CER (%)
Our method	9.16	12.35
Otsu method	12.17	52.13

6 Conclusions

A rough-set theoretic approach for color text segmentation was proposed. The proposed algorithm is a variant of the histogram-based segmentation algorithm in which the graph of roughness measure verses intensity values has been used as the basis of segmentation. The computational complexity of the proposed method is slightly more than the conventional histogram-based thresholding algorithm, but the key point of our approach is that, using the roughness index for selection of peaks and valleys results in more realistic values and thus achieves better segmentation results. The experimental results show the superiority of the algorithm. The proposed approach may have many image processing applications and can be easily extended for the segmentation of multispectral images.

References

1. Dimitrova, N., Zhang, H.J., Shahraray, B., Sezan, I., Zakhor, A., Huang, T.: Applications of video content analysis and retrieval. IEEE Multimedia 9(3), 43–55 (2002)
2. Wang, Y., Liu, Y., Huang, J.C.: Multimedia content analysis using both audio and visual clues. IEEE Signal Process Mag. 17(6), 12–36 (2000)
3. Liehart, R., Wernicke, A.: Localizing and segmenting text in images and videos. IEEE Trans. Circuits Syst. Video Technol. 12(4), 256–268 (2002)
4. Cheng, H.D., Jiang, X.H., Sun, Y., Wang: Color image segmentation: Advances and prospects. Pattern Recognition, 2259–2281 (December 2001)
5. Pawlak, Z.: Some Issues on Rough Sets. In: Peters, J.F., Skowron, A., Grzymała-Busse, J.W., Kostek, B.z., Świniarski, R.W., Szczuka, M.S. (eds.) Transactions on Rough Sets I. LNCS, vol. 3100, pp. 1–58. Springer, Heidelberg (2004)
6. Chen, D., Odobez, J., Bourlard, H.: Text detection and recognition in images and video frames. Pattern Recognition 37, 595–608 (2004)
7. Sato, T., Kanade, T., Hughes, E.K., Smith, M.A.: Video OCR for digital news archive. In: Proceedings of the International Conference on Pattern Recognition, pp. 831–834 (2000)
8. Song, J., Cai, M., Lyu, M.R.: A robust statistic method for classifying color polarity of video text. In: Proceedings of the IEEE International Conference on Acoustics, Speech, Signal Processing, pp. 581–584 (2003)
9. Otsu, N.: A threshold selection method from gray-scale histogram. IEEE Trans. Syst. Man Cybern. 9, 62–66 (1979)

Online Path Planning for UAV Navigation Based on Quantum Particle Swarm Optimization

Jinchao Guo, Junjie Wang, and Guangzhao Cui

Dept. of Electric & Information Engineering, Zhengzhou Institute of Light Industry,
Zhengzhou, China
kingjin0923@gmail.com

Abstract. With regard to modern warfare, the environmental information is changing and it's difficult to obtain the global environmental information in advance, so real-time flight route planning capabilities of unmanned aero vehicles (UAV) is required. Quantum Particle Swarm Optimization (QPSO) is introduced to solve this optimization problem. Meanwhile, According to the threats distribution of terrain obstacles, adversarial defense radar sites and unexpected surface-to-air missile (SAM) sites, Surface of Minimum Risk (SMR) is introduced and used to form the searching space. The objective function for the proposed QPSO is to minimizing traveling time and distance, while exceeding a minimum pre-defined turning radius, without collision with any obstacle in the flying workspace. Quadrinomial and quintic polynomials are used to approach the horizon projection of the 3-D route and this simplifies the original problem to a two dimension optimization problem, thus the complexity of the optimization problem is decreased, efficiency is improved. The simulation results show that this method can meet online path planning.

Keywords: QPSO algorithm, Surface of Minimum Risk, polynomial, online path planning.

1 Introduction

During the last decade, Unmanned Air Vehicles (UAVs) have replaced piloted aircraft in a broad band of missions, showing a high potential for further growing, especially due to the avoidance of human risk in dangerous environments. Most researches are motivated by robotic applications [1]. There already exist methods that produce either 2-D, or 3-D trajectories for guiding mobile robots in known, unknown, or partially known environments. In some cases, neural network based controllers were designed and trained to guide robots in multi-agent robot soccer system [2,3]. In other cases, fuzzy based controllers were used to solve the 2-D mobile robot online navigation problem, with its parameters being optimized in real time, through an evolutionary procedure, such as in [4,5]. Heuristic algorithms are a viable candidate to solve NP-hard problems, including path planning problem, effectively and provide feasible solutions within a short time. The majority of promising Heuristic algorithms including: GA (Genetic Algorithm)[6], PSO (Particle Swarm Optimization)[7], ACO

Y. Wu (Ed.): International Conference on WTCS 2009, AISC 116, pp. 291–302.

(Ant Colony Optimization)[8], SA (Simulated Anneal) and their derivations etc, which all have a solid academic and engineering background. Several UAV related papers address path planning using different methods and the associated algorithms involve, for example, Robust control[9], Gain Scheduling[10], potential field technique [11], Virtual Force [12] etc.

Contrary to the ground based navigation, UAV flight wrong decisions and strategies may easily result in disaster. For this reason, the feasibility of the path line is the main concern. Nevertheless, the high velocities and the flight dynamics impose specific constraints for the smoothness and the curvature of the calculated path line. Different curve smoothing methods are adopted to optimize flying trajectory, such as B-Spline curves based [13], Bezier curve based [14] and polynomial curve based method [15]. When calculation speed, flight dynamics are taken into account, herein, the Quadrinomial and quintic polynomials formulation are adopted.

Because of UAVs' limited sensing ability, gradually constructed polynomial curves are adopted in the online planner. The radar provides data needed for calculating smooth near optimal curves connected to each other, which are naturally fitted to the UAV flight limitations. This work has been motivated by the challenge to implement a dynamic path planning for autonomous UAV navigation and collision avoidance in partially known 3-D rough terrain environments. The problem being solved considers the threats distribution of terrain obstacles, adversarial defense radar sites and unexpected surface-to-air missile (SAM) sites. And Surface of Minimum Risk (SMR) is introduced and used to form the searching space. UAVs are assumed to be equipped with a set of on-board sensors, including radar, global positioning system (GPS), inertial navigation system (INS), and gyroscopes, through which it can sense its surroundings and position. The final destination is known and UAVs must follow an as smooth as possible trajectory (imitating real flight restrictions), planned and re-planned in real-time, given its initial position and initial flight direction. The vehicle is assumed to be a point (its actual size is taken into account by SMR).

The rest of the paper is organized as follows: Section 1 summarizes QPSO fundamentals along with the Point-to-Point Trajectory Representation and cost function design. Scheme of the online planner is presented in Section 2, Simulation results are shown in Section 3, followed by discussion and conclusions in Section 4.

2   Problem Description

In this paper, Quantum Particle Swarm Optimization (QPSO) algorithm was used to construct the local flying trajectory of UAVs based on local information obtained from onboard sensors, so QPSO is introduced firstly. The process of cost function construction is displayed as well. Considering the calculation speed and UAVs' flying dynamics, a quadrinomial and a quintic polynomial are used to model the segments of the trajectory.

2.1   Introduction of QPSO

Particle swarm optimization (PSO) algorithm is developed by Kennedy and Eberhart in the late 90s [16]. Many researchers focused on PSO and had made many modifications on PSO to improve its performance. Inspired by the analysis of convergence of PSO,

Professor Sun et al. studied the individual particle moving in a quantum style and proposed the Quantum Particle Swarm Optimization (QPSO) algorithm [17,18]. In QPSO, particles update its position according to the following formulations:

$$Mbest = \frac{1}{N}\sum_{i=1}^{N} P_{ibest}, \quad (i=1,\cdots N) \tag{1}$$

$$p = r \times P_{ibest} + (1-r) \times p_{gbest}, \quad (i=1,\cdots N) \tag{2}$$

$$x_i(t+1) = p \pm \alpha \bullet \left| Mbest - x_i(t) \right| \bullet \ln(\frac{1}{u}) \tag{3}$$

Where N is the number of particles; D is the problem space dimension; r and u are random scalar; M_{best} is the average position of single best and p is the position between P_{ibest} and P_{gbest}; α is a design parameter called contraction-expansion coefficient [19]. The contraction-expansion coefficient can be written as:

$$\alpha = \alpha_{min} + (\alpha_{max} - \alpha_{min}) \times (Gen_{max} - T)/Gen_{max} \tag{4}$$

α is the only one parameter in QPSO. It determines the convergence speed. Hence, QPSO is simpler and less time-consuming compared with GA and other heuristic algorithms. In this work, chaotic sequences instead of random sequences are used in QPSO to diversify the initial position of QPSO population and improve the QPSO's performance [20]. Due to the non-repetition of chaos, it can carry out overall searches at higher speeds than stochastic ergodic searches that depend on probabili-ties. Due to the ergodic and dynamic properties of Zasl-avskii's map, the QPSO approach chooses chaotic sequ-ences based on Zaslavskii's map to realize initialization. The Zaslavskii's map has the formulation as:

$$\begin{cases} y(t) = \mathrm{mod}\left[y(t-1) + v + az(t), 1 \right] \\ z(t) = \cos(2\pi y(t-1)) + e^{-r} z(t-1) \end{cases} \tag{5}$$

Where mod is the modulus after division, Lyapunov exponent for v = 400, r = 3 and a = 12.6695 are used to attain the attractor, as shown in Fig. 1.

2.2   Point-to-Point Trajectory Representation

According to point-to-point trajectory method, a trajectory consists of several segments with continuous acceleration at the intermediate via point, as shown in Fig. 2.

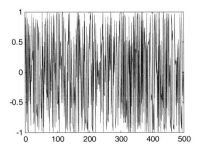

Fig. 1. Attractor of Zaslavskii's map for v = 400, r = 3 and a = 12.6695

294   J. Guo, J. Wang, and G. Cui

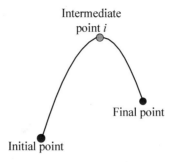

Fig. 2. Intermediate points on the point-to-point path

The position of each intermediate point is supposed to be optimized according to the environment. This is useful especially when there is an obstacle in the working area. Of course, the intermediate points can also be given as particular points that should be passed through.

Here, we adopt the method proposed by S.G. Yue et al. A quadrinomial and a quintic polynomial are used to model the segments of the trajectory [21]. Let us assume there are m_p intermediate via points between the initial and the final points. Between the initial points to m_p intermediate via points, a quadrinomial is used to describe these segments as:

$$p_{i,i+1} = a_{i0} + a_{i1}t_i + a_{i2}t_i^2 + a_{i3}t_i^3 + a_{i4}t_i^4, (i = 0, \cdots, m_p - 1) \tag{6}$$

Where a_{i0} to a_{i4} are constant coefficients and must be conformed to the following constrains:

$$p_i = a_{i0} \tag{7}$$
$$p_{i+1} = a_{i0} + a_{i1}T_i + a_{i2}T_i^2 + a_{i3}T_i^3 + a_{i4}T_i^4 \tag{8}$$
$$\dot{p}_i = a_{i1} \tag{9}$$
$$\dot{p}_{i+1} = a_{i1} + 2a_{i2}T_i + 3a_{i3}T_i^2 + 4a_{i4}T_i^3 \tag{10}$$
$$\ddot{p}_i = 2a_{i2} \tag{11}$$

Where T_i is the execution time from point i to the neighbour intermediate point. The five unknown constant coefficients can be solved as:

$$a_{i0} = p_i \tag{12}$$
$$a_{i1} = \dot{p}_i \tag{13}$$
$$a_{i2} = \ddot{p}_i/2 \tag{14}$$
$$ai3 = (4p_{i+1} - \dot{p}_{i+1}T_i - 4p_i - 3\ddot{p}_iT_i^2)/T_i^3 \tag{15}$$
$$a_{i4} = (\dot{p}_{i+1}T_i - 3p_{i+1} + 3p_i + 2\dot{p}_iT_i + \ddot{p}_iT_i^2/2)/T_i^4 \tag{16}$$

The intermediate point $(i+1)$'s acceleration can be obtained as:

$$\ddot{p}_{i+1} = 2a_{i2} + 6a_{i3}T_i + 12a_{i4}T_i^2 \tag{17}$$

The segment between the intermediate point and the final point can be represented as a quintic polynomial:

$$p_{i,i+1} = b_{i0} + b_{i1}t_i + b_{i2}t_i^2 + b_{i3}t_i^3 + b_{i4}t_i^4 + b_{i5}t_i^5, (i = m_p) \tag{18}$$

Just like a_{i0} to a_{i4}, b_{i0} to b_{i5} can be calculated by the same method.

$$b_{i0} = p_i \tag{19}$$

$$b_{i1} = \dot{p}_i \tag{20}$$

$$b_{i2} = \ddot{p}_i / 2 \tag{21}$$

$$b_{i3} = (20 p_{i+1} - 20 p_i - (8\dot{p}_{i+1} + 12\dot{p}_i)T_i - (3\ddot{p}_i - \ddot{p}_{i+1})T_i^2) / 2T_i^3 \tag{22}$$

$$b_{i4} = (30(p_i - p_{i+1}) + (14\dot{p}_{i+1} + 16\dot{p}_i)T_i + (3\ddot{p}_i - 2\ddot{p}_{i+1})T_i^2) / 2T_i^4 \tag{23}$$

$$b_{i5} = (12 p_{i+1} - 12 p_i - (6\dot{p}_{i+1} + 6\dot{p}_i)T_i - (\ddot{p}_i - \ddot{p}_{i+1})T_i^2) / 2T_i^5 \tag{24}$$

As formulated above, the parameters to be determined are the planar coordination of each intermediate via point ($n \times m_p$ parameters), the velocities of each intermediate point ($n \times m_p$ parameters), the execution time for each segment (m_p+1 parameters), and the posture of the final configuration (n-m).

Considering the calculation expense, only one intermediate via point is selected to construct the trajectory. In this case, there are five parameters should be optimized, which means the solution space are five dimensions.

2.3   Design of the Cost Function

The cost function is an important component of QPSO, which is the so-called object function. In this work, the cost function consists of three indices. All indices are translated into penalty functions to be minimized. Each index is computed individually and is integrated in the cost function evaluation. The fitness function f_f adopted for evaluating the candidate trajectories in flying workspace is defined as:

$$f_f = \beta_1 f_1 + \beta_2 f_2 + \beta_3 f_3 \tag{25}$$

Where β_i ($i = 1,.., 3$) are the weighting factors and they are subjected to the following relationship:

$$\sum_{i=1}^{3} \beta_i = 1 \tag{26}$$

The f_1 index represents the length of the local segment used to provide shorter paths:

$$f_1 = \sum_{i=1}^{n} \sqrt{(x_{i+1} - x_i)^2 + (y_{i+1} - y_i)^2 + (z_{i+1} - z_i)^2} \tag{27}$$

The index f_2 is designed to provide curves with a prescribed minimum curvature radius. This characteristic is essential for a flying vessel. The angle that is determined by two successive discrete segments of the curve (defined by the dots in Fig. 3) is calculated and if less than a prescribed value, a penalty is added to the fourth term of the fitness function.

Fig. 3. Angle between two successive discrete segments

The value of index f_3 depends on the value of a potential between the starting point and the final target. The potential field between the two points is the main driving force for the gradual development of the path line in the online procedure. The potential is similar to the one between a source and a sink, given as :

$$f_3 = \ln\left(\frac{r_2 + c * r_0}{r_1 + c * r_0}\right) \tag{28}$$

Where r_1 is the distance between the last point of the current curve and the starting point, r_2 is the distance between the last point of the current curve and the final destination, r_0 is the distance between the starting and final destination and is a constant and c is a given constant.

A visualization of the corresponding potential field is demonstrated in Fig. 4.

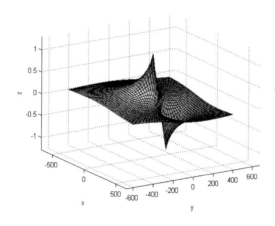

Fig. 4. Potential field of Eq. 28

The minimization of Eq. (25), through the QPSO procedure, results in a set of polynomial control points, which actually represent the desired path.

The pseudo code of QPSO algorithms is presented here.

1)   Construct solution space in the local environment, the horizontal coordinate x is selected manually and the yaw coordinate y is determined by the optimizing process;

2)   Initialize QPSO population, position, ending condition, iteration step and other parameters;

3)   Calculate the coordinate value of the intermediate point;

4)   Calculate the fitness value according to Eq. (25);

5)   Update particle position and speed according to Eq. (1)-(3) and outputting the best solution g_{best};

6)   The end? If so, stopping, otherwise, return back to step 3).

3   Scheme of the Online Route Planning

As previously mentioned, all ground nodes are initially considered unknown. The corresponding algorithm, simulates the radar and checks whether the ground nodes within the radar range are "visible" or not and consequently "known" or not. As the UAV is moving along this segment and until it has traveled about 2/3 of its length, the radar scans the surrounding area, returning a new set of environment information data. The online planner, then, produces a new segment, whose first point is the last point of the previous one and whose last point lies somewhere in the newly scanned area, its position being determined by the QPSO online procedure. The process is repeated until the ending point of the current path line segment lies close to the final destination (Fig. 5). The position at which the algorithm starts to generate the next path line segment (here taken as the 2/3 of the segment length) depends on the radar range, the UAVs velocity and the algorithm computational demands.

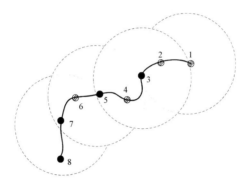

Fig. 5. The sketch of online planning--the curved solid line is the generated path line, formed by 4 successive polynomial segments (1–3, 3–5, 5–7, 7–8). Point 1 is the starting position, while point 8 is the ending one. The dashed circles define the radar-covered area in the position where the QPSO procedure starts to calculate the next polynomial segment. The corresponding circle centers are marked with white dots (points 1, 3, 5, 7)).

The system flowchart is presented in Fig.6.

4   Experimental Results

The online planner has been extensively tested in MATLAB^{TM}. All experiments have been designed in order to search for near optimal path in unknown environment. Before the simulation, the flying space should be defined firstly and the grid map should be constructed accordingly, as shown in Fig.7. As previously mentioned, the final destination is known.

During simulation, two different maps are constructed and two SMRs are calculated respectively. Based on the environmental information obtained by onboard sensors, the path can be optimized by QPSO effectively,   which can be seen from the

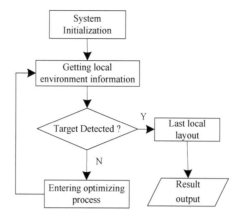

Fig. 6. Flow chart of route planning system

simulation results. The weighting factors β_i ($i = 1,..,3$) are experimentally determined. Based on the two different maps, different weighting factor sets are tested. Case studies show QPSO has good computation precision and distinct fast computation speed compared with GA. Different weighting factors mean different consequence, which also are validated by the results.

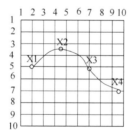

Fig. 7. Plane figure of local map planning

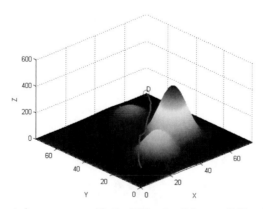

Fig. 8. Route generated on map one with $\alpha_1 = 0.38$, $\alpha_2 = 0.1$, $\alpha_3 = 0.52$, with source (10,10,5) to destination(70,70,5)

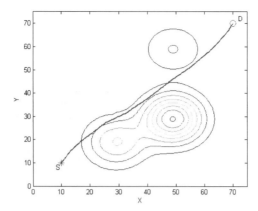

Fig. 9. The contour map corresponding to Fig. 8

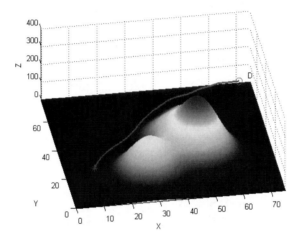

Fig. 10. Route generated on map two with α_1 =0.38, α_2 =0.1, α_3 =0.52, with source (10,20,5) to destination(70,70,5)

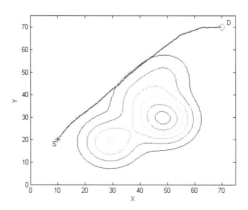

Fig. 11. The contour map corresponding to Fig. 10

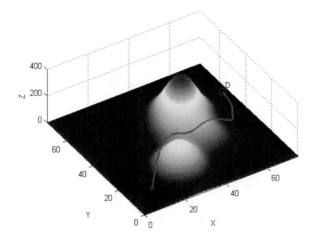

Fig. 12. Route generated on map two with α_1 =0.4, α_2 =0.1, α_3 =0.5, with source (10,10,5) to destination(70,70,5)

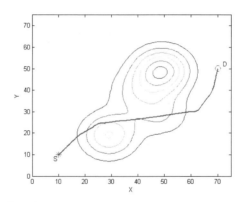

Fig. 13. The contour map corresponding to Fig. 12

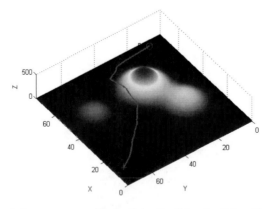

Fig. 14. Route generated on map one with α_1 =0.4, α_2 =0.1, α_3 =0.5, with source (10,70,5) to destination(70,10,5)

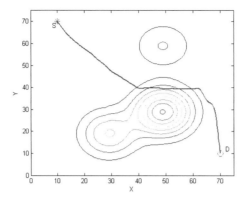

Fig. 15. The contour map corresponding to Fig. 14

5 Conclusions

In the above sections, the problem of point-to-point trajectory planning was studied in detail. Trajectory planning method based on QPSO to minimize traveling length and time expense considering other limitations are presented. Quadrinomial and quintic polynomials are used to describe the segments that connect the initial, intermediate, and final points in UAVs' flying space. Fig. 8, 10, 12, 14 are simulation results under matlab platform and fig. 9, 11, 13, 15 are their corresponding contour map. The results validate the efficiency of the online planning scheme.

Acknowledgments. The authors would like to thank Prof Guangzhao Cui and Doc. Yanfeng Wang who work in Key Lab of Information-based Electrical Appliances of Henan Province. Thanks for their kindness and help in instrument provision. Many experiments are executed in the key lab. This work is supported by the Natural Science Foundation of Henan Province of China under grant No. 0923004-10036 and the doctoral research fund of ZZULI.

References

1. Latombe, J.C.: Robot motion planning. Kluwer Academic Publishers (1991)
2. Jolly, K.G., Ravindran, K.P., Vijayakumar, R., Sreerama Kumar, R.: Intelligent decision making in multi-agent robot soccer system through compounded artificial neural networks. International Journal of Robotics and Autonomous Systems 55(7), 589–596 (2007)
3. Ghatee, M., Mohades, A.: Motion planning in order to optimize the length and clearance applying a Hopfield neural network. Expert Systems with Applications 36(3), 4688–4695 (2009)
4. Vadakkepat, P., Penga, X., Queka, B.K., Leea, T.H.: Evolution of fuzzy behaviours for multi-robotic system. International Journal of Robotics and Autonomous System 55(2), 146–161 (2007)
5. Zamirian, M., Kamyad, A.V., Farahi, M.H.: A novel algorithm for solving optimal path planning problems based on parametrization method and fuzzy aggregation. Physics Letters A 373(38), 3439–3449 (2009)

6. Darintsev, O.V., Migranov, A.B.: Genetic algorithms-based path planning system for the motion of a group of mobile micro-robots. Journal of Computer and Systems Sciences International 6, 493–502 (2007)
7. Adham, A., Phon-Amnuaisuk, S., Kuan-Ho, C.: Navigating a robotic swarm in an uncharted 2D landscape. Applied Soft Computing 10(1), 149–169 (2010)
8. Duan, H.B., Yu, Y.X., Zhang, X.Y., Shao, S.: Three-dimension path planning for UCAV using hybrid meta-heuristic ACO-DE algorithm. Simulation Modelling Practice and Theory 18(8), 1104–1115 (2010)
9. Avanzini, G., Ciniglio, U., de Matteis, G.: Full-Envelope Robust Control of a Shrouded-Fan Unmanned Vehicle. Journal of Guidance, Control and Dynamics 29(2), 435–443 (2006)
10. Fujimori, A., Terui, F., Peter, N.: Flight Control Design of an Unmanned Space Vehicle Using Gain Scheduling. Journal of Guidance, Control, and Dynamics 28(1), 96–105 (2005)
11. Paul, T., et al.: Modelling of UAV formation flight using 3D potential field. Simulation Modelling Practice and Theory 16, 1453–1462 (2008)
12. Dong, Z.N., Zhang, R.L., Chen, Z.J., Zhou, R.: Study on UAV Path Planning Approach Based on Fuzzy Virtual Force. Chinese Journal of Aeronautics 23(3), 341–350 (2010)
13. Nikolos, I.K., Valavanis, K.P.: Evolutionary Algorithm Based Offline/Online Path Planner for UAV Navigation. IEEE Transactions on Systems, Man, And Cybernetics—Part B: Cybernetics 33(6), 898–912 (2003)
14. Jolly, K.G., Sreerama Kumar, R., Vijayakumar, R.: A Bezier curve based path planning in a multi-agent robot soccer system without violating the acceleration limits. International Journal of Robotics and Autonomous Systems 57, 23–33 (2009)
15. Kazem, B.I., Mahdi, A.I., Oudah, A.T.: Motion Planning for a Robot Arm by Using Genetic Algorithm. Jordan Journal of Mechanical and Industrial Engineering 2(3), 131–136 (2008)
16. Kennedy, J., Eberhart, R.C.: Particle swarm optimization. In: Proceedings of IEEE International Conference on Neural Networks, vol. 4, pp. 1942–1948 (1995)
17. Sun, J., Feng, B., Xu, W.B.: Particle swarm optimization with particles having quantum behaviork. In: Proceedings of, Congress on Evolutionary Computation, USA, pp. 325–331 (2004)
18. Xi, M., Sun, J., Xu, W.: Quantum-Behaved Particle Swarm Optimization for designing H infinity structured specified controllers. In: DCABES 2006 Proceedings, vol. 2, pp. 42–46. Shanghai University Press, Shanghai (2006)
19. Sun, J., Xu, W., Feng, B.: Adaptive parameter control for quantum-behaved particle swarm optimization on individual level. In: Proceedings of IEEE International Conference on Systems, Man and Cybernetics, Big Island (HI, USA), pp. 3049–3054 (2005)
20. Coelho, L.D.S.: A quantum particle swarm optimizer with chaotic mutation operator. Chaos, Solitons and Fractals 37, 1409–1418 (2008)
21. Yue, S.G., Henrich, D., Xu, X.L., Toss, S.K.: Point-to-Point Trajectory Planning of Flexible Redundant Robot Manipulators Using Genetic Algorithms. Journal of Robotica 20, 269–280 (2002)

Research and Practice of Teaching Reform of Applied Compound Talents Cultivation System of Environmental Management of the Local University
A Case Study of Wenzhou University

Qi Wang and Jun Li

School of Life and Environmental Sciences, Wenzhou University
Wenzhou, China
wmkfb7897@sina.com

Abstract. Based on the analysis of the market's demand for environmental management talents, the paper explores the cultivation target, curriculum system, educational reform means to improve talents' knowledge and ability for a teaching reform case study of Wenzhou University. The enthusiasm, independency and ability of students were improved through reform. Significant good effects have been obtained in our reform exploration of environmental management.

Keywords: applied compound talents cultivation system, practical teaching curriculum system, environmental management, Wenzhou university, teaching reform, teaching test management system.

1 Introduction

Talents in environmental management allow them to apply for positions in businesses such as wildlife conservation, natural resources, animal population, management positions within those areas, and various policies, and public relations positions. Environmental management has characteristics of humanities, social sciences and applied strong sciences. It is necessary to explore the environmental science and management interdisciplinary teaching and learning reform of environmental management.

Raymond, C. M. et al evaluated the processes and mechanisms available for integrating different types of knowledge for environmental management[1]. Sanchez, E. and Craig, R. proposed teaching effectiveness was increased through collaborative learning activities[2]. The huge demands for environmental management talents bring the continual expansion in the scale of higher education of environmental management. However, during this process, problems appeared that environmental management talents are incapable and relatively surplus, which result from the imprecise positioning of environmental management education's training target and the unscientific setting of courses that lead to the students' failure of acknowledge, ability and quality structure to meet the demands of the market. Based on the analysis

Y. Wu (Ed.): International Conference on WTCS 2009, AISC 116, pp. 303–309.

of the market's demand for environmental management talents, this paper explores the training target of the cultivation target, curriculum systems, educational reform means to improve talents' knowledge and ability for a teaching reform case study of Wenzhou University.

Wenzhou University, which was founded by merging Wenzhou Normal College with Wenzhou University (former) with the approval of the ministry of education. Its main campus is situated in Wenzhou Higher Education Zone. It has a history of 75 years and is a type local University.

2   Applied Compound Talents Cultivation System

The aim of environmental management talents is to cultivate applied compound talents with solid scientific theory and technology, management skills, spirits of humanity. The cultivation system mainly include cultivation target, curriculum system, educational reform means in this study.

2.1   Cultivation Target

To cultivate excellent applied compound talents of environmental economics, firstly in importance, reasonable cultivation target for the local university should be set up. It is important to bring up talent of environmental management through adjusting the training target. It has been register system of environmental protection in China. It is the first issue for running higher education that the cultivation target of the discipline and the criterion. There is an important practical significance for directing the discipline construction and development to study the cultivation target and criterion of environmental economics education discipline seriously, to understand the meanings exactly and to master the basic requirements.

2.2   Curriculum System

For further cultivating the high quality talents with good foundation, wide-range knowledge, strong ability, high quality which has the innovative spirit, avoiding the question of seriously apartment of theory and practice, innovative curriculum system was carried out in the teaching of environmental management.

In the teaching process, we had many lessons through the whole process of teaching and training students to combine theory and practice capabilities to ask colleges students focusing on combining theory and practice, teaching and research combining local problems of environmental management, curricular and extracurricular practices to improve the analysis problem-solving ability and ability to work of them.

In the environmental management teaching contents, set the case topics, in the teaching process, we should pay attention to local features, contact Wenzhou and Zhejiang, the reality of the course. We had many lessons to try to increase the number of cases, the latest research results and applications in the process of teaching reform of environmental management. Environmental cases specifically set up four special cases, such as the coordination of environmental and economic research, we chaired two major tenders Wenzhou circular economy and conservation, quantitative research and countermeasures social aspects, student interest is high. Massoud, M. A. et al assessed the factors influencing the implementation of ISO 14001 Environmental Management System in developing countries taking the Food Industry in Lebanon as a case example[3]. We may take the case for teaching process. The use of student ratings of instructional quality is enhanced by an understanding of the nature of the underlying dimensions[4].

The practical teaching curriculum system of environmental management was shown in the Figure 1.

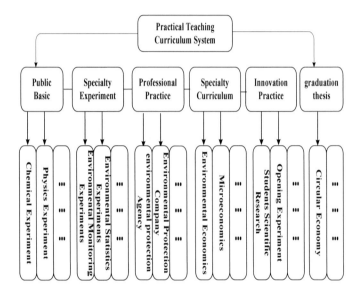

Fig. 1. Practical teaching curriculum system of environmental management

2.3   Educational Reform Means

To improving the course assessment system, teaching effectiveness evaluation should focus on the evaluation of the teaching process, and we should pay attention to the scientific evaluation of teaching content, integrity and teaching links convergence, level of the rate of students to classes, job completion rates, pass rates and test scores.

We adopted appropriate self-learning to participate in teaching, such as environmental management policies analysis which can be called classroom teaching by student-led.

We may make full use of computers in the teaching of environmental management teaching reform. Computer-aided programs such as SPSS software analysis, Microsoft Excel software and manual checking were mixture performed for the analysis of the questionnaire of environmental economics.

We adopted the real project as platform to develop the path way of demonstration in the practice of enterprise education of environmental protection service and taken some measures to local individuals and effectively plan the orientations of enterprise of environmental protection service of undergraduates.

Entrepreneurship education of colleges was built such as establishing the enterprise guiding studio, the training system of enterprise consciousness of environmental protection service and building the training system and entrepreneurial platform system and entrepreneurship policy of enterprise education of environmental protection, etc.

Teaching Test Management System (TTMS) is a Chinese version. The Chinese Interface of TTMS was presented in the Figure 2.

Fig. 2. The Chinese Interface of Teaching Test Management System

TTMS software based on access database by Qi Wang designed has many functions including marking examination papers required, on-line examination, the objective questions automatically corrected, as well as course management and examination management according to the template ideas for the teacher. This enriched test management, improved efficiency and brought great convenience.

Case-based Teaching is an important mean in environmental management teaching process. Case-based teaching is a flexible teaching model. Environmental Kuznets Curves have become standard features in environmental policy which are important contents in environmental management teaching. Fodha, M. and Zaghdoud, O. investigated the relationship between economic growth and pollutant emissions for a small and open developing country, Tunisia[5]. Therefore, we arranged student work of EKC of Wenzhou. The EKC for a case of student work was in Figure 3.

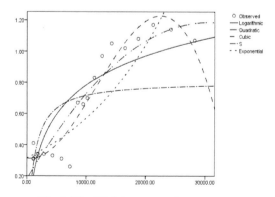

Fig. 3. The EKC for a case of student work

3   Educational Reform Achievements

3.1   Employment

The University's online employment system helps college students find the best fit for career at the University. Educated talents focus on development prospects, income and welfare benefits, as well as the working environment. Through the reform, employment rate improved 40%.

3.2   Entrepreneurship

Entrepreneurship is a major source of employment, economic growth, and innovation, promoting product and service quality, competition, and economic flexibility[6]. The five students in Wenzhou University established their new carbon professional corporation which has the software production. Further, the software for carbon footprint to compute various enterprises was made. Moreover, the group by guide of the authors attended the project of the "Challenge Cup" in Zhejiang, and this resulted in improving the quality of the science and technology and promoting abilities of creativity and enterprise of university's students. Therefore, the reform made a good effect. The Chinese interface of carbon footprint computation for a case of home version was in Figure 4.

Fig. 4. The Chinese interface of carbon footprint compution for a case of home version

3.3   Awards

The authors were get best guidance teachers for university students competition works in Zhejiang. We Guide a student in Zhejiang Province award "Extracurricular academic and scientific works Challenge Cup by 2005" of the first prize, and also the first prize in Wenzhou University. We Guide six students award entrepreneurial Challenge Cup in 2010 the first prize and also first prize for entrepreneurial Challenge Cup in Wenzhou University, and we guided three students award the second prize in energy saving competition in Wenzhou in 2010, and guided some students to publish three papers.

The Chinese award certificate of the project of the "Challenge Cup" in Zhejiang was in Figure 5.

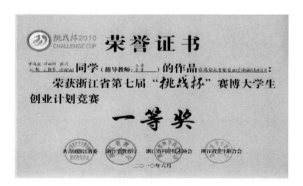

Fig. 5. The Chinese award certificate of the project of the "Challenge Cup" in Zhejiang

3.4   Teaching Effectiveness

Gross, J. et al proposed students' evaluations were strongly influenced by their personal tastes regarding teaching[7]. Therefore, personal tastes in teaching combined with local environmental management practices were focused. We guided student cleaner production audit in Wenzhou to give enterprises the economic, environmental and social benefits. By the reform, we enhanced teaching effectiveness and quality.

4   Conclusion

Environmental management teaching reform aims to make education better for students. This means students should be part of the process. Learn something new every year about the intersection of technology and teaching. In order to ensure that the reform was done successfully, the paper explores some practices of college students, and provides some case studies for the students. After the teaching reform, practical abilities and interests of students are obviously improved. The reform has achieved great successes in research and practice of environmental management teach.

Acknowledgment. This work was supported by the Project of Innovation Experimental Zone for Enterprise Education Talent Cultivation Mode of Wenzhou in the Ministry of Education of China and the Enterprise Education Reform based on Environmental Protection Service Industry of Wenzhou University under Grant 2009JG03.

References

1. Raymond, C.M., Fazey, I., Reed, M.S., Stringer, L.C., Robinson, G.M., Evely, A.C.: Integrating local and scientific knowledge for environmental management. Journal of Environmental Management, London 91, 1766–1777 (2010)
2. Sanchez, E., Craig, R.: Increasing teaching effectiveness through collaborative learning activities. Hortsciene, Alexandria 41, 974–974 (2006)
3. Massoud, M.A., Fayad, R., El-Fadel, M., Kamleh, R.: Drivers, barriers and incentives to implementing environmental management systems in the food industry: A case of Lebanon. Journal of Cleaner Production, Oxford 18, 200–209 (2010)
4. Jackson, D.L., Teal, C.R., Raines, S.J., Nansel, T.R., Force, R.C., Burdsal, C.A.: The dimensions of students' perceptions of teaching effectiveness. Educational and Psychological Measurement, Thousand Oaks 59, 580–596 (1999)
5. Fodha, M., Zaghdoud, O.: Economic growth and pollutant emissions in Tunisia: An empirical analysis of the environmental Kuznets curve. Energy Policy, Oxford 38, 1150–1156 (2010)
6. Hisrich, R., Langan-Fox, J., Grant, S.: Entrepreneurship research and practice - A call to action for psychology. American Psychologist, Washington 62, 575–589 (2007)
7. Gross, J., Lakey, B., Edinger, K., Orehek, E., Heffron, D.: "Person Perception in the College Classroom: Accounting for Taste in Students' Evaluations of Teaching Effectiveness. Journal of Applied Social Psychology, Malden 39, 1609–1638 (2009)

Teaching Reform in the Course of Communication Principles

Fangni Chen

Department of Communication and Electronic Engineering
Zhejiang University of Science and Technology
Hangzhou, China
chaofang333@sohu.com

Abstract. Aiming at improving student learning ability and accelerating the teaching reform, this paper research the difficulty problems in teaching of Communication Principles. Combining with the experiences in our teaching practice on the courses, we put forward some novel opinions of teaching reform involving the selection of textbook, methods of teaching, arrangements of class structure and so on. Teaching practice shows that the learning interest of students and the teaching effects are improved greatly after the teaching reform.

Keywords: Teaching reform, Communication Principles.

1   Introduction

Information transmission is the essence of communication to meet the demands of individuals and society. Communication Principles is one of the most important basis courses of communication engineering, electronic information engineering, automatic control and other similar specialties. This course researches basic concepts, basic principles and basic analysis methods of communication systems and signals [1]. Communication Principles combines theory, engineering with practicality in one course, includes numerous contents, abstract concepts and complicated mathematical reasoning and so on, so it is difficult for teachers to teach and students to learn [2]. With the rapid development of communication technologies, how to inspire the students' interest, how to let the students to understand and grasp the basic theory of communication and how to cultivate the students' analytical and design ability are most important.

In order to accelerate the teaching reform and promote the curriculum development, in this paper some opinions are proposed on the selection of textbook, methods of teaching, arrangements of class structure and so on. Teaching practice shows that the students' initiative and their interest are inspired, desirable teaching effects and students' high praises have been gained.

Y. Wu (Ed.): International Conference on WTCS 2009, AISC 116, pp. 311–316.

2   The Teaching Status of Communication Principles

During our practical teaching on Communication Principles course, we have found that some problems haven't been resolved well, such as conflict of plenty of teaching contents and fewer lessons, some students have weak foundations of the former courses, some students lack of confidence in Communication Principles and so on.

Conflict of Plenty of Teaching Contents and Fewer Lessons

Communication Principles course has plenty of contents which includes communication system model, random signal analysis, channel, digital signal transmission, analog modulation and digital modulation, the best receiver and error control theory or technique, etc.

At the same time as a course combined theory with experiment, it still needs a lot of practice and experiment class hours to help the students strengthen understanding the knowledge point in Communication Principles. At present most universities and colleges exit teaching hours compression phenomenon, so the conflict of plenty of teaching contents and fewer lessons is manifested. How to finish the task of teaching and meanwhile achieved good teaching effect becomes one of the difficulties in teaching this course.

Some Students Have Weak Foundations of the Former Courses

Communication Principles always opened after the curriculums of Higher Mathematics, Electric Circuit Principles, Low Frequency Circuits, Signals and Systems and so on. In another word, these curriculums are the foundation of Communication Principles. Teachers and students of electronic information specialty all know these curriculums are not easy to learn. So in the teaching process of Communication Principles, students always forget the knowledge points or they haven't grasp the knowledge points of these former curriculums yet. Teachers should spend corresponding time to review the knowledge points and that must affect the teaching plan.

Some Students Are Lack of Confidence in Communication Principles

Communication Principles is a strong theory curriculum including many technical terms and concepts, abstract and difficult contents. Teachers must do complex mathematical deduction for make the principles clear enough for students, and that may increases comprehension difficulties. Moreover teaching effect is not working well over the years, high-grade students' horrible description increase the psychological burden of low-grade students so that students lose their interest in this course. Although the students try to master Communication Principles, they will lose their confidence and give up study when they meet some difficulties which they can't overcome easily.

3   Teaching Reform Methods and Results

3.1   Carefully Select the Appropriate Materials

The corresponding teaching contents should be updated with the development of communication technology. Teachers should select materials which can be able to catch up with the development of the current communication technology. For example, after multiple comparison we chose Bernard Sklar's "Digital Communications Fundamentals and Applications, Second Edition" [3] as the textbook, which can satisfy this requirement. At the same time to meet our school's Undergraduate training objectives, we selected some of the important chapters to teach, the main teaching contents are singals and spectra, formatting and baseband modulation, baseband demodulation/detection, bandpass modulation and demodulation/detection, synchronization. The key points of the teaching contents are baseband and bandpass modulation and demodulation. Meanwhile we arrange three times exercise lessons, students can grasp the knowledge point better by the explanation and expansion of the homework.

As students' former courses foundation is weak, the knowledge of Higher Mathematics, Electric Circuit Principles, Signals and Systems courses is not solid, teachers may briefly deduce the theoretical analysis and focus on the principles and conclusions in the class. In this way, students can get out from the complex process of deduction and they may feel the degree of difficulty has been descended. On the other hand it will save a lot of deduction time for teachers and improve the effective utilization of teaching hours.

3.2   Always Emphasize the Concept of Systematic in Process of Teaching

According to the teaching practice through recent years, the point of view which emphasize on analysis of information process module makes the students lack of systems engineering concepts and feel difficult to learn Communication Principles. Therefore, we change that teaching point and always emphasize the systematic concept in process of teaching. In the first class of Communication Principles, we show a typical communication system block diagram as shown in Fig.1.

It will help students establish a communication system concept. Teachers briefly explain the function of every block (unit) and point out the teaching range for students. When teaching course precedes to each chapter i.e. each module, be sure to seize the four-level teaching theory of system - modules - circuit - signal. Guide the students to each module from the system and then to the specific circuit and signal analysis step by step. First introduce the unit module' applications, the location in system, then illustrate the function of the unit module, and then explore the specific circuit, and analyze the signal generation and transmission, and guide students to think about how the basic circuit can meet the system requirements derived from practical circuits. Through this teaching method, students can clearly

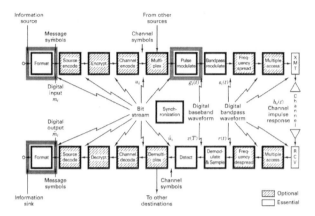

Fig. 1. Typical communication system block diagram

see the complete picture of communication system; get a better understanding of the various sections of communication courses and the connection of them. In the learning process students may summarize and compare the knowledge point more consciously and gradually build the system concept.

3.3   Reform Teaching Methods

With the popularity of multimedia teaching, most colleges and universities now use multimedia teaching methods. Communication Principles should adopt this approach too. The main benefits are as follows: Firstly, pre-produced good courseware can save plenty of class time for teachers writing or drawing on the blackboard. It can expand teaching capacity, accelerate teaching rhythm and facilitate comparison of class contents. Secondly, by adding the super link, the maps, text, voice, animation, simulation software [4] and other forms of embedded multimedia into the courseware, the abstract contents is visualization and vivid and make it easy to understand and accept for students.

For example, digital modulation is the key content of Communication Principles, and the concept is so abstract that the students are difficult to understand. Teachers can add the super link in the courseware, such as using MATLAB software to simulate the communication system with different digital modulation modes and it can easily plot the error performance figures as shown in Fig.2. Through this simulation demonstration, students can clearly see the different error performance of BPSK, QPSK, DPSK and QAM modulation and the evolution as the M-ary number increased. Further more it can motivate the learning interest of students and improve teaching effects.

As the multimedia teaching may proceeds fast than traditional blackboard teaching, it's not easy for students to focus on the key point and to get a deep impression of the lessons. So teachers should pay attention to the integration with the traditional blackboard teaching when utilizing multimedia teaching methods. Writing the outline contents of each lesson, the important circuit structure, formula and conclusions on the blackboard for students may help them a lot.

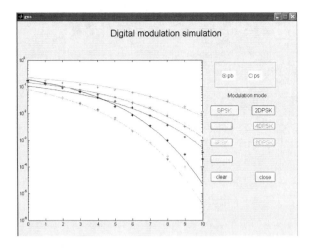

Fig. 2. Performance simulation of digital modulation using MATLAB

3.4   Built a Network Multimedia Teaching Platform

Utilizing the on-line resources and carrying out the network teaching may extend teaching time and space to make up for the in-class lessons shortage. For example teachers can built a network multimedia teaching platform, teaching notice, courseware, homework submission and correction, self-tests, tutorials and other teaching activities can be realized through the network platform. It can greatly promote communications between teachers and students and enhance autonomous learning impetus of students. Also it will be a powerful help and promotion for in-class teaching.

4   Conclusions

In order to accelerate the teaching reform and promote the curriculum development, in this paper some opinions are proposed on the selection of textbook, methods of teaching, arrangements of class structure and so on. Teachers should select materials which can be able to catch up with the development of the current communication technology, always emphasize the systematic concept in process of teaching, be sure to seize the four-level teaching theory of system - modules - circuit – signal, use multimedia teaching methods integrated with the traditional blackboard teaching. The students' initiative and their interest are inspired, desirable teaching effects and students' high praises have been gained in the practical teaching process.

Acknowledgment. This paper was supported by Scientific Research Fund of Zhejiang Provincial Education Department (Grant No. Y200907111).

References

1. Liu, Y., Tan, Z., Lei, G.: Teaching Reform in the Course of Communication Principles. Science & Technology Information 34, 108–110 (2009)
2. Yao, Y., Dong, S., Cao, F., Wu, Y.: Thoughts of Teaching Reform for Communication Principles. In: Proc. 2008 International Symposium on Knowledge Acquisition and Modeling (KAM 2008), December 2008, pp. 467–471. IEEE Press (2008)
3. Sklar, B.: Digital Communications Fundamentals and Applications, 2nd edn. Publishing House of Electronics Industry (June 2006)
4. Herrera, A.: Design of a course of random signals using MATLAB. In: Proc. 26th Annu. Frontiers in Education Conf., Salt Lake City, UT, November 1996, pp. 1219–1222 (1996)

Application Research on Computer Aided Design of Environmental Art Design Teaching

Dan Zhang

College of Arts, Qingdao University of Science & Technology
Qingdao, China
de9510@yeah.net

Abstract. The computer aided instruction(CAI) takes an emerging educational model in university's practice teaching, and the computer aided design will certainly bring new idea, new thought and new manifestation to the teaching. This paper gives analysis and illustratation of CAI of Environmental Art Design Teaching. Using CAI can raise students' skill utilization, and attention on aesthetie ability and the practical ability ; and also strengthens students' theoretical knowledge.

Keywords: Computer aided instruction, environmental art design, teaching, practicality.

1   Introduction

In the days of computer highly popular, the computer aided design has been introduction into curriculum as an emerging less in the university environmental art design teaching[1], and was playing the importment role. It gives university's teaching development and brings the new opportunity, the new task and the new challenge. How to construct a computer aided design curriculum system reasonably, and how to use the computer aided design and teach the specialized theoretical knowledge, and unify the specialized project curriculum effectivly, is a topic which is worth discussing.

2   Content of CAD

The computer aided design is that it uses the computer as the high tech carrier[1], expresses the artistic language and the design concept artistic form the software. Its application scope is vey widely, such as plane design, industrial design, environmental art design, animation design, machine design and so on professional field. In the environmental art design, the main utilization is two-dimensional surface design, the three dimensional modelling software and the light exaggeration software; designer can use them to complete design proposal and performance and the manufacture of architectural design, landscape design, indoor design, demonstration design, furniture design in the environmental art design.

Y. Wu (Ed.): International Conference on WTCS 2009, AISC 116, pp. 317–322.
springerlink.com

In the environmental art design, plane design mainly refers to the two-dimensional space design. And mainly used software is PhotoShop which is presented by Adobe Corporation and AutoCADwhich is presented by Autodesk[2] [3]. And, AutoCAD is mainly used to draw up blueprints and horizontal plan, elevation, sectional drawing, node chart; PhotoShop is very popular in China for image processing. It can be used to complete the adjustment in later period processeing, such as the optical fiber, the light and shade, the degree of saturation, the sense of reality and so on. The three dimensional modelling design by 3D Max[4], sktech up is most typical. We can construct the model by them; then use Lightscape or the Vray-light exaggeration software to carry on light processing and the material quality by past the chart to achieve the simulation real effect. These are the environmental art design college major courses. They have provided the unprecedented art manifestation and the broad creation space for the designer, thus raised the working efficiency.

3   Characteristic of Cad

3.1   Scientific Nature, Technical Nature

The computer aided design is one brand-new design method. In fact it is taking computer as the platform, using design software to complete the artistic work. It covers the graph production, the demonstration and the memory related information technology. And CAD could achive the accurate design[2], build a scientific and technical nature which the people lived newly.

3.2   Rapidity, Convenience

Computer can enhance the plan art work greatly by the new high tech, it means that complete the creativity expression the working efficiency. The computer aided design's utilization make the design drawing, the creation, and the revision become very fast. In the project planning phase for designer creativity, the design plan needs to be massivly revisied. Many software as AutoCAD, PhotoShop, 3D max, sktech up used in the environmental art design, have rework or the historical record function repeatedly; they may restore to the work different period and give the record or the prompt memory; The graph appears frequently by the redundant form, but former handpainting needs to carry on the plan one by one, also has to guarantee that graph absolutely consistent[5], needs to spend the massive time and the energy. The computer can complete duplication function simply and accuratly. For instance, the commercial hotel standard plane plan, design may draw up one as a module, then do the duplication. That saves massive time for designer. This is one that the tradition handpainted is unable to compare with. Moreover we can complete the task such as the drawing brushwork and the line revision, the color, the light and shade, the contrast gradient, the degree of saturation revision, picture revolving, the perspective, the distortion and so on very shortly;before CAD, they are extremely complex and the time-consuming project work[6].

3.3   Practicality

The environmental art design is not only one kind of thought creation, but also requests   the designer to show feeling through the manufacture. The computer aided design is one very strong practical activity, as the expression design idea creativity. It requests the user to have the strong practice ability, to operate, to utilize each kind of charting software nimbly, and to accuratly transmit their own design concept [1].

3.4   Artistic Expressive Force

The artistic form is diverse. Designs from the platform of computer definitely will create more richly, and more concrete artistic work. This kind of work has usability; the designer mainly gives the human one kind of vision the esthetic sense through artistic performance. In the environmental art design, each kind of software can produce colorful visual results. For example: in 3DSMAX, the material quality pastes make the work simulate the object refraction, and the reflection effect in reality, as well as texture[4]. Lightscape and the Vray, the light exaggeration software, may utilize technical exaggeration effect, energy of light transmission, ray tracing and overall situation illumination[7]; In Photoshop,there has artistic effects such as the water color, the gradation, blurring and the filter. According to the need, they can be used to create the richly colorful design work. Through the computer aided design's utilization, the work can be more perfectly developped, achieving effect which the handdrawing can hardly have, so the work can better express designer's creation intention[6].

4   Computer Aided Designs in Environmental Art Design Teaching

4.1   Focus on Students' Esthetic Ability

The computer aided design has the high artistry. It can provide a broader platform through the computer for the designers to create and display. The work created from CAD have not only usability, but also satisfy the people' artistic esthetic demand[2].

   In the university environmental art design's teaching, the computer aided design is an important curriculum. CAD lesson not only gives the student massive computer theoretical knowledge, but also to teach the student to use graphics software. And it focuses on the student esthetic ability artistic quality enhancement. Through the sketch, the color and so on specialized fundamental course students are trained on composition, modelling ability, then students have the solid fine arts foundation of basic skills. In the teaching, techer should pay great attention to the picture primary and secondary distinction, near and far view, the harmonious relations of unity and parts, and the composition full exquisite[6]. The computer aided design complements with the teaching to forrm the basic artistic accomplishment utilization though the specialty fundamental training into the computer aided design. For example: In the sketch training, we can form the light and shade relations understanding to utilize it in the later period of the light exaggeration, that can achive the better processing light chiaroscuro effect; In the handpainted effect chart training, we should pay great attention on the perspective and the proportional relationship, which is also of great help in the computer aided design [2].

4.2   Focus on Theoretical Knowledge Education

At present, in the college art education, the students generally have the problem: their pursue to learn technology constantly, is only to know how to academicaly use computer, but to neglect the specialized theoretical knowledge study integrated with the computer skill[8]. In fact the theory is allways practice instruction, and to a certain extent, the awaress, the new idea, and the creativity originate from the theoretical knowledge. Therefore, in the teaching, we must take the attendance not only in artistic practice, but in the artistic theory and the theory history research at the same time.

The student not only needs to master the specialized design skill, but pay great attention to study the theory knowledge correlated with design. And specialized design skill includs indoor and outdoor envoirment design, demonstration design, furniture design, landscape design, campus design, computer aided design ; the design foundation is art and design theory, it includs the arts historical treatise, design historical treatise and design methodology. The student can get understanding and experience, expand their mentalities of the design, and found the appropriate performance method during studing the Chinese and foreign history of art and outstanding art work. An outstanding art work coveres many contents; regardless of expanding the aspect of knowledge, mastering specialized technique, we can not leave from the most basic theory basic course; these are the design sources, otherwise   it can lead to unclear mentality, the fuzzy concept, and the narrow content. Therefore, in the university teaching, the teacher should pay great attention to students' specialized theoretical knowledge education; because only on this solid foundation, the student can effectively unify the theoretical knowledge and the computer operation skill, for better grasping and utilizing.

4.3   Focus on Students'   Practicality

The computer aided design software is the carrier which displays the design idea, and it needs the designer have the good operation ability. In the university environmental art design teaching, the teacher should focus on the practice, pay great attention to raising students' actual operation ability, in order to enable the students to grasp computer graphics software as the expression design tool. During the teaching, the teacher coach students to operate with design software, make the student form understanding and the feeling during the operation process to the aided design, and strengthen the skill by using. Only the students have the strong ability, after the graduation, can they be competent of designer's work rapidly, can they smoothly complete the role transition between the school and society[8].

Furthermore, the achievement that the environmental art design produces is directly with the social life, and production; and the design achievement needs to accept the market and society's practice and the examination. Therefore, in university's environmental art design teaching, we have to unify the theory and the practice closly. In computer graphics operation's teaching, using the actual projects as the examples, the teachers should unify material which is practically used in indoors and outdoors design, and processing craft and so on to laed the student in the actual design process of full study and the experience. In the computer aided design's teaching process, paying great attention to union of the theory and the practice, is graet to raising the qualified graduates which meet the society demand.

5   Conclusions

Along with the introduction of computer aided design in the university teaching, the software's fast development has impelled specialized tradition educational model, the teaching system's development. Through the reform of university computer aided design teaching,   the teaching activity could be more reasonable, more scientice, and the systemmable. Through correct guidance to the students, teacher can make the student realize the relation between the computer aided design and the specialized design class, and the specialized theory class; and the relation between the computer skill study and the social practice; thus lead stedents to carry on the comprehensive study, enhancing their artistic accomplishment and the practical ability, and becoming the qualified graduates.

References

1. Marcus, C.C., Francis, C.: Design Guidelines for Urban Open Space, pp. 230–238. John Wiley and Sons, Inc.
2. The Photoshop Book for Photographerss Kelby. New Riders Publishing (2003)
3. The AutoCAD book: drawing, modeling, and applications, including release 13 JM Kirkpatrick. Prentice Hall (1996)
4. KL Murdock 3ds Max 5 Bible, 1st edn. John Wiley & Sons, Inc. New York (2002)
5. Fischer, M.A., Luiten, G.T., Aalami, F.: Representing Project (1995)

6. Information and Construction Method Knowledge for Computer-Aided Construction Management. In: CIB Proc. 180, Stanford, CA, August 21-23, pp. 404–414 (1995)
7. Björk, B.-C.: A Unified Method for Modelling Construction Information Building and Environment 27(2), 173–194
8. Issa, R.R.A., Flood, I., O'Brien, W.J.: 4D CAD and Visualization in Construction: developments and applications
9. Christodoulou, S.: Discussion of Feasibility study of 4D CAD in commercial construction". In: Koo, B., Fischer, M. (eds.) J. Constr. Engrg. and Mgmt., ASCE, May/June 2002, pp. 274–275 (2002)

The Computer Calculation of Contact Stress of High Order Contact Surfaces and Gearing

Liu Huran

Zhejiang Univ Science and Technology
Dept. Mechanical Engineering, Zust University
Hangzhou, China
lhy05120@21cn.com

Abstract. At the first, the theoretical basis and calculation formulas have been deduced in this thesis, based on the derivative geometry and the theories of gear meshing. So that the surfaces of the two conjugate teeth have the order of contact as higher as possible. Than, three methods (polynomial, involutes, revision of 2nd order derivative) have been devised and used in the configuration of the profile of the basic rake with high order of contact. Later on, the profiles of the gears have been generated on the gear generator, by the help of NC machining and the modern precise technique has made out, the corresponding cutting tools. Finally, these cutters have generated the new kind of gears. The reducer box has the meshing effect as expected. This paper studies the elastic contact problem in the contact area and contact stress model of local density computing, a set of well resulting cylindrical gears meshing process of calculating the contact stress calculation methods and modular code system. Paper to a pair of aviation for the inverted helical gears is described.

Keywords: gear, profile, curvature.

1   Introduction

According to the theory of contact, if the two conjugate surfaces could realize the concave-convex contact, and the induced curvature equal to zero, the two surfaces will have lowest contact stress.

2   The Condition for the Gap between Tooth Surfaces to Be an Infinite Small of Fourth Order

According to the theories of gear meshing, suppose that the profile of basic rack is: $y = y(x)$, the corresponding profile and normal vector of the gear would be (see Fig.2)

Y. Wu (Ed.): International Conference on WTCS 2009, AISC 116, pp. 323–330.
springerlink.com   © Springer-Verlag Berlin Heidelberg 2012

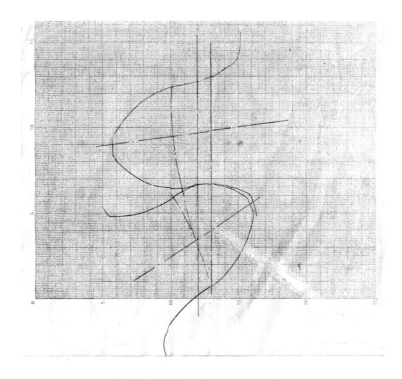

Fig. 1. The high order contact gearing

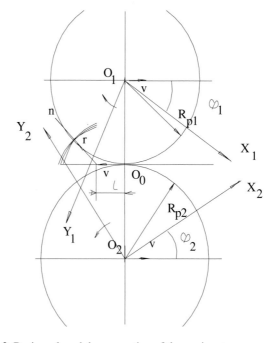

Fig. 2. Basic rack and the generation of the conjugate gears

$$\begin{bmatrix} X \\ Y \end{bmatrix} = \begin{bmatrix} \cos\phi & \sin\phi \\ -\sin\phi & \cos\phi \end{bmatrix}\begin{bmatrix} x \\ y \end{bmatrix} + \begin{bmatrix} \cos\phi & \sin\phi \\ -\sin\phi & \cos\phi \end{bmatrix}\begin{bmatrix} -R_p\phi \\ R_p \end{bmatrix} \tag{3}$$

$$\begin{bmatrix} N_x \\ N_y \end{bmatrix} = \begin{bmatrix} \cos\phi & \sin\phi \\ -\sin\phi & \cos\phi \end{bmatrix}\begin{bmatrix} n_x \\ n_y \end{bmatrix} = \begin{bmatrix} \cos\phi & \sin\phi \\ -\sin\phi & \cos\phi \end{bmatrix}\begin{bmatrix} -y' \\ 1 \end{bmatrix} / \sqrt{y'^2+1} \tag{4}$$

Where: x , y ——the coordinates of the basic rack
nx ,n y ——the normal vector of the basic rack
X,Y ——the coordinates of the gear
NX,NY ——the normal vector of the gear
ϕ ——the angular displacement of the gear

According to the theory of gear meshing, the transverse displacement l of rack and the angular displacement of the gear ϕ must satisfy the following equation:

$$\phi = \frac{l}{R_p} = \frac{x + yy'}{R_p} \tag{5}$$

Where : l ——the transverse displacement
Rp ——the radius of the pitch circle
Equ.3 can be expressed by the form of   Matrix:

$$\{R\} = M\{r\} + M\{\Phi\} \tag{6}$$

Where:

$$\{R\} = \begin{bmatrix} X \\ Y \end{bmatrix}, M = \begin{bmatrix} \cos\phi & \sin\phi \\ -\sin\phi & \cos\phi \end{bmatrix} = \begin{bmatrix} m_{11} & m_{12} \\ m_{21} & m_{22} \end{bmatrix}, \{r\} = \begin{bmatrix} x \\ y \end{bmatrix},$$

$$\{\Phi\} = \begin{bmatrix} -R_p\phi \\ R_p \end{bmatrix}$$

differentiate $\{R\}$ form 1—4th order :

$$\{R\}' = M'\{r\} + M\{r\}' + M'\{\Phi\} + M\{\Phi\}' \tag{7}$$

$$\{R\}'' = M''\{r\} + 2M'\{r\}' + M\{r\}'' + M''\{\Phi\} + 2M'\{\Phi\}' + M\{\Phi\}'' \tag{8}$$

$$\{R\}''' = M'''\{r\} + 3M''\{r\}' + 3M'\{r\}'' + M\{r\}''' $$
$$+ M'''\{\Phi\} + 3M''\{\Phi\}' + 3M'\{\Phi\}'' + M\{\Phi\}''' \tag{9}$$

$$\{R\}^{(4)} = M^{(4)}\{r\} + 4M'''\{r\}' + 6M''\{r\}'' + 4M'\{r\}''' + M\{r\}^{(4)} $$
$$+ M^{(4)}\{\Phi\} + 4M'''\{\Phi\}' + 6M''\{\Phi\}'' + 4M'\{\Phi\}''' + M\{\Phi\}^{(4)} \tag{10}$$

in the above equation, the differentials of $\{r\}$ and $\{\Phi\}$ are:

$$\{r\}'=\begin{bmatrix}1\\y'\end{bmatrix}, \quad \{r\}''=\begin{bmatrix}0\\y''\end{bmatrix}, \quad \{r\}'''=\begin{bmatrix}0\\y'''\end{bmatrix}, \quad \{r\}^{(4)}=\begin{bmatrix}0\\y^{(4)}\end{bmatrix} \tag{11}$$

$$\{\Phi\}=R_p\begin{bmatrix}-\phi\\1\end{bmatrix}, \quad \{\Phi\}'=R_p\begin{bmatrix}-\phi'\\1\end{bmatrix}, \quad \{\Phi\}''=R_p\begin{bmatrix}-\phi''\\1\end{bmatrix},$$

$$\{\Phi\}'''=R_p\begin{bmatrix}-\phi'''\\1\end{bmatrix} \quad \{\Phi\}^{(4)}=R_p\begin{bmatrix}-\phi^{(4)}\\1\end{bmatrix} \tag{12}$$

Since that the original point of coordinate is the intersect point of normal vector and the pitch line of rack. Where: $\phi=\dfrac{l}{R_p}=\dfrac{x+yy'}{R_p}=0$, So that, we have

$$M\big|_{\phi=0}=\begin{bmatrix}1&0\\0&1\end{bmatrix} \tag{13}$$

$$M'\big|_{\phi=0}=\begin{bmatrix}-\phi'\sin\phi&\phi'\cos\phi\\-\phi'\cos\phi&-\phi'\sin\phi\end{bmatrix}_{\phi=0}=\begin{bmatrix}0&\phi'\\-\phi'&0\end{bmatrix} \tag{14}$$

$$M''\big|_{\phi=0}=\begin{bmatrix}-\phi''\sin\phi-\phi'^2\cos\phi&\phi''\cos\phi-\phi'^2\sin\phi\\-\phi''\cos\phi+\phi'^2\sin\phi&-\phi''\sin\phi-\phi'^2\cos\phi\end{bmatrix}_{\phi=0}=\begin{bmatrix}0&\phi''\\-\phi''&0\end{bmatrix} \tag{15}$$

$$\{R\}''=\phi''\begin{bmatrix}y\\-x\end{bmatrix}+2\phi'\begin{bmatrix}y'\\-1\end{bmatrix}+\begin{bmatrix}0\\y''\end{bmatrix}+R_p\begin{bmatrix}0\\2\phi'^2\end{bmatrix}$$

The normal vector of the gear is determined by the normal vector of the rack, so that we have:

$$\begin{bmatrix}N_x\\N_y\end{bmatrix}_{\phi=0}=\begin{bmatrix}1&0\\0&1\end{bmatrix}\begin{bmatrix}-y'\\1\end{bmatrix}/\sqrt{y'^2+1}=\begin{bmatrix}-y'\\1\end{bmatrix}/\sqrt{y'^2+1} \tag{16}$$

The second parameter of the Taylor series of the tooth profile $n\cdot\dfrac{d^2\mathbf{r}}{du^2}$ is: :

$$\{N\}^T\{R\}''=\varphi''[-y'\ \ 1]\begin{bmatrix}y\\-x\end{bmatrix}+2\varphi''[-y'\ \ 1]\begin{bmatrix}y'\\-1\end{bmatrix}+$$

$$[-y'\ \ 1]\begin{bmatrix}0\\y'''\end{bmatrix}+R_p[-y'\ \ 1]\begin{bmatrix}0\\2\varphi'^2\end{bmatrix}$$

$$\phi''(-y'y-x)+2\phi'(-y'^2-1)+y''+2R_p\phi'^2$$

$$=-2\phi'(y'^2+1+2R_p\phi')+y''$$

$$(\because \phi=0, l=x+yy'=0)$$

For gear 1 and gear 2 we have the expressions as following respectively:

$$\{N_1\}^T\{R_1\}''=-2\phi_1'(y'^2+1+2R_{p1}\phi_1')+y'' \quad, \qquad \phi_1=\frac{l}{R_{p1}}=\frac{x+yy'}{R_{p1}} \quad,$$

$$\{N_2\}^T\{R_2\}''=-2\phi_2'(y'^2+1+2R_{p2}\phi_2')+y'' \quad, \qquad \phi_2=\frac{x+yy'}{R_{p2}} \quad,$$

let the second parameter equal zero: $\{N_1\}^T\{R_1\}''-\{N_2\}^T\{R_2\}''=0$ we have :

$$\phi_1'=\frac{l'}{R_{p1}}=0,\ \phi_2'=\frac{l'}{R_{p2}}=0,\ l'=0 \qquad (x+yy')'=0,\ 1+y'^2+y''=0 \quad, (18)$$

The coordinate of the curvature center: $y_c=y+\dfrac{1+y'^2}{y''}=0$, the curvature center

of basic rack must lay on the pitch line.

Some essential findings are:

(1) If $x+yy'=0$, the two surfaces would conjugate with each other, there are no additional requirements for the profile of basic rack. The gap between tooth surfaces at adjacent area of mesh point is an infinite small of the second order. The parameter of the second order can be expressed by: $(x+yy')'$.

(2) If $(x+yy')'=0$, the curvature center of basic rack must lay on the pitch line, the gap between tooth surfaces at adjacent area of mesh point is an infinite small of the third order. The parameter of the third order can be expressed by: $(x+yy')''$.

(3) If $(x+yy')''=0$, The gap between tooth surfaces at adjacent area of mesh point is an infinite small of the fourth order. The parameter of the fourth order can be expressed by $(x+yy')'''$. From the viewpoint of logic it is very regular and reasonable, where y=y(x) is the equation of rack profile.

(4) For gear 1 and gear 2 we have the expressions as following respectively:

$$\{N_1\}^T\{R_1\}^{(4)}=\qquad 4\phi_1''(-y'^2-1)+y^{(4)} \quad,$$

$$\{N_2\}^T\{R_2\}^{(4)}=4\phi_2''(-y'^2-1)+y^{(4)}$$

$$\{N_1\}^T\{R_1\}^{(4)}-\{N_2\}^T\{R_2\}^{(4)}=\qquad 4l'''(-y'^2-1)(\frac{1}{R_{p1}}+\frac{1}{R_{p2}})$$

The gap between tooth surfaces at adjacent area of mesh point is an infinite small of fourth order.

$$\delta=k_4\Delta x^4 \tag{19}$$

3   The Contact Stress of the Tooth Surface with High Order of Contact

The Hertz contact formula suppose that the gap between the two bodies at the adjacent area is the infinite small of second order, $\delta = k_2 \Delta x^2$. While now the gap between the two bodies at the adjacent area is the infinite small of fourth order, $\delta = k_4 \Delta x^4$. The Hertz formula is invalid for this occasion. Adopt the induced relative surfaces and the

elastic abbreviated synthesize insignia $\vartheta = \dfrac{2(1-\gamma_1^2)}{E_1} + \dfrac{2(1-\gamma_2^2)}{E_2}$, the contact problem transformed to another problem: how to calculate the contact stress when a rigged body expressed by equation (19) is pressed into the half infinite great space.

Let us develop the displacement with the Qibchev polynomial. Considering the symmetric of the problem, let the coefficients of the odd order b1,b3....equal to zero.

The programming is following:

```
Private Sub Picture1_Click()
Picture1.Scale (-300, 300)-(300, -300)
Picture1.Cls
Picture1.ForeColor = &H0
Picture1.Line (-300, 0)-(300, 0)
Picture1.Line (0, 300)-(0, -300)
Picture1.Line (100, 100)-(100, 0)
Picture1.Line (-100, 100)-(-100, 0)
Picture1.ForeColor = &H0
L = 20
yL = 1.2
k4 = yL / L / L / L / L
x0 = -L
y0 = k4 * x0 * x0 * x0 * x0
For x = -L To L Step 0.01
y = k4 * x ^ 2 * x ^ 2
Picture1.Line (5 * x0, 5 * y0)-(5 * x, 5 * y)
x0 = x
y0 = y
Next x
a = 8.4
p0 = 2 * k4
x0 = -a
y0 = 0
For x = -a To a Step 0.001
```

y = 500 * p0 * (a * a + 2 * x * x) * Sqr(a * a - x * x)
Picture1.Line (5 * x0, 5 * y0)-(5 * x, 5 * y)
x0 = x
y0 = y
Next x
End Sub
The results:

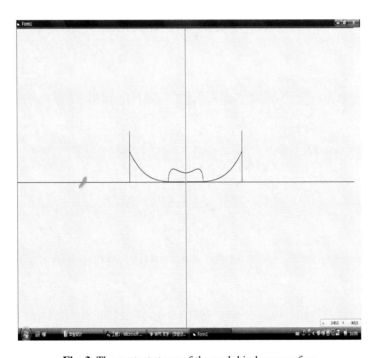

Fig. 3. The contact stress of the such kind gear surface

Acknowledgments. This project is supposed by the natural scientific foundation of China, No.2006-50675235. and the natural scientific foundation of Zhejiang province, China, No. Y106047 and   Y1080093.3

References

1. Komori, T., Arga, Y., Nagata, S.: A new gear profile having zero relative curvature at many contact points. Trans. ASME 12(3), 430–436 (1990)
2. Komori, T.: A new gear profile of relative curvature being zero at contact points. In: Proceeding of International Conference on gearing, China CMCC, pp. 230–236 (1988)
3. Feng, X., Linda, A.W.L.: Study on the design principle of the Logix gear tooth profile and the selection of its inherent parameters. International Journal of Advanced Manufacture Technology 24(11/12), 789–794 (2004)

4. Li, F.: Study for the forming principle of Logix gear tooth profile and its mesh performance. Journal of Xiamen University 36(1), 12–16 (2002)
5. Dalen: Deferential Geometry and Theory of gear mesh (in Chinese)
6. Litvin, L.: Theory of gear mesh and its application
7. Neimann, F.: Machine Element
8. Dooner, D.B.: On the Three Laws of Gearing. Journal of Mechanical Design 124 (December 2002)

Many Strong Limit Theorems for Hidden Markov Models on a Non-homogeneous Tree

Shaohua Jin[1], Jie Lv[1], Zhenyao Fan[2], Shuguang Sun[1], and Yanmin Zhang[1]

[1] Hebei University of Technology, Tianjin, 300130, China
[2] Tangshan College, Tangshan, 063000, China
tanjin154@sina.cn

Abstract. The strong limit theorems is one of the central questions for studying in the International Probability theory. In this paper, some strong limit theorems of random selection System for hidden Markov models indexed by a non-homogeneous tree were obtained by constructing a martingale difference series and using the martingale difference sequences convergence theorem.

Keywords: Markov models, martingale, non-homogeneous tree, random selection, strong limit theorem.

1 Introduction

The strong limit theorems is one of the central questions for studying in the international Probability theory. (c.f. [1] and [2]). In this paper, some strong limit theorems of random selection System for hidden Markov models indexed by a non-homogeneous tree were obtained by constructing a martingale difference series and using the martingale difference sequences convergence theorem.

Let T be an infinite tree, $x \neq y$ are two arbitrarily vertices of T. Then there is only one route from x to $y : x = z_1, z_2, \cdots, z_m = y$, which $z_1, z_2, \cdots, z_m$ are different from each other, and z_i, z_{i+1} are two adjacent vertices. So the distance between x to y is $m-1$. To label the vertices of T, we select one vertex to be "root vertex" and label it by O. If the distance from a vertex to O is n, we say the vertex is on the level n. We also say the root O is on the level 0.

Let T be an infinite tree whose root vertex is O, $\{N_n, n \geq 1\}$ is a positive integer set. If every vertex on level n has N_{n+1} adjacent vertices, we call the tree T is a Bethe tree in broad sense or a Cayley tree in broad sense.

Particularly, to classify the non-negative integer set N, we divide N into several sets as follow:

$$(0) = \{0, m, 2m, 3m, \cdots, nm, \cdots\},$$

$$(1) = \{1, m+1, 2m+1, 3m+1, \cdots, nm+1, \cdots\},$$

Y. Wu (Ed.): International Conference on WTCS 2009, AISC 116, pp. 331–336.
springerlink.com   © Springer-Verlag Berlin Heidelberg 2012

$$(2) = \{2, m+2, 2m+2, 3m+2, \cdots, nm+2, \cdots\},$$

$$\cdots\cdots\cdots\cdots\cdots\cdots\cdots\cdots\cdots\cdots\cdots\cdots\cdots\cdots\cdots,$$

$$(m-1) = \{m-1, 2m-1, \cdots, (n+1)m-1, \cdots\}.$$

When $n \in (i)$, let $N_{n+1} = \alpha_i, i = 0, 1, 2, \cdots, m-1$,

(α_i is a positive integer and not all of them is 1). In this way, we get a particular tree $T_{\alpha_0, \alpha_1, \cdots, \alpha_{m-1}}$. It is convenient for us to denote $T_{\alpha_0, \alpha_1, \cdots, \alpha_{m-1}}$ by T.

Let s, t be two arbitrary vertices of T. If s is located at the only one route from O to t, we denote it by $s \leq t$, and denote the number of edges as $|t|$. Apparently, if $|t| = n$, we say t is on level n. Let s, t be two arbitrary vertices of T, $s \wedge t$ is the vertex farther from O which satisfies $s \wedge t \leq s, s \wedge t \leq t$. Let $t \neq O$, $\bar{t}$ satisfies $\bar{t} \leq t$ and $|\bar{t}| = |t| - 1$, t is called the son of $\bar{t}$.

Definition 1. Let $U = \{1, 2, \cdots, M\}$, $V = \{1, 2, \cdots, N\}$ be two finite sets, $\{X_\sigma, \sigma \in T\}$ and $\{Y_\sigma, \sigma \in T\}$ be random variables defined on probability space $\{\Omega, F, P\}$ and taking values in U and V. Let $q_0 = (q(1), q(2), \cdots)$ be the initial distribution of Markov chains $\{X_\sigma, \sigma \in T\}$ indexed by a non-homogeneous tree. Denote $P(X_{r+1,j} = x_{r+1,j} | X_{r,i} = x_{r,i}) = a_{r,i}(x_{r,i}, x_{r+1,j})$.

Let $\{P_r\}$ be the transition matrices of $\{X_\sigma, \sigma \in T\}$, where

$$P_r = \begin{bmatrix} A_{r,1} & 0 & \cdots & 0 \\ 0 & A_{r,2} & \cdots & 0 \\ 0 & 0 & \cdots & A_{r,N_0 N_1 \cdots N_r} \end{bmatrix},$$

$$A_{r,i} = (a_{r,i}(x_{r,i}, x_{r+1,N_{r+1}(i-1)+1}), a_{r,i}(x_{r,i}, x_{r+1,N_{r+1}(i-1)+2}),$$

$$\cdots a_{r,i}(x_{r,i}, x_{r+1,N_{r+1}i})).$$

Let $P^{(m,n)} = P_{m+1} \cdot P_{m+2} \cdots P_n$.

If $\forall \sigma \in T, \tau_i \in T$ which satisfies $\sigma \wedge \tau_i \leq \bar{\sigma}$, $1 \leq i \leq n$, we have

$$P\left(X_\sigma = x_\sigma \middle| Y_{\bar{\sigma}} = y_{\bar{\sigma}}, X_{\bar{\sigma}} = x_{\bar{\sigma}}, Y_{\tau_i} = y_i, X_{\tau_i} = x_i, 1 \leq i \leq n\right)$$

$$= P\left(X_\sigma = x_\sigma \middle| X_{\bar{\sigma}} = x_{\bar{\sigma}}\right), \quad \forall x_\sigma, x_{\bar{\sigma}}, x_1, \cdots, x_n \in U, y_{\bar{\sigma}}, y_1, \cdots, y_n \in V.$$

$$P\left(Y_\sigma = y_\sigma \middle| X_\sigma = x_\sigma, Y_{\sigma^-} = y_{\sigma^-}, X_{\sigma^-} = x_{\sigma^-}, Y_{\tau_i} = y_i, X_{\tau_i} = x_i, 1 \le i \le n\right)$$

$$= P\left(Y_t = y_t \middle| X_t = x_t\right), \quad \forall x_\sigma, x_{\sigma^-}, x_1, \cdots, x_n \in U, y_\sigma, y_{\sigma^-}, y_1, \cdots, y_n \in V.$$

Then $(X^T, Y^T) = \{(X_\sigma, Y_\sigma), \sigma \in T\}$ is called a non-homogeneous Hidden Markov Model indexed by a non-homogeneous tree T. $\{X_\sigma, \sigma \in T\}$ which can not be observed is called hidden chains. $\{Y_\sigma, \sigma \in T\}$ which can be observed is called observed chains.

2 Some Strong Limit Theorems of Random Selection System for Hidden Markov Models

Let $P(Y_{r,i} = y \mid X_{r,i} = x) = b_{r,i}(x, y)$,

and $B = (b_{r,i}(x, y), 0 \le r \le n-1, 1 \le i \le N_0 N_1 \cdots N_r)$

be the transitional matrix from $\{X_\sigma, \sigma \in T\}$ to $\{Y_\sigma, \sigma \in T\}$.

Lemma 1. Let $(X^T, Y^T) = \{(X_\sigma, Y_\sigma), \sigma \in T\}$ be a non-homogeneous Hidden Markov Model indexed by T defined as above, then

$$P(Y^{T_n} \mid X^{T_n}) = P\left(Y_{0,1} \middle| X_{0,1}\right) \prod_{r=1}^{n} \prod_{i=1}^{N_0 N_1 \cdots N_r} P(Y_{r,i} \middle| X_{r,i}) \tag{1}$$

Lemma 2. Let $(X^T, Y^T) = \{(X_\sigma, Y_\sigma), \sigma \in T\}$ be a non-homogeneous Hidden Markov Model indexed by T defined as above , then

$$P(X^{L_{n+1}}, Y^{L_{n+1}} \mid X^{T_n}, Y^{T_n}) = P(X^{L_{n+1}}, Y^{L_{n+1}} \mid X^{L_n}) \tag{2}$$

Lemma 3. Let $(X^T, Y^T) = \{(X_\sigma, Y_\sigma), \sigma \in T\}$ be a non-homogeneous Hidden Markov Model indexed by T defined as above , $f(x, y, z)$ be real functions defined on $U \times U \times V$, then

$$E[f(X_{r,i}, X_{r+1,j}, Y_{r+1,j}) \mid X^{T_r}, Y^{T_r}] = E[f(X_{r,i}, X_{r+1,j}, Y_{r+1,j}) \mid X_{r,i}]$$

$$(j = N_{r+1}(i-1)+1, \cdots, N_{r+1}i) \tag{3}$$

Let $V_{0,1} = V_{0,1}(X_{0,1})$,

$V_{r,i} = V_{r,i}(X_{0,1}, \cdots, X_{r-1,i}), r = 1, 2, 3, \cdots, i = 1, 2, \cdots, N_1 \cdots N_r$.

where $V_{r,i}$ be a measurable function defined on U^r and taking values in R, $\{V_{r,i}, r = 0,1,\cdots, i = 1,2,\cdots, N_0 N_1 \cdots N_r\}$ will be called predictable sequence of T. If $V_{r,i}$ be a function taking values in $\{0,1\}$ and $V_{0,1} = 1$, $\{V_{r,i}, r = 0,1,\cdots, i = 1,2,\cdots, N_0 N_1 \cdots N_r\}$ will be called random selection system on T.

Theorem 1. Let $(X^T, Y^T) = \{(X_\sigma, Y_\sigma), \sigma \in T\}$ be a non-homogeneous Hidden Markov Model indexed by T defined as above, $f_r(x, y, z)$ be real functions defined on $U \times U \times V$, $\{V_{r,i}, r = 0,1,\cdots, i = 1,2,\cdots, N_0 N_1 \cdots N_r\}$ be random selection functions, $\{a_n, n \in N\}$ be a sequence of real numbers satisfying $a_n \uparrow \infty \, (n \to \infty)$.

If

$$\sum_{n=1}^{\infty} \frac{1}{a_n^2} \sum_{r=0}^{n-1} \sum_{i=1}^{N_0 \cdots N_r} \sum_{j=N_{r+1}(i-1)+1}^{N_{r+1}i} \{V_{r+1,j}$$
$$\cdot E\left[f_{r+1}^2\left(X_{r,i}, X_{r+1,j}, Y_{r+1,j} \big| X_{r,i}\right)\right]\} < +\infty \tag{4}$$

Then

$$\lim_{n \to \infty} \frac{1}{a_n} \sum_{r=0}^{n-1} \sum_{i=1}^{N_0 \cdots N_r} \sum_{j=N_{r+1}(i-1)+1}^{N_{r+1}i} V_{r+1,j}\{f_{r+1}\left(X_{r,i}, X_{r+1,j}, Y_{r+1,j}\right)$$
$$-E\left[f_{r+1}\left(X_{r,i}, X_{r+1,j}, Y_{r+1,j}\right)\big| X_{r,i}\right]\} = 0 \quad a.s. \tag{5}$$

Proof. Let

$$T_{r+1} = f_{r+1}\left(X_{r,i}, X_{r+1,j}, Y_{r+1,j}\right)$$
$$-E\left[f_{r+1}\left(X_{r,i}, X_{r+1,j}, Y_{r+1,j}\right)\big| X_{r,i}\right] \tag{6}$$

We first prove that $\{T_{r+1}, r \geq 0\}$ is a martingale difference series.

Because $E[f_{r+1}\left(X_{r,i}, X_{r+1,j}, Y_{r+1,j}\right)\big| X_{r,i}]$ is measurable on $\sigma(X^{T_r}, Y^{T_r})$, so we have

$$E\{E[f_{r+1}\left(X_{r,i}, X_{r+1,j}, Y_{r+1,j}\right)\big| X_{r,i}]\big|\sigma(X^{T_r}, Y^{T_r})\}$$
$$= E[f_{r+1}\left(X_{r,i}, X_{r+1,j}, Y_{r+1,j}\right)\big| X_{r,i}] \tag{7}$$

For $r \geq 0$, by (6)and (7), we have

$$E[T_{r+1}|\sigma(X^{T_r}, Y^{T_r})]$$

$$= E\{f_{r+1}(X_{r,i}, X_{r+1,j}, Y_{r+1,j})$$

$$-E[f_{r+1}(X_{r,i}, X_{r+1,j}, Y_{r+1,j})|X_{r,i}]|\sigma(X^{T_r}, Y^{T_r})\}$$

$$= E[f_{r+1}(X_{r,i}, X_{r+1,j}, Y_{r+1,j})|X_{r,i}]$$

$$- E[f_{r+1}(X_{r,i}, X_{r+1,j}, Y_{r+1,j})|X_{r,i}]$$

$$= 0 \qquad (8)$$

Let $F_r = \sigma(X^{T_r}, Y^{T_r})$, we have

$$E[T_{r+1}|F_r] = 0.$$

So $\{T_{r+1}, F_r, r \geq 0\}$ is a martingale difference series.

Because $V_{r+1,j}$ is measurable for F_r. So $\{V_{r+1,j}T_{r+1}, F_r, r \geq 0\}$ is also a martingale difference series.

So we have $E[V_{r+1,j}T_{r+1}|F_r] = V_{r+1,j}E[T_{r+1}|F_r] = 0$

Because

$$E[V_{r+1,j}^2 T_{r+1}^2|F_r]$$

$$= V_{r+1,j}^2 E[T_{r+1}^2|F_r]$$

$$= V_{r+1,j}^2 E\{[f_{r+1}(X_{r,i}, X_{r+1,j}, Y_{r+1,j})$$

$$-E[f_{r+1}(X_{r,i}, X_{r+1,j}, Y_{r+1,j})|X_{r,i}]]^2|X_{r,i}\}$$

$$\leq V_{r+1,j}^2 E[f_{r+1}^2(X_{r,i}, X_{r+1,j}, Y_{r+1,j})|X_{r,i}] \qquad (9)$$

so we have

$$\sum_{n=1}^{\infty}\frac{1}{a_n^2}\sum_{r=0}^{n-1}\sum_{i=1}^{N_0 \cdots N_r}\sum_{j=N_{r+1}(i-1)+1}^{N_{r+1}i}E[V_{r+1,j}^2 T_{r+1}^2|F_r]$$

$$\leq \sum_{n=1}^{\infty}\frac{1}{a_n^2}\sum_{r=0}^{n-1}\sum_{i=1}^{N_0 \cdots N_r}\sum_{j=N_{r+1}(i-1)+1}^{N_{r+1}i}\{V_{r+1,j}$$

$$\cdot E[f_{r+1}^2(X_{r,i}, X_{r+1,j}, Y_{r+1,j}|X_{r,i})] < +\infty\} \qquad a.s.$$

$$(10)$$

By (10) and the convergence theorem of martingale difference series (c.f. [3]), we have

$$\lim_{n\to\infty}\frac{1}{a_n}\sum_{r=0}^{n-1}\sum_{i=1}^{N_0\cdots N_r}\sum_{j=N_{r+1}(i-1)+1}^{N_{r+1}i}V_{r+1,j}T_{r+1}=0 \tag{11}$$

By (6) and (11), (5) holds.

Definition 2. Let $\left\{X_{r,i},F_r,r\geq 1,i=N_1N_2\cdots N_r\right\}$

Be a sequence of adaptive and positive random variables, $\left\{V_{r,i},F_r,r\geq 1,i=N_1N_2\cdots N_r\right\}$ be a random selection function,

$$B_{r+1,j}=E\left[X_{r+1,j}\middle|F_r\right].\text{ If }\lim_{n\to\infty}\frac{\displaystyle\sum_{r=0}^{n-1}\sum_{i=1}^{N_0\cdots N_r}\sum_{j=N_{r+1}(i-1)+1}^{N_{r+1}i}V_{r+1,j}X_{r+1,j}}{\displaystyle\sum_{r=0}^{n-1}\sum_{i=1}^{N_0\cdots N_r}\sum_{j=N_{r+1}(i-1)+1}^{N_{r+1}i}V_{r+1,j}B_{r+1,j}}=1 \qquad a.s. \tag{12}$$

Then we call the above ratio is the random equity ratio limit theorem.

Corollary 1. Under the assumption of Theorem 1, we have

$$\lim_{n\to\infty}\frac{\displaystyle\sum_{r=0}^{n-1}\sum_{i=1}^{N_0\cdots N_r}\sum_{j=N_{r+1}(i-1)+1}^{N_{r+1}i}f_{r+1}\left(X_{r,i},X_{r+1,j},Y_{r+1,j}\right)}{\displaystyle\sum_{r=0}^{n-1}\sum_{i=1}^{N_0\cdots N_r}\sum_{j=N_{r+1}(i-1)+1}^{N_{r+1}i}V_{r+1,j}E\left[f_{r+1}\left(X_{r,i},X_{r+1,j},Y_{r+1,j}\right)\middle|X_{r,i}\right]}=1 \tag{13}$$

$$a.s.$$

Proof. Let

$$a_n=\sum_{r=0}^{n-1}\sum_{i=1}^{N_0\cdots N_r}\sum_{j=N_{r+1}(i-1)+1}^{N_{r+1}i}V_{r+1,j}E\left[f_{r+1}\left(X_{r,i},X_{r+1,j},Y_{r+1,j}\right)\middle|X_{r,i}\right]\text{ By Theorem 1, the}$$

conclusion holds.

Acknowledgment. The guidance project of the research and development of science and technology of Hebei province. （No: Z2008308）.

References

1. Jin, S.-H., Lu, J.-G., Wang, N., Jiang, H.-H.: A class of Strong Deviation Theorems for the Sequence of arbitrary Integer-Valued Random Variables. In: Proceedings of the second International conference on Information and computing Science, pp. 193–195. IEEE CS (2009)
2. Liu, W., Yang, W.: The Markov approximation of the sequences of N—Valued variables and a class of small deviation theorems. Stochastic Process. Appl. (1), 117–130 (2000)
3. Doob, J.L.: Stochastic Processes. Wiley, New York (1953)

Many Strong Limit Theorems of the Random Selection System on a Non-homogeneous Tree

Shaohua Jin[1], Ying Tian[1], Shuguang Sun[1], Zhenyao Fan[2], and Yanmin Zhang[1]

[1] Hebei University of Technology, Tianjin, 300130, China
[2] Tangshan College, Tangshan, 063000, China
tanjin154@sina.cn

Abstract. The limit theorems is one of the central questions for studying in the International Probability theory. In this paper, the random selection system in broad sense was introduced to a special kind of non-homogeneous trees, and many limit theorems of the random selection in broad sense for non-homogeneous Markov chains on a special kind of non-homogeneous tree were obtained by constructing a non-negative martingale and using Doob's martingale convergence theorem.

Keywords: Non-homogeneous Markov chains, Martingale, Random selection system in broad sens, Strong Limit Theorem.

1   Introduction

The limit theorems is one of the central questions for studying in the International Probability theory. (c.f. [1] and [2]). In this paper, the random selection system in broad sense was introduced to a special kind of non-homogeneous tree, and many limit theorems of the random selection in broad sense for non-homogeneous Markov chains on a special kind of non-homogeneous tree were obtained by constructing a non-negative martingale.

Let T be an infinite tree, $x \neq y$ be different vertices of the tree, there will be only one path between x and y, $x = z_1, z_2, \cdots, z_m = y$, where $z_1, z_2, \cdots, z_m$ are different vertices and z_i is next to z_{i+1}. So the distance between x and y is $m-1$. To label the vertices of T, we fix any vertex as the "root" and label it by O. If a vertex is n distance from O, we say this vertex is on level n. We also say the root O is on level 0.

Definition 1. Let T be an infinite tree, $\{N_n, n \geq 1\}$ be a countable space, where N_n is a positive integer. If every vertex from level $n(n \geq 0)$ has N_{n+1} sons, the tree T is a *Bethe* tree in broad sense or a *Cayley* tree in broad sense. Particularly, let $N = \{0, 1, 2, \cdots\}$, $m(m \geq 2)$ be an integer, we divide N into several sets as follow:

Y. Wu (Ed.): International Conference on WTCS 2009, AISC 116, pp. 337–344.
springerlink.com   © Springer-Verlag Berlin Heidelberg 2012

$(0) = \{0, m, 2m, \cdots\cdots\}$

$(1) = \{1, m+1, 2m+1, \cdots\cdots\}$

$\cdots\cdots$

$(m-1) = \{m-1, 2m-1, 3m-1, \cdots\cdots\}$

When $n \in (i)$, let $N_{n+1} = \alpha_i$ (α_i is a positive integer and not all of them is 1), $i = 0,1,2,\cdots, m-1$. In this way, we get a particular non-homogeneous tree $T_{\alpha_0,\alpha_1,\cdots,\alpha_{m-1}}$.

Let s,t be two vertices of T. Write $s \leq t$ if s is on the unique path connecting O to t, and $|t|$ be the number of edges on this path. For any two vertices s,t, $s \wedge t$ be the vertex farther from O satisfying $s \wedge t \leq s$ and $s \wedge t \leq t$.

If $t \neq O$, then we let $\bar{t}$ stand for the vertex satisfying $\bar{t} \leq t$ and $|\bar{t}| = |t| - 1$ (we refer to t as a son of $\bar{t}$).

In this paper, $T_{\alpha_0,\alpha_1,\cdots,\alpha_{m-1}}$ is denoted by T. We denote L_n be the set of all vertices on level $n(n \geq 0)$, T_n be the subtree of T containing the vertices from level 0 to level n. We denote $|B|$ be the number of the vertices of the subtree B, $X^B = \{X_t, t \in B\}$, and $s(\sigma)$ be the sons of the vertex σ.

2   Many Strong Limit Theorems on a Special Kind of Non-homogeneous Tree

Definition 2. Let $G = \{0,1,2,\cdots\}$ be a countable state space, $\{X_\sigma, \sigma \in T\}$ be random variables defined on probability space $\{\Omega, F, P_n\}$ and taking values in G. Let

$$p_0 = \{p_0(x), x \in G\} \tag{1}$$

be a probability distribution on G, and

$$P_n = (P_n(y|x)) \quad x, y \in G \tag{2}$$

be random matrices on G^2. If $\forall \sigma \in T$, $\forall \xi_i \in T$ which satisfies $\xi_i \wedge \sigma \leq \bar{\sigma}, 1 \leq i \leq n$, we have

$$P_{i_\sigma}\left(X_\sigma = y \middle| X_{\bar{\sigma}} = x, X_{\xi_i} = x_i, 1 \leq i \leq n\right)$$

$$= P_{i_\sigma}\left(X_\sigma = y \,\middle|\, X_{\stackrel{-}{\sigma}} = x\right) = \left(P_{i_\sigma}\left(y\,|\,x\right)\right), \forall x, y, x_1, \cdots, x_n \in G \qquad (3)$$

and

$$P_0\left(X_{0,1} = x\right) = p_0\left(x\right), \forall x \in G \qquad (4)$$

Then $\{X_\sigma, \sigma \in T\}$ is called G-valued non-homogeneous Markov chains indexed by a non-homogeneous tree with distribution (1) and transition matrices (2).

Let

$$V_0 = V_0(X_0), \; V_{\xi_r} = V_{\xi_r}(X_0, \cdots, X_{\xi_r})$$
$$(\xi_r \in L_r, r = 0, 1, 2, \cdots), \qquad (5)$$

where V_{ξ_r} is a measurable function defined on G^{r+1} and taking values in R. If V_{ξ_r} is a function taking values in $[0, b]$ ($b \in R^+$), then $\{V_{\xi_r}, \xi_r \in L_r, r = 0, 1, 2, \cdots\}$ will be called random selection system in broad sense on T.

Lemma Let $\{X_\sigma, \sigma \in T\}$ be a non-homogeneous Markov chain on T, $\{g_n(x, y), n \geq 1\}$ be a sequence of functions defined on G^2 and taking values in R. Let λ be a constant,

$$F_n(\omega) = \sum_{k=0}^{n-1} \sum_{\sigma \in L_k} \sum_{\tau \in s(\sigma)} V_{\xi_k} g_{k+1}\left(X_\sigma, X_\tau\right) \qquad (6)$$

$$t_n(\lambda, \omega) = e^{\lambda F_n(\omega)} \bigg/ \prod_{k=0}^{n-1} \prod_{\sigma \in L_k} \prod_{\tau \in s(\sigma)} E\left[e^{\lambda V_{\xi_k} g_{k+1}(X_\sigma, X_\tau)} \,\middle|\, X_\sigma\right] \qquad (7)$$

Then $\{t_n(\lambda, \omega), \sigma(X^{T_n}), n \geq 1\}$ is a non-negative martingale.

Proof. By definition 2, we have

$$P\left(x^{T_n}\right) = P\left(X^{T_n} = x^{T_n}\right)$$
$$= p_0\left(x_{0,1}\right) \prod_{k=0}^{n-1} \prod_{\sigma \in L_k} \prod_{\tau \in s(\sigma)} P_k\left(X_\tau \,|\, X_\sigma\right), n \geq 1 \qquad (8)$$

Hence

$$P(X^{L_n} = x^{L_n} \,|\, X^{T_{n-1}}) = \frac{P(x^{T_n})}{P(x^{T_{n-1}})}$$
$$= \prod_{\sigma \in L_{n-1}} \prod_{\tau \in s(\sigma)} P_{n-1}\left(X_\tau \,|\, X_\sigma\right) \qquad (9)$$

By (8),we have

$$
E\left[e^{\lambda \sum\limits_{\sigma \in L_{n-1}} \sum\limits_{\tau \in S(\sigma)} V_{\xi_{n-1}} g_n(X_\sigma, X_\tau)} \middle| \sigma\left(X^{T_{n-1}}\right) \right]
$$

$$
= \sum_{x^{L_n}} e^{\lambda \sum\limits_{\sigma \in L_{n-1}} \sum\limits_{\tau \in S(\sigma)} V_{\xi_{n-1}} g_n(X_\sigma, X_\tau)}
$$

$$
\cdot P\left(X^{L_n} = x^{L_n} \middle| X^{T_{n-1}}\right)
$$

$$
= \prod_{\sigma \in L_{n-1}} \prod_{\tau \in s(\sigma)} E\left[e^{\lambda V_{\xi_{n-1}} g_n(X_\sigma, X_\tau)} \middle| X_\sigma \right] \qquad a.s. \qquad (10)
$$

It is easy to see that

$$
t_n(\lambda, \omega)
$$

$$
= t_{n-1}(\lambda, \omega) \frac{e^{\lambda \sum\limits_{\sigma \in L_{n-1}} \sum\limits_{\tau \in S(\sigma)} V_{\xi_{n-1}} g_n(X_\sigma, X_\tau)}}{\prod\limits_{\sigma \in L_{n-1}} \prod\limits_{\tau \in s(\sigma)} E\left[e^{\lambda V_{\xi_{n-1}} g_n(X_\sigma, X_\tau)} \middle| X_\sigma \right]} \qquad (11)
$$

By (10) and (11), we have

$$
E[t_n(\lambda, \omega) \mid \sigma(X^{T_{n-1}})] = t_{n-1}(\lambda, \omega) \quad a.s.
$$

Thus $\{t_n(\lambda, \omega), \sigma(X^{T_n}), n \geq 1\}$ is a non-negative martingale.

Theorem 1. Let $\{X_\sigma, \sigma \in T\}$ be a non-homogeneous Markov chain indexed by T, random selection system $\{V_{\xi_r}, \xi_r \in L_r, r = 0, 1, 2, \cdots\}$ in broad sense on T is defined by (5). $\{g_n(x, y), n \geq 1\}$ is defined by lemma. Let $\{a_n, n \geq 1\}$ be a sequence of positive random variables. $F_n(\omega)$ is defined by (6). Let

$$
G_n(\omega) = \sum_{k=0}^{n-1} \sum_{\sigma \in L_k} \sum_{\tau \in s(\sigma)} V_{\xi_k} E\left[g_{k+1}(X_\sigma, X_\tau) \mid X_\sigma \right] \qquad (12)
$$

Suppose $\alpha \geq 0$, Let

$$
D(\alpha) = \Big\{ \omega : \lim_{n \to \infty} a_n = \infty,
$$

$$
\limsup_{n \to \infty} \frac{1}{a_n} \sum_{k=0}^{n-1} \sum_{\sigma \in L_k} \sum_{\tau \in s(\sigma)} E\left[e^{\alpha V_{\xi_k} |g_{k+1}(X_\sigma, X_\tau)|} \middle| X_\sigma \right] < \infty \Big\} \qquad (13)
$$

Then

$$\lim_{n\to\infty}\frac{F_n(\omega)-G_n(\omega)}{a_n}=0 \qquad a.e.\ \omega\in D(\alpha) \tag{14}$$

Proof. By lemma, $\left\{\ t_n(\lambda,\omega),\sigma(X^{T_n}),n\ge1\ \right\}$ is a non-negative martingale. By Doob's martingale convergence theorem (c.f. [3]) , we have

$$\lim_{n\to\infty}t_n(\lambda,\omega)=t(\lambda,\omega)<+\infty \quad a.s. \tag{15}$$

By (13) and (15), we have

$$\limsup_{n\to\infty}\frac{1}{a_n}\ln t_n(\lambda,\omega)\le0 \qquad a.e.\ \omega\in D(\alpha) \tag{16}$$

By (6), (7) and 16), we have

$$\limsup_{n\to\infty}\frac{1}{a_n}\left\{\lambda\sum_{k=0}^{n-1}\sum_{\sigma\in L_k}\sum_{\tau\in s(\sigma)}V_{\xi_K}g_{k+1}(X_\sigma,X_\tau)\right.$$

$$\left.-\sum_{k=0}^{n-1}\sum_{\sigma\in L_k}\sum_{\tau\in s(\sigma)}\ln E\left[e^{\lambda V_{\xi_k}g_{k+1}(X_\sigma,X_\tau)}\Big|X_\sigma\right]\right\}\le0 \qquad a.e.\ \omega\in D(\alpha) \tag{17}$$

By (17) , the inequalities

$$\limsup_{n\to\infty}(a_n-b_n)\le d$$
$$\Rightarrow\limsup_{n\to\infty}(a_n-c_n)\le\limsup_{n\to\infty}(b_n-c_n)+d$$

$$\ln x\le x-1 \qquad\qquad (x>0), \qquad\qquad e^x-1-x\le\frac{x^2}{2}e^{|x|} \qquad \text{and}$$

$$\max\left\{x^2e^{-hx},x\ge0\right\}=\frac{4e^{-2}}{h^2}\ (h>0),\ \text{letting}\ 0<|\lambda|<\alpha,\ \text{we have}$$

$$\limsup_{n\to\infty}\frac{1}{a_n}\lambda\left\{\sum_{k=0}^{n-1}\sum_{\sigma\in L_k}\sum_{\tau\in s(\sigma)}V_{\xi_K}g_{k+1}(X_\sigma,X_\tau)\right.$$

$$\left.-\sum_{k=0}^{n-1}\sum_{\sigma\in L_k}\sum_{\tau\in s(\sigma)}V_\xi E\left[g_{k+1}(X_\sigma,X_\tau)|X_\sigma\right]\right\}$$

$$\le\limsup_{n\to\infty}\frac{1}{a_n}\sum_{k=0}^{n-1}\sum_{\sigma\in L_k}\sum_{\tau\in s(\sigma)}\left\{\ln E\left[e^{\lambda V_{\xi_k}g_{k+1}(X_\sigma,X_\tau)}\Big|X_\sigma\right]\right.$$

$$\left.-\lambda V_\xi E\left[g_{k+1}(X_\sigma,X_\tau)|X_\sigma\right]\right\}$$

$$\le\limsup_{n\to\infty}\frac{1}{a_n}\sum_{k=0}^{n-1}\sum_{\sigma\in L_k}\sum_{\tau\in s(\sigma)}\left\{E\left[e^{\lambda V_{\xi_k}g_{k+1}(X_\sigma,X_\tau)}\Big|X_\sigma\right]\right.$$

$$\left.-1-\lambda V_\xi E\left[g_{k+1}(X_\sigma,X_\tau)|X_\sigma\right]\right\}$$

$$\leq \frac{2\lambda^2 e^{-2}}{(|\lambda|-\alpha)^2} \limsup_{n\to\infty} \frac{1}{a_n} \sum_{k=0}^{n-1} \sum_{\sigma\in L_k} \sum_{\tau\in s(\sigma)} E\left[e^{\alpha|V_{\xi_k}g_{k+1}(X_\sigma,X_\tau)|}|X_\sigma\right] \quad (18)$$

Letting $0 < \lambda < \alpha$, dividing two sides of (18) by λ, we have

$$\limsup_{n\to\infty} \frac{1}{a_n}\left\{\sum_{k=0}^{n-1}\sum_{\sigma\in L_k}\sum_{\tau\in s(\sigma)} V_{\xi_K}g_{k+1}(X_\sigma,X_\tau)\right.$$

$$\left.-\sum_{k=0}^{n-1}\sum_{\sigma\in L_k}\sum_{\tau\in s(\sigma)} V_{\xi_k}E\left[g_{k+1}(X_\sigma,X_\tau)|X_\sigma\right]\right\}$$

$$\leq\frac{2\lambda e^{-2}}{(\lambda-\alpha)^2}\limsup_{n\to\infty}\frac{1}{a_n}\sum_{k=0}^{n-1}\sum_{\sigma\in L_k}\sum_{\tau\in s(\sigma)} E\left[e^{\alpha|V_{\xi_k}g_{k+1}(X_\sigma,X_\tau)|}|X_\sigma\right] \quad a.e.\ \omega\in D(\alpha) \quad (19)$$

Letting $\lambda\to 0^+$, we have

$$\limsup_{n\to\infty}\frac{F_n(\omega)-G_n(\omega)}{a_n}\leq 0 \quad a.e.\ \omega\in D(\alpha) \quad (20)$$

Taking $-\alpha < \lambda < 0$ in (18), dividing two sides of (18) by λ, we have

$$\liminf_{n\to\infty}\frac{1}{a_n}\left\{\sum_{k=0}^{n-1}\sum_{\sigma\in L_k}\sum_{\tau\in s(\sigma)} V_{\xi_K}g_{k+1}(X_\sigma,X_\tau)\right.$$

$$\left.-\sum_{k=0}^{n-1}\sum_{\sigma\in L_k}\sum_{\tau\in s(\sigma)} V_{\xi_k}E\left[g_{k+1}(X_\sigma,X_\tau)|X_\sigma\right]\right\}$$

$$\geq\frac{2\lambda e^{-2}}{(\lambda+\alpha)^2}\limsup_{n\to\infty}\frac{1}{a_n}\sum_{k=0}^{n-1}\sum_{\sigma\in L_k}\sum_{\tau\in s(\sigma)} E\left[e^{\alpha|V_{\xi_k}g_{k+1}(X_\sigma,X_\tau)|}|X_\sigma\right] \quad a.e.\ \omega\in D(\alpha) \quad (21)$$

Letting $\lambda\to 0^-$, we have

$$\liminf_{n\to\infty}\frac{F_n(\omega)-G_n(\omega)}{a_n}\geq 0 \quad a.e.\ \omega\in D(\alpha) \quad (22)$$

By (21) and (22), (14) holds. The theorem 1 is proved.

Theorem 2. Let $\{X_\sigma,\sigma\in T\}$ be a non-homogeneous Markov chain indexed by T, random selection system $\{V_{\xi_r},\xi_r\in L_r, r=0,1,2,\cdots\}$ in broad sense on T is defined by (5). $\{g_n(x,y),n\geq 1\}$ is defined by lemma. Let $\{a_n,n\geq 1\}$ be a sequence of positive random variables. $F_n(\omega)$ is defined by (6), $G_n(\omega)$ is defined by (12), Suppose $\alpha\geq 0$, Let

$$D_0(\alpha)=\left\{\omega:\lim_{n\to\infty}a_n=\infty,\right.$$

$$\limsup_{n\to\infty} \frac{1}{a_n} \sum_{k=0}^{n-1} \sum_{\sigma\in L_k} \sum_{\tau\in s(\sigma)} E\left[V_{\xi_i} g_{k+1}^{2} \left(X_{\sigma}, X_{\tau} \right) e^{\alpha V_{\xi_i} |g_{k+1}(X_{\sigma}, X_{\tau})|} \big| X_{\sigma} \right] < \infty \Bigg\} \quad (23)$$

Then

$$\lim_{n\to\infty} \frac{F_n(\omega) - G_n(\omega)}{a_n} = 0 \qquad a.e.\ \omega\in D_0(\alpha) \qquad (24)$$

Proof.   By Theorem 1, we have

$$\limsup_{n\to\infty} \frac{1}{a_n} \lambda \Bigg\{ \sum_{k=0}^{n-1} \sum_{\sigma\in L_k} \sum_{\tau\in s(\sigma)} V_{\xi_K} g_{k+1} \left(X_{\sigma}, X_{\tau} \right)$$

$$- \sum_{k=0}^{n-1} \sum_{\sigma\in L_k} \sum_{\tau\in s(\sigma)} V_{\xi} E\left[g_{k+1} \left(X_{\sigma}, X_{\tau} \right) \big| X_{\sigma} \right] \Bigg\}$$

$$\leq \frac{\lambda^2 b}{2} \limsup_{n\to\infty} \frac{1}{a_n} \sum_{k=0}^{n-1} \sum_{\sigma\in L_k} \sum_{\tau\in s(\sigma)} E\left[V_{\xi_i} g_{k+1}^{2} \left(X_{\sigma}, X_{\tau} \right) e^{\alpha V_{\xi_i} |g_{k+1}(X_{\sigma}, X_{\tau})|} \big| X_{\sigma} \right] \quad a.e.\ \omega\in D_0(\alpha) \quad (25)$$

Letting $0 < \lambda < \alpha$, dividing two sides of (25) by λ, we have

$$\limsup_{n\to\infty} \frac{1}{a_n} \Bigg\{ \sum_{k=0}^{n-1} \sum_{\sigma\in L_k} \sum_{\tau\in s(\sigma)} V_{\xi_K} g_{k+1} \left(X_{\sigma}, X_{\tau} \right)$$

$$- \sum_{k=0}^{n-1} \sum_{\sigma\in L_k} \sum_{\tau\in s(\sigma)} V_{\xi} E\left[g_{k+1} \left(X_{\sigma}, X_{\tau} \right) \big| X_{\sigma} \right] \Bigg\}$$

$$\leq \frac{\lambda b}{2} \limsup_{n\to\infty} \frac{1}{a_n} \sum_{k=0}^{n-1} \sum_{\sigma\in L_k} \sum_{\tau\in s(\sigma)} E\left[V_{\xi_i} g_{k+1}^{2} \left(X_{\sigma}, X_{\tau} \right) e^{\alpha V_{\xi_i} |g_{k+1}(X_{\sigma}, X_{\tau})|} \big| X_{\sigma} \right] \quad a.e.\ \omega\in D_0(\alpha) \quad (26)$$

Letting $\lambda \to 0^+$, we have

$$\limsup_{n\to\infty} \frac{F_n(\omega) - G_n(\omega)}{a_n} \leq 0 \qquad a.e.\ \omega\in D_0(\alpha) \qquad (27)$$

Taking $-\alpha < \lambda < 0$ in (25), dividing two sides of (25) by λ, we   have

$$\liminf_{n\to\infty} \frac{1}{a_n} \Bigg\{ \sum_{k=0}^{n-1} \sum_{\sigma\in L_k} \sum_{\tau\in s(\sigma)} V_{\xi_K} g_{k+1} \left(X_{\sigma}, X_{\tau} \right)$$

$$- \sum_{k=0}^{n-1} \sum_{\sigma\in L_k} \sum_{\tau\in s(\sigma)} V_{\xi} E\left[g_{k+1} \left(X_{\sigma}, X_{\tau} \right) \big| X_{\sigma} \right] \Bigg\}$$

$$\geq \frac{\lambda b}{2} \limsup_{n\to\infty} \frac{1}{a_n} \sum_{k=0}^{n-1} \sum_{\sigma\in L_k} \sum_{\tau\in s(\sigma)} E\left[V_{\xi_i} g_{k+1}^{2} \left(X_{\sigma}, X_{\tau} \right) e^{\alpha V_{\xi_i} |g_{k+1}(X_{\sigma}, X_{\tau})|} \big| X_{\sigma} \right] \quad a.e.\ \omega\in D_0(\alpha) \quad (28)$$

Letting $\lambda \to 0^-$, we have

$$\liminf_{n \to \infty} \frac{F_n(\omega) - G_n(\omega)}{a_n} \geq 0 \quad a.e.\ \omega \in D_0(\alpha) \tag{29}$$

By (27) and (29), (24) holds. The theorem 2 is proved.

Acknowledgment. The guidance project of the research and development of science and technology of Hebei province. (No: Z2008308).

References

1. Jin, S.-H., Lu, J.-G., Wang, N., Jiang, H.-H.: A class of Strong Deviation Theorems for the Sequence of arbitrary Integer-Valued Random Variables. In: Proceedings of the second International conference on Information and computing Science, pp. 193–195. IEEE CS (2009)
2. Liu, W., Yang, W.: The Markov approximation of the sequences of N—Valued variables and a class of small deviation theorems. Stochastic Process. Appl. (1), 117–130 (2000)
3. Doob, J.L.: Stochastic Processes. Wiley, New York (1953)

A Strong Deviation Theorem for Markov Chains Functional Indexed by a Non-homogeneous Tree

Shaohua Jin[1], Yanmin Zhang[1], Yanping Wan[1], Nan Li[1], and Huipeng Zhang[2]

[1] Hebei University of Technology, Tianjin, 300130, China
[2] Institute of Military Traffic, Tianjin, 300161, China
tanjin154@sina.cn

Abstract. The strong deviation theorems is one of the central questions for studying in the International Probability theory. In this paper, a strong deviation theorem for Markov chains functional indexed by a non-homogeneous tree were obtained by constructing a non-negative martingale.

Keywords: Markov chains, martingale, non-homogeneous tree, strong deviation theorem.

1 Introduction

The deviation theorems is one of the central questions for studying in the International Probability theory. (c.f. [1] and [2]). In this paper, a strong deviation theorem for Markov chains functional indexed by a non-homogeneous tree were obtained by constructing a non-negative martingale.

Definition 1. Let T be an infinite tree, $\{N_n, n \geq 1\}$ be a countable space , where N_n is a positive integer. If every vertex from level $n(n \geq 0)$ has N_{n+1} sons, the tree T is a *Bethe* tree in broad sense or a *Cayley* tree in broad sense. Particularly, let $N = \{0, 1, 2, \cdots\}$, $m(m \geq 2)$ be an integer , we divide N into several sets as follow:

$$(0) = \{0, m, 2m, \cdots\cdots\}$$
$$(1) = \{1, m+1, 2m+1, \cdots\cdots\}$$
$$\cdots\cdots$$
$$(m-1) = \{m-1, 2m-1, 3m-1, \cdots\cdots\}$$

When $n \in (i)$, let $N_{n+1} = \alpha_i$ (α_i is a positive integer and not all of them is 1), $i = 0, 1, 2, \cdots, m-1$. In this way, we get a particular non-homogeneous tree $T_{\alpha_0, \alpha_1, \cdots, \alpha_{m-1}}$.

Y. Wu (Ed.): International Conference on WTCS 2009, AISC 116, pp. 345–352.
springerlink.com   © Springer-Verlag Berlin Heidelberg 2012

2   A Strong Deviation Theorem for Markov Chains Functional on a Non-homogeneous Tree

Definition 2. Let $S = \{s_1, s_2, s_3, \cdots\}$ be a countable state space, $\left\{ X_{\xi_k}, \xi_k \in T \right\}$ be random variables defined on probability space $\{\Omega, F, P_n\}$ and taking values in S. Let

$$\left(p(s_1), p(s_2) \cdots \right) \tag{1}$$

be a probability distribution on S, and

$$P_n = \left(P_n(y|x) \right), x, y \in S \tag{2}$$

be random matrices on S^2. If $\sigma_i \in T$ which satisfies $\sigma_i \wedge t \leq \overline{t}$, $1 \leq i \leq n$, we have

$$P_t\left(X_t = y \middle| X_{\underset{t}{-}} = x, X_{\sigma_i} = x_i, 1 \leq i \leq n \right) = P_t\left(X_t = y \middle| X_{\underset{t}{-}} = x \right) = P_t(y|x)$$

$$\forall x, y, x_1, \cdots, x_n \in S \tag{3}$$

and

$$P_0\left(X_{0,1} = x \right) = p_0(x), \forall x \in S \tag{4}$$

Then $\left\{ X_{\xi_k}, \xi_k \in T \right\}$ is called S-valued non-homogeneous Markov chains indexed by a non-homogeneous tree with distribution （1）and transition matrices （2）.

By definition 2, the joint density function of the non-homogeneous Markov chains on a non-homogeneous tree is

$$P(X^{T_n}) = P(X^{T_n} = x^{T_n})$$

$$= p_0(x_{0,1}) \prod_{r=0}^{n-1} \prod_{\xi_r \in L_r} \prod_{\xi_{r+1} \in S(\xi_r)} p_{r+1}(X_{\xi_{r+1}} | X_{\xi_r}) \tag{5}$$

Definition 3. Let P_n be random matrices defined by (2), Q be another probability measure defined on probability space $\{\Omega, F\}$, the joint density function on probability measure Q is

$$Q(X^{T_n}) = Q(X^{T_n} = x^{T_n})$$

$$= q_0(x_{0,1}) \prod_{r=0}^{n-1} \prod_{\xi_r \in L_r} \prod_{\xi_{r+1} \in S(\xi_r)} q_{r+1}(X_{\xi_{r+1}} | X_{\xi_r}) \tag{6}$$

Let

$$r_n(\omega) = \ln \frac{P\left(X^{T_n} \right)}{Q\left(X^{T_n} \right)} \tag{7}$$

and

$$r(\omega) = \limsup_{n \to \infty} \frac{1}{|T_n|} r_n(\omega) \tag{8}$$

Among them, ω is the sample point. $r(\omega)$ is called the sample relative entropy rate of $\{|T_n|, n \geq 0\}$.

Let $\{f_{r+k}(X_{\xi_r}, \cdots, X_{\xi_{r+k}}), r \geq 0\}$ be a sequence of real functions defined on $S^{k+1}, \{a_n, n \geq 0\}$ be a sequence of increasing positive numbers.

Let

$$Y_{r+k}(X_{\xi_r}, \cdots, X_{\xi_{r+k}})$$
$$= \begin{cases} f_{r+k}(X_{\xi_r}, \cdots, X_{\xi_{r+k}}), & \left| f_{r+k}(X_{\xi_r}, \cdots, X_{\xi_{r+k}}) \right| \leq a_{r+k} \\ 0, & \left| f_{r+k}(X_{\xi_r}, \cdots, X_{\xi_{r+k}}) \right| > a_{r+k} \end{cases} \tag{9}$$

$$b_{r+k}(X_{\xi_{r+k}}) = E[f_{r+k}(X_{\xi_r}, \cdots, X_{\xi_{r+k}}) | X_{\xi_{r+k-1}}], r \geq 1 \tag{10}$$

$$b^*_{r+k}(X_{\xi_{r+k}}) = E[Y_{r+k}(X_{\xi_r}, \cdots, X_{\xi_{r+k}}) | X_{\xi_{r+k-1}}], r \geq 1 \tag{11}$$

Then

$$Q_{r+k}\left(\lambda; X_{\xi_{r+k-1}}\right) = E\left\{ \exp\left[\frac{\lambda\left[Y_{r+k}(X_{\xi_r}, \cdots, X_{\xi_{r+k}}) - b^*_{r+k}(X_{\xi_{r+k}}) \right]}{a_{r+k}} \right] | X_{\xi_{r+k-1}} \right\}$$

$$= \sum_{x_{\xi_{r+k}}} \phi_{r+k}\left(\lambda; X_{\xi_{r+k-1}}\right) q_{r+k}\left(X_{\xi_{r+k}} = x_{\xi_{r+k}} | X_{\xi_{r+k-1}}\right) \tag{12}$$

$$q\left(\lambda; X^{T_n}\right) = q_0\left(x_{0,1}\right)$$

$$\prod_{r=0}^{n-k} \prod_{\xi_r \in L_r} \cdots \prod_{\xi_{r+k} \in S(\xi_{r+k-1})} \frac{\phi_{r+k}\left(\lambda; X_{\xi_{r+k-1}}\right) q\left(X_{\xi_{r+k}} | X_{\xi_{r+k-1}}\right)}{Q_{r+k}\left(\lambda; X_{\xi_{r+k-1}}\right)} \tag{13}$$

Where

$$\phi_{r+k}\left(\lambda; X_{\xi_{r+k-1}}\right) = \exp\left[\frac{\lambda\left[Y_{r+k}(X_{\xi_r}, \cdots, X_{\xi_{r+k}}) - b^*_{r+k}(X_{\xi_{r+k}}) \right]}{a_{r+k}} \right]$$

Lemma. Let $\left\{ X_{\xi_k}, \xi_k \in T \right\}$ be a non-homogeneous Markov chain, $P(X^{T_n}), q(\lambda; X^{T_n})$ be defined by (5) and (13). Let

$$t_n(\lambda, \omega) = q\left(\lambda; X^{T_n}\right) / P\left(X^{T_n}\right) \tag{14}$$

Then $\left\{t_{t_n}(\lambda,w),\sigma\left(X^{T_n}\right),n\geq k\right\}$ is a non-negative martingale.

Theorem. Let $\left\{X_{\xi_k},\xi_k\in T\right\}$ be a non-homogeneous Markov chain. Let $\{f_{r+k}(X_{\xi_r},\cdots,X_{\xi_{r+k}}),r\geq 0\}$ be a sequence of real function defined on S^{k+1}. $Y_{r+k}(X_{\xi_r},\cdots,X_{\xi_{r+k}})$, $b_{r+k}(X_{\xi_{r+k}})$ 和 $b^*_{r+k}(X_{\xi_{r+k}})$ are defined as before. $\{\varphi_n(x),n\geq 1\}$ is a sequence of continuous positive even function defined on $(-\infty,+\infty)$,and when $|x|\uparrow$,

$$\frac{\varphi_n(x)}{|x|}\uparrow \text{ and } \frac{\varphi_n(x)}{x^2}\downarrow \tag{17}$$

hold. Let

$$D(c)=\left\{\omega:r(\omega)\leq c\right\} \tag{18}$$

Where c is a non-negative constant.

$$D^*=\left\{\omega:\limsup_{n\to\infty}\frac{1}{|T^n|}\sum_{r=0}^{n-k}\sum_{\xi_r\in L_r}\cdots\sum_{\xi_{r+k}\in S(\xi_{r+k-1})}\right.$$

$$\left. E\left[\frac{\varphi_{r+k}(f_{r+k}(X_{\xi_r},\cdots,X_{\xi_{r+k}}))}{\varphi_{r+k}(a_{r+k})}\Bigg| X_{\xi_{r+k-1}}\right]=\sigma(\omega)<\infty \right. \tag{19}$$

If $\displaystyle\lim_{n\to\infty}\sum_{r=0}^{n-k}\sum_{\xi_r\in L_r}\cdots\sum_{\xi_{r+k}\in S(\xi_{r+k-1})}P\left(\left|f_{r+k}(X_{\xi_r},\cdots,X_{\xi_{r+k}})\right|>a_{r+k}\right)<+\infty \tag{20}$

Then

$$\limsup_{n\to\infty}\frac{1}{|T^n|}\sum_{r=0}^{n-k}\sum_{\xi_r\in L_r}\cdots\sum_{\xi_{r+k}\in S(\xi_{r+k-1})}\left[\frac{f_{r+k}(X_{\xi_r},\cdots,X_{\xi_{r+k}})-b_{r+k}(X_{\xi_{r+k}})}{a_{r+k}}\right]$$

$$\leq\beta(c,\sigma(\omega))+\sigma(\omega) \quad a.e.\omega\in D(c)\bigcap D^* \tag{21}$$

$$\liminf_{n\to\infty}\frac{1}{|T^n|}\sum_{r=0}^{n-k}\sum_{\xi_r\in L_r}\cdots\sum_{\xi_{r+k}\in S(\xi_{r+k-1})}\left[\frac{f_{r+k}(X_{\xi_r},\cdots,X_{\xi_{r+k}})-b_{r+k}(X_{\xi_{r+k}})}{a_{r+k}}\right]$$

$$\geq\alpha(c,\sigma(\omega))-\sigma(\omega) \quad a.e.\omega\in D(c)\bigcap D^* \tag{22}$$

Where

$$\psi(\lambda,x,y)=\frac{x}{\lambda}+\lambda e^{2|\lambda|}y,0\leq x,y\leq+\infty,\lambda\neq 0 \tag{23}$$

$$\alpha(x,y)=\sup\{\psi(\lambda,x,y);\lambda<0\},0\leq x,y\leq+\infty \tag{24}$$

$$\beta(x,y)=\inf\{\psi(\lambda,x,y);\lambda>0\},0\leq x,y\leq+\infty \tag{25}$$

Proof. By (9) , (10) and (11) , we have

$$\frac{f_{r+k}(X_{\xi_r},\cdots,X_{\xi_{r+k}})-b_{r+k}(X_{\xi_{r+k}})}{a_{r+k}}$$

$$=\frac{f_{r+k}(X_{\xi_r},\cdots,X_{\xi_{r+k}})-Y_{r+k}(X_{\xi_r},\cdots,X_{\xi_{r+k}})}{a_{r+k}}$$

$$+\frac{Y_{r+k}(X_{\xi_r},\cdots,X_{\xi_{r+k}})-b^*_{r+k}(X_{\xi_{r+k}})}{a_{r+k}}$$

$$+\frac{b^*_{r+k}(X_{\xi_{r+k-1}})-b_{r+k}(X_{\xi_{r+k}})}{a_{r+k}} \quad (26)$$

By Lemma and *Doob* 's martingale convergence theorem (c.f. [3]), $\exists A(\lambda)\in F, P(A(\lambda))=1$, such that

$$\limsup_{n\to\infty}\frac{1}{|T_n|}\ln t_n(\lambda,\omega)\leq 0 \quad a.e.\omega\in A(\lambda). \quad (27)$$

Hence

$$\limsup_{n\to\infty}\frac{1}{|T^n|}\ln t_n(\lambda,\omega) = \limsup_{n\to\infty}\frac{1}{|T^n|}\ln\left\{\left[q_0(x_{0,1})\prod_{r=0}^{n-k}\prod_{\xi_r\in L_r}\cdots\prod_{\xi_{r+k}\in S(\xi_{r+k-1})}\right.\right.$$

$$\frac{\phi_{r+k}(\lambda;X_{\xi_{r+k-1}})q_{r+k}(X_{\xi_{r+k}}|X_{\xi_{r+k-1}})}{Q_{r+k}(\lambda;X_{\xi_{r+k-1}})}]\cdot\frac{1}{P(X^{T^n})}\right\}$$

$$=\limsup_{n\to\infty}\frac{1}{|T^n|}\ln\left\{\lambda\sum_{r=0}^{n-k}\sum_{\xi_r\in L_r}\cdots\sum_{\xi_{r+k}\in S(\xi_{r+k-1})}\left[\frac{Y_{r+k}(X_{\xi_r},\cdots,X_{\xi_{r+k}})-b^*_{r+k}(X_{\xi_{r+k}})}{a_{r+k}}\right]\right.$$

$$-\sum_{r=0}^{n-k}\sum_{\xi_r\in L_r}\cdots\sum_{\xi_{r+k}\in S(\xi_{r+k-1})}\ln Q_{r+k}(\lambda;X_{\xi_{r+k-1}})-r_n(\omega)\leq 0$$

$$a.e.\omega\in A(\lambda) \quad (28)$$

By (18) ,(28) and the inequality

$$\limsup_{n\to\infty}(a_n-b_n)\leq d \Rightarrow \limsup_{n\to\infty}(a_n-c_n)\leq\limsup_{n\to\infty}(b_n-c_n)+d, \quad (29)$$

we have

$$\limsup_{n\to\infty}\frac{1}{|T^n|}\left\{\lambda\sum_{r=0}^{n-k}\sum_{\xi_r\in L_r}\cdots\sum_{\xi_{r+k}\in S(\xi_{r+k-1})}\right.$$

$$\left.\left[\frac{Y_{r+k}(X_{\xi_r},\cdots,X_{\xi_{r+k}})-b^*_{r+k}(X_{\xi_{r+k}})}{a_{r+k}}\right]\right\}$$

$$\leq \limsup_{n \to \infty} \frac{1}{|T^n|} \sum_{r=0}^{n-k} \sum_{\xi_r \in L_r} \cdots \sum_{\xi_{r+k} \in S(\xi_{r+k-1})} \ln Q_{r+k}\left(\lambda; X_{\xi_{r+k-1}}\right) + c$$

$$a.e.\omega \in A(\lambda) \bigcap D(c) \qquad (30)$$

and

$$E\left[\frac{Y_{r+k}(X_{\xi_r}, \cdots, X_{\xi_{r+k}}) - b^*_{r+k}(X_{\xi_{r+k}})}{a_{r+k}} \mid X_{\xi_{r+k-1}}\right] = 0 \qquad (31)$$

$$\left|\frac{Y_{r+k}(X_{\xi_r}, \cdots, X_{\xi_{r+k}}) - b^*_{r+k}(X_{\xi_{r+k}})}{a_{r+k}}\right| \leq 2 \qquad (32)$$

By the inequality $e^x - 1 - x \leq \frac{1}{2}x^2 e^{|x|}$, we have

$$0 \leq Q_{r+k}\left(\lambda; X_{\xi_{r+k-1}}\right) - 1$$

$$\leq \frac{1}{2}\lambda^2 e^{2|\lambda|} E\left[\left(\frac{Y_{r+k}(X_{\xi_r}, \cdots, X_{\xi_{r+k}})}{a_{r+k}}\right)^2 \mid X_{\xi_{r+k-1}}\right] \qquad (33)$$

By (17), we have

$$\frac{x^2}{(a_{r+k})^2} \leq \frac{\varphi_{r+k}(x)}{\varphi_{r+k}(a_{r+k})} \quad , |x| \leq a_{r+k}$$

then

$$[\frac{Y_{r+k}(X_{\xi_r}, \cdots, X_{\xi_{r+k}})}{a_{r+k}}]^2$$

$$\leq \frac{\varphi_{r+k}(Y_{r+k}(X_{\xi_r}, \cdots, X_{\xi_{r+k}}))}{\varphi_{r+k}(a_{r+k})}$$

$$\leq \frac{\varphi_{r+k}(f_{r+k}(X_{\xi_r}, \cdots, X_{\xi_{r+k}}))}{\varphi_{r+k}(a_{r+k})} \qquad (34)$$

Hence

$$0 \leq \limsup_{n \to \infty} \frac{1}{|T^n|} \sum_{r=0}^{n-k} \sum_{\xi_r \in L_r} \cdots \sum_{\xi_{r+k} \in S(\xi_{r+k-1})} \left[Q_{r+k}\left(\lambda; X_{\xi_{r+k-1}}\right) - 1\right]$$

$$\leq \frac{1}{2}\lambda^2 e^{2|\lambda|}\sigma(\omega) \qquad a.e.\omega \in D^* \qquad (35)$$

By the inequality $0 \le \ln x \le x-1 \quad (x \ge 0)$, we have

$$\limsup_{n\to\infty} \frac{1}{|T^n|} \sum_{r=0}^{n-k} \sum_{\xi_r \in L_r} \cdots \sum_{\xi_{r+k} \in S(\xi_{r+k-1})} \ln Q_{r+k}\left(\lambda; X_{\xi_{r+k-1}}\right)$$

$$\le \frac{1}{2} \lambda^2 e^{2|\lambda|} \sigma(\omega) \quad a.e.\omega \in D^* \tag{36}$$

By (30) and (36), we have

$$\limsup_{n\to\infty} \frac{1}{|T^n|} \left\{ \lambda \sum_{r=0}^{n-k} \sum_{\xi_r \in L_r} \cdots \sum_{\xi_{r+k} \in S(\xi_{r+k-1})} \left[\frac{Y_{r+k}(X_{\xi_r}, \cdots, X_{\xi_{r+k}}) - b'_{r+k}(X_{\xi_{r+k}})}{a_{r+k}} \right] \right\}$$

$$\le \frac{1}{2} \lambda^2 e^{2|\lambda|} \sigma(\omega) + c \quad a.e.\omega \in A(\lambda) \cap D(c) \cap D^* \tag{37}$$

Let R^* is a positive rational number set. Let $A^* = \bigcap_{\lambda \in R^*} A(\lambda)$, then $P(A^*) = 1$, by

(37), we have

$$\limsup_{n\to\infty} \frac{1}{|T^n|} \sum_{r=0}^{n-k} \sum_{\xi_r \in L_r} \cdots \sum_{\xi_{r+k} \in S(\xi_{r+k-1})} \left[\frac{Y_{r+k}(X_{\xi_r}, \cdots, X_{\xi_{r+k}}) - b'_{r+k}(X_{\xi_{r+k}})}{a_{r+k}} \right]$$

$$\le \frac{1}{2} \lambda e^{2|\lambda|} \sigma(\omega) + \frac{c}{\lambda} \quad a.e.\omega \in A^* \cap D(c) \cap D^* \tag{38}$$

Since $\psi(\lambda, x, y)$ continuous for λ, then for $\forall \omega \in A^* \cap D(c) \cap D^*$, $\exists \lambda_n(\omega) \in R^*, n \ge 1$, such that

$$\psi(\lambda_n(\omega), c, \sigma(\omega)) = \beta(c, \sigma(\omega)) \tag{39}$$

By (39) and (40), we have

$$\limsup_{n\to\infty} \frac{1}{|T^n|} \sum_{r=0}^{n-k} \sum_{\xi_r \in L_r} \cdots \sum_{\xi_{r+k} \in S(\xi_{r+k-1})} \left[\frac{Y_{r+k}(X_{\xi_r}, \cdots, X_{\xi_{r+k}}) - b'_{r+k}(X_{\xi_{r+k}})}{a_{r+k}} \right] \le \beta(c, \sigma(\omega))$$

$$a.e.\omega \in A^* \cap D(c) \cap D^* \tag{41}$$

Since $P(A^*) = 1$, then

$$\limsup_{n\to\infty} \frac{1}{|T^n|} \sum_{r=0}^{n-k} \sum_{\xi_r \in L_r} \cdots \sum_{\xi_{r+k} \in S(\xi_{r+k-1})} \left[\frac{Y_{r+k}(X_{\xi_r}, \cdots, X_{\xi_{r+k}}) - b'_{r+k}(X_{\xi_{r+k}})}{a_{r+k}} \right] \le \beta(c, \sigma(\omega))$$

$$a.e. \omega \in D(c) \cap D^* \tag{42}$$

Since $a_n \uparrow$, we have

$$\lim_{n\to\infty} \frac{1}{|T^*|} \sum_{r=0}^{n-k} \sum_{\xi_r \in L_r} \cdots \sum_{\xi_{r+k} \in S(\xi_{r+k-1})} \left[\frac{f_{r+k}(X_{\xi_r}, \cdots, X_{\xi_{r+k}}) - Y_{r+k}(X_{\xi_r}, \cdots, X_{\xi_{r+k}})}{a_{r+k}} \right]$$

$$= 0 \qquad a.e.\ \omega \in D(c) \cap D^* \qquad (43)$$

By (17) and (19), we have

$$\limsup_{n\to\infty} \frac{1}{|T^n|} \sum_{r=0}^{n-k} \sum_{\xi_r \in L_r} \cdots \sum_{\xi_{r+k} \in S(\xi_{r+k-1})} \left[\frac{|b_{r+k}(X_{\xi_{r+k}}) - b^*_{r+k}(X_{\xi_{r+k}})|}{a_{r+k}} \right]$$

$$\leq \sigma(\omega) \qquad\qquad a.e.\ \omega \in D^* \qquad (44)$$

By (26), (42), (43) and (44), (21) holds.
Similarly, we can get the other results. The theorem is proved.

Acknowledgment. The guidance project of the research and development of science and technology of Hebei province. (No: Z2008308) .

References

1. Jin, S.-H., Lu, J.-G., Wang, N., Jiang, H.-H.: A class of Strong Deviation Theorems for the Sequence of arbitrary Integer-Valued Random Variables. In: Proceedings of the second International conference on Information and computing Science, pp. 193–195. IEEE CS (2009)
2. Liu, W., Yang, W.: The Markov approximation of the sequences of N—Valued variables and a class of small deviation theorems. Stochastic Process. Appl. (1), 117–130 (2000)
3. Doob, J.L.: Stochastic Processes. Wiley, New York (1953)

The Innovation Research of Teaching Methods on Higher Education of Management Discipline

Wei Cheng and Hong Jiang

College of Economics and Management
HLJ Bayi Agricultural University
Daqing, China
pnc7899@sina.com

Abstract. Traditional teaching method of management discipline in higher education has not adapted to the information age of development needs. In the new economic background, higher schools and teachers as the teaching subject to improve the teaching effect of higher education and promote the development of subjects, finally improve the students' comprehensive quality, it is necessary to reform the traditional teaching methods. Based on the teaching practice of the management course, this thesis from the perspective of teachers and based on group behavior of students in higher education, putting forward some suggestions in the higher education of management discipline.

Keywords: higher education, teaching method, innovative research, management discipline.

1 Introduction

Teaching method is accelerating transformation with the development of science and technology progress. The research on systemic teaching method has attracted more attention of the whole society. The traditional teaching method requires a revolution in order to reach the teaching goal. Management discipline has strong cross-disciplinary background, so it is necessary to explore new teaching method. Some factors will influence the way of new teaching. This thesis analyzed the current situation of management discipline teaching and learning then putted forward some practical suggestions. [1]

2 Evolution of Teaching Methods

For the education-researchers are at different times and have different perspectives, they make the definition of the "teaching methods" differently. Generally we considered: teaching methods is that students and teachers in order to achieve a common goal to complete a common task of teaching, in the teaching process in use of a term of methods. [2]

Y. Wu (Ed.): International Conference on WTCS 2009, AISC 116, pp. 353–358.
springerlink.com © Springer-Verlag Berlin Heidelberg 2012

Combining the characteristics of regular higher education, teaching methods can be defined as: in a specific environment, teachers and students take a variety of means to achieve the common goal of teaching then take to facilitate the completion of teaching tasks. Teaching method should not only assess the teaching effectiveness, but also improve the overall quality of teachers and students.

Since the teaching method in essence demands a high combination of the teaching activities by teachers and learning activities by students. To achieve the optimal effect of teaching, which doesn't mean the independent instruction of teachers or the independent study of students, but how do those elements combined perfectly to realize the teaching goal. [3] It needs considering of the specific environment that teachers and students are in and the overall competence that they have. The effective teaching method need to be considered that the external environment like policy, economy, culture, technical level and basic level of the whole distinct or the same occupation. And the overall competence has a great influence upon the effects of teaching methods, which consisting lots of patterns, should be concretely chosen with a view of "balance and change". There is no universal accepted ways to distinguish good and bad among the teaching methods. The important thing is to change flexible based on the specific circumstances and to choose the most suited way.

Traditional teaching methods emphasize the view that from the system to look at the whole teaching process and emphasize the separate the roles of teachers and students. According to Babansiji's process-oriented teaching method, the relationship of every teaching element and link should be firstly put into consideration in reaching optimization. "Weston and Granton's teaching methods focus on specific methods of classification. Raska's teaching method ranges between the two methods above. [4] The biggest drawback of the traditional teaching methods is the lack of the specific analysis of the external environment in which teachers and students stay, and only emphasizing the role of teachers and students separation.

3   Characteristics of Management Discipline

In the current cross-disciplinary background, management disciplines have become one of the most populated cross-disciplinary sciences. New subjects appeal fast and discipline of crossover and subdivide become the biggest obstacle in the management disciplines learning process. [5] This phenomenon put forward higher request for the teachers and students' quality. The main characteristics of management disciplines are as follows:

3.1   Emphasize the Learning of Public Basic Course

To the students, they can study further into their major basic course and combine their interests in choosing professional elective only when they learn the public basic course well. These are the three steps to the ideal assumption of the learning model for the students majoring in management. For example, one can only learn professional curriculum like management science, operations research, management economics etc, and study further into the principles and theories of comprehensive evaluation method when he/she learns higher math well. One won't have to start from the very beginning when he/she study industrial technology economics or agricultural technology economics if he/she learns technical economics well. The students should pay more attention to the public basic course learning for management is a inter discipline. So the students' early formed misconception must be corrected because the pubic basic course learning is of the most importance. As for the teachers, they must have a solid grounding in basic courses, which is the prerequisite to ensure an effective teaching.

3.2   Cultivate Versatile Talents

Attention should be paid to the education of inter-disciplinary talent. For a long time, the function of higher education was to develop high-level specialists. It was a high-level specialized education. Thereof, universities offered specialized education in accordance with the related professional emphasis. Students took up occupations in terms of their majors after graduation. This became a basic operation model of higher education. In the past 50 years, this model has fostered over 10,000,000 professionals for the society. It greatly facilitated social economy. Its contribution was indelible. However, current society is inclined to the talents who feature inter-discipline and the integration of skill. It is an inevitable trend in the requirement of the society on human resource. Inter-disciplinary talents are not simple specialized talents, nor are they simple all-round persons. They are the persons who have prominent talent in one area with comprehensive abilities in other fields. For example, marketing major students should be equipped with not only marketing knowledge, but also medical treatment, agricultural products and other industries' special knowledge. It makes the students adapt themselves to the sales of medical treatment and agricultural products more easily.

3.3   The Practice Is Eessencial

Theories are in a particular social circumstance, summing up a large number of community development practices. Management is a high academic technology, and also a high practical art. Sometimes, the theory is too strong but will affect the students learning. Some classical theories rarely the actual business activities in the enterprise has been applied, while the theoretical front there are strong limitations. For management, there is no universal applied theory. The most important in the theory of the learning process is to understand the different way of thinking. Theoretical study is only a part of the learning process. Babansiji in accordance with Raska's third class teaching method: "Check and teaching effectiveness of the method of self-examination" for students to participate in practice, based on the theory to solve practical problems is a very effective teaching method. To enable students to understand business activities in a short time, most of the schools lead student to enterprise and take practice. Many practical courses are a mere formality, just lead students to go visit the corporate. Practical teaching is important, but the subjective and objective conditions do not allow students have more opportunities to practice. This restricted curriculum integration badly. So, management discipline has to adopt innovative research, changing the traditional method, and puts forward the new idea. In accordance with the constraint conditions, teaching objectives should put forward some new teaching methods.

4   The Main Factors of Innovation Teaching Methods and the Concrete Measures

4.1   The Main Factors of Iinnovation Teaching Methods

- *Stage of Social Development*

Education is under certain social environment. Some factors within the environment could influence the teaching method and effectiveness. For example, lecture with some ancillary methods is still the most popular way. Multimedia, network, white board, visual classroom and visual environment are new trends. With the increase of education investment, as a result of government pays more attention to the education, great changes have been occurred in teaching method. Thanks to the advance of science and technology and economic growth. It laid foundation of teaching method innovation.

- *Quality and Subjective Tendency of Teacher*

Teachers in higher education obtained teacher qualification through related training. It is the basic requirement of a teacher. The education standard could be jeopardized by the absence of high-quality teachers. Un-normal college graduates lack of related practice and experience in classroom control and group psychoanalysis. In teaching, some teachers are inclined to lecture;+ others adopt interaction between students and teachers. Subjective tendency is an individual action with great differentiation. It plays a big role in the teaching method selection and teaching effectiveness.

- *Quality and subjective tendency of student*

After the college expansion plan in 1999, elite education turns to mass education. Under this background, it is necessary to classify the students. Teaching students in accordance of their aptitude is no longer a slogan. It has been involved in practical teaching. Teachers should have the ability to distinguish the differences of students coming from different area with different major, grade and gender. Different teaching method should be applied to different.

4.2   The Concrete Measures of Innovative Teaching Methods

- *Launch Unitary and Continuous Teacher Training courses*

The teachers' quality directly affects teaching effectiveness. If condition is allowed, universities should organize and launch pre-post training, pre-class training and teaching evaluation system for the teachers. Different teachers could achieve different teaching effect, even if they take the same course and syllabus. To achieve the best teaching effect, we should spare on effort on the communication among same subject teachers and between students and teachers, teaching normalization training, exchanging ideas with first-class high school teachers and the betterment of teaching method.

- *Research on Classified Students*

Comprehensive quality of students can differ drastically. Not all the students can adapt themselves to a certain teaching method. It is impossible to offer tailor-made course for each individual student. That is why we should classify students in terms of their personalities, scores and genders and research their sensitiveness towards different teaching method.

- *Materialize the Innovative Teaching Methods*

Practice is the sole criterion for testing truth. Innovation is not necessarily practical. Innovative teaching method should be evaluated by teaching effect. Innovation should be encouraged and tested in small scale. Teaching effect should be evaluated on the basis of long-turn observation and extensive consultation. For example, for agricultural-forestry economy and management students, their project thesis could be check in the form of investigation report. Students can finalize their professional emphasis and finish their investigation reports under teacher's guidance during the social practice in farms arranged by universities. In our school in this 5 years test period, this method received positive feedback from both students and teachers. It can be spreader to the similar majors in other universities. Our country features large quantities in universities with numerous internal students. Government should roll out stimulus measure and distribute special fund to boost the bringing forward, practice and popularization of the innovative teaching method and enhance the development of institution of higher learning to better contribute to the progress of the society.

5   The Conclutions

At present, under the condition of higher education in the management training process, we no longer constrained by some teaching methods. Teaching students according to their quality and decide to choose the most optimism teaching method is allowed. Teachers and students should improve themselves simultaneously during the learning process. This is not only beneficial to the promotion of higher education from traditional education mode to innovative education mode and cultivates both the students and the teachers in their comprehensive quality. Finally the innovative education mode can accelerate the integrated development of the whole society.

References

1. Wang, W.: The science of higher education, 1st edn. Fuzhou& fujian education press (2001)
2. Li, B.: Teaching theory. People's education press, Beijing (1999)
3. Zhaohong: University of research-based teaching and teaching methods reform. Journal of Higher Education Research A247, 55–57, 36–38 (2006)
4. Yu, G.: The green education in the teaching process. Journal of Environment and Society A315, 25–26 (2000)
5. Wang, X.: The theory and practice of higher finance and economic education. Zhejiang University Press (2006)

College Students' Career Planning Education Should Be Valued and Enhanced

Luo Jianguo[1] and He Maoyan[2]

[1] Mechanical and Electrical Engineering Department, North China Institute of Science and Technology, East Surbub of Beijing, China
[2] Labour Union, North China Institute of Science and Technology, East Surbub of Beijing, China
lchjiang03@163.com

Abstract. The college students' career planning education/guidance for the students in school to enhance their employability and future career development are of great significance. In this paper, students need for career guidance and career planning about how to implement the guidance of university students in terms of university students in colleges and universities and to guide the work of career planning were discussed.

Keywords: College Students, Career Planning, Education.

1   Introduction

Career planning is also called career design, which refers to a person's career objective and subjective factors analysis. It summarizes and measures to determine a person's career goals and life goals, and goals to a reasonable arrangement, with the feasibility, adaptability and so on. College Students Career Planning is aimed to help college students in self-understanding, understanding of the basis of society, to determine the direction of career development from the practical and social needs of its own proceeding, the development of university study and the overall goals and short-term goals and the steps in life. They will try their best to plan for future career development.

2   College Students Need Career Guidance

The growing number of university graduates, job pressures. China's colleges and universities enroll every year, in 2009, a total of 6.11 million university graduates. The employment of university students has become a very important issue for the government and universities. Chinese Premier Wen Jiabao, in January 7, chaired a State Council executive meeting. He deployed and said it is essential for college graduates to find a job. The Conference confirms the strengthening of the employment of college graduates into seven measures. The State Council, in January 19, 2009, issued a "State Council General on the strengthening of universities and colleges to inform the work of graduate employment ". 2009 Government Work Report of the State Council pointed out that" trying our best to promote employment and to promote the employment of college graduates in a prominent position ". In April 2, the State Council convened a national average employment of college graduates teleconference meeting to further deploy this issue.

Government aimed to increasing jobs and promoted the employment of university students. However, not every student can find a suitable job. The solving of this problem depends not only on the number of jobs, but also on the quality of their own.

Gap between college students' quality and social needs. According to the survey group of "China's Employment Problems of College Students" in Renmin University of China, the results of the survey of 600 units adopted in 2004 in Beijing, Guangzhou and Shanghai show that the first five indicators employers in the design 15 units in recruiting college students consider most are: (1) professional knowledge and skills; (2) professionalism; (3) strong desire to learn, a high plasticity; (4) communication and coordination capacity; (5) basic problem-solving abilities. The five indicators among the 14 lacing factors for graduates in the employers' opinion are: (1) professionalism; (2) the basic problem-solving skills; (3), to overcome difficulty under pressure; (4) relevant work or internship experience; (5) communication and coordination capacity. [5] The results of this survey show that college students own a far cry from the quality and social needs.

College students need for personal growth. Our recent survey of students in our college, for "no clear objectives and in a daze", respectively, 17.7%; "very consistent", 24.1% of students said the situation is with their "basic line", 32.2% of the people choose "ambiguous" and 18.7% said the situation is with their "basic non-compliances"; only 7.3% said the situation is "very consistent" with him. In other words, 41.8% of the students are aware of their lack of clear objectives in life.

For most students in high school, their only goal is to enter university. They lost their way when the goal is gained. They do not understand themselves, and do not know what they can do. Even they are not sure about their own interests, personality, values. Their knowledge is not enough for their professional and future career. Lacking of self-target, resulting in many students impetuous attitude, lazy behavior, not studying carefully and even indulging in network, muddling along. These not only affect their stage of development at the university, but also will make them to face employment difficulties.

College students' awareness of career planning is not strong. A survey for students in Beijing's major colleges of cultural economics pointed out that 62% of them have no career plan, 33% not clear, only 5% with a clear one [6] .

For most college students, their parents do everything for them during the process of their growing, from kindergarten to primary, junior, high school. Few of them take initiatives to explore and think their own path in life development issues independently. After entering the colleges, they still live in a passive way and do not try to understand the society, the occupation, and analyze themselves actively. During the job search process, they depend on schools and parents. They lack job skills and do not understand the employment policy, the environment and they are poor in adapting. In addition, their business awareness and entrepreneurial capacity is insufficient.

Lag of college students' career guidance. From the survey in www.askform.cn Network, from August 4, 2008 to May 20, 2009, the employing units believe that before the graduates go to work, their disadvantages in vocational education are: career planning guidance, 50%, professional ethics and education, 50%.

Based on the above said, we think that society needs good professional preparation of university graduates and university students need career planning. College Students Career Planning for students in school learning and future career development are of great significance. Firstly, it can give students a correct understanding of their own personality traits, the value of their own accurate positioning; secondly, it enables students to define their ideal career and development goals, and find more suited to their career path development; thirdly, it can encourage the students to explore their own potential advantages, enhance the employability and competitiveness of students; fourthly, it can make them adapt to changes better and enhance the students' ability of adaptability. Students need the guidance of career planning, so colleges should pay more attention to strengthen the career planning guidance.

3   Implementation of College Students Career Planning Guidance

The formation of college students' career planning guidance institutions and teachers. Schools should set up a special college student's career planning guidance agencies, which should be leaded by a competent school leader, school workers, the Ministry of Employment and various departments such as sector-specific organization and implementation.

Career planning should include both full-time and part-time teachers.

Full-time teachers are responsible for providing students with individual and group counseling, career development courses or seminars set up to carry out investigation and research career development of students. By this way, they can develop students' career development plans and training programs and organize the students with social practice, and to provide theoretical guidance for the activities, and actively Faculty professional development training and other education staff. Therefore, the full-time teachers must be proficient in career development theory, have targeted individuals, groups, counseling and consulting skills, and have the knowledge and skills in psychological assessment to understand the industry and career information, with knowledge of the elements of professional ethics and related legal knowledge. Full-time teachers must be highly rigorous, and be familiar with systematic career counseling training, career design, professional psychological adjustment, employment, market forecast, personnel training and the quality of human resources management. Besides, they should have a strong expertise. To implement sexual education and counseling according to the characteristics of the students. To meet these qualifications, they must be a holder of teacher evaluation and career guidance qualification certificates.

Part-time teachers shall be subject to professional development training for the teaching profession, with the basic professional theoretical knowledge and practical operational skills. It is best to hold vocational guidance teacher credentials. Counselors should become the backbone of part-time teachers.

First, college students and the instructors are responsible for all aspects of life guidance. Therefore, they have more opportunities to communicate with the students, so they know generally every student's personal interests and hobbies. Therefore, they can better advise the students to locate their own values. It is easy to carry out Mentor type of individual counseling and in-depth understanding of the specific circumstances of the students. For those students in dire need of professional counseling, they can give advices for appropriate professional functions.

Second, the counselors usually work at a fixed faculty. They are more familiar with their majors, the counterpart of the occupation, employers, and other aspects of recruitment conditions. With this understanding, they can accurately guide students to assess the professional needs assessment and to determine their own career aspirations and development goals.

Third, in general, college counselors will accompany the students through the whole college lives, so they can give the students a full professional career planning, guidance and personality combined with guidance at any time to help students solve various problems arising at different stages.

Opening courses on career planning. First, courses, primarily to enable students to create program awareness, they are the beginning of college students career planning, which usually lower grades for university students in particular. They know fewer careers planning, being lack of understanding of university life. Often for a very long period of time, the use of university resources confused them. They experienced many setbacks that could have been avoided. Through "college students' career planning", "college students career planning and career development" and other courses of study to guide students to understand and accept the career planning. The course focused on the popularization of knowledge of career planning, career planning to introduce students with the basics of basic theories, professional tests, job analysis, self-analysis, career planning principles, methods, professional ideals, professional ethics education, occupational choice ability, professional resistance capacity, entrepreneurial ability, and to develop students' awareness of career development and capacity.

Second, special counseling sessions for students in career planning and development problems are needed, as well as organizing seminars, lectures, exchanges or advisory Council. All these are designed to help students solve the specific career planning practice problems, including clearing up doubts, and guiding them to establish their own development goals in life, and inspiring them to study hard, to achieve career aspirations. Special counseling has the advantage of being problem-oriented, of a clear purpose and effectiveness prominence, especially those counseling sessions initiated by student, in which condition, students will participate actively and the counseling effects can be fully guaranteed.

Training career planning guide.

(1) to guide students with a clear sense of career planning, resulting in the planning awareness.

A sense of career planning is a good foundation and starting point for the creation of courses on career planning functions, which is to guide students to create the planning awareness for the study of the course. For those whose awareness is still not strong, teachers can talk with them through mentoring, and further strengthen their awareness of career planning.

(2) to guide students to conduct a comprehensive self-assessment and assessment of career needs.

Self-assessment is a process of scientific, comprehensive, thorough anatomy, aims to understanding ourselves. Students' self-assessment means learning about their self-values, personality, hobbies, abilities and temperament and so on. Career planning mentors guide students with the self-assessment. Their main means of evaluation work by the manners for qualified personnel, using of modern psychology, management science and related disciplines of research results. Through psychological tests, scenario simulation and other means of human ability, personality characteristics and other factors are measured. A scientific evaluation is got by measuring the students' demand and the organizational characteristics. The students' quality status, development potential, personality characteristics and other psychological characteristics will be a professional evaluation to determine the direction of persons.

Occupational assessment is to enable students to learn the social needs about their major. In other words, that is to understand what they can do in the future, as well as what is required of such employment of qualified personnel. Before the assessment, guidance teachers should focus on education and the normal use of opportunities to communicate with and guide students to understand the nature of the profession as well as what kind of quality and ability they should have. Besides, the teachers should also point out the general direction of future development. Conducting professional assessment, they need also consider the following factors, such as career development prospects, professional work environment, the economy and the jobs of non-financial remuneration.

(3) to guide students to establish career goals.

From a psychological point of view, the process of human behavior is essentially caused by the objective needs and, thus, motivation, causing a positive behavioral pattern to achieve desired goals. Goal is to induce an important driving force of human behavior. The primary prerequisite for success is to have clearly defined objectives, and it is the same with the career planning. Career goals can be determined on the basis of needs assessment in the self-assessment.

A positive and a clear career goal of the behavior have a strong appeal for college students, not only inspire students to study at the university stage, but also inspire him to struggle for their cognitive development, innovation and creation, as well as proactively adapt to society. It has a strong dynamic effect for them to adapt to society.

Each student should establish a suitable system for their own career development goals, including one's own life goals, a four-year objectives, the objectives for each semester or academic year.

The right goals, put it simply is a person level match. Personality characteristics and job requirements are required to match the individual's ability and professional expertise. To guide students to find a good fit between the personal qualities and social needs, without momentum effect by the impact of interest or a trend, choose to give full play to their talents, to maximize value and realize their career development goals.

The correct target is a continuous improvement goal. The choice of career development goals for college students is hard to be accomplished overnight, which needs continuous improvements with the objective and subjective conditions for change needs. Mentors should guide and assist students in practice, changes with the environment based on their own and take the initiatives to adapt to new situations, timely adjust and improve their career development goals.

(4) to guide students to develop career goals to achieve the implementation of the program.

At present, four-year undergraduate academic structure is generally divided as the following: the first grade is to adapt to stage, the second and third year is to stabilize the stage of development, the fourth grade for the sprint phase. Specialist school system is generally a three-year, with the second year to stabilize the stage of development. Students in career planning program development usually completed in adapting in the initial phase.

First of all, during the process of the students developing specific programs, teachers should remind students to consider their own ability, not being blindly over-ambitious.

Secondly, students must seize the most important content during the process. Anyone's energy is limited, so he can not achieve good results, but easily worth the candle if he focuses on several areas at the same time. At this time, guidance teachers should guide students to develop specific programs to seize key objectives. At each stage, there will be a core objective, while others are designed to target the development of this core objective. Therefore, the guidance teachers should guide students to determine the order to achieve other goals according to the core objective for the importance of the capacity of its own conditions.

(5) The implementation of programs to guide students to make adjustments and amendments.

Students will encounter many problems in the process of implementation of the program. At this time, the guide teachers should give effective guidance to them.

First, the program implementation will take a quite long time, and this requires students should have the continued determination and strong willpower. Therefore, in this process, some students can not adhere instead of giving up easily. So, the guide teachers will always have to pay attention to the specific circumstances of the students in difficulty in the implementation of the program and encourage them to develop persistence and perseverance.

Second, during the implement of the program, sometimes because of the change of their own capacity and the surrounding environment, they can not achieve the desired results. Some students are not good at analysis and adjustment, are still in accordance with the program have been unrealistic to implement. At this time, guidance teachers should guide students to adjust new programs according to the re-assessment and evaluation of results got by analyzing their own abilities and the surrounding environment, by doing so to avoid the students doing more exercise in vain.

Third, in the process of implementation of the program, students will easily lead to some negative frame of mind. For example, when there is a gap between the actual results with their targets, some students are become afraid of difficulties. While if the actual process is shorter than anticipated, some students prone to self-sufficient state of mind. At this time, guidance teachers should pay attention to the mentality of the students, to change the negative attitude of students and to build students' self-confidence, humility and other positive frame of mind.

Finally, it is necessary to conduct regular review and sum up the implementation of the program. The positive points will get praise, while make up the negative ones. Appropriate adjusts should be made accordingly on the next plan of action.

College students' career guidance approach.College students' career guidance can be group counseling may also be individual counseling. Group counseling means that teachers have those students who have similar problems or the same needs as a team, and then members of the group inspire, discuss, solve problems, and achieve their goals under the guidance of the organization. Through group counseling approach to mobilize participation in peer activities, it can often have a multiplier effect. Group counseling can not only help students being conducive to career planning and development to solve the problems encountered, but also help to develop their interpersonal skills, cooperation, ability and awareness to help others.

Individual counseling refers to the individual counseling for individual students.

Students' career awareness, career expectations, career interests, career development path are of various kinds, so career planning and development should meet individual requirements. At the same time, students' career aspirations, career aspirations include some personal privacy, by group counseling it easily leads students to conceal their true thoughts, which is not conducive to the achievement of counseling effects. In addition, the individual problems of individual students, groups, guidance counseling increased the cost of the students is not conducive to the rational use of time and energy. In one-way, individual counseling for the individual problems of individual students is direct and specific, but the common implementation of policies aiming at groups increases the cost of education. Although individual counseling is an interaction between teachers and students, but it lacks of inspiration, encouragement and interaction among students. Therefore, group counseling and individual counseling will be used in combination on the analysis of specific practical problems.

4   Conclusion

In short, a college student career planning guidance is not only career development needs of university students, but also is conducive for the community colleges and universities to send more high-qualified personnel. It must be noted that college students' career planning guidance is a complex and ongoing task, which requires a high degree of attention and all-round organization and implementation of many colleges and universities.

References

1. Dong, W.Q., Tan, Z.C.: Students career planning. Northwestern Polytechnical University Press, Lanzhou (2007)
2. Min, C.: University of career development and management. Fudan University Publish House, Shanghai (2008)
3. Xie, B.Q., Virginia: Career planning and employment counseling. South China University of Technology Publishing House, Guangzhou (2008)
4. Wang, Y.X.: Career planning and practice guidance. Northwestern University Press, Lanzhou (2007)
5. The Chinese People's University, China's Employment Problems of College Students, Task Force, You are favored by employers who do. Journal of China Education 10, 8–8 (2004)
6. Leung, K.S., Yan, Y.: Career management employment of university graduates encounter difficulties. Journal of China Youth 7, 19–21 (2005)

Exploratory Analysis of the Substance of Work Security in Harmonious Society

Luo Jianguo[1] and He Maoyan[2]

[1] Mechanical and Electrical Engineering Department, North China Institute of Science and Technology, East Surbub of Beijing, China
[2] Labour Union, North China Institute of Science and Technology, East Surbub of Beijing, China
lchjiang03@163.con

Abstract. Work security is the indispensable composition of harmonious society. From the viewpoints of history, practice, as well as philosophy, actual facts ,deep analysis and discuss are carried out in this paper. Finally we come to an important conclusion that the basic gist of work security in constructing harmonious society consists of law and rules, technology and resources. The research and analysis also tell us that the three factors above integrated together closely and developing by relying on each other are the necessary trend and the essential content.

Keywords: Harmonious Society, Law and rule, Technology, Resource, work security.

1 Introduction

Since the Sixteenth Congress of the Party, the state has formulated a series of strategic development plans, and these plans depict us a grand blueprint of how to build socialism in this new century and new stage. The "people-oriented" concept of scientific development, building the socialism as a harmonious society, building a conservation-oriented society, as the era masterpieces, they demonstrate to the world the ability and determination of the new generation of Chinese leaders in administrating the construction of the socialism modernization. If we want to achieve the country's prosperity and people's well-being, we must ensure work security and living security. While work security is the only effective way of all the material wealth and spiritual wealth accumulation, thereby, guaranteeing work security is the premise and assurance of building a harmonious society. In the fruitful yesterday is the case, today is the case, and in the promising future will also be the case. The harmonious society what we wish is a stable and orderly society with the effective co-ordination of the interests of all aspects, with the constant innovation and fullfillment of social management system. Specifically speaking, it is a society of democracy and the rule of law, justice, sincerity, amity, vitality, stability and order, man and nature living in harmony. To make people everywhere feel democracy, fairness, being cared about, being trusted,

Y. Wu (Ed.): International Conference on WTCS 2009, AISC 116, pp. 367–372.
springerlink.com © Springer-Verlag Berlin Heidelberg 2012

energetic, orderly, security ... ,they must lay their desires, efforts, rewards on the base of understanding, help, contribution, must achieve their goals in a dynamic process of regulation, guarantee, adjustment and through the collaboration of state, collectives and individuals. while the system is the strong security and operational guidelines of the smooth realization of this process, and technology is the magic weapon and tool for the effective enforcement of this process, and resources is the motivation and mind of the smooth realization of this process.

2   Requirements for Work Security in Harmonious Society

The proposal to build a harmonious society is not accidental. With the sustainable and rapid economic growth, per capita GDP has reached 1000 U.S. dollars. At the same time, the diversity of various interests relationships within the community, the complexity of human relationships, relationship between human and nature becoming increasingly strained, more and more conflicts resulting from those conditions, also social risk is also growing. In other words, a variety of disharmony in Chinese society has become increasingly prominent. All of these not only affect the construction of the well-off society, but also contrary to the purpose of the Chinese Communist Party and the socialist in nature. It is obvious that building a harmonious society is actually the objective requirements of China's social development. At the same time, it also shows the deepening of our understanding of socialism, i.e., socialism is not only a democratic and affluent society, but also the sublimation of coordinated development in all aspects of society. The characteristics of a harmonious society is: first, enhance the creative vitality of society as a whole by mobilizing all positive factors; secondly, maintain social justice through the coordination of various interests; third, creating a good social atmosphere to form a good interpersonal environment; fourth, safeguard social stability through the strengthening of democracy and the rule of law; fifth, addressing the relationship between man and nature to ensure sustainable development. Specific to security production, what content should it have?

First of all, perfect work security laws and regulations, complete security quality performance standards, safe and effective prevention, operations and aid rules, fair and transparent security supervision management and disposal system, all of those are the priority of government, business, security scientific and technical personnel, at the same time they constitute the first requirement for work security of harmonious society. Although in recent years, the CPC Central Committee and the State Council have been adopted a series of important security measures, as long as having a brief look at the history of the legal system of production security will we be aware of the institutional problems where exist. With the regulations and standards stated on the Internet about security production by The State Administration of Work Security, we can briefly sum up the formulation and promulgation time of the current rules and regulations and standards for the relevant departments and fields, a table made of as follows

Table 1. Issue time of Law and Rules

relevant departments or fields	The earliest time (years)	the latest time (years)
electric power	1987	1997
chemicals	1983	1996
machinery	1985	1992
building materials	1956	1995
transportation	1979	1998
mining	1980	1996
coal mines	1983	1999
health	1956	1998
stress	1957	1991
other industries	1956	1997
vocational training	1989	1996

From the above Table 1 we can see: the earliest adopted time of the formulation and promulgation of the current regulations and standards was in 1956.Nearly half a century has passed, they still play a role for two reasons: First, these within the industry equipment and management models are still the early days, which can still be applied now; Second, the rules and regulations no longer comply with the development of the industry requirements, so they required to be modified or re-enacted. Whichever the cause, it reflects the characteristics that does not meet requirements of the times. In particular, New China experienced a decade of the Cultural Revolution, and later introduced the predominant reform and opening up, in 2000 she joined WTO, and so have been, is being or will be giving a substantial impact on the development of Chinese society. The times need advanced legal institution and system, and China need the legal institution and system which meet the actual situation in China and keep up with the times. As the basis link for socialism construction, China's security production must have a solid system security and institutional guarantees.

Second, reform archaic technology, invest vigorously in developing a more secure and reliable innovative technology, promote the coverage of security technology. With technology and system integration, producing more security management and practice of talent is the state, enterprises, educational institutions and each workers long-term task, and it is also an another requirement for security production in harmonious society. Seeing from national security production situation, the number of accidents and the total number of deaths was rising before 2003, and a slight decrease since 2003 (ignoring the impact of SARS). Rapid economic development, road criss-crossing, and travel spreading bring challenges to the security production. In 2003 than in 2002 the total number of accidents fell by 11% ,and the number of deaths dropped 1.9%; in 2004 compared to 2003, the number of accidents dropped 15.7%,and   the number of deaths dropped 0.2%.There are deep-seated reasons and shallow reasons for accident-prone and serious security situation; there are historical reasons and the reasons for the development. Concrete can be summed up as follows: First, with rapid economic growth, technological level and equipment growth in relatively backward. Rapid economic growth has brought challenges to the security production. Second, there are inadequate investment for technical updating and transformation, and security production is difficult to gain the assurance of being synchronized with the international. In particular, coal, heavy chemical industry due to security debts, according to statistics, the state-owned coal mine security statistics 505 million outstanding loans, and there are about one-third of the equipment to be eliminated, inadequate investment in security as long as inadequate investment in basic industries. Third, the industry technical standard management weaken, and enterprise management landslide. China is still in the early stage of industrialization, and will achieve industrialization by 2020. We are not post-industrial country, and China is not the United States, Japan, so we should start from the national conditions. Over the years industry management did not keep pace with the world, and industry technical standards, design specifications were not changed for some years, the executive or the provisions of the original 7, 8 years ago. The era of progress, science and technology in development, with the backward procedures, standards, we cannot build factories, enterprises with advanced level to the times, and it is also impossible to produce internationally competitive products. Industry technological breakthrough is also lack, such as the formation mechanism of coal gas, gas control means, austria gray flood, electrostatic hazards, high-pressure blowout and so on, and many issues remains unresolved. Fourth, there is embarrassment between the security technology training and transfer of rural labor force. Today we have already 100 million 3 million rural labor force transmission, which is a manifestation of social progress. There are another 100 million peasants waiting for labor, what work do they come to do? Bitter, hurt, tired, dangerous work, from the types of work speaking, coal, urban construction, environmental protection, health, if ever an accident of these industries happened, we will owe it to the quality of migrant workers. Complain is unfair and we cannot complain our peasant brothers. Who is going to train them? Who is training anyway? With the absence of culture, how to enable them to understand and master the security knowledge and skills? No one to train and no one have been trained are often the answer, and only the bad news of death tells us the findings. Therefore, we should conduct "people-oriented" job training, adapt the technology equipment, so that workers and the working tool become more reliable.

Moreover, changing the structural model of resource exploration and amusement, building resources recycling, saving security production model, actively seeking clean and environmental renewable alternative sources. Through the improvement of system and technology innovation to achieve sustainability of the resources is the society as a whole shared responsibility, and it is also the last requirement for work security in harmonious society, realizing the laborers - working tools - all elements of the object of labor security. In the past 50 years, we have achieved world-noted achievements during the construction of new China, but all of these are based on high energy consumption, high pollution, high input, high-emission, low-return, low-earnings as models and prices. In one hand, our country spend 31% of the world's coal, 29% of the steel, 8% oil, 45% of the cement, creating the world's 4% of GDP. Of course, as a developing country, we should have a process. However, we must acknowledge that, the environment and resources can no longer support, and if this mode of growth not transited to relying on technological progress and the people's quality, it is not able to support any longer.

Output of 1.95 billion tons of coal last year, there were 1.2 billion tons with security security capacity and the remaining 700 million tons were insufficient of security security capacity. But the pattern of economic growth cannot change in one year or two, it is a deep-seated problems, and the CPC Central Committee and State Council have attached great importance, so we must make determined efforts to solve this problem. On the other hand, China's security production conditions cannot compare to other countries. Compared to the United States and developed western countries, even to developing countries there are many disadvantages. Accounted for 95% of the mining shaft, open pit is very rarely. Accounted for 46% of high-gas mines, spontaneous combustion of coal seam easily accounted for half, and the conditions are relatively poor. China is a major coal producer century, and now the average depth of mine is 420 meters, and extending down to 20 meters per year, so coal miners will pay more labor and hardship. We gain million tons coal losing 3 people, and the United States is 0.03, even Poland and South Africa is 0.3. China's coal production accounts for 31% of the world, but accounted 79% of the world's total death toll from coal mine. Situation are grim, tasks are arduous.

Finally, we also have problems in security supervision, numbers of documents, meeting, inspections have tried, but it boils down: supervision is not solid; law enforcement is not strict; liability failure to implement; the measures are not in function. It is difficult to imagine that we can build advanced level of security production manufacturers, produce first-class products, create a brilliant performance in the difficult operating environment with the backward system, procedures, standards, and backward techniques and equipment, with the workers who are lack of security awareness and skills.

3   Realization of Work Security in Harmonious Society

To achieve the harmonious society work security must be based on for the basis and the starting point, but to achieve work security we must start   from the holding up of the three-legged tripod of work security,that is to say ,from the system, technology, resources, three long-term efforts to achieve security, stability, effectiveness. To this end, we should do the following work.

(1) improving the security management system specification, promoting the security and quality standards. At all levels of government and enterprises should actively develop and improve production security laws and regulations, refine and improve the technical specifications, quality standards and management system, standardize work processes, operating procedures and operational processes, and be organized to build standard, upgrade the standard and compliance work, try efforts to achieve the controlling and being,in controlled of security production.

(2) further clarify the responsibility of governments and enterprises, increase investment of security, promote the scientific and technological progress of security production, and increase security education and training,let work security personnel education, training, selection, etc. into the talent development program units to improve the security level of specialization of production management team in order to meet the needs of enterprise security production.

(3) austerity, promote the development cycle, and actively explore new resources and optional resources,strengthen the supervision and monitoring of the corporate resources for production amusement, strengthen the government's macro-control capacity,establish the declaration of emergency resources and limiting mechanism, establish a strategy for emergency resources reserve mechanism,strengthen international trade and   cooperation to increase the source and stability of resources, configure and regulation the exploration and use of resourses within a reasonable basis of guiding the industrial structure.

(4) Establish and complete security supervision and management mechanism, and constantly deepen the security special rectification, establish security production accountability system, strengthen law enforcement efforts and penalties, build production security's system - education -punishment as the trinity of universal regulatory mechanism.

4   Conclusion

The construction of harmonious Society was put forward on the foresight and profound vision with Comrade Hu Jintao as general secretary of the new central leadership and government introducing a "people-oriented, comprehensive, coordinated and sustainable scientific development concept".Harmony means "to co-ordinate appropriate and well-balanced",and society means "a community constituted by a certain economic base and superstructure, also known as social form." With Comrade Jiang Zemin for the work security and insight:"hidden hazard is more dangerous than fire,and prevention is better than disaster relief, and responsibility is extremely heavy".We re-integrated history, reality, philosophy, the perspective of the specific evidence and then obtained the conclusion in simple terms that the system, technology, resources as three elements are three legs in supporting the three-legged tripod of work security, and on this basis,an analytical study on how to achieve in harmonious society. so as to lay a solid theoretical foundation for China's socialism modernization drive from one victory to another.

References

1. Xu, S.: The construction and exploration of China's security culture. Sichuan Science Press, Chengdu (1994)
2. Jin, L., Xu, D.S., Luo, Y.: China's 21st Century Strategy for Disaster Reduction. Henan University Press, Kaifeng (1998)
3. Wang, T.J.: The establishment of environmental resource industries-lead the road of harmonious development between human and nature. Journal of China Audit 16, 85–87 (2003)
4. Yang, X.L.: Harmony of human and nature with sustainable development. Journal of Shandong University of Technology (Social Science Edition) 5, 32–39 (2003)
5. Xie, Z.H.: Take the road of recycling economy to achieve sustainable production and consumption. Journal of Environmental Protection 3, 3–4 (2003)
6. Xu, D.S.: Deep thoughts to security and security culture. Journal of Labor Security and Health 9, 28–30 (2001)

An Adaptive Immune Genetic Algorithm and Its Application

Gang Shi [1,2], Jia Ma [2], and Yuanwei Jing [2]

[1] School of Information Science and Engineering, Northeastern University, Shenyang, China
[2] Shenyang Institute of Automation, Chinese Academy of Sciences, Shenyang, China
keesri@gmail.com

Abstract. To avoid premature and guarantee the diversity of the population, an adaptive immune genetic algorithm(AIGA) is proposed to solve these problems. In this method, the AIGA flow structure is presented via combining the immune regulating mechanism and the genetic algorithm. Experimental results showed that the proposed AIGA can rise above efficiently such difficulties of SGA as precocious convergence and poor local search ability and provide well the global converging ability to enhance both global convergency and convergence rate, thus solving effectively the flexible job-shop scheduling problem (FJSP).

Keywords: flexible job-shop scheduling problem, immune genetic algorithm, immune operator, adaptive strategyvaccine.

1 Introduction

IGA (Immune Genetic Algorithm) is a improved algorithm based on biology immune mechanism, which is the combination of immune principle and SGA (Simple Genetic Algorithm) [1]. Immune algorithm is an effective global optimization algorithm, which adds the immune operations. And in some degree restrains the degradation in the optimizing progress and the late evolution stage. So, introducing the immune operator into GA can increase the algorithm overall performance. In the meantime it can restrains the unit degradation in the optimizing process by using the information characteristics, and increase the algorithm global convergence [2].

This article puts forward the improved IGA, which can not only add immune operators and also apply self-adaption strategy to overlap and variation operation, namely Adaptive Immunity Genetic Algorithm. By applying the algorithm to the flexible job-shop schedule example, the result shows the algorithm is effective and feasible [3].

2 The Model

The FJSP description is: N different processing sequence workpiece are finished on M machines .Each workpiece is processed by the same machine more than one time,

Y. Wu (Ed.): International Conference on WTCS 2009, AISC 116, pp. 373–380.

and each process can't be enable to interrupt .We use U to denote each piece J contains nj processes, and each process order can't be enable to change. O_{ij} denotes J workpiece i process, which can be processed on any machine with processing capacity. $p_{i,j,k}$ denotes the time needed in process O_{ij} dealt by machine M. Let T denotes as follows:

$$T = \left\{ p_{i,j,k} \middle| 1 \leq j \leq N; 1 \leq i \leq n_j; 1 \leq k \leq M \right\}$$, N denotes workpieces, M denotes machines. Please see the table 1,which is a processing time of FJSP.

Table 1. Processing time of FJSP

		M_1	M_2	M_3	M_4
J_1	$O_{1,1}$	1	3	4	1
	$O_{2,1}$	2	6	2	1
	$O_{3,1}$	3	5	3	7
J_2	$O_{1,2}$	4	1	1	4
	$O_{2,2}$	2	3	8	3
	$O_{3,2}$	7	1	2	2
J_3	$O_{1,3}$	8	5	3	5
	$O_{2,3}$	3	5	8	1

In the FJSP, the assumptions as follow should be satisfied:

Every machine should be available at t=0,and every job may be started at t=0;

In given time one process can only be adopted by one time, another process can be scheduled after finishing the process, this is called source restriction;

Every job process only can be machined as scheduled and this is called precedence constraints.

The FJSP optimization goal reached in this paper is to find a most optimized schedule scheme satisfied the described precedence constraints and source restriction to make the makespan shortest.

3 Job-Shop Scheduling of Adaptive Immune Genetic Algorithm

Inspired by the immune system, an self-adaptive immune genetic algorithm was put forward, which can keep the original selection, crossover and mutation genetic operators. Premising the characteristics, change the crossover in different stages of genetic evolution, keep the excellent individual with genetic code, promote the optimal solution [4] .

1) Antigens recognition

Antigens recognition is to input the optimal designed objective function as an antigen immune genetic calculation. For the FJSP, the optimal designed index may be the performance indicator of the finished time, such as the maximum completion time; may be performance indicator of delivery: the maximum delay time; may be the performance indicator: the average stock for processing work packages [5-6]. This article adopt the maximum completion time as the FJSP optimal index.

2) Population initialization

According to the characteristics of problem solving, produce the antibody group. The reasonable design of antibody encoding mechanism has a great impact on IGA quality and benefit.   The chromosomes legitimacy , feasibility , effectiveness and the integrity of the solution space representation must be considered when immune coding .This article adopts expression method based on the process, namely to assign the same sign to all the part, and then to explain according to the appearance order in the give chromosomes. If three parts, each workpiece have three procedure, chromosome "122313132 ", the first 1 "denotes the first piece first procedure, the second "1" denotes the first workpiece the second process, the rest may be deduced by analogy [7].

3) Immune choice

a) Affinity evaluation valued operator.

Affinity denotes the combination intensity between immune cells and antigen, similar with IG fitness. Affinity evaluation valued operator usually is a function aff(x). The input for the function is antibody unit, and the output is the evaluated result. The affinity evaluation is related to the concrete problem, for the different optimal problems it is should define affinity evaluation function according to the characteristic of the problem under the premise of understanding the problem's substance. The article's affinity is defined as the max finished time [8].

b) Antibody concentration evaluation operator

Antibody concentration den(x) denotes the antibody population diversity good or bad. The antibodies concentration is high ,which means the similar units exit in large, so the search for optimization is focused on a sector in the available interzone, it goes against to the global optimization. So the optimal algorithm should restrain the high concentration units and keep the individual diversity. Antibody concentration is usually defined as:

$$den\left(ab_i\right) = \frac{1}{N}\sum_{j=0}^{N-1} aff\left(ab_i, ab_j\right) \tag{1}$$

Thereinto, N denotes population scale; ab_i denotes population i antibody; $aff(ab_i, ab_j)$ denotes the affinity between antibody i and antibody j.

$$aff\left(ab_i, ab_j\right) = \frac{1}{1 + H_{i,j}\left(2\right)} \tag{2}$$

Thereinto $H_{i,j}(2)$ denotes the average information entropy of the antibody group comprised of antibody ab_i and antibody ab_j, which is defined as :

$$H_{i,j}\left(2\right) = \frac{1}{L}\sum_{k=0}^{L-1} H_{i,j,k}\left(2\right) \tag{3}$$

Thereinto L denotes the code length, $H_{i,j,k}(N)$ denotes the information entropy on the k gene of the antibody group comprised of antibody ab_i and antibody ab_j, which is defined as :

$$H_{i,j,k}(2) = \sum_{n=0}^{S-1} -p_{n,k} \log p_{n,k} \tag{4}$$

In this formula S the allele quantity for every dimension in the discrete codes , such as binary system S=2, $p_{n,k}$ denotes the probability of the K dimension n allelic in the antibody group comprised of antibody ab_i and antibody ab_j.[9] .

c) Motivation efficiency calculation operator
Antibody motivation efficiency sim(x) is the final evaluation for the antibody quality , which should be considered the antibody affinity and antibody concentration. Usually high affinity low concentration antibodies will get more motivation efficiency [4]. The calculation of motivation efficiency usually can be got by using the evaluation for the antibody affinity and antibody concentration by simple mathematics, such as:

$$sim(ab_i) = aff(ab_i) \cdot e^{-a \cdot den(ab_i)} \tag{5}$$

And $sim(ab_i)$ denotes antibody ab_i motivation efficiency ; a and b is calculating parameter.

d) Immune optional operator
Immune optional operator Ts is based on antibody motivation efficiency to make sure which antibodies are chosen to clone operation. The higher motivation efficiency antibodies have better quality and easier to be chosen to clone operation, which can go on local search in more valuable search zone in search space. Immune optional operator are defined as:

$$T_s(ab_i) = \begin{cases} 1, & sim(ab_i) \geq T \\ 0, & sim(ab_i) < T \end{cases} \tag{6}$$

Thereinto I denotes motivation efficiency threshold.

4) Clone operation
Clone operator Tc denotes duplicating the antibody units chosen by immune choice operator. Clone operator is defined as:

$$T_c(ab_i) = clone(ab_i) \tag{7}$$

In this formula $clone(ab_i)$ denotes the assemblage comprised of m_i clones the same with ab_i ; m_i antibody cloning quantity.

5) Self-adaptive crossover and variation
a) Population average information entropy

$$H(N) = \frac{1}{L} \sum_{k=0}^{L-1} H_j(N) \tag{8}$$

$H_j(N)$ denotes gene j information entropy:

$$H_j(N) = \sum_{n=0}^{S-1} -p_{i,j} \log_2^{p_{i,j}} \tag{9}$$

$p_{i,j}$ denotes the probability that i signal appearance on loca, namely :

$$p_{i,j} = \text{(the total quantity i signals on j) } /N$$

b) Population similarity

$$A(N) = \frac{1}{1 + H(N)} \tag{10}$$

A(N) represent the whole population similarity, $A(N) \in (0,1)$,The larger $A(N)$ is , the similarity of the population higher, vice versa. When $A(N) = 1$, each antibody in the population is totally the same.

c) Self-adaptive strategy
Crossover probability p_c and variation probability p_m is the two important parameters in IGA, no matter huge or small ,which will effect the algorithm convergency rate directly. It is difficult to find an optimum suitable for every problem to the different optimal problem. Self-adaptive strategy is able to make the p_c and p_m change voluntarily following the population diversity. When the population unit fitness tends to the same or tends to local optimum, p_c and p_m increase. When the fitness tends to dispersion p_c and p_m decrease. p_c and p_m adjusts by the following formula.

$$p_c = e^{2(A(N)-1)} \tag{11}$$

$$p_m = 0.1e^{2(A(N)-1)} \tag{12}$$

6) Vaccine extraction
Supposed each generation keeps K_b fitness optimal antibodies, the latest K (current generation included) generation totally keeps antibody group comprise of $K * K_b$ optimal antibodies, every gene loci of each antibody has $k_1, k_2, \cdots, k_s$ five selectable signs. The probability that i is k_i is $\quad p_{i,j} = \frac{1}{K * K_b} \sum_{j=1}^{K*K_b} a_j \quad$,when $g(i) = k_j$, $a_j = 1$, if not , $a_j = 0 \cdot g(i)$ is the i allele sign in the population. The allele maximum probability (namely $p_i, i = 1, 2, \cdots, L$) k_j is taken as the allele vaccine section, thereby to extract the vaccine $H = (h_1, h_2, \cdots, h_L)$.

7) Vaccine inoculation
Selecting the antibodies needed inoculation from the father generation, to get the new immune units which are replaced by one or more gene segment according to the roulette, thus forming the more excellent group. The gene segment selection way is as following: from section 3.6, vaccines are denotes $H = (h_1, h_2, \cdots, h_L)$, supposed

$$q_i = \frac{p_i}{\sum_{j=1}^{L} p_j} (i = 1, 2, \cdots, L)$$

, correspond with h_i, the p_i is same with section 3.6.

According to principle of gambling wheel, q_i partitioned the whole disk. The probability that every vaccine is chosen depends on q_i.

8) Clone restrained operator

Clone restrained operator choose the units operated following the mentioned above, restrain the low affinity units, retain the high affinity units, form the new antibody group. In the cloning process, the original antibodies and clone antibodies, by crossover variation and vaccination,then form a temporary antibodies group X_{ti} , cloning restrained operation will keep the highest affinity units, restrains other antibodies. Cloning restrained operator may be defined as the following:

$$T_r \left(X_{ti} \right) = ab_i^*$$ (13)

ab_i denote the highest affinity antibodies in the gather X_{ti} , aff(ab_i)=max{aff(ab_k, $ab_k \in X_{ti}$}. Because of cloning, the original antibodies in variation operator operation is excellent units, and cloning restrained operator temporary antibodies group includes the father original antibodies, so the artificial IA operator implicates optimal units reservation scheme [8].

9) Population refresh operator

Population refresh operator T_d refresh the low motivation units, cancel this antibodies from antibodies group and generate the new antibodies randomly ,which is useful for the antibody diversity and searching for the new feasible solution space.   Population refresh operator is defined as the following:

$$T_d \left(X_n \right) = X_n^*$$ (14)

X_n is the low motivation antibody gather, X_n^* is the new antibody gather randomly, and gather X_n^* and gather X_n has the same scale.

4   Simulation

Because many factors will be considered in the actual production process, scheduling problem is also very complex, so in different production system, JSP exist some differences. Taking a manufacturing processing workshops as the example, simplify the production data , design a simulation   with 7 workpiece, 6 processes , a total of 11 machines to complete assignment. The specific workpiece, processes, machinery and processing information are in table 2, 3, and 4 following. In table 2, 1 denotes the processes that can be completed by machines and 0 denotes not.

Table 2. Working procedure-machine relationship

	Turning M1	Turning M2	Boring M3	Milling M4	Milling M5	MC M6	MC M7	MC M8	Grinding M9	Grinding M10	Planning M11
Turning	1	1	0	0	0	1	1	1	0	0	0
Milling	0	0	0	1	1	1	1	1	0	0	0
Planning	0	0	0	0	0	0	0	0	0	0	1
Grinding	0	0	0	0	0	0	0	0	1	1	0
Drilling	0	0	0	0	0	1	1	1	0	0	0
Boring	0	0	1	0	0	0	0	0	0	0	0

Table 3. Workpieces procedure information

Work-pieces	working procedure					
	1	2	3	4	5	6
J1	Turning	Milling	Planning	Boring	Drilling	Grinding
J2	Milling	Grinding	Boring	Drilling	Turning	Planning
J3	Boring	Drilling	Turning	Planning	Grinding	Milling
J4	Grinding	Planning	Drilling	Turning	Milling	Boring
J5	Planning	Turning	Grinding	Milling	Boring	Drilling
J6	Turning	Boring	Milling	Grinding	Drilling	Planning
J7	Drilling	Boring	Grinding	Milling	Planning	Turning

Table 4. Working procedure over time

		J1	J2	J3	J4	J5	J6	J7
Turning	M1	4	6	4	7	9	3	3
	M2	3	5	3	8	8	5	2
	M6	4	7	7	5	7	7	4
	M7	5	6	6	6	6	8	5
	M8	5	5	5	4	5	8	4
Milling	M4	3	5	8	4	3	9	4
	M5	2	6	7	3	5	10	6
	M6	5	9	4	5	7	8	7
	M7	4	8	6	6	5	8	6
	M8	6	7	5	5	5	6	8
Planning	M11	4	5	3	7	3	4	5
Grinding	M9	7	3	4	8	6	8	6
	M10	6	5	4	7	4	6	4
Drilling	M6	4	6	5	4	9	3	5
	M7	6	4	4	6	7	4	6
	M8	5	7	6	5	8	3	5
Boring	M3	5	3	3	6	4	6	7

Table 5. Simulation parameters for three algorithms

Algorithms	population quantity	crossover probability	variation probability	vaccination probability	Cloning size	generations
GA	200	0.9	0.1			300
IGA	200	0.9	0.1	0.2		300
AIGA	200	0.9	0.1	0.2	30	300

This article adopt the GA, IGA and AIGA to dispatch the example separately , the parameters for each algorithm is just as table 5. From the result, the SA-IGA property introduced in this article is better than GA and IGA. SA-IGA not only can keep the population diversity, accelerate the evolution speed, but also avoid the early convergent phenomenon in the evolution process.

5   Conclusion

This article analyzes the GA principle and convergence and puts forward the GA improved strategy introduces immune operator and self-adaptive strategy, forms the SA-IGA and applied it to the FJSP. Immune operator can prevent degeneration phenomenon in the population units re-crossover and re-variation process, self-adaptive strategy keeps the population diversity and algorithm convergence [10]. The result shows AIGA can resolve the FJSP quickly and effectively.

References

1. Holsapple, W., Jacob, V.S., Pakath, R., et al.: A genetics-based hybrid scheduler for generating static schedules in flexible manufacturing contexts. IEEE Transaction on Systems, Man and Cybernetics 23(4), 953–972 (1993)
2. Huang, S.: Enhancement of thermal unit commitment using immune algorithms based optimization approaches. Electrical Power and Energy System 21, 245–252 (1999)
3. Jiao, L., Wang, L.: A novel genetic algorithm based on immunity. IEEE Transactions on Systems, Man,and Cybernetics 30(5), 552–561 (2000)
4. Xiao, R., Wang, L.: Artificial immune system: the principle, model, analysis and forecast. Computer Journal 25(12), 1281–1293 (2002)
5. Srinivas, M., Patnaik, L.M.: Adaptive probabilities of crossover and mutation in genetic algorithms. IEEE Transactions on Systems, Man and Cybernetics 24(4), 656–665 (1994)
6. Kacem, I., Hammadi, S., Borne, P.: Pareto-optimality approach for flexible job-shop scheduling problems: Hybridization of evolutionary algorithms and fuzzy logic. Mathematics and Computers in Simulation 60, 245–276 (2002)
7. Najid, N.M., Stephane, D.P., Zaidat, A.: Modified simulated annealing method for flexible job shop scheduling problem. In: Proceedings of the IEEE International Conference on Systems Man and Cybernetics, pp. 89–94. IEEE, NJ (2002)
8. Dasgupta, D.: Artificial Immune Systems and Their Applications. Springer, Heidelberg (1999)
9. Ho, N.B., Tay, J.C.: GENACE: An efficient cultural algorithm for solving the flexible job-shop problem. In: Proceedings of the IEEE Congress on Evolutionary Computation, pp. 1759–1766 (2004)
10. Hunt, J.E., Cooke, D.E.: Learning using an artificial immune system. Journal of Network and Computer Application 19(4), 189–212 (1996)

Phone Mobile Learning Model and Platform in the Construction of Research[*]

Xiong Guojing

School of Economy and Management
Nanchang University
Nanchang, Jiangxi Province, China
jngji@sohu.com

Abstract. This paper analyzes the phone-mobile-model learning of the opening education, concentrate on the study of phone-mobile-model learning and the construction of the phone mobile learning platform. The purpose is to promote and enrich student's learning for the phone mobile learning that can meet the needs of students to learn better, promote the opening education to serve society better.

Keywords: M-learning, Mobile phone, Model, Platform.

1 Introduction

The data statist iced by Ministry of Industry and Information Technology of China (MIIT) shows that the number of mobile phone users which accounts for 47.3% of the population which has reached 633.84 million in china by November 2008.The number of mobile phone users reached 3 billion, and china account for 21% of global users. Mobile phone is no longer a simply tool for query and transmission message but become an important tool for mobile learning by communication systems, Bluetooth and 3G video phones.

M-learning characterized by realization of "Anyone, Anytime, Anywhere, Any style" (4A) under the freedom of learning is a new stage of development of the remote education. Mobile Learning relies on the relatively mature wireless mobile networks, the Internet and multimedia technology. In this case, students and teachers can achieve interactive teaching activities more conveniently and flexibly by using mobile devices (such as wireless Internet access for portable computers, PDA, mobile phone, etc.) and moving the teaching server. Phone learning possesses all the characteristics of remote learning, including flexibility, portability, interactivity, personalization, pervasive and so on. As long as they realize their phone devices wireless connectivity, they will be able to learn freely. The development of mobile learning will enable students to study more freely in the remote education. From the performance of mobile phone, it can display and play the text, MP3, pictures, etc, run a simple interactive learning software; also can play video, run a strong interactive multimedia

---

[*] This work is supported by the Jiangxi Province Education Science Foundation Grant # 08YB206.

Y. Wu (Ed.): International Conference on WTCS 2009, AISC 116, pp. 381–386.
springerlink.com © Springer-Verlag Berlin Heidelberg 2012

learning software which meets the needs of the learning of handheld devices. Besides calling, mobile phone possesses the most features of PDA, especially the Function of personal information management and wireless data communications based on the browser and e-mail, to meet the needs of handheld devices learning. It has functions of short message service (SMS), sending and receiving e-mail, Internet browsing, information management, a variety of commercial applications, communications device (Web phone web-telephone), multimedia playback editing features, opening and editing office documents.

2   Phone Mobile Learning Model

2.1   Phone Mobile Learning Goes into the Online Access Model from the Short Message of Cellular Phone

In the past, mobile phone was used for SMS and calls. Learning based on SMS is the simplest, most efficient format. Through the short message, learners, between learners and teachers, between learners and the Internet server can be achieved between the limited characters of the transmission.   To take advantage of this way, learners can be realized through a wireless mobile networks and the Internet communication between and complete a certain teaching activities. Short messages are generally divided into three categories: SMS (Short Messaging System, Short Message Service), EMS (Extension Messaging System, Enhanced Message Service), MMS (Multimedia Messaging System, Multimedia Messaging Service).   As to the short message, its data communication is interrupted and can not be real-time connectivity, so interactivity is very poor.   With the development of Internet and its 3G technology, online access to learning emerges.

Learner can access to the Internet and WAP according to their own needs, i.e. at any time, at any place by mobile phone, to access to support mobile learning educational site for learning.Learners can browse a lot of information resources, including databases, digital libraries, broadcast media, and also can use online applications and software libraries and other software resources.

2.2   The Technical Barriers of Phone Mobile Learning to Be Overcome

Currently, 3G networks are not generally applied and the charges for mobile communications are higher. All of these add substantial economic burden to learners. Moreover, the expression of short message is limited, generally only for text, video and audio as well as the animated way. And the content is very monotonous. Limited by the phone features, Phone can not effectively reflect the learning content in time. The speed of phone mobile learning network is very slow, which limited the online learning.

The bandwidth of network is difficult to ensure ready access to mobile learning server, and the traffic restriction affect the effectiveness of phone mobile learning.

The monotonousness, lack of learning resources, simple content, poor interactivities etc, have direct impact on the mobile phone initiative.

The technology of mobile learning achievement is based on mobile internet technologies, such as WAP, Bluetooth, GPRS, UMTS technology, mobile terminal. Large-size color screen mobile phone technology matured and the rapid spread, high-capacity flash memory applications and the price is low, the use of mobile expansion card slot free to expand storage space, Mobile processor performance, computing faster, enough to complete the more complex computing tasks, which can overcome the multi-media learning software running on mobile phones technical obstacles.

2.3   Phone Mobile Learning Model

1) Problem-based learning model. Problem-based Learning (Problem-Based Learning, referred to as PBL) is received widely attention in recent years as a teaching method, which emphasizes the study set in complex and meaningful question's situations. By letting learners work together to solve real - of (authentic) problems, to learn the science behind the question implicated in the knowledge, learners can form the problem-solving skills and self-study (self-directed learning) capabilities. Because of simple operation and carrying convenient, learning mobile phone is very suitable for problem-based learning model. For example, when you forget how to write a word or you do not know how to implementation of an action, and at this time the computer can not be opened or the internet can not be accessed, in this case, you can solve the problem by using the phone itself the Knowledge Base or mobile communication network.

2) Resource-based learning model. Resource-based learning is a kind of learning, in which learners achieve the curriculum aim & information culture aim based on the development & utilization of various learning resources so that they can renew & develop by themselves.

Resource-based learning is learner-centered for learners can take active part in learning with a great quantity of resources available, learning places changeable and learning time flexible. As the learning resources around learners and achieves that learning is everywhere.

3) Learning model with the aid of network & mobile terminal. There exists some limitations in the model based on SMS & online access; however, with advantages combined the model with the aid of network & mobile terminal can extend strong points of the former one. Mobile phone learners can learn by online browsing, i.e. network permitting, they can browse learning materials online and communicate and even cooperate with tutors and other learners. Furthermore, mobile phone gets connected with M-learning platform so that learning materials can be downloaded from server terminal and stored in the phone for studies. To accelerate the speed, other such terminal devices as pc could also be used for downloading the information from the server.

3   The Construction of M-Learning Platform

At present, learning software based on the platform of mobile phone are rare, which mainly concern e-books, mp3 and short videos based on mobile phone. In addition, the contents are monotonous and there is no featured handheld learning platform based on mobile phone. Therefore, the development of mobile learning platform is the inevitable trend of development of mobile learning.

3.1   Technical Support of Phone M-Learning Platform

From two ways of realization of M-learning, the technical based on mobile Internet technology as WAP, GPRS, Bluetooth, UMTS and the realization tool is mobile phone. Internet is the backbone of learning mobile phone ,for learning resources can be downloaded from the Internet so as to form the seamless platform for the interaction between phone & Internet.

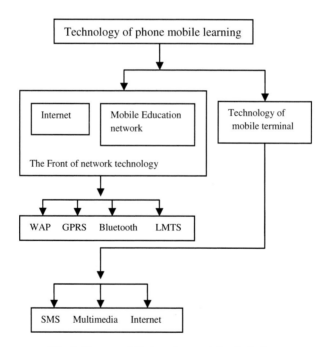

Fig. 1. Phone mobile learning model technical

1) Network support. How to carry out M-learning based on mobile phone mainly relies on Internet & mobile education network. As a kind of education media, mobile education network, used to send and receive information from mobile station and Internet, consists of several bases and is a branch of entire mobile network. Moreover, the seamless connection between mobile station and Internet will be achieved by air interface.

2) Support of mobile terminal technology. Mobile phones become more powerful in that they can Display and play text, pictures and videos, run simple interactive learning software, log in system sites for M-learning, and even interact with the system to gain more resources.

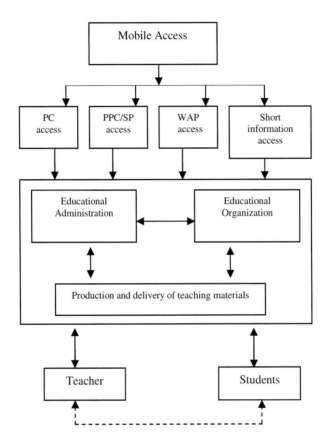

Fig. 2. Phone mobile learning platform

3.2   Construction of Platform for Phone M-Learning

1) platform for mobile SMS interaction. Platform for mobile SMS interaction is that SMS applications are embedded in the processes of teaching organizations and academic management in the environment supported by education of wireless mobile network. Learners interact with academic managers and teachers by mobile phones and wireless network to achieve such Organizational functions of teaching as Distribution of educational resources, test of Teaching effectiveness, Online discussion Q & A, Activity-Based Management. Thus academic managers and teachers can track, gather statistics of, and analyze the teaching processes and results whenever and wherever necessary.

2) Platform for Production and distribution of multimedia courseware. Platform for Production and distribution of multimedia courseware is the core of m-learning mode based on mobile phone. The courseware suitable for computer can be converted into mobile courseware with various models and various sizes by Converters in the platform. Mobile courseware should be Dapper so as to satisfy learners' needs, the needs of the efficient use of screen and the battery in Handsets, and then delivered to network platform to provide maximum support to m-learning based on mobile phone.

4   Conclusions

As a form of M-learning, phone m-learning features portability, mobility, autonomy. It can provide a learning environment at any time at any place so that fixed open education will no longer be conducted only when computer is available. One of important aims of learning is to gain stable and flexible knowledge and then put them into application in the new environment. Phone m-learning enables learners to interact with the context and rich context information helps to enhance the significance of learning. As a result, learners can construct their knowledge more easily and deeply.

Since 2001, mobile education has drawn much attention worldwide and Series of researches have been done to m-education both at home and abroad, including technology support, system Architecture, realization of platform, production of learning materials. Mobile technology equips phone m-learning with just-in-time learning contents for phones can be used to search relevant information when learners are travelling or studying outside. Mobile learning software should be vivid and lively so as to make learners gain better results in limited time and Multimedia Interactive Learning, is the one and the only one source into this purpose.

References

1. Keegan, D.: Mobile Learning:the Next Generation of Learning. Open Education Research (6), 22–27 (2004)
2. Latchman, H.A., Salzmann, C., Gillet, D., Bouzekri, H.: Information Technolog Enhanced Learning in Distance and Conventional Education. IEEE Transactions on Education 42(4) (November 1999)
3. Hu, H., Ren, Y.: Legitimate Peripheral Participation and Mobile Learning Community. China Educational Technology (9), 9–12 (2006) (in Chinese)
4. Ran, M., Zhou, Z.: The Learning Model based on Mobile Terminal. China Information Technology Education (3), 9–20 (2008) (in Chinese)
5. Xie, Y., Yu, L.: The status of m-learning development and Application model. Software guide (2), 6–8 (2006) (in Chinese)
6. Zhang, H.: The research of mobile education based on WAP. Journal of Shanxi Normal University (Social Science Edition) (5), 34 (2007) (in Chinese)
7. Zhu, S., Zou, X.: The Construction and Application of Instructional Design Mode in the Mobile Learning. Modern Educational Technology (10), 69–72 (2008) (in Chinese)

Research on Cognitive-Based KLSKT Teaching View and KM Teaching Theory

Bingru Yang[1], Nan Ma[1,2], and Hong Bao[2]

[1] School of Information Engineering, University of Science and Technology Beijing
Beijing, China
[2] College of Information, Beijing Union University, Beijing, China
bb74123@sohu.com

Abstract. In this paper, we propose a new teaching model including a cognitive-based KLSKT teaching view and a KM teaching theory considering characteristics of computer courses. This model is based on underlying logic of mind process, employments steps of pumping point-connection-netting-extension-modeling-Embedded form, forms overall intellectual framework and logic diagram of knowledge elements through micro structure and macro formation to firmly grasp knowledge. The practical study shows that the method effectively improves the efficiency and effectiveness of study.

Keywords: Cognitive, KLSKT teaching view, KM teaching theory.

1 Introduction

Modern science and technology has made rapid development in recent decades and made more brilliant performance than before resulting in "knowledge explosion", "information explosion" of new pattern. In this case, contradiction between the limited time of human learning and richness of knowledge becomes more and more sharp. Effective way to resolve this contradiction is not mechanically repeat the teaching of knowledge of their predecessors who proceed from the most primitive ideas; but uses a shorter time, the right way to get understanding of the knowledge "leap" , i.e., we are required to use a higher point of view, organize contents systematically.

It is an inevitable trend to take the internal logical structure of knowledge as the core of course content organization medium, i.e., we regard logical structure based on ontology as the heart of teaching; and the idea based on core of logical structure of knowledge can be analyzed by the ideological theory of the structure (internal level) approach. Structure of the paper is organized as follows: The first part describes the KLSKT teaching view, KM teaching method and correlation between them; the second section elaborates the operating mechanism of KLSKT teaching view and KM teaching theory; third part of the paper expounds process of pumping point-connection-netting-extension-modeling for practical science and engineering teaching in higher education. Also, we set forth integration of teaching within an interactive "thin - thick - thin" model. we take   C + + program as an example of teaching

Y. Wu (Ed.): International Conference on WTCS 2009, AISC 116, pp. 387–393.
springerlink.com   © Springer-Verlag Berlin Heidelberg 2012

practice and statistical results show the advance of KLSKT teaching view and KM teaching theory.

2   KLSKT Teaching View and KM Teaching Theory

2.1   KLSKT Teaching View

KLSKT teaching view is a teaching ideology and teaching ideas which refers to the establishment of logical structure, theoretical framework and the internal relationship in accordance with the internal logic of the theoretical knowledge and adheres to the "establishment first, then filling, then induction" and "less and precise" principle. Also, it constructs the teaching content with the multi-level logical structure based on the laws of cognitive psychology and information processing theory.

2.2   KM Teaching Theory

KM teaching is the acronym of knowledge logic structure and the mind mapping [1]. Mind mapping is created by Tony Buzan, showing a natural expression of divergent thinking process where learners can improve thinking ability in mind maps. Thinking mind map adopts a radiation thinking approach which can not only provides an accurate and quick learning methods and tools, but produces surprising results when used in the creative divergence and convergence, project planning, problem solving and analysis, meeting management, etc.

KM integrates knowledge of the logical structure and mind mapping, and it puts macro idea based on logical structure of knowledge into effect which aims to seem the students as principal part   such that they can learn quickly to improve teaching quality [2].

2.3   Relevance of KLSKT and KM

KLSKT teaching view and KM teaching theory are complementary in macro idea and micro implementation which requires us to research KLSKT teaching view and KM teaching theory by use of scientific ideas and methods. KLSKT teaching view is the guiding ideology and KM teaching theory is specific teaching methods under the guidance of KLSKT teaching view. The macro-thinking which is based on the logical structure of knowledge blended into the microscopic interpretation of teaching ideas makes the teaching process come up with reality, rather than empty words. The implementation steps of KM teaching theory fully demonstrate the specific cognitive processes with the guidance of KLSKT teaching view such that it is up to the theoretical level of teaching and has a profound connotation and meaning.

3 Operating Mechanism of KLSKT Teaching View and KM Teaching Theory

3.1 Operating Mechanism

Our theory starts from cognitive science, takes the logical structure of knowledge as the core, integrates the mind map and forms fusion teaching mechanism after implementation of micro knowledge and macro deduction. Under the guidance of the syllabus teachers based on the internal logic of theoretical development, establish the guiding ideology which give priority to teaching of the logical structure of knowledge, theoretical framework and the internal relationship as well as the thinking activities. Also, they adhere to "establishment first, then filling, then induction" and "less and precise" principle.

3.2 Teaching Content

As for the teaching content, we construct the logical system of Knowledge in the multi-level structure in accordance with the laws of psychology and information processing theory to. And in teaching methods, we focus on deepening the application level and make full use of inspiration function induced by the mind mapping such that obtain leap in knowledge.

The core of KM is a deductive system based on the macro level and micro level which are complementary in teaching content. In terms of the macro level: the main expression of the logical structure of knowledge is logical structure diagram (abbreviated as KLSG), where we can get the overall architecture of knowledge systems which characterize the intrinsic link between subsystems, concept of proposition (theorem), deduction, proof, problem solving, class of (species relations) and other internal links within the subsystems. In terms of the micro level: mind mapping integrates the concept (concept map), proof, problem solving, and other sectors to characterize their specific, detailed, dynamic, and development characteristics of the logical form and logical deduction, reveal the proof of concept formation and problem solving ideas and reveal the progressive refinement process.

The comprehensively integrated, multi-layer hierarchical structure of knowledge systems which takes KLSG as the main body integrated with the mind mapping method is the essence of innovative teaching methods. In summary, KLSKT teaching view and KM teaching theory form the operating mechanism with the logical structure of knowledge in macro integrated with mind mapping spread in micro.

What summarized in the knowledge of the system can be given in the form of chart structure which has a clear structure. Moreover, it can give us a macro picture of knowledge and clarify the clues in overall knowledge such that the students are easy to remember and use what they learn. In teaching the teachers first teach "skeleton" and "structure", then filling in the key, difficult and critical points. The secondary side of points can be given to the students to study by themselves to fully activate their initiative. Finally, the teachers teach "overall structure" followed by summarization.

4 Construction Pumping Point-Connection-Netting-Extension-Modeling- Embedded Form during Higher Engineering Education and Integration of Teaching Law of "thin - thick - thin" Model. "

4.1 Pumping Point

Generally analyze the theoretical system by unit-chapter-book one by one. And we take out concept, theorem, law, theory of knowledge in every part, discarding those minor or side things.

4.2 Connection

In the process, the first analysis of local profiles with its expansion then the last of the whole. In the content, we should seek two kinds of elements: The first is the concept, theorem, rule, theory and the intrinsic link between them; The second is the main thread throughout all parts of concept, theorem, rule, called "knowledge chain."

4.3 Netting

Deepen the structure with multi-level from simple to complex level and from concrete to abstract level based on intrinsic link between knowledge and constantly enriched theory. Pay attention to knowledge in the horizontal and vertical direction in order to form the "knowledge networks."

4.4 Expansion

Based on the previous frame of knowledge, along each "context" to develop an extension, we will add details into every part(explain the key, difficult and critical knowledge points, others studied by students themselves), to expand and increase the overall state of knowledge. At this time what the students learned is not original material and the intellectual content of a repeat list in the book, but the high point of view, a solid knowledge of probability model. In this way knowledge students learn has a "space" structure, rather than simple show with "flat" structure. From the epistemological point of view, knowledge is the spiral of time.

4.5 Modeling

The teaching process can be divided into three levels, the first three is known as A-level, expanded as B-level, model called the C-level. Different from the learning process with thick - thin – thick approach we clearly propose the process of interactive teaching with thin - thick – thin approach.

4.6 Embedded Form

In analyzing the structure of knowledge, appraisal issues, grasping the intrinsic connection between knowledge points and forming a broad association with the creative thinking activities, mind mapping plays an important role. We can develop properly the existing mind mapping method to form incomplete radiation pattern, the "forest" -based structure and empowered tree structure, generalizing the existing thinking and results analysis.

Its implementation can be described in the following flow chart:

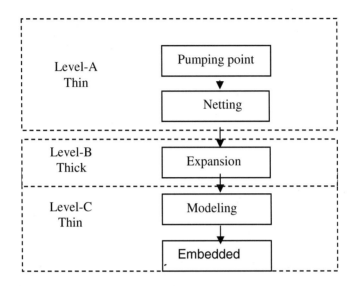

Fig. 1. flow chart of KM implementation

Teachers can use KM method in teaching with the guidance of the flow chart in Figure 1. Also, teachers organize what they teach in the class as KM chart described above such that the students can understand what they learn clearly, which in turn will help organize classroom teaching combined with induction teaching and heuristic teaching, making the classroom focus on the important contents developed systematically.

5   Teaching Practice

Practice teaching results show the significance of KM compared with the case without use of KM.

5.1   Practice of "C + + Program Design" Course

For C + + programming courses, the logic mind diagram are described as figure 2 after process of pumping point-connection-netting-extension-modeling- Embedded form.

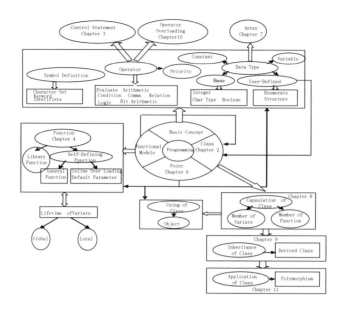

Fig. 2. logical Structure of C++

5.2   Evaluation

To evaluate results of teaching we compare the correct ratio of teaching content in the classroom with the case after some intervals. We take chapters II and III as examples. The former has selective questions and filling in the blanks which may be referred to as objective questions while the latter has filling in the blanks and programming. Test time is just after class, and the data are obtained by use of KM and not. Comparative results are shown in Table 1.

Table 1. Form of test result

		Chapter 2	Chapter 3	
		Objective questions	Objective questions	Programming problem
Without KM	Class 051	41.65%	48.93%	39.45%
	Class 052	40.34%	45.76%	38.33%
	Class 053	38.83%	42.32%	39.02%
	Class 054	39.95%	40.98%	35.67%
With KM	Class 071	48.76%	54.65%	40.31%
	Class 072	52.56%	53.32%	42.32%
	Class 073	53.34%	55.78%	45.34%
	Class 074	54.45%	58.67%	46.17%

As can be seen from table above, use of KM can improve correct rate of mastery of knowledge, especially in programming learning efficiency such that students can improve programming capabilities. Comparison between the data shows students can master knowledge longer with KM method.

6   Conclusion

In this paper, we propose a cognitive-based KLSKT teaching view and KM teaching theory for great contradiction between the limited human learning and richness of knowledge accumulation. Curriculum practice shows that the method effectively improves the efficiency of students' learning. The method has been implemented in C + + teaching, and its basic principle can be applied to other computers courses in the teaching process.

References

1. Yang, B.R., Zhang, T.H.: A Research of Pedagogy of KM in Science and Engineering Courses. Modern University Education 4, 83–85 (2006)
2. Zhang, T.H., Xie, Y.H., Yang, B.R.: KM in the "Computer Culture" course Teaching. Education for Chinese After-School 12, 331 (2009)

The Construction and Application of the Three-Dimensional Teaching Resources of University PE Based on the Modern Education Technology

Xu Feng

Zhe Jiang Gong Shang University, Hangzhou, China
xfeng250@163.com

Abstract. This paper discusses the three-dimensional teaching resources of university physical education from its contents and features. It presents the scheme and application of the course construction. This three-dimensional physical teaching is based on the modern education technology, which has lots of advantages. It can reduce teachers' workload, achieve great convenience for students to inquiry the results, arrange tests, and guarantee the fairness of examinations. On the other hand, it can properly guide and arrange for the professional sports teams, and adjust the players' physiological and psychological condition.

Keywords: three-dimensiona, informationization physical education, construction, application .

1   Introduction

The basic approach to carry on the information technology education and realize the education informationization is to construct the three-dimensional teaching resources. These years in the process of the university PE teaching reform, there come into being a large scale of new teaching materials and the corresponding digitized teaching resources, which have been able to reflect the new research results of course construction and course development, to manifest the idea of modern education, to utilize the modern education technology. However, the overall construction of the three-dimensional teaching resources is still in the exploration stage of improvement. Therefore, we must strengthen the construction and application of the PE teaching resources while teaching. In doing so we can realize the informationization of physical education, fundamentally change the educational thoughts and concepts , reform educational model, thus improve the quality and effectiveness of physical education.

2   Concept, Contents and Features of the Three-Dimensional Teaching Resources

The three-dimensional teaching resources is based on the modern education concepts, the networking technology platform and the traditional paper teaching material,

Y. Wu (Ed.): International Conference on WTCS 2009, AISC 116, pp. 395–400.

centering on the subject curriculum, which is the collection of multi-media,multi-form, multi-purpose, multi-level of teaching resources and a variety of corresponding teaching publications. It is the symbol of the modern teaching of this digital era. It is also an important approach to realize educational informationization, networking and optimize teaching resources' allocation. The three-dimensional teaching resources consists of three aspects: the three-dimensional teaching package, the teaching resource pool, and the course Website or specialized Website. It can also be divided into three levels. The three-dimensional teaching resources has such features as digitized processing, optical storage, multimedia display, network transmission, seriation of learning resources , intelligentized teaching process, etc. Such features can meet the teaching and learning needs of diversity, individuation and practicality.

3   The Construction of the Three-Dimensional Teaching Resources

3.1   The Three-Dimensional PE Teaching Package

The construction of the three-dimensional teaching resources is teaching content-based, focusing on developing teaching materials of theoretical knowledge and skill raise required for physical education, teacher s' reference books, study guides, skill instruction books, test bank, etc. It also develops PowerPoint courseware for teachers' teaching and students' reviewing, CAI courseware for solving both important and difficult problems, and network self-learning and self-training courses, etc. Meanwhile, the teaching design should be integrated, combining teaching content, teaching methods and means properly, presenting with paper books, audio-visual and electronic publications, three-dimensional products. And these teaching resources of related contents and different forms should be repeated, emphasized, overlapped, complemented, coordinated, in order to form a teaching resource design, as to improve teaching quality.

3.2   The PE Teaching Resource Pool

After completing the construction of the three-dimensional PE teaching package, a learning system should be developed aided by the computer application software system, which combines the teaching resource management system, the target system, database of the teaching package, advanced programming languages, and the media production software. This learning system conveys the theoretical knowledge in the form of text, graph, image, audio, video, animation, streaming media and other materials. These knowledge is in a tree structure arranged in accordance with the curriculum knowledge points, which is convenient for teachers and students to search. The system also provide PE teaching with simulation practice convenient for students to learn sport skills. To complete the whole construction of the teaching resource pool, it is necessary to improve the management system, so as to facilitate teachers and students to query and use.

3.3   The University PE Websites

The university sports Websites should be developed according to its features, student-oriented, taking website as the gateway to construct and practice. The construction sites should include teaching sites, extra-curricular counselling sites, sports knowledge sites, and several other parts (Figure 1). Teaching sites include teaching management system, classroom teaching content system, test system, evaluation system, feedback system and so on. Extra-curricular counselling sites include extra-curricular exercise knowledge, club activity, campus sports competition and training, equipment conditions and so on. Sports knowledge sites includes all kinds of sports website links, sports academic website links and so on. The purpose of this Website construction is to combine traditional teaching resources and activities with digitized learning resources and cyberspace under the informationization education environment. Teachers and students can learn sports knowledge and experience with social experience in campus, in classroom, on the net, through various communication and daily activities, thus realizing high quality teaching.

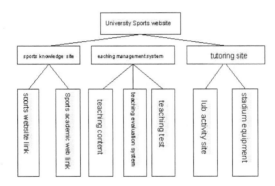

Fig. 1. The University PE Websites

4   Realization of the Three-Dimensional Teaching Resources

4.1   The Design Principles of the System

1) Advanced Technology
This system is constructed based on the B/Sstructure mode, adopting multimedia technology, streaming media technology, Windows API technology, ADO, database technology and web development technologies. It also equips new technologies of the current large-scale commercial website design, such as upload component file and non-upload component file. Thus the advancement of this system is guaranteed.

2) Standard Resources
The resources used for system design are corresponding to one or several standard formats, so that the majority computer users do not need to install any software or

plug-in. Moreover, the system provides with all kinds of softwares and plug-ins to help minority users to solve the browse problem.

3) Diversity Resources
The resources pool is constructed by teacher users and administrators comparatively, together by student users' recommendation, which guarantees the diversity and quality of the resources.

4) Scientific Browsing and Search
The PE teaching resources pool is divided into several individual pools according to modular design, such as chart gallery, document pool, video pool, audio pool, courseware pool and so on. These resources are organized according to the resource type and provide retrieval. On the other hand, taking the advantage of the network, the system offers another retrieval with the index of subject. These two kinds of retrieval complement with each other, narrowing the searching scope and improving the searching efficiency.

4.2   The Composition of the System

1) Component Modules
The system is composed of system management module, user management module, resource management module and the user browsing module.

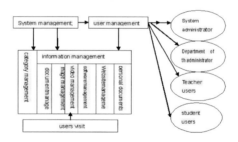

Fig. 2. System diagram map

2) System Management Module
The system management module mainly contains the user management and category management. The user management includes users' verification and logging off, password initialization, permissions allocation and so on. The category management includes module management and classification management of teaching resources.

3) User Management Module
Users can be divided into four ranks: system administrator, faculty office administrator, teacher user, student user. Administrators can verify users, log off users and initialize user's password, while users can modify personal information and password.

4) Resource Management Module
Users can add to and maintain the resources upon registering. Firstly, he should select the corresponding resource module, taking all the resources of this module. Then he

can add resources, modify or delete the resources issued by himself or the subordinate users. It is convenient to select   resources because there is classification mark in the attribute. Teacher users can only issue resources related to their teaching and research. While both teacher and student users can recommend resources to any faculty office. And these resources should be verified, transferred or deleted by the related faculty office.

4.3   Key Technologies and Their Implementation

1) Video Resources on Demand
The modern education technology center has accumulated over the years a large number of video resources of physical education, sports events and information, which is irreplaceable and of   historical value and teaching value. Thus the key point of this research is how to make teachers and students process easy access to these video resources.

The video resources saved in the   modern education technology center are in high quality formats like mpeg4, vi and the like, which can be broadcast and also made into DVD, VCD. However, these video formats cannot be used on demand on the net. Streaming media is a realization of real-time play of video, animation, audio and other multimedia resources, whose format supports media resources' streaming and playback. Therefore we use Windows Media's streaming delivery system to realize the adjustment of the multimedia data streams and to transmit streaming media, meeting the demand for network.

Windows Media's streaming delivery system consists of Windows Media Encoder, Windows Media Service and Windows Media Player. In order to meet the different needs of network load conditions, it is recommended to encode by bit rate of 282kbps、340kbps、548kbps、764kbps、1128kbps during format conversion.

The streaming video resource mainly uses the conponent upload way. It saves video file to the Windows Media Service in a specified directory. This directory is set in the disk array connected to the server. Other related information of the video file are also stored in the database. For extra-large video which is unable to be uploaded, users can split it into several sections and upload them one by one. This extra-large video can be broadcastedcontinuously in a playlist.

2) Courseware On-line Browsing
Various kinds of coursewares can be divided into single document courseware and multi-document courseware on the basis of format.

a.Single Document Courseware
While Powerpoint and flash documents can be broadcasted on line, others not. Users have to download those documents to the local computer and play them using related software. IIS automatically refuses to offer download many kinds of documents in special formats in order to protect the Server. Out of the same concern, this system permits documents uploading in several certain kinds of formats, i.e. documents with suffix of ppt, pps, swf or rar. Other format documents should be uploaded after being packaged into a rar file. Considering the standardization, the system does not support other compression and packing methods, such as winzip.

b.Multi-document Courseware
Multi-document courseware should be uploaded after being packaged into the rar file and users can operate the courseware after downloading. Multi-document

coursewares include network courseware, which can be browsed online with the default boot file. Since the multi-document courseware must be packaged to upload, it must be decompressed on the Server to achieve online browsing. The system uses VB unzip command to call WinRar, decompressing rar documents on the Server, recording the decompressed file directory to the database. Users can broadcast the courseware online by accessing to the file directory.

5   The Application of the Three-Dimensional Teaching Resources

The construction of the three-dimensional teaching resources of physical education lags behind. The way of traditional physical education is that teachers teach sports knowledge and skills while students blindly imitate and practice. Teachers' teaching philosophy, students' learning habits and service quality of teaching can not meet the requirements of the modern information technology education. Therefore, we should take effective measures to promote teachers and students to use the three-dimensional teaching resources to enhance the sports teaching.

5.1   To Improve the Students' and Teachers' Comprehensive Use of the Three-Dimensional Teaching Resources

Firstly, teachers and students should improve their ability to get information effectively, to evaluate information critically, to absorb, store and extract information effectively. Secondly, teachers should improve their ability to use Powerpoint and CAI coursewares and to teach in the multimedia forms. Thirdly, students should improve their ability to use network resources, to study and exchange sports knowledge effectively and independently. In this way, we can improve the informationization of physical education.

5.2   The Three-Dimensional Service

The application of three-dimensional teaching resources need to be supported by the three-dimensional service. Firstly, the department of the physical education should offer dynamic service continuing to the whole semester to satisfy students' need of the sports teaching resources. Secondly, the information service system of physical education provide teachers and students multi-class, multi-format and multi-stage teaching services, achieving the three-dimensional service format. Finally, to achieve the three-dimensional service client, any teaching institution or individual can access to the rich sports teaching resources through various means.

References

1. Zhu, Z.: Modern Education Technology with information-based education. Education Science Press, Beijing (2003)
2. Wang, Y., Liu, J.: Modern Assessment of. Heilongjiang People's Publishing House, Harbin (1995)
3. Xue-nong, Qing, D., Wen-ling: Optimizing multimedia design. Guangdong Higher Education, Guangzhou (1996)
4. Xu, W.: Physical Education Teaching Reform of Sports and Science 11, 75–76, 80 (2001)

Study on Structural Instability of Large Crawler Crane Boom Structure

Wang Anlin, Jiang Tao, Dong Yaning, and Ling Fei

College of Mechanical Engineering
Tongji University
Shanghai, China
Kitty_lwy@yeah.net

Abstract. Taking a certain large crawler crane boom as research object, the local buckling critical load is calculated by finite element analysis, and a full-range analysis for the boom structure is conducted. The analysis method of boom structure instability propagation process is established, and the true load response of actual structure is reconstructed. The results show that the dynamic effect is obvious after the occurrence of local instability accompanied with dynamic hopping. The research results of this paper have certain reference value to make measures for preventing the instability propagation, and also have guiding significance for the large boom structure design.

Keywords: Large crawler crane boom structure, eigenvalue analysis, structural instability, nonlinear analysis.

1 Introduction

Engineering structure or component will be in a balanced state under certain loads, constraints and the influence of other factors. However it will deviate from its equilibrium position under the effect of any small external disturbance. The balanced state is stable if it can automatically revert to the initial equilibrium position after the disturbance is removed. Otherwise, the initial equilibrium state is unstable. The behavior that the engineering structure or component changed from an equilibrium position to another equilibrium position because of the instable balance form is called buckling, or known as the instability [1].

For a structure system subjected to external effect, we not only concern its critical load when the local buckling occurring, but also concern the instability behavior when the load exceeds the critical load, which is post-buckling behavior. The dynamic effect is very obvious when the local instability accompanied with dynamic hopping happens, so it is necessary and important to make a full-rang analysis for the structure. However, the traditional static analysis can not represent the true load response of the actual structure because it greatly simplifies the whole analysis process. Many scholars and researchers have done a lot of work on the buckling analysis of large span arch bridge, and shell structures [2-4]. The paper adopts this

Y. Wu (Ed.): International Conference on WTCS 2009, AISC 116, pp. 401–406.

method in the construction machinery field, considering the dynamic response; we put forward the analytical method of instability propagation process for large crawler crane boom structure, and use the international common finite element software-ANSYS to analyze the loading patterns, local instability form's influence on the instability propagation process.

2   Eigenvalue Buckling Analysis of Large Crawler Crane Boom Structure

Eigenvalue buckling analysis as one kind of linear analysis is generally used for predicting the theoretic buckling strength (that is, bifurcation point) of the ideal elastic structure. Because it doesn't consider any non-linear and initial defects, the calculation speed is particularly fast. Prior to conducting the non-linear buckling analysis, eigenvalue buckling analysis can be used for obtaining the buckling shape quickly.

We define the structure stress stiffness matrix $[K_\sigma]$ as stress stiffness matrix under unit external load, and the load multiplier is λ, the external load is $[R]$, the displacement matrix is $[D]$. Under linear conditions, $[K_\sigma]$ and usual stiffness matrix $[K]$ aren't displacement function. Under baseline conditions, the mechanical balance equation of the structure is defined as follow:

$$([K]+\lambda[K_\sigma])[D]=[R] \tag{1}$$

The structure will reach a new equilibrium state with the effect of external load. Defining the virtual displacement matrix as $[\overline{D}]$, we can get the new mechanical balance equation:

$$([K]+\lambda[K_\sigma])[D+\overline{D}]=[R] \tag{2}$$

According to the above two equation, we obtain the following equation:

$$([K]+\lambda[K_\sigma])[\overline{D}]=0 \tag{3}$$

Equation (3) is used for solving the structure's eigenvalue λ and the displacement eigenvalue vector $[\overline{D}]$. Bifurcation point critical load P_0 can be obtained from λ multiplying load. If the initial load is unit load, we know that the λ is P_0. The displacement eigenvalue vector $[\overline{D}]$ denotes buckling shape which also is called buckling mode. In common buckling analysis, the first eigenvalue and eigenvalue vector are important.

We take a certain large crawler crane boom shown in Fig.1 as study object. The bottom is under hinged constraints, and the top is supported by two steel wire ropes. It can withstand the maximum load 10e6kg under working conditions.

Fig. 1. One certain crawler crane boom structure drawing

Importing the drawing into finite element software, by setting the parameters, the previous 4 order modal (that is, buckling mode frequency) is obtained. It is shown in Fig.2.

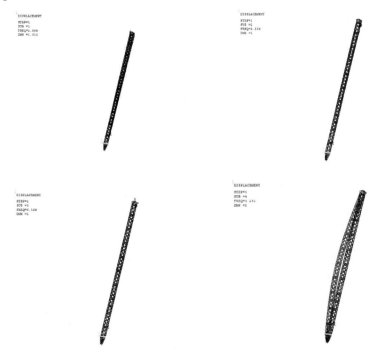

Fig. 2. Eigenvalue buckling analysis (the previous 4 order modal)

Because the eigenvalue buckling load can be denoted by load×buckling first order eigenvalue, and the boom structure's first order eigenvalue is 1.667 according to Fig.2, we get the expression of the eigenvalue buckling load as: $F \times g \times 1.667$ Where F denotes the applied load, g denotes the gravity acceleration.

However, in the real engineering practice, before the applied load reaches the theoretical elastic buckling load, the boom structure will become unstable because of its material defects and non-linear characteristics (see Fig. 3). Therefore, eigenvalue buckling analysis's result is non-conservative (Calculated critical load is low), and it can't be used in the real engineering analysis. In the actual structure design and calculation of critical buckling load, it's necessary to carry out nonlinear stability analysis of the crawler crane boom structure.

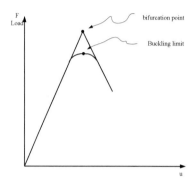

Fig. 3. Linear load- deform cure of eigenvaule analysis

3 Nonlinear Stability Analysis of the Large Crawler Crane Boom Structure

3.1 Nonlinear Stability Analysis Method

Through the gradual increment of applied load, nonlinear stability analysis obtains the critical load which begins to make the structure unstable. Eigenvalue buckling load which is calculated in above part1 is the expected upper limit of the nonlinear buckling load, which can be used as a given initial load for nonlinear buckling analysis. Before the gradual increments load reaches the eigenvalue load, the nonlinear solving should be divergent. Critical load which makes nonlinear solving divergent is named as nonlinear stability load.

The large crawler crane boom is a truss structure, whose nonlinearity mainly refers to the phenomena of truss axis deviating with the load pressure line under the effect of load. It is unavoidable because the pressure line is changeable during the boom structure under working conditions. And the geometric nonlinearity of the boom structure belongs to elastic deformation field.

Boom nonlinear equilibrium equation is as follow:

$$([K_0]+[K_L]+[K_\sigma])\{\delta\} = \{F\} \tag{4}$$

Where $[K_0]$ denotes small displacement elastic stiffness matrix, $[K_L]$ denotes large displacement matrix; $[K_\sigma]$ denotes initial stress matrix; $\{F\}$ denotes equivalent node load; $\{\delta\}$ denotes node displacement; $[K_L]$ and $[K_\sigma]$ are the function of $\{\delta\}$.

We use load increment method to solve (4). Load gradually increases from zero to $\lambda_i\{F\}$. The $\lambda_i\{F\}$ value that makes $\{\delta\}$ begin to diverge is the boom stability limit load. The nonlinear equations are written in common as follow:

$$[K(\delta)]\{\delta\} - \lambda\{F\} = \{0\} \tag{5}$$

where λ is the load factor, given by:

$$0 = \lambda_0 < \lambda_1 < \lambda_2 < \cdots < \lambda_i < \lambda \tag{6}$$

Under normal circumstances, the large boom structure's stability factor $\lambda \geq 10$ when it reaches its limit bearing capacity. For safety purposes, we set $\lambda = 100$. We apply the self-modified Euler method during linearizing (5), and the limit load value of the boom structure is between the load diverging and the front convergent load. If incremental steps are more enough, we can take the front step as the convergent load, which can avoid complex calculations.

3.2 Nonlinear Stability Analysis of Large Boom Structure

Setting the geometrical defect factor of the boom structure as 0.01, the structure in Fig. 1 is imported into the finite element software for nonlinear stability analysis. During the analysis, the arc length method is adopted to predict the approximate buckling load. When the applied load nearly reaches to the desired value of the critical buckling load, the load increments should be small enough. If the increment is excessive, the accurate predicted buckling load can't be obtained. Opening the dichotomy and automatic time step items is better for avoiding this problem.

Taking a certain top node of the boom structure to analyze, the load-time curve can be got (shown in Fig.4). From the Fig.4, we know that the boom structure becomes unstable when the load factor increases to 2.67.

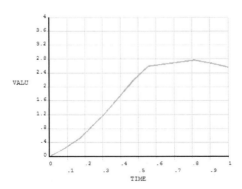

Fig. 4. Load-time curve of a certain top node of the boom structure

Fig.5 shows the bottom support reverse force. According to the Fig.5, when the load factor (time) increase to 1.2, the support reverse force begins to fluctuate, which represents the applied load can't exceed the 1.2 times of the maximum load due to the constraints of the boom size and shape

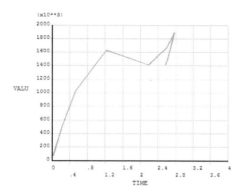

Fig. 5. Load-time curve of the bottom support reverse force

4   Conclusion

Eigenvalue buckling analysis and nonlinear stability analysis of a large crawler crane boom structure are developed in this paper. It can be concluded that:

（1）Carrying out eigenvalue buckling analysis on the large crawler crane boom structure, the previous 4 order modal are extracted. The first order eigenvalue supplies the initial value for the sequent nonlinear analysis.

（2）Developing nonlinear stability analysis on the crane boom structure, the nonlinear dynamic response indicators are obtained.

（3）The buckling analysis method is introduced to the field of construction machinery's boom structure dynamics, which provides a thought and method for the design of large boom structure and instability propagation research.

References

1. Shang, X., Qiu, F., Zhao, H.: ANSYS Structure Finite Element Analysis method and Examples of Application. China Water Power Press, Beijing (2008)
2. Cui, J., Wang, J., Sun, B.: Nonlinear buckling analysis for large span concrete filled steel tube arc bridge. Journal of Harbin Institute of Technology 35(07), 876–878 (2003)
3. Liu, J., Zhou, J., Su, G.: Long-span Steel Arch Bridge Dynamic Characteristics and Stability Analysis. Railway Construction Technology 2, 93–96 (2008)
4. Xiao, J., Ma, K.: Geometrically nonlinear stability analysis of prestressed partial single layer shallow reticulated shells. Spatial Structures 9(02), 31–37 (2003)

Research of Neural Networks in Image Anomaly Identification Application

Xianmin Wei

Computer and Communication Engineering School of Weifang University
Weifang, China
s2005100153@21cn.com

Abstract. Images anomaly refers to a relatively constant image suddenly appeared continuous monitoring of abnormal signs, its online identification is widely used in public security, environment, transportation, production process monitoring. This paper studies online identification algorithms and implementation techniques of image anomaly. and proposes the learning algorithm based on BP neural network for anomly identification.

Keywords: image processing, image analysis, image comparison, BP neural network.

1 Image Pre-processing Algorithm and Implementation

In the 24-bit true-color image is on pre-processing, the image is to be converted first. One way is to convert images to the appropriate 256 grayscale image, with one byte representing one pixel of the image. After converting images to grayscale images, based on the need to balance the gray scale of the images, the aim is to increase the dynamic range of pixel gray value to achieve the effect of contrast-enhanced images as a whole , thus facilitating the comparison images.

Image processing flow shown in Figure 1.

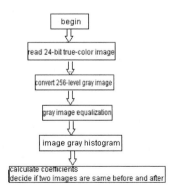

Fig. 1. Image processing flow

Y. Wu (Ed.): International Conference on WTCS 2009, AISC 116, pp. 407–413.
springerlink.com

24-bit true-color image
A BMP file is generally divided into 4 parts:Bitmap file header, bitmap information header, palette, the actual image data.Typically, for true-color image, each pixel is directly expressed with R, G, B these three components bytes, instead of the palette technology, so the value of R, G, B component data is taken directly from the image . In general, BMP files data order is from bottom to top and from left to right. First read from the file is the bottom line of image pixels to the left first, then the second left pixel., ..., followed by the penultimate row on the left of the first pixel, the second left pixel, ..., and so on, and finally get the top line of the last pixel. For true-color map, image data is the actual R, G, B value, from left to right is B, G, R value.

The initial image convert into 256 grayscale images
Each pixel of grayscale image has the brightness value from 0 (black) to 255 (white), totally 256 gray-scale.

1) grayscale diagram
To express grayscale diagram, it is necessary to quantify the brightness value. Usually divided into 256 levels from 0 to 255, 0 is the most dark (all black), 255 is the brightest (all white).

However, BMP file format has not the concept of grayscale, but it can easily be expressed into a grayscale with BMP file. Method is to use 256-color palette, but this palette is a bit special, each of the RGB values are the same, that is RGB value from (0,0,0), (1,1,1) has been to (255,255,255). Among them, (0,0,0) is all black, (255,255,255) is all white, in the middle is gray. This gray color diagram is to represented with 256-color diagram. For R = G = B colors, using them into YIQ or YUV color system conversion formula, it can be seen in the color components are 0, that is, no color information.

2) true-color bitmap into a grayscale
To convert 24-bit true-color bitmap into grayscale, you must first calculate the corresponding gray value of each color. Grayscale and RGB colors have the following correspondence:

$$Y = 0.299R + 0.587G + 0.114B \qquad (1)$$

Then, RGB values of each image pixel is replaced with the corresponding gray value, thus 24-bit true-color image is converted into gray scale images.

Fig. 2. 24-bit true color image   **Fig. 3.** Converted gray image

Gray image equalization
Gray image equalization is sometimes also called histogram equalization, the aim is to increase the dynamic range of pixel gray values to achieve the overall image contrast enhancement effect.

3) Algorithm Design
In accordance with the definition of the image probability density function (PDF, normalized to unit area histogram):

$$P (x) = H (x) / A_0 \qquad (2)$$

Where, x is gray-scale, H(x) is for the histogram, A0 the area of the image.

To Set before converting the image probability density function is Pr(r), the converted image probability density function is Ps(s). Where r, s respectively gray level before and after conversion, the conversion function s = f (r).
By the probability of knowledge, we can get:

$$Ps(s) = Pr(r)dr/ds \qquad (3)$$

In this way, if to make probability density function of the converted image as 1 (ie the histogram is uniform), you must meet:

$$Pr(r) = ds/dr$$

Intergral of r on both sides of the equation, as follows can be obtained:

$$s = f(r) = \int P_r(u)du = \int H(u)du / A_0 \qquad (4)$$
$$(u = 0 \sim r)$$

This conversion formula is called the cumulative distribution function (CDF) of the image.

The above formula is derived after normalization, and for the not normalized situation, as long as multiplied by the maximum gray value (DMax, the grayscale is 255). Gray equalization conversion formula is:

$$D_B = f(x) = D_{MAX} \bullet \int H(u)du / A_0 \qquad (5)$$
$$(u = 0 \sim x)$$

For discrete images, the conversion formula:

$$D_B = f(x) = D_{MAX} / A_0 \bullet \sum H_i \qquad (6)$$
$$(i = 0 \sim x)$$

Where x is the gray level, Hi is the i-level gray pixels.

2   Image Feature Extraction

In digital image processing, one of the most simple and most useful tools is histogram. It summarizes the contents of an image's gray level, any one image histogram may include considerable information, some types of images can be completely described by its histogram.

2.1   Gray Histogram

Gray histogram is a function of gray level, it reflects occurrence frequency of an image pixel of each gray level. It is an important feature of the image and reflects the image gray scale distribution.

Gray histogram is also another way of definition: Suppose a continuous image defined by the function D (x, y), it changes grayscale smoothly from the center to the edge of the high gray-level to the low-gray value. We can choose a gray-scale D1, then a contour is defined, the contour line connects all the points in the image with the same gray level of D1. Contour obtained forms the closed curve with gray level greater than or equal D1 surrounding region. To call in a continuous image by contours with a gray-scale D to surround all the area as the threshold area function A (D). Histogram can be defined as:

$$H(D)=\lim(A(D)-A(D+\Delta D))/\Delta D=-dA(D)/dD(\Delta D \to 0) \qquad (7)$$

From the above it can be concluded: a continuous image histogram is negative derivative of its area function. The emergence of a negative sign is due to D in the increase as DA(D) in the decrease. If a two-dimensional image as a random variable, then the area function is equivalent to the cumulative distribution function, while the histogram is equivalent to the probability density function.

For discrete functions, fixed ΔD is 1, then the above formula becomes:

$$H(D)=A(D)-A(D+1) \qquad (8)$$

2.2   Draw the Image Histogram

The gray-scale is as horizontal coordinates, the frequency of gray levels is vertical coordinates, rendering the relationship between the frequency with gray-scale map is the histogram. When histogram is in the picture, first to select gray level with most pixels as the highest benchmark, the other gray levels are drwan in an appropriate ratio.

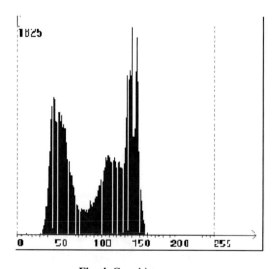

Fig. 4. Gray histogram

3   Anomaly On-Line Identification Algorithm and Implementation

After the image pre-processing on the initial BMP files, the next work is to compare the image. There are many image comparison algorithm, the simplest one is the image histogram for a rough comparison, before and after the two images by calculating the mean variance of gray histogram to understand in general the difference degree of two images. But this method results in the actual implementation is not very satisfactory, There are two main reasons:

(1) Histogram does not reflect the specific location of the image pixels, even worse is two completely different images may correspond with a histogram;
(2) Image the number of pixels will have an impact.

To sum up above conditions, using another method, which calculat the correlation of two images, its basic principles is through the calculation of correlation function to determine the similarity of images.

Algorithm Design
Given two functions f(t) and g(t), their cross-correlation function is defined as:

$$R\ (v)\ =f\ (t)\ \cdot g\ (-t)\ =\int f\ (t)\ \cdot g\ (t+v)\ dt$$

$$(-\infty< t <+\infty) \tag{9}$$

For discrete images, we can use the following correlation function for similarity measure:

$$R=\sum f\ (t)\ g\ (t)\ /\ (sqrt\ (\sum f2\ (t)\ )\ \cdot sqrt\ (\sum g2\ (t)\ )\ ) \tag{10}$$

Where the function f(t) and g(t) respectively a same property of the two images to be compared, R is the correlation coefficient of two images.

According to Schwarz inequality, we can know in the above of $0<R\leq 1$, and only in the ratio of f(t)/g(t) is constant it can be max value of 1. Larger R, the higher the degree of similarity of two images.

In the time of comparing the two images, using the image gray histogram, as to histogram getting rid of location features of pixels, this comparison may be wrong during the time. Here one method is to provide first the image is divided into several parts, then they are in the image preprocessing according to a few steps of above, and finally, the corresponding part of the two images to compare, as long as a portion of which is not the same, you can view these two images is inconsistent.

4   Based on BP Neural Network Learning Algorithm for Image Anomaly Identification

The basic idea of BP algorithm is that the learning process is composed of positive communication of the signal and back propagation process of error. When forward propagation, input samples transfer from the input layer, through the hidden layer processing, transmission to the output layer. If the actual output of the output layer does not match the expected output, then transfer to the error back propagation stage.

Error back propagation is the output error by some form through hidden for layer to layer back-propagation, and error are assessment for all units on each layer, so to gain error signal from each layer unit, this error signal is correction weight basis of each unit. As this signal in forward propagation with error back-propagation, each layer weight adjustment process is carried out of the cycle. Weights continuously adjustment process is network learning is and training process. This process continues until the network output error is reduced to an acceptable level, or to a predetermined number of the study times.

BP algorithm using feedforward neural network is by far the most widely used neural network, in multilayer feedforward network applications, as shown in Figure 5, the application of single hidden layer network is most common. Generally accustomed to a single hidden layer feedforward network known as the three-layer feedforward network, or sensor, so-called three-tier, including the input layer, hidden layer and output layer.

When the network design is completed, to apply the design value to train. Training for all samples to run one cycle positively and one time weight modifies negatively as one time training. During the training sample data set to be used repeatedly, but each getting data is not a fixed sequence. Usually needs thousands of times to train a network.

Network performance mainly depends on whether good or bad generalization ability is, generalization ability test can not be carried out on the training set data, and use test data outside the training set to be tested. General practice is available to collect samples randomly divided into two parts: the training set, the other as the test set. If the network error on the training set was very small, while error of test set sampleswas great, to indicate that the network has been trained to over-match, so generalization ability is poor.

When hidden nodes are fixed in some cases, in order to obtain good generalization ability, there is a best training times. To illustrate this, training will be alternating between training and testing, each training to record a training mean square error, then keep network weight value unchanged, with test data positively to run, recording test MSE. Using these two error data can plot the curves in Figure 6 of two mean square error with training times.

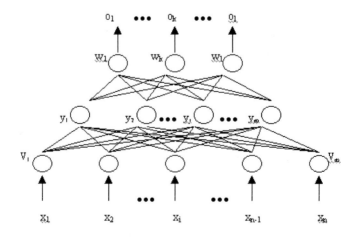

Fig. 5. A single hidden layer BP network

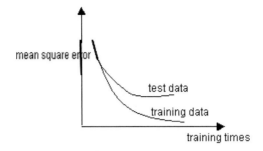

Fig. 6. Mean square error curve with the training times

5   Conclusion

From the error curve can be seen before a certain number of times in training, with the training increasing, the two error curves fallen. When the number of times is over the training, the training error continues to decrease while the test error is starting to rise. Therefore, this training times shall be the best training times, if stopped training called insufficient training, after which is called overtraining. Based on the above BP algorithm basic idea, when we compare more images, could use a similar algorithm. The images first divided into two groups: one group was similar to the image, the other group was not similar to the first group images; then let computer literacy, the output compared with the expected results to determine a boundary range, so long as meet the range of two or more similar images, otherwise dissimilar.

References

1. Bin, H.: Visual C + + digital image processing. Posts & Telecom Press (April 2001)
2. Castleman, K.R.: Digital Image Processing, Tsinghua University Press (February 2002)
3. Han, L.: Artificial neural network theory, design and application. Chemical Industry Press (January 2000)
4. Zhang, Y.: Image processing and analysis. Tsinghua University Press (February 1999)
5. Jia, Y.: Computer image processing and analysis. Wuhan University Press (September 2006)

Research of Dynamic Buffer Pre-allocation Method Based Network Congestion Control

Xianmin Wei

Computer and Communication Engineering School of Weifang University
Weifang, China
s2005100153@21cn.com

Abstract. In this paper, contrary to defects of the traditional average distribution pre-buffer to resolve network congestion control strategy, the paper proposes a dynamic buffer pre-allocation improvement strategy, according to the size of the network entrance traffic to adjust all the entrance buffer size in real-time, and to carry on simulation The simulation results show that the strategy can better solve the network congestion problem than the traditional buffer pre-distribution method.

Keywords: dynamic buffer pre-allocation method, network congestion control.

1   Introduction

In the past ten years, with Internet's rapid growth, network congestion problem has become more and more serious, the root causes of network congestion is that the network load by users is greater than the capacity of network resources and processing power, its performance is, data package delay increase, the number discarded increase the upper application system performance degrade and so. Effective congestion control is the key to ensure Internet stability. Traditionally, with the increasing network capacity, congestion will be reduced. However, recent studies have shown that with increasing network capacity, but the stability of congestion control problem is getting worse, when network capacity increases to a certain limit, the existing congestion control system will be unstable.

Congestion is the positive feedback phenomenon resulting from network link failure or severe overload on circuit devices. Internet system structure determines the congestion happening possibility because in no prior consultation and admission control mechanism of resource share network, there must be some IP groups reach router simultaneously, and to expect the possibility of forward from the same output port. Obviously, not all groups can be handled by the same time, we must have a service order, the cache is to provide a protected method for the waiting for services groups, but when the cache space is exhausted, the router only dropping packets.

Y. Wu (Ed.): International Conference on WTCS 2009, AISC 116, pp. 415–420.
springerlink.com

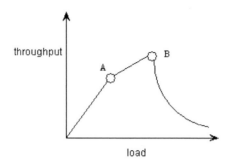

Fig. 1. The relationship between network load and throughput

Figure 1 shows the relationship between the load of computer network and throughput. When the load is smaller, a linear relationship is between the load and the throughput; After thoughput reaching A point, as the load increases, the throughput gradually has smaller increments; when the load across the B point, the throughput has dropped drastically. Near A point is usually called the congestion avoidance interval; Between A point and B point is the congestion recovery interval; Outside B point is the congestion collapse range. For the maximum use of resources, networking status in mild congestion should be ideal, but it also increases the possibility of sliding into congestion and collapse. When the network is in state of congestion and collapse, the small load increment will cause the effective network throughput declined sharply. Packet loss rate is higher, end to end delay increases, the whole system may even crash, to cause partial or total paralysis of the network, seriously affecting network performance and service quality. Therefore, congestion control mechanisms must be used to constrain and limit.

2   Dynamic Buffer Pre-allocation Method

From the point of view control theory, the computer network is a large system, so many methods of control theory can be used to analyze the mechanism of computer network congestion, to design congestion control method. Using control theory to analyze computer network systems, the key is to establish a mathematical model describing the network conditions. The complexity of current computer networks is very larger, to build a model accurately to describeis very difficult, this paper will use more intersection vehicles passing model to describe the model of network congestion.

We know that in a busy city intersection, although the peak of traffic congestion in the traffic situation is very serious, but waiting for each traffic direction is a different number of vehicles, the traditional red and green light supply passage time equal to the each other direction of approach. The more apparent is in some direction having more vehicles, some direction not having vehicles in all directions, such phenomena does not relieve traffic congestion problems. Later, red-green light is for the improvement, according to statistical laws, the travel time of each direction was adjusted, so more serious traffic congestion, travel time longer the direction, this efficiency for traffic congestion is sloved effectively. This way there is a problem, the number of vehicles passing each direction is a random variable widely, and only through real-time testing before we know the exact number, and real-time detection is also a lag process, it occurs after vehicle congestion. Therefore, in order to effectively solve the traffic congestion, we must advance forecast the number of vehicles in each direction, adjust the direction of real-time dynamic travel time.

Network congestion is similar with intersection traffic congestion, and similar to the principle of solution, its difference is even more complex network model, the network entrance number is much larger than the amount of vehicles crossing direction. According to traffic congestion control principle through the intersection, to set the network entry number is M, the total buffer pre-allocate capacity is N , the average distribution of each network entrance buffer capacity is ma, at the n sampling time the entry i the data packet capacity is Mni, the etwork entrance traffic change percentage is η1 and η2 at sampling moment n and n-1, $0<η1< η2<1$, buffer dynamic adjustment coefficient is kni,the network system satisfies the following formula:

$$\sum m_i = M \tag{1}$$

$$m_a = M / N \tag{2}$$

If adopting the following adjustment control strategy, to n sampling moment, when

$$1 \le \frac{M_{ni} - M_{(n-1)i}}{M_{(n-1)i}} \tag{3}$$

$k_{(n+1)i}$ is not changed. When

$$\frac{M_{ni} - M_{(n-1)i}}{M_{(n-1)i}} \le 0 \tag{4}$$

$k_{(n+1)i}$ is not changed. When

$$\eta_1 \le \frac{M_{ni} - M_{(n-1)i}}{M_{(n-1)i}} \le \eta_2 \tag{5}$$

$k_{(n+1)i}$ is not changed. When

$$0 \prec \frac{M_{ni} - M_{(n-1)i}}{M_{(n-1)i}} \prec \eta_1 \tag{6}$$

$k_{(n+1)i} = k_{ni} \times (1 - \eta_1)$. When

$$\eta_2 \langle \frac{M_{ni} - M_{(n-1)i}}{M_{(n-1)i}} \langle 1 \tag{7}$$

$$k_{(n+1)i} = k_{ni} \times (2 - \eta_2)$$
,

$$M_{(n+1)i} = M_{ni} \times k_{(n+1)i} \qquad (8)$$

Where the value of $\eta 1$ and $\eta 2$ is determined according to the statistics of network traffic or network administrator experience. For (3) and (4) cases, that is a very small probability of unexpected events, the stability of the adjustment coefficient remains unchanged.

3   Simulation

According to the algorithm designed in this paper, in order to better reflect the advantages of network congestion control improved by dynamic buffer pre-allocation method, to use MATLAB to develop results simulation software of the dynamic buffer pre-distribution control method for simulation analysis procedures. Program source code is as follows:

```
clear all
pass=0;
wpass=0;
n1=0.2; % Coefficient 1
n2=0.5; % Coefficient 2
randnmuber=800; % number of random
mem=1000; % size of pre-buffer in average allocation
ma=0.8*mem;
for i=1:randnmuber
it(i)=mem*rand(1);
k(i)=1;
end
m=it;
for i=1:(randnmuber-1)
if it(i)>ma
pass=pass+1;
end
if i==1
if (it(i)-ma)>=ma
k(i+1)=1;
elseif (it(i)-ma)<=0
k(i+1)=1;
elseif n1<=(it(i)-ma)<=n2
k(i+1)=1;
elseif 0<(it(i)-ma)<n1
k(i+1)=1*(1-n1);
else
k(i+1)=1*(2-n2);
end
m(i+1)=it(i+1)*k(i+1);
```

```
end
if (it(i)-ma)>=ma
k(i+1)=k(i);
elseif (it(i)-ma)<=0
k(i+1)=k(i);
elseif n1<=(it(i)-ma)<=n2
k(i+1)=k(i);
elseif 0<(it(i)-ma)<n1
k(i+1)=k(i)*(1-n1);
else
k(i+1)=k(i)*(2-n2);
end
m(i+1)=it(i+1)*k(i+1);
if m(i+1)>ma
wpass=wpass+1;
end
end
wpass/pass
```

The simulation results shown in Figure 2. Figure 2 is a network congestion control curve at the entrance, the horizontal coordinate is on behalf of the data at entrance to be transported, while the vertical coordinate is the control effect ratio curve of the sampling dynamic allocation buffer with average distribution buffer, and it can be seen from the figure that effects using dynamic allocation buffer allocation is better 4.8 times or so of average distribution buffer for relieving network congestion.

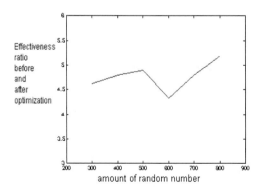

Fig. 2. At the entrance of a network congestion control effects curve

4   Conclusion

In this paper, contrary to the defects of traditional average distribution pre-buffer to resolve network congestion control strategy, a dynamic allocation pre- buffer improvement strategy is proposed, and its simulation results show that this improved strategy can better solve the congestion than traditional buffer pre-distribution method.for the network congestion problem.

5   Application and Conclusion

This paper presents an effective solution for digital text watermark, this shows in a blank text messages secretly embedded in is entirely possible. This program can be widely used in text of Unicode or other format, especially for the many cases requiring the protection of content and ownership and high-security of content exchange is very appropriate, while in large enterprises and government agencies also have broad application prospects.

References

1. Wang, Y., Cao, C.: Based on active network congestion control. Computer Engineering and Science 24, 45–48 (2002)
2. Ren, M., Wang, W.: Algorithm for optimal control theory based on network congestion control. Electrical Engineering 20(5), 123–125 (2003)

Security Analysis on Exam System Based on Network

Li Guofang[1], Li Guohong[2], Qi Yubin[3], Wang Tao[3], and Zou Ping[3]

[1] College of M&E Engineering, Hebei Normal University of Science and Technology,
Qinhuangdao, China
[2] College of Finance and Economics, Hebei Normal University of Science and Technology,
Qinhuangdao, China
[3] College of E&A, Hebei Normal University of Science and
Technology Qinhuangdao, China
lhy05120@21cn.com

Abstract. Different from the traditional exam, the network exam has more complicated security problem. The security over the whole process of general network exam is analyzed. The security of network exam is considered in the organization and management of exam, the seriousness of exam, system accident and protocol and data transmission. Some measurements to improve network exam security are present, such as personal identification, secondary login, time synchronization and restoration of exam. New flow chart of network exam is designed with security measurements. A network exam system based on above security analyses is developed and has got a good result in application with high safety. The security methods can provide a basis for the design of network exam system.

Keywords: Network exam system, security, exam bank, exam organization and management.

1 Introduction

Network exam has many advantages including objectivism, accuracy, quickness, convenience for later treatment; therefore the exams based on network (LAN, Internet) are widely used. From the insight of technology, the function for different exam systems is largely identical with each other, for example including such parts as maintenance of questions, paper-producing and score-marking. The security problems in network exam are much more complex than those in traditional examinations, and they deal with paper form, ID identification, exam timing and secondary login. These are the points to be discussed in the thesis.

2 Characteristics of Network Exam System

2.1 The Advantages and Disadvantages of Network Exam

The advantages of network exam focused on the followings:
   The exam scale is changeable, and the number of examinee can be random.

Y. Wu (Ed.): International Conference on WTCS 2009, AISC 116, pp. 421–427.
springerlink.com

The examination time can be unfixed, and examinee can take the exam whenever he comes. There are no duplicated exam questions.

Automatic score-marking and the score can be marked as soon as the exam is over.

Lightens the load of the teachers, reduces errors in marking score and in recording the marks.

Automatically deals with the examination results, such as the highest mark, lowest mark, average and score distribution and root-mean-square deviation.

The main disadvantages of network exam system are that exam question types are restricted only to objective questions, such as single-choice, multiple-choice, blank-filling and true/false exam questions.

2.2   Processing Flowchart of Network Exam System

Figure 1 is the processing flowchart of general network exam system.

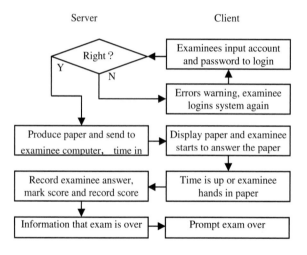

Fig. 1. Process flow chart of network exam system

When network exam began, the examinee entered his account and password, which was given to every examinee respectively by the supervisor before exam, to login to exam server. When login successfully, the server produces the paper and documents it, then to hand it out to the examinee. As the paper is transmitted to the examinee computer, it is displayed in question number sequence, and then the students began to answer the paper. When the exam time is up or when the student is willing to hand in his paper, the examinee computer will transmit the paper up to server. The server will record the answers and mark score to achieve the result, and then write it to database. Finally the server gives off the signal that the exam is over.

3   Organization and Management of Network Exam

3.1   Exam Organization

New technology based on computer can guarantee the exam process secure, yet as a non-technology secure method, strict invigilation is essential to network examination. Such human factors as exam-taker, entrainment books cheat, cheat each other tips, cannot be distinguished and prevented only by computers. So the network exam need concentrate the students to a specified computer room and specified seats, and then the specialized supervisor checks identity and maintains exam discipline.

3.2   The Position of the Examinees

To avoid exam-taker, the students should take the exam at the computer position specified by supervisors. Whether he is in his position can be guaranteed by supervisor inspecting, or by server judging whether some account logins from the corresponding student computer.

3.3   Exam Question Types

Network exam paper had better give priority to objective questions, such as multiple-choice, single-choice, true/false examination and blank-filling questions, while subjective questions, such as short-answer questions and essay questions do not fit in it. The purpose is primarily to quickly mark the paper. Still some subjective items can be manually marked by teachers after exam, yet the exam period has been lengthened and also the potential safety hazard increases.

3.4   Types of Exam Paper

AB paper types are intended during traditional exam to prevent from plagiarizing answers created by others. In network exam, random paper-producing method is used, as examinees are located near another, and the computer are set without no strict rules. This requires that there are enough questions designed in question bank and the questions are equal in difficulty.

4   Security Associated with the Exam Seriousness

4.1   The Position of the Question Bank

In order to improve the security of the system, the question bank should be saved in server, rather than appeared in any examinee computers.   Moreover the questions should be converted and saved in encrypted message form, not in plaintext form. The question bank cannot be in the same folder as the exam system. The best way to control question bank is with SQL-Server or ODBC. This can help prevent question bank from hidden danger as the exam system position is exposed.

4.2   The Security of the Server

To assure the server to work smoothly and stable, antivirus firewall and hacker software should be installed in the server. During exam process, unnecessary ports should be closed and only those ports to be used and related to the exam are kept open. If the exam is on LAN, the physical connection of the LAN with the Internet should be shut off.

4.3   Storage and Display of the Paper

During the course of the exam, it is on the server to produce paper and save it, which includes standard answers for the use of marking score. Then it reaches the examinee computer through network. However, the examinee computer just does display the paper in the exam system, and performs decryption conversion on the paper, rather than stores the papers in the auxiliary storage.

4.4   Exam Timing

Unified exam timing is to be offered. As the examinee logins the exam system, the server is used to realize time synchronization with the examinee computers, then the exam times are simultaneously recorded on both server and examinee computers. The examinee computer uses the accumulative time, and is not associated with the current date and time.

5   Security Problems Resulting from System Accidents

5.1   System Abnormality

System abnormality means system crashes, sudden power break or lock up in server or examinee computers. During the exam, the possibility that such abnormality is on examinee computer meets a few bigger. To solve the problem, when the examinee switches the questions by clicking next question or former one, all the answers on the examinee computer upload to the server and the server will record them. When the examinee computer is abnormal, the server should alarm real-time to inform supervisor of performing corresponding treatment, so as to avoid interrupt the exam for a long time.

5.2   Secondary Login

Secondary login refers to that the examinee logins the system again, both by inputting the examinee's account and password and by the supervisor's inputting secondary login password when the examinee computer is abnormal. This method can avoid students to frequently use secondary login for the purpose of delaying time.

When he logins the network, the system sends the former answered paper and the exam time to the examinee computer according to the examinee account. Thus comes to the exam circumstance and exam resumes.

6   Security Measurements of Network Protocol

6.1   Data Encryption Transmitting

System controls present in visual programming languages are used in designing the whole system.Those controls directly support TCP or UDP protocol, and in general developers no longer control the application level protocol. To enforce the security of the exam system, consideration should be taken to convert the data sent on the network according to certain regulations, without changing TCP itself. As a result, it is the cipher text that is transmitted on the network, and can defend the attack of sniffing effectively. Meanwhile, auxiliary control messages, such as handshake and synchronization, should be added at the beginning and the end of the data string, to prevent forgery attack. Hence improve the system security.

6.2   Examinee Computer

In those systems, which is designed based on B/S, the exam is performed on IE Browser in the examinee computer. There are disadvantages in security existed in general browsers. To enforce the security of the system, C/S should be used to develop exam system, that is, developing exam service prefers specialized system to Web information distribution systems, such as IIS. Examinee computer utilizes special Client Software to improve the system security.

7   Processing Flow Chart of Network Exam Based on More Security

According to the above analyses, various aspects of the security associated with network exam should be taken account into design, and new processing flow chart is illustrated in Figure 2.

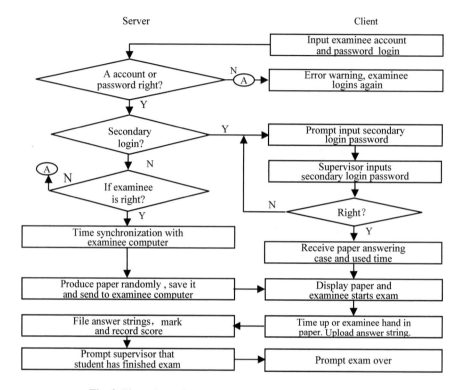

Fig. 2. Flow chart of network exam system based on security

Compared with Figure 1, Figure 2 increases the following methods, the examinee secondary login, judgment that the examinee computer is correct, system time synchronization and return to normal exam after abnormal system.

Based on above security analyses, we designed a network exam system on C/S with VB.Net. The system has got a good result in application with high safety when using.

Acknowledgments. Author thanks for the supporting from Education Science Plan of Hebei Province (grant No. 06030096) and Education Reform Fund of Hebei Normal University of Science and Technology (grant No.0802, "Teaching method reform of Engineering Thermodynamics").

References

1. Fu, C.: Advances computerized Adaptive Testing. Journal of Modern Information, 61–64, January 1 (2005), doi:CNKI:SUN:XDQB. 0.2005-01-018
2. Ren, J., Zhu, L.: The Design of Network Test System. Journal of Henan Mechanical and Electrical Engineering College 17(6), 119–121 (2009), doi:CNKI:UN:HNJD.2009-06-044

3. Lin, J., Yan, H., Wu, B.: Theory analysis on Computerized Adaptive Test. Jouranl of Taiyuan University of Technology 35(2), 221–224 (2004), doi:CNKI:SUN:TYGY.0.2004-02-031

4. Yu, H., Xin, G.: Network test system development and implementation. Journal of Jilin Teachers   Institute   of   Engineering   and   Technology 23(2),   85–86   (2007), doi:CNKI:SUN:JLGC.0.2007-12-032

5. Fu, X., Zou, B.: The Research on Multi-level Management Model of Grade Security in Examination System. Science Technology and Engineer 31(24), 219–221 (2005), doi: CNKI:SUN:KXJS.0.2005-06-011

Application of CAI in the Teaching of Physical Education in College

Tao Shi[1], Ligang Ma[2], and Zhijie Xing[3]

[1] School of PE, Handan College, Handan, China
[2] Department of Public PE, Hebei University of Engineering, Handan, China
[3] Sport Group, HanShan Experimental Elementary School, Handan, China
shisi60@yahoo.cn

Abstract. With the rapid development of science, Computer Assisted Instruction (CAI) is being applied in the teaching of physical education in college, as well as other modern education technologies, which largely promotes the development of teaching theory and practice in college education. Physical training serves as an important component of quality education and should be equipped with modern technologies of education to improve the quality and effect basic on its characteristic.

Keywords: CAI, college, teaching of physical education, network.

1 Introduction

CAI system and Multimedia technology, which are the most remarkable research fruits and crystal of human wisdoms in the education of 20th century, have become the main components of education science and modern education technology and have very positive significance in breaking traditional education concept. CAI means "Computer Aided (or Assisted) Instruction" and is an approach of education through the interaction between student and computer which serves as the teaching medium. With the development of modern technology, the way of gaining knowledge has been greatly changed and people also raised higher requirement for education, so CAI is gradually showing its importance in advanced education. How to combine various modern education media with traditional education media organically, especially CAI and multimedia technology, to meet the requirement for PE teaching in modern times, is the important subject teaching staffs of physical education must face.

2 Function of CAI in PE Teaching in College

2.1 Help with Teacher's Instruction and Demonstration

CAI technology is able to give prominence to the key and difficulty of technical motions in the PE teaching, especially for some items of high difficulty in which students are required to complete a series of complicated motions in a minute and

Y. Wu (Ed.): International Conference on WTCS 2009, AISC 116, pp. 429–434.
springerlink.com © Springer-Verlag Berlin Heidelberg 2012

technical requirements are so harsh. Students are often hard to see clearly these motions finished in a moment and cannot set up a complete motion image rapidly. Teacher's repeated demonstration and instruction may mislead students. We can make multimedia courseware with computer and modern education technology to show the technical parts, which are hard to demonstrate by teacher, with slow motion, freeze-frame and replay combined with teacher's instruction, so that students can see clearly the motion and technical detail in every moment and set up motion image faster and more comprehensively. This way gives special shoot to the difficulty and key of motion, improves teaching effect and shortens teaching process.

2.2   Help with the Learning Interest of Students

The multimedia courseware optimizes teaching content, method and practice and its expression is very intuitive and vivid. The courseware adds artistic expression and strong impression to teaching content and makes teaching and learning funny by diagram, image, animation, music, glinting, color and letterform. Diagram is for summarizing, concluding and comparing; imagine can raise the class atmosphere, attract the attention of students and be also applied to the training of technical skills. CAI and multimedia technology improves the combination of teaching and learning and fully exerts the advantage of students' cognitive system, and thereby improves the interest in sport activities and helps them set up the concept of lifetime sports.

2.3   Help the Teacher Correct the Fault Action of Students by Comparison

To reduce or avoid the mistake in exercise, it is better to point out common mistakes and their causes while demonstrating a correction action. By modern education technology, we can present both correct and error actions in a courseware, so that students can think and compare while watching and avoid making many common mistakes. This way enables students to master an action rapidly and trains the ability to observe and analyze.

2.4   Help Improve Teaching Efficiency and Reform Teaching Structure

Multimedia technology can release the teacher from repeated labor to a certain extent and enables the teacher to play active and dominant role from a higher level. Meanwhile, in the research of multimedia courseware and exploration in CAI and multimedia teaching method, the teacher can constantly improve theoretical level and knowledge system, and thereby strengthen the teacher's enthusiasm and creativity in teaching and teaching efficiency.

2.5   Help to Accelerate the Update of Knowledge and Ability

Convenience and changeability are the characteristic of CAI and multimedia technology which can have the content of courseware changed anytime tracking domestic and overseas advanced sport technology, teaching method and approach, and thereby abandon the outdated teaching theory of traditional books which were compiled decades ago to let students master as much latest knowledge as possible.

2.6   Help to Change Teaching Concept

The application of multimedia technology to physical education is a new approach and opens up the new land for PE teaching. According to the feedbacks of students in class, teacher's giving lecture by modern approaches can develop and train students' potential, and stimulate their curiosity for new knowledge, desire to explore and imitative to think while expanding the scope of knowledge, in order to promote the comprehensive development of quality, which is hard to reach by traditional teaching in class. CAI and multimedia technology provide advantageous technical foundation and conditions for the reform of PE teaching.

3   Influencing Factors of CAI in PE Teaching in College

3.1   Hardware Establishment

CAI is supported by computer, network and multimedia. Without necessary hardware facilities like proper computer, projector, multimedia room and so on, CAI in PE teaching cannot exist. As for now, in some developed areas, there are already abundant multimedia teachers and campus network has already become the hotspot in the technical environment construction of modern education, but in some colleges, campus network only stays in document delivery, multimedia resources are only used by the single machines within a workstation, campus network is still in its early form, and hardware environment limits the development of CAI.

3.2   Supporting Software for PE Teaching

Development of supporting software is a professional field and requires the cooperation of computer professionals and PE teaching staffs and is the integration of multiple knowledge and technologies. Firstly, affected by exam-oriented education and economic benefit, the supporting software for PE teaching is still in short; secondly, for some existing supporting software, such as "computer arrangement and score system of sports", "student's constitution evaluation and monitoring system" and so on. As PE teachers are lack of related computer knowledge or use motivation, these software can't show their functions; in return, this restricts the enthusiasm of software development and holds back the application of computer assisted instruction.

3.3   Teaching Ideology and Concept

Computer assisted PE education is what information times require from PE teaching and the combination of modern education technology and PE teaching. We cannot deny the fact that many teachers hold the outdated teaching thought that I can work well and live well without the knowledge of computer. Changing the traditional teaching ideology and concept of teachers and improving the academic quality has become the important factor to affect the application and promotion of CAI.

3.4   Attentions of Governors

In some developed areas, schools have made some related researches and explorations in the CAI of physical education. This can promote the application of CAI to a certain extent and show a new kind of assistant approach to physical education; however, the effort only stays in "discussion, demonstration and observation" level and little effort has been made to apply it to the practice of teaching to show its function. Theoretical research is still in short. This has much to do with the attention of governors.

4   Courseware Structure and Organization

4.1   Courseware Structure

Define abbreviations and acronyms the first time they are used in the text, even after they have been defined in the abstract. Abbreviations such as IEEE, SI, MKS, CGS, sc, dc, and rms do not have to be defined. Do not use abbreviations in the title or heads unless they are unavoidable.

The structural design of multimedia items is different from traditional program design because it should work with several kinds of media and reasonably determine information structure and overall arrangement, which is especially important. Therefore, to reach the expected effect, besides abundant graphs, images, elegant music, instruction and harmonious color arrangement, the overall structural design is also very critical. In creating CAI courseware, program structure and overall arrangement is nonlinear tree structure. With navigation map as inquiry center, related subjects can be switched to each other providing the information of other levels and setting up a pyramid-shaped information structure. The CAI courseware with tree structure has an obvious hierarchical structure and inheritance relation and the connection between lower node and upper node is unique. This structure is the best one to show the affiliation between each component and provide users overall concept and clear thought. It can avoid the limitation of sequence structure as well as avoiding the "get lost" defect of net structure.

4.2   Content Organization of Courseware

The CAI courseware of physical education is usually composed by 5 subjects which independently express teaching content and are somewhat interactive. Summary: this part briefly describes the technology and skill of modern sports by dynamic video and synchronization commentary; stimulates learners' feeling toward this sport and shows the general picture of sport technology and application in practice to set up necessary foundation for learning by intuitive video effect. Technical teaching: this part gradually shows learns each technical action step by step with "Follow Me" method and through combination of dynamic video demonstration and synchronization commentary. This part presents standard actions, gives brief and to-the-point instructions and includes slow-play and replay modes to give profound understandings to learners. Image browse: shows learners some wonderful images. With delightful background and pleasant background music, learners can browse between pages which not only enables learns to simulate the technical actions of high difficulty, but also can stimulate the enthusiasm of learning. Text reading: this part mainly includes reading materials and shows to learners comprehensive, particular and systemic skills, methods, characteristics and basic principles of technical actions. It can provide abundant and sufficient knowledge information and teaching materials for learners. To avoid the baldness in reading text, the software can include explanation, background music or concomitant music. Question explanation: this part equips the courseware with shared features of student and teacher, including question explanation and teaching aid. Question explanation is basic on the key questions often asked by learners in learning and mastering sport technology according to years of teaching experiences and designed by "answer-question" mode to help learners solve confusion; teaching aid shows the mistakes teachers are likely to make in giving instruction and demonstration during technical teaching process, corrections, notes, main training methods, examples and so on for the reference of teachers.

5   Suggestions

A.   Teachers should understand the goal of computer assisted instruction and must try to achieve this goal in making courseware and giving lessons. Teachers should make full use of advantages of computer assisted instruction, not just for using it. Efforts should be made to combine the leading role of teachers with computer assisted instruction to improve teaching effect.

B.   Students should understand the main goal of CAI is to learn faster and better. If some students cannot master some part of training well in class, they can review and repeatedly learn the content of this part by teaching software, which also embodies the advantages of CAI. But, we should also avoid the phenomenon that some students turn a deaf ear to teacher in class and copy the courseware to learn by themselves after class.

C.   Efforts should be made to strengthen network construction to meet the requirement of CAI for basic equipment. In software, to reduce repeated labor and improve the use efficiency of teaching software, schools should set up a network platform to enable the sharing of software resources of physical education around school.

6   Conclusion

Education always moves forward in the way that the old is replaced by the new. CAI has already become the necessary trend of technical development of education. Development, promotion and application of CAI courseware will necessarily reform traditional education structure and teaching method of physical education. College is the development base of modern education technology. We should promote the development and promotion of CAI to the strategy of PE development, actively create conditions, research, develop and apply CAI in PE teaching. PE teachers should make CAI courseware by themselves and reform traditional teaching method, which is of great significance.

References

1. Zhaoli, P.E.: Information Technology Application and Development. Journal of Beijing Sport University (02), 145–147 (2008)
2. Song, P.: Brief Analysis on the CAI in College. Journal of Guangxi Normal College (edition of philosophy and social science) (10), 124–125 (2007)
3. Zhang, G., Pu, Z.: Primary Exploration in the Reform of CAI in College. Journal of Sichuan Normal College: social science edition (1), 37–40 (2003)
4. Gu, X.: Strategic Conception of Development and Application of PE Resources. Journal of Physical Education (4), 98–100 (2003)

Study on Practical Teaching System and Its Operation Modes of Materials Science and Engineering

Kegao Liu, Bin Xu, and Lei Shi

Shandong Jianzhu University, School of Materials Science and Engineering
Jinan, China
luca.m126@gmail.com

Abstract. The practice education reform is an important part of professional development and quality projects of university running. It has a very important role for the cultivation of creative thinking and practice ability. The research status of practice teaching in the undergraduate stage is summarized and it introduces the details of practice teaching systems of materials speciality in domestic universities. The suggestions and proposed measures of improving the practice teaching system were put forward and the reforming objectives of the practical teaching system reform were proposed. It suggested that the modular management, segmenting assessment, skills competition modes can be implemented in practice teaching.

Keywords: Materials science, practice teaching, operation modes.

1   Introduction

The practice education reform is an important part of professional development and quality projects of university running. In recent years accompanying with rapid economical development, specially with the development of mechanical industry at home, the overseas many manufacturing industries come into China for further expanding. As the students specialized in materials specialities may find jobs not only in machinery industry but also in similar industries. So the employment rates of students specialized in materials specialties has always been high. In many schools, the employment rates of materials speciality are over the first few ranks. The employment rate of the materials specialty in our school has been higher than those of other specialities in recent years. The employment rates of materials science and engineering in our school are shown in Table 1. However, with the 2009 arrival of the financial crisis it has seriously affected the manufacturing industry, thus affecting the employment rate of students. Its reasons are that the employing demand of enterprises declines and the employment standard threshold is raised. When the students are looking for a job, the most prominent things reflected are their bad practical skills, poor comprehensive ability to combine theoretical knowledge and practical skills and lack of practical experiences for application. It puts forwards some questions about teaching for college. What parts need to be improved in the current practice of

Y. Wu (Ed.): International Conference on WTCS 2009, AISC 116, pp. 435–440.
springerlink.com

teaching system, so as to it better promotes the students' practical level and meets the needs of student employment. Therefore, it requires that we carefully and comprehensively analyze the current status of the practice teaching system of materials specialities, then focusing on the training objectives of materials science and engineering, reform the practice teaching system, establish the practice teaching system independent of theoretical teaching system relatively, optimize the content of teaching and operation modes, urge students to enhance their ability to exercise and improve the practical level, which is of great significance to improve the comprehensive qualities of students [1-3].

Table 1. The statistics of employment rate and graduates

comtent years	2005	2006	2007
The rate of graduates	40.2%	15.4%	18.9%
first-time employment rate of graduates	91.4%	85.5%	84.2%
comtent years	2008	2009	2010
The rate of graduates	19.4%	17.4%	19.4%
first-time employment rate of graduates	86.3%	59.6%	96.3%

2   The Status of Domestic and Foreign Research about Practice Teaching

The professional teaching system for materials speciality has been used for many years, it has been formed a mature setting system. This is corresponded with materials development in economy society. The metal material science, inorganic non-metallic and polymer science are important branches of materials disciplines. The development of socio-economic has always played a tremendous role in promoting development of material science. The development of materials science and engineering has enormous impact on social progress. Materials science and engineering has abundant connotations with not only traditional metal, ceramics and other structural materials, but also a variety of functional materials. Since the materials science is a direct practical science and technology, the demand of socio-economic development has always huge impetus and traction to research and development of materials. As materials application is getting more and more widespread, and seeps into various professions, many fields are closely related with material preparation, properties and application and so on, it enables materials to become the foundation of machinery, electron, chemical industry, building, energy, biology, metallurgy, transportation, information science and technology and other industries. These related disciplines are developing crosswise. Therefore, overlapping of materials discipline with other disciplines is an important character of materials discipline development. It requires that when set the professional curriculum or teaching practice, it is necessary to pay attention to students' ability to fully develop, and to focus different professional orientations of students studied different specialities. Before the founding of the nation, for meeting the social need and exploring materials resources, materials education in our country mainly trained talented persons for mining and metallurgy; After the founding of new china, due to great lack of engineering talent, it mainly cultured engineering personnel for industrial design, construction, operation and other business professional works in engineering; Since reform and opening, along with the development of economy, society and science, materials science and engineering began to blur their boundaries it has more internal links and similarities among several materials. Therefore, domestic many colleges held on the chance to develop disciplines domains and overlapping and new disciplines research, to break down the original professional boundaries gradually and strengthen the mutual seepages and relations of two, third-level special disciplines and to renew course contents and reform mentality; In the 1990s, along with switching of the economic system from the planned economy to the market economy in our country, for broadening the speciality caliber, the education ministry had made the adjustment to materials sepecialities of undergraduates, which further impelled the reform and development of material science and engineering education in our country. In recent years, China has been active in the fields of materials science, new concepts, new ideas and new methods emerged unceasingly, the novel materials have had considerable progress in theory and technology, the materials class professionals must have comprehensive knowledge of material science and engineering about materials preparation and processing, composition and structure, performance and application [3-8].

Various professional training objectives of materials science and engineering are different for different specialized directions, so this is also reflected in the professional training program and model. With continuous development of China's modernization materials science and engineering disciplines is developing to a discipline. So the professional personnel with the basis of a single material speciality can not meet the needs of society. The social demand for materials speciality has been changed from the original need of single technical personnel to the personnel with multi-specification to adapt to different productivity levels and different industry departments. It requires that the students should have the comprehensive elementary knowledge and can be competent for changing different positions. Thus it puts new requests to the undergraduate teaching of materials speciality. It is the key to investigate cultivation of comprehensive quality, practical ability, innovative spirit of students specialized on materials science and engineering. In the new situation it is the key how to improve the practice level of the students for promoting their employment [8-11].

3   The Practice Teaching System of Domestic Materials Speciality

At present the domestic professional practice system mainly includes the following several aspects. The first is experimental, including course experiments, which are lack of comprehensive experiments, most of them are demonstration-type experiments. Most comprehensive experiments need a week or more time to be completed. Although the laboratory equipments are already sufficient for students, but because the too many students are difficult to be managed and assessed, which causes the teaching effect to be ordinary. Therefore, the operation modes of practice teaching are also needed improving. The second is the course design. Although this type requests to achieve one topic for every 4 people as far as possible, but it is still has questions of management. There are too many students to realize one topic for one person. Originally common design by group promotes communication and collaboration among students. But it is difficult to avoid the same design among students and to guarantee that all students can positively think and practise. Therefore it also needs to explore the innovation in the inspection assessment way. The third is practice, including decentralized and centralized practice. First of all, it is the metalworking practice for the students majored in engineering, and then the cognition, professional and graduation practices are set. Although the types of practice designed are relatively complete, but in recent years the relationship between schools and enterprises is changing. Since the strength and means of support practice are also changed it leads to be difficult to achieve good practice results. Another aspect of improving practice ability is open experiments and science-technology innovation activities. At this aspect the quantities of open experiments need increase. So that it can provide a platform of practice and innovation for more students. The fourth is graduation thesis or design. For this aspect the difference is very large along with school characteristics, for example, the graduation thesis, for research-oriented universities it is primarily theoretical, for teaching-oriented universities it is summary paper type. While for the teaching-research engineering university it is graduation thesis or design with different proportions for different school.

Due to the big caliber of materials science and engineering there are many courses for each specialized direction. From student employment interview processes and real feedback after graduation it was found that practical skills are the weakest link in comprehensive professional quality. The practice section in the training plans running currently should stress on the professional fields of technology. How can adjust the practice teaching system, the content and operation modes, which is an important problem to insure further enhancing the practice ability of students. It is significant for overall development, potential and adaptability in jobs of students.

4   Suggestions and Measures to Improve Practice Teaching

4.1   Settings and Optimization of the Practice Teaching System

At present main practice courses are metalworking, professional, production and graduate practices. This aspect needs to fully investigate the setting situation of same specialized practice category in domestic schools. The proportion of several types of practices and the effect and experiences of practices should be investigated. Thereby according to this information the time, category and term distribution of different practices can be optimized.

4.2　The Content Reform of Practice Teaching

In the aspect of practice courses the key points are curriculum experiments and synthesis experiment weeks. The practice course contents should be optimized according to basic business requirements of new social development situation. It should focus on strengthening the practical teaching links with more practical applications for promoting employment. This requires investigating requirements of employers.

4.3　The in Operation Modes of Practice Teaching

This includes management approaches of the practice links, incentive mechanisms and evaluation ways of students. It is the key to explore the operating modes according to different experimental teaching contents. It can be discussed to set different types of experiments such as the design-experiment, open experiment and demonstration experiment. For the demonstration experiments, which often cause the effect of a cursory look. Therefore, these experiments are most needed to be improved. A series of designed experiments and open labs should be set as far as possible under the existing conditions. It makes the students change from passive to active to enhance the experimental effect. For this aspect it should consider assessment methods and incentive mechanisms to students. Therefore it needs to discuss the reasonable assessment means in experiment processes and positive management modes. Similarly, the same assessment means and incentive mechanisms can be adopted in the course designs.

4.4　Discussion about Teaching Supervision and Encouragement on Teaching Innovation

This aspect mainly depends on information feedback by teachers and students. The teaching and research staff room and academic supervisors should strengthen inspection and guidance to the experimental courses and fully consider the recommendations of teachers and students to take specific assessment measures.

5　Reform Objectives of Practice Teaching

(1) Optimizing the practice teaching system of materials science and engineering for the foundation of new version of personnel training plans.

(2) Adjusting the relevant contents of various experimental links; Putting forward their basic requirements and ideas; Establishing the evaluation system of the effect of practice teaching links in order to constantly optimize the practical education, which can set up a dynamic platform for the common progress of teachers and students.

(3) Establishing the operation modes and management systems as well as the assessment and incentive mechanisms for main practice courses.

6   Summary

Experimental courses are quite easy to carry on and to achieve better results if the school is equipped with sufficient experimental teaching equipments. Therefore, this part is easier to improve the effect than the practices in factories, mines and enterprises. Therefore the key is that it needs to construct the operation modes and assessment systems for solving course experiments, experimental weeks and open experiments. It can adopt module management for different types of practice teaching and adopt advancing grade system in operable practices. By the modular management in practice teaching it enables students to improve practice ability step by step. For the teaching practices it can adopt the skills competition ways to explore the key management specifications. Anway it suggests that the modular management, segmenting examination and skills competition modes can be adopted in practice teaching.

Acknowledgment. This work was supported by the teaching and research project funded in 2009 of Shandong jianzhu University.

References

1. Zhao, Y.: Construction and implement of the experimental teaching system of higher education in the new circumstances. Chemical Higher Education (4), 90–99 (2003)
2. Xie, J.L., He, F.: Talking on engineering education of materials speciality from the demand of society. Polytechnic Higher Educational Study 26(4), 49–51 (2007)
3. Wu, H.B., Zhang, L.M.: Education reform research on materials science and engineering, pp. 1–20. Wuhan University of Science and Technology Press (2001)
4. Song, X.P., Hao, H.Q.: Exploration on the construction of first-level materials science and engineering. Higher Engineering Education Research (1), 9–12 (1997)
5. Li, H.D., Shi, C.X.: The present situation of Chinese material development and main Strategies stepped into new century, pp. 1–10. Shandong Science and Technology Press (2003)
6. The teaching steering committee ministry of education of materials science and engineering, the development strategy study of material science and Engineering, http://www.edu.cn/yjbg_10009/20100601/t20100601_481172.shtml
7. Luo, J., Zeng, Y., Zhang, W.: Preliminary probe into practical teaching system about material shaping and controlling engineering specialty. Journal of Chongqing University of Science and Technology (social sciences edition) (4), 137–139 (2006)
8. Ji, Z.H.: An exploration to practice teaching system establishment of materials science and engineering. Research Studies on Foundry Equipment, 52–54 (May 2003)
9. Li, X.L., Cao, X.X., He, X.F.: Study on the practice education system of Materials Science and engineering. Light industry education of China (1), 49–50 (2009)
10. Nian, S.: Research into the practical teaching of metal material engineering specialty. Journal of Anhui University of Technology (social sciences) 22(1), 96–98 (2005)
11. Shi, J.H., Wang, J.J., Sha, G.Y.: Multi-level combination of production, study and research about metal materials engineering speciality. Journal of Shenyang Institute of Aeronautical Engineering 26(z1), 101–105 (2009)

A Hash-Chain Based Anonymous Incentive Mechanism for Mobile Ad Hoc Networks

Yuan Zhong, Jianguo Hao, Xuejun Zhuo, and Yiqi Dai

Department of Computer Science and Technology, Tsinghua University, Beijing, China
tanygq@sogou.com

Abstract. In the routing protocols for mobile ad hoc networks (MANETs), both route setup and data transfer rely on the collaboration of participant nodes. The existing node incentive mechanisms for MANETs do not consider the privacy protection of the participant nodes; therefore, they cannot be applied to the network environments with high privacy protection requirements. This paper proposes an anonymous incentive mechanism (AIM) by introducing the hash-chain based micro-payment mechanism to the DSR protocol. The proposed mechanism realizes the real-time stimulation to the participant nodes while preserving their anonymities in the routing. In contrast to the existing incentive mechanisms in MANETs, AIM protects the anonymity of nodes with reasonable overheads, and can withstand the common security attacks in the process of routing.

Keywords: Mobile ad hoc networks, node incentive, anonymous, micro-payment, hash chain.

1 Introduction

The mobile ad hoc networks (MANETs) are the wireless networks that do not have communication infrastructures. Therefore, the routing in MANETs depends on the mutual cooperation of nodes. The common routing protocols for MANETs are designed under the assumption that all nodes cooperate in route setup and data forwarding. However, in the actual MANETs, selfish nodes may reduce data forwarding for other nodes as much as possible to save energy and bandwidth, and this can result in the above routing protocol becoming inapplicable in the network environments with selfish nodes.

To stimulate selfish nodes to cooperate in the routing, the researchers have proposed many node-collaborative incentive mechanisms in recent years. These mechanisms are divided into two categories: reputation-based [1,2,3] and credit-based [4,5,6,7,8] incentive mechanism. The SPRITE proposed in [7] is a typical reputation-based mechanism, but it requires the online CCS to calculate the reputation value, and the destination node's participation when the forwarding node obtaining its reputation value. Based on the efficiency and security of hash chain, some researchers, such as the authors of [4], introduced the hash-chain based micro-payment [9] into the collaborative incentive mechanism to improve the efficiency. However, the "cashing

Y. Wu (Ed.): International Conference on WTCS 2009, AISC 116, pp. 441–450.

incentive value" in [4] depends on the safe channel between the forwarding nodes and the trusted authority, which is actually difficult to realize in actual MANETs.

Moreover, because of the openness of the wireless channel in MANETs, the participant node's privacy information, such as its identity and position, is susceptible to eavesdropping of the attackers. Since all existing incentive mechanisms aim for maximizing the network's performance without taking into account the privacy protection of nodes, it is necessary to introduce a node incentive mechanism with the privacy protection to nodes. Because meeting the demands of nodes' privacy protection definitely raises the complexity of the incentive mechanism, and increases computation and communication overheads, it is a challenge to design an efficient anonymous incentive mechanism for the routing protocols in MANETs.

In this paper, a credit-based node incentive mechanism, Anonymous Incentive Mechanism (AIM), is proposed to manage this challenge. AIM introduces an anonymous micro-payment mechanism to realize the source node's real -time payment to the forwarding node during route setup and data forwarding while protecting the nodes' anonymities in the routing. The micro-payment mechanism therein contained profits from [10]. Our mechanism can not only stimulate the forwarding behaviors of nodes in routing efficiently and guarantee the performance of the routing protocol, but also protect the privacy of nodes. Compared with the existing node incentive mechanisms for the routing in MANETs, AIM has the following advantages: i) It uses the high computation efficiency of hash chain and does not rely on any on-line Trusted Third Party (TTP), and is an efficient real-time incentive mechanism; ii) With the irreversibility and non-repudiation of hash chain, it can withstand the common security attacks and guarantee the security of the incentive mechanism; iii) It can protect the node's anonymity by a pseudonym mechanism integrated into the hash-chain based micro-payment mechanism.

2   Mechanism Overview

A typical process of the AIM mechanism includes the following entities: a source node (S), a series of forwarding node (R_i), a destination node (D), and a Trusted Third Party (TTP). The incentive flow is shown as Fig. 1.

AIM is based on the basic DSR [11] protocol. In AIM, TTP manages the credit accounts of all nodes in the network. At the initialization stage, each node added into the network passes accreditation from the TTP and becomes a legal node to obtain the initial values of the account. A series of pseudonyms is produced at each node. A pseudonym is enabled when each route is launched. Only TTP and the node itself can open these pseudonyms. The node interacts with TTP once after each time slot (such as one day). The credit value in the node account obtains the TTP stamp on several hash

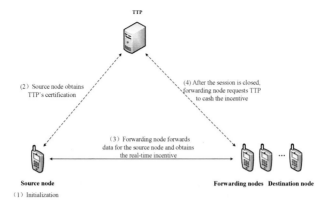

Fig. 1. The incentive flow of AIM

chains. Each stamp corresponds to one hash chain. When the node has data to send to another node, the source node initiates the routing request on the destination node. In route setup, the forwarding node proposes the reward necessary for data forwarding in the answer message of the route. The source node chooses the lowest quotation or the shortest route based on the answer message of the route. At the inception of data forwarding, the source node sends a stamp to each forwarding node, it adds the hash value to the forwarding node while sending the subsequent packets, and it sends the incentive value for each forwarding node using the hash chain. When receiving subsequent packets and hash values, the forwarding node uses a stamp to authenticate the validity of the hash values. At one time slot after routing, the forwarding node applies cash Hash values to TTP. TTP adds the corresponding credit value to its account after validation. In the network, the node increases the credit value of its account only by forwarding data to the node. Only with enough credit values does the node make other nodes forward data to itself. This mechanism produces an effective stimulus on the data forwarding of the node to improve network performance. Table 1 describes the notations used in our mechanism.

Table 1. Notation Description

Px	Pseudonym of node x
Sig_x	Signature with the private key of the node x
PK_x	Public key of node x
SK_x	Private key of node x
$H_i(*)$	Hash function that generates the ith hash-chain
Wi_0	Hash chain's length
M	Hash value
$stamp_i$	TTP's signature for the ith hash-chain

3   Anonymous Incentive Mechanism

According to the routing process of DSR, AIM is divided into five stages: initialization, route setup, data forwarding, route maintenance, and incentive cashing.

3.1   Initialization

After a node enters into the network and passes TTP accreditation, the node establishes a credit account in TTP. The initial value of each account is α_0. The node uses its private key to generate a series of pseudonyms (P_{A_i}). Only the node itself and TTP can open these pseudonyms. Each time a route is opened, the node enables a new pseudonym.

The node interacts with TTP once after a time slot. Node S generates n (the number) hash chains. The length of each hash chain is m.

$$W_{i_0} = h_i^m (W_{i_m}), i = 1, \ldots, n$$

Node S needs to store the vector matrix of n (number) hash chains.

To obtain the TTP signature of n hash chains (stamp), the source node uses its pseudonym P_S to send a message encrypted by TTP's public key. The message includes the hash anchor of the hash chain (W_{i_0}), the length of the hash chain (m), and P_S's signature.

$$S \rightarrow TTP: \{ [(W_{1_0}, W_{2_0} \ldots W_{n_0}), m] \, sig_{P_S} \} \, pk_{TTP} \circ$$

Upon receipt of the message, TTP uses its private key to open the message, validates the signature of the source code, and calculates the amount of payment needed by the source code based on the length (m) and number (n) of hash chains submitted by the source node.

Macropayment=mn

TTP requests the account balance β of node S. If $\beta \geq Macropayment$, TTP sends n stamps encrypted by S's public key to node S. Each stamp includes the hash anchor(W_{i_0}) and the time stamp (TS_i) of the hash chain, signed by TTP's private key.

$$TTP \rightarrow S: \{ stamp_1, stamp_2, \ldots stamp_n \} pk_s.$$

where $stamp_i = \{ W_{i_0}, TS_i \} \, sig_{TTP}$.

TTP reduces a suitable amount (*Macropayment*) from the account of node S.

3.2   Route Setup

The route's response message increases the incentive value field, indicating the incentive value needed by each forwarding node in forwarding one datagram. The DSR protocol is divided into two stages: route discovery and maintenance. In the route discovery stage, when node S has data to forward to the destination node D, and its route buffer does not reach D's route, the source node uses the pseudonym P_S to broadcast the routing request. Each node receiving the routing request checks if it appears in the address list. If yes, it discards the packet; if not, the middle node adds its address to the address list and uses the pseudonym to broadcast the route's request message. When the route's request message reaches the destination node, the destination node reverses the list of routing address and sends back the route request message. When the answer message reaches each middle node along the reversed route list, the middle node transfers the incentive needed by each packet and attaches it onto the route answer message. When the route answer message reaches the source node, the source node stores this route in the route's buffer. The source node chooses the best or the cheapest route from the route buffer queue based on the billing information sent back from the middle node.

Below is the information flow for the route setup stage.

$$S \rightarrow * : \{P_s, (D)pk_D\} \ ;$$

$$R_1 \rightarrow * : \{P_s, (D)pk_D, P_{R_1}\} \ ;$$

$$R_2 \rightarrow * : \{P_s, (D)pk_D, P_{R_1} P_{R_2}\} \ ;$$

$$\cdots\cdots$$

$$R_n \rightarrow * : \{P_s, (D)pk_D, P_{R_1} P_{R_2} \ldots P_{R_n}\} \ ;$$

$$D \rightarrow R_n : \{P_s, P_D, P_{R_1} P_{R_2} \ldots P_{R_n}\} \ ;$$

$$R_n \rightarrow R_{n-1} : \{P_s, P_D, P_{R_1} P_{R_2} \ldots P_{R_n}, (P_{R_n}, b_n)\} \ ;$$

$$\cdots\cdots$$

$$R_2 \rightarrow R_1 : \{P_s, P_D, P_{R_1} P_{R_2} \ldots P_{R_n}, (P_{R_n}, b_n) \ldots (P_{R_2}, b_2)\} \ ;$$

$$R_1 \rightarrow S : \{P_s, P_D, P_{R_1} P_{R_2} \ldots P_{R_n}, (P_{R_n}, b_n) \ldots (P_{R_2}, b_2), (P_{R_1}, b_1)\} \ .$$

Where $\{R_1, R_2, \ldots, R_n\}$ is the forwarder in the route, $\{P_{R1}, P_{R2}, \ldots, P_{R3}\}$ is the pseudonym of each forwarding node, and $\{b_1, b_2, \ldots, b_n\}$ is the reward proposed by each forwarding node needed for forwarding each packet.

3.3   Data Forwarding

In forwarding the earliest packets, the source node sends the payment instruction C_i^S to node R_i. C_i^S includes the stamp, pseudonym P_S of the source node, pseudonym P_{R_i} of the forwarding node, and digital signature Sig_S of the source node. C_i^S is the certificate that the forwarding node issues when applying to TTP for cashing the incentive after the routing is completed. After each forwarding node is received, the C_i^S belonging to itself is retained, and the signature is validated. After passing the validation, it forwards the packet and gets ready to receive subsequent packets and to pay the hash value.

Below is the data flow of the packet's forwarding process.

$$S \rightarrow R_1 : \{\mathrm{Msg}_0, \ C_1^S, \ C_2^S, \ldots C_n^S\}$$

$$R_1 \rightarrow R_2 : \{\mathrm{Msg}_0, \ C_2^S, \ldots C_n^S\}$$

$$\cdots\cdots$$

$$R_n \rightarrow D : \{\mathrm{Msg}_0\}$$

whereas $C_i^S : \{\mathrm{stamp}_i, \ P_S, P_{R_i}\} Sig_S$.

The source node forwards subsequent packets, adding hash value Wi_j to each forwarding node. Wi_j is j^{th} hash value for i^{th} hash chain.

$$S\rightarrow*: \{Msg_j, (W_{1_j}, P_{R_1}), (W_{2_j}, P_{R_2}), \dots, (W_{n_j}, P_{R_n})\}$$

The forwarding node always stores the last Hash value received. When new hash values are received, it needs only one hash computation $H(W_{i_j}) \overset{?}{=} W_{i_{j-1}}$ to judge the legality of the hash value.

The process of paying real-time incentive during data forwarding is shown as Fig. 2. When a hash chain assigned to a forwarding node is used up, the source node forwards a new stamp to this forwarding node, thereby enabling a new hash chain.

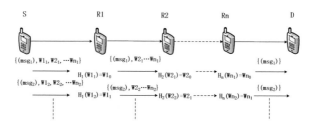

Fig. 2. Paying real-time incentive during data forwarding

3.4   Routing Maintenance

Due to the strong mobility of the node in MANET, the network topography changes from time to time. When a central node moves to a far place or when electricity is used up, resulting in the route's interruption, the source node finds a new route in the routing buffer or initiates a new route discovery process in order to find the new route leading to the destination node. At this point, the source node does not need to contact TTP and only needs to send new stamps to the new forwarding node in order to stimulate the new middle node to continue data forwarding and the node stimulus.

3.5   Incentive Cashing

After each route is completed, the forwarding node contacts TTP after an account settlement cycle (T) and applies the hash value to the cash. The node sends to TTP the highest hash value ($W_{i_{highest}}$) and the payment instruction C_i^A it receives.

$$R_i \rightarrow TTP: \{W_{i_{highest}}, C_i^A\}.$$

After receiving the message, TTP validates the C_i^A's digital signature. C_i^A includes the source node and TTP's digital signature. To judge whether the hash value is the legal hash value that node Ri needs to obtain, TTP validates the digital signature twice. After passing the validation twice, TTP adds the relevant amount to the node's account, issues a receipt, and sends it to the forwarding node.

TTP→R_i: {receipt$_i$}.

With regard to the source node, TTP also puts the remaining hash value into the account of the source node.

In cashing incentive values, TTP runs the validation of the digital signature twice to ensure safety in cashing hash values, increasing the overhead of the TTP validation to a certain degree. Due to the fact that the interaction between the node and TTP is offline, this does not increase the overhead calculation and communication of network nodes during online validation.

4 Performance Analysis

The node incentive mechanism in this paper adopts an efficient hash-chain based micro-payment mechanism. The mechanism's real-time nature is reflected as follows: the route response message is added with the hash values needed to forward the packet, and the forwarding node is stimulated when forwarding the data packet. TTP finishes the hash chain signature for the source node and cashes the incentive for the forwarding node offline. The node produces the n hash chains and obtains the TTP signature by batches; when the route changes, the source node immediately sends a new stamp to the new forwarding node instead of establishing a connection with TTP.

AIM uses an efficient hash-chain based micro-payment mechanism to stimulate the forwarding node, which validates the incentive value through the hash computation. The number of digital signatures and validations is the main computation overhead. In the process of withdrawing a large sum and paying by several small sums, TTP does not need to participate in the incentive each time. Therefore, this reduces communication traffic, the number of encryptions, or the signature of key information, and improves efficiency while guaranteeing safety.

We analyze the computation efficiency of AIM by comparing the number of digital signature, the validation signature, and hash calculations in the key messages with Express. We assume that the source node sends p (the number) packets in one routing, passes through n (the number) forwarding nodes, generates the digital signature (G) once, validates the digital signature (V) once, and runs the public key encryption or decryption (E) once and the hash computation (H) once.

In Express, the source node conducts n (the number) digital signatures when the route starts, whereas the forwarding node conducts the digital signature once. The number of digital signatures and validation signatures is $n(G+V)$. When forwarding each packet, the source node runs hash computation npH once, whereas the forwarding node validation runs hash calculation $2npH$ twice. To generate receipts, the source node conducts one hash computation npH on each forwarding node/packet. The total computation of the Express node is $O_E=n(G+V)+5npH$.

In AIM, the node's overhead computation is brought by the incentive and anonymity mechanism because the node obtains a TTP stamp on each hash chain during initialization. When sending several earliest packets, the source code adds the stamp of each forwarding node and the digital signature of the source node; the time of the signature is nG. Upon receipt of the message, the forwarding node validates the signature for nV (times). When forwarding subsequent packets, the forwarding node conducts hash computation once upon receipt of the hash value. The incentive is cashed after the routing is completed. The forwarding node is cashed by TTP with the highest hash value and the corresponding stamp. Its safety is guaranteed by the validation of the digital signature by TTP without the need to increase the node's overhead computation. To guarantee the anonymity of the destination node, the source node uses the destination node's public key pk_D to encrypt the destination node's ID when forwarding the route request message. Upon receipt of the message, the forwarding node needs to run "n" (times) decryption computation, that is, nE. Hence, the total computations for AIM nodes is $O_A=n(G+V)+npH+nE$. In AIM, the pseudonym mechanism is introduced in the overhead computation to trade for privacy protection on the nodes.

5 Anonymous and Secuiry Analysis

Anonymity Analysis

Anonymity means that the active attacker does not know the true identity of the other nodes in the route, and that the forwarding node only knows its own route anonymity and not the true identity of the other nodes in the route. This mechanism adopts the pseudonym to realize the anonymous node incentive. When the node is added to the network, and the private key is generated to produce a series of pseudonyms, only the TTP and the node itself can open these pseudonyms. When the source node initiates a route request, a new pseudonym is enabled. After the middle node receives the route request of a certain node for the first time, a new pseudonym is enabled. Hence, this mechanism also realizes the anonymity of the node identity; any forwarding node does not know the true identity of the other nodes in the route. Thus, even if the attacker intercepts the message, it still cannot link the source node and the destination node because it will not be able to know the true identity of any node in the route.

5.1 Security Analysis

AIM can defend the common attacks in the node incentive process, such as interception of payment instruction, and replay attack, and is a secure incentive mechanism.

1) Interception of payment instruction
When the source node sends the payment instruction to the forwarding node, other malicious nodes might intercept the message and use C_i^S to claim the forwarded

reward falsely. Since there are the digital signature of source node and the pseudonym of forwarding node R$_i$ in C_i^S, the malicious nodes cannot tamper with C_i^S. Therefore, it is impossible for the malicious node to let PPT transfer the payment to its account.

2) Replay attack
Replay attack refers to the attacking behavior wherein the attacker repeatedly sends the same cashing message in an attempt to increase its own account value repeatedly. In AIM, TTP puts the time stamp into each stamp of every hash chain, and can prevent the replay attack effectively.

6 Conclusion

To protect the anonymity of the participant nodes in the routing incentive mechanism for MANETs, this paper proposes AIM, an anonymous real-time node incentive mechanism for MANETs. AIM uses a hash-chain based micro-payment mechanism to conduct the real-time stimulation to forwarding nodes. The introduction of the pseudonym mechanism into this incentive mechanism realizes the privacy protection of nodes. We give a quantitative comparison with the AIM mechanism and a current incentive mechanism EXPRESS, and the results indicate that AIM can protect the anonymity of the participant nodes with the reasonable overheads. Through anonymity and security analysis, we conclude that AIM is an anonymous and security node incentive mechanism. However, although TTP is designed to be offline in AIM, which means it need not to participate in every routing incentive, other overhead still may bring TTP to become the bottleneck of performance and security. So, our future work will around the optimization of TTP to solve this problem.

References

1. Buchegger, S., Le Boudec, J.-Y.: Performance Analysis of the CONFIDANT Protocol (Cooperation Of Nodes-Fairness In Dynamic Ad-hoc NeTworks). In: Proc.3rd Symp. Mobile Ad Hoc Networking and Computing (MobiHoc 2002), pp. 226–236. ACM Press (2002)
2. Bansal, S., Baker, M.: Observation-based cooperation enforcement in ad hoc networks. Technical Paper, Computer Science Department, Stanford University (July 2003)
3. Liu, Y., Yang, Y.R.: Reputation propagation and agreement in mobile ad-hoc networks. In: Proc. IEEE Wireless Communications and Networking Conference (WCNC), New Orleans, LA (March 2003)
4. Janzadeh, H., Fayazbakhsh, K., Dehghan, M., Fallah, M.: A Secure Credit-Based Cooperation Stimulating Mechanism for MANETs Using Hash Chains. Future Generation Computer Systems 25(8), 926–934 (2009)
5. Hu, Y., Johnson, D.B., Perrig, A.: SEAD: Secure efficient distance vector routing for mobile wireless ad hoc networks. In: Fourth IEEE Workshop on Mobile Computing Systems and Applications, WMCSA 2002 (2002)
6. Buttyan, L., Hubaux, J.P.: Stimulating cooperation in self-organizing mobile ad hoc networks. ACM Journal for Mobile Networks (MONET), special issue on Mobile Ad Hoc Networks (2002)

7. Zhong, S., Yang, R., Chen, J.: Sprite: A Simple, Cheat-Proof, Credit-Based System for Mobile Ad Hoc Networks. In: Proc. INFOCOM 2003, March 2003, pp. 1987–1997 (2003)
8. Marbach, P., Qiu, Y.: Cooperation in wireless ad hoc networks: A market-based approach. IEEE/ACM Transactions on Networking 13(6), 1325–1338 (2005)
9. Jakobsson, M., Hubaux, J.P., Buttyan, L.: A micropayment scheme encouraging collaboration in multi-hop cellular networks. In: Proc. of Financial Crypto 2003, La Guadeloupe (January 2003)
10. Tewari, H., O'Mahony, D.: Multiparty micropayments for ad hoc networks. In: Proc. IEEE Wireless Communications and Networking Conf., New Orleans, LA, pp. 2033–2040 (2003)
11. Johnson, D.B., Maltz, D.A.: Dynamic Source Routing in Ad Hoc Wireless Networks, pp. 153–181. Kluwer Academic (1996)

Unsupervised Texture Segmentation Based on Redundant Wavelet Transform

Guitang Wang, Wenjuan Liu,
Ruihuang Wang, Xiaowu Huang, and Feng Wang

School of Information Engineering, Guangdong University of Technology,
Guangzhou, P. R. China
zgh6338@sogou.com

Abstract. The algorithm of Redundant Wavelet Transform (RWT) and laws texture measurement is proposed and applied to image segmentation. Based on the characteristics of the indentation images, this article uses texture features to extract the indentation silhouette from the point view of texture segmentation. We adopt Redundant Wavelet Transform and laws texture measurement algorithm to describe the texture characteristics of the indentation image, forming a n-dimensional feature vector, introducing texture features smoothing algorithm based on quadrant to smooth the features. Finally we combine with the improved k-means clustering algorithm to get texture segmentation result. The experiment demonstrates that in the material Vickers hardness image segmentation the proposed algorithm was significantly effective and robust.

Keywords: RWT; laws texture measurement; improved k-means clustering algorithm; Texture segmentation.

1 Introduction

The texture classification and the division question have been the focal points which people pay attention, involving so many research areas such as pattern recognition, application mathematics, statistics, nerve physiology and neural network. The textural property extraction is the key to do image texture description, classifying and division successfully, because texture feature influences following processing quality directly. Formerly analysis of Vickers hardness indentation was mainly based on the above-mentioned method. Textural property is seldom used to do indentation extraction. Multi-resolution Wavelet is used to analyze sampling signals of Vickers indentation and least squares method is used to fit indentation edge line, thus extract the micro-hardness indentation edge, as in [1]. Gray level co-occurrence matrix, fractal and the K-means cluster algorithm is used to do indentation texture division, its segmentation effect is perfect, but the algorithm running time is not very ideal. Therefore, this article proposed that redundant wavelet transformation, the Laws texture measure analyze the K-means cluster algorithm is used to do texture segmentation, which obtained the same effect as in [2], but the speed actually enhanced

Y. Wu (Ed.): International Conference on WTCS 2009, AISC 116, pp. 451–456.

obviously. Wavelet transform is applied to standard or rule texture images, but it's ineffective for the nature images with complex background, because noise interference or texture pixels are not similar everywhere in some region. Besides, its computation is very great. The algorithm optimization is the question which we should solve

2   Textural Features Extraction Based on Redundant Wavelet Transformation

Most of the texture feature extraction methods are heuristic, which are mainly the methodological and experimental research at present. The majority methods did experiment and confirmation on the texture with differentiate in vision and images with relatively simple texture edge. There are so few research on many kinds of texture types including complex boundary or   indistinguishable   texture in vision.

$N \times N$ two dimensional original image I_l^0 uses Mallat algorithm to do L-pyramid Wavelet Decomposition and obtains $3L+1$ sub images. I_l^k is used to represent the k-th sub image of the level l decomposition. As a result o, these sub images often correspond to middle and high frequency component of the original image and its corresponding frequency range is wide. Therefore, every point value is very different from each other even if they are the same texture region. So, it's not appropriate that wavelet coefficients extracted directly from the corresponding pixels in the image constitute eigenvectors of the pixels.

In order to obtain relatively stable eigenvalue of each point, Laws proposed energy norm-a feature extraction method. This method is used to obtain mean square deviation in standard window of each wavelet sub-image. Mean square deviation replaces the wavelet coefficient and is regarded as pixel eigenvector, that is:

$$e(i, j) = \frac{1}{(2n+1)} \sum_{x=i-n}^{i+n} \sum_{y=i-n}^{i+n} \left| s(x, y) - \bar{s}(x, y) \right| \tag{1}$$

The size of the window is decided by n. $s(x, y)$ represents wavelet coefficients of the initial image of and $\bar{s}(x, y)$ represents the mean value of wavelet coefficients in the window centered by point (x, y). Mean square deviation corresponding to each pixel is calculated point-by-point, then we get $F(x, y)$, a eigenvector whose number of dimension is $3N+1$. Here N represents the number of layers of wavelet decomposition. This is a texture feature extraction method-energy measurement proposed by Laws based on the pyramid of wavelet transform Pyramid Wavelet transform is non-redundant, small calculation, less memory required. But when integral translation happens to the image, which corresponding wavelet coefficients have large changes and this variability affects texture features accurate extraction and.

Therefore, this article uses a 53 integer wavelet redundant wavelet [3] instead of pyramids to decompose the original image and then extract energy features as the corresponding eigenvectors by means of formula. Because the redundant Wavelet canceled the next sampling, so Wavelet coefficients images with various resolution have the same size as original image, which is very convenient for each pixel grayscale values corresponding to Wavelet coefficients one-to-one.

3 Smooth Texture Boundary

Texture boundary treatment is a key issue of texture segmentation. Mean square deviation is used to obtain texture energy measurement, which brings the obvious error along the regional boundaries and make the calculated texture feature deviate from the desired value. The resolution of this issue can be attributed to how to smooth noise without blurring the edges.

A texture features smoothing algorithm based on quadrant [6] is used in this paper. The basic idea of this algorithm is: each pixel of the characteristics image is regarded as the center respectively to establish four neighborhood sub windows. If a certain window has neighboring pixels belonging to a different texture regions, or contain pixel texture even belong to the same region, but the differences between these pixels are larger and the window corresponding variance is larger. So we select a window with minimum variance and calculate the mean or median as eigenvalue of the pixel.

Concrete step of the algorithm are:

(1) On the bases of the level-1 decomposition obtains average value and the variance of four w×w neighboring sub-window images region centered on (x, y), that is:

$$m_{l,k}^1(x, y) = \frac{1}{w^2} \sum_{i=0}^{w-1} \sum_{i=0}^{w-1} e_l^k(x+i, y-j)$$

$$v_{l,k}^1(x, y) = \frac{1}{w^2} \sum_{i=0}^{w-1} \sum_{i=0}^{w-1} [e_l^k(x+i, y-j) - m_{l,k}^1(x, y)]^2$$

$$m_{l,k}^2(x, y) = \frac{1}{w^2} \sum_{i=0}^{w-1} \sum_{i=0}^{w-1} e_l^k(x+i, y-j)$$

$$v_{l,k}^2(x, y) = \frac{1}{w^2} \sum_{i=0}^{w-1} \sum_{i=0}^{w-1} [e_l^k(x+i, y-j) - m_{l,k}^2(x, y)]^2$$

$$m_{l,k}^3(x, y) = \frac{1}{w^2} \sum_{i=0}^{w-1} \sum_{i=0}^{w-1} e_l^k(x+i, y-j)$$

$$v_{l,k}^3(x, y) = \frac{1}{w^2} \sum_{i=0}^{w-1} \sum_{i=0}^{w-1} [e_l^k(x+i, y-j) - m_{l,k}^3(x, y)]^2$$

$$m_{l,k}^4(x, y) = \frac{1}{w^2} \sum_{i=0}^{w-1} \sum_{i=0}^{w-1} e_l^k(x+i, y-j)$$

$$v_{l,k}^4(x,y) = \frac{1}{w^2} \sum_{i=0}^{w-1} \sum_{i=0}^{w-1} [e_l^k(x+i, y-j) - m_{l,k}^4(x,y)]^2$$

In the four neighboring sub window image areas, the minimum variance is:

$$v_{min}^k(x,y) = \min\{v_{l,k}^e(x,y), e = 1,2,3,4\}$$

(2) Selecting the window with minimum variance, the mean of the window is regarded as eigenvalues of the (x, y) pixels.

4   Improved K-Means Method

K-means principle is choosing S centers of initial distance to calculate the distance between each sample and the S centers and discover the minimum distance then samples belong to the nearest cluster center. Revising the central point value to be average value of all samples and calculating the distance between each sample and the S centers is to classify again. Revising the new central point, which will stop until the new distance center is equal to the previous time central point.

K-means algorithm is able to get better visual effects, but the iterative algorithm K-means has an obvious flaw that the running time is excessively long. So K-means algorithm is needed to make some improvements. Tanimoto coefficient is regarded as the category determination standard. That is, Tanimoto measurement, the formula is:

$$d_t(x_i, x_j) = \frac{x_i^T x_j}{x_i^T x_j + x_j^T x_i - x_i^T x_j}$$

Each pixel contains nine texture vectors from Laws texture measure. Vectors are regarded as a sample and use the improved K-means clustering algorithm [5]. Improved K-means method: a group of energy measurement which is the maximum in the image and distance is greater than the threshold T (T is optional) are regarded as cluster centers. Concrete steps are:

(1) The initial cluster centers are selected;

1) Occurrences of each texture measure is counted and the sequence is ordered in descending by A_i($i = 0, 1 \cdots, N-1$)

2) Assuming that there are already m measurement categories

3) if distance between measure A_i($i \geq m$) and m measures is greater than the threshold T, then add it to the class, that is $A_m = A_i$, then go to step 4), otherwise the next measurement value A_{i+1} repeat step 3).

4) If a known class has k classes, then the selection of initial cluster centers C is complete, go to step (2), choose A_{i+1} from in the sequence and go to step 3)

(2) In accordance with the minimum distance, each measure will be mapped to the corresponding S

(3) A new cluster centers C' is obtained at the t-th iteration.

(4) In accordance with the minimum distance, Ai is mapped to the corresponding cluster S'

(5) if the maximum distance between cluster centers obtained by two adjacent iteration corresponding to the measurement $\max_{i=1\cdots n} \|S - S'\| \le T_1$, the program ends.

5   Experimental Results and Discussion

Experiments have been performed on the Vickers hardness indentation image using the proposed algorithm. The result of the approach based on morphology filter is fig. 2 while the result of the method based on our proposed algorithm is shown in fig.3. Another result of the method based on fractal dimension and co-occurrence matrix algorithm is shown in fig.3. Comparing the results in fig. 2 and fig. 3, we can find that the effect of segmentation only based on morphology filter is not very satisfactory. Besides, some of the background image pixels still appear in the image after segmentation, which are still not completely separated from the target object. Although the result based on morphology filter has not fully meet the requirement of segmentation, on the other hand, we can certify that the background and objective are two different kinds of texture. According to fig. 3, it can be seen that our method separates the background and objective of the indentation image beautifully and accurately attains the indentation image contours. Comparing the results in fig.3 and fig. 4, the results are the same but run time of our method is much shorter than reference [2].

Recently comparisons mainly are computation complexity of texture feature extraction, separation degree of extracted texture feature, accuracy of texture classification and segmentation. It's lack of research on performance appraisal standard of textural feature extraction method, which leads the comparison work among each method to be very difficult to do and it rarely has persuasive   power.

Fig. 1. Original Vickers hardness indentation images   **Fig. 2.** Based on Redundant Wavelet Transform (RWT)

Fig. 3. Based on our proposed algorithm

Fig. 4. Based on fractal dimension and co-occurrence matrix algorithm

6 Conclusions

Although texture description method based on the wavelet obtained very good research but the filter group's choice question still waited for solving. Wavelet transform is applied to standard or rule texture images, but it's ineffective for the nature images with complex background, because noise interference or texture pixels are not similar everywhere in some region. Besides, its computation is very great. Laws texture survey is easy to understand, but the successor researches are few. We could not find the aspect research work in each kind of reference and the periodical magazine and the application is limited. The innovation of this article is as follows. The algorithm of Redundant Wavelet Transform (RWT) and laws texture measurement is proposed and applied to image segmentation. A new segmentation method for material Vickers hardness indentation image has been proposed. The experiment demonstrates that in the material Vickers hardness image segmentation the proposed algorithm is significantly effective and robust. Because fractal and wavelet do studies on nonlinear problem essentially whose content involves details of the object, which is related to object similarity. Therefore combination of fractal and wavelet will be the important developing direction in the texture analysis

References

1. Wu, L., Zhou, Q., Deng, Y., Zhu, M.: Automatically Analyzing The Image of Vickers Hardness Test Using Wavelet. Chinese Mechanical Engineering, pt.15, 498–500
2. Wang, G., Zhu, J., Cao, P.: Application of Fractal Dimension and Co-occurrence Matrices Algorithm in Material Vickers Hardness Image Segmentation
3. Lu, L.: Research on Texture Segmentation Method Based on Wavelet transformation. Master Degree Paper of HeBei Univercity of Technology, 26–27
4. Wang, L., He, D.C.: A new statistical approach for texture analysis. Photogrammetric Engineering and Remote Sensing 56(1), 61–66 (1990)
5. Zhou, X., Tu, H.: Image segmentation algorithm based on improvement K-means cluster 29(5), 258–265 (2007)
6. Donitson, P.P.: Quantitative Evaluation of Edge Preserving Noise-Smoothing Filter. In: Geoscience and Remote Sensing Symposium, vol. 3, pp. 1590–1591

Oriented to Small and Medium Enterprises of Applied Undergraduate of Electronic Commerce Course System

Hantian Wei

Software Institute, NanChang University, NanChang, JiangXi, China
amituofo60@sohu.com

Abstract. There is a great gap of e-commerce talent in China, and a low employment rate of e-commerce undergraduates, which calls for reform of teaching models in e-commerce education. Based on the development of small and medium enterprises (SMEs) in China, and analyse of e-commerce applications, the paper summarized categories of e-business needs and skills. According to the combination of Undergraduate personnel training, it is proposed that plot the direction of elective modules, and develop a hierarchical curriculum structure, teaching methods and practice of training evaluation programs, to avoid the past wrong teaching method of extensive but not extractive.

Keywords: E-commerce, application personnel, training model, knowledge structure.

1 Introduction

Electronic commerce is one kind of new business model in information age, which provides a new opportunity and external environment to construct core competitiveness for enterprises. It plays an important role to reduce costs, expand market, capture chance, share information, and enhance their competitiveness. According to statistics, there are about 500,000 SMEs use e-commerce as regularly in 2005. They search business opportunities through the Internet, and ultimately realize the transaction amounted to 300 billion by a variety of ways. According to survey data released by iResearch, the survival rate of SMEs who use e-commerce is 5 times higher than traditional enterprise offline business. From the demand for qualified personnel the next two years, including e-commerce Assistant or Commissioner, e-commerce executives, e-commerce manager, e-commerce sales, e-commerce engineers, and sales professionals, sales executives, sales managers, sales director, and etc. particular the compound people who understand operations and highly educated will have a increased requirement. But e-commerce graduate employment rate is far lower than the national employment rate of university graduates. A survey found that in recent years the employment rate of graduates only 20%, compared to the average 47% of employment. Strong demand for e-commerce expertise and low employment rate formed a pair of sharp contradictions. It is not difficult to find that the orientation and teaching models have some problems in a number of universities.

Y. Wu (Ed.): International Conference on WTCS 2009, AISC 116, pp. 457–463.
springerlink.com

E-commerce is a nascent and constantly developed new area. Both domestic and international have exploring e-commerce field of education to the problem that how to adapt to the market, and train the person with e-business ability.

The "Social Blue Book" published by Chinese Academy of Social Sciences shows that private enterprises and self-employed is the major employer of university graduates. On 445,000 sample survey among 2007 graduates, Academy of Social Sciences found that 36.2% of undergraduates and 57.1% of the vocational or junior college students employed by private enterprises. From the scale, the SMEs below 300 people employ less than 48% of university graduates. The universality of electronic commerce made its professional location in the different schools varies greatly. Currently the professional construction and the curriculum of e-commerce have dramatic differences in some universities and colleges, and talent training e-commerce is extremely vague. The extensive and volatilization of the person needed by enterprises has affected the professional training college on the e-commerce location, also affect the employment of electronic commerce graduates. This is not beneficial to the professional development of electronic commerce. Therefore, more and more enterprises will be concerned about e-commerce. Person with the e-commerce has a larger demand, which should prompt the students to master business and technology in college, to meet the future market.

2   E-Commerce Needs of SMEs

2.1   Training for SMEs' Goal

In China, small and medium processing enterprises are less than 2,000 workers, or sales at 300 million. According to statistics, the numbers of SMEs accounted for 99%.of the total number of Chinese enterprises, and create around sixty percent of China's total economic output and nearly half of tax revenue, and eighty percent jobs.

The SMEs who need for professional e-commerce may not have an e-commerce department, or a professional e-commerce does not necessarily work in e-commerce sector. They may work in the client department, marketing department, or IT department. In other words, the actual e-commerce companies need professionals that understand the specific business e-commerce personnel or technical staff, but not only know how to e-commerce expertise.

2.2   Demand Characteristics of Jobs in SMEs

To cultivate the applied talent people who has good professional ethics, professionalism, innovation, team awareness, professional e-commerce required familiarity with laws and regulations, mental health, master of e-commerce website design and management and application rules, with the necessary electronic business knowledge and application. They can engaged in network marketing and planning, online transactions, electronic payment, logistics, EDI processing operations,

E-Commerce Platform work and management of e—Commerce. So the main body of e-business talent, which is characterized by fluent in modern business, with sufficient technical knowledge of e-commerce, can operate in the area of electronic commerce. Business-type e-commerce professionals are widely applicable. They have a good grounding in business theory, familiar with the basic use of computers and the Internet technology, proficiency in e-commerce environment, business operation, such as online shops, online procurement , Internet banking, online transactions; grasp of marketing, consumer psychology, business etiquette, customer relationship management, international trade, logistics management, economic law and other knowledge. Through the survey, the demand of e-commerce companies generally divided into three categories: e-commerce technology professionals, e-commerce marketing professionals, e-commerce management professionals.

1) E-commerce technology professionals
The work required to design and develops websites, web animation, the network database maintenance, website routine maintenance, familiar with the web site promotion techniques to understand the commonly used network marketing, operations and so are familiar with CRM and ERP. The corresponding jobs are: web development, web design, website maintenance, corporate ERP technical consultants. These are the largest demand of the current job market.

2) E-Commerce Marketing professionals
The work required to proficiency in marketing through network, familiar with the network marketing skills, strong communication skills and writing skills, ability to have network marketing plan. Corresponding positions are: network marketing personnel, call center managers, clerks, the demand for qualified personnel in this area are also high.

3) E-commerce Management professionals
The management professionals are the strategy personnel in the pyramid, who are general leadership of the enterprise, or who have high professional quality, communication skills, and strong oral and written communication skills and are able to use e-commerce tools skilled personnel to carry out trading activities. The two levels above can't be trained directly in school, but continue to accumulate in the workplace, and exercise with grew up.

3   Reform of Curriculum System

According to the e-commerce specialization training ideas, e-commerce system professional courses should consider the professional basic courses, e-commerce core curriculums, professional skills curriculums, professional elective curriculums.

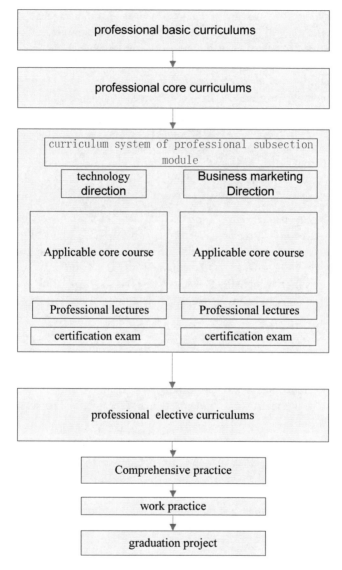

Fig. 1. Structure of curriculums system

3.1   Professional Basic Curriculums

The basic courses in Electronic Commerce should be the relevant business and technology courses, such as: management, finance, accounting, financial management, management information systems, ERP, CRM, SCM, operations research, statistics and so on. These courses must be able to lay the foundation for learning.

3.2   Professional Core Curriculums

Core curriculum must be open to all e-commerce professional courses. Such as Introduction to e-business subjects, e-commerce management, e-commerce engineering, modern logistics management, computer network, network marketing, electronic payment, e-commerce security, and etc.. E-commerce core curriculum should enable students to understand the whole picture of e-commerce, master the basic concepts, basic techniques and basic methods.

3.3   Curriculum System of Professional Subsection Module

E-commerce is a complex interdisciplinary, as a single e-commerce professional, contains too many elements. It is difficult to include all aspects of electronic commerce in the four years of undergraduate courses. Actual business really needs is to understand the specific business e-commerce professionals or technical personnel, but not a single e-commerce expertise. So according the categories above, it should be set professional elective modules, sub-commerce direction. The two main kinds are e-commerce technology and e-commerce marketing. If the enterprise knows structure and professional skills clear, the different understanding of e-commerce cultivation between enterprise and school can be avoided, and students will not feel confused.

1) Curriculum setup of the two classes
Technology direction Course includes Computer Network, database technology, e-commerce platform development, e-commerce web design and web site management, network security;
    Business marketing Direction Course includes business English, online customer service and management, online advertising sales and production, public relations practice, international trade, logistics management, e-commerce.

2) Combine the teaching and certification exam
Encourage students to participate in the relevant e-commerce certification exam, that the certified person has basic knowledge of e-commerce, but also the ability to operate third-party e-commerce platform, familiar with network marketing, with the corresponding ability of the network business.

3) Professional elective curriculums system
The elective courses of E-commerce can be combined with the general student professional interest, such as team management, business environment and culture, business management, business management, e-commerce and legal side of Theoretical Course, aimed at expanding the professional knowledge of students and professional development of potential mining. Various e-commerce professionals can also be broken down according to the actual need to identify specific professional orientation; professional elective courses can be reflected on this.

3.4   Practice

1) Comprehensive practice
According to the e-commerce professional specialization training ideas, e-commerce system of professional courses in the practice session to set the problem more easily

solved. Subdivide it into several professional e-commerce expertises, the various e-commerce specialty practice on the part of the requirements needed to be clearer.

For example, the direction of application, which will pay more attention to web-based e-commerce technology and information systems skills, is part of the urgent practical need to do. The website and e-commerce platform for the design and development requirements of each student, the appropriate course of a semester, and a large design can be arranged. The other e-commerce marketing profession needs more companies to do the actual depth research related.

It is clearly that E- commerce require compared high Laboratory conditions. At present, there are more than 300 undergraduate colleges and universities set up e-business professional, but many schools have no e-commerce engineering talents necessary experimental conditions. According to the e-commerce professional specialization training ideas, the schools, whose e-commerce project without the necessary experimental conditions for personnel training, will run by amendment of specific e-business professional personnel training orientation.

2) Work practice

Practice, professional practice and graduate practice are important to train the professionals in e-commerce. Combination of business practice, enhance the management of these practice areas are especially important too. Requiring students to visit and investigation of practical business applications of information technology and e-commerce specific problems, graduate practice require students to combine practical business or practical issues to do research for some time after the completion of thesis.

4   Conclusion

In recent years, with the rapid growth of global e-commerce, Electronic Commerce drastic development, bring serious shortage of e-business talent. As the Internet user increase rate reach 100% per year, China is expected to need two million professionals in the next 10 years. E-commerce, as the high-tech fields business, should be paid more and more attention. The nature of e-commerce business is commerce innovation based on technology and technical innovation oriented to business. E-commerce professional education itself is an innovative service, and will continue to require innovative. Therefore, this paper presents e-commerce for small and medium enterprises Undergraduate Major. The fundamental purpose of e-commerce professional education is to train community urgently needs to e-commerce professionals. In addition to undergraduate professional e-commerce, the Chinese universities can set up e-commerce and e-commerce Master of Engineering Management MBA, training senior personnel of e-commerce businesses need.

References

1. Nambisan, S.: How to prepare tomorrow's technologists for global networks of innovation. Communications of the ACM 48(5), 29–31 (2007)
2. Flieder, K.: Integrating the virtual world: Teaching EAI and e-Business Integration at Universities of Applied Sciences. In: Conference ICL 2007 (2007)

3. Flieder, K.: Virtual continent in the classroom: Teaching e-Business Integration Issues with Web-based Technologies. In: Conference ICL 2007 (2007)
4. Weaver, A.C.: Experience with Teaching Electronic Commerce Frontiers in Education, 2005. In: Proceedings 35th Annual Conference, FIE 2005, October 19-22 (2005)
5. Liu, S., Tang, M.: An effective tool for e-commerce teaching and learning. International Journal of Information and Operations Management Education 2(1), 103–115 (2002)
6. Susser, B.: Teaching e-commerce Web page evaluation and design: a pilot study using tourism destination sites. Computers & Education 47(4), 399–413 (2006)
7. Journal of Computing Sciences in Colleges 23(1), 14th (October 2007)
8. Zhou, Z.: Teaching e-commerce in an information systems program: striking a balance between business models and implementations. In: Annual CCSC Midwestern Conference and Papers of the Sixteenth Annual CCSC Rocky Mountain Conference, pp. 231–240

Influence of New Media Technologies on College Students' Ideological and Political Education

Wang Jia and Dai Yanjun

School of Marxism, Dalian University of Technology, Dalian, China
pkrysl@yahoo.cn

Abstract. New media technology (NMT) has been widely accepted by college students, which incur tremendous impact on the growth and development of college students. Ideological and political education (IPE) should take advantage of the development opportunities and take measures to cope with the challenges caused by new media technologies, such as establishing the students' principal position in the education process, integrating the advantage of the traditional medium of IPE and the NMT, paying more attention to the guiding roles of opinion leader to peers, organizing various campus activities in the virtual world.

Keywords: new media technology (NMT), ideological and political education (IPE), college students.

1 Introduction

New media refers to the communication patterns of using such channels as digital technology, network technology, by the Internet, broadband local area network, wireless communication networks, satellite etc., with computers, mobile phones, digital televisions and other devices, to provide information and entertainment for users. With the wide spread and a large amount of applications of blog, podcasts and so on, the Internet is becoming a freedom forum for exchanging information and expressing the public opinion. New media communication has three characteristics: ① Fuzziness of communication levels: each users can be the information sender, so it is "all for all" social communication; ②Generalization of communication relation: instantaneous communication instruments allow people to communicate and exchange their ideas without the restrictions of time and region, which breaking the traditional communication barriers of society, culture and mentality; ③ Diversification of communication subject: transmitting relationship has changed in new media era, and the communication subject is no longer unique, so different social groups and community sectors may express their own opinion through new medias.

New media technology (NMT) has been generally accepted by college students, which incur tremendous impacts on the college students' communication ways, thoughts and personality formation. Transmission characteristics of new media technologies not only offer the development opportunities for IPE, but also bring

Y. Wu (Ed.): International Conference on WTCS 2009, AISC 116, pp. 465–471.
springerlink.com

many new challenges. IPE should change educational ideas, innovative the medium of IPE, and expand the space of IPE to deal with the new circumstances and new problems caused by new media technologies for IPE.

2   Specific Manifestation and Reasons of the Influence on College Students by NMT

The new media will become the important propagation force to promote the changes of social patterns, and it will shape new cultural form. College students are in the formative stage of world, life and value views, which would be enormously influenced by new media.

2.1   The Specific Manifestation of the Impact of NMT on College Students

First of all, college students' communication ways change a lot. By means of BBS, E-MAIL, QQ, blog, short message service and other new medias, students feel more relax and convenient in their interpersonal communication, and they feel more free to solve their problems than the expectations. On the Internet, the views exchange between two sides adopt more anonymous way to reduce the interference from social factors or other persons, it is conducive to protect personal privacy and expression freedom and contribute to exchange ideas and express affection effectively, so the Internet has become the ideal choice for students to express their ideas and voice their opinions. The proportion of virtual space activities is increasing in college students' common lives, which become a pressure relief valve for their intense study and life. Nevertheless, at the same time, the openness and virtual of new media may break away college students from the constraints of real human relations and morals, and some students easily indulge their behaviors and forget their social responsibilities. Some students are interested in virtual communication and alienate from the real human communication, resulting in interpersonal obstacles, and then causing a evasive psychological tendency.

Secondly, the rapid development of new media changed the students' learning and thinking ways. The traditional learning way mostly refers to students' classroom learning under the guidance of teachers. But in the new media age, students can use the Internet to acquire a lot of educational resources for interactive learning. New learning methods help students develop the habits of exploring the unknown world initiatively, increasing exploration experience and accepting new knowledge and technology independently. Confronting with the information from new mass media, because of their limitations of knowledge, experience and the ability of understanding and thinking, the students easily tend to think partially, so their identification ability is usually less effective, which is in urgent need of proper guidance.

Thirdly, new media has important influence on the formation of students' characteristics. In the new media era, it is obvious for students' desire to pursue and express their personality. Having more sufficient opportunities and rights, it is essential condition for individuals to promote their personalities. In this sense, new media satisfied the need of the contemporary college students' personalization, and provided more choice for them, which made every college student have free access to the information production and dissemination. But the new media have some negative effects to the students who have not form a stable pattern of life and values: the lack of rational judgments and strong curiosity would make students accept outside information passively, which would reduce their own thinking scope; the immature of theoretical systems and thinking ways would easily cause university students losing themselves and confusing their values in pluralistic environment; the disorder of new media information would weaken the college students' sense of responsibility and make them pursuit absolute individual freedom excessively.

2.2   The Underlying Causes of Acceptance of NMT by College Students

"The 24th China Internet Development Statistics Report" which was released by China Internet Network Information Center showed that 90.5% of college students use the Internet as an important information channel. College student widely accepted new media have three reasons:

Firstly, the students of knowledge and the technical requirements of new media are in line. Being in the phase of knowledge acquisition, 90s' students are more familiar with the network technology than the other ages.

Secondly, the college students' motivation of using new media and their needs of psychological development are corresponding. The university stage is designated as pre-adults in developmental psychology, and the students face 10 development tasks, such as learning, interpersonal communication and career planning, so they often acquire the valuable information actively. The Internet is not only a access to information for the students, but also a important channel to learn social values and improve their interpersonal skills.

Finally, the students' elite consciousness, their pursuit of individual personality and the spread feature of new media are in line. New Media Communication is no longer a government-led and top-down linear spread pattern, but a multi-center and non-linear spread pattern. College students realize the possibility of becoming the "opinion leader" of a "new view class", and thus stimulating the young people's elite consciousness and the pursuit of individuality.

3   The Opportunities and Challenges of IPE with Features of New Media Technology

3.1   Take Advantage of NMT to Build New Platform for the Development of Ideological and Political Education

First of all, NMT is colorful and three-dimensional, which provides IPE with the unprecedented dynamic and imagery feature. The features of new media technology(NMT) are text-image and audio-visual integration, which could reproduce the historical scenario, illustrate relative procedures, demonstrate data clearly, transform the simple input form of IPE into integration of input and guide, and thus enhance the persuasive and appealing feature of IPE.

Secondly, NMT is open and interactive, providing quick and timely feature for IPE. One of the NMT features is boundlessness, which enable a more frequent, wider, and thorough communication, and thus enrich the carriers of ideological and political education. E-mail, BBS, MSN, QQ, BLOG and mobiles, such carriers could enable the educator and the receiver to exchange ideas and share opinions remotely, which could realize remote education, resolving doubts and confusion and sharing authority.

Thirdly, the NMT is anonymity and hidden, which guarantees truth for IPE. To some extent, NMT shortens the psychological distance of social communication, eliminates the fear of first impression, and enable people to communicate freely. For ideological and political educators, it is possible to detect the students' real opinion. By SM, MSN and QQ, college students can express negative emotions with awkward feelings for face to face interview, and it provide a reasonable possibility for target-orientedly resolving their practical issues.

Finally, NMT is interpersonal and multivariant, which provides college students' IPE an unprecedented multi-point correspondence. "In the Internet age, everyone can become channel of information, become the main subject of opinion expression. There is a vivid metaphor that in front of everyone there is a microphone." Mode of transmission from the "one to many" to the "many to many" breaks the one-way mode of information flow from a single source to other directions, showing multifaceted, extensive, in-depth propagation characteristics for IPE.

3.2   The Challenge of the College Students' IPE under the Conditions of NMT

The diversity of NMT promotes the college students' IPE to change its educational concept and develop its educational theories. The students express preferences to the new media technologies, so IPE should enrich the educational content, change educational concepts and develop educational carriers. Especially the teaching methods should respect the principal position of college students and meet their behavior habit and psychological characteristics. Presently, newspaper, magazine and other traditional media are cyberizing progressively, and the ideological and political educators often comment on some events with "blog" which the students concerned. These are the exploration of the diversified educational methods.

The information disorder from NMT requires IPE to strengthen positive guidance and monitor public opinion. NMT enrich information which is also messy, so thus it could easily lead to false and bad information spreading. The students in the growth stage have novelty mind, and they tend to reject and doubt authoritative information, which make it easier for students to take an interest in false or harmful information. Few students use new media technologies to self-hype, which challenge the baseline of social ethics and values. Therefore, the ideological and political educators should strengthen the guidance of public opinion under the new media environment, including standardizing clearly what is right and what is wrong.

The equality of subject and object of new media technologies makes IPE insist on people-oriented thought. Under the conditions of NMT, the interpersonal communication shows hidden and private feature strongly. Especially in the QQ, MSN, Blog and other communication environments, the informal way of expression is not only to reflect the true feelings, but also avoid some psychological readiness caused by the "face to face, words by words" communication. The subject and object have achieved equality and mutual trust of mentality and superficiality in the specific environment. If you want to get students' really opinion and achieve effective educational effect, you must have a equal attitude to communicate and eliminate psychological barriers.

The rapid update of NMT requires ideological and political educators to improve their work. The update rate of the operation technologies, the external forms and the communication pattern of new media are very quickly. These changes require new task to the ideological and political educators constantly, including changing their educational ideas, using new information technology, improving their communication and adjusting their attitude. With the development of science and technology, NMT will be more frequently updated. Therefore, the ideological and political educators must be constantly adaptive to new challenges, adjust their work and innovate their method.

4   Measurs of College Students' IPE under the Condition of NMT

Firstly, changing the concept of IPE. With the help of NMT, the college students are able to grow a stronger ability of self-awareness, and a higher lever of self-acceptance. The IPE should obey the principles of ease-oriented education when facing with the wrong, immature point of view of college students. Ideological and political educators must enhance their understanding of NMT and the abilities of using NMT to conduct ideological and political work. In the aspect of understanding NMT, ideological and political educators must keep the pace with college students. While the code of conduct of using NMT, ideological and political educators must walk in front of college students as an example.

Secondly, integrating various IPE carriers. Traditional campus carriers of cultural activities are newspaper, blackboard, campus broadcasting and cultural facilities. They can not be replaced by NMT for a relatively long time as they still have the characters of clear orientation and strong public credibility. To merge the carriers of IPE is to maintain the base of traditional advantage and to spread and develop the traditional advantage by using NMT, to correct the traditional advantage and improve both advantages mutually. Henan University made a useful exploration. The first mobile newspaper of Communist Youth League in campus was born in November 9, 2008.Youth of Henan University (mobile version) were based on the original newspaper 'Youth of Henan University' and the internet version of it, using multi-media broadcasting technology. The IT broadcasting platform was build by Chinese Communist Youth League of Henan University and China Mobile Communication Henan branch. The newspaper were sent to the students and teachers of Henan University in form of multimedia message on every Monday with contents about the theoretical study, group organizations, campus cultural activities, scientific and technological innovation. The mobile newspaper is a bold innovation of building new IPE carriers in Henan University.

Thirdly, extending the IPE space. With the development of new media, a new view class is emerging in China - a class focus on current events and express opinions online, and college students are the major part of this class. The central figures are opinion leader in the new view class 'who have a quite influence, thus requires IPE workers to spread their influence in NMT to be initiative.

Fourthly, organizing diversify virtual campus activities. Social activities of college students become more spontaneous from passive, scattered from centralized, paid from free, social from professional. Ideological and political educators should recognize the student group change and adjust the educational policy in time. They should consciously convert the orientation of campus website from broadcasting to service, respect fully the spirit of independent innovation of students, and communicate by equal attitude with the students to win their trust. IPE should help students to discover, organize and manage knowledge. Ideological and political educators should clean the barriers in the student life in a style of guide not mould and be equal with students to learn each other.

5   Conclusions

With the rapid development of NMT, new media has increasingly affected people's lives. This paper researched in the perspective of IPE and got the following conclusions:

NMT meet the college students' need of pursuit elite consciousness and showing their personalities. Especially the college students are more familiar with Internet technology than other social groups, so new media technologies are widely accepted by them.

NMT have both positive and negative impact on the college students, mainly concluding three aspects: being keen on virtual communication escaping from the real life; using the interactive learning methods which need proper guidance; the pursuit of developing personality and showing themselves, which would weaken their responsibility.

NMT bring some development opportunities for the IPE, mainly the following four aspects: The text-image and audio-visual integration features of NMT enhance the persuasive and appealing feature of IPE; their open and interactive features provide quick and timely feature for IPE; their anonymity feature guarantees interaction for IPE; their multiple features provide multi-point correspondence for IPE.

NMT bring a lot of challenges to IPE, mainly concluding the following four aspects: The diverse technologies promotes IPE to develop its carries and enrich its educational theories; the disorder of information require IPE strengthen positive guidance and monitor public opinion; the equality of subject and object makes IPE insist on people-oriented thought; the rapid update of NMT requires ideological and political educators to improve their work.

IPE should take active measures to cope with the challenges of NMT: The IPE should obey the ease-oriented education principles and make college students become the object of education; it should merge the advantage of traditional and new carriers; it should strengthen the guide role of "opinion leader" to the peers; it should organize diversify virtual campus activities.

Acknowledgment. The study was co-supported by the Social Science Fund of Liaoning Province (Grant No.L09BDJ030) and the Fundamental Research Funds for the Central Universities (Grant No.DUT10RW108).

References

1. Feng, G.: Research on the Function of the Ideological and Political Education of New Media Technologies. Journal of Beijing Education: Moral Education 6, 5–9 (2009)
2. Wang, X.: New challenge of Ideological and Political Education of 90s' college students from New Media. Journal of Ideological & Political Education 1, 71–74 (2010)
3. Jiang, E.: The College Students' Ideological and Political Education in the New Media Environment. Journal of the Theory Front of Institutions of Higher Learning 6, 54–56 (2009)
4. Feng, Z.: The Internet's Influence of the Ideological & moral Construction and the Corresponding Strategies. Journal of China Higher Education Research, 38–43 (May 2003)
5. Zhang, Z., Zhang, J.: More Views More Force and More Innovation: Some Questions of Resource Allocation of New Media in Social Transformation. Journal of Modern Communication 2, 56–60 (2008)

A Comprehensive and Efficient NAT Traversal Scheme on SIP Signaling

Wang Yue and Cheng Bo

The State Key Lab of Networking and Switching
Beijing University of Posts and Telecommunications
BeiJing, China
szlily2005@sina.com

Abstract. Due to the characteristics of SIP protocol, various NAT network has become a stumbling block for the extensive application on SIP. However, several existing NAT traversal solutions are not suitable for all the NAT networks, or some solutions significantly affect the instantaneity during the interaction of SIP, which is due to that they either need us to modify the existing equipments in the network, or have limitations on their practical application. Therefore, according to the features of SIP protocol and various traversal protocols, this paper proposes a comprehensive and efficient traversal method. It is based on extension protocol of SIP, STUN protocol, and TURN protocol. In detail, this method applies the extension protocol of SIP to the traversal of SIP signaling. For streaming media in asymmetric NAT network, it uses STUN protocol. What's more, TURN protocol is adopted for the traversal of streaming media in symmetric network. Practice has proved that the SIP soft-phone integrated with this method traverses all kinds of network.

Keywords: SIP, NAT Traversal, STUN, TURN, asymmetric NAT, symmetric NAT.

1 Introduction

With the furthering development and maturation of the next generation network, Researchers have paid close attention to the technology of VOIP. Currently, there are two major technical specifications applied to VOIP, which are H.323 protocol of ITU and SIP protocol of IETF. Traditional telephone signaling mode is adopted to the H.323 protocol because of its convenience for the connection with PSTN, which is why it is used relatively widely. However, other standards and ideas in the protocols on the Internet are referenced in SIP protocol. Therefore, SIP protocol has many features such as simplification, flexibility, compatibility, expandability, and so on, which H.323 protocol doesn't have, so it has become a new furthering development direction for VOIP.

Rapid expansion of the internet has led to a crisis that the space of global IPv4 address will be exhausted, which results in that NAT becomes an important technology to solve the crisis. According to mapping feature of NAT, it can be divided

Y. Wu (Ed.): International Conference on WTCS 2009, AISC 116, pp. 473–480.

into two categories, which are asymmetric NAT and symmetric NAT. The main features of asymmetric NAT is that as long as the packets are sent from the same internal address and port, NAT will map it to the same external address and port, even if the destination address was changed. However, once either the destination address or port is changed, the symmetric NAT will map to a new IP address and port. The SIP header and body contains the communication address, so it will not lead to establish a normal session if one part of the communication is in private network. The information of IP addresses in SIP signaling has become an obstruction from the traversal of NAT. Based on this, a comprehensive and efficient traversal method is proposed in this paper, which makes SIP signaling and media traverse all kinds of NAT.

2   Sip Soft-Phone Framework

Even if IPv6 technology has matured, private networks of many big companies still use the way of NAT to connect into the Internet. So the module of NAT traversal is necessary for the SIP softphone.

The Framework of the SIP softphone is made up of four modules, that are the SIP UA module, the STUN/TURN Client module, the JMF module, and the audio and video codec module. SIP UA is mainly responsible for the handling of SIP signaling, which is a core mudule to establish a SIP session. STUN/TURN Client is a module of NAT traversal in SIP softphone, which provides effective information for the SIP UA. JMF is a media framework module based on java, which is used to capture audio and video stream from the equipment, and call the codec to process audio and video stream, finally send RTP stream to the other party of the communication. The audio and video codec is mainly responsible for encoding and decoding.

Once the SIP soft-phone starts up, SIP UA will begin to monitor the port used to send SIP signaling. When it receives a request of invitation to a SIP session, it will call JMF to obtain the local ports of audio and video when the SDP is created, and make JMF capture audio and video stream. At this time if the SIP soft-phone is in the private network, SIP UA will then call the interface of STUN/TURN client module to obtain the public addresses of audio and video used to communication in this session. If the SIP soft-phone is in asymmetric NAT network, the interactive interface will use STUN client to create a STUN message, and extract the mapping NAT address from the response for STUN request. However, the interactive interface user TURN client to create TURN message, and extract the relay address for audio and video stream in the symmetric NAT network. SIP UA will fill out the valid address in the SDP, which insures that the SIP soft-phone establish a correct connection for audio and video. After the communication of SIP signaling, JMF begins to encapsulate, send, and receive audio and video stream. When the session is completed, SIP UA will notify JMF to stop sending media stream and releasing the ports listened.

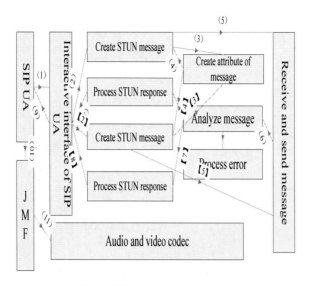

Fig. 1. SIP Soft-phone framework

3   Implementation of the Nat Traversal Scheme

Due to the complexity of the network, the SIP soft-phone can't make SIP signaling and media stream traverse all kinds of network only through applying one kind of traversal methods. Thus, according to the characteristics of SIP signaling, in this method, we organically combine three ways that are the simplified STUN protocol, TURN protocol and extension protocol, which efficiently completes NAT traversal for SIP signaling and media stream. It can be divided into three processes that are learning the type of NAT, establishing SIP session and maintaining the ports for the media stream.

Learning the type of NAT is the cornerstone of the whole traversal process, which provides an evidence for the decision of the second process. The integration of SIP, STUN and TURN signaling is the most critical process, so whether SIP signaling and media stream can traversal NAT or not depends on the effectiveness of the information obtained in this process. The maintaining process of the media ports is essential to the whole traversal process, because even if the second process obtains effective information, the media stream still can't be received successfully without this process, which is due to the time limitation of NAT and the firewall.

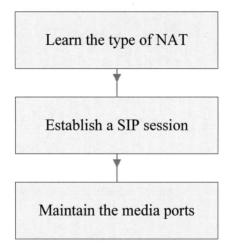

Fig. 2. Three process of NAT Traversal

3.1   The First Process for Learning the Type of Nat

Before sending SIP register messages, SIP UA must confirm which network the SIP soft-phone is in, so the SIP soft-phone should learn the type of network first. SIP UA will call STUN Client to detect the network environment. First of all, STUN client sends a STUN request message to IP1 and PORT1 of STUN server. If the return address is the same with the local address, it illustrates that the network is public and then it stops sending STUN request. While, it proves the SIP soft-phone is in private network if not, and then STUN client continues to send a STUN request message to IP1 and PORT2 of STUN server. If the return address in this response is the same with the address in the first response, it shows that the type of network is asymmetric NAT and the useStun flag should be set to 'true'. Otherwise, it illustrates that the type of network is symmetric NAT and the useTurn flag should be set to 'true'".

3.2   The Second Process for Establishing a Sip Session

SIP signaling includes the SIP message header and the SIP message body. The SIP message header contains the local IP address and the port for establishing a signaling session. The SDP in SIP message body carries the IP addresses and the port for receiving media stream of both side during the communication. If the SIP message header of one part brings the private network address, the other part will send a response to that private address according to this SIP message header. However, the signaling to private address can not be routed in the Internet, which leads not to establish a normal session. To ensure that the session is established in real time, the SIP signaling without the SDP don't need the assistance of STUN and TURN signaling, but need to apply the extension protocol of SIP. When one part in communication needs to notify the media address, the SIP signaling must include a SDP. The SIP UA will modify the private address in SDP via STUN or TURN signaling.

If the SIP soft-phone wants to communicate with the SIP application server, it should not only register in the application server first, but also confirm that the address of registration is public network address. As the registration message is the SIP message without SDP, it applies the extension protocol of SIP to interact with the SIP application server. The SIP UA will send a registration message, which is added a RPORT flag to the via field and the other field of which are still filled with the local address and port. When the SIP application server detects that the registration message received contains RPORT flag in via filed, it will extract the source IP address and port, then fill them in RPORT and RECEIVED filed of the return message, and register them as the address of SIP UA. As there is time limitation during the address translation, the SIP UA must send a registration message every 60 seconds in order to maintain and update the registration address. Only through this way, it can ensure that the SIP application server can find the SIP client in private network through the registered address in any time.

In the process of establishing a session, the SIP application server will send an "invite" request to the SIP UA according to its registration address. The SIP UA reply a "180 ring" message after receiving the " invite "request, and then send a " 200 OK" message with a SDP to tell the audio and video address to the SIP application server. When constructing a SDP, it will check the useStun and useTurn flags. If useStun flag is 'true', the SIP UA separately sends STUN requests by audio and video ports locally listened. Then SIP UA obtains corresponding export NAT address of local audio and video ports from the responses to those requests. If useTurn flag is 'true', the SIP UA separately sends TURN requests by audio and video ports locally listened to acquire the relay address for media stream. Fill the export address or the relay address in SDP, and at the same time extract the address from the "to" filed of "invite" request and fill it in the "contact" filed of the "200 Ok" message. As a result, the interaction of SIP session can be ensured to carry out smoothly.

3.3   The Third Process for Maintaining Media Port

The ports usually used to receive and send media are not the same, in this case NAT and firewall will prevent the audio and video stream from entry into this network. Because they will block those packets from the outer hosts which the hosts in private network have never sent messages to. They think this behavior could break the security of the private network. In order to communicate normally, the SIP soft-phone must open up a hole for media stream on NAT and firewall. This method starts up two threads which bind local audio and video ports for sending a simple UDP packet every 60 seconds. These packets won't influence receiving real media ports whose purpose is to open up a hole on NAT and firewall. When the session ends, it will end the two threads. If the method adopts TURN protocol, it still sends 'refresh' message to TURN server using the local media ports, which make TURN server release resources allocated to the TURN client.

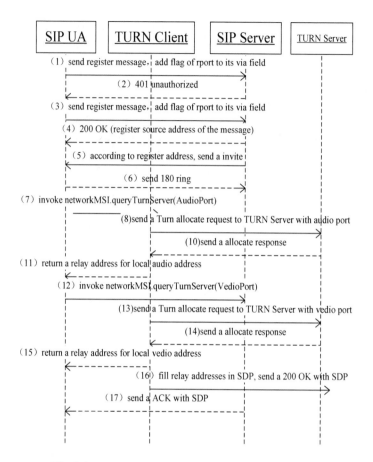

Fig. 3. Message interaction in asymmetric NAT network

4   Application

This method has been applied in the web application software terminal of the multi-media conferencing system. When a conference is to be created, the application sever will send request to all of the participants, in order to invite users in all kinds of network to join the conference and start audio and video conference. The two invitations for yaya whose IP address is 10.13.17.23 and bupt13 with IP address 192.168.1.87 are shown in the following graph. In detail, yaya is in a non-symmetrical NAT network, using STUN traverse the NAT. While bupt13 is in symmetric NAT network, and it uses TURN to traverse the NAT. During the conference, the participants have completed the SIP signaling normal interactions, the correct audio receiver and display. 59.64.156.153 is the address of relay Server.

What's more, it has passed through the test in the campus intranet, internal OA Unicom network, residential broadband and other private networks. It has been proved that the method can ensure the SIP signaling and the normal reception of audio and video.

202.106.171.102	10.13.17.23	SIP	Request: INVITE sip:yaya@10.13.17.23:6060
10.13.17.23	202.106.171.102	SIP	Status: 180 Ringing
10.13.17.23	202.106.171.102	SIP	Status: 180 Ringing
10.13.17.23	202.106.171.102	SIP	Status: 180 Ringing
10.13.17.23	202.106.171.102	SIP	Status: 180 Ringing
202.106.171.103	10.13.17.23	STUN	Message: Binding Response
202.106.171.103	10.13.17.23	STUN	Message: Binding Response
10.13.17.23	202.106.171.102	SIP/SDP	Status: 200 OK, with session description
202.106.171.102	10.13.17.23	SIP/SDP	Request: ACK sip:yaya@10.13.17.23:6060;transport=ud

Fig. 4. Message interaction in asymmetric NAT network

192.168.1.87	202.106.171.102	SIP/SDP	Status: 200 OK, with session descriptior
192.168.1.87	202.106.171.102	SIP/SDP	Status: 200 OK, with session descriptior
202.106.171.102	192.168.1.87	SIP/SDP	Request: ACK sip:bupt13@59.64.158.243:55

```
bytes on wire, 1074 bytes captured)
Elitegro_9a:f9:d6 (00:14:2a:9a:f9:d6), Dst: Netgear_03:3e:aa (00:1b:2f:03:3e:aa)
, Src: 192.168.1.87 (192.168.1.87), Dst: 202.106.171.102 (202.106.171.102)
>tocol, Src Port: sip (5060), Dst Port: sip (5060)
>n Protocol
:P/2.0 200 OK
```

```
·iption Protocol
:cription Protocol version (v): 0
:or, Session Id (o): bupt13 0 0 IN IP4 59.64.156.153
ie (s): -
Information (c): IN IP4 59.64.156.153
ption, active time (t): 0 0
·iption, name and address (m): audio 59907 RTP/AVP 0 8 97 3 5 4
bute (a): rtpmap:0 PCMU/8000
bute (a): rtpmap:8 PCMA/8000
bute (a): rtpmap:97 ILBC/8000
bute (a): rtpmap:3 GSM/8000
bute (a): rtpmap:5 DVI4_8000/8000
bute (a): rtpmap:4 G723/8000
bute (a): fmtp:4 packetization-mode=1
·iption, name and address (m): video 61909 RTP/AVP 99 34 26 31
:nformation (b): AS:640
bute (a): rtpmap:99 H264/90000
bute (a): fmtp:99 packetization-mode=1
bute (a): fmtp:34 QCIF=1 CIF=1 MaxBR=1960
bute (a): rtpmap:34 H263/90000
bute (a): rtpmap:26 JPEG/90000
bute (a): fmtp:26 QCIF=1 CIF=1 MaxBR=1960
```

Fig. 5. 200 OK message in symmetric NAT network

5   Conclusion

A comprehensive and efficient traversal method is proposed in this paper. It is based on extension protocol of SIP, STUN protocol, and TURN protocol. In detail, this method applies the extension protocol of SIP to the traversal of SIP signaling. For streaming media in asymmetric NAT network, it uses STUN protocol. What's more, TURN protocol is adopted for the traversal of streaming media in symmetric network. Although this method can traverse all kinds of NAT, it still adopt the way of relaying the media stream. It is inevitable to increase time delay of media stream in this way.

Therefore, The next step is to improve the way in symmetric NAT and try to detect ports via STUN method in order to stop from relaying the media stream.

Acknowledgment. This project is supported by the National Natural Science Foundation of China under Grant No. 60432010 and No.60872051; 973 project (Universal trustworthy Network and pervasive Services) under Grant No. 2007CB307100

References

1. Huang, C.-L., Hwang, S.-H.: The Asymmetric NAT and Its Traversal Method. In: IFIP International Conference on Wireless and Optical Communications Networks, WOCN 2009 (2009)
2. Rosenberg, Weinberger, Huitema, Mahy: STUN- Simple Traversal of UDP Through NATs. IETF Draft, RFC-3489
3. Rosenberg, J., Huitema, C., Mahy, R.: Traversal Using Relay NAT (TURN). IETF Draft draft-ietf-behave-turn-16
4. Rosenberg, J., et al.: SIP: Session Initiation Protocol. RFC 3261 (June 2002)
5. Rosenberg, J., Schulzrinne, H.: An Extension to the Session Initiation Protocol (SIP). RFC358 (August 2003)
6. Srisuresh, P., Holdrege, M.: IP Network Address Translator (NAT) Terminology and Considerations. Request for Comments 2663 (1999)
7. Gou, X., Jin, W.: Multi-Agent System for Multimedia Communications Traversing NAT/Firewall in Next Generation Networks. In: CNSR 2004, May 2004, pp. 99–104 (2004)
8. Midcom, W.G., Takeda, Y.: Symmetric NAT Traversal Using STUN. draft-takeda-symmetric-nat-traversal-00.txt, IETF (2003-2006)
9. Huang, T.-C., Shieh, C.-K., Lai, W.-H., Miao, Y.-B.: Smart Tunnel Union for NAT Traversal. In: SNCA 2005, pp. 227–231 (2005)
10. Gou, X., Jin, W.: Multi-Agent System for Multimedia Communications Traversing NAT/Firewall in Next Generation Networks. In: CNSR 2004, May 2004, pp. 99–100 (2004)

Study on College Scientific Research Capability Evaluation System Based on Neural Network

Huizhong Xie[1], Zhigang Ji[2], and Lingling Si[1]

[1] Handan College
[2] Hebei University of Engineering
380282773@qq.com

Abstract. According to the fuzziness of assessment system about university scientific research capability, this paper puts forward to a comprehensive judgment method basis on improved discredited Hopfield neural network to establish assessment model and studies with examples. It indicates that it is a practical and simple assessment model with a higher resolution and actual assessment results and it offers scientific reference for recognizing university scientific research capability, self-diagnosis and adjustment as well as promoting university ranking.

Keywords: Neural Network, scientific research capability of university, evaluation.

1 Introduction

Scientific research capability has become an important indicator to measure a university comprehensive strength. How to accurately evaluate the research capability of universities has brought serious concern by governments, enterprises and universities. At present, there are many methods used to evaluate university research capability [1-6], but most of them have shortcomings such as tedious, time-lags and so on, and can not avoid the interference coming from subjective factors. How to quickly and accurately get an objective evaluation of research capability for large number of colleges and universities is a problem needed to be solved urgently. This paper uses a improved Hopfield neural networks to evaluate the research capability of universities. Result shows that our comprehensive evaluation model has higher resolution; the calculation is simple and easy to handle by computer, etc. have a high reference and application value.

2 Establishment Evaluation Index System of University Scientific Research Capability

2.1 Basic Principles to Construct the Evaluation Index System [7]

1) Principle of systematization.
According to the ideology of system theories, university scientific research capability evaluation index system should reveal the complete picture of the university scientific

Y. Wu (Ed.): International Conference on WTCS 2009, AISC 116, pp. 481–487.

research capability more comprehensively and systematically as possible as it can do to prevent overgeneralization. At the same time, we should also obtain the key points and select correct index which can reflect the nature of the university scientific research capability to prevent deviations. On the basis of this principle and the author's understanding about the connotation of the university scientific research capability, this paper will demonstrate university scientific research capability by the four criteria of scientific research teams, research equipments and environment, research outputs, research and management capability, and then explain the four criteria respectively by the index assemblages which may rely on each other, emphasize different points and reflect the capability systematically and completely as the assessment of evaluating the university scientific research capability.

2) Principle of science
The main purpose to evaluate the university scientific research capability is to get a comprehensive evaluating result, rather than assessing all aspects. When selecting indexes, we should highlight as far as possible the comprehensive function of indexes. We should use more analyzing and evaluating indexes, less descriptive indexes to distinguish between the general technological, economic statistics.

3) Principle of synthesizing absolute indexes & relative indexes "hard" indexes & "soft" indexes
Absolute indexes reflect total volume and size, relative indexes are used to describe speed of response and rate; hard" indexes can be quantified and obtained from actual data's, so they can be handled easily; "soft" indexes can reflect more aspects related to the university scientific research capability better and have more powerful ability to provide a measure method in strategic point of view. These indexes complement each other to more accurately reflect the actual situation.

4) Principle of comparability
This principle asks us to choose evaluation indexes which have clear meaning and one voice. Moreover these indexes should consistent with the requisition of domestic identified indexes in order to ensure the rationality, impartiality and objectivity of the evaluation.

2.2   Construction of Evaluation Index System

According to the above principles and the author's understanding about the connotation of the university scientific research capability, this paper establishes the following evaluation index system of the university scientific research capability (shown in fig.1).

3   Establishment of the Model Based on Discrete Hopfield Neural Network

3.1   Thoughts of Design

Design balanced spots of discrete Hopfield neural network with several evaluated factors which shall be corresponding to typical classified levels. Among which the

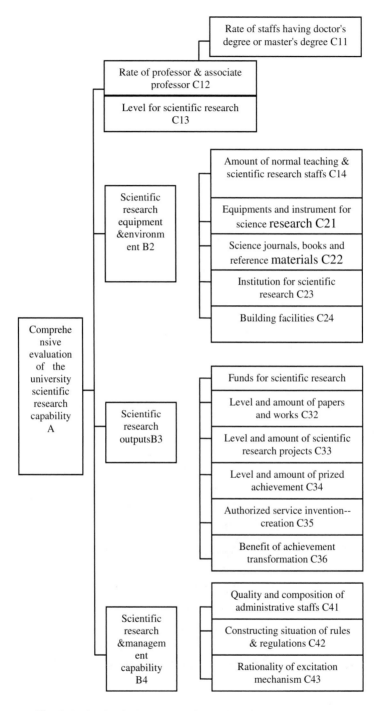

Fig. 1. Evaluation index system of the university research capability

university scientific research capability can be divided into five grades: very good ĉ), comparatively good (Ċ), average (ċ), comparatively bad(Č),very bad(č).

The learning process of Hopfield neural network just is the process those typical classified evaluated indexes approaching to the balance spots of discrete Hopfield neural network ,and the balance spots will be reserved by Hopfield neural network as various typical classified evaluated indexes when the learning process accomplishes .When inputting university evaluating indexes which are waiting to be classified , the Hopfield neural network will approach to some storage balance spot by utilizing the ability associating with memory in imagination. If this state stands still, the balance spot which is corresponding to will be the classified level.

3.2   Designing Steps

a) To calculate the average of total evaluation factors which come from colleges having the same level and regard them as the ideal evaluation factors for every level.

b) Coding those ideal evaluation factors getting from step 1 ,and treat them as the balanced spot of Hopfield neural network .The rules are :when the code is greater than or equal to the factor at some level ,the state of corresponding neuron is set up to be "1" ,or it will be "-1".

c) Coding every evaluation factor of college waiting to be classified by the same way as step 2.

d) To establish Hopfield neural network;

e) Simulation testing result analysis .

Shown as the following figure:

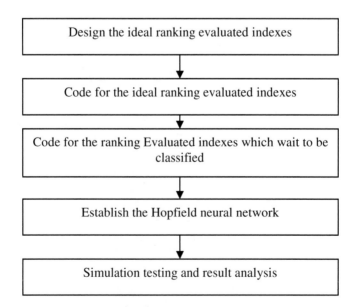

Fig. 2. Designing steps of evaluation model based on discrete Hopfield Neural network

3.3   Improved Discrete Hopfield Neural Network

Through the pre-treatment for sample of memory to change the weights of the network we can improve network performance [8-9]. At first we make Schmidt orthogonalization to samples of memory. The Schmidt orthogonalization formula is as follows:

Presumes the memorial sample respectively as ∂_1 , ∂_2 , . . . ∂_r

$$\beta_1 = \alpha_1$$

$$\beta_2 = \alpha_2 - \frac{[\beta_1, \alpha_2]}{[\beta_1, \beta_1]} \beta_1 ;$$

. . . .

$$\beta_r = \alpha_r - \frac{[\beta_1, \alpha_r]}{[\beta_1, \beta_1]} \beta_1 - \frac{[\beta_2, \alpha_r]}{[\beta_2, \beta_2]} \beta_2 - \ldots - \frac{[\beta_{r-1}, \alpha_r]}{[\beta_{r-1}, \beta_{r-1}]} \beta_{r-1},$$

After orthogonalization and according the Hebb rule to design weight, if we put into a certain model and through several iteration, net-work will arrive to steady which is the result of net recognition.

Evaluation model function provided by MATLAB neural network toolbox [10]. The assessment process can be realized by using MATLAB language to program. The process contains three parts: modeling, training and simulation.

4   Empirical and Result Analysis

There are so many factors to influence university scientific research capability that we can not use them all. Then we just pick out 11 correspondingly more important elements among those as the assessment index ,they are team of scientific research (x1)、 scientific research basement(x2)、 scientific technology and knowledge with corresponding support(information materials books) (x3) 、 funds of scientific research (x4)、 management of scientific research (x5)、 capability of information receipt and manufacturing(x6)、 capability of knowledge accumulation and technical storage(x7) 、 capability of scientific research and technology innovation(x8) 、 capability of knowledge releasing (x9)、 capability of adjustment by itself(x10)as well as ability lf of scientific decision (x11) and so on.

Using the data's of above 11 assessment indexes from 20 universities to build assessment model of university scientific research capability, we valuate several universities' scientific research capability and judge which level they belong to.

As Fig.3 showing, it is the result through empirical analysis by Matlab process.

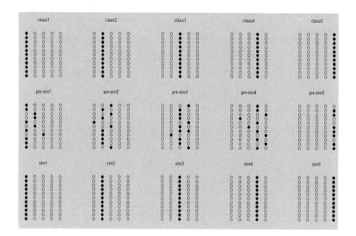

Fig. 3. Results of simulation testing

It implies from Fig.3 that it can correctly conclude a classified assessment on universities' scientific research capability through improved Hopfield neural network simulation.

Discussion on classification effect:

When there is a an quite significant coexistence of strengths and weakness which means some factors with higher marks and some lower at an university, the Hopfield neural network won't get the accurately classification. It's in consistent with the result from Expert marking method. As Fig. 4 showing

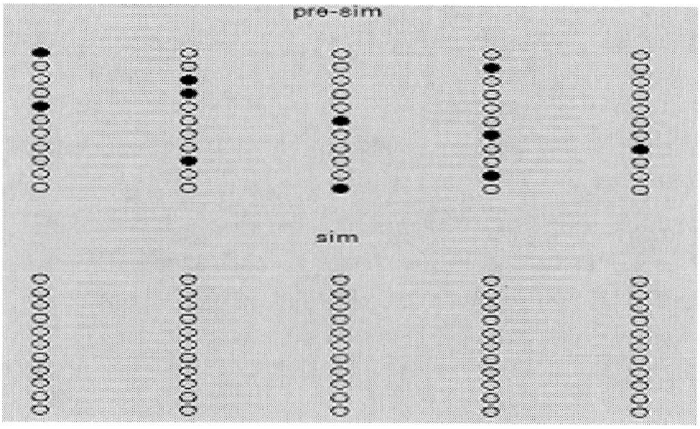

Fig. 4. Results of un-classification

5   Conclusion

This paper introduces an evaluation model for the university scientific research capability based on improved discrete Hopfield neural network, which can provides scientific reference for colleges and universities to identify the status of their own research capability enhance scientific research capability and improve their ranks. The model makes concepts of evaluation for scientific research indexes into quantitative data and use them as a Hopfield neural network input, use comprehensive evaluation results as output. Using Matlab to make an empirical analysis we found that if use this model to evaluate university scientific research capability, it can not only overcome the disorder causing by subjective factors of evaluated subject appearing the evaluation process but also can receive a satisfactory evaluating results, with broad applicability.

References

1. Zhang, J.: The project on the Evaluated Criterion system for the Teachers' ability of the scientific research in the University. Heilongjiang Researches on Higher Education (5), 101–103 (2006) (in Chinese)
2. Zhiqiang, M., Qiang, M., Yongyue, Z.: Discussion on establishment of achievements appraisal indexes system about styles of "teaching &researching" teachers in universities. China Economist (03), 98 (2008) (in Chinese)
3. Zhu, W.: The establishment of systematized evaluation indexes to university scientific research capacity. Journal of Anhui University of Technology and Science (9) (2003) (in Chinese)
4. Li, Y., Wu, H.: Study of the developmental evaluation of teacher index system of higher education institutions. Science –Technology and Management (05), 144–146 (2006) (in Chinese)
5. Wang, J., Hu, Z., Xiao, J.: Research into the Assessment Index system of University Teachers' Level in Scientific Research. Journal of Hunan institute of Engineering (Social Science Edition) (04), 32–34 (2005) (in Chinese)
6. Xu, X., Han, M., Wang, W.: Probe into the Fuzzy comprehensive Appraisal of Teachers" performance Research-oriented Universities. Agricultural Education China (04), 32–34 (2005) (in Chinese)
7. Xu, M., Dai, C., Hu, B., et al.: Research into the University scientific research capacity based on indistinctive mathematic theory. Science and Technology Management Research (08), 185–187 (2006) (in Chinese)
8. Xu, H., Yang, Z., Jing, H.: The research of Hopfield networks associative memory exterior products designed weight. China Computer & Network (3), 83–87 (2006) (in Chinese)
9. Ma, X., Tian, B.: Analysis on Discrete and continuous hopfield neural networks applications. Computer Simulation 20(8), 64–66 (2003) (in Chinese)
10. Yang, S.: Identify of model and count of intelligence -—Matlab achieve technique. Electronic Industry Press, Beijing (in Chinese)

The Design of Scientific Research Collaboration System Based on Web

Huizhong Xie, Lingling Si, and Dejun Qiao

HanDan College
380282773@qq.com

Abstract. Web-based scientific research collaboration system is in accordance with requirement of university scientific research; through analyzing the richness of communication media, it has selected scientific research collaboration system to be as the communication media, has analyzes and designed the system function structure, then to analyze system implementation and develop system through three-layer applicative structure, system security strategy, identity authentication mechanism, Web security and backup strategy. System program has reasonable structure, complete function, friendly interface, and it is easy to control and use; system has the strong expansibility, portability, security and allsideness of data.

Keywords: Web, scientific research, collaboration system.

1   Introduction

It is the important thoughts of modern university that combining teaching with scientific research and making them both important[1]. Informatization of scientific research is one core content of university digital campus construction. The target of university scientific research is that using campus Web and internet to promote sharing of resources and equipments, speeding up transmission of scientific research information, advancing international academic communication, developing research on Web in common; and using Web to advance transfer from the latest scientific research achievement transfer into teaching area; as well as industrialization and marketization of scientific research achievement; then it will greatly increase the scientific research's innovation and impact [2].

Taking scientific research by Web collaboration can fully use the existing resources advantage, organize high-level and experienced researchers with cross-disciplinary and cross-region to research project in common. Combining with existing university situation, aiming at the scientific research course, researching on richness of team communication media, it has designed system functional module and architecture, and has developed Web-based scientific research collaboration system, so to strengthen the ability of collaboration for scientific research, promote informatization of scientific research.

Y. Wu (Ed.): International Conference on WTCS 2009, AISC 116, pp. 489–496.
springerlink.com   © Springer-Verlag Berlin Heidelberg 2012

2   Necessity of Scientific Research Collaboration

Scientific research collaboration is cooperation and collaboration of scientific research; in order to innovate knowledge better and faster, produce more economic and social value.

2.1   Requirement for Self-Development of Scientific Research

- At present, scientific research subjects are getting more complex, research courses are complex, more relevance, multi-level, holistic emergence and non-linear. Scientific research subjects refer to broad area and have strong professional and requirement of integrating special technology with comprehensive knowledge. So if it takes scientific research to fulfill the facing task and solve existing problem, it will need to collect power of each area, use complex, comprehensive and integrative ideas and systems engineering, as well as the way of collaboration.
- For researchers, existing social information is very large, scientific research specialization and technical difficulty is very large, individual has limit ability to deal with knowledge and is impossible to master all the knowledge, it should combine with all specialties and researchers from all areas to organize the scientific research collaboration team to take scientific research.
- In order to produce and innovate knowledge, individual need organization supply well surroundings, platform and support. Innovation ability of organization is higher than individual, it is important that organization supplies well surroundings to individual, therefore, it is necessary for scientific research to show cooperative group mode.

2.2   Analysis on Market Economy Demand of Scientific Research Collaboration

- Low transformation rate of university scientific research achievements, which usually less than 10%; the main reason is that researcher always starts from self-science or basic technology, and neglects market requirement and economic value when taking scientific research and technical innovation. The main reason why researchers have not the stronger market economy consciousness is that they are insensitive to market economy information. So researchers should deep into social problems and enterprise production to capture the information of social requirement and market opportunity. Researchers should have the market consciousness to create more economic value for society.
- University scientific research needs capital support and technical experiment production platform. Science found always focuses on vertical basic research, and the lateral support for large market requirement and good social benefit is insufficiency. The fund of scientific research is limit, so researchers should get more capital support from society and market. A lot of capital support of

overseas university comes from enterprises. Researchers should cooperate broadly to pursuit win-win collaboration.

- For the former study on scientific research collaboration, one from the point of knowledge production and innovation, that is the construction and management of research team; one from the point of knowledge application, which focuses on the production of production, learn and research, and is restricted to example analysis. However, analyzing scientific research collaboration from knowledge activity system, it can be found that scientific research collaboration is not only itself, but also includes knowledge production like scientific research and technical innovation, as well as application step of market economy orientation of scientific research achievements, it is a systematic course, so it should start from point of system to make the whole harmonious and cycle beneficial.

- With regard to scientific research collaboration, firstly, government and enterprise provide capital, technology and equipments; researchers take scientific research collaboration combing with their advantages, namely, this is knowledge production course. It uses the former scientific achievement to innovate; then through enterprise and government, new achievement shows its market value under market mechanism, serves society, then it produces social and economic benefit; through market feedback mechanism and government guiding mechanism, university transmits knowledge, and it enters into scientific research collaboration system again[3], which as shown in figure 1.

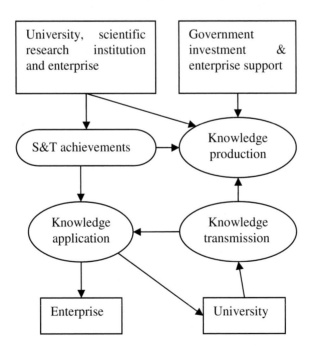

Fig. 1. Scientific research collaboration system

3   Richness and Choice of Communication Media for Scientific Research Team

3.1   Analysis on Richness of Communication Media

According to the media richness theory of Daft & Lengel[4-5], media richness is the communication capability supplied by media in a certain period, it can be judged with instant feedback capability, transferring multi-clue, drawing individual attention and diversity of language expression. Daft & Lengel shows that richness and poverty of media are technology's natural objective attribute; if a media could ensure instant feedback, use video/audio channel and communication with natural language, it will be richness.

According to these four rules, different media has different richness; the order of media richness from high to low is as the followings: face to face→ video conferencing→ Web platform or system→ telephone→ E-mail or written files (letter and memo) → report form and announcement. Thereinto, face to face is the most richest, which has the highest speed of feedback, the represented information can be corrected instantly, and can pass multi-clue like language and nonverbal (gesticulation, tone, intonation and expression); and does not like electronic communication, it could not be confined and affected by technical factors (bandwidth and reliability of hardware). Video conferencing is much richer, which has characteristics of synchronization, multi-clue, natural video /audio channel and instant feedback; Web platform or system is richer, which has characteristics of synchronization, multi-clue, natural audio channel, written information and instant feedback; telephone is rich, which has characteristics of synchronization, less clue, natural audio channel and instant feedback; E-mail or written file is low richness, which has characteristics of asynchronization, single dimensional clue, written information and time-delayed feedback; report form and announcement are poor, which has characteristics of asynchronization, single transmission and written information. As shown in table 1.

Table 1. Commutation Media & Richness

Media	Richness	Richness Characteristic
Face to face	Most richest	synchronization, multi-clue, natural language, instant feedback
video conferencing	Much richer	synchronization, multi-clue, natural language / audio channel and instant feedback
Web platform or system	Richer	synchronization, multi-clue, audio channel, written information, instant feedback
Telephone	Rich	synchronization, less clue, natural audio channel and instant feedback
E-mail or written file	Low richness	asynchronization, single dimensional clue, written information and time-delayed feedback
Report form and announcement	Poor	asynchronization, single transmission and written

3.2   Choice of Communication Media

Scientific research team is a temporary with "centering on project", members take interaction crossing border (time, place, organization and culture) in order to fulfill the common target [6]. Member is independent on time and space; during inner team, they usually communicate with face to face and combine with communication based on electronic technology; using their knowledge and specialty, members cooperate with each other to solve problems of project.

Compare with traditional team, scientific research team has obvious distributed characteristic, which mainly shows as the followings:

- Members come from many areas, which usually distribution with crossing-area, even crossing time zone.
- Groups of member are vague, members may be come from different group, group broader of team are undefined, and it is a temporary team with "centering on project".
- Communication among members need the support of technical media, it mainly carry out by electronic mode.

When team taking scientific research, there are more than 50% scientific research include complex communication course. However, due to the inherent distribution of scientific research team, face to face has been restricted by time and space, and it to become scarce resource, team even could not supply the face to face communication to members, then communicating by electric technology is a key to operation and success for team, but video conferencing is confined and affected by technical factors (bandwidth and reliability of hardware), therefore, Web platform or system is the preferred communication media for scientific research collaboration.

4   System Design

4.1   Target of System Design

Scientific research collaboration system is an application system based on Web, which managing and operating scientific subject. For the subject, the important content for management is plan, task, progress, file and resource allocation; according to the whole course of scientific research subject, system should support the management and collaboration of online subject, as well as digital composition file management; in the scope and period of scientific research subject, under fixed constraint conditions, in order to achieve optimal target, according to the scale of subject, the principal can plan, organize, command, control and cooperate the whole course, and also can organize the scientific team, advance communication and collaboration among members under the flexible and high security surrounding, to fulfill files communication, version management and scientific research flow management. That is to say, through scientific research collaboration system, the subject principal can achieve plan design, progress tracking, quality supervision, successful file management and asset equipment management, to fulfill the purpose of instant feedback of subject information, instant monitoring of research course, recording the whole task flow, at the same time it has accumulating store of files and achievements to supply experiment and data; so to fulfill the whole course management of scientific research.

4.2   System Function Design

Combining with actual situation and according to customer's different requirement, system includes three functional modules like member module of subject team, subject team leader module and system manager module.

1) Member module of subject team: It has the functions of managing individual information, understanding and carrying out the task of subject plan, as well as communicating information, etc. Members of subject team manage information of them in order to provide convenience for statistical analysis on different customers; understand the task of subject plan and carry out the appointed task, write the implementation of task at any moment; keep in touch with members timely.

2) Subject team leader module: It has the functions of user management, plan design, task appointment, progress tracking and task's checking and confirmation, etc. Subject team leaders can manage member of this subject team; and also has functions of designing the implementation plan of this subject, appointing and allocating task to members of subject team, tacking the progress, checking and confirming the implementation of members; as well as has the functions of information distribution, real time information, online and inserting discussion.

3) System manager module: It has the functions of system initiation, system control, subject team management and subject information's management, statistical analysis and inquiry, etc. Before the Web scientific research collaboration, manager should take system initiation, set system style and visibility of module function, as well as manage subject team.

5   Implementation of Scientific Research Collaboration System

5.1   Three-Layer Applicative Structure

Combing with actual situation, the scientific research collaboration system has been developed into three-layer applicative structure based on B/S (Browser/Server, namely, user interface, intermediate layer and database[7-8] . Use ASP to develop the user interface, users submit request by Web browser or special user program; use Window 2003+IIS as Web application server to deal with business logic; database uses SQL Server 200 for stored data. Three-layer applicative structure lays the user interface and user logic on different dealing layer, user accesses database only through intermediate application layer, these provide convenience to control the server access and also ensure the security of database.

5.2   System Security Strategy

Scientific research collaboration system refers to the important information like subject and communication among members, so the database should has high security. Program uses Script Encoder to encrypt the ASP page layout, which can effectively prevent ASP source code from leaking; program uses Session object of ASP to take registration verification, which can effectively prevent unregistered user from skipping user interface to directly enter into intermediate layer, and password cheat[9]. It can control the hierarchical access of subject team member, subject team leader and system manager through combining the Windows 2003+IIS application server setting and SQL Server 2000 database server setting.

5.3   Identity Authentication Mechanism

Scientific research collaboration system lay the user interface and user logic on different dealing layer by three-layer applicative structure. User accesses database only through intermediate application layer, and it uses identity authentication technology to authenticate the identity of subject team member, team leader and system manager; submit individual information and relevant receipt through user interface server, after has been authenticated by server, the identity authentication has been completed. Web identity authentication mechanism not only can ensure team members, team leader and system manager conveniently use the system input and obtain information; but also can ensure disabled users shall not access into the system.

5.4   Web Security and Backup Strategy

Through setting server and firewall before server, scientific research collaboration system logically insulates inhouse network from the outside. All the connections should be checked, and only has been authorized can access the application and database server. Database will be heavily threatened when computer system has fault due to many factors like soft hardware system, surrounding; then it uses SQL Server 2000 backup strategy to take daily physical and logic backup for database[10].

6   Conclusion

It analyzes the necessity of scientific research collaboration from its development and demand of market economy; it selects scientific research collaboration system as the communication media of scientific research collaboration by analyzing the richness of communication media and combining with situation, and also shows system's design and implementation.

References

1. Min, W.: Reflections on the Development of World Class Universities. Peking University Education Review (3) (2003)
2. Han, X.: Design of On-line Scientific Research Supporting Platform in Digital Campus. Distance Education in China (3) (2006)

3. Li, H.: Study of Mode, Issues and Policy for University Scientific Research Collaboration. Science & Technology and Economy (1) (2007)
4. Daft, R.L., Lenge, R.H.: Information richness: a new approach tomanagerial behavior and orgnization design. Research in Organizational Behavior (6) (1984)
5. Daft, R.L., Lenge, R.H.: Organizational information requirements, media richness and structural design. Management Science (32) (1986)
6. Ye, F.: The Research for Dynamic Harmonizing and Communicating Strategy Based on Distribution Scientific Research Team. Science Research Management (7) (2002)
7. Huang, T.: Management Information System. China Higher Education Press, Peking (2000)
8. Sa, S.: An Introduction to Database System. China Higher Education Press, Peking (2002)
9. Cheng Liang, A.S.P.: Network Programming Instances. Post & Telecom Press, Peking (2001)
10. Zhang, X.: SQL Server 2000 Management and Application System Development. Post & Telecom Press, Peking (2002)

The Modeling and Transmission Simulation of the Hardened and Grinded Worm Gear

Liu Hu Ran

Zhejiang University Science Technology, China
rwgn87937@sina.com

Abstract. In order to enhance the worm drive the stability and saves the metallic material, proposed one new conjugate forming method and the new worm bearing adjuster worm gear processing method. Is opposite in the traditional formed method, the new formed method is line contact meshing. Infers through the Archimedes worm bearing adjuster makes this worm bearing adjuster's grinding wheel's outline coordinate formula. Again using the VB procedure and the variable value scope, in the definite grinding wheel outline each spot coordinate, and the obtained each spot on the coordinate paper will express that thus obtains grinding wheel's outline shape. Difference tradition worm gear processing method, this article imagination grinding wheel in worm bearing adjuster interior, thus infers this grinding wheel's outline coordinate formula. This grinding wheel namely to process worm gear's grinding wheel. Obtains this wheel shape method and obtains the above grinding wheel's method to be the same.

In brief, understands and knows the gear meshing principle, the VB programming, aspect and so on worm bearing adjuster worm gear tooth profile line independences and the overall design. The design key point is may rub the hard tooth face worm bearing adjuster worm gear to gnaw the vanishing line design; The analysis key point is to may rub the hard tooth face worm bearing adjuster worm gear's meshing performance study.

Keywords: Conjugation, worm, gear.

1   Introduction

Research of the new kind of meshing is an important research direction of mechanics and manufacture. At present, we have only few methods. Such as the Oliver first method, the Oliver second method and the F. Litven's method. This paper presents a new kind of conjugate method. The new method is of great significant both in theory and in practice. In order to save nonferrous metal and to increase the strength of the worm gear, scholars at home and abroad suggested producing the worm gear with steel and iron. In other words, both the surface of worm and worm gear are hardened and grinded. So that it is necessary to probe the new way of conjugation.

At present we have only few methods. Such as the Oliver first method, the Oliver second method and the F. Litven's method. Up to now, there has been no method of conjugation presented by Chinese. In the history of mechanics, during the resent more than 20 year, no one found a new conjugate method. This paper presents a new kind of conjugate method, is of great significant both in theory and in practice.

Y. Wu (Ed.): International Conference on WTCS 2009, AISC 116, pp. 497–504.
springerlink.com © Springer-Verlag Berlin Heidelberg 2012

In order to save nonferrous metal and to increase the strength of the worm gear, scholars at home and abroad suggested producing the worm gear with steel and iron. In other words, both the surface of worm and worm gear are hardened and grinded. So that it is necessary to probe the new way of conjugation. For the purpose to realize this kind of worm transmission with worm gear hardened and grinded, we must find a new kind of conjugate method.

2 The Oretical Basis for the New Kind of Conjugation

Suppose that we have an auxiliary surface ΣF, which has line contact with an existing surface $\Sigma 1$, moves together with surface $\Sigma 1$, and envelop it. Let the surface ΣF serve as the generating surface. When the surface $\Sigma 1$ moves according to the regulation, ΣF will form a conjugate surface $\Sigma 2$. The contact line between $\Sigma 1$ and ΣF is $\alpha\alpha$, the contact line between $\Sigma 2$ and ΣF is $\beta\beta$. If $\beta\beta$ and $\alpha\alpha$ intersect at point Pi, when ΣF is removed, $\Sigma 1$ and $\Sigma 2$ will realize the point conjugate contact at the point Pi, As shown in Fig.1 and Fig.2.

The background of the new conjugation: employ a disk-grinding wheel that internal contacts with the worm helical surface, and imaginary remained in the body of the worm. Let worm revolves, the grinding wheel moves transparently. On the grinding wheel, the contact line between the worm and the grinding wheel is always the same line. In the same movement, grinding wheel will generate the surface of the worm gear. Therefore the worm will conjugate with the worm gear with point contact.

As is known for every one, when a helical surface revolves, the disk grinder will move transparently. On the grinding wheel, the contact line between the worm and the grinding wheel is always the same line. We imagine that there is a disk located in the internal of the body of the worm, contact with the worm, and attached with the worm. When worm rotates, the grinder moves transparently. On the grinder, the contact line between the worm and the grinder is always the same line. Since that when worm rotates, the worm gear should rotate correspondingly as well. In the same relative movement grinder will generate the worm gear. Between the worm and worm gear there must be a contact line as well. Both of the contact lines are on the surface of the grinder. The former contact line is the common contact line between the grinder and the worm; the later contact line is the common contact line between the grinder and the worm gear. If the two contact lines have intersection, then the intersect point must be the common point between the worm and the worm gear. Remove the imaginary grinder; the worm and worm gear will realize the point contact conjugation. It should be emphasized is that: the grinder is imaginary and located in the inertial of the solid body of the worm. What should be emphasized again is the mini difference from the Oliver second method. In the Oliver second method, the auxiliary surface is fixed to the first surface.

3 The Proof of the New Conjugate Method

Theorem: If surface $\Sigma 1$ and surface ΣF are conjugate surfaces with line contact conjugation, and the contact line on the surface ΣF is a constant line, if the surface ΣF, in the same time meshing with surface $\Sigma 1$, generate a new surface $\Sigma 2$, then, surface $\Sigma 1$ and surface $\Sigma 2$ will point contact conjugate with each other.

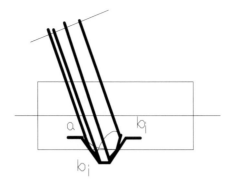

Fig. 1. The new kind of conjugate method

4   The Grinder Was Imaginary in the Internal of the Solidity of the Worm

In the solidity of the worm, the section profile of the grinder contacting with the worm is shown in the Fig.2. The contact condition of the grinder and the worm:

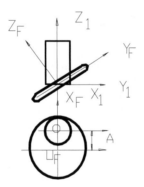

Fig. 2. The grinder contact with the worm

$$z_1 n_{x1} + Actg\gamma n_{y1} + (A - x_1 + pctg\gamma)n_{z1} = 0 \qquad (4.1)$$

The equation of the worm and the normal vector:

$$x_1 = u\cos\alpha\cos\theta$$
$$y_1 = u\cos\alpha\sin\theta$$
$$z_1 = u\sin\alpha + p\theta$$

$$(4.2)$$

$$n_{x1} = p\cos\alpha\sin\theta - u\cos\alpha\sin\alpha\cos\theta$$
$$n_{y1} = -p\cos\alpha\cos\theta - u\cos\alpha\sin\alpha\sin\theta$$
$$n_{z1} = u\cos^2\alpha \tag{4.3}$$

Solve above equations, we can find the equation of contact line. Transform the contact line into the system fixed to grinder:

$$x_F = x_1 + A$$
$$y_F = y_1\cos\gamma + z_1\sin\gamma$$
$$z_F = -y_1\sin\gamma + z_1\cos\gamma \tag{4.4}$$

$$R = \sqrt{x^2 + y^2}$$

$$z_F = z_F \tag{4.5}$$

5   Take the Archimedes Worm as Example

In the practical practice, the grinding wheel can by no means into the internal of the solidity of another body. We can only imagine that. Suppose that the grinder is located in the internal of solidity of the worm body, we can calculate the profile of the grinder according to the contact condition. From the meshing principle of the worm and worm gear, we know that this grinder is in fact the grinder to machine the worm gear. Let us imagine the situation when grinder inter the internal of the solidity of the worm:

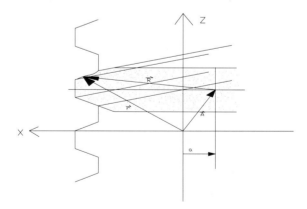

Fig. 3. The grinder is located in the internal of solidity of the worm body

Establish a system as shown in the Fig.3., we can know that:

$$\vec{r} = x\vec{i} + y\vec{j} + z\vec{k}$$

$$\vec{A} = -a\vec{i} + \left(\frac{dtg\alpha}{2} + \frac{\pi mn}{4}\right)\vec{k}$$

Since: $\vec{R} + \vec{A} = \vec{r}$

So that:

$$\vec{R} = \vec{r} + \vec{A}$$

$$= \left(x\vec{i} + y\vec{j} + z\vec{k}\right) + a\vec{i} - \left(\frac{dtg\alpha}{2} + \frac{\pi mn}{4}\right)\vec{k}$$

$$= \left(x + a\right)\vec{i} + y\vec{j} + \left(z - \frac{dtg\alpha}{2} - \frac{\pi mn}{4}\right)\vec{k}$$

In the equation: R—the radius of the grinder; r—the radius of the worm; x, y, z—the coordinates of any point on the contact line.

According to the contact condition we can get:

$$\left(x + a\right)n_x + yn_y + \left(z - \frac{dtg\alpha}{2} - \frac{\pi mn}{4}\right)n_z = 0$$

The above equation is deduced under the circumference that the grinder is not inclining. In practical machining the grinder should incline an angle, suppose that the incline angle is Σ. By deduction, we may have the following equation:

$$\left(-x - a\right)n_y + \left(-x - a\right)n_z tg\,\Sigma + \left[y + \left(z - \frac{dtg\alpha}{2} - \frac{\pi mn}{4}\right)\right]n_x = 0$$

The above can be abbreviated as: $f = 0$ $\qquad\qquad$ (5.1)

By this equation we can find all the points on the grinder that satisfy the condition. Connecting all of these points, we can get the sectional profile of the grinder. So as determine the shape of the grinder. The basic radius of the grinder (the radius with which the grinder contact the worm at the pitch cylinder) can be determined by the equivalent curvature of the pitch radius of the worm.

$$\rho = \frac{r}{\cos^2 \gamma}$$

6   The Data of the Imaginary the Grinding Wheel Which Is Collated in the Solidity of Worm

Table.1   the data of the profile of the grinder

```
Private Sub Picture1_Click()
Open "C:\Myfile3.dat" For Output As 3
sig = 5.71
a = -250
For u = 15.2 To 24.8 Step 0.5        /
For thi = -20 To 20 Step 0.5       /
x = u * Cos(20 / 180 * 3.14) * Cos(thi / 180 * 3.14)
y = u * Cos(20 / 180 * 3.14) * Sin(thi / 180 * 3.14)
z = u * Sin(20 / 180 * 3.14) + 2 * thi / 180 * 3.14
nx = 2 * Cos(20 / 180 * 3.14) * Sin(thi / 180 * 3.14) - u * Cos(thi / 180 * 3.14) *
Cos(20 / 180 * 3.14) * Sin(20 / 180 * 3.14)
ny = -2 * Cos(20 / 180 * 3.14) * Cos(thi / 180 * 3.14) - u * Sin(thi / 180 * 3.14) *
Cos(20 / 180 * 3.14) * Sin(20 / 180 * 3.14)
nz = u * Cos(20 / 180 * 3.14) * Cos(20 / 180 * 3.14)
f = (a - x) * (ny + nz * Tan(sig / 180 * 3.14)) / nx + y + (z - 10.419) *
Tan(sig/180*3.14)                       /*/
X0 = a - x
Y0 = -y * Cos(sig / 180 * 3.14) - z * Sin(sig / 180 * 3.14)
Z0 = -y * Sin(sig / 180 * 3.14) + z * Cos(sig / 180 * 3.14)
R = Sqr(X0 * X0 + Y0 * Y0)
Z0 = Z0
Print #3, u, thi, R, Z0, f
Print u, thi, R, Z0, f
Next thi
Next u
Close #3
End Sub
```

The explanation to the above data: the u and θ is the surface parameters of the worm. According to equation (2.6) by change u and θ the whole surface of the worm will be covered. From equation (2.6) we can find its solution. Then take it into equation (2.5), we can find the sectional profile of the grinding wheel.

From the data listed above, we know that the equation has solutions. According to the data, we can draw the sectional profile of the grinding wheel. Before the grinding wheel configured, we should test that if the grinder can contain the worm or not. The normal curvature of the grinder should be larger than the normal curvature of the worm at the contact point. Suppose that the radius of the worm of one point on the worm is R, according to the Muriel theorem:

$$k = \frac{k_t}{\sin \eta} = \frac{1}{R \sin \eta} \le k_w$$

7 The Modeling and Transmission Simulation of the Hardened and Grinded Worm Gear

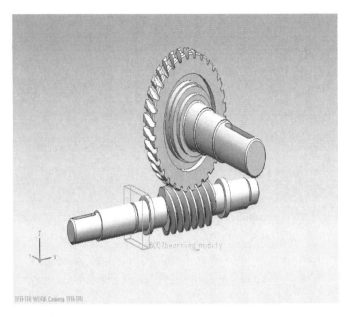

Fig. 4. Transmission Simulation of the Hardened and Grinded Worm Gear

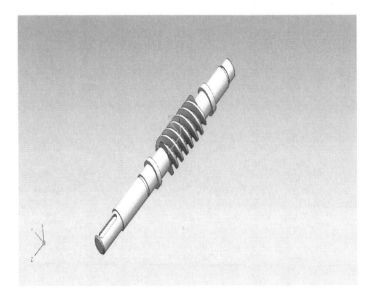

Fig. 5. The Modeling of the Hardened and Grinded Worm

Fig. 6. The Modeling of the Hardened and Grinded Worm gear

8   Conclusions

This project is supposed by the natural scientific foundation of China, No.2006-50675235. and the natural scientific foundation of Zhejiang province, China, No. Y106047 and Y1080093.3 The difference between new worm transmission and the old are listed bellow:

The material: ferrous metal /noun- ferrous metal

The tooth surface: hardened /noun-hardened

The machining method: grinding/ hobing

The machining tool: grinding wheel/ hob

The contact condition: point contact / line contact

The profile of the tooth surface: the envelope of a rotary surface/the envelope of a helical surface.

References

1. Neumann, F.: Elsveir Press, London (1990)
2. Komori, T., Arga, Y., Nagata, S.: Trans. ASME 12(3), 430–436 (1990)
3. Komori, T.: Proceeding of International Conference on gearing, China CMCC, pp. 230–236 (1988)
4. Feng, X., Linda, A.W.L.: International Journal of Advanced Manufacture Technology 24(11/12), 789–794 (2004)
5. Li, F.: Journal of Xiamen University 36(1), 12–16 (2002)
6. Wu, D.: Deferential Geometry and Theory of gear mesh. Science Press China (in Chinese)
7. Litvin, L.: Theory of gear mesh and its application. Mechanical Industrial Press, China (1984)
8. Dooner, D.B.: Journal of Mechanical Design 124 (December 2002)
9. Di, P.: Alternative, Mechanism and Machine Theory (2005)

Research on the Science and Technology Innovation Ability of Undergraduates Based on the Fuzzy Evaluation Approach

Chen Lidong, Ma Shuying[*], Shi Lei, Feng Lizhen, Zheng Lixin, Li Guofang,
and Zhang Liang

College of Mechanical and Electronic Engineering
Hebei Normal University of Science and Technology
Qinhuangdao, China
lfdj175@sogou.com

Abstract. Based on establishing scientific and rational science and technology innovation ability evaluating index system of the university student, a fuzzy comprehensive evaluation approach of the university student innovation ability synthetic evaluation has been established through establishing and weighing the evaluation index set, the index weight, the reviews set, the membership matrix and fuzzy comprehensive evaluation vectors. The method has overcome the shortcoming which the sole assessment method exists. And verifies through example, we provided one kind of newly and feasible method for synthetic evaluating innovation ability of university students.

Keywords: science and technology innovation ability, evaluation method, evaluation index, undergraduates.

1 Introduction

With the development of technology and increased competition, science and technology innovation plays increasingly prominent even decisive role on the competitiveness of a country. College students as the backbone of innovation and reinforcements of building innovative country carry the historical mission of national and social development. Personal training is a fundamental task of universities whether can train and bring up a number of high-level and high-quality personnel is a measure of an important indicator of the high-level education. At present, more and more universities in China pay much attention to strengthening the scientific and technological innovation ability, and they consciously develop some technological innovation activities for college students, such as organizing students participating in the research activities, scientific and technological competitions, or teachers' topic researches etc.. The activities can not only simulate the students' interest and curiosity to exploit their own creative spirit, and practice their ability to work independently, but also make the

---

[*] Corresponding author.

Y. Wu (Ed.): International Conference on WTCS 2009, AISC 116, pp. 505–512.

students familiar with the advanced concepts of modern engineering design, work flow, production process and solve the general problem, which will lay a certain foundation for the future work [1]. However, the current research on student innovation mainly focused on the institutionalization level of creating the students' innovative ability and the training methods and approaches; most university didn't have a unified and reasonable evaluation on these activities on the achievements and technological innovation ability. Developing innovation capability evaluation for the college students can provide a workable basis for objectively investigating the undergraduates' innovation ability, in addition, it can promote their innovation initiative and creativity, encourage students to active innovation. The evaluation also can help to optimize the ability structure of college students, broaden the students' creative ability, and well improve the education to achieve the objective of fostering innovative talents [2]. Therefore, it has important significance to scientific evaluate the students' innovation ability and guide their all-round.

2   Technological Innovation Ability Evaluation Index System of the University Students

Innovation ability refers to the ability of breaking the restraint of the old ideas and old patterns, solving new problems and getting new results and opening up new prospects. Innovation ability and innovation spirits are the fundamental conditions of creating innovation talent; they are the necessary targets for training innovative talents in colleges and universities. There are many component factors of scientific and technological innovation capability of undergraduates, and its structure is complex. Therefore, only when we establish an evaluation index system based on different angles and levels, it can accurately reflect the students' scientific innovation level [3]. In this paper, based on our investigation, we divide the evaluation index system into five first level indexes (innovation knowledge base, innovation learning ability, innovative thinking, innovative skills and innovations achievements), 18 evaluation indexes (2nd level index). The comprehensive evaluation index system of innovative ability is shown in Figure 1.

2.1   Innovation Knowledge Base

Innovation knowledge base is security conditions for innovation ability evaluation, including professional knowledge base, adjacent field knowledge, professional practice knowledge and skills, psychology knowledge base, knowledge base of creating science and thinking science of knowledge base and so on [4]. The index can be reflected by the level of the basic knowledge, the level of professional knowledge, the level of cross knowledge and the level of innovation knowledge. Among them, the level of the basic knowledge and the level of professional knowledge basic knowledge can be inspected by the traditional approach; the level of cross knowledge and the level of innovation knowledge are the important indicators to measure innovation knowledge base, we should respectively inspect the range of the corresponding knowledge and the generosity of theoretical knowledge.

Innovation Learning Ability

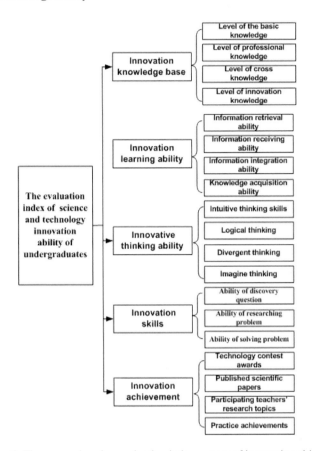

Fig. 1. The comprehensive evaluation index system of innovative ability

Innovative learning is that students should be independent thinking combined with practice on the basis of former knowledge, and they can integrate their knowledge and put forward some new ideas and new methods under the guidance of teachers, but not superstition book knowledge [5]. Innovative learning ability means that the students, after having mastered knowledge and developed intellectual, can obtain their integrated practical skills through education training and the practice session. So we let the information retrieval ability, information receiving ability, information integration ability and knowledge acquisition ability as the indexes of the innovation learning ability.

2.2   Innovative Thinking Ability

The science sense of inventor is crucial to innovational practice. It is necessity to who want to innovative to be creative. The innovative thinking is one core substance of what make up of creation. The innovative thinking ability is that people can propose some creative ideas or new methods during the process of thinking and solving problems, which is the thinking base of innovation and creativeness [6]. The indicators include intuitive thinking skills, logical thinking, divergent thinking and imagine thinking of four secondary indicators.

2.3   Innovation Skills

Innovation skills refers to the action abilities of the innovators' behavioral skills, including new information processing capacity, operation capacity, ability to master and apply innovative techniques, and so on.

2.4   Innovation Achievement

Technological innovation achievement mainly refers to the technology contest awards (including the "Challenge Cup" Business Plan Competition and Competition of Undergraduate's Science and Technology Works after school, etc.), participating teachers' research topics, and responsible or liable for the Students Technology Innovation Fund projects, and in the practice they obtain some achievements. In addition, the achievements can also be the extracurricular works completed by the students independently, patents and published scientific papers and so on. To assess this indicator is because technological innovation is the most important aspects of innovative practice.

3   Undergraduates' Evaluation Method of Scientific and Technological Innovation Ability

The undergraduates' innovation ability has its own ambiguity and uncertainty, and the evaluation indexes have qualitative and quantitative evaluation of points. The traditional evaluation method has a disadvantage of single evaluation results. In this paper, a fuzzy comprehensive evaluation approach is introduced, which can provide quantification from people's qualitative results, and it will make the evaluation effect have credibility [3].

3.1   Design Evaluation Index Set (the Universe U) [3]

The major factors of Evaluation on the students' creative ability embody in the innovation knowledge base U_1, the innovation learning ability U_2, the innovative thinking U_3, the innovative skills U_4 and the innovations achievements U_5, which are the first level factors, namely $U = (U_1, U_2, U_3, U_4, U_5)$. In which the influencing factors of the innovation knowledge base U_1 are selected as the level of the basic knowledge U_{11}, the level of professional knowledge U_{12}, the level of cross knowledge U_{13}, and the level of innovation knowledge U_{14}, namely $U_1 = (U_{11}, U_{12}, U_{13}, U_{14})$; the influencing factors of the innovative learning ability are selected as the information retrieval ability U_{21}, information receiving ability, U_{22} information integration ability U_{23} and knowledge acquisition ability U_{24}, namely $U_2 = (U_{21}, U_{22}, U_{23}, U_{24})$; the innovative thinking ability selects intuitive thinking skills U_{31}, logical thinking U_{32}, divergent thinking U33 and imagine thinking U_{34} as its factors, that is $U_3 = (U_{31}, U_{32}, U_{33}, U_{34})$; the influencing factors of the innovative skills are information processing capacity U_{41}, operation capacity U_{42}, ability to master and apply innovative techniques U_{43}, namely $U_4 = (U_{41}, U_{42}, U_{43})$; the factors of innovation achievement are selected as science and technology contest awards U_{51}, published papers U_{52}, participating research topics U_{53}, completing the technology works U_{54}, namely the $U_5 = (U_{51}, U_{52}, U_{53}, U_{54})$.

3.2   Design Index Weight of Science and Technology Innovation

Weight refers to the influence degree of a certain index factor to the students' technological innovation ability. The size of weight value is related to influence degree, and the weight value is between 0 and 1, the sum of the weight value of the evaluation factor is 1. Usually there are many methods to determine the weights; in this paper we adopt an improved method of the analyzing level, which is an integrated technical approach of combining most experts' experience and subjective judgments determination. The way can naturally satisfy the compliance requirements, eliminating the consistency test. According to the method we can get the weight coefficient vector of the first and the second level indexes, namely $X=(w_1, w_2, w_3, w_4)$; $X_1=(w_{11}, w_{12}, w_{13}, w_{14})$; $X_2=(w_{21}, w_{22}, w_{23}, w_{24})$; $X_3=(w_{31}, w_{32}, w_{33}, w_{34})$; $X4=(w_{41}, w_{42}, w_{43})$; $X_5=(w_{51}, w_{52}, w_{53}, w_{54})$.

3.3   Design Reviews Set E

Reviews set is to point out the standard of the levels of the evaluation factors, and gives numerical results of the reviews set, that is a standard satisfaction degree vector W. We set the reviews set be E = (strong, relatively strong, general, bad), that is, W = (100, 85, 70, 60).

3.4   The Membership Matrix and Fuzzy Comprehensive Evaluation Vector

After the former steps, using expert score method, we can determine the membership vector of each sub domain of the second indicators and form the membership matrix. Then we can calculate the fuzzy comprehensive evaluation vector of the sub domain Y_i,

$$Y_i = X_i \text{ o } A_i \ (i = 1, 2, 3, 4).$$

Where A_i is the membership vector of each sub domain.

3.5   Calculating the Integrated Assessment Calculation Score

After calculating Yi and normalizing Y, we can get matrix A. Then according Y = X o A, we can get Y'. After that, we can obtain the integrated assessment calculation score, $u = W'(Y')^T$.

$$U = \begin{pmatrix} 1 & \dfrac{1}{2} & \dfrac{1}{4} & \dfrac{1}{5} & \dfrac{1}{3} \\ 2 & 1 & \dfrac{1}{5} & \dfrac{1}{3} & \dfrac{1}{4} \\ 4 & 5 & 1 & 2 & 3 \\ 5 & 3 & \dfrac{1}{2} & 1 & \dfrac{1}{2} \\ 3 & 4 & \dfrac{1}{3} & 2 & 1 \end{pmatrix}$$

4 Evaluation Examples

With discussing the comprehensive evaluation method for many indexes, the paper constructs an index system to evaluate the science and technology innovation ability of the student from agricultural machinery specialty in Hebei Normal University of Science and Technology. According to the evaluation method, we select 6 relevant experts, and then the experts present each index score based on the student's comprehensive evaluation results, teacher's evaluation, expert evaluation, and innovation awards and so on. After that, based on the 9 scale method [3], innovation index weight coefficient is obtained, as shown in Table 1.

By the weight coefficient and the Innovation Index calculation methods, the integrated assessment calculation scores are:

$$u = W'(Y')^{T} = 78.564$$
$$u_1 = W'(Y1')^{T} = 83.352$$
$$u_2 = W'(Y2')^{T} = 81.134$$
$$u_3 = W'(Y3')^{T} = 78.355$$
$$u_4 = W'(Y4')^{T} = 79.866$$
$$u_5 = W'(Y5')^{T} = 78.425$$

From the evaluation results we can get a conclusion, that is, the student's creative ability is relatively strong; his innovation knowledge base is more solid; his learning ability is strong; his innovation thinking and innovation skills should be further improved. By the above method, the science and technology innovative ability of students can be valued, and on the basis of the results, we can make a deep analysis and adopt some appropriate countermeasures.

Table 1. The weight coefficient of innovation index

First level evaluation index	Weight coefficient	Second level evaluation index	Weight coefficient
U$_1$	0.0674	U$_{11}$	0.0770
		U$_{12}$	0.1542
		U$_{13}$	0.3128
		U$_{14}$	0.4560
U$_1$	0.0674	U$_{11}$	0.0770
		U$_{12}$	0.1542
		U$_{13}$	0.3128
		U$_{14}$	0.4560
U$_2$	0.1365	U$_{21}$	0.1024
		U$_{22}$	0.2130
		U$_{23}$	0.4056
		U$_{24}$	0.2790
U$_3$	0.2547	U$_{31}$	0.1048
		U$_{32}$	0.0526
		U$_{33}$	0.3164
		U$_{34}$	0.5262
U$_4$	0.3352	U$_{41}$	0.1284
		U$_{42}$	0.3795
		U$_{43}$	0.4921
U$_5$	0.2062	U$_{51}$	0.2256
		U$_{52}$	0.1543
		U$_{53}$	0.2210
		U$_{54}$	0.3991

5   Conclusions

(1) The undergraduates' science and technology innovation evaluation method introduced in this paper has a stronger pertinence. The main evaluation object of the method is science and engineering college students. If the method will be applied to the social science students, it needs to add some appropriate evaluation indexes.

(2) Innovation is a complex dynamic process, and the evaluation method should be applied in different grades according to the student enhancing innovation ability.

Acknowledgment. This research is supported by the 2010 teaching research program of Hebei Normal University of Science and Technology (grant No. JYYB201007). Author also thanks for the supporting from Education Science Plan of Hebei Province (grant No. 06030096).

References

1. Feng, Y., Wang, M.L., Zhang, Y.F., et al.: Research on the evaluation system of college students capacity of scientific and technological innovation. Journal of Nanjing Institute of Technology (Social Science Edition), Nanjing 7, 64–68 (2007)

2. Cao, Y.Y.: Research on the index system construction of innovative capability of college students. Wuhan University of Technology Master Paper (2008)
3. Wang, J.Q., Cao, Y.Y.: Comprehensive evaluation of innovative capability of college students. Journal of WUT, Wuhan 29(8), 133–137 (2007)
4. Shang, Y.M.: The establishment of the foundation of innovative knowledge for the college student. inJournal of Henan Vocation-Technical Teachers University (Vocational Education Edition), Xinxiang (3), 83–84 (2004)
5. Song, C.S.: Innovative Learning and Training Students' Creation. Heilongjiang People's Publishing House, Harbin (2005)
6. Liu, W.P.: Innovation Thinking. Zhejiang People's Publishing House, Hangzhou (1999)

The Data Analysis of Unceasing Axle Dynamic Track Scale

Zhang Wen-ai and Jia Run-zhen

Taiyuan University of Technology, Taiyuan, Chinese
vaigei@tom.com

Abstract. The testing data's analysis is the key factor of influencing the metrical accuracy of unceasing axle dynamic track scale. This thesis used Hilbert-Huang transform to analyze the detection signal, adopted experience mode decomposition arithmetic, and simulated through MATLAB. Then put the maximum value of residuals to curve fitting four times. Then the average curve fitting's value is what we seek. Simulation shows that this method can greatly improve the measurement accuracy of dynamic track scale, and heavier trains on the measurement of relatively high precision.

Keywords: Dynamic track scale, Hilbert-Huang Transform, EMD.

1 Introduction

From the current situation of domestic and foreign research, wheel axis scanning measurement-type track Weighing is more suitable for high-speed dynamic Unbalanced Load Detection. This article thorough analytical study wheel weight detection signal's characteristic and the data processing method and carried on the simulation confirmation using the MATLAB digital simulation technology to it.

2 Train's Vibration Characteristic

The unceasing axle dynamic track scale's sensor output signal is the weighing signal with each kind of unwanted signal's combination. The unwanted signal is making up by many kinds of frequency fluctuation signals and the impulse signal. Therefore the system mechanical vibrations characteristic analysis is the overall system design foundation. In system vibration characteristic rough analysis, we simplify the system as two degree-of-freedom vibrating systems composed by the train and track. The train and the track may simplify separately to a system have quality m, damping C and the elasticity coefficient K. As shown in Figure 2.1 [1] [2] [3] [4]

Y. Wu (Ed.): International Conference on WTCS 2009, AISC 116, pp. 513–519.
springerlink.com © Springer-Verlag Berlin Heidelberg 2012

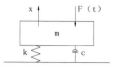

Fig. 2.1. System molded relief map

Normally, as the train quality is bigger than the track's, therefore the train oscillating component's frequency is lower, the oscillation amplitude is bigger; the track oscillating component's frequency is higher, the oscillation amplitude is smaller. [5] For the 20 tons compartments, its base frequency is 7~10Hz; For 100 tons compartments, its base frequency is 2.5~4Hz. The train it self's oscillation amplitude is related to the train speed, the compartment rigidity condition and the axle seam transition situation. Weighing the axle to lay down directly on the 50m long's reinforced concrete monolithic concrete track bed, so its rigidity is biger; the base frequency is higher than the weighing platform's. Usually the weighing platform's base frequency is 20~40Hz.

Based on the weighing sensor's principle of work, its output signal u (t) is closely related to the sensor's micro displacement x (t). So we can get the composition of u (t):

$$U\ (t) = p\ (t) + a\ (t)$$

Steady-state output:

$$P\ (t) = \begin{cases} p_0 & 0 \leq t \leq \tau \\ 0 & \text{Other} \end{cases}$$

In the formula: τ- The time that the axle travel in the weighing track measurement's area.

p_0 -The weighing sensor outputs steady-state value which the wheel weight produces.

$$a\ (t) = \sum_{i=1}^{j} a_i\ (t)$$

$$a_i(t) = \begin{cases} m_i \sin(2\pi f_i t + \theta_i) & 0 \leq t \leq \tau \\ 0 & \text{Other} \end{cases}$$

In the formula: m_i - oscillation amplitude; f_i - vibration frequency; θ_i - stochastic initial phase angle.

a (t) is the fluctuation component which caused by the train and track's two kind of vibration condition : f_1=2.5~10Hz, f_2 > 40Hz.

As in the real measure, it will inevitably exist some interference factors, thus actually when the train through the weighing zone, sensor's output signal is:

$$u\ (t) = p(t) + a(t) + \beta(t)$$

$$\beta(t) = \sum_{j=1}^{k} \beta_j(t)$$

$\beta(t)$ – is a variety of combinations of disturbance.

In the survey it uses the round scanning detection mode, therefore each sensor's output signal u(t) is the train through weighing axle's wheel weight detection signal. From the above mathematical model's analysis we can get the conclusion:

- When dynamic weighing, train's vibration may be regarded as withouting damping free vibration. Its vibration frequency is related to train's turning frame spring and the train's quality.
- When dynamic weighing, weighing axle's vibration includes the transition process and the stable state process. Transition process's attenuation oscillation frequency, attenuation extent's magnitude and speed are related to weighing axle's inherent frequency and damping ratio ξ. A direct component, a sine AC component and the disturbance ingredient constituted the stable state process. And the sine AC component's frequency is related to the train's characteristic.
- Sensor output signal includes three parts: steady state value that the wheel weight produces, and the fluctuation disturbance ingredient and the external interference ingredient which are caused by both the train and the weighing axle's vibration state. [6]

3   Data Analysis and Disposal

3.1   The Train's Dynamic Signal Simulation and Experimental Verification

According to the above statement, weighing sensor's input signal is: weight steady signal (Pulse Generator),the train it self's oscillation disturbance signal (Wave), other harmonic interference signal(Wave1 、 Wave2),and random disturbance(Random Number). And the train it self's oscillation disturbance frequency is 2.5~10Hz, the oscillation amplitude is the weight steady value's 10%. The harmonic interference frequency is 45~60Hz, the oscillation amplitude is the weight steady value's 5%. The scope of random disturbance's A/D value scope is 0~200. So based on the platform of MATLAB simulation, we can establish the train's dynamic detection signal simulation model as shown in Figure 3.1.

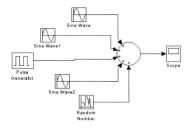

Fig. 3.1. Train Signal Simulation Model for Dynamic Detection

3.2   Hilbert-Huang Transform

Taiwan Academia Sonic academician Norden E. Huang et al. proposed Hilbert-Huang Transform. [7][8][9]. This method's processing object is the non-stable state and the misalignment signal.

Fourier Transform is decomposes a signal to infinite many sines and the cosine wave analyze data, while Hilbert-Huang Transform analyzes data by decomposes a signal to several approximate cosine wave's signal and a tendency function.

Hilbert-Huang Transform has the following merit and the demerit:

Merit:

1) It has avoided the complex mathematics operation.
2) It may analyze the frequency along with the time variation signal.
3) It is more suitable for Climate analysis; Economical data and so on has the tendency analysis.
4) It may discover a function's tendency.

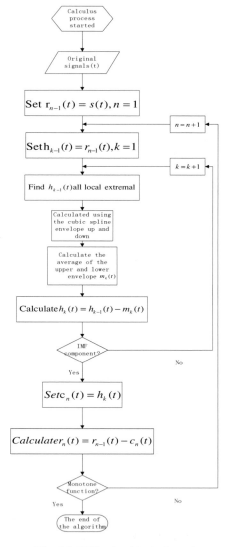

Fig. 3.2. EMD algorithm's flow chart

Demerit:

1) It lacks for rigorous physics significance.
2) It needs the complex recursion and the operation time must be longer than STFT.
3) Hilbert transform can unnecessarily correctly compute the Intrinsic Mode Function's instantaneous frequency.
4) It is unable to use FFT.
5) It is only when exceptional case did Hilbert-Huang Transform can be quickly commutated.

Establishes IMF is to satisfy Hilbert transform's pre-processing regarding to the instantaneous frequency's limiting condition. However the majority of materials are not IMF, in order to solve non-linear and the non-stationary material then decomposes IMF meets difficulty, therefore developed EMD. EMD algorithm's flow chart as shown in Figure 3.2.

3.3   Data Analysis

We are mainly apply it regarding the Hilbert transformation to be able to extract a function the stable state value. According to the front analysis, we obtains the primitive data as shown in Figure 3.3. In the chart we take the sensor input signal's stable state value is 102000kg.

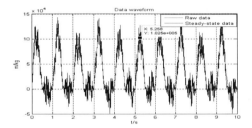

Fig. 3.3. Primary data

The stable value which after the primary data carries on the experience modality decomposes is smoother. In the field we use eight weighing sensors carry on the survey, so we have eight chart 3.3 such profile. If uses the round axis of scan measurement mode to calculate each vehicle, then generally the freight vehicle has two turning frame and four axes, then every four stable value's wave ridge data may think that is a compartment's weight data, neighboring four wave ridge data just is neighboring compartment's weight data. We select train's axis heavy stable state value is 102t and the 10.2t. As shown in Figure 3.4.

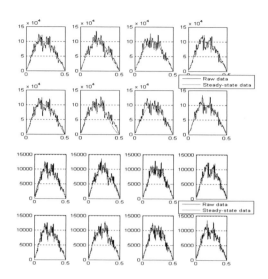

Fig. 3.4. Eight sensors to identical round heavy data profile

According to the stable state extreme value, we carried on four time multinomial curve fitting as shown in Figure 3.5.

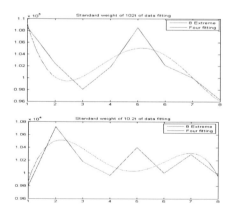

Fig. 3.5. Makes 4 curve fittings to the stable state maximum value

Then we put the curve which after fitting on average:

When the Mint-weight m = 102t, the mean value $\overline{m_1}$ = 102.023t. Relative error δ_1 = 0.02%.

When the Mint-weight m = 10.2t, the mean value $\overline{m_1}$ = 10.225t. Relative error δ_1 = 0.2%.

From the computed result, when the weighing weight is heavier the precision is relatively higher and the computed result conforms to "People's Republic of China Measurement Examination Regulations JJG-90" 0.5 level dynamic examines weighs the accuracy the request. Table I is each sensor's stable state data maximum value

Table 1. Sensor's stable state maximum values

	Sensor Number							
Standard Weight	1	2	3	4	5	6	7	8
102000kg	108538.30kg	102408.15kg	98048.71kg	101825.22kg	108536.65kg	102139.06kg	100033.69kg	96332.70kg
10200kg	9799.87kg	10724.07kg	10178.60kg	9963.56kg	10398.42kg	10000.51kg	10292.97kg	9973.56kg

4   Conclusion

The paper has studied the unceasing axle train weighing system's vibration characteristic, proposed carries on processing using the Hilbert-Huang transform pair train's dynamic data, and after processing the stable state data maximum value carries on 4 curve fittings, finally the curve averages is the value for the train weight. The MATLAB simulation proves this method to be able the effective guarantee measuring accuracy. And when the weight measured is big the measuring accuracy is relatively high.

References

1. Li, H.: Dynamic Weighing Computer Detection System: MS Thesis, Liaoning Technical University, Shenyang (2001)
2. Wang, X.-F., Zhou: Dynamic Weighing with your simulation test. Automation Instrumentation (17), 37–41 (1996)
3. Min, G.L., Wei, S.: e said. Measurement Press, Beijing (1982)
4. Xiao, J.: Movement of dual-use and application of the electronic railway scale. Industrial measurement (3), 34–35 (2001)
5. Lu, J.-W., Liu, J.: Dynamic track scale weight control and signal processing. Lectures (5), 59–61 (2002)
6. Zhang, H.: Dynamic monitoring of the train set features ultra-biased signal analysis and processing of MS Thesis, Central South University (2004)
7. Huang, N.E., Shen, Z., Long, S.R.: A new view of nonlinear water waves: the Hilbert spectrum. Ann. Rev. Fluid Mech. 31, 417–457 (1999)
8. Huang, N.E.: The empirical mode de-composition and the Hilbert spectrum for nonlinear and nonstationary time series analysis. Proc. R Soc. Lond. A (454), 903–995 (1998)
9. Hilbert - Huang Transform Wikipedia,
   `http://zh.wikipedia.org/zh-cn/%E5%B8%8C%E7%88%BE%E4%BC%AF%E7%89%B9-%E9%BB%83%E8%BD%89%E6%8F%9B#.E7.B6.93.E9.A9.97.E6.A8.A1.E6.85.8B.E5.88.86.E8.A7.A3.28EMD.29`

Computer Curriculum Development Based on Professional Group

Jinpeng Tang and Linglin Li

School of Traffic Information, Hunan Communication Polytechnic, Changsha, China
djy182@tom.com

Abstract. Relying on a professional group of traffic safety and intelligent control, to use real project as the carrier in computer curriculum development. The paper described the process of project carrier selection, cutting and transformation. Then we elaborated on different professional teachers in the same professional group how to jointly develop courses. Finally, the learning scenarios of the course are determined. This curriculum development approach promotes the professional group development, helping to train computer science students that not only understand technology and familiar business.

Keywords: Professional Group, Computer Science, Curriculum Development, Learning Scenarios.

1 Introduction

The ideal of professional group had already appeared in 90 mid-20th centuries [1], and some vocational institutions to study and practice of self. In 2006, China National Model Higher Vocational School clearly required to construct professional groups. Therefore, professional group of construction [2] is extensive attention by the vocational colleges.

Hunan communication polytechnic, who aimed at developing compound talents, core professional driving other professional development for the purpose ,formed traffic safety and intelligent control professional group. This professional group was Chinese national demonstration professional in 2007, it including a traffic control professional and computer. In this paper, this professional group as a platform, we use the real business of the project traffic control industry carrier to develop computer professional core courses - ". Net Web Application Design". This can be beneficial to the professional groups to play the advantages of functional, to enhance professionalism among teachers of different communication and collaboration, and helping to train not only understands technology and familiar business computer science students [3].

2 Curriculum Development Process

2.1 Teachers Requirements

Members in curriculum development team come from the traffic safety and intelligent control professional group. Professional group of teachers in collaborative curriculum

Y. Wu (Ed.): International Conference on WTCS 2009, AISC 116, pp. 521–526.

development, that traffic control professional teacher guide business, and computer teacher guidance technology. This is important internal conditions that make the smooth implementation of curriculum development.

2.2   Project Selection

In the process of curriculum development, project selection is the very important .If the selected project is too simple, it can not reflect the core content of courses; if too difficult, students lose confidence in learning. Therefore, this project case must be comprehensive (fully included the knowledge of courses), reflecting the typical working process, biodegradable (entire course of the project better to decompose to various teaching situations, and again by the teaching situation broken down into individual work activities), for the specific business (as software development is oriented to the business for a specific industry).

In the all kinds of projects, we selected a real project-"XX Port management platform" which reflects the characteristics of the transportation industry, and includes the core curriculum. We use this project as the main line of development process, implementation of the whole teaching process.

This project has the overall coverage (i.e., covering the core knowledge points, such as: master pages, themes, skin, ADO.net programming, Web controls). But for teaching, this project has many shortcomings that is its large scale, technically demanding and more complex business.

Therefore, this project was cut for teaching. First reduce the size of the project, then part of the business knowledge to be simplified: to retain the typical business, such as dangerous goods management, user management, accounting fees, for teaching process. The work is done by the computer science teachers in collaboration with the traffic control professional teachers, as far as possible without stripping out of business operations.

Projects is selected by a professional group of teachers, it is also in line with the work process of systematic curriculum development methods: project carrier for the curriculum development come from the actual work, but should be higher than the actual work [4]. It should not just a playback of a particular work process. It should follow the teaching law that from simple to complex, from concrete to abstract.

2.3   Business Knowledge Development

Software development is based on the actual business, so for software developers that is one essential quality to understand the real industry business. Through the study of business-oriented program helps students develop business knowledge.

(1) Way of Business Knowledge Development
In the curriculum development process, first students learn to identify business entities and relationship between businesses entities. Such as ports, docks, berths, affiliated companies, ships, equipment, goods, dangerous goods, the relations between business entities and so on. It is only possible to understand these business entities to understand tables and corresponding relationship in the database table. Also enable students to understand the entity-relationship design is essentially derived from the business entity of business requirements that is further optimized. We designed

teaching methods to be adapted to business: role-playing. The whole class is divided into several teams; each team plays a business entity. Traffic control professional teachers according to the established classics business processes guide every team acting as the role has to do.

(2) Identify Business

Business content is determined by the three officers: computer science teacher, traffic control professional teachers and industry experts. The process of determining business knowledge is as follows:

To discuss business related use case of the requirement documents.

To determine the business knowledge to be taught by discussing and three iterations:

- The first iteration, traffic control professional teachers submitted business knowledge document according to requirements document
- The second iteration, the project developer (Computer teacher) in conjunction with traffic control professional teachers utilized use case to analysis businesses level.
- The third iteration, the teaching managers and industry experts examine the businesses.

Business is divided into three categories: peripheral business, core business, general business.

Core business: it is closely with the business process of the project. Do not understand this business, then the project can not be completed. Such as dangerous goods information maintenance: must to conduct regular maintenance of dangerous goods comply with the name of dangerous goods, property, storage, transportation, emergency and other information issued by authorities, expert bodies, industry bodies enacted. Conduct regular maintenance of dangerous goods: dangerous goods property modification, dangerous goods salvage, dangerous goods addition. To achieve this business, students have to study GB6944 "dangerous goods classification and code", GB12268-90 "List of dangerous goods" and other Chinese national standards. In addition, processes such as traffic revenue is belonged to the project's core business.

General business: it is referring to foundation business knowledge which including meaning of business entities and the composition of business entities.

Surrounding business: business knowledge that nothing to do with the project development. Such as: Management operations in the dangerous goods, emergency treatment for different classes of dangerous goods emergency response capacity of auxiliary information provided, including: alarm and communications methods, channels and means of communication, emergency evacuation methods and the establishment of alert areas, on-site first aid to help aid means Leak, fire fighting. Emergency treatment is a very important business, but its processes are not automated in the realization of this project is to deal with, so after discussed, we classified it as a peripheral business, students need to know that this business, in teaching and we will not provide detailed references for this business.

Through three iterations, the required business knowledge of teaching is confirmed and they will be integrated into the relevant learning scenarios.

2.4   Learning Scenarios

".Net Web Application Design" learning field is composed by a number of learning scenarios. Each learning scenario use one or several forms of teaching carrier to achieve (reflect) the work process (project carrier used in this article); it is used to describe a software development work scene. The learning scenarios are designed as following .

Scene 1 : Construct .NET development platform, create a "port management platform" solution, analysis of project requirements. Task: To understand project requirements; understand requirements document; understand even flow of the use case diagram; Understand Entity Relationship Diagram; Create project by visual studios; to establish static pages processing. Skills Objectives: To understand the use case diagrams and the basic event flow. According to basic event flow and alternative event flow analysis of user requirements; To understand the meaning of the data dictionary; Familiar with Vs.net development environment. Business knowledge Objectives: Understanding the meaning of business entities and relationship. Such as ports, docks, berths, affiliated companies, ships, equipment, goods, dangerous goods, and so, the meaning of business entities and their subordinate relationship. Understanding data items of the various business entities, (data dictionary), focus on understanding the dangerous goods data table.

Scene 2: According to user requirements, design web pages, unify the pages style. Task: Design pages of port management module; Design the relevant pages of member management module; Design the relevant pages of port charge module; Design the master page of management side, design theme and skin combined with the CSS. Skills Objectives: Vs.net development environment; master pages, content pages; cascading style sheet; site map; configuration of Web.config ; HTML tags; server controls; standard control; HTML Control; navigation controls; CSS; theme. Business knowledge Objectives: Collecting the port business processes, the material business entities required to use (pictures, scanned documents, and Flash animation) (each project team independently complete the job based on scene 1. Traffic control professional teachers can act as a client representative role review the appropriateness of material submitted by project teams).

Scene 3: According to requirements validation of all data entered by the page. Task: Check all data entry; Check weather the Web control for data input is appropriate (by the computer professional teacher approval);Check whether the s terms for business data is properly. (Reviewed by the traffic control professional teachers); According to the document put forward various demands validation of input data required by the Traffic Control professional teacher review; cording to verify the work required to complete client authentication. (Related to the database server authentication for the time being do not do). Skills Objectives: Required validation controls, compare validation controls, regular expression validation controls, range validation controls, custom validation controls, and validation summary controls; using the client-side validation script (JavaScript). Business knowledge objectives: Identify all kinds of business reports, such as administrative inspection table for loading and unloading storage (Dangerous Goods), handling storage (grocery) administrative checklist, "port operation license" e-file, handling storage (bulk cargo) administrative inspection table ,handling storage (container)

Chief checklist, facilities management Corporate for shipping terminals (anchorage, buoys) checklist, passenger service administrative checklist, assessment classification regulation summary, evaluation reference, the meaning of evaluation projects.

Scene 4: access back-end database. Task: The port authority management, port management, port enterprise management, port enterprise information display; To generate fees data, print fees report, send fees data; Issued operating licenses of dangerous goods; Approving operations of dangerous goods; search, add new members in member management. Skills Objectives: data source controls; data-bound controls; data source controls with parameters of; binding syntax; common events of standard control; indirect events; selecting data controls; access database using sql statement ;access database using stored procedures ; access back-end database with the ObjectDataSource through stored procedures.

Scene 5: to achieve the rights management. Task: To create roles; assign permissions for the role; realize members of the property, data, passwords, rights management. Skills Objectives: roles, members; page-level authorization; Creating user profiles; operating the system configuration file- Web.config; read and write Web.config with code. Business knowledge Objectives: Understand the systems organizational structure; According to the organizational structure to determine the role of categories (review).

Scene 6: Implement the system reuse and RSS. Task: 1. Using RSS technology allows information to synchronize with the external network (waterway network); Call domestic container weighing data packets using the packets from Web Service interface. Skills Objectives: Practicing to create custom controls; Release and call WebService; achieved Rss using Xml. Business knowledge Objectives: To understand the knowledge management system, complete documentation collection, including: laws and regulations, industry regulations, professional literature, news and other content; Domestic Container Load data packets process.

Scene 7: Building three-tier web application. Task: The reconstruction of the original system from two-tier applications to the three-tier applications;. To achieve the dangerous goods management in the form of three-tier structure. Skills objectives: Multi-layer structure; to achieve multi-layer structure in a solution; Business Logic Layer; Data Access Layer, ORM.

Scene 8: System testing, release, configuration and debugging. Task: To do unit testing with Nunit; to do performance testing with LoadRunner; Adjust the system performance; Release port management platform. Skills objectives: Performance testing; unit test;. Net Web application publishing and configuration (Web.config); performance tuning; To manage project using SVN. Business knowledge objectives: Traffic Control Traffic Control professional students (end-user role playing) in accordance with user requirements specification for Alpha testing, and complete test reports.

In the above, we treaded a cut the traffic control industry real projects as the carrier develop the learning scenario, each learning scenario is made up of a series of practical tasks in project development, and that this task has covered the core knowledge and skills of the course. More features are, the tasks will also cover a number of business knowledge of traffic control industry. In this course involving: the basic knowledge of water transport, business knowledge about dangerous goods management and accounting fees. This knowledge has always been part of college tradition of professional. Therefore, in the teaching process, teachers in professional group may do collaborative teaching according to the skill knowledge and business knowledge. This contributes to the professional group in the depth of integration of different professional.

3   Conclusion

The paper put forward, in the same professional group, to fully utilize complementary of traffic control professional and computer professional to develop courses. The collaborative approach for take advantage of their own professional advantage and professional group teacher's competitive advantage, enable different professions in the same professional group to achieve a deep level cooperation. At the same time, it is conducive to train not only understands technology and familiar business computer science students. This way of curriculum development also provides a reference for other vocational colleges to develop curriculum against background of professional groups.

Acknowledgment. This work is supported by Education Research Project of China Institute of Communications Education(NO. 1002-108), school-level research projects of Hunan Communication Polytechnic(NO.HJY09-1010), school-level research projects of Hunan Communication Polytechnic(NO.HJY08-0906)

References

1. Xing, W., Liang, C.: Module model to cluster strengthen the management of vocational high school. Tianjin Education 1, 22–24 (2000)
2. Mei, Y.: Higher professional group intensive construction. Education Development 5, 68–71 (2006)
3. Lu, L.: From the Perspective of the Software System Requirements. Programmer 2, 91–93 (2006)
4. Xu, H.: Work Process-Oriented Vocational Education. Education and Vocational 4, 5–8 (2008)

Research on the Teaching Skill of DSP Application Technology

Xizhong Lou, Yanmin Chen, Xiuming Wang, and Bo Hong

College of Information Engineering, China Jiliang University, Hangzhou 310018, China
sniafar@gmail.com

Abstract. DSP Application Technology is an important professional elective course of the electronic information and communication engineering. This course is a part of the Key Course Construction Plan in China Jiliang University. A series of reform measures and innovative practices such as teaching contents, teaching method, teaching skills and teacher's ability are introduced. As the rapid development of the digital signal processing technologies, the FPGA technologies and the embedded CPU technologies, it is a useful exploration to consider these technologies together. It is a good practice on the training the students to fulfill the requirements of the society and the companies. The students are more interesting in DSP Application Technology than before by introducing these methods.

Keywords: DSP Application Technology, Teaching Contents, Teaching Skills.

1   Introduction

"DSP Application Technology" is a very important and useful professional elective course for the students of electronic information and communication engineering, automatic Control, mechanical engineering and electronics engineering. It is a follow-up course for students after mastering digital logic circuits, digital signal processing, microprocessor or Microcomputer Principle and Interface. The traditional teaching methods for the DSP Application Technology are mainly focus on the explanation how the DSPs' chip works. It is hard for the student to have a good overview of the course. And most of the student is easy to give up when they find that they can not find interesting things in this course.

Our group asks students to participate in the research-based teaching practice and researches projects undertaken by teachers in order to achieve substantial results. These kinds of methods let the students know how a practical DSP system works, which ignited the interesting of the students and strengthened the ability of theory with practice. The project works well with the supporting of the training mode and teaching reform achievements, which has been carried out for many years.

2   "DSP Application Technology" Teaching Methods Investigate

2.1   The Development of Technology Requires Adjusting Teaching Contents

With the development of society, economy and technology, DSP and the embedded applications system are developing quickly. Old models of DSP chips continued to be

Y. Wu (Ed.): International Conference on WTCS 2009, AISC 116, pp. 527–531.

eliminated and new models of DSP chips are emerged in large numbers. At the same time, new technology, new methods and new tools are invented time to time. These force the teachers to learn more things. The teachers must meet the higher requirements for this course. In particular, the latest DSP technology commonly uses the traditional DSP processor core with common purpose processor core, such as DaVinci Series DSP chips. In general, the TI DaVinci series of DSP chips have one TI DSP core and one ARM processor IP core. And the Multi-core DSP chips are designed more and more. So the DSP technologies and embedded technologies begin to fuse with each other. At the same time, as the development of programmable technology, FPGA(Field-programmable gate array) and CPLD(Complex programmable logic device) also have made rapid progress. Many FPGA Chips now integrate the traditional DSP cores, which become a powerful digital signal processing competitors. These trends are shown in Figure 1. Traditional DSP device with only one DSP core will share less market from now on.

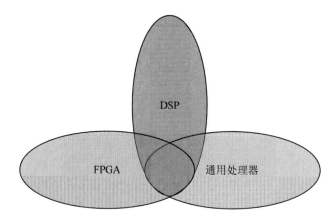

Fig. 1. The Development trend of DSP、 FPGA and common purpose processor

The rapid development of technology requires adjusting teaching contents in time. Then it will keep up with technology progresses. Therefore, as to the teaching contents, it can't be limited to the original TI 5000 series DSP, 6000 series DSP and single DSP core architectures. We should adopt the latest DSP devices for teaching, which is on behalf of DSP technology development. At the same time we should break the fence among the course of DSP and the course of embedded processor and the course of FPGA. These three courses should be unified reformed. It is a good idea to combine these three courses into one. Then the close relationship among these three courses are clear to anybody. The selection of application solution is base on the application its self, not depends on fixed device.

2.2   Students Psychological Changes Ask to Improve Teaching Skills

DSP application technology is usually an elective course. The hours of the course (class hours plus experimental hours) are 36 to 42. During the class hours, all of the hours have been exhausted after finishing DSP architectures, addressing modes, assembly language programming design, DSP peripheral applications and 3 or 4 experiments. The students have no more time to learn much more important practical programming techniques such as the applications of DSP algorithms.

At the beginning, many students knew that the course is very useful, so they selected it. But after several courses, many of them gave up because the course is not only more difficult than microprocessor, but also very boring. In particular, the time of the course is always in the senior semester. Many students were busy in preparing for the entrance examination for the graduate, the civil service examination, or under training and hunting jobs. Therefore they are more likely to ask for leave. Then the situation will be worse. This kind of courses such as DSP application is more difficult to keep up with when the students left the class for several times. Therefore the students will give up the course unconsciously. However, it is found that students generally are more diligent in the experimental hours, because each experiment must be checked and accepted. In the acceptance process, the teachers will ask questions about the experimental procedure, the coding, so the students will try to seek answers actively, read related materials details, understand the procedures well. Therefore, we believe that it is more effective to reduce the amount of class lessons appropriately, and add the experimental hours. Because it offers students more opportunities to coding to solve the practical problems, they will have more interests and pressure to learn. And most important they like this way.

2.3   Changes in Teaching Contents and Teaching Skills Result More and Higher Requirements to the Teachers

"DSP Applications Technology" is a very practical course, which relates to software design and hardware design. The requirements to the teachers are relatively high. With the rapid development of DSP technology, teachers must continue keeping up with the pace of technological development, and continue learning new skills. DSP technology is now combining with programmable logic devices (such as the CPLD / FPGA) and embedded processors technology (such as ARM). These kinds of combination let people easy to find a low-cost, high performance application solution. On the other hand, the traditional DSP solutions have shown its lack more over. It is necessary to have a good understanding and grasp the programmable logic device, ARM processor for the teachers. So the requirements for the teachers are greatly improved. We let the teachers go out to visit other schools to keep in touch with teachers of other schools. At the same time, We let the teachers go to some companies to learn the practical technologies. And it is necessary to train the teachers from time to time to help the teachers to improve themselves. And we have done these works. All most all of the teacher are participate some kind of training. That made the teachers more confident when they face the students.

2.4   Enhance the Students Interesting by More Extracurricular Practice

Because of the limited hours of the "DSP Applications Technology" course, it is impossible to give the students enough hands-on opportunities in class or experiment. However most of the students who have spare time may participate in the research projects of teachers. It can ignite the interests of the students. It can combine what they have learnt in class with practical application. It makes students quickly grasp technical knowledge of DSP applications through the training. For example we let some students to participate in the project of water monitoring. The students can take the initiative to learn more from others, and do some works, and achieve some good results. When they got result, they were very happy to do this. We also allow students to participate in extra-projects, such as smart car competition. Involved in these projects, DSP technology will be applied to the actual project. Finally, the students who actively participate in these projects and activities usually have more opportunity to hunt a good job.

3   Conclusion

With rapid development of DSP technology, several useful teaching reform explorations on the course of DSP application technology has been carried out, such as the adjustment of teaching content, the update of teaching skills, the teachers' improvement in capabilities and the students' participation in extracurricular practice. A good result has been achieved as shown in Table 1. In table 1, till 2008 the students' satisfaction with the course is under 33%, and the average final examination score is 67. The first year(2009) after we reform the teaching contents and teaching skills, the satisfaction was improve to 46%, and the average final examination score is 73. The second year (2010) we trained the teacher and improve they abilities in DSP technology and let more students do extracurricular practice, the figure became 57% and 76. Also all the teacher are more satisfied with the student, because less students were escaped from the class.

Table 1. The result of the teaching reforming

	Students' satisfaction (%)	Average final examination score	Teachers satisfaction (%)
~2008	33%	67	38%
2009	46%	73	55%
2010	57%	76	75%

Among all the factors, we think the improvement of teachers' teaching and research ability is a prerequisite. Adjustments on the teaching content and innovation on the teaching skills are the guaranty of teaching effect. The students' participation in research projects and extracurricular activities of science and technology is a further enhancement of the teaching effect. All of these measures increases the students' interest and improves their practical ability.

Acknowledgment. The project is supported by the Scientific Research Fund of Zhejiang Provincial Education Department (Y200805880) and the Key Course Construction Fund of China Jiliang University.

References

1. Xu, S., Cong, W., Sheng, Q.: Retrieval teaching practice. Journal of Tianjin Manager College, Tianjin 13(5), 61–62 (2007)

2. Bai, J., Zhang, X.: Research on the teaching practice for the course of the digital circuit and logic design. Journal of EEE, Nanjing 29, 69–71 (2007)
3. Yue, Y., Zhang, Z., Tan, T.: Information Theory course teaching reform and practice. Science and Technology of Guangxi University, Nanning 32, 226–228 (2007)
4. Yu, W.: Reform NC teaching methods to train students creative and practical ability. Dianda Ligong, Shenyang 232(3), 37–38 (2007)
5. Dou, Y., Shi, Q.-Z.: Research on computer foundation teaching system of non-computer major. Research in Teaching 32(1), 67–70 (2009)
6. Li, J., Ni, Y., Chang, Y., Wang, M.: Analysis on the status of education in computer system course in University. Higher Education and Research 11, 5–7 (2008)
7. Computer Science. Curriculum 2008: An Interim Revision of CS (2001), http://www.acm.org/education/curricula/ComputerScience2008.pdf

Research on Mine Universities Automation Curriculums

Meiying Qiao and Zheng Zheng

School of Electrical Engineering and Automation, Henan Polytechnic University,
Jiaozuo, China
zhengleng@sohu.com

Abstract. According to the regional economic characteristics and actual industry background, combining with automatic professional features, this paper firstly has analyzed shortage of original automation curriculum system. Moreover, the new curriculum reform has been proposed in the light of the guiding principles of new cultivation program. Finally, the specific implementation of automation curriculum reform has been described for four modules which include direction mode, general cultivation, professional curriculum set and teaching practice. The author has illustrated necessity and significance of each reform in detail, and a feasible curriculum system is ultimately given. In the new situation, the curriculum reform possesses some reference value for designing professional curriculum of automation in Mine University.

Keywords: Coal mine, Automation Curriculums, Cultivation program.

1 Introduction

With the growth of energy demand and automation technology of coal mine improved, the automation professionals needs for coal industry is further increased. Our school was the first national coal institutions (to be established in 1909), therefore raising automation talent for coal mine industry, is its responsibility. Due to economic situations of coal industry change better as well as other profession's employment pressure increases, our automation graduates starts to develop themselves in the coal mine. However, student's knowledge structure and their choosing profession scope in the future are all decided by curriculum system. Therefore laying down good professional courses, the students can not only improve the employment rate, but ensure school rapid and healthy development.

However, the coal industry is the high risk and hard industry, and universities in the employment selection process are bidirectional. These factors resulted in a pair of contradictions. If curriculum designed only considers the coal industry, many students who will not engage in this vocation, would face a difficult choice at graduation. If the curriculum designed does not think over the coal industry, on the one hand, the automation students could not be competent the coal mine job; on the other hand, school resources will not be able to be fully utilized.

The automation specialty is a wide aperture specialty, nearly faces the various trades and occupations. At present, employment pressure of atomization students mainly

Y. Wu (Ed.): International Conference on WTCS 2009, AISC 116, pp. 533–539.
springerlink.com

comes from two aspects in our school: first, this pressure comes from the automation specialty of Research University, moreover it also faces many other specialty which possess characteristic of the automation specialty. How to make the automation specialty survival and develop must be considered problem by our professional teachers. Curriculum had been further reformed and innovated when cultivation program of the automation specialty is revised and discussed in 2010 years.

According to actual situation, integrating the local economical characteristic, the original cultivation program had been analyzed and generalized. Meanwhile, many deficiencies also had been pointed out in revision process. The new Cultivation program had been worked out based on feedback of automation graduate and demand information of employing enterprise.

2   Analysis on the Existing Problem of Original Curriculums

2.1   Inconsistent of Training Objectives and Curriculum

Curriculums are not strictly corresponds to the object of professional training in original cultivation program because there are several training direction, but they did not map out specific curriculums in connection with the different direction. So the professional curriculums which provided by the graduate student in the job search process will not reflect automation characteristics of our school. On the one hand, this problem makes the students lack confidence for their learning professional knowledge. On the other hand, it also possibly lead the employing enterprise not better understand our automation students who only are considered a branch of the control and engineering. The automation training of our university is not understood so comprehensive by the enterprise that they are often partially comprehend on the professional study and employment.

2.2   Theory Courses and Practice Courses Is not Closely

Curriculum setting can not integrate the theoretical knowledge with the practical skills in the original cultivation program. This main reason is that position of automation specialty in our university was not cleared in the old cultivation program. It makes the specialty lack clear thought in professional development. In curriculum setting, it is so irregular as to much advantage run off contrast with the other subjects with automating color. These reason also led characteristics of the automation can be better reflected by graduates in the job search process, so the original advantage will be lost.

2.3   Practice Is Relatively Weak

The automation specialty is an emphasis on professional practice, and its engineering value can be realized only in practice. At the same time, the practice also can enhance the students understanding of basic theoretical knowledge, so it not only can make students lay the foundation for further study, but also stimulate their interest in learning. However, because of teaching resources and practice base resource constraints, the theoretical teaching hours is more, and the practice is less in the curriculum setting process. Therefore, students who had been trained under the curriculum system of existing professional cultivation program will lack of practical operational ability and practical engineering analysis ability.

3   The Laying Down Principle of a New Curriculum System

3.1   Outstanding Professional Characteristics

Automation feature of our school is to train compound applied talents who will possess stronger actual engineering capabilities and some of studying capacity according to the background of the school culture and economic characteristics of the region. Therefore, comprehensive and engineering application research must be highlighted in the curriculum laying down process .The students must grasp some of Communications technology strong electrical knowledge   and foundation courses in related profession except possess   strong theoretical knowledge. That is to say that school needed to foster compound professional talents who not only understand the theory, but understand technology. Therefore, automation of our school ought to highlight the characteristics of school development according to their own development and resources integration. However, the characteristic of a profession is embodied through laying down the reasonable curriculum.

3.2   Compatible with the Industry's Needs

Today, China is one of big energy consumption country, and the main energy comes from coal. The coal mining rate is demanded higher along with the economic gradual development. It is an only way that the automation of coal mining level be improved to achieve this requirement. Therefore, requirements of automation graduates increase more in recent years than that of in previous years in the coal industry with the coal development. However, due to the employing situation gets stern in recent years, and it makes more and more students choose the coal industry to work and realizes their own value.   Based on the current situation and own advantage, automation talents of the coal mine are increased in our school. Henan is one of big coal production province in China, furthermore our school is adjoin to Jincheng of Shanxi province, so Enterprises related the coal mine are many surroundings the university. Therefore, in the new curriculum system formulation process, the objective which depends on the coal industry and radiates other professions is proposed.

3.3   Optimization of the Curriculum Structure

Comprehensive knowledge is demanded especially high for automation of the coal industry. It not only requires students to possess some theoretical knowledge as to lay the foundation for the follow-up learning, but requires students to know how to deal with weak electricity and strong electricity. Simultaneously, knowledge of explosion-proof, real-time data collected, field monitoring and tracking also must be understood to the students of automation. Therefore, these reason demand professional curriculum be made reasonable arrangements according to their mutual advantage. So the complementary of advantage specialty will be formed to optimize their own disciplines and cultivate talents to meet the social needs. Thus, theories and practices are complementary each other in the course structure and achieve an organic combination. Holistic and complementary have been formed in laying down process.

4   Specific Reform Measures on the Curriculum System

4.1   The Reform Measures on Different Direction Training

Since automation is a wide scope professional, if student's culture were specifically concentrated on the coal mine automation, they would be bound to many limitations in employment process in the future. On the other hand, our automation is classified to coal institutions, so this is the main duty to train automation students for the coal industry. At present, more and more automation students are likely to work in the coal mine because of the pressure of employment. In line with statistical analysis from graduated students, when they have been the coal mine for long time, the students can find there are many basic curriculums on the coal industry not studied in the university so as to they can not be quickly familiar with the work environment. In order to make the automation students can soon meet the requirement of coal industry and other industry in the future, a clear automation training reform was provided in 2010 years. First, general education of professional curriculums which must be studied by all undergraduate students   is set in the first and second grade. The third year is training course for the basis professional curriculums.   These curriculums will be classified in detail according to different professional direction from the latter period of the third year to later. Coal mine automation, electrical drives automation and industrial process automation are proposed in the light of actual teaching conditions of school and regional economic background of Henan Province. Three directions are not same in choice of curriculums, and they are complementary, but also different from each other. For example, there are" coal capital equipment control", "digital Mines and Technology" and "Introduction to mining" in curriculums set of Automatic Orientation etc. "Modern communication speed", "Field Bus and Industrial Network" and "electrical equipment fault diagnosis technology" can be selected by the students of   Electric Drive Automation. However, curriculums selected become "process safety management and inspection equipment", "Field Bus and Industrial Network" and "process control engineering design and implementation" in the industrial process automation.

4.2   Reform Measures of General Education

With the development of network technology the university's campus is influenced by various types of information from the network and the student's life is impacted by every aspect. So, the university education also withstands tremendous pressures and challenges. How to cultivate useful talents who possess good health, sound mind and a particular expertise should be thought about by university education. According to the lack of human knowledge for engineering students, general education platform is set to guide healthy and happy of students in our school. Many elective general education courses are provided based on compulsory general education in the original plan. They include the following categories: Humanities and social sciences, natural sciences, engineering technology and art classes. Relational courses include:

"Contemporary World Economy and Politics", "Student Career and Development Planning," "Sports and Health", "University of Aesthetic Education", "Mental Health Education", "innovations", "Mathematical Modeling "" College Chinese "and so on. These courses will be repeatedly set in four years, and students may elect them according to their own ability and time. In order to implement, the students have to at least select one credit in each category to make these courses can play their own role. But for the public art classes, the art of restrictive taking elective courses, they are demanded at least 2 credits. such as "Introduction to Art," "Music Appreciation", "Art Appreciation", "Dance Appreciation," "Film appreciation", "Calligraphy Appreciation", "Drama Appreciation" and "Music Appreciation" .

All kinds of Elective curriculums increased in General Education platform, on one hand, it could broaden the students' horizons, and help them set correct world view and values. On the other hand, the character deficiencies of engineering majors can be made up in the learning process. So the overall quality of university engineering students can be improved, and it plays a role to training comprehensive talent for our school.

4.3   Reform Measures of the Professional Curriculums

Some professional curriculums whose respectabilities are higher had been canceled, and some professional Curriculums had been restructured and consolidated in the curriculums setting process. For example, "Computer Principles" and "SCM principles" are combined a course as "Microcomputer Principle and interface technology ", while knowledge of microcomputer can be used one week course designed to enable students to master. The original "Computer Graphics" course removed in the new cultivation program, and its content will be set on the part of other several curriculums and the final graduation project. Through the training in these areas, students will eventually master these basic computer technologies. "Enterprise for Distribution Technology", "operational Research" and "Signals and Systems Analysis" and other courses are increased. These set main taken into account the complex industrial spot which requires the automation students not only possessing the weak electricity knowledge, but understanding the strong electricity principle and knowledge of signal and systematic optimization in control aspect. These reforms make the students more meet the personnel development and industrial needs.

In order to further optimizing the curriculum, since graduate enrollment and the pressure of employment situation in recent years, more and more students choose to further study after graduated from the university. "Modern Control Theory" had been moved up from the seventh semester to the sixth semester and its nature had been come from elective into compulsory in order to enable students to set goals earlier and earlier to make preparations. In order to do some preparatory work for direction of the seventh semester, some elective courses are set in   the sixth semester, such as "Control Motor", "Introduction to Artificial Intelligence", "Navigation Guidance and Control", "Power Electronics and New Energy Technologies", "DSP Theory and Applications," "VHDL Technology and Application".

4.4   Reform of Practice Teaching

Automation is a practice-oriented professional and the specialized knowledge can be mastered only in reality studies. Therefore, the whole study process is a theory - practice - again theory - and then practices the process, and most of the time theory and practice are complementary and simultaneous. In addition to traditional metalworking, electronics technology practice and Electrical Engineering Practice in the freshman and sophomore years, students begin perceptual practice and Microcomputer Principle and Interface Technology curriculum design from the fifth semester. In particular, from the original course design and simulation demonstration to achieve the minimum CPU into the nature of the physical process control system, which greatly enhanced the students' ability creativity and understanding to simple microprocessor and electronics.

Production practice will be moved up from the seventh semester to the sixth semester in the new curriculum. On the one hand, the main reason is that the students had completed basic professional courses, which is not only able to make the students to analyze the basic principles, but also deepen their understanding about the theoretical knowledge in spot, meanwhile, it will lay the foundation for the next semester directional training. On the other hand, it is more convenient that the students contact practice location with various business in this period. Otherwise, many enterprises also are employing people in this time, and they hope expanding their influence through selecting excellent students to their enterprise from our school. According to some practical situation, our school had established a number of long-term practice bases with some enterprise which possessed high automation level, such as the cable plant, the power plant, the HuaFei electronic and electrical factory and so on.

Meanwhile, taking geographical advantages, our automation had also established long-term practice bases with the coal mine enterprise surrounding school, such as Pingdingshan Coal Group, Jiaozuo Coal Group and Jincheng Coal Group., some students practice in theses bases during the production practice. Practices make the students further be close to the enterprise scene, which will lay a good foundation for directional cultivate in the seventh term.

Except for these traditions practice given in above, an innovative module for the student's elective is added on the basis of directional training modules mentioned in the seventh term, which is to further develop their comprehensive abilities. Such as innovative experiment, research training and paper published, technological inventions, science and technology festival.

It quires the students at least gain four credit, and each part is two credit in these modules. Students also can gain credit through participating SRTP, Electronic Design Contest, Freescale's Smart car racing, the National Mathematical Modeling Competition in school and provincial or national. Our students have already participated in the large-scale national event every year for a long time, but only did minority teachers and students who interested in contest participate in previous events. Therefore, these were not specifically arranged in cultivation program to correctly manage. Now, in order to extending influence and coverage of this practice to the students, and specifically managing, the practice reform is provide, which can stimulate the enthusiasm of students and innovation. Our students had firstly obtained a national patent whose name is "a slow road with power energy storage" under guidance of this thought.

5   Conclusion

Our take full advantage of the school's resources, according to the characteristics and background of the school, developed a new curriculum. In this paper the curriculum of automation specialty is used as an example to illustrate reform and innovation under the new situation. On the one hand the school characteristics have been retained in the new curriculum system; on the other hand the new training mode increased employment opportunities for students. In the new model, the curriculum was carried out the reform and innovation accordingly in all aspects. New curriculum structure not only take into account the schools actual situation in, but also enhance the school's employment rate, while increasing the students to continue in-depth study possibility. Meanwhile, the new curriculum reforms have some reference value for other coal specialty.

Acknowledgment. Financial support for this work, provided by Teaching Reform Fund of Henan Polytechnic University (No. 2007JGD28) is gratefully acknowledged.

Double Languages Teaching Research of Oracle Database Technology

Hejie Chen and Biyu Qin

Department of Publishing and Management
Beijing Institute of Graphic Communication
Beijing, China
x.ychen@yahoo.com.cn

Abstract. The double languages teaching is an important part in higher education. This paper presented the double languages teaching experiences of Oracle database of information management and Information system speciality at the Beijing Institute of Graphic Communication. The paper firstly introduced double languages teaching objective and double language content of Oracle database. Some important questions during teaching process are described. Then the paper mainly introduced that we adopted some effective pedagogical methods to improve double language teaching effect. Because Oracle is a CS course, the paper described experimentation arrangement. Finally, some students' evaluations are presented. The formal evaluation demonstrated a very effective and realistic learning experience for the students.

Keywords: Oracle double languages teaching, pedagogical methods, experimentation arrangement, a case study approach, a problem-based approach, a ladder teaching method.

1 Introduction

Over past years, culture originality industry of China rapidly increased. And that, press industry held 70% culture originality industry. According to the General administration of Press and Publication of The P. R. China Report, china published 233971 kinds' books and 6408 million volumes in 2006[1]. Great changes have taken place in the publishing with the remarkable development of new system capabilities, new technologies, and new forms of media. The person not only read all kind of contents from traditional books, but also from new media which includes CD_ROM, digital library, the networks and mobile services. We can find the following scenes:

1) An undergraduate is downloading some papers of computer science from digital library.
2) When a girl studied to paint, she saw a painting CD_ROM.
3) If the writer make use of World Wide Web, the book can be read by mans of all over the world.
4) When the man or the woman is walking, he or she is reading the news through mobile telephone.

Y. Wu (Ed.): International Conference on WTCS 2009, AISC 116, pp. 541–548.
springerlink.com © Springer-Verlag Berlin Heidelberg 2012

The press industry is utilizing the computer technology now. The Digital Publishing, a new the press industry concept, has presented. Some people consider that Digital Publishing is a serial press activity that utilized digital technology [2].Other person presents that all information with unify of binary system code store in light, disk etc. Some computer devices or similar equipments finish information process and translation [3].These demands influenced talent culturing objective and contents of the high education.

Beijing Institute of Graphic Communication specially train talented person of publishing industry in china. The department of Publishing and Management emphasized the publishing talent cultivating. The bachelor degree of management department was launched by the Beijing Institute of Graphic Communication in September 1990[4].At present, the department of Publishing and Management offer five specialities: marketing speciality; information management and Information system speciality; financial management speciality; edit and publishing speciality; publishing advertisement speciality. Oracle database is a double languages course of information management and Information system speciality.

Students of information management and Information system speciality master to analyze and design of information system. Students have a good grasp of recourse exploitation, management and utilization. Students are familiar with the production and operation management and information management. Students are engaged in information management and information systems analysis, design, implementation, management and evaluation in publishing or other industry. Some differences occurred between information management and Information system speciality and computer application speciality.

- Information management and Information system speciality emphasizes the students' analysis, design and maintenance abilities.
- Computer application speciality emphasizes facility ability except analysis, design and maintenance abilities.

The students learned Oracle database in the seventh semester. Teaching languages are English and Chinese. The teaching objective of the course improved database management and science English ability.

Section 2 introduced database course group information. We also introduced teaching objective and contents of Oracle database. Some important questions are described. Section 3 introduced some effective pedagogical tools through teaching practices. Oracle database emphasizes hands-on capacity, while the hands-on abilities rely mainly on curricular experimentation and the comprehensive experimentation. Therefore, Section 4 presented the experimental arrangement. Section 5 described some students evaluated our Oracle database teaching works. At last, some conclusions and future works is introduced.

2 Double Languages Teaching Actuality of Oracle Database

2.1 Base Information Introduction

The students learned Oracle database in the seventh semester. Teaching languages are English and Chinese. Its previous courses included database theory, database comprehensive experimentation, and Java program language. These courses focused on the relational database system and program language application.

- Database theory emphasized the theoretical and practical aspects of database application. The course content covered data models; relational data model; schema normalization and integrity constraints; query processing; recovery; isolation and consistency; and security.
- The database comprehensive experimentation cultivated analysis and design ability of the project. The teacher provides some projects of database design. Two students composed a team, every team selected a project. Every project involve requirements analysis, conceptual design, relational database mapping and prototyping, and database system implementation using a given database software.
- Java program language finished management information system.

Teaching item of Oracle database improved the large database software application ability. The students will read plentiful science English papers. Because the students graduated, some base speciality English must study.

2.2   Existing Problems

Because English level of our students is low. How to arranged Chinese and English in teaching process?   Other question is that we didn't find good double languages book at present. Oracle database is very complex. It is difficult to select teaching contents. Traditional pedagogical methods can't improve application ability. This paper presented some pedagogical methods to improve the students' abilities.

3   Some Pedagogical Methods Application

Because Oracle database software content is very complex, double languages teaching has self-character. It is important that used compatible pedagogical methods. In order to increase practice ability, self-study ability and innovation ability of students, we have adopted a variety of pedagogical methods in the actual teaching process. These methods mobilize initiative and enthusiasm of students to learn knowledge and encourage the students to independently think.

3.1   A Problem-Based Approach

Problem-based learning is an instructional method that is said to provide students with knowledge suitable for problem solving [5]. In fact, we use the problem-based lecture approach and the problem-based learning approach. The former, the teacher asks a question and translates the knowledge surrounding the question. The later, the students propose a solution according the question.

   The problem-based approach is main pedagogical method. For example, the balance of finite vocabulary and course knowledge is main difficulty of double languages teaching. However, we found that the problem-based approach can solve this question. We used English and Chinese to describe the problem. Students not only thought solve method, but also found the vocabulary and phrase.

3.2   A Cooperative Learning Approach

A cooperative learning is an instructional technique that requires students to work together in small, fixed groups on a structured learning task [6]. It can train positive interdependence and individual accountability. We usually adopt the cooperative learning approach to finish a project. For example, the teacher provided some projects of Oracle database application. 2-3 students composed a team, every team selected a project. Every project involved requirements analysis, database design, and database system implementation using an Oracle database software and database management. Every students finished different contents. At the same time, students are required to actively participate in each phase. The students assume different roles in each phase to allow them to experience different leadership responsibilities. A cooperative learning approach can improve self-study ability, innovation ability and team cooperation spirit of students.

3.3   A Case Study Approach

When some theories are very complicated or difficult to understand, we use the case study approach. Case study is the study of the particularity and complexity of a single case, coming to understand its activity within important circumstances [7]. Selecting a case standard is that is easy to understand and comprehensively explanation this problem. The teacher looked for the detail of interaction with its context during teaching. Because the teaching time is finite, we do some video. The video based on the case study.

3.4   A Ladder Teaching Method

A knowledge unit teaching adopted the ladder teaching method.

- Firstly, some vocabularies and a problem are introduced.
- Secondly, the teacher showed correct operations.
- Thirdly, the students imitated these operations.
- Fourthly, the students solved the similar questions.
- Finally, the teacher and the students summarized the knowledge unit.

4   Experimentation Arrangement

The key aim of experimentations is to ensure that students operate Oracle software. The other aim is to ensure that students make an in-depth understanding theoretical content through operating software. Experimentation of information management and Information system speciality are subdivided into three categories according to different specialties. They are the curricular experimentation, the comprehensive experimentation and emulation experimentation (Fig.1).

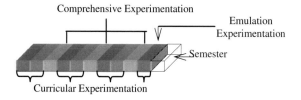

Fig. 1. Experience arrangement

Experimentation of Oracle technology is mainly the curricular experimentation, because the course hours are limited. But the emulation experimentation trained the Oracle database. In the following, we introduce experimentation content and Oracle course experimentation content.

4.1   The Curricular Experimentation

The curricular experimentations were allocated within the 16-week semester. The curricular experimentation items of Oracle deeply understand theory knowledge and master software operation.

Because Oracle academic content and operation are tremendous and complex, 12-hours can't satisfy need. But the students can practice core operations, a number of software operations videos are provided to students for extracurricular time.

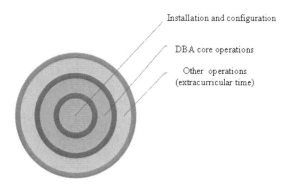

Fig. 2. The curricular experimentation arrangement

4.2   The Comprehensive Experimentation

Between 16-week semesters, our Institute arranged 5-week semester, which is called the small semester. Some courses are launched 1-week or 2-week to finish project. Oracle hasn't the comprehensive experimentation.

4.3   The Emulation Experimentation

Before graduation design start, the emulation experimentation is arranged. The students are trained to realize core questions of publishing industry. The students master the Oracle software, which included database analysis, database design, database recovery, database security and other questions.

5   Evaluation

Because Oracle database rapidly developed and Oracle database is widely used in various industries, Oracle teaching has taken on an increasingly important significance and prominence. Since the launch of Oracle technology in information management and Information system speciality in the 2006 academic year, course content has been revised many times. A key requirement is to culture students' application ability of database and science English. To achieve this assessment, an evaluative questionnaire is developed and given to the students. The results from this questionnaire are presented in table 1 in which students were asked to score each question in the range of 0 to 100 and where the satisfaction level is an average of these scores for each question. A maximum score given by all students for a given question would result in a satisfaction level of 100% for that question. Comparative satisfactions exhibited in this questionnaire. Students acknowledge that database and program language courses improve computer courses with the highest score. Students have also generally acknowledged that the courses arrangement is reasonable and the courses have helped understanding of Oracle database theory and application ability of science English. Similarly, students acknowledge that the experimentations arranged effective and improved practical hand-on ability. Finally, the courses have assisted the students in being able to work in groups.

Table 1. Oracle Technology Evaluation

Questions	Satisfaction
This course has helped my understanding of the theoretical knowledge	85%
This coourse has helped my application ability of science english	85%
This course has helped my practical hand-on ability	80%
This course has educated independent thinking ability	80%
The number of class hours arrange reasonable.	70%
Experimentations improved my practical hand-on ability	80%
The Comprehensive Experimentations arranged reasonable and effective.	80%
This course has helped my ability to work in groups.	75%
The course was well organized and effective.	85%

The evaluative questionnaire also provided students with an opportunity to provide additional comments about the three general questions listed in Table 2.For question 1, students acknowledged strong satisfactions, and main themes presented as follows: improvement the database application ability, masterly using of software, working in the group and reading ability of science English.   For question 2, a number of issues, and even some contradictory questions arise. Main themes presented as follows: to increase/decrease the English spoken time, to extend/shorten curricular experimentation time, to deepen/reduce the theoretical contents, to increase/decrease English reading. For question, some students hoped that college provides the practical chance of companies, which are using their studying software.

Table 2. General Questions

General Questions
What aspects of the course did you enjoy the most?
What aspects of the course did you dislike the most?
Tell us how could improve the students Oracle database ability and science english?

6   Conclusions

With computer and communication rapid development, computer technology are applied every industry. On the other hand, students are demanded high computer technology by the enterprise. Database technology is an important computer technology. At the same time, international communion required science English. So, it is important to find effectively cultivating two aspects abilities. This paper described double languages computer course information of information management and Information system speciality at the Beijing Institute of Graphic Communication, Beijing, China. Some questions existed in the teaching process. The paper also introduces several effective pedagogical methods in order to increase practice ability, self-study ability and innovation ability of students. Experimentation is an important ability cultivating means, which train students a high level of application ability. So this paper introduces experimentation arrangement. A formal evaluation has been carried out and demonstrated a very effective and realistic learning experience for the students.

References

1. http://www.gapp.gov.cn/cms/cms/website/zhrmghgxwcbzsww/layout3/indexb.jsp?channelId=493&infoId=448190&siteId=21
2. Li, Z.: Analyzing Digital Publishing Concept. China Publishing Journal, 11–14 (December 2006)

3. Yi, W.: Digital Publishing, Which are your choice? Science and Publishing, 6–8 (May 2006)
4. http://www.bigc.edu.cn/jgsz/jgsz1/
5. Linge, N., Parsons, D.: Problem-Based Learning as an Effective To Teaching Computer Network Design. IEEE Transactions on Education 49(1), 5–10 (2006)
6. Dietrich, S.W., Urban, S.D.: A Cooperative Learning Approach to Database Group Pjects: Integrating Theory and Practice. IEEE Transaction on Education 41(4) (November 1998)
7. Eisenhardt, M., Kathleen: Building theories from case study research. Academy of Management Review 14(4), 532–550 (1989)

The Realization of 3-D Garment Art Design in CAD Software

Liu Jun

Guilin College of Aerospace Technology
liuengineering@TOM.COM

Abstract. In this article, base on operating experience and current technology, we raise that establish 3-D table first and base on determination of 3-D table structural points and structural lines, acquire virtual dress by defining related interval values. We also concluded some effective and practical methods for 3-D garment style generation. Through fabrication and display of 3-D style, it enables designers or customers to operate interactively in a good visible environment, and view real-time design results intuitively and precisely in any time. From 2-D to 3-D, garment style designs become easier and make it possible to realize automatic style graph to 2-D clipped sheets transformation.

Keywords: Garment CAD, Human Body modeling, 3D.

1   Introduction

The Chinese garment industry began to utilize garment CAD technology from 1990s, early computer aided garment designs are plotted in 2-D mode. The current global garment industry has been undergone a transformation from traditional design and fabrication mode to a Hi-tech way. Digitized design and fabrication has become an important trend [1]. Garment products digitization is a practice of utilizing 3-D digitized design in garment design and production process. This technology is becoming a focal point for garment companies and it is an important practice for local companies to gear to international standards. That's why, in recent years, 3-D garment style display is regarded as the best way for style designing and it becomes more and more popular. Never the less, 3-D design has to base on 2-D drawings. Effective 2-D/3-D transformation then becomes the focal points in garment CAD design research.

2   Techniques and Process of Garment Style Design

Ordinary garment style 2-D graph are plotted with the help of professional graphic software such as PHOTOSHOP and ILLUSTRATOR to make accurate and standard style graph. Styles are illustrated by lines. These graphs are generated by fitting lines and curves of the original design and can be saved as vector or non-vector graphs. Although they are able to display features of the design, they are still in a 2-D manner, and can not unveil the essential features of the design including the structure,

Y. Wu (Ed.): International Conference on WTCS 2009, AISC 116, pp. 549–554.

slack or tight, materials etc. They are not suitable for further adjustments and processes. In other words, the 2-D design work inclines to be rigid rather than more understandable.

Currently, the focal point of garment CAD research is the 3-D style display. This is also an important section of garment CAD work chain. It is decided by many factors, such as design, data input, size, establishment of human body model, 2-D virtual sewing and the combination of 3-D human body and 3-D dress model. At this stage, we are able to see detailed 3-D dress style display. At the same time, the design and fabrication work comes to a substantial phase. 3-D display plays an important role. The feasibility of the style can be testified here. Base on original 2-D design, by means of 3-D table modeling, 3-D dress modeling, 3-D clothes parts unfolding, 2-D styles are transferred to 3-D ones.

3-D garment style designs are carried out on 3-D table. Thus, before the 3-D CAD work, table modeling on computers must be done. During the 3-D design work, static and dynamic dimensions of human body must be considered. We must consider the elasticity of human body tissues, tolerance for breathing, and movements. In short, we must consider the degree of slackness. So, base on available 3-D table, in the modeling phase, freedom of human body movements must be considered. At the same time, 3-D clothes parts are formed through the formation of structural lines. As 3-D clothes parts created, according to complicity of garment style, the first approaching and unfolding of curved surface can be considered, that is to unfold the entire dress contour. Then it comes to the partial curve unfolding.

The 3-D CAD technologies are applied in dress design. We use improved curve parts method to do the fitting of dress curve surfaces, combined with standard 3-D table data points, increase appropriate interval values to acquire shape-defining points. Base on this, we put forward a technical route which fulfills requirements of 3-D effects display and integrates features of the entire system.

3   2-D to 3-D Transformation Process

3.1   Preparation of 3-D Table

Normally, a 3-D table is established by scanning data points on 3-D human body, determining 18 key curves of 3-D table. Each curve consists of 13 points. 18*13 key points are determined by pre-process. In order to improve its smoothness, the original data are interpolated to get 86*85 orderly arranged points. In this way the 3-D model meets accuracy requirements. Connect acquired points per sequence as Bezier curved surface points to get 3-D table surface. Never the less, there are some problems with the use of 3-D scanning: 1. data acquired are actually not continuous; 2.enormous data; 3. redundant data caused by overlapped scanning [2]. That's why we need processed human body surface as collision imitation geometry model. Data are processed with human body scanning software. It can transfer data, edit raw scan data, and rebuild Triangle mash, merge different scanned parts etc. A displayed continuous high resolution surface is not continuous in fact. Since the high density of points which forms the surface, it looks like that. While, as we zoom in to some extent, the lines become saw tooth like. There is a new human body modeling method base on human body measurement. Through interactive amendment on 21 anthropometry parameters, customize personalized human model. (1)Adopt normal 3-D human body model(standard model). (2) Acquire new skeleton model while doing surface model editing, and keep the two models matching with each other. The new human body model can be directly driven up in virtual environment. (3) Controlling model figure with 21 measured human body parameters makes it possible to create more delicate, plenty, and comprehensive human models [3]. 3-D tables created in this way are easily driven, the models formed are more delicate, plenty, and comprehensive, easier to be popularized and meet customized requirements.

3.2   Foundation of 3-D Style Formation

Normally, traditional style designs are based on 2-D plane. Although some current style design software claim to have 3-D effects, they are all 2-D based. Imitate real drape with gridlines, pad materials on dress according to curved surface. Retain original texture and shadow. Such display effects are not genuine 3-D effects. The foundation of 3-D application is 3-D data. In garment CAD, they are 3-D points which forms curved planes.

Ones the 3-D tables are ready, 3-D style can be generated from 2-D design. 2-D style graphs provide designers design idea and intuitional structural features. Ones dress contour, style, proportion characters are determined in 2-D graphs, 3-D transformation can be conducted through steps shown in Table 1.

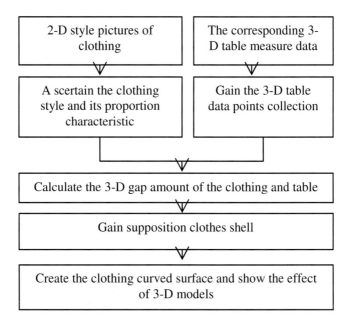

Fig. 1. Flow chart on transforming 2-D style into 3-D style

In real world, when people ware dresses, relevant structural lines and structural points on dress have their counterparts on human body, in other words, they cover on corresponding lines and points on human body. While between dress and body, there is what we previously mentioned slack degree which enables convenient body movements, to be more precise, the interval values, as we figure out interval value between dress and body structural points, put a virtual dress on 3-D table, make adjustments base on it, it won't be difficult to get needed 3-D dress style.

3-D table data points are normally obtained by 3-D scanning. Dress shape-defining points can also be acquired. Characters of 3-D contour can be determined from 3-D contour values. Base on this, new side face interval values are introduced to each 3-D table section. As other new interval values are calculated, new shape-defining points are determined. In this process, Interval values of different part are not necessarily identical. Since there are some key supporting points between dress and body, such as shoulder points, chest point, hip point and elbow point. While for other parts other then supporting points, since the irregularity of human body, interval values can't be regarded identically. For example, intervals in middle of waist and side of waist are different. To make 3-D dresses look more real, these interval values must be determined according to actual condition. Final dress effects will be created through dress emulating process. Physically based 3-D dress emulation realizes static and dynamic dressing emulation. It enables designer and customer to view effects in all directions. They can make choice and adjustment on dress styles through real-time interactions. These functions will no doubt have wide application prospect, such as dress customization base on internet, product sale and virtual display etc.

3.3 Surface Modeling Method

Base on original points congregation, obtain curved surface points through graphical algorithm, and then base on point congregation and known points (3-D points),through connecting these points form several simple triangle, fitting 3-D curved surface. Assign number to known points as V0, V1, V2□and so on. Connect neighboring points, form simple triangle. By calling basic plotting function, plot triangle curved surface, join curved surfaces, 3-D curved surface can be created. When 3-D model curved surface are divided into extremely small portions, curved surface can be substituted by small planes. In this method, as the grids are divided finer, we get smother curved surface. It applies to point congregation model in various techniques [4]. Curved surface created by this triangle splicing method are lifelike smooth and beautiful. Its static effects approach real dresses as graph 1 shows.

Fig. 2. Curved surface effect generated by joining triangles together, engine of HE XUAN SHI YI

So we must use triangle grid to solve this problem. Triangle grid is a grid structure formed by triangle connected points. Since every 3 points always form a triangle, a kind of model surfaces which are relatively distinct, continuous and stable can be defined according to this. During operation, needed model surface can be created with different triangle grid density. We can reduce grid triangle number, optimize triangle shape without making big changes on model surface. Thus, we are able to create relatively small numbered model whose surface structures are even. It is more convenient to create 3-D dress surface with triangle grid file then with original data.

Never the less, dress materials are normally made of soft substance. Unlike human body, changes in atmosphere constantly lead to change in material physical properties. This leads to surface shape changes and makes the imitation difficult. In this field, dress material physical modeling method is the focal point and main difficulty in our research. As we study 3-D dress style□we can only control displayed 3-D dress style in a relatively stable state, then conduct 3-D modeling.

It is certain that as the dressing form varies, interval values vary, followed by reshape of dress model. We have to rebuild model. This will pose difficulties to us.

With current technology, we are not able to build up models which reflect random drape changes on material surface caused by dynamic change. That's why, in many cases, 3-D dress styles displayed are tend to be rigid.

Moreover, to realize garment CAD 2-D to 3-D transformation, we must solve technical problems such as displaying textile texture and dynamics, 3-D rebuilding, lifelike and flexible curved surface shaping, and transforming 3-D design model into 2-D dress sheet. Other than this, how to make adjustment on 3-D dress style in real time is a project we need to complete.

4   Conclusion

As customers raise higher requirements on garment quality, fitness and personality, popularized 3-D garment technology and girded garment design have become necessary method for Chinese garment industry to gear to international standards [5]. From 2-D to 3-D, garment style designs become easier and more intuitive and make it possible to realize automatic style graph to 2-D clipped sheets transformation. Base on 3-D style effects, designers are able to make adjustments on clipped sheets. Afterwards, physical models are introduced to conduct virtual sewing to get final 3-D dressing effects. The entire route is a highly intelligent process. The criteria for garment technology become much lower.

References

1. Huang, Y.: The combination of computer technology and the garment industry. Journal of Suzhou Institute of Silk Textile Technology (4), 78–80 (2000)

2. Xing, Q., Zhang, X., Li, Y.: Research on Virtual Fashion Show. Journal of Nantong Textile Vocational Technology College (3), 1-5 (2005)
3. Mao, T., Wang, Z.: An Efficient Method for Customizing Individual 3D Virtual Human Body. Journal of Computer-Aided Design & Computer Graphics (10), 2191-2195 (2005)
4. Hu, J., Geng, Z., Qian, S.: Study On 3D Pattern Simulation of Garment Based On 2D Pattern. Control & Automation (1-3), 95-97 (2006)
5. Zhou, L.: Digital Technology—The Inevitable Tendency of Clothing Industry in 21st Century. Shandong Textile Science & Technology (1), 38–40 (2004)

Research and Improvement of Linux Real-Time Performance

Yang Yang[1] and LingLing Hua[2]

[1] Computer Department, North China Institute of Science and Technology,
Beijing, China
[2] Basic Department, North China Institute of Science and Technology, Beijing, China
`shui.yang@sohu.com`

Abstract. Linux, as a general operating system, is not suitable for real-time applications. In order to solve these problems, this paper puts forward the optimized for Linux clock accuracy of a kind of improvement strategies, and experiment verifies the improved system. Through the experiment, improvement scheme can effectively improve the accuracy of real-time clock. The time delay millisecond timescale reduced to microsecond. Especially in high load conditions, after the improvement of the kernel stability in low delay working condition, conform to requirements.

Keywords: Linux, Interrupt, Real-time, Operating system.

1   Introduction

Linux as a general network operating system, the main consideration is scheduling of fairness and job throughput, reflected in real-time, Linux mainly adopted two measures: one is to put all the tasks in the running state in a queue hung up on different tasks, in its mission control block with a "policy" attribute to determine its scheduling strategy, and the real-time task and real-time tasks to stricter requirements. Real-time tasks using "SCHED-FIFO" scheduling in a scheduling operation is completed. After the mechanism, using "softirq" mechanism Linux, this method in solving the problem of missing, interruption has improved Linux real-time.

2   Standard Linux Kernel Restricting Factors of Real-Time Performance

Nonetheless, real-time performance is still an obstacle of embedded system developing for Linux, In the following aspects: Linux system clock for precision clock 1Hz, too coarse; When a task through the system calls into the kernel mode runtime, It is not to be preempted real-time tasks, and this leads to the uncertainty of the execution time; The lack of effective real-time scheduling mechanism and scheduling algorithm. [1]

Y. Wu (Ed.): International Conference on WTCS 2009, AISC 116, pp. 555–559.
springerlink.com

To solve these problems, such as "RT-Linux", "Kurt-Linux", "RED-Linux", these system have realized some improvement methods, but these methods or more complex, or just to solve some of the problems, according to the general characteristics of embedded system tasks, and put forward some practical feasible improvement methods.

3   Linux Real-Time Performance Improvement

3.1   To Improve the System Clock Accuracy

In order to improve the open-source Linux system real-time performance, puts forward some solutions. Such as, "KURT-Linux" , "RT-Linux" , "Monta-Vista Linux" etc.

The University of Kansas developed KURT - Linux. It changes the Clock interrupt mode will be fixed frequency Clock chip set for a single trigger mode. Use Pentium CPU architectures provide Time Stamp machine tracking system, the core of the Clock Time scale are based on TSC accuracy, can achieve the delicate grade. The plan was proposed in this paper for KURT ideas. Different is, provides a with precision calibration of core real-time clock processing system, this system and standard Linux core clock parallel operation. The system can not only improve the stability and efficiency, and independent core clock easy maintenance.

2.6.16 Linux kernel USES "jiffies" as a timer, but in low frequency for accuracy, 1000Hz, only reached 1 ms. In order to improve the accuracy and precision clock can use a clock interrupt source record since the last time, since after thus Linux kernel can provide higher accuracy than beats cycle to determine the current time.

This work mode can use the following pseudo-code description:

```
timer_interrupt():
hi_res_base=read_timesource()
xtime+=NSECS_PER_TICK+ntp_adjustment
gettimeofday():
now=read_timesource()
return xtime+cycles2ns(now-hi_res_base)
```

X86 architecture CPU TSC provide high precision scale. This scheme, system using TSC increased in a high precision timer, need to maintain two related with the clock interrupt requests queue: The system clock interrupt requests queue and HRT interrupt request queue. They are respectively responsible for respective interrupt service routine. [2]

Real-time data structure definition of timer shown below:

```
Type def structure timer
{heap queue,
The real-time timer expires time,
Real-time tasks, the periodic real-time tasks for 0,
The timer process of corresponding CPU,
The timer expires shall implement the service functions,
```

Service function of parameters,
}
The following is the key data structure hrt-timer:
Structure hrt-timer
{Structure rb-node node;
ktime-t sub-expires;
enum hrt-timer-state state; /* Timer state * /
int(* function)(structure hrt-timer*);
Structure hrtimer-base* base; /*Timer standard */
#ifdefCONFIG-HRT-REQ
intmode; /*Timer mode* /
Structure list-head cb-entry; /* Function queue * /
#endif};

Linux need maintenance a clock source abstraction layer, timing system through the abstraction layer over the past one time distance calculated the time now. In order to calculate the passage of time, the timing of the system hardware abstraction, and in general timing code through a pointer to the selection of the corresponding source as a time clock hardware. The clock can be used as a source of simple structure, said: [3]

```
struct clock_source{
    char *name ;
    int rating ;
    cycle_t (*read)(void *data) ;
    int mask ;
    int mult,shift ;
    }
Write a read clocks source function:
cycle_t simple_clock_counter_read(void)
    {
    cycle_t ret = readl(simple_clock_ptr);
    return ret;
    }
```

Use the register_clocksource (&simple_clock), simple _clock can be registered to the system, then we can use simple_clock in timer system.

3.2   Real-Time Scheduling Algorithm Module Design

In retrofit scheme is introduced based on the two common real-time scheduling algorithm of priority: Rate Monotonic Analysis (RM) and Earliest Deadline Firs (EDF).

RM scheduling algorithm is a kind of typical static no-preemptive priority scheduling algorithm, is also the most classic of real-time research method. RM scheduler data structure as follows: [4-5]

In the header files "sched .h".Add the macro definition: RM
create RM

In process control block "time" by structure - increases the task scheduling strategy RM attributes:

```
Structure RMS-structure {
Unsigned long period;
Unsigned long ready-time;
Unsigned long service-time;
Unsigned long time-serviced;
};
```

This structure to join "time" before, need structure - first initialize:

```
Rm
{
init-Rm. period=-1;
init-Rm. ready-time=0;
init-Rm. service-time=1;
init-Rm. time-serviced=0;
}
```

Also called deadline driven EDF scheduling algorithm (DDS), is a kind of dynamic scheduling algorithm. The EDF scheduler data structure as follows:

In header file "sched.h", add the EDF macro definition:

#define EDF;

Add in the process control block EDF scheduling strategy adopted the task of properties:

```
Structure EDF-structure
{Unsigned long deadline;
Unsigned long period;
Unsigned long ready-time;
Unsigned long service-time;
Unsigned long time-serviced ;};
```

4   System Testing and Conclusion

All of the following procedures for the test environment is CPU: Pentium 4 2.0GHz, 2048MB memory, 250GB Hard disk. According to the definition, interrupt response time, using testing at different moments of numerical method of reading timer plan. The basic idea is to test procedures, assume certain time to process need sleep for some events (such as I/O operation ended), because the precision clock system, so actual sleep problems and expectations of sleep time will have certain difference. The clock system precision, the difference is smaller. Procedures for 1,000 times, each time sleep cycle count 1 second and final average minimum and maximum, the sleep time delay. [6-7]

The program pseudo-code are as follows:

```
do{
clock_gettime(now)        /* Get system current time */
nanosleep(1000*USEC_PER_SEC) /* Sleep for 1 second */
clock_gettime(next) /* Get to sleep after 1 second of time */
```

diff = calcdiff(now+1000, next)
/* Actual sleep time is calculated with the expectation of sleep time, Unit for microseconds */
if (diff < stat->min)
stat->min = diff   /*Minimum of time delay sleep */
if (diff > stat->max)
stat->max = diff   /*Maximum of time delay sleep */
stat->avg += (double) diff / * Average sleep time delay */
} while (loop<1000)

Test results such as shown in table 1 and table 2:

Table 1. The system time delays, No load conditions

The kernel	Maximum Difference (μs)	Minimum Difference (μs)	Average Difference (μs)
Standard 2.6 kernel	1011	2035	1530
The kernel after improving	10	23	15

Table 2. The system time delays, Existing load conditions

The kernel	Maximum Difference (μs)	Minimum Difference (μs)	Average Difference (μs)
Standard 2.6 kernel	2332	10341	5447
The kernel after improving	12	51	22

From the test results can be seen: after the improvement, the kernel delay before have significantly reduced, delay millisecond timescale from below microsecond.

Especially in high load conditions, more obvious advantage. The delay in Standard 2.6 kernel has greatly increased, at least, most 130.66% increased 408.16%, and average delay 256.01% increased, and improved the kernel delay is relatively stable in low latency, keep state.

References

[1] Ding, C., Zhang, Y.-L.: Improvement of Real-Time Performance of Linux 2.6 Schedule Algorithm. Journal of Universityof JiNan 22(04), 362–365 (2008)
[2] Li, B., Li, Z.-W.: Analysis of Linux Real-timeMechanism. Computer Technology and Development 17(09), 41–44 (2007)
[3] Wang, B.J., Wang, Z.G.: Uniprocessor static priorityscheduling with limited priority levels. Journal of Software 17(03), 602–610 (2006)
[4] Wang, Y.J., Chen, Q.P.: On schedule ability test of rate monotonic and its extendible algorithms. Journal of Software 15(06), 799–814 (2004)
[5] Xing, J.S., Liu, J.X., Wang, Y.J.: Schedule ability test performancean alysis of rate monotonic algorithm and its extended ones. Journalof Computer Research and Development 42(11), 2025–2032 (2005)
[6] Kar, R.P.: Implementing the rhealstone real-time benchmark. Dr.Dobbs Journal 15(04), 46–55 (2000)
[7] Kaashoek, M.: Interrupt and task scheduling of rt-linux. Embedded Systems Programming 20(8), 1456–1471 (2003)

The Study on Product Art Design Decoration Style

Liu Jun

Guilin College of Aerospace Technology
liuengineering@TOM.COM

Abstract. In this paper, a preliminary discussion on new decoration style is conducted. Since the time for this study is short, this paper is insufficient in depth and width. The follow-up study will focus on the problem how to enhance the application of decorating elements in our local culture, thereby enabling product design to reflect regional culture and national & cultural characteristics better.

Keywords: New Decoration Style, Product Design.

1 Introduction

From the patterns on ancient Egyptian mummies to those on "Dressing" sofa made by Moooi, and from the laces on medieval noble clothing to the flower patterns on the metal surface of Nokia mobiles, decoration is sentimentally expressing the inspiration of designers and has strong characteristics of humanism & naturalism. Decoration is not static: at the end of the 20th century, a new decoration style integrating many styles brought innovation to design, giving birth to some designs full of imagination. These designs don't stress the traditional "glam" and "bling" but use decoration to embellish their details, resulting in unexpected uniqueness and a new trend. "In the current situation we are seeing a slowing down of rigor and a return to a more pleasing design, more suited to domesticity", said Gillo Dorfles, an Italian art critic [1]. In Milan Design Week 2009, the designs with new decoration style accounted for a quarter of all the exhibits. In addition to the basic practical functions, these new designs are giving off a warm taste of home, thus being in vivid contrast against the minimal exhibits that are rational and cold.

2 The Formation of New Decoration Style

The development of modern decorative design has undergone many representative historical periods. During Renaissance, humanism was taken as the guiding ideology, the artistic forms in ancient Greece & Rome were chosen as references, and a lot of decorative forms of art that were in line with the features of that era appeared. As time went by, the hypocritical, complex and highly decorative Baroque style and Rococo style were pursued by the royal family, aristocrats, churches and rich men, which made the two styles a symbol of status and wealth until the advent of industrial revolution. As we know, industrial revolution brought people to the age of modern

Y. Wu (Ed.): International Conference on WTCS 2009, AISC 116, pp. 561–567.
springerlink.com   © Springer-Verlag Berlin Heidelberg 2012

industry. The designers headed by William Morris triggered the Arts and Crafts Movement, in which they stressed the traditional, natural and delicate Gothic style and adopted the unvarnished patterns of plants or animals to form the unvarnished decoration style that is quite different from the luxurious style. Also, the Arts and Crafts Movement gave birth to another design movement – Art Nouveau Movement. With expensive materials, designers took beautiful stems, leaves, flowers and delicate women as the basic patterns and made various luxurious articles manually for the bourgeoisie. However, the luxurious and tired style of decorative design was soon replaced by the Art Deco Movement that combines handicraft with industrial production and integrates the Eastern and Western culture. [2]

The development of decorative design stopped after the outbreak of World War II. Due to a serious shortage of materials and the turning of manpower & machines to the manufacture of war-related products, designers had to focus on the practicality of products, as they could no longer find extra materials and manpower for decoration. At the same time, the new materials for wars were developed significantly, and gradually the modernism style emphasizing practicality and opposing decoration became the mainstream of design, and this style had been popular for nearly a century.

According to Adolf Loos in his "Ornament and Crime", decoration can only cause waste to social resources, time and money, and will cause regression to our society like pestilence in human's development. The viewpoints of some modernism design pioneers, such as Adolf Loos and Ludwig Mies van der Rohe, etc. represented the thinking of designers who were willing to assume obligations for the socialist ideals. They hoped that their products were in line with the taste and consumption capacity of the public and far from those nobles & bounders' specific articles that could reflect their status and wealth. In their opinion, the general public who were busy in pursuit of food and clothing didn't need the articles with patterns of flowers or beautiful ladies, so they attached more importance to function and hoped that people could buy cheap and practical articles. At that time, such reality was determined by the social productive forces. [3]

The modernism design, which culminated between 1920s and 1930s, stresses function and the requirements of the public and pursues the clean and rational style of design. However, it also makes the buildings or products dull, listless and inhumane. Under such circumstances, people all live in the same box-like houses and use the same articles. Yes, such life is efficient but lacks of emotion. This situation hadn't been changed until 1960s ~ 1970s.

After the advent of media age and information-based society, media and the social culture have been inducing consumers to pursue personality and fashion, while the designers are trying to meet consumers' demands for fashion in various ways. Astute designers began to apply new decoration methods in their design, and this is not a simple repetition of the traditional decoration but a way to highlight the personality of design by decoration as well as the significance of design to people, thereby arousing people's poetic response[4], and then forming the new decoration style.

3   New Decoration Style in Product Design

A new decoration style is not a pure style of design but a synthesis of many styles, and its form and theme of expression in product design are not absolutely the same.

3.1   Detail Decoration in Minimal Form

Until 1990s, the minimal style has been playing the leading role in product design, while the products are rational and practical but not distinctive. Currently, designers like highlighting personality in their design, and they prefer to express themselves with bold colors and abundant patterns. The minimal style hasn't been abandoned by designers but integrated with decoration to form a new aspect – the combination of minimal appearance with beautiful colors and abundant patterns. While maintaining the purity of modeling, designers maximize the role of decoration elements by boldly applying new technologies and materials, and they make their design vivid and lively by means of comparison. Some excellent designs, such as HB Group (UK)'s "Bling Console Tables", Hella Jongerius (NL)'s "Ps Jonsberg" vase (Fig. 1) and Hannes Wettstein (CH)'s "Jackson" sofa, reflect the perfect combination of minimal-style products with detail decoration. Compared with the products produced automatically, the hand-made products have more human touch and value due to the uniqueness of every product, though there are significant differences among different individuals. Some wise designers would rather conduct detail decoration manually, such as Hella Jongerius (NL)'s "Nymphenuburg Sketches" Plates (Fig. 2). Some other designers even deliberately maintain and exaggerated hand-made signs to indicate the origins of the products by the meticulously designed "flaws", thereby highlighting the personality of products and making them different from the industrial products.

Fig. 1. "Ps Jonsberg" vase by Hella Jongerius

Fig. 2. "Nymphenuburg Sketches" Plates by Hella Jongerius

3.2   Romantic Fantasy and Classic Feelings

Romantic fantasy is an eternal theme of decoration. The design style of Tord Boontje (NL) is romantic and poetic – "I see myself as a storyteller". With the help of computer-based cutting techniques, he converts the natural elements, such as flowers, grass, swan, reindeer and snowflakes, into a fairy tale. Some of his products, such as the "Fairy Tail-Midsummer" lamp (Fig. 3) and "Think Of You" vase cover, are warmly favored. The similar designs include Joris Laarman (NL)'s "Heatwave" Radiator (Fig. 4), Danny Venlet (BE)'s "D2V2" Lamp, and Nahoko Koyama (JP)'s "Delight" Lampshade, etc, in which the designers also adopted the romantic elements from nature. The designers filled the products with artistic temperament by referring to the decoration methods in the 17-19 centuries and taking the patterned vines and flowers as the major decorating elements.

Fig. 3. "Fairy Tail-Midsummer" lamp by Tord Boontje

Fig. 4. "Heatwave" Radiator by Joris Laarman

Lace is a common element in clothing design, and it's now widely used in product design due to its inherent romance. By use of some computer software, designers are designing the lace patterns with more distinctive characteristics of the times, and they can create the products with strong decorative value by means of laser cutting. Lace element is not subject to geographical constraints and can easily accepted by the public, so its elegant figure often appear in the design of modern products, such as jewelries, furniture, lamps and decorations, etc. People can taste the feminine style in Ronan & Erwan Bouroullec (FR)'s "Algues" Interior decoration and Marcel Wanders (NL)'s "Crochet Chair" Armchair (Fig. 5).

Fig. 5. "Crochet Chair" Armchair by Marcel Wanders

Decorating products with classical elements has become a new trend, and designers are extracting classical signs and integrating the traditional culture into modern product design. Some works, such as Philippe Starck (FR)'s "Victoria Ghost Chair" armchair, Jasper van Grootel (NL)'s "Plastic Fantastic Swarovski" Dining chair (Fig. 6) and Pieter Jamart (BE)'s "Louis III" sofa, are innovated products based on the classical products, and the designers re-highlight the classic works by extracting the main visual elements thereinto and using the modern concepts, materials and techniques, etc. Actually, these designs have a touch of irony while arousing people's retro feeling. In addition, the abundant patterns and forms in Baroque and Rococo Period have become an inexhaustible source of inspiration for designers. In William Brand (NL) & Annet van Egmond (NL)'s "Candles And Spirits" Lamp (Fig. 7), the designers boldly adopt the complex digital-designed decorative elements to create an impressionistic palace-style sense of extravagance.

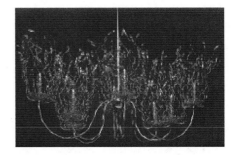

Fig. 6. "Plastic Fantastic Swarovski" Dining chair by Jasper van Grootel

Fig. 7. "Candles And Spirits" Lamp by William Brand & Annet van Egmond

3.3   Integrated Features of New Decoration Style in Product Design

By analyzing the themes, forms, materials and techniques of new decoration style in product design, we summed up the integrated features of new decoration style in product design, as shown in Tab. 1.

Table 1. Integrated features of new decoration style in product design

Features of product style	Originate in the 17th ~ 19th Century, and reflect the romantic feelings of naturalism; the products are usually given the characteristics of artworks, and the mash-up method is often used; local decoration details and the overall sense of form are stressed in design, and both visual stimulation and dramatic effect are emphasized. Classical decoration is widely used.
Expression form of product	Most of the design themes are related to romantic fantasy and classical feelings; a lot of natural elements (e.g. flowers, trees, grass, birds, worms and animals, etc.) are used to generate decorating patterns in the manner of modern graphic design; bright colors are adopted boldly; the products often have minimal appearance, and a visual balance is maintained between abundant decoration details and the minimal product appearance.
Materials / processing techniques	The commonly used materials include stainless steel, feldspar porcelain, polymers, plastic, wood, MDF, composites, metals, glass, marble, felt, resin, paper, leather, cotton, rubber, silver.[1]The processing methods include water cutting, laser cutting, etching, milling, stereo-lithography, selective laser sintering (SLS), 3D printing, screen print, laminated object modeling (LOM),fused deposition modeling (FDM),quick Polymerization of synthetic materials, CAD (computer Aided Design), CAM (computer aided manufacturing). [1]
Coverage of product design	The application of design is strong, covering the fields of jewelry, carpets, furniture, home accessories, cutlery, glasses, etc.

4   Conclusions

Generally, the inspiration for design with new decoration style comes from natural elements and classical elements. The decorating elements extracted from nature have the strongest vitality, and they can help the urban people living in houses made with steel and concrete return to nature, thereby obtaining valuable spiritual comfort. On the other hand, the introduction of classical elements has also blurred history and reality and aroused people's retro feeling. The verbose traditional decoration may make people stressful, while the products with new decoration style can help to regulate mood and highlight personality. [5]

With distinctive humanistic features, decoration can reflect the cultural perspective and values of designers. Compared to the exotic culture, designers are better at using the local decorating elements. So far, new decoration styles have laid a positive impact on the product designers in China. The BenQ 5250C scanner was awarded the 2004 IF Product Design Prize, which rests not only with its slim body and excellent performance but also with the decorating patterns of Chinese calligraphy on its top. For another example, Delux exhibited some blue-white porcelain series laptops in

Computex 2009, and such laptops were quite outstanding among all the laptops due to the application of blue-white porcelain patterns in the design of laptop shells.

As technology advances, the cost to apply decoration in products is becoming increasingly lower, and decoration doesn't mean a waste any more, thus being no longer a crime. On the contrary, decoration can make a product more vivid and the communication between designers and users more direct. Like vines, decoration is freely spreading in people's lives, and people can no longer live normally without decoration.

References

1. Bucquoye, M.M., Storm, D.V.D.: Forms with Fantasy, p. 9, 94, 95. Prgeone, Singapore (2007)
2. Wang, S.: A History of Modern Design, pp. 88–105. China Youth Press, Beijing (2002)
3. Bayley, S., Carner, P.: Twentieth-century Style and Design, pp. 65–67. Sichuan People's Publishing House, Chengdu (2000)
4. Diani, M.: The Immaterial Society, p. 4. Sichuan People's Publishing House, Chengdu (2000)
5. Lin, Z.: The Revival of Art Deco design. Product Design (6), 36 (2006)

Professional Accounting Professional Training Model Application in Practice

Liu Ye

Siping Professional College
gqliuwhu62@sohu.com

Abstract. The professional quality is the sum of the knowledge, abilities and professional values competent to be a job and the core of Higher Vocational Education. This post needs to explore the basis of the accounting profession Accounting Majors professional quality content, and described the practice of culture in the Accounting students have taken the application of professional quality models.

Keywords: Accounting Professional, Professional Quality, Training Mode, Application.

1 Content of Computerized Accounting Professional Quality

Vocational education training shoulders for the community, "high-quality, skilled" talent of the important task of Accounting students should have what professional quality? To address this issue on our school graduates and graduates of the past two years where the unit conducted a follow-up survey and questionnaires, through sorting, analysis, summarized in the quality of computerized accounting professional career include[1]:

1.1 Positive Values, Good Professional Ethics

Accounting is dealing with career and money, which requires the accounting officer must establish the correct values, to maintain a normal state of mind, maintaining the balance of mind, and with good behavior, correct treatment of fame and fortune, be honest and self-discipline, law-abiding individual interests must be subordinate to social interests and national interests.

1.2 Refining Professional Skills

Professional skills is the most basic capabilities of accounting personnel is a basic condition for the completion of accounting work.

1) Accounting Professional Judgement
January 1, 2007 started the implementation of new accounting standards, some business enterprises in the processing of Chu Li does not provide specific methods, but to require the accounting officer in Mianlin uncertainty Qingkuang O'clock, Genju

Y. Wu (Ed.): International Conference on WTCS 2009, AISC 116, pp. 569–576.

accounting standards, Institution requirements, combining the businesses of operating environment and characteristics of professional knowledge and experience of enterprise financial services of the accounting treatment and final financial statements to be taken by the principles, methods, etc. to judge and choose. In this case, the same question to judge the results of different workers may not be as natural to the business impact and results are also different, therefore the occupational sense of accounting has become the most important professional skills[2].

2) Accounting capacity
Accounting business accounting capability is the most basic capabilities of accounting personnel, the specific design capabilities include accounting, accounting business daily processing capacity, property inventory capacity, tax accounting, reporting capabilities, cost accounting capabilities, reporting and analysis capabilities.

3) The capacity of financial management
Any enterprise, its objectives are to develop business bigger, stronger, able to achieve this goal, affected by many factors, but its core is the business of financial management issues, specifically including financial planning and control, financial report analysis, decision-making capacity[3].

4) Ability to collect and analyze information
According to statistics, the information needed in the management of enterprises, 70% or even more from the accounting department can be said to affect the accounting information is a critical factor in decision-making, thus, asked the accounting staff must have quick response to information, organize , analytical ability, and thus for the enterprise managers and relevant departments to provide useful and reliable financial information[4].

5) Accounting capacity
Because of the widespread use of special computer was online tax system, the applications enable business-to-capacity computerized accounting requirements has become a prerequisite for appointment of accounting personnel's, Bing Chu Yu important of Wei Zhi, computerized accounting Nengli including its Xitong set , daily operational computerized processing, maintenance, capacity capabilities.

6) Sensitive force to number
Accounting numbers is mainly accomplished by the accounting staff every day and figures to deal with a number to record the occurrence of economic operations, with digital to analyze problems, to solve problems with numbers, and different numbers and they represent different economic and business , represents a different meaning, representing different results, which requires the accounting officer of the figures to be particularly sensitive to the value seen in particular the number should be written down in their minds, whenever they were taken, so do accounting work for will play a multiplier effect[5].

1.3   Good Language, Communication Skills

Accounting as a service position, with all aspects of dealing with people both inside and outside units, must learn to communicate with others, deal with the relationship between the various stakeholders, such as the annual inspection for industry, tax registration and tax returns, bank financing involved, with so better external auditing, accounting staff of these jobs require strong language communication skills.

1.4   Teamwork

Today's society, with the knowledge economy era, all kinds of knowledge, technology continues to introduce new competition has become increasingly intense, more diverse social needs, and let people work in the face learning situation and environment is extremely complex. In many cases, rely solely on individual ability has been difficult to independently deal with complex issues and achieve effective results. All of which need people to constituent groups, and further interdependence between members Requirements, mutual Guanlian and joint cooperation, the establishment of cooperative team to grasp the complex issues and make the necessary Hangdongxiediao, development team and the response Neng Li Xie Zuo, Yi Kao team the power of miracles[6].

1.5   Life-Long Learning, Innovative Spirit

Lifelong Learning in the 21st century rules of survival is to adapt to each individual's own career development and the only way. A knowledge acquired in school life, according to statistics only account for 5% of the required knowledge, most knowledge outside the school environment to get to enrich themselves become a necessity; Further continued economic development, social change has many aspects, new situations and new problems, new business after another, accounting standards and economic laws and regulations has continued to make the appropriate changes and adjustments that require accountants to be with the times, constantly learning, updating professional knowledge and, based on changes in economic activity, Accordingly, the financial and accounting innovation model, to adapt to the needs of work, to adapt to social development[7].

1.6   Ability to Adapt

Students learn the end of ten years of life, into society, due to environmental changes, many people feel too much different from past, many students are not feeling well, and some students also created a greater psychological pressure, to a tremendous impact on work and life[8]. This requires that students must have the environment to meet the edge and can Suizhe surrounding changes Erxun Su adjust our ways, thinking habit, thinking of the knowledge of the living environment to adapt to Xianyou actively work to adapt to the changing Zhu Dong environment.

1.7   Risk Prevention

Society is rapidly changing, the risk exists objectively. Enterprises to survive in society, in the financing of fiscal and investment decision-making is bound to face the risks, how to predict the risk, risk diversification, risk aversion has become an essential financial management professional quality.

2   Computerized Accounting Students Professional Quality Training Model Application in Practice

Over the years, we have been insisting the students ability-based, in practice, and practice of continuous bold reforms formed a multi-faceted professional quality training mode.

2.1   Demand Orientation Professional Jobs Professional Accounting Training Objectives

Related businesses in the region through a survey of accounting jobs, combined with the characteristics of vocational talents, defined "according to local economic development needs, front-line positions for corporate accounting work, student-based, computerized accounting profession essential to basic vocational skills and overall capacity of the base point, with strong practical training, applications and creative accounting professional personnel "training objectives.

2.2   Dynamic of Talent

First of all, to go to professional schools in the Steering Committee of experts on accounting standards work, professional competence and professional accounting curriculum and some specific issues, to seek business experts, and business experts on personnel training program content and implementation of the Ke Xingxing to discuss justification; then, the personnel training program implementation process, many times companies held special seminars, in-depth corporate accounting department for accounting positions Demanding professional competence and conduct a survey and recommendations according to business experts, survey data and students post practice feedback, timely revision of personnel training programs.

2.3   Deepen the Educational Reform, the Formation of Sub-sectors, Sub-level Personnel Training Mode

- According to accounting in different industries and different job requirements, implemented by industry sub-modular teaching and training posts. Targeted selection of production and circulation of commodities and other industries and cashier 4-6, materials, cost accounting positions 8-11 months, by industry and job specifications and standards, to determine the teaching content of each module and the teaching methods adopted and so on. Modular separation of points by teaching and training posts, reducing the accounting industry to work with the distance.
- To construct a competence-based, basic skills, professional training at different levels as the core of the training pattern.

Accounting expertise will be divided into three levels:

The first level of vocational basic skills training. Including: office automation skills; oral and written language skills; point of note, abacus and other basic skills. According to the basic skills standards, through training, you can equip students with basic accounting capabilities needed to post.

Second, the level of accounting skill training. Divided into five main modules: the job of coordination and communication skills; settlement and accounts processing; business recognition and measurement capabilities; records and accounting reporting capabilities; day to day management and control, and professional skills development into their curriculum teaching, extra-curricular the whole process of activities and social activities.

Third, the development of innovative ability level. On Improving the work of students in the future re-creation of job skills and potential development capacity. Include: creative development capabilities; thinking changed requesting different capacities; life-long learning; independent planning capacity; business planning capacity module.

To achieve these objectives, we get "four cardinal principles", that is: every day for a speech contest; adhere to a weekly computer operations training; adhere to a professional once a month basic skills test; adhere to a professional every semester simulation training (or training), and to ensure that the training, the continuity and effectiveness of practice.

2.4   Curriculum System Oriented by Work Process

According to the accounting professional job groups, analyze their typical work tasks that correspond to typical tasks to complete vocational ability, combined with occupational skills standards, corresponding areas of action summarized, and finally converted to fields of study courses.

In accordance with the design, computerized accounting professionals identify the typical career positions Gongzuorenwu You: Ri Chang Shen He filled in documents and accounts processing, payments Clearing, tax filing, profit calculation of the cost of Preparing the Financial Statements Fenxi, accounting, computer daily Ye Wu treatment 6, combined with professional competency requirements, identified 60 areas for action, and finally converted into "accounting basis", "Corporate Tax Practice", "cost accounting practice" and other six learning areas of core courses and 20 specialized courses.

2.5   Adhere to Combining Work with Study, to Strengthen Cooperation with Enterprises, Deepening the Curriculum Reform, Professional Ability for Students to Create the Conditions

1) Strengthen cooperation between schools and enterprises, to explore a "three-three" teaching mode.
"3": One teacher out of school to the business of teaching base for practice, training, enhance the experience; two students out of the classroom to the corporate understanding of internship; three students to leave school before graduation, to practice, post practice teaching base or candidates posts[9].

"Three": The first company to enter the school part-time teaching specialists; Second, the corporate accounting cases, examples into school to enrich teaching content; Third, the different sectors of accounting experts come in, a professional steering committee to help plan professional development.

2) Speed up the pace of curriculum reform
Accounting Professional and business together, "Financial Accounting" and "Accounting" and other curriculum development and building work. And business people to study the course content, teaching methods, assessment methods, preparation of training materials together. For example: we combine the requirement of accounting professional job, and business experts together to discuss and change the "Financial Accounting" course of testing methods to remove the traditional beginning of a book examination of the manner by highlighting the practical skills assessment, that is, the settings from the books to handle business documents, keep accounts, report preparation, corporate accounting practice used in all business, and require students to conduct field operations, the entire examination process is always the students exposure to the environment of accounting positions, reducing the student and the distance between the actual accounting positions[10].

2.6   Use Multiple Teaching Methods

According to the teaching profession were the characteristics of different courses to take a "case teaching", "task-driven" and many other teaching methods. One representative is:

1) The role of experience law
Transmission in the accounting documents, the daily business processes such as teaching content to students, with handling personnel play a business, manufacturing a single staff, audit staff, cashier, bookkeeping personnel role, through mutual business simulation, to equip students with the accounting positions in the actual division of work and interface, accounting and related business relationship, and how to play in the accounting oversight functions and so on.

2) Extracurricular activities, learning through racing law.
Professional Accounting students often carry out the basic skills for all to participate in professional competitions, contests, prizes set up groups and individuals, the end of each semester to conduct a table on the competition Jiang, Bing Jiang results of the competition and the scholarships and student assessment Xiang annual Appraising Contact. Activities have greatly stimulated and increased enthusiasm for learning, playing through the contest to promote learning.

2.7   Implement the "1 +2" Certificate Management Regulations

1 refers to the "graduation certificate", 2 means "accounting qualification certificate" and "Basic Skills Professional Certificate"
    To strengthen the professional ability of accounting students, we ask students to receive diplomas at the same time must have "qualification certificates" and the school, "the basic skills of the professional certificate" two skills certificates.
    "Professional Certificate in Basic Skills" is a unique school Accounting outstanding professional skills of students in basic skills training as a school professional certificate, which includes the abacus skills, the digital word, the word accounting-specific writing skills, Counting skills, character entry skills Financial practical writing skills such as assessment content.

2.8   Through the Introduction of External and Corporate Practice, Exercise and Other Measures to Establish a High-Quality "Double" Teacher Team

At present the professional full-time teachers all have a dual qualification, of which 4 business consultant, corporate trainer and a person from the company's three part-time teachers, long-time "business accounting", "Tax Practice" course and practice links teaching task, as the practical ability of students to play its unique features.

2.9   Strengthen Training Base Outside the Building Practice for Students, Providing Favorable Conditions for Employment

Accounting is a very practical profession, should have enough practice teaching activities, or difficult for students to master basic accounting theory and basic skills. To protect the practice, practice practice teaching and other smooth, on the one hand, we construct or expand the accounting manual simulation, computerized accounting training room two schools, the purchase of advanced teaching equipment and genuine financial software, and accumulated large enterprises and from the training is constantly updated data, the most important day of our training room open to students, full training platform for students; the other hand, the past few years, we use a variety of way, comprehensive and stable set up outside a dozen practice base for students to create the conditions for field work, but also provide employment opportunities for students.

Acknowledgment. This work is a part of the Eleventh Five" social science research project "The Research and practice of Accounting Majors professional quality training mode " ([2008] No. 405).

References

1. Page, S., Meerabeau, L.: Hierarchies of evidence and hierarchies of education: reflections on a multiprofessional education initiative. Learning in Health and Social Care 3(3), 118–218 (2004)
2. Payne, M.: Social Work Theories and Reflective Practice. In: Dominelli, L., Payne, M., Adams, R. (eds.) Social Work: Themes, Issues & Critical Debates, 2nd edn., Palgrave / Open University, Basingstoke (2002)
3. Redfern, M.: The Report of the Royal Liverpool Children's Inquiry. The Stationery Office, London (2001)
4. Schon, D.: The Reflective Practitioner: How Professionals Think in Action. Ashgate, Brookfield (1998)
5. Smith, J.: Death Disguised: The First Report of the Shipman Inquiry. HMSO, London (2002)
6. Smith, S., Roberts, P.: An investigation of occupational therapy and physiotherapy roles in a community setting. International Journal of Therapy and Rehabilitation 12(1), 21–29 (2005)

7. Tate, S.: Using critical reflection as a teaching tool. In: Tate, S., Sills, M. (eds.) The Development of Critical Reflection In the Health Professions. Occasional Paper No. 4. (2004), http://www.healthacademy.ac.uk/publications/occasional
8. Wenger, E.: Communities of Practice. Cambridge University Press, Cambridge (1998)
9. Wenger, E., McDermott, R., Snyder, W.: Cultivating Communities of Practice. Harvard Business School Press, Boston (2002)
10. Yelloly, M., Henkel, M. (eds.): Learning and Teaching in Social Work: Towards Reflective Practice. Jessica Kingsley, London (1995)

The Study on Digital Art Based on Science and Technology

Chen Su

Guilin University of Electronic Technology
chen.xiaojing@yahoo.cn

Abstract. It is difficult to resist the allure and charm in modem information technology. But it is certain that we must admit that digital art is also art. Digital art works can not make the rules of technology bigger than the rules of art; it can not modify and disguise the shallow performance of the concept of art with the coat of the technological digital form which weakens the charm of digital art and the interaction tension in the spirit and emotional between works and audience.

Keywords: Digital Art, Art, Art Form.

1   The Definition and Background of Digit

Digital art refers to the art with various forms of independent aesthetic value and with the basic features of interactive and the use of the network media produced by a variety of digital and information technology

Since the 20th century, it is undeniable that the development of technology has changed many aspects of the world, including art. Two powerful forces are promoting the innovation of art: First is the change of concept originated from quantum mechanics and relativity of the natural sciences and other thoughts such as modern philosophy and social science and Postmodernism. The second is the changes of scientific development tools originated from the innovation of the media and network technology. The effects of these on the innovation of art are done through media. Fresh artistic experience of art creation and trends of social life are compiled by the media too. Therefore, the change of media is one of the most far-reaching historical events. And the newly-emerging digital media not only provides an unprecedented means of communication which enhances the innovation of art forms and reflect s the needs of social life for scientific theory and social thoughts but also improves the innovation of art research in concepts and methods. Humankind is in an era of rapid high-tech development. And the accumulated knowledge of the last 30 years is equivalent to the sum of the past two millennia. We must acknowledge that the development of scientific technology has cultivated a new land for human civilization.

The digital art we are familiar with has not been like this; it is evolved gradually to meet the multidimensional needs of the audience with the development of social forms. The separation of science and art is a phenomenon arising from the particular

Y. Wu (Ed.): International Conference on WTCS 2009, AISC 116, pp. 577–581.
springerlink.com   © Springer-Verlag Berlin Heidelberg 2012

stage of human civilization. This separation not only means different areas of activity, vocational skills and knowledge, but also means different outlook to the world.

Arts and Sciences are certain different as two ways to see the world. For example, it is recognized that art is a form of ideology, but it is a controversy to say that science is an ideological form. However, the differences between arts and sciences do not affect the penetration and integration of them. Art can not do without rational support, intervention and help. Artistic imagination can become orderly and fee, art can become aesthetic emotion; artistic creation can become co-rule and fee with a rational, scientific reason in particular. This reason is provided in large part by the development of science and intellectual structure. Art can be prosperous and highly-developed by the intervention and support of scientific.

2   Introduction

With the evolution of human being, art was bor. For thousands of years, Creative art Inspiration has been stimulating by the advantage of technology. "Science and art are the two sides of coin." Tsung-Dao Lee says, the famous Physicist in the world, won the Nobel Prize with Zhenning Yang, also the famous Physicist in 1957. That means that Arts and Sciences are the same and can not be separated each other. Nowadays, computer technology and network technology are developing rapidly, which makes technology a realistic tool and communication vector for artistic creation.

The faster development of science and technology is, the more obviously Digital Art shows us its special art form. Art is to adapt to the needs of people, born and evaluated In certain social conditions. Digital art, as the emerging form of Mass Art, depends on the development of digital media and the Internet to spread. Because it' s no question that the art works which displayed in font of audiences are more influential than that not. Digital Art is really different from the traditional art in communication, which has very strong features of the post-modem art.

3   Digital Art Is a New Art Form Based on Science and Technology

It's widely accepted that Digital A is a kind of new art, which is based on modem. Today, we can see clearly the distinctions between the forms of Digital A and of others. Such as literature, paintings, music, dance, although these arts also depend on technology, for examples, Literature with application of the printing technology, paintings with application of the perspective theory, music (musical instruments) with application of manufacturing technology. But it cannot be said that they are the direct products of technology, because technology itself is external and acquired to them. On the other hand, literature, paintings, music and dance can exist without these so-called technologies, while Digital A can't. Digital A, we know that, is a new art form, established new media which is the carrier, dependence and means of it. Therefore, Digital A concentrates science and technology. The birth and development of digital art is the direct application of results in science and technology revolution in the field of art. It marked the beginning of human art into a new era.

4   Science and Technology Development Is the Basis on the Existence Digital Art

Digital art is the close combination of information communication technology and new art ideas. With digital technology and new media as the technical conditions, it is supported by multi-disciplinary integration with the media of Graphic images, humanistic ideas and scientific technology. It has a double attributes of science and humanity. It is a newly-emerging arts that links by science and culture.

Digital a is generated with the development of scientific technology and culture media industry which becomes a new field of human creation. It provides an unlimited potentiality and high efficiency for artistic creation and enriches its contents, expands the space through computer technology, such as input, storage and process.

Digital art is an art form that is difficult to deliver by the performance of the traditional art forms. It is a newly -emerging art that relies on the first productivity - scientific technology.

Digital art does not change the nature of art. It is an extension form of expression of the traditional art. It turns the art language expressed by traditional arts into a new art form that can be understood easily by using high-tech means. It draws the distance between the artist and the audience, becoming a strong integrated vector and makes it an independent art form.

Digital A differs from other arts in the elements of instruction. Such as words in language art, colors and kinds of stone in formative art, and body in performance art. As a new Category of art, artistic expression forms of Digital A goes far beyond the scope of traditional art. Besides the visual factors, such as image, picture, word, Digital A also has a voice, images, audio, interactive and so on in it. And its use of these basic visual elements is not like other types of art, by the direct processing, treatment, but with a decisive technical tools - modem high-tech tools - to deal with. As a result, thanks the tool itself is involved in the process of artistic creation, Digital A marked with a technical branding, rather than a single skill. Technology creates unprecedented new forms and new feeling of arts. Therefore, Digital A has a strong technology-based color art form within a basis on science and technology. And it is a comprehensive holistic systematic new concept of art style, and a set of artistic creation and digital technology in one, which is relative to science and technology each other. It's true that Digital art will increasingly become an important pat of the wholly art world.

5   Influence to Digital Art with the Development of Science and Technology

It has in-depth, intrinsically relationship between the number of arts and science and technology. First of all, the development of modem science and technology stimulates the scientific passion of human beings, develop the human consciousness and scientific thinking, and promote people's desire and attention to the scientific and technological achievements. Consequently, science and technology creates the external conditions of application in all areas. Secondly, for the emergence of categories for arts, science and technology provides a direct, necessary technological base. More importantly, the technical characteristics and rules produced by modern science and technology play an important role in artistic norms and shaping.

Digital art is the product, when development scientific and technological goes to a certain stage. Because, as the technological advance in digital technology as an ar digital art, technology constitutes the premise of its existence and indispensable conditions. Digital art depends on the scientific and technological achievements of directly, which makes technical decisions in the basic characteristics of ar. Undoubtedly, technology had a tremendous impact on the arts.

Science and technology, a strong creative ability and their ideology, and direct manipulation to concept, is bound to affect and change the ideology of art. So, Digital A is one outstanding example. As it's dependence on technical means, Digital A becomes an usage and display of current technological capabilities, and a understandable form of the technical characteristics. Technological means of Digital A weakens the purpose of art in itself. And the combination of a and technology has changed the original meaning of art. Difference between Digital Art and traditional arts, or the unique nature of digital art, lies in the changes between aims and means. Recently , this obvious transformation from human-to in order to mater (technology)-based, establishes one of the most important prerequisite for Digital Art, which provides all the basic characteristics of type, shape, style and substance to Digital Art, as one kind of arts.

With today's rapid development in information technology, the technical content in digital art is much higher than ever previous art forms. However, we should clearly see that this phenomenon will abate, once the technology development tends to homogenization, So, this is not avant-garde awareness. In fact, it is a traditional concept "technology frst" of continuity in the industrial society.

6   Conclusion

At present, we criticized the lack of digital art works of humanistic care. At same time, we also affirm the value of digital art features. Because digital art makes a modern sense of cultural transmission in a global scale through a variety of media, leads to a new concept of ethics and human spirit using Digital art, and poses a new challenges for the living style in modern human society. It fully shows that digital art permeated with cultural connotations with support of technology. Artists continue to learn the essence of excellent culture from the personal integrity and intellectual level. I fully shows that digital art permeated with cultural connotations with support of technology. Artists continue to lea the essence of excellent culture from the personal integrity and intellectual level. Meanwhile, digital artists can use digital media to do multi-dimensional creation with the combination of digital a in a market.

References

1. Yang, E., Mei, B.: Art. People Publishing Press, Beijing (2001)
2. Mark, D.: Non-material Society–Design Culture and Technology of Post-industrial World. Sichuan People's Publishing Press, Sichuan (1998)

3. Dannal: Philosophy Art, Fulei. Social Sciences Academic Press Tianjing (2004)
4. Negroponte: Being Digital. Hainan Publishing Press, Haikou (1997)
5. Tian, S.: Network Communication. Science Press, Beijing (2001)
6. Lu, X.: Huze, Situation - Dialogue of Chinese Contemporry Art. lincheng Publish Press, Beijing (2002)
7. Wang, S.: Modem Design History in the World. China Youth Publish Press, Beijing (2002)

Improved Intelligent Answering System Research and Design

Chuan Liang

Computer Science College
Sichuan University
Chengdu, China
KaiLiang011@126.com

Abstract. As the developing of Internet, Intelligent Question Answering System is playing an important role in the network education. In the current, Intelligent Question Answering System has some problems such as low recall and precision .In this paper, we use maximum matching algorithm with improved lexicon, we also use a kind of similarity calculation method with statements based on the weight vector feature model. So we greatly improve the efficiency and accuracy of Intelligent Answering System.

Keywords: Distance Education, Intelligent Question Answering System, Improved Lexicon, Chinese Word Segmentation, Similarity Calculation, Weight.

1   Introduction

With the development of internet, traditional education is gradually developing to the network. It combines current teaching ideas, teaching mode and the educational system, realizes the change in the mode of teaching, overcomes the traditional education in time and space constraints, makes high-quality education resources have been more extensive spread. One need only access to the Internet, you can participate learning anytime and anywhere, more and more people participate in the network education. But now network education still has some issues to be resolved, For example: in real life education and teaching, students are free to ask questions to the teacher and teacher can answer face to face. But in network education, the teacher can't answer so many questions in time. And if answering constantly repeated questions, it is not efficient for the teacher's work. Therefore a convenient and efficient intelligent question answering system is very important for network education. The Intelligent of Intelligent Question Answering System is reflected in:

1. Permit the natural language to describe the problem
2. Answering automatically by a computer
3. System can give a specific answer.

Y. Wu (Ed.): International Conference on WTCS 2009, AISC 116, pp. 583–589.
springerlink.com   © Springer-Verlag Berlin Heidelberg 2012

2   System Model

System has the following modules: authentication Module, problem input module, Chinese word processing module, statement matches module, teachers answering module.

Background database including: Words Library, Question Library, Answer library.

(1)Authentication Module: Intelligent Question Answering System has access control, only after the user log on, he can question to the system.

(2)Problem input module: Students enter the problem in the browser's text box, and then submit the problem to the system by form.

(3)Chinese word processing module: the Chinese word extraction is much more difficult than English, so Chinese word segmentation is a key step in Intelligent Question Answering System. We use maximum matching with improved lexicon process words, so it can improve efficiency, sub-word length limit even ambiguity.

(4)Statement matches module: After the issue of word, the problem need compare similarity with the problem in Answer Library. In this paper, we use the weight-based similarity measure; it is more efficient and accurate than common similarity calculations measure.

(5)Teachers answering module: If the system can't find the match answer, the system put the question to the teacher.

3   Key Technology and Algorithms Intelligent Question Answering System

3.1   Traditional Chinese Word Segmentation

Traditional Chinese Word Segmentation methods often can be divided into two common methods. They are understanding syncopation and machine matching.

1) Understanding Syncopation Method
Understanding syncopation method's word segmentation system consists of lexicon, knowledge base and inference engine parts. Lexicon store vocabulary entry. Knowledge Base store formal language rules, grammar and Linguists' knowledge and experience in the process of segmentation and inferring. Reasoning mechanism use large amounts of data and knowledge which provided by lexicon and knowledge base, imitate Linguists' logic thinking process, realize automatic segmentation. This is actually an automatic segmentation expert system. This kind of system has too much overhead. In addition to theoretical difficulties, this system also has the problem of system complexity and implementation difficulties.

2) Machine Matching Method
Machine Matching Method is mainly based on the principle of string matching. When matching, there are no parsing, no semantic analysis, only Mechanically Matching and comparison. It is based on a large enough vocabulary, and use a certain treatment strategy to compare string in the text with words in the lexicon one by one; if matching success, the string is a word. Common Word Segmentation method has

positive maximum matching method, reverse maximum matching, minimum cut algorithm, maximum matching word segmentation algorithm is easy to implement, but it has many obvious defects:

a) Length limitation: At first, the maximum matching method must set an initial value of the matching word length. If word length is too long, efficiency will be relatively low and if word length is too short, long word will be wrong.

b) Low efficiency: Even if the word length can be set to a relatively short, such as word length of 5, when the length of our large number word is 2, at least 3 times the matching algorithm is wasted.

c) Maximum matching may be not a expected segmentation approach: Maximum matching method is based on the idea of finding the largest matching words, but sometimes expect the maximum matching words, may only need part of the word.

Based on the above analysis, we propose a solution to improve the efficiency of segmentation algorithms, word length limit and even ambiguity handling.

3.2   Improved Chinese Word Segmentation

Building lexicon is the first step, in order to improve Maximum Matching Algorithm in efficiency and limit of the word length even ambiguity, it must have lexicon to match the words. It needs to reform lexicon and make lexicon more suitable for matching word. Broke up the word in relational database by word and store them in the Hierarchical Database. The following is an example of hierarchical lexicon, as shown in:

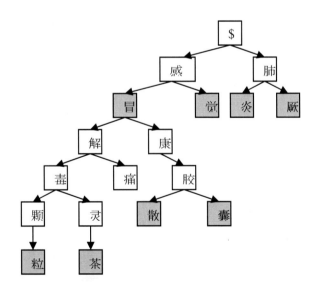

Fig. 1. Hierarchical lexicon sample

Gray words means that string on the tree can be composed a single word. For example "感冒" can be found in the lexicon ,all the gray word is the terminator character. The white word mean the string on the tree can't compose a word, for example, "感冒解" is not a real word. After lexicon modified, any sentence would be broken down into a word, then match the tree structure word, the length of the word become a tree height, every match become a tree traversal.

The specific steps of word segmentation:

(1)   First of all, the full text will be divided into sentences by punctuation marks.

(2)   Traversal the sentence in the tree structure, if you experience white word then continue; if you experience gray terminator, we found that the word is a complete WROD, so you can put the word as a WORD.

(3)   From the next word, continue to do Step (2), so recycle, at last word segmentation would be finished.

3.3   Similarity Calculation Based on Weigh

In the intelligent answering system, sentence processing use Vector Space Model, it is the most commonly used information retrieval model, then calculate the similarity between vectors. At last, one or several candidate answers ranking front will return to the user.

Calculation method of weighted semantic similarity between sentences:

Suppose eigenvector of two sentences are A and B;

$$A=(x_1, x_2, \ldots, x_n)$$

$$B=(y_1, y_2, \ldots, y_m)$$

$x_1, x_2, \ldots, x_n$ are words extracted from sentence A, $y_1, y_2, \ldots, y_m$ are words extracted from sentence A

(1)Structural similarity matrix M_{AB}.

$$M_{AB} = \begin{bmatrix} X_1Y_1 & X_2Y_2 & \cdots & X Y_m \\ \vdots & \vdots & \ddots & \vdots \\ X_nY_1 & X_nY_2 & \cdots & X_nY_m \end{bmatrix}$$

X_iY_j represent words X_i's and words Y_j's semantic relevancy

(2)Calculation of weight vector

Professional words should set higher weight, it will set by relevant professionals. The common words' weight is mainly on the basis of obtaining statistics. Frequency is divided into absolute frequency and relative frequency. Absolute frequency is frequency of occurrence in the text. Relative frequency is normalization frequency. Calculation method using general TF - IDF formula :

$$W(t, \vec{q}) =$$

$$\in \frac{tf(tf(t,\vec{q})) \times \log(\frac{N}{n_i} + 0.01)}{\sqrt{\sum_{t \in \vec{q}} \left[tf(tf(t,\vec{q})) \times \log(\frac{N}{n_i} + 0.01) \right]^2}}$$

$W(t, \vec{q})$ is the weight of t in question $\vec{q}$, tf $(t, \vec{q})$ is frequency of in question $\vec{q}$, N is the total number of questions in answering system, n_t is the number of question which appear t in question set, the denominator is normalized molecules.

(3)Matrix M_{AB} compression to one-dimensional, such as formula

Set $m_1, m_2, \dots, m_n$ represent $x_1, x_2, \dots, x_n$ weight.

Weighted Semantic Similarity (AB)

$$= \frac{\sum_{i=1}^{i=n}(w_i * (\max(x_i, y_i), j \in [1, m]))}{\sum_{i=1}^{i=n} w_i}$$

That is, get the maximum of each row of the Matrix, then weighted sum of these maximum. Using the same method can be obtained:

Weighted Semantic Similarity (BA).

(4)Calculation of A and B sentence semantic similarity formula.

Weighted Semantic Similarity (|AB|) =1/2(Weighted Semantic Similarity (AB) + Weighted Semantic Similarity (BA))

This algorithm to join the weight difference algorithm of words, make semantic similarity calculation   more reasonable.

4   Technology Implementation

The system is implemented by the Java EE development technology. We use some application Framework such as Struts 2, Spring 3.0 and MyBatis. We use Mysql database.

Apache Struts is a free open-source framework for creating Java web applications. Web applications based on JavaServer Pages sometimes commingle database code, page design code, and control flow code. In practice, we find that unless these concerns are separated, larger applications become difficult to maintain. One way to separate concerns in a software application is to use Model-View-Controller (MVC) architecture. The Model represents the business or database code, the View represents the page design code, and the Controller represents the navigational code. The Struts framework is designed to help developers create web applications that utilize a MVC architecture.

Spring is a popular and widely deployed open source framework that helps developers build high quality applications faster. Spring provides a consistent programming and configuration model that is well understood and used by millions of developers worldwide. Unlike the traditional Java EE platform, Spring provides a range of capabilities for creating enterprise Java, rich web, and enterprise integration applications that can be consumed in a lightweight, a-la-carte manner.

The MyBatis data mapper framework makes it easier to use a relational database with object-oriented applications. MyBatis couples objects with stored procedures or SQL statements using a XML descriptor. Simplicity is the biggest advantage of the MyBatis data mapper over object relational mapping tools.

MySQL is a small relational database management system. Currently MySQL is widely used in small and medium website. Because of small size, high speed, low total cost of ownership, in particular the characteristics of open source, many small and medium sized websites chose the MySQL database in order to reduce total cost of ownership.

Students log in through the browser, and then input the problem. The problem is submitted to the business layer. Using the improved Chinese word Segmentation technology, we are to sentence segmentation. After segmentation, we judge the similarity of the questions, and then matching answer will be return to students. If there is no matching answer, the problem will be submitted to teachers.

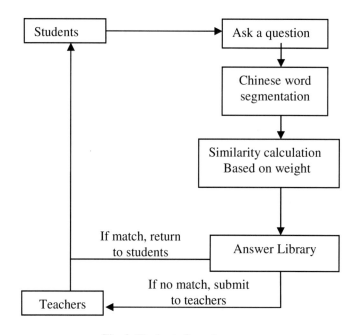

Fig. 2. The basic flow chart

5   Conclusion

This paper make a comprehensive analysis and introduction in Intelligent Question Answering System, after searching a lot of relevant information, we use maximum matching algorithm with improved lexicon and weight-based similarity measure, it make Intelligent Answering System efficiency and accuracy improve greatly. Experimental results show that users can quickly search for intelligent answering system to your answer. Relative to the general Intelligent Question Answering System, the system efficiency has improved by using effective weighted similarity algorithm and the improved lexicon, but there are still many deficiencies.

References

1. Qi, W.: An improved algorithm for Chinese word segmentation. Huangshi Institute of Technology 23(4), 24–25 (2007)
2. Kang, W., Yang, Z.-Q.: Similarity calculation in intelligent answering system and its application. Computer Technology and Development 20(2), 71–74 (2010)
3. Wang, H.Q., Zou, Z.: Internet-based teaching network engineering intelligent answering system. Science Information (33), 148–149 (2008)
4. Yan, H., Yan, D.: ntelligent Question Answering System Research and Design. Computer Development and Applications (5), 12–14 (2010)

The Research on the Application of Multimedia Technology in the Teaching of Professional Sport Judge

Lei Xu, Wei Zhang, and Xiujie Ma

School of PE
Handan College
Handan China
xusongxiaorong@sina.com

Abstract. According to the social requirement for basketball judge, we reformed the teaching program, teaching content and method, exam content and mode of basketball course in physical education and closely combined theoretical teaching with in-site practice creating internship to students, improve students' application abilities as a judge in basketball game, increase the passage rate of class-2 basketball judge certification, and provide a good platform for students.

Keywords: Multimedia Technology, Physical Education, Judgment Teaching.

1 Introduction

Basketball judgment teaching is an important component of basketball teaching in a sport college and its teaching quality may directly affect students' understanding about basketball and competition level. We should locate the difficulties and key points of training in view of the training of basketball judge. Application of the advanced teaching approach, multimedia technology, enables learners to do selective and effective learning according to demand and achieve double effects with only half of effort required; by the simulation of microcomputer, we can give simulated tests and trainings, learning performance can be fed back in time, and enthusiasm of students can be fully mobilized to improve the learning performance. From the current research, the study on the teaching of basketball judgment is only centralized in traditional media (lantern slide, projection and electroacoustic system) while the study on the application of multimedia combined with network technology is very rare. For that, this paper offered a discussion.

2 Training of Judgment Abilities and Existing Problems in Basketball Teaching

2.1 Training of Judgment Abilites

1) Imagine Training
Image training is the most common training way of mental skills in sport field and requires students to repeatedly imagine some kind of sport or sport scene under the

Y. Wu (Ed.): International Conference on WTCS 2009, AISC 116, pp. 591–596.

guidance of implication. A teacher should help students set up the integral set of concept and action image of basketball game. For example, teacher can express some illegal action with brief words and then asks students to memorize with the same words, improve and consolidate related action image with the words.

2) Simulation Training

Simulation training aims to enable students to do repeated exercises to deal with various possible situations or problems in a game and prepare students for various problems in working as a judge during a game. At first, students can complete action modeling by simulating teacher's posture, as well as watching game video and simulating the whistling, gesture and manner of modern international judges. Teacher can also design common problems in a game to let students learn to control the situation and solve the problems.

3) Self-Suggestion Training

Self-suggestion is the process to impose effect on the psychology of students by stimulus, like language, and then control their behavior. This kind of training mainly aims to enhance students' confidence in working as a judge. As students in basketball major are common students, though they have received more systemic trainings than other students, they are still far from the level of a professional judge in the understanding about rules and in-site skills and the control on the scene is not complete resulting in lack of confidence in working as a judge. Therefore, teachers should give more encouragements to students in usual time and improve their confidence with self-suggestion training.

2.2   Existing Problems in the Training of Judgment Abilities

1) Gap between theory and practice

The most common teaching method of judgment teaching is the learning of judgment theory first and then the application of practice. As the practice lessons of judgment only take a very small proportion of basketball teaching, teachers have to insert the knowledge of judgment into normal basketball lessons. At present, there are more than 30 students in each class and each student gets very limited time for practice, and many students think everything will be fine as long as all the rules and judging methods are memorized, but in fact they have low practical abilities and are hard to be competent for the basic work of a judge.

2) Lack of opportunities of practice

School environment and the restrictions of social rules and mechanisms decrease the opportunities of practice and the knowledge learnt in class will be forgotten very soon for lack of practice. The training and improvement of judging abilities should be through practice and the teaching without practice is incomplete.

3) Single Teaching Method

At present, many PE teachers of colleges don't pay enough attention to intuitionistic multimedia courseware and are lack of assistant equipments for technical analysis and strategic study. Though some colleges are equipped with multimedia teaching equipments, video materials cannot keep pace with teaching process and teachers are lack of assistant equipments for technical analysis and strategic study, especially the video materials recording the gesture of basketball judges, movement and regulations, and students are lack of perceptual knowledge about the work of judge.

3   Significance of Multimedia Technology in Judgment Teaching

3.1   Improve the Efficiency of Memory

Cognition psychology proves that human's perception starts with sense organs and people collect all kinds of information by vision, hearing and touch, and then process these messages and form memory, thinking and imagination basic on analysis and comprehensive judgment. The cognition process of the youth is with imaginal memory and thinking as main forms. Cooperation of multiple receptors is an important condition to improve the efficiency of memory. The way of multimedia fits the cognition characteristic of human. Multimedia technology presents the information by image, text, sound and video after computer process and can fully stimulate various organs of learners to improve learning effect and efficiency.

3.2   Improve the Efficiency of Basketball Theoretical Teaching

The adoption of vivid and visual teaching mode by multimedia technology is the approach to improve students' interest in basketball theoretical teaching and teaching performance. In the teaching of basketball judgment, dull notes and tedious instructions often make students feel extremely bored. Those static pictures or models, which are exhibited now and then, are also hard to make students understand the complicated dynamic process of body activity. Multimedia technology can show it by real scene or animation. For example, for the related introduction about the history of basketball rule, regulations and judgment rules, we can insert some images and animations with texts to enable students to become clear at a glance. Besides, the interactive function of multimedia enables students to be involved in teaching and convert passive learning to positive learning , and thereby largely mobilize the enthusiasm of students in learning. This will be of great help for improving learning quality of students, learning environment and enhancing learning performance.

3.3   Helpful to Improve Case Analytical Abilities

In a basketball game, a judge often has to face various problems. Sometimes a judge will make a false judgment due to insufficient understanding about rules or misunderstanding, which may result in the dissatisfaction of players. Therefore, case analysis has become a key subject in the theoretical examination for judge.

In the past judgment teaching, a teacher only read or wrote down the blackboard for students to learn. Even though a student offered a correct answer, we couldn't tell he/she had fully understood. This way deviates from the original target of training. CAI multimedia teaching can present both texts and game videos with teacher's instruction, so that students can understand at a glance and remember it in heart.

3.4   Helpful to Improve In-site Commanding Abilities of Students

High quality basketball games are very fierce and raise more and more requirements for the judge. Sometimes the body contact between players happens in a moment, but a judge will be in a difficult situation to make a correct judgment for lack of experience. The opportunity to be a judge in a high-level game is very limited and students are impossible to gain the experience from many games in a short term. But, multimedia teaching software can solve this difficulty. Teachers can make a video set of controversial game segments and give instructions to help students accumulate experiences. In schools with sufficient conditions, teachers can shoot a whole high-class game. In multimedia teaching, students feel like being personally in the scene while watching how a judge makes decision in the video. This way can largely improve students' in-site commanding abilities no matter from psychology or flexibility.

3.5   Enrich Teaching Resources and Improve Teachers' Quality

During teaching, the search and editing of teaching resources will greatly enrich PE teaching resources and build up the foundation for multimedia teaching and network teaching in the future. Meanwhile, this new teaching mode will bring many changes to teaching, for example, changes in teaching approaches and methods, ways of giving lessons, relationship between teachers and students, combination of teacher's experience with teaching contents, and so on. These are largely different from traditional teaching ways. Computer-assisted instruction raises higher requirements for the quality of teachers. Therefore, through it, teachers can improve their own qualities to adapt to social and technology development and complete teaching assignments better.

4   Suggestions

4.1   Liberate Mind and Specify Direction

Place the construction of multimedia at the first place of PE development, make practical and feasible development plans, gain sufficient attentions from school leaders, and carry out multimedia construction under the attentions of school leaders.

4.2   Improve Understanding and Set Up the Basic

Comprehensively improve the understanding about multimedia technology in PE teachers and staffs, enhance the consciousness of its assistance to education by modern approaches, attach importance to the training and learning of young teachers, and set up the basic and make good preparation for the opening of multimedia teaching.

4.3   Transverse Cooperation to Exert Advantages

The fabrication of multimedia courseware for physical education is an integrative project covering professional skills and computer skills. Transverse cooperation in the development of multimedia courseware is a short cut to multimedia teaching, so it is critical to exert the advantages of cooperation between PE staffs and professionals in the development of media teaching.

4.4   Key Investments to Guarantee the Implementation

The basic construction, development and application of multimedia requires certain investment and equipments to guarantee basic working environment and conditions.

4.5   Attentions from Leaders and Proper Policies

The development and application of multimedia courseware consume large amount of human resources and time, so related leaders should offer certain preferential policies in achievement evaluation, discourse study, title assessment and teaching assignment to guarantee the smooth implementation of the project.

5   Conclusion

1) In the teaching of basketball judgment for students of PE education department, application of multimedia teaching method caters for the requirement of teaching reform and helps to improve students' abilities as a basketball judge and flexibility in the scene. It has a specific teaching goal and easy to control and master.

2) In multimedia teaching, the teaching content and target of each phase are single and feedback is in time; students and teachers can communicate and exchange information timely and constantly improve judging skills of basketball. Multimedia teaching for the training of basketball judge is the optimization of current training methods and represents the direction of education reform.

References

1. Wen, J., et al.: Experimental Study on Optimizing the training of Basketball Judge by Modern Information Technology. Journal of Hubei Institute For Nationalities (natural science edition) (December 2007)

2. Xie, L., et al.: Training Basketball Judge by Multimedia Teaching Approaches. Liaoning Sport Technology (October 2002)
3. Yuan, T., Xi, Y.: Application of Computer-Assisted Instruction (CAI) in Basketball Course. Journal of Anhui Normal University (natural science edition) (July 2007)
4. Ma, L., et al: Development and Application of CAI Courseware for Basketball Judge Training. Journal of Shanghai Sport College (November 2001)
5. Liu, F., et al: Cultivation of Basketball Judging Abilities of Students in PE Major. Journal of Chongqing Industrial and Commercial University (natural science edition) (February 2009)
6. Li, D.: Experimental Study on the Application of Multimedia CAI Teaching in the Training of Basketball Judge in College. Journal of Sport Technology (December 2009)
7. Ren, T.: Experimental Study on the Application of Multimedia CAI Teaching Technology in Public Physical Education of College. Journal of Sport Technology (2007)

Feasibility Analysis of Yoga Course in Public PE Course of College

Wei Zhang, Lei Xu, and Weimin Guo

School of PE, Handan College, Handan China
xusongxiaorong@sina.com

Abstract. The Current PE teaching of college attaches more and more importance to the cultivation of student's lifetime sport awareness and ability and particularly emphasizes the introduction and development of an uprising project. Yoga, as an ancient sport, has stood out from so many sports with its unique charm and is favored by more and more people. This paper discussed the feasibility of yoga as a course of public PE course in college by interviews and questionnaires, and provides reference for the reform of PE teaching in college.

Keywords: College, yoga course, feasibility.

1   Introduction

There are more and more reports about the mental problems of college students in recent years and it is very urgent to establish a course which can relieve the mental pressure of college students. Sport can yet be regarded as a good way to reduce pressure. Yoga, which roots in the practice method of the ancient Indian a few thousand years ago, has its unique characteristics from other sports: it can coordinate the functions of every part of body and make people healthier; from higher level, it can cultivate people's focus, peace, calm and objective mind, heart and nature to achieve mental and physical health. It is an excellent sport suitable for college students. Therefore, the study on the feasibility of setting yoga as a course of PE in college has important significance of enriching teaching content of higher education, promoting the reform and development, and cultivating students' healthy awareness.

2   Target

To define the target of course is an important premise to guarantee the smooth implementation of yoga course. The setting can give the following three main effects:

2.1   Strengthen the Physical Quality of College Students

Traditional yoga postures include stretch, press, wrest, pull, bend and so on. Through the learning in proper sequence, students can have more powerful muscle, have

Y. Wu (Ed.): International Conference on WTCS 2009, AISC 116, pp. 597–603.

physical capacity and bone intensity improved, and strengthen the feasibility and toughness of joints. Yoga can accelerate blood circulation to eliminate the toxin from body and promote metabolism, and make muscle more flexible. Yoga can not only maintain the health of nerve system, but also help those abnormal nerves return to normal status. For example, due to various pressures, college students often have endocrine disorder while bow and fish poses can help to adjust the function of endocrine and prevent it from falling into abnormal status.

2.2   Improve the Mental Health of College Students

As college students have dynamic energy, rich emotion, large emotional fluctuations as well as various external pressures from study and environment, if the pressure can't be released timely and correctly, spirit will be in long-term tension which is likely to result in anxiety, schizophrenia and other mental diseases, and even dangerous behaviors to society. Yoga exercise can help to eliminate fretful and complicated emotions and return to a kind of peaceful and comfortable status. Various yoga exercises aim to help people maintain physical health, but also bring the peace to one's heart. Also, to maximize the benefit from practicing these poses, one must have a peaceful and quiet mood to face these exercises. The breathing technique of yoga can adjust autonomic nerve, control heart rate, and ease tension, especially the complete yoga breathing which can smooth internal nerve, relax nervous body and mind, bring peace and sedateness and shine the charm of life. Meditation of yoga is the essence of the whole system and can help with mental focus, controlling self-awareness and adjusting the capacity of mind and body, and make one calm down without panic before changes, and keep a kind of peaceful and pleasant status [1].

2.3   Improve the Moral Cultivation of College Students

When mentioning yoga, people always relate it with the improvement of physical and mental health. Of course, it also fits the guiding concept of PE teaching "health first", but few people will mention its function in personal moral cultivation. At present, many college students are the "only-child" in the family and grow up with the excessive love and care from parents. They often develop selfish personality and always start with their own interest and consider about themselves first without caring about others' interests and feeling. Such a reality brings great difficulties to our harmonious society. Of the main yoga systems, there are two worthy of our attentions, practice yoga and service yoga. The practice yoga (or working yoga) is the system of unselfish activities or work; service yoga refers to the devotion to the absolute whole. During the practice of yoga, when we bring loving care and unselfish devotion into it, it can not only improve physical health and inner happiness in return, but also sublime spirit and morality. In life, when we devote our caring heart to the ones who need help, we can get great happiness in heart which in return promotes the physical health. In the "Gift of Yoga" written by Ms. Zhang Huilan, the "mother of yoga in China", her instructor wrote: "'the root provides nutrients to make tree bear sweet fruits, but it never asks for a return; we need to know our unselfish devotion will bring endless happiness in return.' This is the yoga spirit of Zhang Huilan" [2].

3   Feasibility Analysis

By literature, questionnaire and mathematical statistics, we investigated and studied the 17 colleges in Hebei province and sampled 608 students as research subjects whose majors are not PE education. According to the goal of research, we designed questionnaire and verified its validity and credibility to make sure it fits the requirement of statistical research. 608 questionnaires were sent and 587 effective questionnaires were collected, effective rate being 96.55%.

3.1   The Opening of Yoga Course Is Required by the Teaching Reform of New Curriculum

"Decision of State Department about Deepening Education Reform and Promoting Quality Education around the Nation" explicitly points out that "a healthy body is the fundamental premise for the youth to serve the nation and people and the embodiment of dynamic vigor of Chinese nation. The school education should set "health first" as guiding thought and practically strengthen PE course to help students form the good habit of persisting in physical training." The new PE curriculum standards gives colleges larger space in course setting and raises higher requirements for college education, especially in the selection of teaching materials which requires us to start with the multiple functions of PE, integrate the fitness, education and entertainment functions of PE, and constantly get rid of the stale and bring forth the fresh, and establish colorful contents in PE course. "Yoga, as a kind of fashionable, healthy and elegant fitness method, is introduced into college to help students relieve the pressure from study and life, figure perfect shape, and meanwhile return to peace, confidence and health. Yoga benefits both physical and mental health, and chases for the perfect integration of body and spirit. "[3] Therefore, yoga is popular among college students, especially the girls. If yoga is set as a college course, it not only benefits the mental and physical health of students, but also can provide a new mode for the reform of college courses.

3.2   Yoga Helps Students Set Up the Awareness of Lifetime Physical Training

"The National Physical Fitness Program" points out that national physical fitness program face all national people, especially the youth." College students are a centralized group and also the group with conditions and advantages of participating in physical fitness. College students are in an uprising and growing stage, the best time for self-perfection and individual socialization through education, so the PE teaching in college has a profound significance in cultivating students' abilities of independent training and formation of life-time physical-training concept and establishment of national fitness awareness [4]. During the 4 years' study in college, it is very necessary for college students to master one or two fitness items which can benefit their lifetime. The practice teaching of yoga just meets this requirement. Yoga course includes abundant contents which are easy to learn and practical, and only a few simple actions can bring fitness, beauty and shape figuring effects; humanized action design and affordable practicing method can subject students to a relaxing and comfortable atmosphere and thereby easily recognized by most people. Besides, yoga practice is not restricted by the condition of field, time, equipment, age, gender, body status or training level, so it is very suitable to be a life-time fitness item for students.

3.3   The Opening of Yoga Course Is the Urgent Demand of College Students

1) Students' interest in yoga
Interest is the internal drive of learning. It is created and developed upon mental demand, the main factor to directly affect learning enthusiasm and consciousness, and the key part to guarantee an expected effect.

Table 1. Interest in Yoga N=587

	Like	So-So	Dislike
Number	540	31	16
Percentage	91.99	5.28	2.73

Through the comprehensive investigation on students, we found that 91.99% of students are interested in yoga. Thus, it can be seen that most of students have much interest in yoga. The table indicates there is urgent demand for yoga course among students. The teaching of yoga course enriches the form of PE course while traditional sport activities cannot fully meet the requirement of students who require the kind of sport form which is fashionable, new and up-to-date. Yoga is a kind of elegant activity in civilized society with the integration of entertainment, sport and art into one and this just caters for students' physiological and mental characteristics and cultural requirement. It can inject a stream of fresh air to campus, add bright colors to leisure life and improve the cultural taste of entertainment. What is more important is that it can promote the improvement of basic activity capacity and physical quality of students, effectively train students' strength, flexibility, resistance, sensitivity and ability to balance, and achieve the function of losing weight, fitness, disease prevention and healing while figuring shape. It fits the principle of mental and physical development and caters for the demand of modern society.

2) Motivation
Motivation can reflect one's learning object, determine the learning attitude of individual and provide basics for teachers in arranging teaching materials.

We can see from table 2 that 49% of students take yoga course to change appearance and modify imperfect figure, and make a fit shape; 18.7% of students want to improve body capacity and physical quality; 15.3% of classmates want to adjust the tension of learning, release pressure and relax by learning yoga; 12.8% of students want to improve artistic accumulation through learning and fully embody the perfect temperament and elegant style of modern youth; some of them think it is a kind of fashion and want to follow the fashion with blurry motivation. Most of students think yoga is the hottest fitness method in modern society and want to truly get to know it by learning. However, for financial conditions, students cannot afford the training in gymnasium, so the yoga course in college will certainly be popular and have the support of teachers and students.

Table 2. Motivation to Learn Yoga N=587

Motivation	Number	%	Sequence
Fashion	19	3.24	5
Fitness	110	18.74	2
Figure Shaping	288	49.06	1
Reduce Pressure	90	15.33	3
Artistic Cultivation	75	12.78	4
Others	5	0.85	6

3) Students' perception and approach to yoga

Table 3. Students' Perception about Yoga N=587

Item	Well-Known	A little or just heard of	Don't care
Number	226	302	59
Percentage	38.50	51.45	10.05

Through investigation, we found that 38.50% of students were well-known about yoga and 51.45% of students just knew a little or heard of yoga. From the data, we can see that students have insufficient understanding about yoga and their understanding just stays at the surface. In general, though students don't know about yoga course very well and are just at starting point, most of students hold positive and recognized attitude toward yoga.

Table 4. Approach to Know Yoga N=587

Channel	TV Network	Books	Social Promotion	Friends	In Class
Number	167	50	315	38	17
Percentage	28.44	8.5	53.66	6.5	2.9

From table 4, the approach to gain information, we can see that most of students are dependent to social publicity which stimulates the interest, internet, books and communication with friends to get to know yoga further; some of them start to know about yoga from books and internet by themselves and few of them know it from school. At present, our teaching still stops in traditional items.

4   Suggestions

4.1   Set Yoga Course in College

Firstly, we should strengthen publicity and invite experts for knowledge lecture to let more students know about yoga and improve their understanding; hold regular training classes, organize students to watch materials of yoga and improve students' understanding and admiration level by audio-visual aids. In spare time, we should set after-class instruction of yoga and hold some performances and activities about yoga to let students experience the health and happiness yoga brings. Besides, we also need to cultivate some sport actives to mobilize the enthusiasm of students and form an active and good external environment.

4.2   Improve Professional Qualities of Faculty

The setting of yoga course in college conforms to the development of leisure sports and caters for the requirement of national fitness. To carry out yoga teaching better, we should give trainings to teachers and provide opportunities to communicate with domestic and overseas excellent tutors and hold related academic exchange to perfect the theoretical and technical system of yoga teaching. Teachers should not only study teaching materials, explore, constantly develop innovative ideas and improve professional level, but also need to understand the truth of yoga and master its essence.

4.3   Make Full Use of Site Condition of School

Such a kind of fitness method, which can benefit the body and spirit, doesn't need any additional appliance and is not restricted by ground condition like ball games. We should make full use of current conditions of school, such as body room and exercise room, and make them as yoga classroom. Some internal rooms of college are not equipped with the conditions required by professional yoga lessons. Then, we can choose a clean field with fresh air, lay cushions on, place audio equipments and select proper teaching materials to start. This way can also achieve expected teaching effect.

5   Conclusion

1) The great interest of students in yoga indicates the urgent requirement of students for yoga course in college and sets up solid foundation for the introduction of yoga.

2) The clear motivation of students in learning yoga tells that yoga is one of the requirements of students to chase for high-class sports and also reflects modern college students' view toward health, sports and value, and aesthetic taste.

3) The setting of yoga course in college not only caters for the trend of modern sports and meets the mental and physiological requirements of students, but also can improve the physical quality of students, adjust psychological status and shape a health and elegant figure. It is required by the reform in curriculum and has certain practical significance for comprehensive quality education and cultivating students' life-time sport awareness. Therefore, the introduction of yoga into college education

fits the reform of teaching, conforms to the trend of PE teaching toward lifetime use, fitness, entertainment and mass and it is also feasible and necessary to do that.

References

1. Bai, Z., Zhang, H.: Yoga—Qi Gong and Meditation, vol. 374. People's Sports Publishing House, Beijing (2005)
2. Zhang, H.: Gift of Yoga, vol. 1. People's Sports Publishing House, Beijing (2006)
3. Liu, Y., et al.: Study on the Practice of Yoga Course in College. Journal of Capital Sport College 20(3), 82 (2008)
4. Du, H., et al.: Feasibility Analysis of Setting Yoga as Selective Courses in College. Journal of Chang Ji College (2), 63 (2007)
5. Wen, J., et al.: Experimental Study on Optimizing the training of Basketball Judge by Modern Information Technology. Journal of Hubei Institute For Nationalities (natural science edition) (December 2007)

Application of Audio-Visual Instruction in the PE Teaching of Deaf-Mute School

Zhansuo Liang and Xiuhua Feng

School of Physical Education, Handan College, Handan, China
J.J_Bai@yahoo.cn

Abstract. Audio-visual instruction is a kind of advanced modern teaching approach and one of the important methods in PE teaching. The application of audio-visual instruction in the physical education of deaf-mute school is very important in cultivating students' interest in sports, promoting the formation of sport technology, skill and good habit and moral education. The paper pointed out the matters we should pay attention to in audio-visual instruction (ASI) and gave suggestions.

Keywords: Audio-Visual Instruction (ASI), deaf-mute school, PE teaching, application.

1 Introduction

Audio-visual instruction is the teaching approach basic on modern teaching approaches, such as slide projector, TV, radio, video and network and a kind of modern teaching method enabling students to master basic knowledge, technology and skill better and faster. The modernization of teaching approach is the important symbol of education modernization. Modern information technology is being applied more and more widely in education and each subject to different extent. It will fully play its function in physical education, the subject integrating knowledge instruction, skill training and physical exercise. No matter for theoretical teaching or skill lesson in physical education, the role of modern information technology in PE teaching is being recognized by more and more PE teachers. According to survey, the main existing problems in the PE teaching of deaf-mute school are the lack of qualified teachers and the small proportion of professional PE teachers in faculty; most PE teachers are not equipped with due technology level or demonstration ability to deliver standard actions, which directly weakens the learning performance of deaf-mute students who can only gain information by vision. Meanwhile, a failed demonstration will affect the image of a teacher in the heart of students. Besides, there are other problems in the PE teaching of deaf-mute school, such as limitation of field and appliance, dull content, single teaching method and approach, lack of learning interest, inactive class atmosphere and so on, which seriously affects teaching quality. Therefore, PE teachers always prefer to arrange the lesson they like and are good at, in which dull contents and single teaching approaches make inactive

Y. Wu (Ed.): International Conference on WTCS 2009, AISC 116, pp. 605–610.

atmosphere in class and affect students' learning habit. With the aging of PE teachers, their sport abilities will certainly decline, so in choosing teaching contents, they will avoid those which are hard to show and technically-demanded, which affects the comprehensive development of students. In the teaching materials of physical education, there are many technical actions, such as soaring up, acceleration, overturning and so on which will be very hard for students to see clearly in moments and set up a complete action image in such a short time. In this situation, teachers can only demonstrate repeatedly and the outcome will be at the cost of teaching process. ASI can largely make up this defect, so it is very necessary to apply this advanced teaching approach in deaf-mute school. So, how to reasonably and correctly apply this advanced teaching approach to the teaching in deaf-mute school will become an important topic of our research.

2   Application of ASI in the Physical Education of Deaf-Mute School

2.1   Promote the Conversion of Student and Teacher

According to the requirement of modern teaching concept, a teacher is not only an initiator of knowledge, but also should be the guide and partner of students in learning; students are required to become the owner of learning from a passive "acceptor". A proper environment is required in teaching process to give students the opportunity to experience by themselves and independently seek for what they want, and train the abilities of independent and explorative learning; teachers are also required to organize, guide, help and supervise students in learning. The modern teaching approach, i.e. ASI, can serve as a convenient approach to achieve this goal.

2.2   ASI Helps to Show the Key Points and Difficulties of Technical Actions and Is an Effective Approach to Improve Teaching Effect

Physiologists hold that the early period of establishing action image is led by visual appearance; the nerve centers in pallium receive the stimulus and excitement from hearing and vision, tidy up the information and then gradually form a clear and complete image. According to the survey and statistics of UNESCO (United Nations Educational Scientific and Cultural Organization), people can only remember 15% of oral information, 25% of visual information and 65% of information carried by both hearing and vision. In the PE teaching of deaf-mute school, students can only simulate and learn by vision, but the deaf-mute school is lack of qualified teachers and the proportion of professional PE teachers is very small in faculty, and most teachers are not equipped with due technology level and demonstration abilities and are hard to give standard demonstration, which directly affects learning effect. ASI technology is a kind of new assistant teaching approach to optimize the teaching process with bright images, vivid pictures and changeful animation effects. ASI courseware simplifies the key points and difficulties in teaching. During teaching, teachers can present the technical parts to students, which are hard to be described through demonstration, by animation, video or freeze-frame and students can see every technical detail clearly and set up a comprehensive image more rapidly. ASI deepens understanding, shortens generalization process and has obvious effect in helping students master learning content rapidly and improving teaching effect. For example, in jumping instruction, we can make jumping technology into lantern slides to help with explaining each action, such as run-up, take-off, actions in the air and landing and students will understand the connection and function between them and thereby master the main points of action. ASI can show the key points and difficulties of teaching and comparison between wrong and right and students will master the main points and actions after the proper and timely instruction of teachers.

2.3   Promote the Reform of Teaching Concept and Method

The traditional mode in the PE teaching of deaf-mute school is "teacher gives instruction and student exercises". With the development of modern technology, the center of modern education is shifting from class, textbook and teacher to student and emphasizes students' active exploration into knowledge, discovery and organization of significance of knowledge. The required learning environment of modern education is strongly supported by the achievements of latest modern information technology which promotes the reform of teaching ideology in various schools. Meanwhile, ASI also plays an active role in the reform of education method and changes the traditional "feeding" teaching mode, and provides conditions and basics for the creation of new teaching mode, and gives teachers wider space and creative area in teaching process.

2.4   ASI Can Create the Environment of Happy Learning and Increase Learning Interest

According to educational psychology, learning motivation is a kind of internal drive to directly put students into learning and its most active component is cognitive interest. Learning motivation and interest are the main sources and mental factors of learning enthusiasm. It not only affects the learning of students, but also is an important target of school education. In the PE teaching of deaf-mute school, students are lack of the interest for learning due to the restriction of teacher's qualification, field and equipment, single teaching method and approach and dull content. ASI can completely change the traditional teaching mode of "teachers do the talking while students do exercise". By advanced teaching approaches and powerful functions of computer, it can impose effect on the psychology of students by ray, color, figure and light, meet their dynamic desire for knowledge and curiosity, and stimulate their learning interest. Teachers can intuitively and vividly reflect the dynamic changes of actions by image, animation, video and other media, attract the vision of students and enlighten the mind of students in order to stimulate their learning motivation. By ASI technologies, teachers can draw the attention of students to pictures and animations and enable them to enter into a good learning environment with pleasant mode. Therefore, the application of ASI can make students learn happily and actively. Besides, ASI can facilitate instruction, demonstration, operation, assessment, feedback and the optimization of other aspects in PE teaching with its convenient interaction and sharing. If by ASI game software for assistant teaching, we can set a game environment at will under certain rules and easily show the combination of basic technologies. With vivid animations and powerful 3D video recording of software, students will show great interests. In this way, students will become active in PE learning, take initiative to participate in sports, make sports a kind of life necessity and gradually form the habit of taking exercise actively.

2.5   Carry Out Moral Education Better

Moral education is one of the important missions of PE teaching in deaf-mute school. We can arrange students to watch the videos or TV programs about the tough training of athletes, especially handicapped athletes, and some important international games in which the athletes' efforts for the honor of nation, tenacious and persistent spirit, and the scenes of accepting an award, playing national anthem, lifting national flag and the patriotic enthusiasm of cheering squads will deeply impress every student in the scene. By timely instruction and education, we can cultivate children's patriotism, collectivism and the spirit of overcoming difficulties and fighting bravely, and those wonderful segments of sports and the spirit shown also arouse the drive and sense of success in them in "happy sports", gradually improve deaf-mute children's interest in sports, and make them form the habit of "lifetime sports" penetrating sports into their life.

3   Matters That Should Be Given Attention in ASI

3.1   The Application of ASI Should Have a Specific Target and Pay Attention to Effect

According to the teaching content of a term, teachers should let students watch some related wonderful fragments first in class to improve their learning interest. In a technical class, teachers should arrange students to learn and watch the filmstrip and video of technical actions, and then explain the key points and difficulties to deepen the impression of image in students and set up solid foundation for mastering technical actions.

3.2   Make Preparation before Class and Seize on Teaching Session

Before playing teaching materials, teachers should raise some questions for students and tell them to think while watching; use slow-play and freeze-frame to help with explanation to make students understand each technical part.

3.3   Organize Discussion to Enhance Understanding

After watching teaching materials, teachers should organize a discussion for students to let them judge how well they perform in technical actions, and then give revisions. Teachers should give active and timely instruction, help to solve the questions before class, and strengthen the understanding about technical actions and formation of image.

3.4   ASI Is an Important Assistant Approach, Not the Main Teaching Approach of PE Teaching

As long as we exert the advantages of ASI and hold a correct view about the practical value of traditional teaching approach, we can put it into reasonable and correct use to achieve better teaching effect and improve the quality of PE class.

4   Suggestions

1) PE teachers should improve understanding and transform ideology. Some teachers, especially old teachers, still stay in original understanding and are lack of the sense of responsibility without understanding the importance of ASI. For this reason, our PE teachers should update their own knowledge, change ideology and reasonably arrange practice class and theoretical class.

2) PE teachers should improve the ability of fabricating courseware. The innovative, changeful and funny courseware can be matched with large amount of cartoons, pictures, video clips as well as comparisons between right and wrong to present the whole process of action to students clearly, completely and precisely. To achieve this target, PE teachers are required to improve the ability of fabrication courseware which is composed by abundant professional theories, computer skills and innovative abilities.

3) Support of leaders and large investment. At present, the hardware of deaf-mute schools is insufficient, especially ASI equipments and related teaching materials. Investment must be increased to purchase some necessary hardware and software of ASI, such as video recorders, pictures, lantern slides, discs and so on. Meanwhile, schools can make some materials and clips by equipments and related departments are also expected to fabricate and issue some audio-visual products for PE teaching to make ASI widely used in PE teaching.

5   Conclusion

ASI is an advanced modern teaching method and important assistant approach in PE teaching. Its application in the PE teaching of deaf-mute schools can help to show the key points and difficulties of technical actions and improve teaching quality; it plays an important role in cultivating students' interest in sports, promoting the formation of sport technology and skill, forming a good sport habit and starting moral education. As long as we exert the advantages of ASI and hold a correct view about the practical value of traditional teaching approach, we can put it into reasonable and correct use to achieve better teaching effect and improve the quality of PE class.

References

1. Wang, H., Zhu, J.: Application of ASI in PE and Health Class. Academic Journal of Chemical Advanced School, Lianyungang (14), 39–40 (2001)

2. Wang, G., Wu, T.: Application of Computer-Assisted Teaching in PE Teaching Reform. Physical Education and Science 21(122), 59–61 (2000)
3. Zhu, L.: Application of ASI Approach in PE Teaching. Academic Journal of Beijing Sport University 28(6), 825–826 (2005)
4. Zhou, X.Y., Wang, F.C., Cao, H.: Application of Multimedia CAI Technology in PE Teaching. Academic Journal of Beijing Sport University (02), 245–246 (2000)
5. Zhou, H.: Discussion of Multimedia Computer-Assisted PE Teaching [J]. Chinese Modern Education Equipments (09), 5–7 (2005)

One Consistency Model of Collaborative Learning for Distance Education

Wang Xiaohua and Li Tianze

Network and Further Education College, Xi'an University of Electronic Science and Technology,
Xi'an, Shaanxi, China
w.l.l_so@qq.com

Abstract. Modern distance education is teaching activities based on the Internet and terminals, Collaborative learning is an important way to improve the efficacy of distance learning, It is Required that in-depth research of consistency maintenance, collaborative awareness and other key technologies based on collaborative learning semantic, So high efficient, natural interaction and customized learning will be achieved. This paper analysis Model and Algorithm of collaborative learning Consistency about the Distance Education, By this model, a specific algorithms based on knowledge point structure will be Build to maintain the consistency.

Keywords: CSCL, knowledge points, Operational Transformation, consistency.

1   Introduction

Modern distance education is teaching activities based on the Internet and terminals, real-time remote virtual teaching environment mainly include remote classroom teaching, Tutor, group study and discussion etc., Must be achieved through the interaction between teachers and students to a barrier-free, smooth teaching and learning， And collaborative learning is an important way to improve the efficacy of distance learning. But a major factor of restricting the development and application of collaborative learning is that the collaborative learning model, collaborative learning perception, personalized replica consistency, concurrency control based collaborative learning semantic and other key technologies are not yet in-depth study to achieve adequate and efficient, natural interactive, personalized learning in supporting collaborative learning system (CSCL). some key algorithm Need to study for distributed real-time collaborative learning system applied in the modern distance education. This paper presents the analysis of Consistency and Algorithm for collaborative learning under the Distance Education. Through improved data structure, the unique operational transformation algorithm based on knowledge point is formed to maintain consistency. collaborative learning systems based on Real-time collaborative editing system generally use the half-copied, distributed architecture, So the consistency of collaborative learning system is very important. Its solution relate to semantic of application. some consistency models

Y. Wu (Ed.): International Conference on WTCS 2009, AISC 116, pp. 611–617.
springerlink.com   © Springer-Verlag Berlin Heidelberg 2012

and algorithms are present for different types of collaborative editing systems. Typically It have operational transformation, multi-version control, locking methods. For distance education, learning model based knowledge point need to design new data structures of knowledge point, And the algorithms of consistency maintenance based on a new data structure will be made. It is Aimed ambiguity, causal conflict, the intention conflict. Specifically complex object and the data based on knowledge points in real-time collaborative learning how to maintain consistency, which requires to improve the existing collaborative consistency algorithm. Firstly design the new data structure based knowledge points, Secondly improving the locking mechanism for dealing with conflict, it present specific operational transformation algorithms to maintain consistency on Collaborative learning. This paper discusses the model of consistency based on knowledge point structure on collaborative learning.

2   Consistency Model and Algorithm

2.1   Model

Complex objects based Knowledge point can be Classified from Manifestations: Text, graphics, images, video, audio, various embedded objects(PPT, data tables, xml, etc.), various databases, Therefore, in virtual real-time distributed collaborative learning required to create a new multi-level architecture and data model, goals of effective collaborative learning, collaborative awareness and other will be achieved To meet the wide area network environment. Under WAN environment, the characteristics of collaborative learning are as follows:

1、the network environment is unstable, cause congestion, jitter and other issues, Lead to real-time systems face more difficulties in the response time。

2、access to resource-constrained, Due to wide area network environment, so learning resources can not be passed on to all students, Only adaptive extraction and mapping.

3、compatibility must be considered because students use a heterogeneous platform.

4、the difficulties of collaborative awareness, semantic layer description Must be extracted from the structure of knowledge point.

5、adjusting the model and description of knowledge points to support the dynamic, distributed data storage.

6、the control of consistency and concurrent achieve by Hierarchy.

For the above characteristics, in wide area network environment, the consistency model of collaborative learning based on knowledge point structure is as follows:

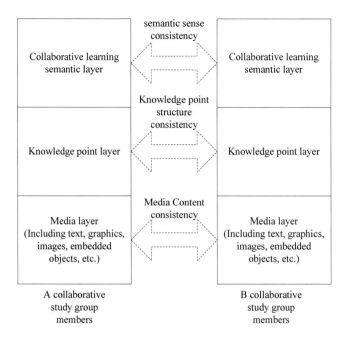

Fig. 1. Consistency Model

2.2   Description of Knowledge Points

In the collaborative learning process, Usually the structure of knowledge points can be divided into meta-knowledge, knowledge points, and knowledge modules (courses, thematic discussions, collaboration, answering the theme, etc.), knowledge (professional, cross-professional, etc.), discipline. Knowledge point is expressed as a 9 per group (ID, Name, Description, Keywords, Type, Applicability, Importance, Difficulty, Extension). Among those ID Mean the knowledge-point mark composed of two parts: Creator + timestamp; Name is the name of knowledge points, or the title of meta-knowledge point; Description is Description Department of the knowledge points which make a   brief description about knowledge points by the form of text or other media content , it may be a file links to the actual storage location of the content; Keywords is the keywords set of knowledge point which can be used to retrieve this knowledge point; Type is the type of knowledge points, range of (parent node, leading point set, reference set, support point set, child node set, threshold), After learning this knowledge point, while the test results point not less than the threshold student Can leave this knowledge point. if parent node is empty this knowledge point is the root (can not be discipline, etc.), if child node set is empty this knowledge point is the meta-knowledge. Applicability is the application of knowledge points, show that the knowledge points suit for which types of students; Importance is the importance of knowledge points, show the importance of this knowledge points, the range is [0,1] which present the greater the value the more important point that the knowledge; Difficulty is the difficulty of knowledge points, reflecting the difficulty of the knowledge points, the range of [0,1], larger values indicated greater difficulty;

Extension is expanding content about knowledge point by the publisher of knowledge points, such as creative note of knowledge point by professional teachers.

Analysis of collaborative learning process, the initial System of personalized curriculum knowledge is good template for a pre-determined (or sponsors such as teachers or head learning build according to the static principle), in the process of learning all members of the group Dynamic Construct the knowledge system, which is collaborative learning. Then the operation to knowledge point generally include view, insert, delete, create, edit, replace, merge, split, practice. about these view, insert, delete, create, edit are basic operation, others are combined operations, combined operations can be formed by a number of basic operations.

Knowledge points can be polymerized by a number of knowledge points, between the knowledge points the knowledge structure form tree structure. Relationship of level between the knowledge points can be obtained by the division of knowledge in accordance with the horizontal structure and vertical structure. The relationship between different knowledge points can be divided into parent-child relations, brotherly relations, dependencies relationship, parallelism relationship, reference relationship and free relationship.

2.3   Consistency of Knowledge Point Architecture

Operational transformation algorithm based on knowledge point is based on a total order to carry out, it first leads to all kinds of operations relationships:

Definition 1: Causality. Given any two operations Oa and Ob which Apart located at sites I and J , Oa and Ob have a causal relationship (denoted Oa $\to$ Ob), if and only if Oa and Ob have the following three conditions: ① I = J and the operation Oa took place before Ob ; ② I $\neq$ J and the implementation of operation Oa in the site j is before the production operation Ob; ③ operation Ox exist, and there are Oa $\to$ Ox and Ox $\to$ Ob.

Definition 2: dependency / parallel relations. Given any two operations Oa and Ob, Oa and Ob that have dependencies, if and only if Oa and Ob meet Oa $\to$ Ob or Ob $\to$ Oa; said Oa and Ob with non-dependent (parallel relationship, denoted Oa $\parallel$ Ob), if and only if Oa and Ob not meet Oa $\to$ Ob, not to meet the Ob $\to$ Oa. Definition 2 is actually another alternative description about definition 1 .operations for the existence of dependencies can follow the causal sequence of operations to be ordered, for non-dependent operations can be further classified.

Definition 3: The Consistency model of Knowledge point architecture. If you meet the following three conditions, the knowledge points system of collaborative learning is Consistency: 1. The end result Consistency: After performing the same set of operations, a copy of the knowledge point system all sites share is consistent; 2. Causality to maintain: any two operation Oa and Ob, if Oa-> Ob, then all sites Oa implement before Ob ; 3. intention to maintain: for any operation O, O in the local and other sites have the same effect of implementation with the intention of O; and if Ox exist and Ox $\parallel$ O, then the implementation of the results of Ox do not affect each with the implementation of the results of O.

Definition 4: time vector. Set N is the number of collaborative learning in the site, the site label were 0, ···, N. each site the time vector SV has a N-component. Initially, SV [i] = 0, i ∈ (0, ···, N-1). After the site I perform an operation, SV [i]: = SV [i] +1. after Implementation of any operation in the local site, it will be done that operation and the local time state vector SV sent to the remote site.

Definition 5: The global order relation ">." Given two operation were generated by sites I and J , SV1 and SV2 is the corresponding time vector. O1 in the global order in ahead of O2, denoted as O1> O2, if and only if: Sum (SV1) <Sum (SV2) or I <J and

Sum (SV1) = Sum (SV2).in which $Sum(SVj) = \sum_{i=1}^{n} SVj[i]$ .For any two operations O1,

O2, if O1> O2, the O1 is the first order operation to O2, O2 is the subsequent operations to O1. All operations can determine the unique relationship between the global order, global order and causal order is Consistency. If the site execute according to the operation of the global order, If the site execute operation according to the global order , or for conflict operation do operational transformation to make the same effective of operation Executive to the global order , so in collaborative learning system the ultimate Structure of knowledge point is consistent in accordance with the 3 consistency requirements of definition 3.   Accordingly, we propose a new algorithm of operational transformation based on knowledge point. through operational transformation for Conflict operation before the implementation (usually including the conflict between father and son relation, dependency conflicts, operation conflict of the same knowledge point, etc.), then Execute the operation after transformation to achieve the consistency of knowledge point architecture in collaborative learning system.

2.4   Consistency of Knowledge Point Architecture and Algorithm of Operational Transformation

Definition 6: The background structure of knowledge point BC. While operation O produced The structure of knowledge point called The background structure of knowledge point of the operating O, denoted by BC [O]. Background structure of knowledge point is the context of knowledge point operation to record all attributes and relationships for the knowledge points in the current structure of knowledge point, this article discusses the structure of knowledge point based on tree structure. Recorded the initial structure of knowledge points as BC0. the results structure of the knowledge point obtained after the implementation of Operation O in the background structure of knowledge point, denoted as BR [O]. For any operation O, obviously BR [O] = BC [O] + O. "+" Indicates the role of O in BC [O]. On a set of operations: HB = [01,02, ..., On], BR [On] = BC0 + O1 +02 + ... + On.

Definition 7: background structure of knowledge point equivalence.   for the operation produced on two sites: O1, O2, if the BC [O1] = BC [O2], the background structure of knowledge point of the O1 and 02 is equivalent to, that for the O1 ≒ 02.

Performed operations in accordance with the global order has been saved in the history queue HB, a site has performed operations O1, O2 ,..., On, recorded as HB = [O1, O2 ,..., On], and there is O1> O2> ....> On. Knowledge point operations O can Express by

Costudy (PS), PS description that in the background structure of knowledge point BC meta-knowledge point have be inserted, deleted, created, edited. In the collaborative learning system, generally when started a collaborative learning task, all learners in the terminal click the link of the knowledge point, access to specific content of the knowledge points to complete the collaborative learning. Each terminal on the background maintain a copy of the background structure tree of knowledge point, for the local operation, results of operations do not write in the background structure of knowledge point, but in a temporary operation until the operation is complete, then calling sIOPT algorithm do operational transformation and update the background structure of knowledge point. For remote operations, a direct call sIOPT algorithm is to operational transformation and update the background structure of knowledge point after the operation received; As the causal relations and the global order is Consistency, so the use of operational transformation causality do not operate in the order of the causal sequence, also can be maintained the effect Consistency of implementation.

Definition 8: operator <: Determine the global order relations: The Global (Oa, Ob). Determine the global order relation between operations Oa and Ob.

Algorithm 1. sIOPT algorithm
/ / in HB there are n-point operations of knowledge point have been implemented
HB: 01,02, ..., 0n
Ok: the remote operation received by this site
BC: f the current background structure of knowledge point of this site BC = BR [On]
sIOPT (Ok) (
/ / Package   operations of knowledge point belong to operations Ok , in Package (Ok) consistency operations of specific media files is processing , the paper not be discussed.
Ok1 = Package (Ok)
/ / Ok1 is the subsequent operations of On
If Ok1 <0n Then
/ / Ok1 added to the HB tail
HB. AddTail (Ok1)
/ / Executive Ok1 and Update the background structure of knowledge point
BC = BC + Ok1
Return
End If
j = n
/ / Ok1 is pre-order operation of Oj
While (0j <Ok1)
j --
End While
Ok1 '= Ok1 / / do operational transformation to all subsequent operations of Ok1
For m = j +1 To n
Ok1 '= IT (Ok1', 0m)
End For
/ /   Ok1 Will inserted in front of 0j +1 in HB
HB.InsertBefore (Ok1, 0j +1)

/ / Update the current structure of background knowledge points, show the implementation results

BC = BC + Ok1 '

)

Definition 9: operational transformation of knowledge point IT (O1, O2): the operation O1, O2, if O1> O2, and O1 $\risingdotseq$ O2, O1 do operational transformation of O2, O1 '= IT (O1,O2), makes results between the implementation of the O2 ,O1 'and the implementation of O1, O2 are consistent, namely: BC [O1] + O1 +O2 = BC [O2] +O2 +O1' [6].

3   Summarizes

This paper discuss how it do consistency maintenance about the tree of knowledge and the specific media file, when the group members complete a collaborative learning task in collaborative learning based on the tree structure of knowledge point. Consistency model of collaborative learning is proposed based on operational transformation of knowledge point. it is useful for the following research of collaborative learning.

References

1. Jia, W., Kun, Q.: The study of relationship between knowledge point in network courses by the view of knowledge. Continuing Education Studies (2008)
2. Zhang, P.A.: Structured representation way of knowledge in network courses. Computer Education (2008)
3. Zhu, Z.: Personalized Curriculum Organization Method Based on Knowledge Topic Ontology. Computer Science 36(12) (2009)
4. Liao, B.: Survey of Operational Transformation Algorithms in Real-time Computer-supported Cooperative Work. Computer Research and Development 44(2) (2007)
5. Xiao, K.: An Operation Formalization of Knowledge Points in Course-ware Computer Science. Computer Science 35(1) (2008)
6. Xu, X.H.: The study of a number of issues for adapted real-time collaborative editing systems, PhD thesis (2005)

The Design and Implement of SKPLC SFC Programming System

Zhao Feng, Sun Xiang, Cheng GuangHe, and Ren XuCai

Shandong Provincial Key Laboratory of Computer Network, ShanDong Computer Science Center , 19 Keyuan Road, JiNan, ShanDong Province, China
zhaoqiansun1213@163.com

Abstract. This article introduces in details the structure of Sequential function chart and the function of every element in SFC and the functions of SFC software. We design the graphic SFC editor and all elements in SFC, We also define the built-in ladder and instruction list and the structure of SFC file. Finally in this paper it is shown how to convert the SFC file into (IL) instruction list and how to translate instruction list (IL) into intermediate code, and how to download the intermediate code to PLC and PLC interpretive execution.

Keywords: PLC, SFC, Programming System, Instruction List, IEC61131-3.

1   Introduction

Programmable logic controllers (PLCs) are widely used for real-time control of automated industrial processes [1]. They are well suited for discrete event control because they are able to directly implement the control sequence specified by means of standard languages such as Ladder, Instruction List, Function Block Diagram, Sequential Function Chart, and Grafcet defined in the IEC 61131-3. Sequential function chart (SFC) is among the most popular design methods of automated control systems and used to capture the sequence of operations executed by the system's control software.

Sequential function chart programming makes use of both function blocks, or steps, and conditional transitions between the steps to graphically represent the control system [2]. Each function block or step represents a control action in the control sequence. Control passes from one step to the next through the conditional transitions. A SFC can represent the sequence flow of the control logic graphically, SFC shows a programming ideas and represents a control language to the structured and modular development trend, Therefore the development of SFC software has important practical significance.

From Fig.1 we can see that A SFC is composed of initial Step, steps, transitions, links and actions [3].

Step: A step indicates a state in a set of sequences. An action of a step depends on the action accompanied with the step. A step has the logical strates ON and OFF. It is described as a box. Particularly the initial step is described as a double box.

Y. Wu (Ed.): International Conference on WTCS 2009, AISC 116, pp. 619–624.
springerlink.com   © Springer-Verlag Berlin Heidelberg 2012

Transition: When the previous step state is ON and the transitional condition is satisfied, the previous step state is changed to OFF and the following step state is changed to ON. A transition is described as a horizontal bar.

Link: A link is a vertical line which represents the the flow proceeds connection between steps generally from up to down.

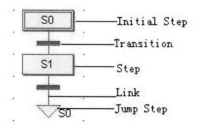

Fig. 1. The Structure of Sequential function chart

Action: Actions may be accompanied with a step, and it is executed when the step state is changed to ON.

There are two types of divergence in a SFC, which are called as "the simultaneous sequence divergence" and "the optional sequence divergence".

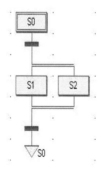

Fig. 2. The simultaneous sequence divergence

The simultaneous sequence divergence is depicted by a pair of parallel horizontal lines in Fig.2. Multiple steps are simultaneously activated by firing one transition. When all last steps for simultaneous sequences are active, the following transition can fire.

In the optional sequence divergence, if conditions of multiple transitions are satisfied without the description of priority, then the transitions fires.

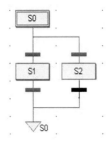

Fig. 3. The optional sequence divergence

2 The Main Function of SFC Software

SFC programming system not only can directly be sequential function chart display to users, allowing users to easily edit and flexible operation, but also be able to compile sequential function chart program, and eventually converted into target sequence function chart code is downloaded to the PLC, therefore SFC software should have the following functions[2]:

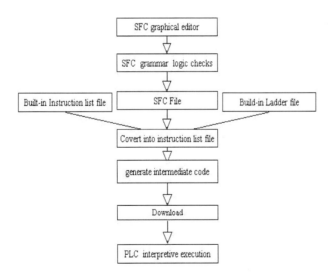

Fig. 4. The Main function of SFC software

1) The graphical display for SFC language elements. According to IEC61131-3, SFC language elements include: steps, transitions, action, single sequence divergence, simultaneous sequence divergence and optional sequence divergence. For every language element are set corresponding graphic symbols. Programming system

should can graphics displayed in the interface according to the size and appearance of various element, should also be able to dynamically adjust the operation of the user graph size, relative position, color and font attributes.

2) Flexible editing. Programming system provides users with easy editing functions: delete, cut, copy, paste, search, replace, undo, redo and positioning and so on.

3) Embedded ladder or instruction list for the action of every step and the transitional conditions of the transition.

4) Compiler. Programming system provides fast compile function, through the compilation of the sequential function chart program for syntax, logic for inspection, and the results feedback to the user. And The SFC File will be compiled and generated object code associated with PLC.

5) Download. SFC Programming system provides serial communication functions, it can download the object code to the PLC from the PC.

3   The Implement of SFC Software

3.1   The Design of SFC Elements

We designed the user interface SFC editor, first of all we use the Canvas property of Image control, The surface of   Image control divide into 64X48 the same size of grid, appears as a fixed-size rectangular region. Each grid has a fixed size and location of the coordinates X, Y (the ranks of the grid number). Each grid is only allowed to put one and only one element. We designed the following elements in the table 1[4][5].

Table 1. All the elements

Number	Name	Graphics
1	Initial Step	
2	General Step	
3	Transtion	
4	Vertical extension	
5	Level extension	
6	The head of optional divergence	
7	The bottom of optional divergence	
8	The head of Simultaneous divergence	
9	The bottom of Simultaneous divergence	
10	Jump Step	

3.2   SFC File Preservation

SFC file save format is defined as the line of text file. The first line describes the SFC version of the basic information. The rest of the line describes the structure of their SFC and the information of various graphic elements.

The second and subsequent line corresponds to the interface on the SFC editing the contents of the corresponding row.

The file structure of every element in one line:

line num	the first element data	the second element data	...	the last element Data

A description of each element:

type num	Column coordinate number	Step number	the pointer of build-in program

3.3   Embedded Ladder and Instruction List

We need built in ladder or instrution list for the action in the initial step and the general step in the SFC. When this step state is active then its built-in instruction list or ladder is executed. The transitional condition also need built-in transition conditions, When the previous step state is ON and the transitional condition is satisfied, the previous step state is changed to OFF and the following step state is changed to ON. All the built-in ladders are stored in the (*. sld) file, and all the built-in instruction lists are stored in the (*. sst) file.

3.4   Convert into the Instruction List File

The SFC file and the built-in ladder file(*. sld) and the built-in instruction list file (*. sst) will be converted into instruction list file(*. stl). But the instruction list need append the STL instruction, STL SN is equivalent to a master logic switch, only the step SN is active then the action of the step SN is scanned every program cycle.

The initial step of the conversion: an initial step before add the following statement before build-in instruction list.

```
LD BM   //Only in the first scan program cycle ,BM is ON
SET Si   // Set the initial step Si to active
STL Si   // add the 'STL Si' instruction list
......   // follwed by the built-in instruction list
```

General steps of the conversion: add 'STL SN' before build-in instruction list.

```
STL SN //add the 'STL SN' instruction list
......       // follwed by the built-in instruction List
```

The transitional conditions of conversion: When the previous step Sn state is ON and the transitional condition is satisfied, the previous step Sn state is changed to OFF and   the following step Sm state is changed to ON, So add 'CLR Sn'and 'SET Sm' after the built-in instruction list.

```
......       // the built-in instruction List
CLR Sn    //add the 'CLR Sn' instruction list
SET Sm    //add the 'SET Sm 'instruction list
```

Jump step of conversion: When the previous step Sn state is ON and the transitional condition is satisfied, the previous step Sn state is changed to OFF and the following step is Jump step Sm, So the conversion of the Jump Step Sm are 'CLR Sn'and 'SET Sm'.

CLR Sn   / /add the 'CLR Sn' instruction list
SET Sm / /add the 'SET Sm'instruction list

3.5   Instruction List into Intermediate Code and Download

SKPLC programming system has two programming language: Ladder diagram and Instruction List. The ladder file must convert into the instruction list file to generate the intermediate code. Therefore the two programming languages all translate the instruction list into intermediate code. The intermediate code is downloaded to the PLC from PC and PLC interpretive execution. Thus we can convert the sequential function chart file into instruction list file, so we   can directly call the function of the instruction list translate into intermediate code to generates intermediate code, download the intermediate code to PLC from PC through the serial communication function and PLC interpretive execution.

4   Conclusion

In this paper, we have designed all elements required in the SFC according to IEC61131-3 standard and the successful implementation of a user-friendly, easy to use and easy to maintain Sequential Function Chart programming system. The stability and humanity of the SFC programming system has been unanimously praised by users.

References

1. Konaka, E., Suzuki, T.: Optimization of Sensor Parameters in Programmable Logic Controller via Mixed Integer Programming. In: International Conference on Control Applications, Taipei,Taiwan, pp. 866–871 (2004)
2. He, M.F., Wang, Z.A.: Design and Implement of PLC SFC programming system. Industrial Control Computer 21(3), 78–81 (2008)
3. Nakamura, S.: Study on a Transformation Method of Ladder Diagram into Sequential Function Chart on the Basis of Linear Programming Technique, pp. 473–478. IEEE, Yokohmia National University, Japan (1997)
4. Liang, G., Bai, Y.: Design and Development of Graph System in SFC Configuration Software. Control and Instruments in Chemical Industry 32(1), 29–32 (2005)
5. Fan, S., Zhang, B.: A Graph System Based on SFC Configuration Software. Computer Simulation 24(8), 272–274 (2007)

Influence of Netizens' Involvement in Virtual Games of SNS Website on Website Loyalty

An Empirical Research Based on the Application of QQ Farm to Tencent Space

Zhang Jun, Peng Xiao-Jia, and Wang Fang

College of Business, Hunan Agriculture University
Changsha, China
zhangxin22@live.com

Abstract. This paper studies the game of QQ Farm and analyzes the influence of netizens' participation in SNS websites virtual games on the loyalty to the websites by adopting empirical study method. The author suggests four methods for SNS websites to cultivate the loyalty of netizens, which conclude providing the netizens with higher perceptive value of game experience, lowering gam ecost,strengthening game environment maintenance and improving self-credibility.

Keywords: Virtual Games, SNS websites, perceptive value, neitizens' loyalty.

1   Introduction

Until June 30,2009, the Chinese net citizen have reached 338 millions and the popularity rate amounted to 25.5%. There has been large proportion of people using internet for entertainment, information acquisition and communication. and the ratio of the SNS websites such as internet forum and BBS has reached 30.4%.In 2009,SNS subscribers have added up to 103 million in community applications. With the increasing development of SNS websites, there are lots of choices for SNS internet users, it is easier for them to be transferred from one choice to another.Therefore how to maintain their loyalty has become the main concern of the SNS websites.

The Farm Game,which is based on the SNS platform and developed by the team Five Minutes, is very popular among the netizens and profound influence.After the Tencent Company which has 860 millions registered accounts introduced the Farm Game in April, 2009, this game has yielded a revenue of 50 millions per month for Tencent.In order to find out whether the net citizens' participating in the virtual games has effect on their loyalty to the websites, this paper studied the significance of the Farm Game to the net citizens.The author choosed some online users of the Tencent zone and sent out more than 150 electronic questionnaires which are regarded as the data support for the analysis. Finally based on through some research on the related literature, this thesis put forward a supposed model and uses the retrogressive method to analyze and verify this supposed model.

Y. Wu (Ed.): International Conference on WTCS 2009, AISC 116, pp. 625–631.

2   Relevant Theory and Hypotheses

2.1   Relevant Theory

From a sociological point of view, Griffiths (1996) analyzed the data and discovered the negative aspects of online games, the consequences both medical and psychological brought by the overfun and indulgence, the aggressive games. Among current achievements, the theoretical model put forward by choi & kim (2004) is related with player-loyalty, and it giving a full explanation of the reason for the continuous addiction to online games in terms of the concepts of customer loyalty, personal interaction and social interaction .

Robert B. Woodruff (1992) has presented customer value refers to the perception of purchasing intentions and preferences of the evaluation given by the customers based on their feedback to the quality and characteristic of products as well as their willing to purchase. Kotler (1967) has stated "customer gapping value", identified the gap between the whole value customers obtain form the product (including services) and the full cost customer has to pay to get the product. Professor Dong Daihai (2000) holds the view that customer value is the comparison between the efficiency of the product that the customer has attained in the process of purchasing and using and the cost he has to pay, which is simply abbreviated as: $V = U / C$.

Corstjens (2000) considered that the concept of website loyalty is in fact the extension of the traditional concept of brand loyalty to the area of online shopping. Oliver (1999) maintains that the formation of customer loyalty is concealed in the attitude development structure, and finally revealed through the purchasing behavior. He defines customer loyalty as the deep commitment of customers to consume certain favored product or service in future periods, and thus resulting the repetitive purchasing of the same brand, regardless of the changes of market conditions. The degree of customer loyalty is measured by both behavior and attitude according to the following indexes: the intention of repetitive purchasing, recommending the company or brand to others, price tolerance and intention of customer cross-buying, etc.

Many researches deepened the customer loyalty theory, but the content and perspectives are very limited, which is contradictory to the urgent need of theory guidance for the current rapid development of online virtual game industry. It is of great realistic meaning to introduce the customer value theory in the exploration of the relationship between the participation of netizens in online games and netizen loyalty, because it is a useful attempt to fill the theory vacancies and resolve the real contradiction.

2.2   Hypotheses

Combined the features of online games, the relevant theories of customer value and website loyalty, the author develops a perceiving value model for netizens' involvement in virtual games in SNS website, as shown in Figure 1. The adoption of the statistical description, reliability, validity analysis and regression approach is applied to the hypothetical model validation with spss15.0.

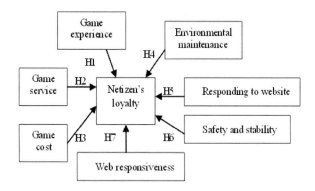

Fig. 1. The influence of netizens' perceiving value on netizens' loyalty

H1: Perceived Game experience will have a positive effect on Netizen's loyalty.

H2: Perceived Game serive will have a positive effect on Netizen's loyalty.

H3: Perceived Game cost will have a positive effect on Netizen's loyalty.

H4: Perceived Environmental maintenance will have a positive effect on Netizen's loyalty.

H5: Perceived Responding to websit will have a positive effect on Netizen's loyalty.

H6: Perceived Safety and stability will have a positive effect on Netizen's loyalty.

H7: Perceived Web responsiveness will have a positive effect on Netizen's loyalty.

3   Emperical Alalysis

3.1   Data Collection

The survey gathers 154 questionnaires obtained through the network in and adopts the Likert scale 5-point to demonstrate the different attitudes: agree very much (1 point), agree (2 points), uncertain (3 points), do not agree (4 points) and do not agree very much (5 points).And there are three measures of game-perceiving value, including the game experience (GE), game service (GS), the game costs (GC). The perceiving value of SNS's operation can be measured from the following aspects: the environmental maintenance (EM), website response (WR), safety and stability (SS), credibility (WP). Netizen's loyalty clings to the view that any netizen holds a positive feedback (UL1) to the games in the website, even if there are other similar websites providing the same game, the netizen will still choose the same website (UL2). The questionnaire is composed of seven independent variables and dependent variables, mounting to a total of 28 measures, and the number of questionnaires meets the requirements of statistical analysis."

3.2   Empirical Results

The values of Cronbach alpha are 0.806 (game experience), 0.721 (game service), 0.835 (environmental maintenance), 0.800 (Responsiveness), 0.740 (security), 0.798(credibility),which suggest that the items associated with each factor are closely related to each other, the Cronbach's α value of Game cost and user loyalty are respectively 0.688 and 0.692. (Gefen, Straub and Boudreau 2000).the internal consistency and the stability of the variables of the questionnaire are fairly good, and the questionnaire has quite good reliability. In the result of SPSS, KMO for each variable is equal to or more than 0.50 and the significance of Barlett Test of Sphericity is 0.00 which is less than 0.01, it meets the general standard. The Factor Loadings of eight variables is over 0.6. And in general social study, if the Factor Loading is more than 0.4, the variable is valid. The strength degree of each variable is more than 50%, so it meets the standard for general social study which requires the degree to be more than 30%. Therefore, the variables are all valid.

variables	Measurement index	Std. Deviation	Cronbach's α
Game experience (GE)	GE1	0.984	0.806
	GE2	0.810	
	GE3	0.762	
	GE4	0.911	
	GE5	0.864	
Game service (GS)	GS1	0.841	0.721
	GS2	0.789	
	GS3	0.899	
Game cost (GC)	GC1	0.990	0.688
	GC2	0.875	
	GC3	0.850	
Environmental maintenance (EM)	EM1	0.988	0.835
	EM2	0.855	
responsiveness (WR)	WR1	0.911	0.800
	WR2	0.964	
stability (SS)	SS1	0.818	0.740
	SS2	0.939	
credibility (WP)	WP1	0.808	0.798
	WP2	0.711	
User loyalty (UL)	UL1	0.917	0.692
	UL2	0.881	

Fig. 2. Reliability test of variables and measurement index of the questionnaire

We can come to a regression equation as shown in Figure 3: $WY=0.146+0.355GE+0.144GC+0.163EM+0.25WP$.

It shows that the game service is 0.334、 the responsiveness is 0.432 and the complete stability is 0.367. According to Liu, D H, Zhao, Y & Li, N(2008), sig values less than 0.1 suggest a good fit of the hypothesize model. so the correlation co-efficiency is not significant. In other words, the influence of game service、 responsiveness and safeness stability on loyalty of netizens is not significant. H2、 H5 and H6 are false. Otherwise, influence of gaming experience and credibility is the most while the cost of environment maintenance and games are less. In order to test the significance of the regression equation, the author goes on to F-test. Its F is equal to 1.943 and sig is 0.000 which is less than significance α (0.01), so the regression equation is significant.

Coefficients[a]

Model		Unstandardized Coefficients B	Unstandardized Coefficients Std. Error	Standardized Coefficients Beta	t	Sig.
1	(Constant)	.146	.235		.621	.536
	(8-12)	.355	.132	.299	2.694	.008
	(13-15)	-.116	.120	-.101	-.970	.334
	(16-18)	.144	.089	.135	1.613	.052
	(19-20)	.163	.077	.187	2.126	.035
	(21-22)	6.423E-02	.081	.071	.788	.432
	(23-24)	7.390E-02	.082	.075	.905	.367
	(25-26)	.250	.099	.224	2.529	.013

a. Dependent Variable: (29-30)

Fig. 3. Regression-table

According to the analysis above, the hypothesis should be modified. (Figure 4).the modified model is more concise and the freedom is higher.

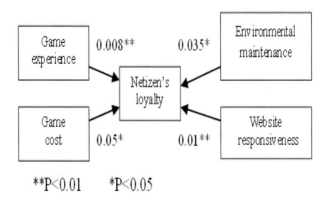

Fig. 4. Empirical Results

3.3   Management Implementation

SNS website, as one of the important applications in the times of web 2.0,has become the most popular topics and network-application highlight,where netizens have access to the formation of perceptual evaluation of the value and preferences form different attitudes, leading to the corresponding appropriate behaviors.These behaviors include enhancement of involvement, increase of frequency of visits to the site,and give referrence of this web to others as well as propaganda. The exploratory factor analysis and reliability analysis show that the individual measures and indices are relevant and reliable.

In other words, such constructs as game experience、 website responsiveness、 environmental maintenance and game cost have significant impact on netizen's loyalty.

Firstly, The enhancement of netizens' involvement in games and the reduce of game cost remain the most efficient way to stregthen netizens' loyalty from the perspective of the virtual games.

Scendly, The emphasis on maintaining game environment and upgrading the credibility of website remain an effective way to boost the loyalty of netizens from the perspective of SNS website. It may be quite safe for the website of SNS to build netizen's loyalty from the above two aspects.

4   Conclusion

This research analyses the relationship between netizens' involvement in virtual games in the website of SNS and loyalty to website based on empirical research on the application of QQ game farm and Tencent space .The results suggest that game experience、 website responsiveness、 environmental maintenance and game cost influence netizen's loyalty.Due to some constraints, however, in choice of variables, the author only selected seven independent variables for analysis and research, without taking other scenarios adjustment factors for consideration, thus, the conclusion has some limitations to some extent; in sample collection, the samples are constrained both in range and number, which can not represent the overall state of the virtual game for netizens' involvement in SNS website. So this paper is a constituent of the whole research on the loyalty of SNS website, more and more experts and scholars are expecting to join in the academic group to add amendment and supplement to the further study on both the theory and practice of the SNS website.

References

1. Clerk Maxwell, J.: China Internet Network Information Center 2009, Statistical Report on Internet Development in China, 3rd edn. A Treatise on Electricity and Magnetism, vol. 2, pp. 68–73. Oxford, Clarendon (1892)
2. Choi, D., Kim, J.: Why People Continue to Play Online Games: in Search of Critical Design Factors to Increase Customer Loyalty to Online Contents. Cyber Psychology & Behavior 7, 11–24 (2004)

3. Corstjens, M., Lai, M.: Building Store Loyalty Through Store Brands. Journal of Marketing Research 37(3), 281–292 (2000)
4. Dong, D.H., Quan, X.Y., Qu, X.F.: Theory of Customer Value and Its Formation. Journal of Dalian University of Technology 20(4), 18–21 (1999)
5. Griffiths, M.D.: Computer Game Playing in Children and adolescents: A Review of the Literature. In: Gill, T. (ed.) Electronic Children: How Children Are Responding to the Information Revolution, pp. 44–58. National Children's Bureau, London (1996)
6. Jahanna, Veronica: Customer Loyalty on Content-Oriented Web Site. Service Market 18, 175–186 (2004)
7. Kotler, P., Wang, Y.(trans.): Marketing Management, p. 392. Truth & Wisdom Press, Shanghai (2009)
8. Liu, D.H., Zhao, Y., Li, N.: SPSS 15.0 Statistical Analysis: from Rudiments to Mastery. Tsinghua University Press, Beijing (2008)
9. Clerk Maxwell, J.: China Internet Network Information Center 2009, Statistical Report on Internet Development in China, 3rd edn. A Treatise on Electricity and Magnetism, vol. 2, pp. 68–73. Oxford, Clarendon (1892)
10. Woodruff, R.B.: Delivering Value to Consumers: Implications for Strategy Development and Implementation. In: American Marketing Association Winter Educator's Conference Proceedings, pp. 209–216 (1992)

Design for Item Bank in Exam System Based on Network

Li Guohong[1], Wang Xiaodong[1], Qi Yubin[2], and Li Guofang[3]

[1] College of Finance and Economics, Hebei Normal University of Science and Technology, Qinhuangdao, China
[2] College of E&A, Hebei Normal University of Science and Technology, Qinhuangdao, China
[3] College of M&E Engineering, Hebei Normal University of Science and Technology, Qinhuangdao, China
liwangmy@163.com

Abstract. The paper analyzed the item bank of the exam system from the aspects of reliability and validity of questions, database structure and database maintenance. The paper presents the following points: the validity should be emphasized; the technical indicators in the database structure should be diverse to fully reflect the knowledge points; the system should provide a more extensive database maintenance method for the convenience of the administrator to operate.

Keywords: Exam System, Item Bank, Reliability And Validity, Structure Design.

1 Introduction

Exam system based on network (LAN, Internet) is objective and accurate, efficient and easy to post processing; therefore it is widely used in various examinations. Item bank is at the center in the exam system. This article will focus on description the organization form, structure, and management of the item bank.

2 Overview of EXAM System

The main advantages of network-based examination system are as follows:

Examinations may vary in size, and the number of candidates can be arbitrary.

Examination time can be uncertain, candidates can pick examination, and examination questions are not repetitional.

Through automatic scoring, the examination results can be given immediately after the examination.

It can reduce the workload of teachers, also reduce marking, recording and other errors.

Test scores can be analyzed automatically, such as the highest points and lowest points, average score, score distribution, standard deviation and so on.

The main disadvantages of the examination system are as below: limited testing questions, in general it can be only tested by objective items, such as single-choice,

Y. Wu (Ed.): International Conference on WTCS 2009, AISC 116, pp. 633–639.
springerlink.com   © Springer-Verlag Berlin Heidelberg 2012

multiple-choice questions, true-false. Subjective items can not be marked by the computer automatically.

3   The Reliability and Validity Analysis of Questions in the Item Bank

The reliability of questions refers to the dependability of the system. The test score includes two parts, the one is real score, and the other is the false score because of the stochastic error. Of course, the more errors are, the higher reliability is. Validity is the effectiveness of test results that means the extent of the test can measure.

There is a relationship of the unity of opposites between the reliability and validity of the test question. They are interdependent, if there is no reliability, the validity is impossible. If no validity, reliability is meaningless. But in the course of the specific composing test paper, the reliability and validity are often mutually exclusive. Relatively speaking, reliability of the data is easy to access, while validity is elusive. If we ensure the reliability, the validity is likely to be ignored, in contrast, the pursuit of its validity, it's difficult to ensure reliability.

Specifically, the contradiction between reliability and validity of the test questions lies in the following aspects.

A.

Standardized test questions are no longer used in testing the system of knowledge only, but focus on testing practical ability. However, ability can not be accurately quantified with number. In this case, if the paper is designed to a quantitative model, then the reliability is high, the validity is low. If the paper is designed to a qualitative model, then the validity is high, the reliability is low. The proficiency test basically belongs to qualitative test, while achievement test is essentially quantitative type.

B.

The limited options, or "true" and "false" in the standardized test is difficult to reflect the level of mastery and application of the knowledge. It is difficult to reflect the extent to which the testee can apply the knowledge in a flexible way in practice and therefore it is difficult to determine the validity of the test .

C.

The complex and integrated knowledge hierarchy is divided into various components as test points through the examination system. It is beneficial to come to an objective and clear test result and to help improve reliability. However this division is against the whole concept of knowledge and not to meet the validity requirement.

D.

Knowledge is multidimensional, not one-dimensional. The examination should be rich in content which includes terminologies, concepts, and knowledge points and so on. The content has the two major aspects, theory and practice. The basic knowledge and extend knowledge constitute the examination. It is difficult for any paper to reflect the multidimensional nature of these areas, thus validity will be affected.

E.

The knowledge is dynamic. It is a process to master the knowledge for students, not just an outcome. Subject knowledge system itself is also at the development and improvement process. Test can coagulate knowledge, it will make the subject development still, and this does not meet the validity requirement.

How to handle the conflict between the reliability and validity? The common approach in the current examination is: the reliability will be ensured first as a key point on the premise of that the validity should be considered as much as possible about the important exam. Subjective test items should be increased appropriate in the exam. If the test have an effect on a relatively small area, it should be focus on ensuring the validity than the reliability. Meanwhile Taking into account the influence on the test to the teaching and quality training of students, the more important test questions should be designed more effective. It should be carried out though enriching the question type.

Table 1. Database Structure

No.	Field Name	Type and Width(bit)	Note
1	Type	Character, 2	Item Type
2	Score	Numeric, 4	Score
3	Chapter	Character, 4	Chapter
4	Level	Character, 1	Degree of difficulty
5	Used	Numeric, 3	Frequency of using
6	Major	Character, 2	Major of testee
7	BelongTo	Character, 2	Knowledge point
8	Flag	Character, 1	Sign of using
9	Item	Character, 250	Question
10	OptionA	Character, 250	Option A
11	OptionB	Character, 250	Option B
12	OptionC	Character, 250	Option C
13	OptionD	Character ,250	Option D
14	Answer	Character, 250	Answer

Item bank is the core of the examination system. The quality of item bank is directly related to the reliability and validity of test results. The larger the capacity of item bank is, the lower the exam duplication rate is. In general, the reliability and validity will be better. But the capacity of item bank is not the more the better, there is a degree. If the item bank is too much, then the duplication of the content will increase and the ability of papers covering the contents will reduce. It should be tagged different weights for the different difficulty of exam questions respectively, which are not only the basis of question designing, but also the basis for marking.

4   Structure Analysis of Item Bank

Nowadays the computerized exam mainly bases on standardized tests. Exam questions are mainly single-choice questions, multiple-choice questions, true or false and blankfilling four types. Questions are saved in the item bank, which are composed by two kinds of indictors that are question ontology indictor and question supplementary indictor. It will be described separately below.

4.1   Database System

Database is divided into three types according to the model, that is: relational database, hierarchical database and network model database. At present, relational database is applied widely. View from the organization of questions, you can use the relational database. From the point of view of common degree, relational database software products are Microsoft Access, SQL-Server, Microsoft FoxPro, etc.. Considered the convenience of programming control, system support and system integration procedures, if the test database is not big, Microsoft Access database can be used.

4.2   The Question Ontology Indicator

The question ontology indicator is the question and the answer of question. Such as the choice question, the question ontology indicators include the question, the options and the answer. The question ontology indicator is the most basic part of the item bank. In designing the structure of the database, we can set the question, option A, option B, option C, option D and answer the six fields. Setting the four option fields is to easily control the format in the formation of the test paper.

4.3   The Question Supplementary Indiactor

Besides the question itself, there are some technical indicators to play a supporting role to the test paper. Such as the type of questions, score, chapter, degree of difficulty, frequency of occurrence, the major of testee and signs of using. The more abundant the supplementary indicators are, the more flexible the controllability of paper formation is. For example, after these supplementary fields are used, you can form test paper according to chapter (mid-term exam, final exam, chapter test, etc.) or major.

4.4   A Suggested Structure

A suggested structure of test item bank system is as tabled below.

The meaning of the important fields in the table is as follows.

Item Type: it is used to identify the type of questions, such as "01" means the Single-Choice, "02," means the multiple-choice, "03," means the short answer, "04" means blankfilling, etc..

Chapter: it means that the section where question is, the chapter can be represented by the first two numbers, and the section can be represented by the last two numbers.

Degree of difficulty: it can be set by teachers. The questions are divided into a total 10(0 to 9) according to the difficulty, which will be flexibility to use in the formation of the test paper.

Frequency of using: it is used to record how many times the item has been used in the formation of paper;

Major of testee: the various majors are used to represent by the 2 bit codes.

Knowledge Points: Teacher divides the course content in terms of knowledge point, and mark the knowledge point of each question in the item bank.

Sign of using: when a test paper forms, the question is used or not.

Options: for the considerations of the paper format post-control, the four options should be saved in a separate field.

5   Item Bank Operation

5.1   Classification of Item Bank Operation

The operation of the item bank can be divided into item bank maintenance and examination using two categories.

Item bank maintenance is under the instruction of teachers from the course content to analysis various technical indicators of questions, and add, delete or modify the questions in the item bank. In the operating process, if there is operation which is potential threat to data integrity of the item bank, there must be a warning mechanism for the operator to confirm. Furthermore, there must be an operate undo mechanism for the operator the opportunity to avoid the misuse.

Examination using is defined as the using to the test item bank of test system in organizing a test. At this point test system obtains control to the item bank, and generates a paper randomly by a certain strategy. In the process of paper generating, some tag information will temporarily make in the item bank, after the paper forms, the tag information should be removed.

5.2   Operation Model of Question Bank

Question bank maintenance operation shown in Figure 1, that is, after the teacher passes the authentication; he can perform operations to the database like adding, deleting and modifying.

The process of using item bank in the test is shown in Figure 2.

In the examination process, the server must first start the test service, and then candidates log on the system on the test machine. After they successfully pass the authentication, the test service system forms the test paper according to certain logic. The logic can be according to the chapter, the type, the knowledge point, the difficulty level, etc.. The more the technical indicators of the questions are, the more the logic which you can choose to set the paper is. In addition, it is also possible to form the complex logic papers in according to several indicators.

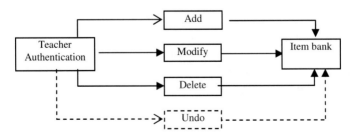

Fig. 1. Item bank maintenance operation

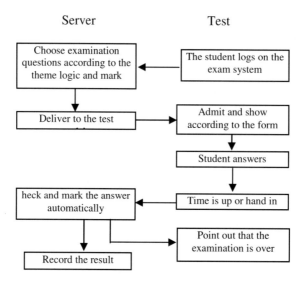

Server   Test

Fig. 2. Using item bank in an exam

Selected questions are marked "selected" at the specific field in the item bank. When a complete paper is formed, the paper will be sent to the candidate's computer by sequence according to the "selected" mark. The test questions are displayed on the candidate's machines in a specific format, and then the students begin to write the answer. When the test time is over or students actively submit papers, the candidate machine system will upload the candidates' answer information to the server, the examination server will automatically score according to standard answer, and record to the database.

Test item bank is the core of the examination system, and the database structure design will directly affect the functions and efficiency of examination system. In database design, we must fully analyze reliability and validity of questions to construct various types and large number of candidate questions. The question technical indicators should be designed as much as possible to enrich the test paper model in structure designing. The item bank is used in accordance with system policy from two aspects of the server and the candidate machine in order.

Acknowledgments. Author thanks for the supporting from Education Science Plan of Hebei Province (grant No. 06030096) and Education Reform Fund of Hebei Normal University of Science and Technology (grant No. 0802, "Teaching method reform of Engineering Thermodynamics").

References

1. Ren, J., Zhu, L.: The Design of Network Test System. Journal of Henan Mechanical and Electrical Engineering College 17(6), 119–121 (2009), doi:CNKI:UN:HNJD.2009-06-044
2. Lin, J., Yan, H., Wu, B.: Theory analysis on Computerized Adaptive Test. Jouranl of Taiyuan University of Technology 35(2), 221–224 (2004), doi.CNKI:SUN:TYGY.0.2004-02-031
3. Fu, X., Zou, B.: The Research on Multi-level Management Model of Grade Security in Examination System. Science Technology and Engineer 31(24), 219–221 (2005), doi. CNKI:SUN:KXJS.0.2005-06-011

Exploration and Analysis on Some Related Issues to E-Learning in the Environment of Computer Network Technology

Wu Peng and Xie Jingjing

Computing Center, Henan University, Kaifeng, China
S.W_Tian@hainan.net

Abstract. The development of computer network technology is exerting a significant impact on education which is also on a way to change, and new teaching models are being constructed. Computer network provides technical support and broad development platform for E-learning, while E-learning teaching model is changing the traditional educational ideas and methods. In recent years, domestic universities initiate E-learning teaching mode, enforcing information construction of education. This paper will explore and analyze some related issues to E-learning in the environment of computer network technology.

Keywords: E-learning, environment, computer network technology, information technology.

1   Introduction

E-learning (Electronic Learning) is proposed firstly by American scholar Jay Cross, who argued that the focus of E-learning was not on "E" but "learning" and the technology was not the point but the technology application in learning level. US Department of Education point out that "E-learning" is an approach to education, including new communication mechanisms and the interactions between people. These new mechanisms refer to computer network, multi-media, websites on professional content, information search, electronic libraries, distant learning and online classes and so on. E-learning means to use the network technology to deliver various kinds of solutions for increasing knowledge and improving learning efficiency. E-learning is "the use of network to promote learning, including the design, delivery, retrieval of learning content and the management in learning experience as well as the exchange between learning communities.", while the "digital learning" refers to any formal or informal sharing activities that apply digital technologies. Any digital technology used in teaching process, such as asynchronous self-learning materials, synchronous online virtual classroom, online learning based on Internet or local network, single machine computer-aid training, audio documents, audio disks, audio tape, video tape, satellite broadcasting, interactive TV, etc, All of which can be called E-learning. E-learning stems from Electronic Learning, and it is translated in Chinese as "digital learning", "electronic learning", "network

Y. Wu (Ed.): International Conference on WTCS 2009, AISC 116, pp. 641–645.

learning" and so on. Different translations indicate different opinions: 1.emphasize learning based on Internet; 2.emphasize digitalization; 3.empahsize the combination of digital content and network resources. All of these opinions attach importance to digital technology and using technologies to change and guide education. Some scholars explain that "E-Learning is the learning and teaching activities mainly done through Internet, and it fully utilizes the learning **environment** with brand new communication mechanisms and rich resources which are provided by modern information technology to achieve a whole new learning approach." We stress E-Learning is a new kind of learning methods, therefore there is a special "network education" in advanced education. The widespread of "network education" reflects people's expectation or a state of mind to consider E-Learning as a special approach to education.

2   Research and Analysis on Some Related Issues to E-Learning Based on Computer Network Environment

E-learning is one of the most important parts of E-Education. The target of E-learning is to create a kind of new learning style through the ideal learning environment provided by modern information technology in order to thoroughly change traditional teaching structure and the essence of education, and then train a large number of creative talents. The specific teaching principles of E-Learning education are shown as follows: creating the real-life situation and emphasizing self-learning, collaborative learning and discovery learning. Compared with the traditional teaching mode, E-learning has presented many new features. The objects faced by E-Learning own mass, personalized features, therefore, E-Learning will add self-learning contents, conditions, resources, methods and activities into the education system in order to meet the learner's individual learning requirements. The research approach to E-Learning is a kind of systematic method. Research and explore the essential relationship between various elements of teaching and learning or between factors and the entirety, and then comprehensively consider the relationship between them and make all elements integrated dynamically in order to achieve the intended purpose of teaching and learning. E-Learning can keep the learning content updated timely and continuously. Besides, in the long term it can make various learning resources including learning materials keep updated and related to technology, which will endow those resources more value. Because of the Internet, the distance between people, people and learning resources has been largely shortened. Through the Internet, we can easily access to information though it's thousands of miles away, communicate with people from all over the world, master the latest technological development trend at the very first time and stand on the forefront of the times. In the traditional education and training, teachers should develop training materials, arrange training venues, organize examination, logistics, and then after announced the end of the training they immediately plough into the next round preparation. Adopting the solution of E-Learning can cut down the cycle and make us work in a real-time mode, while this doesn't mean the rigorous training program be no longer applicable---maybe it's still the best solution, but for today's increasingly rapid working pace, the time consuming by learning itself has exceeded the time that can be dominated by individual and groups. If we want to keep up with surrounding changes, the advanced teaching and information technology maybe a must.

The import of E-Learning internet learning platform can largely improve the flexibility of teaching plans' implementation. At present, domestic higher education has commonly combined teaching and practice. In the learning phase, students should complete the learning of theoretical knowledge and basic technology with the focus on the theory and technology. In the practical phase, based on the practical requirements of jobs that students will engage after graduation, students will carry out practice related to the relative skill and other practical training in the relative positions in school base, laboratories or school-enterprise. In practice, students can make use of flexibility and quick updating features of E-learning to make adjustment to teaching plans in accordance with their' actual situation and social development demands. E-learning almost completely breaks the constraint of teaching space and time, and enhances the flexibility of both teaching and learning. As it's a kind of internet teaching, in theory E-learning can fully utilize the teaching resources of excellent teachers, which is undoubtedly very important to those fields short in teaching resources. If students cannot come to school in the practice phase, E-learning can be used to conduct the self-learning. Students can accept training any time at any place, besides, the communication to teacher and technicians of enterprises through this platform, the theory and practical skills can be combined perfectly. E-learning makes the education socialized and activated. In this information age, new knowledge and new things emerge in large numbers any time at any places, so we must change the learning orientation from one-time learning to life learning, while the Internet has provided a strong support for this change. This is an education socializing and informationizing process. In the next few years, education will walk out from school to home, communities, rural areas and any place that has been universalized in information technology. Internet will become a real school without walls and the network learning will become an integral part in our life, which will all consist of the happiness in everyday life.

E-Learning combines text images and sound together, which has largely improved the effectiveness of education training. The diversity and readability of learning materials, and the interactivity of learning mode is one of the characteristics of college students' learning. Through the establishment of effective platform of education training, internet education training can provide learning resources and environment which is a blend of text, image and sound. For example, IT platform can help students know the latest education training information; the research platform of teaching cases can help students conduct teaching case research based on rich web lesson resources (including teaching cases, design lesson plans, courseware, etc.), and share learning experience through autonomous exploration or cooperative research; teaching video can help students use digital technology to make analysis and evaluation on the classroom teaching example. It can also provide network training courses to help students carry out academic and practical learning research. While the exchange and inter-dynamic platform will help students conduct the thematic exploration, exchange of thoughts, opinions and cooperative learning on the hot or difficult issues through online forums BBS, blog, E-mail, MSN and other ways. Besides, digital learning styles can integrate image, video, audio and other technologies into learning, which will fully adjust to learners' various feelings, give compound stimulus on the brain in order to adapt to laws of learning and greatly improve the learning efficiency. In addition, due to the network's advantage of resource sharing, through the mode of online course, students do not need to pay training fee, moreover, they can upload their own excellent learning resources on the web site, and others can choose the appropriate content to achieve the learning resource sharing. Therefore, this kind of learning methods can help students increase knowledge and broaden horizons, especially for those backward areas.

In traditional classroom teaching, most teachers have no chance to fully exchange ideas with each student, and there are various reasons that many students dare not to have a face-to-face communications with teachers. E-Learning has changed all of that. On the Internet, learners can not only download teachers' handouts, assignments, and other reference materials, but ask questions from thousands of miles away and discuss and comment on the content learned from classroom with other students on the net, which thus mobilizes the enthusiasm of learning. E-Learning endows learners with teamwork spirit, strong communication skills and the ability to discover and deal with problems, which are the focus of higher education. It is found that to adopt collaboration learning strategy tends to achieve a multiplier effect, and thus more effective when involving occasions of high-level cognitive learning. In the network, commonly used collaboration methods are discussion, competition, cooperation, partnership and role-playing and many other different modes. It is safe to say that collaborative learning is the best embodiment of network characteristics and also one of the most conductive teaching modes to train the quality of talents. Researches show that learning methods based on E-Learning allow students a more in-depth discussion and engagement, who are able to apply teaching and communication technology to realize a variety of interactive and collaborative environment. If experts and students enter the same environment, then the learning effects are even better than a traditional small class' interactive effects. Online interactions can provide case studies, description of scene, demonstrations, discussion groups, instructors, temporary project group, chat rooms, practice tests, E-mail, BBS, tips, guides, FAQs, and practice guides and other tools. Another study discovers that online students are more willing to communicate with others than those of traditional classes. The population effect generated by students in network is more active than that led by teachers. Thus, E-Learning enables students to become evaluation subject, an education method for students to understand themselves and educate themselves and a way of feedback for teachers to improve teaching and complete innovative educational objectives, which helps to exert the guiding function of education evaluation, lead teachers to teach creatively and students to study creatively and make possible and maneuverable the evaluation experiment that enhances students' innovative spirit and practical ability.

3   Conclusion

E-Learning, supported by computer network technology, simply speaking is online learning or network learning, that is, to build Internet platform in education field where students connect to Internet through PC, and it is a brand new learning method through Internet. Of course, this method cannot be separated from the completely new learning environment made up by multi-media network resources, network learning community as well as network technology platform. The network learning environment put together a large amount of data, files, programs, educational software, discussion group, news group and so on, forming a highly integrated comprehensive resource base. As the rapid development of higher education informationization, E-learning is changing people educational and learning method at an alarming rate. However, with the receding of the first round of research and practice boom, people gradually return to their sense. E-Learning method has rich multi-media resources, convenient collaborative exchange, friendly interaction and other unique advantages, but it cannot replace the teachers' classes completely, which is lack of the deep involvement of teachers and whose learning effects are not as satisfactory as expected. How to fully represent E-Learning's active participation, how to give full play to the guiding role of the teachers or experts as well as how to properly utilize relevant technology environment to achieve a better result have become a common concern and issues that require further thinking and research.

References

1. Xu, H.-Y., Feng, Y.: Design and implementation on E_Learning system based on Agent and integrated with study context. Computer Engineering and Design 02 (2009)
2. Wu, F., Wu, B., Shen, Z.: Technical Difficulties and Characteristics of Next Generation e-learning Platform. Open Education Research 01 (2009)
3. Zhu, Z.-Z., Wu, Z.-F., Wu, K.-G., Zhou, S.-B.: E-learning Services Discovery Algorithm Based on User Satisfaction. Computer Engineering 03 (2009)
4. Chen, G., Chen, M.-X.: The Design of Individual Interaction Based on E-learning Performance. Modern Educational Technology 02 (2009)
5. Wu, D., Zhao, S., Yang, X., Zhang, Y.: A Study on E-Learning Resource Cataloging and Coding Strategy. Distance Education in China 01 (2009)

The Data Transfer and Data Security Scheme in a College Students' Overall Quality Evaluation System

Ying Zhao[1], Qiang Ren[2], and Zhan Li[1]

[1] Department of Electronic Information, Teachers College of Beijing Union University,
Beijing, China
[2] Beijing Computer Technology and Application Institute, Beijing, China
ZhaohuiRen025@163.com

Abstract. A college students' overall quality evaluation system was designed and developed based on computer and mobile blended networks. Users could do evaluation works anywhere at any time with the use of this system. Mobile office became a reality and it makes work more convenient and more efficient. HTTP protocol was adopted to transfer data between Midlet and Servlet and it makes the mobile client program to be device independent. In addition, ECDSA (Elliptic Curve Digital Signature Algorithm) was implemented in this system, and it ensures the authenticity and integrity of the data needed by the evaluation. Thus, evaluation work can be carried out objectively, fairly and efficiently by the use of this system. This paper described the data transfer scheme and the data security scheme of this system in detail.

Keywords: Wireless data transfer, data authenticity, data integrity, ECDSA, overall quality evaluation.

1 Introduction

We have carried out overall quality evaluation among our college students for years. And it is an important measure to guide students to develop their morality, intelligence and physical fitness in balance. But, all the evaluation works, from data collecting, sorting and verifying to the score calculating, publicizing and publishing, are almost handled manually by far. Teachers are suffering from the heavy and Error-prone works. What's worse, the necessary data needed by the evaluation comes from deferent departments. And the departments are so distributed that it is difficult to collect data from them.

Since computer network has been very popular in our college, we could develop a college students' overall quality evaluation system based on computer network. With the system, all the data could be collected using computer network and score could be calculated automatically by computer. It will greatly reduce the teachers' workload, improve the efficiency and reduce error rate. In the other hand, as we all known that the data transfer capability of 3G mobile wireless network has been improved greatly as well as the 3G mobile performance. So it is also possible for us to expand the system to 3G mobile terminals. If teachers in the distributed departments could upload

Y. Wu (Ed.): International Conference on WTCS 2009, AISC 116, pp. 647–655.
springerlink.com   © Springer-Verlag Berlin Heidelberg 2012

data or visit and use the system through their mobiles, they will certainly feel very convenient to do the evaluation work.

But how to verify the sender's identity and detect whether or not the data was modified after it was send out, if the data is colleted using computer network or mobile network? It is essential to the system, because the evaluation is objective and fair only if the data is authentic and integrate. In addition, how to transfer data between mobile and web site? It is also important to the system, because users could visit and use this system through mobile only if mobile could communicate with the web site. This paper will analyze and introduce the solutions of these two problems.

2   Data Transfer Scheme

2.1   Architecture of the System

Nowadays, J2ME has been supported by major mobile manufacturers because of its good portability, deployment flexibility and supporting advanced communication protocol such as TCP, UDP, etc. So we developed our system using Java and the architecture of the system was designed as shown in Figure 1. It is a Client/Server and Browser/Server blended architecture, where the Client/Server architecture was designed for mobile users and the Browser/Server architecture was for computer users. Both of mobile users and computer users shared the same Web server. Mobile users should download, install and run a client program which is a Midlet program in order to visit and use this system. While for computer users, they can visit and use this system only by use of a Web browser. This architecture helps to realize the seamless integration of computer network and mobile network, so that users could visit and use this system through either computer or mobile. Consequently, evaluation works become more flexible and more convenient.

On the server, the three-layer architecture idea was adopted. JSP and Servlet are on the representation layer. They will receive requests from computer users or mobile users, and then pass the requests to the business logic layer. JavaBean is on the business logic layer. It will visit the MySQL database and handle the requests coming from the representation layer according to particular business logics, and then give responses to the users through representation layer.

2.2   Data Transfer between Midlet and Servlet

We developed our mobile client program using J2ME. The MIDP in J2ME specified three network communication protocols which are HTTP, Socket and DataGram. MIDP also specified that HTTP should be supported by all mobile information devices. In order to make our mobile client program to be device independent, we adopt HTTP protocol to realize the data transfer between Midlet and Servlet.

HTTP specified two kinds of data transfer modes which are POST and GET. In GET mode, data will be transferred as a part of the URL, while in POST mode data will be transferred as a bit stream. Therefore, the data size is limited in GET mode while it is unlimited in POST mode. What's more, only text data could be transferred in GET mode while the data format could be freewill in POST mode. Since the size of data to be transferred in our system is arbitrary, we adopt POST mode.

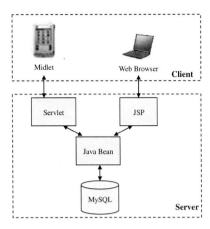

Fig. 1. Architecture of the System

The data transfer procedure between Midlet and Servlet in our system is shown in Figure 2. In order to transfer and display Chinese characters correctly, we created a DataInputStream and a DataOutputStream between Midlet and Servlet, and then read data from or write data to the stream respectively by the use of readUTF( ) method and writeUTF( ) method.

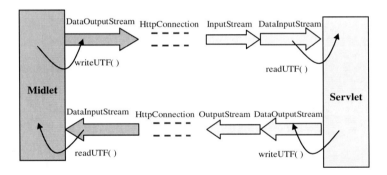

Fig. 2. Data Transfer Between *Midlet* and *Servlet*

3   Data Security Scheme

3.1   Basic Procedure of Digital Signature

We adopted digital signature technology to ensure the authenticity and integrity of the data which is collected by the system. A digital signature is basically a way to ensure that an electronic document is authentic. Authentic means that you know who created the document and you know that it has not been altered in any way since that person created it.

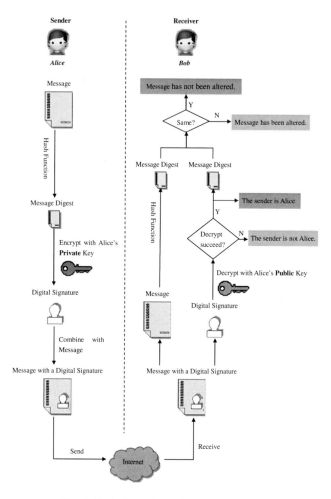

Fig. 3. Basic Procedure of Digital Signature

The basic procedure of digital signature is shown in Figure 3, and it includes a signature generation process and signature verification process. Almost all digital signature algorithms are based on public key cryptosystem and have a public key and a private key. The private key is used in the signature generation process and should be remain secret by the owner. The public key is used in the signature verification process and it need not be kept secret, but its integrity must be maintained. Anyone can verify a correctly signed message using the public key.

3.2 Comparison of Different Digital Signature Algorithms

Digital signatures rely on certain types of encryption to ensure authentication. There are a lot of encryption algorithms to realize digital signature, the famous of which are RSA, DSA (Digital Signature Algorithm), ECDSA, etc. Particularly, ECDSA is the elliptic curve analog of DSA.

1) Security Bases: RSA, DSA and ECDSA algorithms base their security on different mathematical problem. For example, RSA algorithm bases its security on the intractability of the integer factorization problem, while DSA algorithm bases on the intractability of the discrete logarithm problem in a finite field, and ECDSA bases on the intractability of the Elliptic Curve Discrete Logarithm Problem (ECDLP).

2) Security Strength: Unlike the ordinary discrete logarithm problem and the integer factorization problem, no subexponential-time algorithm is known for the ECDLP, which means ECDLP is fully exponential-time. For this reason, the strength-per-key-bit is substantially greater in an algorithm that uses elliptic curve. Table 1 compares the key sizes needed for equivalent strength security in ECDSA comparing with RSA and DSA. From Table 1 we know that at the same strength security, the key length of ECDSA is much shorter than RSA/DSA. In other words, strength-per-key-bit of ECDSA is much stronger than RSA/DSA.

Table 1. The Comparison of Security Strength

RSA/DSA Key Size (in bits)	ECDSA Key Size (in bits)	Time to Break (in MIPS-Years)
1024	160	10^{11}
5120	320	10^{36}
21000	600	10^{78}
120000	1200	10^{168}

Menezes and Jurisic, in their paper [11], said that to achieve reasonable security, a 1024-bit key size would have to be used in a RSA system, while 160-bit key size should be sufficient for ECDSA.

3) Key Size and Signature Size: The signature size of ECDSA is the same as DSA which equals 4t bits, where t is the security level measured in bits, For example, at a security level of 80 bits, meaning an attacker requires about the equivalent of about 280 signature generations to find the private key, the signature size for both ECDSA and DSA is about 320 bits. Whereas the signature size of RSA is much longer than ECDSA. Table 2 shows the different signature size for different algorithm to achieve reasonable security.

Table 2. The Comparison of Key Size and Signature Size

Algorithm	Public Key Size (in bits)	Signature Size (in bits)
RSA	1024	1024
DSA	1024	320
ECDSA	160	320

4) Computational Efficiency: ECDSA is a high efficient algorithm. Certicom, a Canadian company has been studying and promoting the ECC (Elliptic Curve Cryptography) system since the early 1980's. Some of their results of fast implementations of ECDSA compared to RSA are given in Table 3.

Table 3. The Comparison of Efficiency

Function	ECDSA 160-bit (in ms)	RSA 1024-bit (in ms)
Key Generation	3.8	4708.3
Sign	3.0	228.4
Verify	10.7	12.7

From the above comparison, we know that ECDSA has the highest security. And it could achieve the equivalent strength security with much shorter key size and much shorter signature size, which means that ECDSA has lower requirement for storage space and bandwidth. What's more, ECDSA is more efficiency.

In our system, mobile's storage capability and processor speed as well as the bandwidth are all limited. So these advantages might make ECDSA an attractive choice for our system's digital signature implementation.

3.3   Implementation of ECDSA in Java

According to the above analysis, ECDSA is the most suitable algorithm for our system's digital signature. And in order to blend computer network and mobile network so that users could visit our system though either mobile or computer, Java is an attractive choice for our system's implementation. But Java provides very poor support for digital signature. Java 6, which is the latest version of Java, only supports RSA and DSA. In other words, ECDSA is not supported by Java 6. Fortunately, there is an open source library which is name *Bouncy Castle*. *Bouncy Castle* supports many kinds of digital signature algorithm (e.g. SHA1withECDSA, SHA224withECDSA, SHA256withECDSA, SHA384withECDSA, SHA512withECDSA, etc.). SHA1, SHA224, SHA256, SHA384 and SHA512 represent Hash Functions which belong to SHA family. SHA (Secure Hash Algorithm) is the successor of MD (Message Digest). The message digest length generated by SHA is much longer than that of MD. So the security is stronger than that of MD. SHA224, SHA256, SHA384 and SHA512 are mainly deferent in the length of message digest and the number behind each 'SHA' represents to the length of the digest.

Firstly, we should download the latest *Bouncy Castle* libraries such as *bcprov-jdk16-143.jar* and *bcprov-ext-jdk16-143.jar*. And then add them into our project.

Secondly, we should import the necessary package into our java program file. e.g.

import java.security.*;
import java.security.interfaces.*;
import java.security.spec.*;
Finally, we should create Bouncy Castle Provider. e.g.
Security.addProvider(new BouncyCastleProvider());

After doing that, we can implement ECDSA in Java with the use of *Bouncy Castle* Libraries. The implementation procedure is shown in Figure 4. The signature generate implementation is similar to that of signature verification. The distinct difference between is the initialization of the Signature instance. In the signature procedure, the Signature instance should be initialized with the sender's private key by the use of *initSign( )*method. While, in the verification procedure, it should be initialized with

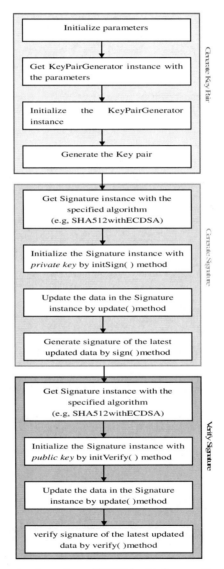

Fig. 4. Implementation Procedure of ECDSA with Bouncy Castle Libraries

the sender's public key by the use of *initVerify( )*method. The other difference is that signature is generated by the *sign( )* method, while it's verification is executed by *verify( )* method.

4   Conclusion

Overall quality evaluation is an effective measure to guide college students to develop their qualities in balance. Our college students' overall quality evaluation system was based on computer and mobile blended network. Users could visit and make use of this system through either computers or mobiles. It means that users could do the overall quality evaluation works anywhere at any time. Thus, mobile office becomes a reality and the evaluation works will be done more conveniently and efficiently. What's more, a high security, low requirement for storage capability and bandwidth, and high efficient digital signature which is named ECDSA was realized in this system. This mechanism will ensure the authenticity and integrity of the data which is collected from different departments with the use of this system. So the evaluation results will be ensured to be objective and fair.

The key pairs used for digital signature in this system were distributed and managed manually now. We should find a good way to distribute, storage and manage the key pairs automatically and securely in the future.

Acknowledgment. We would like to acknowledge the contribution of any people to this article. And the department of electronic information has provided us an excellent environment to develop this system. And we also would like to thank our families for their love and support.

References

1. Jun, L., Xi, Y., Hui, Z.: J2ME Mobile Software Program Design, pp. 121–141. China Water Power Press, Beijing (2009)
2. Electorinc Publication: Wikipedia, Elliptic Curve DSA (July 2010), http://en.wikipedia.org/wiki/
3. Johnson, D., Menezes, A., Vanstone, S.: The Elliptic Curve Digital Signature Algorithm (ECDSA). International Journal of Information Security 1, 36–63 (2001)
4. Qiu, Q.: A pplication of ECC on D igital Signature. Journal of Wuhan University of Technology 26, 78–80 (2004)
5. Zhou, S.S., Dong, P., Su, L.: A New Digital Signature Project Based on Elliptic Curve Cryptography. Application Research of Computers 9, 147–148 (2005)
6. FIPS PUB 186-3, Digital Signature Standard(DSS), Information Technology Laboratory, National Institute of Standards and Technology (2009), http://csrc.nist.gov/publications/fips/
7. Kapoor, V., Abraham, V.S., Singh, R.: Elliptic Curve Cryptography. ACM Ubiquity 9 (May 2008)

8. Johnson, D., Menezes, A.: The Elliptic Curve Digital Signature Algorithm (ECDSA), Technical Report CORR 99-34, Dept. of C&O, University of Waterloo, Canada (February 2000), http://www.cacr.math.uwaterloo.ca
9. Howgrave-Graham, N.A., Smart, N.P.: Lattice Attacks on Digital Signature Schemes. Designs, Codes and Cryptography 23(3), 283–290 (2001)
10. Jurisic, A., Menezes, A.J.: Elliptic Curves and Cryptography. Dr. Bobb's Journal (1997)
11. Dong, L.: The art of Encryption and Decryption about Java, pp. 111–115,167-168, 311-318. Machinery Industry Press, Beijing (2010)

The Affect of History Cultural Art in Art Education

Chen Su

Guilin University of Electronic Technology
chen.xiaojing@yahoo.cn

Abstract. Hezhe ethnic group folk arts is a kind of important culture heritage that belongs to the entire human, we should realize the valuable culture heritage's enormous meanings when we use it for the art design in the future. The most important, we should let young generation knowledge and experience personally the fascination of the culture heritage, and make them create some modern art masterpiece that is featured of Chinese folk culture.

Keywords: Cultural art, Art teaching theory, Teaching theory course.

1 Introduction

The art teaching theory is the course that set up for developing the primary and junior school teacher. Today in overall propulsion education for all-round development, combine the new foundation education "The Standard of the Course" the school and region can compile the textbook of the school or local special feature according to their own characteristic plait; to adapt the need of students from different regions. Knowing that the teaching theory course joints the national minority cultural is viable.

Hezhe ethnic group is an old nation which has formed own special culture through the long history. The art of pattern is very heavy and complicated. Hezhe ethnic group folk arts is a kind of important culture heritage that belongs to the entire human, we should realize the valuable culture heritage's enormous meanings when we use it for the art design in the future. So it's necessary to study this "no word epic". For inheriting and developing the cultural art of the Hezhe ethnic group, the art teaching theory course joins the contents of Hezhe. The Hezhe ethnic group lives in the beautiful and rich Sajiang plain China. In several thousand years, they created and developed the rich and splendid race cultural art. Owning to this, the writer taught the cultural art of the Hezhe ethnic group in the course, being a kind of common sense and investigation.

2 It Is Necessity to Melt the Cultural Art of the Hezhe Ethnic Group as the Agrestic Teaching Material Plait into the Art Teaching Theory Course

The traditional cultural and civil arts of the clan of Hezhe and Nanai are very ancient but abundant. It has the heavy race breathing and local special feature. Their cultural art is mainly emphasis in the dress: the fish fur-lined jacket, the birch skin bucket, the fish skin wallet etc; construction of the residents: Cuoluozi, Diyinzi, Mukeleng, Majiazi etc; in Saman teaching tools and implements: Shenou, Shenqun, Shenxue, Shenshoutao etc. as shown in figure 1, figure2, figure 3, figure 4, and figure 5. All of these emerge the intelligent wisdom of the clansman of Hezhe and Nanai, and have important aesthetic value. These abundant inheritance cultural art constituted the essence of cultural art of the Hezhe and Nanai, Which is an indivisibility part of our country race art and traditional cultural.

The new teaching theory being implemented, the teaching should use the art textbook creatively. According to the students, the characteristics of the school and the region, the teachers choose the content of the course, reorganizing and recreating, and make vivid use of the local nature and the cultural resources, to improve the content of the art teaching theory with the cultural art of the Hezhe ethnic group.

Fig. 1. Clothes of Hezhe ethnic group in the late nineteenth century

Fig. 2. Shaman priest and assistant of Hezhe ethnic group in the late nineteenth century

Fig. 3. Hezhe ethnic group acting out folk drama in the late nineteenth century

Fig. 4. Hezhe ethnic group on dog sledge in the late nineteenth century

Fig. 5. Painting on back of a fish-skin garment of Hezhe ethnic group in the late nineteenth century

The new art textbook of junior and senior school is very difficult to reflect the feature of the local place. The condition of the teaching college is dissimilar. The teaching conditions are also dislike. So there exist the problems how to choose and supply the new supplementary teaching material. Considering the agrestic teaching material should be a certain proportion in art teaching. It is viable to melt the cultural art of the Hezhe ethnic group into the art teaching theory course. In regard to the art education, the students should use the textbook contains the cultural art of the Hezhe ethnic group as the agrestic teaching material that can combine the art education with the native area humanities, the geography and economies construction. We should make full use of the cultural art that was created by the ancestors' diligence and intelligence precipitate through history. It is the comprehensive reflection of the ability of the students follow to the education post after the graduation. Make the students full understand the Hezhe clan race art. Accept the influence of the Hezhe clan race art. Raise the aesthetic sense and aesthetic judgment of the student.

3   Collect the Data of the Cultural Art of the Hezhe Ethnic Group, and Make Preparation for the Students to Write Compile the Teaching Plan and Try to Speak in the Class

Make use of the time outside the class; organize the student to visit the museum, shown in figure 6. Make the student initial understand the living condition and product condition of the Hezhe ethnic groups man .Pay more attention to the clothes they wear and living condition, sketch the painting on the fish skin clothing and birch skin there and then. Go and see the history ambition office, the file building, go to the news agency and television stations to check the relevant data and engrave the CD record. Call on the student to enjoy, observe and learn the data and picture and record. Through see, comment, argument three stage, the students aware to accept the influence of the cultural art of the Hezhe ethnic group. Make good sensitive faculty and reasonableness foundation for the student to teach in the class.

Fig. 6. Investigation in museme of Hezhe ethnic group

4   The Implement Process of the Cultural Art Melt into the Art Course

4.1   Carry Out the Mode Diagram

Under the full understanding of the student in teaching college guide according to the teaching method, compose plan, teaching implement, the teaching links such as evaluation and the information feedback etc, melt the cultural art of Hezhe ethnic group into art teaching theory course. The concrete implement process is as figure 7.

1) The course of appreciation: First of all, have the appreciate lessons combined with ancient culture and the arts of Hezhe ethnic group to make the teaching of appreciate lessons more practically oriented. And have arranged the students visited the museum with questions again, enlightened the students at the right time to make them aware that:" the Parent cultures of Chinese Ethnic Minorities is a folk art with primary and widespread, museums and other places were the convergence point that carrying the mother culture." Then teachers and students make programs together, and collect and sort out data on the Internet for application. The trial teaching process of appreciate lessons could be divided into four stages, the import stage of introducing pictures to the subject, the analytical understanding stages to stimulate students interest in learning, the phase of new lesson to explain of explaining the role of the conversion actually, the summing up stage of the distillation of materials and the summary of theory. Through the four phases of the implementation, a wonderful lesson was accomplished successfully. In the atmosphere of the teachers and students to learn together, cooperation, and explore,   Shown in Figure 8.

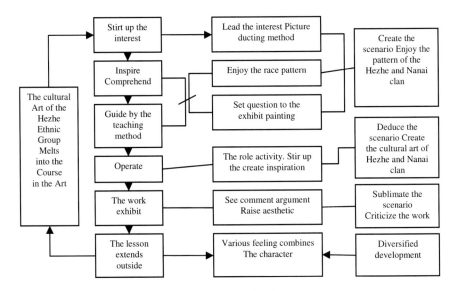

Fig. 7. Flow chart of teaching implement process

Fig. 8. Appreciate fish-skin art of Hezhe ethnic group

2) Painting Lesson: Hezhe ethnic group have motifs from ancient patterns . In the utensils and accessories of the Hezhe ethnic group , there were a large number of symbols , the motifs of plants , animals, moon and stars , as well as the two sides continuous and four-row pattern that abstracted from the concrete . This course is open to students to experience the exposition skills in the field of modeling the performance . The author enlighten the students from both aesthetic and spiritual of the Hezhe to enable students to swim in the design world . On this basis, seizing the opportunity to join the cultural transmission of knowledge ,to let the teachers that will be on the three feet know that they are entrusted with the responsibility of the Historical Heritage ,and the young people they taught will also be the core of cultural heritage in the future . This course aims to enable the local culture be extended and sublimated in the transmission from generation to generation. Trial Teaching teachers in the curriculum not only for the actual operation, but also carried out the extension links of extra-curricular , to make this class both deep and practical significance.

3) The lesson of handwork: In this lesson , the Trial Teaching teachers apply the motifs of Hezhe ethnic group to make fish-skin painting, and use the didactics of teaching with practice in the class , which is that the new teaching methods of teaching combined with practice . In the trial Teaching course , students made crafts with ethnic characteristics, as shown in figure 9, and figure 10. The author introduced the micro-teaching into class and recorded the whole process of Trial Teaching. After-school repeatedly watch video , take the local materials compiled by the Trial Teaching Teachers to be modified, supplemented and perfected again.

4) Activity Course: The activity course is in the area of Exploration , and they are also summary and theoretical sublimation of the few lessons before . To achieve this purpose and Interpretation of the situation to put those present exposure in the native culture . In the background music of ship song of Wusuli river, some students dressed in costumes of Hezhe ethnic group dance with a variety of Hezhe ethnic group living utensils (homemade) in hand--------a successful art class is shown before us.

Fig. 9. Show of fish-skin painting of student's handworks

Fig. 10. Show of fish-skin cuttting of student's handworks

4.2   Carry Out the Process and Feedback

According to the basic characteristics of the art teaching material of the junior and senior school, choose the dual orient teaching method that speaking and practicing, namely the teacher teach and the student try to speak, speaking with practing, practicing with speaking, combine the teaching and practicing teaching method. Because the agrestic teaching material is an old compile teaching material. There maybe some shortage or mistakes. For the complement and perfect of the self compile teaching material, can convince the special subject colloquium after lesson, to hear the opinion extensively. Make use of the history, geography, humanities, art etc. Various course of knowledge, to rectify the wrong spot in time, and add and perfect. Make the successful art teaching theory course emerge in front of the students.

Look back the whole process the course implement and the extend lessons the visit and appreciate outside the class, the analysis of the data of the picture of Hezhe ethnic group initially, especially the highly praise to the pattern arts and the Hezhe ethnic group. They all mean to describe the fine life of the Hezhe ethnic group with their paintbrush. The writer receives the unexpected result from it.

5   Make Full Use of the Geography Advantage, Protect and Develop the Precious Inheritance of the Cultural Art of Hezhe Ethnic Group

Know to all, the Hezhe ethnic group live in the high latitude region. The special cultural art is formed in a particular environment, also reflecting the living idea of the Hezhe ethnic group in the particular region. It is saying that, it is very necessary to develop and protect the cultural art of the Hezhe ethnic group in our province. It is viable to carry on tour development in our two countries by using of their inheritance cultural art. It will help to improve the cultural art of the Hezhe ethnic group. Being the synthetic high college in the Sajiang plan, we have the responsibility to put the cultural art of the Hezhe ethnic group as agrestic teaching material into the art teaching theory course. Let the teacher and student realize and understand the national cultural art from first step to the overall understanding. Draw lessons from and develop the cultural art statue of the national minority. Thus make this work become our school and Russia brothers' colleges to promote solidity. The common progress and prosperous that make.

6   Conclusion

Hezhe ethnic group folk arts is a kind of traditional plastic art, she has various formative patterns, and multifarious styles, which agglomerates wisdom and distillates of Chinese. Hezhe ethnic group folk arts has some special characteristic in some facets-----modeling, inditing, the connotations of folk culture, all which are valuable information for the art design and creation, it is worth of deeply studying and investigation.

References

1. Ling, C.: Hezhe Clan of Downstream of Songhuajiang River. Shanghai Arts Press (1990) (in Chinese)
2. Ministry of Education of the People's Republic of China. The Course Standard of Art. People Education Press (2003) (in Chinese)

Dialectical Thinking of the Evolution of Software Development Technology

Lu Huijuan[1,2], Tang Wenbin[1], and Guan Wei[1]

[1] College of Information Engineering, China Jiliang University, Hangzhou 310018, China
[2] College of Information and Electrical Engineering, China University of Mining and Technology, Xuzhou 221008, China
luu_luu@sina.cn

Abstract. In order to enhance the guide of project development for students in practice of teaching, and to make software development more scientific and effective, this paper, in combination with the research orientation - the practice of computer software development, uses the dialectical perspective to discuss the historical change of object, subject and Technology of software development technology. It concludes that software models are changing from mechanical theory to organic theory, main body of technological from machine-centered to human's activity-centered and that humanized human-machine interfaces of software development are becoming a trend. So we summarize that the education of computer science and technology should keep up with the pace of software technology development, and emphasis on professional development.

Keywords: Dialectical, software development, evolution, object-oriented, structured, model.

1   Introduction

With the continuous progress of society and rapid development of computer science, the applications of computer and software become more and more wide and thorough. Software and hardware depend on each other mutually; software is a series of computer data and instruction set with a specific order, including procedure, data and related documents,   constituted by the program module. A complete programming process may be divided into the following three steps:

(1) Analyzing questions and constructing models;
(2) Designing algorithms and describing the process;
(3) Programming, testing and running

Simply, it can be summarized as "model—algorithm—program".

Software development technology develops software with computer language and its three essential factors are: technical object, technical main body and technical craft.

With the continuous development of hardware, software technology is gradually enriched and improved.

Y. Wu (Ed.): International Conference on WTCS 2009, AISC 116, pp. 665–671.
springerlink.com   © Springer-Verlag Berlin Heidelberg 2012

From late-1960s to the mid-70s, the structured programming technique and the software engineering concepts were presented.

In the period of mid-1970s to the 80s, computer auxiliary software engineering (CASE) became the research hot spot, object-oriented (OO) technology gradually started popularly.

In the 90s, object-oriented technology became the mainstream of software development technology. Software process and software process improvement also became the research hot spot, focusing on software reuse and software component technology.

This paper focuses on the historical changes of software development technology, and some characteristics of software development technology.

2   Views of Nature Changes: From MECHANICAL THEORY TO ORGANIC THEORY

To review the development history of software technology, one of core technologies is basic model of software. Three essential factors closely related with software technology development are computer platform, human thought patterns and question essential feature.

The structured methodology that software early used believes all things, no matter atom, electron, mountains and rivers, the hosts of heaven, animals and plants and so on, are constituted by changeable events or processes. These events or the processes depend on each other mutually, transform mutually, and form the unification world. But, later, object-oriented methodology argues that the world is composed of the different level object, each kind of object has the respective internal behavior and the law of motion, the interaction of different object constitutes the different systems.

2.1   Comparison of Technical Ideas

The main idea of structured method is pursuing accuracy step by step. It resolves complicated system into simple system. Three basic structures, sequence, choice and cycle, can constitute modules which any computer can deal with.

The object-oriented method is simulating habit thinking mode of people, thinking that the objective world consists of all kinds of object and that moreover complex object consists of relatively simple object. Object is the main body of dealing with, letting object and object's operating abstraction be a new data types—class, which synthesizes function abstraction and data abstractions by using the technology of information hiding—encapsulation.

2.2   Differences between the Basic Concepts

The most commonly used concepts in structured methodology are system flow diagram, data flow diagram, modular, etc. System flow diagram depicts the flows of various components in the system information (programs, files, databases, etc.). Data flow diagram is a tool of depicting the logical system. It does not involve physical elements, but depicts the information flow and process in the system. Modular separates the program into independent named modules. It is consistent with the law of humanity to solve general problem, which decomposes a problem into a number of smaller problems so that the overall complexity will be lower.

The most commonly used concepts in object-oriented methodology are object, class, encapsulation, inheritance, message and method. Object is the aggregation of special attributes (data) and behavior (method). Class is the aggregation of objects with the same or similar properties and methods. Inheritance is the mechanism that automatically shares the methods and data in class, subclass and objects. Encapsulation mechanism is a kind of information hiding technology. Users can only see the information in the object encapsulation interface and the inside of object is hidden to the users. Another purpose of the encapsulation is to separate the users of the object from the complexity internal algorithms.

2.3   Differences between Programming Methods

The development of structural program design is to control the complexity of program developing process and the implementation process, its design method focuses on process, 'Programming = Algorithm + Data Structure' is its basic guiding principle. It has a strict theoretical basis in structural program design. Dijkstra, the founder of structured methodology, discussed three mathematical reasoning methods that support structural program design. The methods are enumeration, inference and abstract.

Object-oriented method is to control the complexity result from system requirement variable. 'Programming = object + class + inheritance' is the basic principle of object-oriented programming. Relationship between objects merely transmits the news mutually. Basic concepts of object-oriented method, such as class, object, inheritance, have formal descriptions to a certain extent, but generally speaking, they lack a unified and strict mathematical theory foundation.

2.4   Differences in the Process of Software Development

The process of Software Development is the one that maps the problem of vital concern to the computers by using computer language. It can be described as: real world → modeling → programming → solving by computers.

In the structural design, the programmer can get a process-oriented model after analyzing the problem domain. The implementation process can be described as: real world → flow diagram → process-oriented language → implementation and solution. Structured approach has three major shortcomings: first, there is no unified model to contact all the stages; second, changes in late period are difficult; third, it cannot support the technology reuse.

In the object-oriented method design, the implementation process can be described as: real world → class diagrams → object-oriented language → implementation and solution. In the object-oriented software design, the concept of object is filled in the development process with the analysis, design and coding. The relationship between objects and the processes are the common medium to express in the various stages. The focus of development shifts from the coding to the analysis and from function-centric to data-centric. The iteration and the seamless in the development process make reusing more natural.

3 Practice Shift: The Activities from Machine to Human

Structured methodology was all the rage in the 1970s. Since the mid-80s of the 20th century, it gradually had been replaced by object-oriented methods. The starting point and the basic principles of object-oriented methodology mimic the human way of thinking habits as far as possible, so that software development methods and processes can get close to those of human in understanding the world and solving the problem, and replace functional decomposition with object decomposition.

3.1 Change from Machine-Centered to Affair-Centered

Object-oriented method can construct a better software system relatively easily. Structured approach has the structural characteristics of Von Neumann Machine in essence, namely process-oriented, that is, the "process" and "operation" are the center to construct the system and design program. The outcome of such thinking has poor reusability, and costs much for maintenance. The reason is that people must know what an object is firstly before feeling the changes happened to this object, namely process. Structured methodology is closer to the physical world of computer. Structured approach emphasizes the point that software should be designed according to the characteristics of the computer itself. To write a program, we must consider the memory, stack, pointer, IO address, know the process how the computer realizes this program, and then use the command to complete one by one. Since the 1980s, this situation had been greatly improved after object-oriented approach got popular. Object-oriented approach can be said to come from the development of structured methods, so it inherited a lot of successful experience of structured methods, such as: modularity, data abstraction, strong cohesion and weak coupling.

3.2 Establishment of Human-Cented Activities: Decentralized Post-Industrial Feature

The main target of software design is to achieve good structure, make the development of software with good structure and evolution. The key technologies are to abandon the large central control system and solve complex problems with a lot of spontaneous competition and combination of simple components.

Decentralized structure performances are more prominent on the Internet. At present, software system based on computer hardware platform is experiencing the transition from centralized and closed platforms to open platforms,   network-oriented computing environment is changing from Client/Server to Client/Cluster gradually and toward Client/Network and Client/Virtual environment. The Internet itself has a very loose structure. It is almost impossible to establish a centralized type of control system on the Internet. Therefore, as the founder of the Internet Tim Berners-Lee said: "In the early stages of the world wide web's development, the people who tried to belittle the World Wide Web point out that the network can't become an organized library in the perfect order for ever. Without a central database and tree structure, we can't be sure to find whichever information. These statements are not wrong. But the expression ability of World Wide Web system provides an enormous amount of information. People would think the tools such as search engine were not reality completely ten years ago. Now a complete index   has been compiled for a large amount of information on the World Wide Web ". Of course, the World Wide Web information is not complete, but it can provide a lot of useful information.

4   Humanistic Steering: Humanistic Tendency

In computer software, over the years, there is a debate to the software technology exploring: towards strictly formal (mathematical axiomatization) way, loose informal or semi-formal way?

With the development of science and technology, and the social need enhancement, software industry has also had some prominent problems, such as the software scale promotion unceasingly, the developing price getting higher and higher, maintaining difficulty and so on. These problems have decided the future of software technique developing trend, which requires software to have the good compatibility, openness, scale flexibility and reuse, etc.

Object-oriented technology is now getting mature day by day, its advantage in software development will beyond the structured approach in the future without doubt. Object-oriented technology will reduce the complexity of software development and reduce the threshold of software development. Although, formal methods can solve the errors in the software design, informal methods are more able to accommodate all kinds of people to join a team of software design thus pushing the development of software, which is the software industry's future trend and what we expect.

Object-oriented methodology eliminates the gap between the formal and non-formal methods, which reflects dialectical unity of the formal and non-formal approaches. The reasons are as follows: on the one hand, any kind of programming language including object-oriented programming language such as Smalltalk80, C++ and java is a formal language itself., this programming language design itself is very strictly following the formal methods and requires deep professional foundation and a good mathematical basis; on the other hand, users of these languages can use formal, semi-formal or non-formal methods. Application software developers can write programs like piling up building blocks due to the good design of these programming languages, which makes them to be non-professional programmers. So it is said that training a structured program designer is much more difficult than training an object-oriented programmer.

Object-oriented methodology gives people a free development space. It divides computer software developers into three types. The first kind is computer experts that have profound theoretical knowledge and ground-breaking insights in certain areas. The second kind is computer professionals that have good professional quality and can write software development tools according to professional theory. The third kind is the application software developers that develop specific application software. The former two kinds of people mainly use formal methods while the third mainly uses semi-formal or non-formal methods. Because object-oriented methodology eliminates the gap between the formal and non-formal methods, three kinds of people can cooperate effectively and promote the prosperity of the computer software industry.

Structured methodology emerges with software crises and object-oriented methodology first appeared in the field of operations research. Structured methodology and object-oriented methodology both have their own advantages. They were the products in the particular stage of computer software development history, and promoted the development of software engineering in varying degrees. The purpose of software engineering is to develop products that customer needs and minimize the cost at the same time. So we must choose a suitable design method according to actual needs and make full use of advantages of object-oriented and structured methods.

The applications of structured and object-oriented methods have been beyond the field of computer hardware and software.And they have became a broad general methods,for example, the structured cabling system. There will be more software development methods with the development of the time.From a deeper level of meaning, technical method always generate in practice. Each self-contained method are adapted to the needs of social production,and it will spread to other areas of engineering technology when it reach to a certain extent. Computer software technology can promote the development of other engineering.

5   Conclusions

Over the past decades, software technology has experienced a series of important changes and development. The main line of development is: software models are changing from mechanical theory to organic theory, main body of technological from machine-centered to human's activity-centered and that humanized human-machine interfaces of software development are becoming a trend. The rapid development of network technology brings much more demand and challenges to software technology. The dialectical analysis of evolution of software develepment techniques will help us deepen our understanding and knowledge of historical development, reminding us from another level that the education of Computer Science and Technology should keep up with the pace of software development, emphasize on course reform, and improve key major construction.

Acknowledgments. This work was supported by Computer Science and Technology provincial key major construction funds from   Department of Education of Zhejiang Province and Department of Higher Education Foundation ([2008] No.71) and Special Issues Higher Education Association Foundation of China (No. 2010CX161).

References

1. Wittgenstein, L., He, S.-J.: Tractatus Logico-philosophicus. The Commercial Press, BeiJing (2005)
2. Hu, X.-H.: The Philosophy of Software Technology. Fudan University, Shanghai (2008)
3. Yang, F.-Q., Lu, J., Met, H.: Net structure, software technology system: a system structure for center of approach. Science in China (Series E:Information Sciences) 2008 (2006)
4. Yang, F.-Q., Met, H., Lu, J., Jin, Z., et al.: Some Discussion on the Development of Software Technology. Acta Electronica Sinica 26(9), 515–1115 (2003)

5. Hong, G., Mao, X.-G.: Aspect-Oriented. Software Development: Philosophy and Observatione Computer Engineering & Science 29(11), 94–96 (2007)
6. Wang, F.-X.: Development of Computer Software Engineering Studies. Computer Knowledge and Technology 5(19), 5150–5152 (2009)
7. Liu, G.: The Philosophy of Information: A New Paradigm for the Philosophy of Science and Technology. Journal of Dialectics of Nature (4), 23–28 (2004)
8. Glenn Brookshear, J.: Computer Science: An Overview Liu Yi, Feng Kun, Xu Jianqiao, Translator, 9th edn. The People's Posts and Telecommunications Press (2006)

Study on Information Literacy Education of American State University and Its Enlightenment on the Related Education Development of Chinese Local Colleges

Wu Peng and Yin Ke

Computing Center, Henan University, Kaifeng, China
w.s_lin@yahoo.com.cn

Abstract. American State University not only reforms the university, but also changes the American society. Its appearance and development achieve the miracle of the popularization of American higher education. The analysis and study on the basic situation of the information literacy education of State Universities can find referential methods for the related education development of Chinese local universities.

Keywords: Information Technology, Information Literacy, Education.

1 Introduction

Paul Zurkowski, the chairman of American information industry association in 1974, first put forward the concept of "information literacy" in the report "The Information Service Environment, Relationships and Priorities" named by American National Commission on Libraries and Information Science. In this report, Paul Zurkowski proposed to enact an   information literacy program to realize information literacy in the next ten years. He believes that information literacy is the ability of individual to apply the information tools and information sources to solve problems. Later on, the discussion of the information literacy starts to begin, especially in the field of education. Seen from the gradual improvement process of the definition, information literacy is a comprehensive concept, its basic content includes: it should have information consciousness, use information sources effectively, think information critically, mix the useful information into their own knowledge systems, can identify various types of information initiatively, obtain the needed information to evaluate and analyze, have the ability to develop and disseminate information, etc.

2 Analysis and Study on the Basic Situation of the Information Literacy Education of American State University

The information literacy education receives more attention in America, and early in 1983, the American information scientists Horton believed that the education departments should open information literacy course to raise people's using ability of online database, communication service, electronic post, data analysis and library network. In 1987, American Library Association (ALA) established the information literacy education committee. In 1990, the American higher education commission established "the evaluation outline of the result of information literacy education", and in 1996, it also established "the function framework of information literacy education in general education program". American education ministry believed: "This knowledge has become one of the basic skills of a person, such as the skills of reading, writing, ciphering, etc." From 1994 to 1995, American thoroughly investigated the situation of information literacy education for the national 3236 universities. "Project 2061" was formally started in 1985, which put forward the ideas of the information technology and the integration of each discipline. In the late 1980s, college students' information literacy education was formally incorporated into teaching syllabus as a course. During this time, the college information literacy education developed vigorously, and achieved the fruitful results. Until now, the American information literacy education is guided by ripe theories, studied by special organization structure, and also supported by the relevant policies. In the forum of national information literacy, resource department lists the sites and projects of the information literacy education in number of American State University. According to the analysis and research on these experiment items and concluding their successful experience, which has the important significance for the implementation of the information literacy education of our local institutions. Now, we only analyze and study on some programs with great influence and representative in American state universities.

TILT (Texas Information Literacy Tutorial) is the more implementation project of American college information literacy education, which was continually developed, renewed and used till now. This project is supported and funded by the digital library of the system of American Texas State University, and was designed and developed by Austin, who is the branch school of the system of Texas State University, which is an earlier college information literacy education programs based on web. In 2001, the project proposed a TILT open publication license, as long as the other party complies with this license, then they can use the services of TILT program and have stronger universal characteristics. The entire project includes three main function modules, which are respectively selection module, search module and evaluation module. In addition, it also includes the ABC knowledge of Internet and the library tour guide. In the function module of the ABC knowledge of Internet, it mainly introduces some basic knowledge of the Internet and some people's misunderstanding of Internet. The function module of library tour guide is mainly the introduction of the collection distribution and public services of the library of Texas State University, and provides the corresponding hyperlink. Now I will introduce three main function modules of the project, the contents of their main functions are shown as follows: select function module. ① the introduction of several major information sources. It mainly introduces the concept, characteristics and specific examples of several major information sources; the major principles and common methods of the selection of information source; the offer of interactive exercises. ② the introduction of the libraries and the website. It includes the characteristics of the website and library; the main builder of project.③ the introduction of the related knowledge of periodical index. It includes the concept of periodical index; the method of finding periodical index; periodical index is the reason of better information source. ④ the distinctions between general periodicals and academic periodicals. It includes the main features of general periodicals and academic periodicals; the introduction of the application of general periodicals and academic periodicals. Retrieve function module. ① the introduction of brainstorming. It includes the basic procedures of brainstorming; the offer of interactive exercises. ② the introduction of database retrieval. It includes the definition and classification of database; how to select the appropriate database. Common retrieval methods: subject retrieval, keyword retrieval, author retrieval, combination retrieval, etc; interactive retrieval practice. ③ how to use websites and search engine. It includes the navigation introduction of the project website; the introduction of search engine; how to find information by using search engine. Evaluate function module. ① the definition of information sources. It includes how to determine the information of the printed text data; the offer of interactive exercises; how to determine network information. ② the evaluation of information. It points out that different resource has different evaluation, especially should pay attention to the evaluation of network information resources; the indicator of evaluation; the offer of interactive exercises. ③ the introduction of academic references. It includes the definition of reference and notes its specific principles; the format of reference; the method to prevent plagiarism.

The information competence project of California State University (CSU Information Competence Project) is a more typical example of the American college multi-campus model. The main intent of this project is to create and develop a kind of information literacy education model with guidance and popularization, aiming to help the students of California State University system to evaluate and exchange study by finding and using information. In 1996, California State University system information competence working group held experts and scholars meeting about "study on using information resources to promote active learning, case studies and the related methods" based on school-wide. In June, 1996, one of the five branch schools received about 70 thousand dollars to endow to develop this training program of information capacity education based on multi-campus, in order to develop the materials with interactive teaching guidance to train the information literacy abilities of the five branch schools in the future, and then to promote and implement in all the 22 branch schools. With full cooperation and division of labor, the five branch schools jointly develop this interactive information capacity model information which applies to the related information literacy courses of library and the professional courses of other disciplines, and its main contents includes: (1) The specific courses of information capacity, which was developed by the branch school Cal Poly San Luis Obispo. The backbone contents of specific courses include: the multimedia form presentation of course content, the active learning and related practice of classroom cooperation, the learning manual of electronic text form and the related distance learning content. (2) Put the specific content of the above backbone courses in the independent page of library website in order to teach and test courses. (3) Create the communication center of information capacity resources, all the 22 branch schools can share the information resources and research results related to information capacity through network. (4) Determine the information capacity requirements of each discipline through analysis and study, the cooperation of Cal Poly Pomona branch school and Fullerton branch school, and then to eventually establish the information capacity requirements of disciplines, such as the business, engineering, mathematics, agriculture, construction, etc. The multi-campus model is one of the important achievements of the information literacy capacity education project of California State University system. In the later years, each branch schools of its subordinates jointly develop and study for this project, its main work includes the information ability training of the community college and secondary school student, the information ability training of students in different grades, the information ability training of the disciplines' different students, information ability training of teachers, etc. Till 2002, California State University system has developed 49 projects related to information capacity, which has made great gains. The information literacy Initiative of New York State University (SUNY information literacy Initiative) is put forward by SUNY Council on Library Directors in 1996, the purpose of this project are to identify the information literacy capacity level that the students of New York State University should have, and to design and develop the project entity which is suitable for the information literacy capacity training of all the students. For the concrete implementation of this project, the curator council of New York State University Library particularly has set up two institutions, one is information literacy initiative committee, and the other is the special working group of network information literacy. The main tasks of the special working group of network information literacy have three items: develop and design an information literacy education curriculum, which should be good for learners to train and improve their information literacy capacity; design and develop the specific methods for promoting and implementing this course; design and develop the evaluation mechanism and methods of the learners' learning effect. In order to implement this initiative, Ulster County community college provides a web-based information literacy education courses LIB111 for New York State University, students can learn this course through remote learning mode and obtain the corresponding credits.

San Francisco State University Library has set three types of courses: One is the education of library users; second is to set up online advancement of student information skills for students, which is an interactive teaching software that the students can learn randomly; third is to open the lecture about the improvement of the students' research skills and the enhancement of the information capacity for teachers.

3   Conclusions

The information literacy education of American State University are highly effective, and achieved great success, in which, there are many aspects we should learn and borrow. Now, I will order some experiences and thoughts as follows: if the information literacy education wants to develop rapidly, it must make the various circles realize the importance of information literacy education. China's local colleges and universities should learn from the advanced experience of American State University, make full use of all types of media, publicize information literacy education in various forms, and enhance the awareness and understanding of various circles for information literacy education. One of the very important reason for the development of Information Literacy Education in American State University is that it pays much more attention to theoretical research, forms a full set of theory systems which conforms to their national practical situation, which is verified in the development and implementation of several projects, and it greatly plays the promoting effect on the guidance of the theory to practice. China and the national academic organizations related to information literacy education and its major branch should change the situation of lacking communication between each other, and enhance exchange and cooperation in information literacy education. It can consider establishing the alliance like American (NFIL), or the alliance of (local) college information literacy education, making full use of complementary effect on academic, professional, field, etc. It should develop series of activities jointly and coherently, aiming to promote the rapid and healthy development of our national information literacy education.

References

1. Guan, R., Xue, Y.: The inspiration o the development of our national local universities of United States State University (26) (2009)
2. For a clever country: information literacy diffusion in the 21st century [EB/OL] (March 19, 2009),
   http://www.library.unisa.edu.au/about/papers/clever.pdf
3. American Library Association [EB/OL] (March 19, 2009), http://www.ala.org/
4. Related Information Literacy Web Sites [EB/OL] (March 19, 2009), http://www.infolit.org/sites.html
5. Fowler, C.S., Dupuis, E.A.: What have we done? TILT's impact on our instruction program. Reference Services Review (4) (2000)
6. Texas Information Literacy Tutorial: TILT (University of Texas) [EB/OL] (March 19, 2009), http://tilt.lib.utsystem.edu/
7. TILT resources [EB/OL] (March 19, 2009),
   http://tilt.lib.utsystem.edu/resources/index.html
8. TILT Open Publication License [EB/OL] (March 19, 2009),
   http://tilt.lib.utsystem.edu/yourtilt/agreement.html
9. Clay, S.T., Harlan, S., Swanson, J.: Mystery to mastery: the CSU Information Competence Project. Research Strategies (17) (2000)
10. CSUN's Information Competence Program [EB/OL] (March 19, 2009),
    http://library.csun.edu/Research_Assistance/infocomp.html
11. Information Competence [EB/OL] (March 19, 2009),
    http://library.csun.edu/susan.curzon/infocmp.html
12. CSU Information Competence Projects [EB/OL] (March 19, 2009),
    http://library.csun.edu/susan.curzon/infocmp.html
13. SUNY Connect: Information Literacy (State University of New York) [EB/OL] (March 19, 2009), http://www.sunyconnect.suny.edu/ili/iliover.htm
14. Bernnard, D.F., Jacobson, T.E.: The committee that worked: Developing an information literacy course by group process. Research Strategies (18) (2001)
15. National Forum on Information Literacy [EB/OL] (March 19, 2009),
    http://www.infolit.org/

The Optimization of Professional Sporting Management System Based on Gene Algorithm

Tang Guiqian

Guilin University of Electronic Technology
tangwang112@yeah.net

Abstract. This paper implements a single system to serve different sports league schedule, and it has a huge difference from other papers that a single system with a single sports league. This system that serves two or more sports league can not be adjusted the fixed parameters. It can only be considered to set the general direction, and can not customized its parameters by every league. Although the schedule can be arranged with a variety of sports leagues, it may not be better than single system for the single sports league. Therefore, how to customization parameters of different sports leagues by single system is the direction of future research. Another drawback of this system is that it arranges the schedule without the Continuous battles, such as the MLB's three or four Continuous battles etc., this restriction is an issue to be explored in the future.

Keywords: Optimization, Sporting Management, Gene Algorithm.

1 Introduction

1.1 Foreword

It is laborious to arrange the schedule for all kinds of sports. Overseas professional contests, such as NBA, NHL, MLB ... etc., usually face this problem. There are targets needed to be reached, however, the sporting schedule is constituted ad allotted under limited resource. It is easy to cause the complicated problem and take much time to find the answer. Obviously, it is hard to make the schedule for complete season in a short time. The planner has to consider the regulation for the league and the teams, such as the number of games for a weekend, the fairness of allotting the stadiums, the postponement of rainy days, and so on. The conditions are limited and complex and the schedule management costs a lot of time to settle.

1.2 Research Motivation

To probe the problems of planning a schedule for all professional sport leagues, at the same time, we develop a good system for planning the schedule according to heuristic algorithm by the benefit, speedy calculating of the computers. At one

Y. Wu (Ed.): International Conference on WTCS 2009, AISC 116, pp. 679–687.

hand, it can reduce the burden for the planner, ad bring some statistic reports to sport leagues administrator. On the other hand, it also can reduce move cost and lower expenses.

The scope of this study is about planning schedule and is defined to how to plan the route. We take the NBA as an example to construct an effective schedule which meets the league rules according to heuristic algorithm.

After collecting the references, we proceed to gather the information and to research it deeply for this case. We design the patter for planning the schedules, and organize the related statistic reports about the management. Finally, we estimate the results and analyze the targets. We expect to help NBA to establish the schedule decision support system. In the mean time, it will help other sport leagues schedulers to make the decisions.

2   Literature Review

In the following contents, we will introduce the forms of the problem of planning schedule and classification; next, we will explain how to find the answer for all kinds of problems. Finally, we will compare to all of solutions.

2.1   Sports Management Issues

Many sport leagues such as baseball, ice hockey, basketball ... etc., are going to face this problem in a round robin tournament. If there is a competition for two teams, and the home team will ask that they want to have several competitions in the fixed period. The most important thing in a round robin tournament is to let every team in the schedule compete with each other for one time. So, if the six teams have five times competitions, every team will compete with each other in a week.

The professional sports leagues develop vigorously in the foreign counties. Because of extensive territory and high-participated for the people, there is an obvious system for home team and guest team. The savants in the foreign countries use as full-number scheme to plan the schedule in the full regular season for ACC. This kind of problem is classified to round robin tournament; there are different calculation step and procedure to find the answer for different kinds of sport types.

Macaroon [5] uses ILOG to resolve the problem of the schedule for NBA. He has a good achievement from stereotyping the leagues' schedule limit. Since the difficulty of management is related to the number of teams and usually there are more than ten teams in the leagues from the foreign courtiers, it makes the problem more complicated. Most of savants develop heuristics algorithm to find the answer. They hope for having reasonable time to meet schedule answer.

Russell and Leung [7] planned the schedule for the professional baseball leagues because they developed the minimum total cost for travel ad two phase's math. Table 2. is the comparison to the records of schedule problem.

Table 1. Literature review

Author (Year)	Consideration	solution
Bean and Bridge(1980)[I]	The shortest Traveling cost	Traveling salesman
Russell (1994)[7]	The shortest Traveling cost	Mathematical programming
Costa (1995)[2]	Combine GA ad T A and develop a new algorithm	Gene algorithm& Tabu Search
McAloon (1997)[5]	Modeling schedule constraint	Constraint programming
Nembauser (1998)[6]	Home Away Rest	integer programming
Henz (2004)[4]	Round Robin touament	integer programming

3   System Architecture

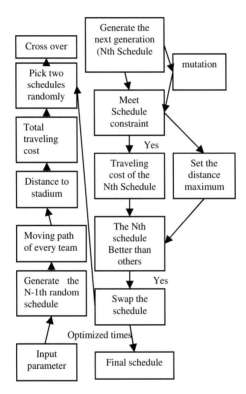

Fig. 1. Algorithm steps

Figure 1 shows the structure of schedule algorithm. A brief introduction on this structure will be given in this chapter.

Step1: Input the parameters, such as the number of teams leagues ad divisions. Input constraint conditions: such as the duration of seasons and optimized times of schedule.

Step2: Generate an N-I random schedule, read the database of distance to stadium and calculate the total traveling cost.

Step 3: Based on the Wheel Theory, to choose two schedules as the swapping genes, and generate the Nth game schedule. Note that the possibility of this particular game schedule to mutate is extremely small.

Step 4: Verify whether the game schedule meets the constraint conditions, which were set up at the first step, after the swapping (mutating). If it does, calculate the total traveling cost of the game; if it does not, set the distance of the game to the maximum, which is the ultimate punishment condition.

Step 5: Identify if the total cost of the Nth schedule is less than that of other game schedules. If yes, swap the two game schedules; if not, no action to be taken at this step.

Step 6: Based on the optimized times constraints set in step 1, repeat step 3 to step 5. The schedules generated in step 3 will overwrite the Nth schedule that are ruled out in step 3 to step 5.

Step 7: Output the complete schedules into a ".txt " file.

3.1   Parameters Setting

The schedules planned by this study does not limited to any single sports leagues .This system is applicable for all sports leagues that are of the same characters. Therefore, the parameter settings should be flexible, so as to provide schedule planners multiple choices of settings.

This study, considering the schedule planner's view and satisfaction of schedule rules, lists all the Hard Constraints and Soft Constraints as follows:

1) Hard Constraints
P: Controls schedule optimized times, which is the terminating condition for the algorithm.

HI-H4: Control the number of teams in each league division.

H5-H7: Control the number of games for the host (guest) teams, for example, if H6 = 1, the system is to arrange every team to have a game against all the other teams of the league at each team's hosting venues.

H9-HIO: Control the state and end dates of the complete seasons for each team

HII-HI2: Control the dates that no games can be held on, such as major public holidays and festivals.

2) Sof Constraints
S 1-S3 Control the unbalanced games that are needed to be increased. For example, if S3=6, each team has 6 games in total against the other teams in the same area.

The team will be the hosting team for 3 of the 6 games and the guest team for the remaining 3 games. S4 controls the day intervals for any two teams battle at the same venues.

3.2   Random Schedule Initialization

This study uses the gene algorithm to plan schedule. Before the genes crossover of the schedule, an initial schedule is required as the initial population for the genetic algorithm. Management is a complex combination of NP-hard. If we use the greedy method or the shortest path algorithm to generate the initial population, the program may fall into infinite loop ad be usable to generate the initial population because of the hard constraint requirements. In order to avoid the above problems, the schedules of this study are randomly generated. Although the schedules are generated randomly, all the constraint conditions are to be verified. When a schedule conflicts with constraint conditions, the particular schedule should be re-generated until it meets the constraint conditions.

As shown in fig 2, the populations randomly picks gene 4 for mutating. Gene 4 verifies the mutating possibilities, which are swap 1 - swap 5. Next, verifies the fitness (X) and picks swap 2, which is the best mutating population. To apply this mutating approach onto the schedule, each day of the schedule is a gene, and the number of the days is the number(N) of the genes. A gene is randomly picked from the population, and the number of swapping approaches is (N-1). Start the swapping from the 1st gene until the $(N-1)^{th}$ gene, and record the schedules' total traveling costs generated from each swapping, and pick the one that has the lowest total traveling cost.

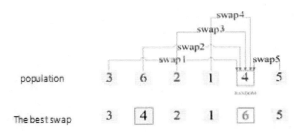

Fig. 2. Mutation

3.3   Soft Constraint and Schedule Swapping

If only one of the statistics is lower than old generation, it would judge the times of the soft constraints violating to determine which one is punished. The punish parameter is all violating times multiplied 0.1 % of total moving distance. All situations are showed as table 2.

Table 2. Ruled out conditions

P:Parent C:Chiid	Moving distance	Times o Constrain violation.	Punished "alUM Moving distance	swap
Case 1	C<P	C<P		y
ease2	C>P	C>P		N
Case3	C<P	C>P	C<P	y
Case4	C<P	C>P	C>P	N
Case5	C>P	C<P	C<P	y
Case6	C>P	C<P	c>p	N

4   Results

The following sections will describe, two different professional sports, NBA and NHL season rules, how to design the parameters in the management system of different professional sports and the output NBA schedule in this study make a comparative with an NBA official 2008-2009 schedule by the traveling cost.

4.1   Experimental Result

After encoding, check the distance between the stadiums with goggle map ,and execute the management system. After a lot of test ,after calculating 200,000 times, the target value convergence to a fixed value, so we identify the termination conditions that is calculated 200,000 times. In terms of calculating with Brute Force algorithms need to be more than 2 million times .

Table 3 values obtained for calculating each mutation rate.

Table 3. Mutating result

Rate X100	1%	0.11%	0.1%	0.09%	0.06%	0.05%
2	245317	286511	24117	281951	305519	3062302
4	233594	2715263	2308291	2660271	2889232	297477
6	23823	259813	2282411	25390	282408	285896
8	2270	2523308	2240249	2489555	2771919	2725
10	226539	2453595	22175	24357	27321	26918
12	225097	239114	220918	2391819	2691109	264950
14	2257422	235740	218950	2331038	265726	26037
16	2257422	232631	2186216	2317253	2603241	2562160
18	Z25718	229222	218251	2298144	2577457	25322
20	225718	227333	218251	22585	25574	249727

After testing all kinds of mutation rate, the mutation rate 0.1% which makes the target value, 2,182,501 is the best. When calculating over 18 million times, the target value have reached convergence value.

The target value of mutations rate 1 % have reached convergence value when calculating over 14 million times. Although the target value is fewer than any other values, the speed of population mutating is too fast resulting in conferencing is also too fast.

Due to the mutation rate 0.05% is too low, there may be too many material gene which can not mutate to good genes, leading to target convergence too slowly. It also takes time for a lot of calculus.

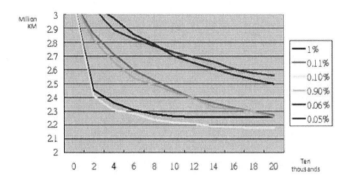

Fig. 3. Comparison for the mutation rate

Figure 3 Comparison for the mutation rate. Each curve from Figure 3 that the biggest difference is the updating rate of target value before 60,000 calculating times. In the second half of the scope, the curve is not much different. Shown in 0.05 and 0.06 mutation rate, due to excessive low, the result of the calculations about tens of thousands of times are meaningless, and later for 20 million, the target value updated too slow.

Although the mutation rate of 1 % make a good target value in the earlier period, the target value convergence is too fast to the updating times which is too low in the second half of the scope.

4.2   Analysis

According to the result from foregoing character, it proved that mutation at 0.1 was the applicable parameter for this system. This study took schedule at 2008-2009 of the America Basketball Association (NBA) for example. It evaluated the total moving distance to show that whether the distance from teams was proper than official schedule or not.

The following as table 4 reveals the calculation of mutation at 0.1 for the system: the decreasing range of goal was about 100 million kilometers from the first count to the 2,000,000th, and the goal, 2,193,379 kilometers was shorter than the official, 2,197,119 kilometers at 135000t time. The goal, 2,182,501 kilometers couldn't be renew at 180000th time, it was the best target.

Table 4. Mutation at 0.1

Execute Times	Moving distance	Execute Times	Moving distance	Execute Times	Moving distance	Execute Times	Execute Times
1	3223883	50000	2293888	100000	2217568	150000	2187366
5000	2682862	55000	2289614	105000	2217545	155000	2186628
10000	2551357	60000	2282411	110000	2217568	160000	2186216
15000	2463916	65000	2270123	115000	2112699	165000	2186216
10000	2411784	70000	3353510	120000	2209186	170000	2184127
25000	2384308	75000	2250126	125000	2208529	175000	2182913
30000	2347492	80000	2240249	130000	2206898	180000	2182501
35000	2311927	85000	2224916	135000	2193379	185000	2182501
40000	2308291	90000	2224916	140000	2189500	190000	2182501
45000	2299275	95000	2217568	145000	2187366	185000	2182501

5   Conclusion and Discussion

Many foreign sports schedule, such as NBA, NHL, MLB, etc., encounters the problems of schedule arrangements. The subject of this study is planning the schedule based on the shortest traveling cost. Therefore, when used to land large counties, the result is much significant.

The system in this research is able to arrange two or more sports league schedule. We hope that this study could referenced by people who want to study about the problems.

The induction of specific findings of this study is as following. The study ranks schedule model in the test volume and design, NBA, NHL, MLB Alliance of regulations and competition system and other aspects related to understand. We also consider the rules of current staff management, and organize the operation method which satisfies arranging the schedules made of this study. The schedules satisfy all character of every league, and it shows that the genetic algorithm of this study solving the management problems effectively.

References

1. Bean, J.C., Birge, J.R.: Reducing Travelling Costs and Player Fatigue in the National Basketball Association. Interfaces 10, 98–102 (1980)
2. Costa, D.: An Evolutionary Tabu Search Algorithm and the NHL Management Problem. INFOR. Ottawa 33(3), 161–179 (1995)
3. Cooper, T.B., Kingston, J.H.: Complexity of Timetabling Construction Problems, vol. 17, pp. 183–295 (1996)
4. Henz, M.: Global Constraints for Round Robin Touament Management. European Joual of Operational Research 153, 92–101 (2004)
5. McAloon, K., Tretkoff, C., Wetzel, G.: "Sports League Management. In: Proceedings of the 1997 ILOG Optimization Suite International Users' Conference, Paris (July 1997)

6. Nemhauser, G.L., Trick, M.A.: Management a Major College Basketball Conference. Operations Research 46(1), 1–8 (1998)
7. Russell, R.A., Leung, J.M.Y.: Devising a Cost Effective Management for a Basketball League. Operations Research 42(4), 612–625 (1994)
8. Saltzman, R.M., Bradford, R.M.: Optimal Realignments of the Teams in the National Football League. European Journal of Operational Research 93, 469–475 (1996)

Large-Scale Sports Events Optimization Management Based on Value Chain

Tang Guiqian

Guilin University of Electronic Technology
tangwang112@yeah.net

Abstract. Large-scale sports event is a complicated and systematical project involving all the sectors of the society, such as politics, society, economy, and culture. Meanwhile, there are the top demands of city infrastructure and other public products and services when holding the large-scale sports events. Obviously, it is the government that has the capability to supply the public products. Therefore, government plays a leading role and has a dominant function in the management system of large-scale sports events.

Keywords: Large-scale sports events, Management, Technology innovation.

1 Aggregation-Diffusion Effect of Technology Application in Large-Scale Sports Events

The technology diffusion firstly has the effect on the development of people's awareness. While people watching or getting to know about the large-scale sports events, they will get the information about the technology application by different kinds of media, which will help them develop and enhance their scientific and innovative awareness. Besides, it has the effect on overall enhancement of national technology innovation capability. For example, China has brought a large amount of manpower, material and financial resources, and all kinds of technology achievements together to make innovations in order to promote the High-tech Olympic Games strategy, which has not only expanded the innovative fields but also shortened the period of technology innovation, furthermore, enhanced China's innovation capability. Last but not the least, it has the enormous influence on the development of the technology industry: on one hand, the achievements aggregated through large-scale sports events will certainly and directly be known and understood with the heavy exposure of the media by the technology corporations, which is good to break the technology barriers between industries and immediately raise the overall technology level and increase the innovative capability of the technology industry; on the other hand, the aggregation of different kinds of technologies will objectively accelerate the aggregation between technologies, form a new model of technology innovation, and enhance the innovative capability in all the technology fields.

The essence of the technology aggregation in large-scale sports events is about the kind of special innovation model and the technology transfer process. According to the latest research results of technology innovation and transfer by Chinese and

Y. Wu (Ed.): International Conference on WTCS 2009, AISC 116, pp. 689–695.
springerlink.com
© Springer-Verlag Berlin Heidelberg 2012

foreign scholars, there is an outstanding social effect in the process of transferring technology. Roger (1995) thought that technology transfer is a general phenomenon of developing the technology, a process of developing and changing the social structure and function. Hereby, technology diffusion will effectively push the development of technology industry and speed up the development of economic society. The large-scale sports events have a more and more powerful influence on the society by the increasing magnification of the media, and even become one of the important social events which have a profound effect on the development of international societies at a certain stage. After doing the researches, scholars believe that every large-scale technology diffusion will bring the many economic and social benefits with the help of modern media, and has the great effect on the international society development.

2   Technology Aggregation-Diffusion Model in Large-Scale Sports Events

2.1   Technology Aggregation-Diffusion Path in Large-Scale Sports Events

Based on my researches on Beijing High-tech Olympic Games Model, the study of technology model in other large-scale international sports events is highlighted in this paper to prove the universality of the previous research results which indicate that the technology aggregation-diffusion path widely and objectively exist in large-scale sports events, in which there are two kinds of technology development path, that is, the aggregation and the diffusion path.

The first kind of path is the path of technology aggregation. In the large-scale sports events, technology aggregation path is a complex path which is made up of three sub-path, which involves the technology aggregation through the market, the government, and the society. The aggregation through the market is an aggregating process of implementing the economic value of technology which is based on the demand and help of the market. Most of the technology corporations take an active part in the process of holding large-scale sports events and set up a specialized system of technology management for the events. The aggregation by the government is that government purposefully guides and promotes the aggregation of the technology power to the field of large-scale sports events in order to achieve the goals of enhancing the technology innovation capacity and provide better service to the society through technology. The High-tech Olympic Project actively pushed by Beijing Olympic Games is a typical example of the aggregation through the government, which is that China's government plays an important and representative role in holding the large-scale sports events and expanding the technology aggregation effect. The aggregation through the society is that some of the social groups promote the technology aggregation actively to implement their own purposes in the light of their respective value orientation. Some independent social organizations including international and domestic organizations carry out some activities to accelerate technology innovation and reasonably achieve technology value with the purpose of the organizations, which is the third sub-path of technology aggregation in large-scale sports events.

While, two technology aggregation path can be summarized from the function implement point of view, which are the path of technology innovation and technology research & development, and the path of technology application aggregation. Technology innovation and technology research & development indicate that the innovation and research projects are carried out to solve the problems based on the demands of large-scale sports events, the economic society development, and the environment value increment and protection. The main task of innovation and research & development is to rationally promote the conduction and increase the value of the technology in the three aspects discussed above. The aggregation path of technology application is rationally setting up the aggregation path of the technology for sport and the basic technology and environment technology in holding the large-scale sports events, and the strategic path of technology development based on the construction of the technology chain and the industry cluster, which provides the fundamental guarantee for completing the specific task of the technology.

The second kind of path is the path of technology diffusion. The essence of establishing technology diffusion path is focused on how to make technology achievements transformed to the society, that is, how to set up the path of technology industrialization and technology publicity.

Setting up the path of technology industrialization is trying to spread the research results and technology achievements in the market mechanism, that is to say, the market tries to promote the industrialization and international development of the technology with large-scale sports events. At present, initiated by holding Beijing Olympic Games, there are many research results from some Chinese scholars on how to accelerate China's technology industrialization with the help of hosting the Olympic Games, among which a project(2006) the National Natural Science Fundation of China "The Impact on Technology Industry Development of Hi-tech Olympics" hosted by Professor Huang Lu-cheng is a representative research with its essential study about the possibilities of advancing technology industry development based on the Olympic market rules.

Setting up the path of technology publicity is aiming at spreading the research achievements through the aggregation by the government among the social public area, that is, technology is used to solve the problems of the people's livelihood, raise the people's science quality, and provide the service to the public. In recent years, that technology provides service to the people's livelihood has gradually become one of the hot topics that some of the Chinese scholars have paid attention to. Some scholars have proposed the idea of technology application in the people's livelihood, and positively give the suggestions to build the technology service system. As we all know, the essence of technology creation and development is to solve the problems of human beings' own existence and development, and to ensure the harmonious development among human beings, nature and environment. There are still much more debates on some academic topics such as the technology value in the academic field, and the topics about the basis of technology value production, the connotation of technology value, the paths of the technology value achievement, and the ontological and practical value of technology are still the hot topics in the academic discussions, however, the idea of the function of technology in promoting the harmonious development among human beings, nature and environment has been widely accepted by all the sectors of the society. Therefore, a great many technology achievements aggregated through the large-scale sports events must be encouraged to spread in order to increase the value of technology in the process of conduction in the society. It is the successful hosting of Beijing Olympic Games and the rapid spread of a large number of technology achievements aggregated through the Olympic Games that set up the persuasive example in this respect. With the opportunity of hosting the Olympic Games, Beijing has reached the goal of the development of technology publicity with five strategies including strategies of digital technology, new construction technology, new material technology, environment protection technology, and biomedicine technology.

2.2   The Structure of Technology Aggregation-Diffusion Model in Large-Scale Sport Events

In this paper, it presents the technology aggregation-diffusion model based on the study of the effects and path of technology aggregation-diffusion in large-scale sports events.

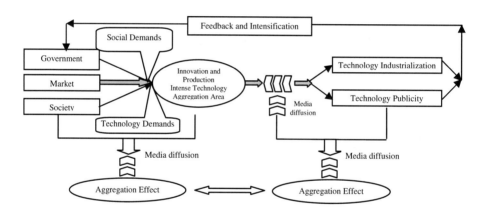

Fig. 1. Technology Value Chain in Large-scale Sports Events

From the figure, we can illustrate the technology aggregation-diffusion model in the following three perspectives: (1) the technology driven by social and its own demands are brought together in three path through the market, the government, and the society to the field of large-scale sports events; (2) the aggregated technologies which form an intense technology aggregation area have made a lot of technology innovations at high levels, got many technology achievements, and brought the aggregation effect; (3) the technology achievements are spread in all the sectors of the society through technology industrialization and technology publicity, which has brought an obvious diffusion effect. Generally speaking, the diffusion of technology achievements is still implemented through some more concrete path. For instance, the diffusion of industrialization of technology achievements in Beijing Olympic Games is carried out and spread in the path through industrialization of the technology in sports, industrialization of the technology infrastructure construction of the Olympic Games, industrialization of the environment technology applied in the Olympic

Games, and the strategic technology industrialization formed by the technology industry chain and cluster. While, the publicity diffusion of the technology achievements more reflects that technology provides the service to the social public benefits, which is proved by the " Technology achievements of the Olympic Games in the Field of Environment Protection and Food Security Seminar".

It is obvious that modern media has played a significant role in the process of technology aggregation and diffusion. Since large-scale sports events have become one of the important social events deeply influencing the international society, the media's focus and reports on sports have objectively had the intense aggregation and diffusion effects which become more and more highlighted owing to the increasing magnification of the reports on every part and field of holding the sports events, and especially on the technologies applied in the events.

2.3 Technology Aggregation-Diffusion Mechanism in Large-Scale Sports Events

The study of technology development mechanism in large-scale sports events is prevalent in the studies of technology industrialization and international development caused by the process of China's promoting High-tech Olympic strategy. For instance, Fang Fu-qian, as well as other scholars did the researches based on the life cycle theory, systematically analyzed the life cycle evolution of the technology cluster of the Olympic Games and its impact on the establishment of regional innovation system, and tried to explain the operation mechanism and fundamental rules of the development of technology industrial cluster in large-scale sports events.

These research results provide more information for our further study of the operation mechanism of technology aggregation-diffusion model. Considering the process of technology aggregation and diffusion as a process of the flow of value, it is easy for us to draw a conclusion that it is a fundamental model of technology value aggregation-diffusion development with the conduction and increment of technology value as its driving forces. In other words, technology aggregation and diffusion is a special process of conducting and increasing the technology value, for it is built in the basic structure of the value chain.

At present, there are numerous research results of the value chain. The analysis and the study of the value chain has been the hot topic since first described by Michael Porter, Professor of Harvard Business School, in his 1985 best-seller, the competitive advantage of nations. The traditional theories about the value chain mostly focused on the study of an enterprise's production process, even if on the international value chain which still pay much attention to a enterprise's economic production, but they still provide a new analysis paradigm for the study of technology value chain. When doing the researches on technology value chain, scholars start their study with the discussion of the technology innovation process, which is because that technology innovation is always developed in the field of corporations, but the more important is that it is the driving force of the development of modern corporations.

The two levels of value conduction are achieved in this mechanism, which are the increment of technology's own value and the use value carried by it. Based on its own demands and the social demands, technology is applied to provide relevant products to carry out the operation of the value chain, which will contribute to the realization of the two levels of conduction.

The adding of technology's own value is based on technology system's characteristic of self-reinforcement. After reviewing the technology system in history and present, we find that the fundamental reason why technology is the great economic driving force of the society is that it is always able to build a relevantly dependent and efficient self-operating system according to its tasks and goals to be completed and reached. Technology elements coordinate, support and depend on each other to promote the overall technology efficiency. Simply speaking, though the technology system is always initiated and maintained by human, an important element of keeping the system's self-operating efficiency is to eliminate the interference from the human factors as possible as we can, owing to which technology has gained the authority in almost the same model in all the sectors of the society. Technology also has the intrinsic demand of self value-adding in large-scale sports events. When we set up a technology system in the field of large-scale sports events, such as the judging system of sports by the optoelectronic technology, there is a continuous upgrading demand of the system generated by the technology system itself. As a result, the process of technology innovation has objectively become the fundamental measure of the system's self-improvement, by which technology value is increased.

The process of increasing the use value carried by technology shows that different social stakeholders have the right to ask for their own interest. (1) From the government point of view, we can see that the technology value must provide great support to the fast development of social health. In large-scale sports events, the government has fully realized that technology plays a significant role in raising the public service level, which is applicable in China as well as the countries with different political systems. So, the increment of technology value in the public has provided the guarantee to social development. (2) From the market point of view, we can know that the increasing technology innovations will bring enormous economic profits, ensure the successful operation of producing the value in the corporations and industry, and enhance the core competence of the corporations and industry, in which their interest plays an essential role. It is obvious that the technology value is finally achieved through technology industrialization, for the input of each node in the innovation value chain includes the financial, manpower, material, and information resources of technology. The output of the technology achievements in each node has different characteristics. If the output of the former node could not be used as the input of the coming one, the value of this node would not be achieved. From the value chain point of view, it shows that the value of innovation value chain will be achieved as long as the technology achievements of the end node are industrialized. (3) From the society point of view, we can find that the demands of technology from the society is also based on the satisfaction of the increasing social development in order to ensure the sustainable development of human beings, which means that the increment of technology value contribute more to achieving the harmonious development among human being, nature and environment, which have more contributions on the value-adding of the humanistic value.

3   Government Responsibilities on Technology Aggregation-Diffusion Model in Large-Scale Sports Events

The technology aggregation and diffusion in large-scale sports events are made up of a series of technology innovations which require the necessary situations of the society and the technology, and also have the vivid feature of the public products. Besides, it is the government that can complete the aggregation of technologies from every sector of the society. Thereby, the establishment and operation of technology aggregation-diffusion model in large-scale sports events need to be completed by the lead of the government as well as the active participation of all parts of the society.

The government has the responsibilities to take some necessary measures to set up the technology aggregation-diffusion model in large-scale sports events and positively promote the operation of this model, which will guarantee the successful conduction and increment of the technology value in the events. It should actively integrate the technologies of all the sectors of the society and ensure the smooth operation of this model so as to achieve the maximum increment of the technology value, which is to reach the maximum increment of the values of the society, the economy, and the environment. With the opportunity of holding large-scale sports events, the government should set up the fundamental mechanism of aggregating the technologies in such short time while making technology innovations in a large scale; meanwhile, it should build the fundamental model of promoting the industrialization and publicity of technology achievements, encourage the market and the society to spread these achievements and increase the technology value.

References

1. Dong, C.-S., Yu, X., Wang, J.: Inventory of Beijing Olympic Games: Value for the Hi-tech Olympics from National Strategy. Journal of Shenyang Sport University (1), 5–8 (2009)
2. Metcalfe, J., Georghiou, L.: Equilibrium and evolutionary foundations of technology policy. OECD. STI Review, 75–100 (1998)
3. Dong, C.-S., Xing, H.-B., Wang, J.: Three key questions to hi-tech Olympics impel our country science and technology industrial production. Studies in Science of Science (2), 65–69 (2007)
4. Fang, F.-Q., Li, X.-Z.: The Characteristics of Scientific and Technological Cluster of Olympics and the Systematic Circle. Social Science of Beijing (3), 87–92 (2008)
5. Porter, M.E.: The competitive advantage of nations. The Free Press, New York (1985)

The Study on Sport Stadium Management Teaching Based on Virtual Reality

Liu Xiaomin

Attached Middle School of Guilin University of Electronic Technology
xiao.m.li@sohu.com

Abstract. By combination of modeling, computer simulation, 3D animation and virtual reality technology, practice teaching of the sports stadium management can provide sophisticated interactive features to achieve good teaching training effects without any additional expensive equipment. The virtual reality system enables users to take full share experience of a stadium and a new technical method is provided to the practice teaching of sports stadiums management .It is the pursuing objective and trend in the future of practice teaching.

Keywords: Sports stadium management, 3D modeling, virtual reality.

1   Introduction

In the construction of a modern international metropolis, sport plays an important role. Many major cities in the world are building large sports venues and facilities and actively bidding for major sporting events in order to establish a good image of their own [1]. The construction and operation of the sports stadium, especially intelligent and digital sports stadium management have put forward higher requirements.

Sports stadium management is a collection of multi-disciplinary basic theory and research methods to form a comprehensive course on the cultivation of stadium management. It must be linked with other disciplines and the comparison of knowledge to grasp the main points of the discipline in order to cultivate the application talents with the ability to use the theory and practical operation.

The teaching of sports stadium management is currently in the traditional way that mainly uses text, pictures, video and other two-dimensional display method. The practical skills training and interactive scenarios of teaching are difficult to achieve due to space constraints. The use of modern educational technology combined with sports stadium management expertise required by experimental teaching is imperative.

2   Application of Virtual Reality Technology

Virtual reality technology (referred to as VR technology) [2] is a computer-generated and high-level human-machine interaction system that constitutes not only a visual

Y. Wu (Ed.): International Conference on WTCS 2009, AISC 116, pp. 697–704.
springerlink.com

experience, but also hearing, touching, sense of smelling. The operator, using specialized equipment (such as helmets, touch gloves, computers, etc.) may be in a particular environment to observe, touch, operate, inspect and other test, the "virtual reality" of the flu effectively expand their cognitive domains. The greatest feature of VR technology is as follows: The operator can use a "natural" way to interact with the virtual environments, while in the past human beings apart from personal experiences (on-site-visit) can only be passive (text, pictures, movies, television, etc.).

VR technology has gradually been applied in urban planning, interior design, real estate development, industrial simulation, military simulation, heritage, electronic games, Web3d / products / still life show, roads and bridges, science education and other fields.

3   The Research on the Application of Virtual Reality Technology in the Practice Teaching of Sports Stadium Management

3.1   The Problems in Stadium Management Teaching

The management of sports complexes can be divided into three types according to the period: the special period of the Games (or the large-scale performances and other activities); general period that used as fitness activity areas; closed period. Management of general period focused on the venue booking, membership management, coaching, financial management and other things, is part of the scope of management information systems; maintenance is the major work during the closed period. Therefore, this paper has placed emphasis in the management practice teaching activities during the special period, that is, period of large-scale activities or sports games.

According to special period in the areas of stadium management, practice of teaching may include three directions: display and promote the venue, facilities control practices of the venue and risk control practices. But in actual teaching process, there are problems in these three directions:

- Venues display and marketing

Due to the constraints of space and time, it is impossible for students to inspect venues anywhere in the real sports venues. When introducing the structure, historical, cultural and the activities of a stadium, text, pictures, video, etc are always used. These methods lack of interactive and intuitive feel poor; and the promotion of venues, here mainly refers to venues and surrounding areas of advertising resources to promote, also rely mainly on oral, written description and effects diagram to represent, can only provide a static partial visual experience. Three-dimensional animation being a stronger dynamic three-dimensional expression also does not have real-time interactivity. So if the proposed program changes, it costs days or even a few weeks to see results [3].

- Practice of venues device control

For the intelligent management of sports venues, especially management and maintenance of the equipment, due to complex and diverse facilities, the students have little chance to go to the scene and have direct knowledge of all of the information including building automation, fire protection, monitoring, security and many other facilities. In the process of event, organizer will not rashly to enable students to do the operations and practice in order to avoid irreparable mistakes. In traditional teaching, students can only learn through abstract symbols of the venues for a large number of devices and their operation. It is dull and difficult to master. In actual application it is very difficult to operate the equipments, so these teaching methods are less applicable.

- Risk Control Practice

Sports stadium as a high-density venue needs to meet not only the high demand of activity, but also the safe and orderly operation of the facility, especially in the emergency evacuation of personnel, traffic order and protection. When unexpected events, it becomes one of the important issues that how to design treatment programs to effectively reduce losses and control risk. The current teaching of management of sports facilities has pay more attention to this problem, but generally limited to theoretical teaching, and there are still difficult in how to enable students to understand the critical knowledge in-depth through the practice session and cultivate the ability to deal with emergencies.

The problem of stadium management teaching practice has become a problem of sports stadium teaching that needs to be addressed urgently. So the virtual reality technology is used to improve the traditional teaching method and model. The use of virtual reality technology with the main body immersion, application of the universal nature, the current sensitivity, training and interactive teaching methods can be well positioned to meet the requirements of experimental teaching and the natural scenes of computer interaction [4].

3.2   Framework of Virtual Sports Stadium Management Experimental System

A stadium management practice is designed in this paper. The system can be divided into three modules: display and promotion module, device controlling and training module, and risk controlling and training module.

The display and promotion module includes such subsystem as venues roaming, cultural shows and advertising resources to promote; device controlling and training module includes the building automation, fire protection, monitoring, security; risk management and training module includes emergency evacuation simulation and vehicle scheduling simulation subsystems [5]. The building automation subsystem also includes air-condition, fresh air, variation power distribution, lighting, elevators and chillers subsystems. The security subsystem includes burglar alarms, access control, parking, ticket-sale, patrol and other subsystems. System architecture diagram is shown in Fig 1.

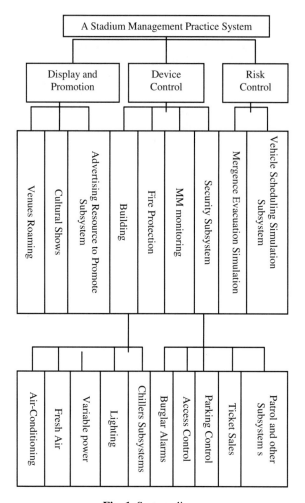

Fig. 1. System diagram

3.3 Function Design of Virtual Sports Stadium Management System

- Display and promotion of venues

1) Roaming: A realistic model library of sports venues is created by using the modeling technique. Users can roam in the venue and change the height, angle, location, and route for a random operation. The 3D scene model is entirely based on real-world, real-time and interactive roaming, feeling walking among the scene change and animation effects, with the support of three-dimensional imaging and multimedia effects, there is no constraint of time and space and it is a more realistic and detailed feelings for the venue as the real environment [3].

2) Cultural shows: Virtual stadiums experiment system can enhance the effects using digital technology and realistically reproduce important historical artifacts, information, apparatus, characters and scenes. The user can share the experience of its territory without physical situations and increase interactivity according to the user's personal interest in selective display.

3) Stadium advertising resources promoting: the function of the core modules of this subsystem servicing for sports venues promotion. In the experimental system, you can organize activities according to the actual situation. The first step is to select the venue according to the event and its size; and then complete the excavation, design and value integration of the venue advertising resources. So students can choose for indoor and outdoor environment to observe the site exposure points and dig out the regions where it can be set up ads both inside and outside the venue according to the actual features and exposure points. The type and form of the advertisements can be set. The specific effects of advertising (including the color, content, size and position) can be also set up and the final sports stadiums advertising design can be previewed. When sports venues and advertising design are complete, 3D models are generated and can be used in roaming and interacting. The 3D models can be exhibited to the sponsors and other related departments so that they can see an extremely realistic three-dimensional effect without going to the actual stadium which helps to their decision-making.

- Device Control

The device models of the experimental system have the same size of real equipments and the models can be viewed from multiple angles and multiple levels. The location, interconnection and other relations between the equipments can also be obtained from the virtual model. When operating the equipment in a virtual scene, the operation is exactly like the real one. Students can simulate the actual operation effects in a virtual control interface under the teachers' demand such as selecting, viewing, setting the device information, lighting, air conditioning, monitoring. The corresponding adjustments can be made to achieve the best effects according to the setting results of the virtual reality scene, which helps students to quickly master the device management skills [6].

- Risk Control

Experimental system has been built using virtual three-dimensional model of the stadium, it can be a clear reproduction, including the outdoor environment (parking lot, the main entrance road junctions) and the interior of the channel layout, facilities, fire signs, indicating the location of its exports can be arbitrarily adjusted evacuation routes the width and layout of obstacles and so on, can be based on different initial data to simulate different emergency scenarios, different personnel density, distribution and behavior; on this based on evacuation simulation, realization of complex roaming scenarios, simulated unexpected situation, the evacuation process, to escape training and fire training.   It also can also be the vehicle scheduling analog control to develop a reasonable scheduling strategy.

Virtual sports stadium management and promotion system is an integrated system in which each sub-module can run independently. In the practice teaching, the corresponding training module can be loaded and activated according to the knowledge points. And it can also be used as a complete example to examine the student's overall grasp degree.

3.4   Critical Technology and Application Examples

- System critical technology

The system's critical technology includes mainly two types, that is modeling and the underlying application. Modeling system based on different applications can also use different methods. The demonstrating of the model class (such as artifacts, utensils, etc.) can be used panoramic image- generation technology that uses cameras to acquire the real images of the model in order to get the discrete images or continuous video as the basic data. The panoramic image and their spatial association are generated to establish an immersive and interactive information environment after image processing. Thus the modeling process is completed [7]. Another type of geometric model is created by using the three-dimensional modeling software (3ds max is used in this system). The sub-modules including complex model, advertising model, equipment model, device control interface model, etc. were established to form classified model bases. In the course of the construction of the model, the composition model of point, line, and surface number should be strictly controlled because when the solid model is imported from 3ds max into the system by the underlying process, the number of model's points is limited. If the number exceeds the limit, models will appear the phenomena of distorted face and point merging which leads to the result that the imported model is incorrect, and thus the interactive function cannot be added to the model.

The underlying system can be optional Direct3D, OpenGL, Virtools or integrated shader programs. The system has been completed by using OpenGL as the underlying development tool to complete the interactive function.

- Application Example

The following is the example of stadium roaming and advertising resources promoting in a virtual experiment system of sports stadium.

An outdoor stadium-roaming interface is shown in Fig 2. The roaming is divided into an overall automatically roaming and the one by users' will.

The main interface of an outdoor stadium advertising resources promoting is shown in Fig 3, which includes venue selection, site construction, advertisement zoning and advertising settings submenu. In sub-menu, different venues in the material base can be selected, such as the soccer field, basketball, volleyball hall, etc. Site building includes not only the fixed facilities in the venue, but also other facilities such as booths, temporary stands, etc. according to the necessary of activities. Advertising locations are classified and calibrated in different colors to meet with different levels of sponsorship's needs according to their exposure values. The advertising setting includes the type, color, content and location of advertisings. When all the settings are finished, the advertising resources can be packaged and saved in order to be exhibited to the sponsors and organizer.

Fig. 2. Outdoors Stadium-roaming Interface

Fig. 3. Advertising Resources to Promote Interface

4   Conclusion

The virtual reality technology is used to construct a practice teaching system of sport stadium management according the present situation and its existing problems. The basic structure, main functions of the virtual stadium management teaching system and its advantages in the practice teaching are introduced. The critical technologies and application examples of this system are also showed in this paper.

References

1. Chen, W.: Research on management of large gyms and stadiums in Beijing. Journal of Capital Normal University 6, 115–120 (1998) (in Chinese)
2. Wang, C., Gao, W.: The Theory, Implementation and Appliance of Virtual Reality, pp. 56–80. Qinghua University Press, Beijing (1996)

3. Zhu, N., Zou, Y.: Appliance of Virtual Reality in Architecture Design. Journal of Beijing University of Civil Engineering and Architecture 24, 34–36 (2008)
4. Cheng, D.: Apply Virtual Reality Technology to Improve the Quality Of Experimental Teaching In tourism management. Education Exploration 24, 134–135 (2008) (in Chinese)
5. Wei, X., Hang, C.: Discussion on the Design & Construction of Building Intelligence System in National Indoor Stadium. Electrical Technology of Intelligent Buildings 2, 23–25 (2008) (in Chinese)
6. Sheng, M., Hu, R., Huang, X.-T.: Prospect and Application of Virtual Reality for Stage Control. Entertainment Technology 5, 47–49 (2008)
7. Zhang, Y., Liu, J., Mu, Y.: Analysis and Design of City Tour ism Information System Based on WebGIS and Virtual Reality Technology. Aeronautical Computing Technique 38, 77–80 (2008)

Study on College Students Sports Tourism Participation Factors

Liu Xiaomin

Attached Middle School of Guilin University of Electronic Technology
xiao.m.li@sohu.com

Abstract. The conclusion shows there has been obviously distinct in gender ad grade about the degree of sports tourism participation; The average degree of participation of Guilin University of Electronic Technology college students is higher than that of nation ones, and the tends of participation of Guilin University of Electronic Technology college students in grades is the same as nation ones; There has been positive affected relationship between the degree of sports tourism participation and cognitive behavior, behavior motive, behavior habit, however, the situation is opposite in subjective criteria ad te pressure on the environment.

Keywords: Factors, sports tourism participation, College students.

1 Introduction

College students are a special group in society. They have some ability to self-life, a relatively related time, more risk-taking ad forward-thinking travel, which have become a potential tourist source markets. In this article, the writers surveyed the sports tourism participation degree of Guilin University of Electronic Technology College ad compare with the average level of Chinese college students objective is tat prove the difference between Guilin University of Electronic Technology and nation. What is more, the article also analyses the influence factors on the degree of sports tourism participation from College students. The purpose is that provide theory information for related department to acknowledge the college student's tourism markets.

2 Research Target and Methods

2.1 Research Target

In this article, all students registered of Guilin University of Electronic Technology have been chosen in this research.

2.2 Research Methods

- Alot of current sports tourism literatures and books on college students have been read ad studied.

Y. Wu (Ed.): International Conference on WTCS 2009, AISC 116, pp. 705–711.
springerlink.com © Springer-Verlag Berlin Heidelberg 2012

- I accordance with the basic requirements of the sociology, psychology and statistical analysis to design and distribution the "Questionnaire on College students participated in sports tourism from Tongren university". The questionnaire was designed by researchers and tested with higher reliability and validity by specialists .The questionnaire collected Physical Activity Participation in the main quantitative indicators .It was divided into three pars tat made of the basic situation of respondents and sports tourism participation level and the influence factors to participation. And the influence factors to participation were made of 20 items and 5 factors by behavior cognition, behavior motivation, subjective standards, behavior habit ad the pressure on the environment. What's more, each factors include 4 projects which Correlation Coefficient were 0.77, 0.72, O. 79, 0.71 and 0.74. In this article , used cluster sampling ad stratified random sapling methods to distributed 500 questionnaire ad 480 were recovered. The recover rate was 96.0%, of which 364 were effective volumes, it effective rate was 75.8%.
- The testing data have been analyzed by the software Excel 2003 and SPSS 17.0. Mainly Descriptive Statistics X, the Weighted Case, the Relevant Analysis.

3   Results and Analysis

3.1   College Student Sports Tourism Participation Comparison of Deferent Demographic Statistic in Tongen University

It can be seen in the statistics of table 1 that the distribution of the degree of the college students male and female who participate in sports tourism in Guilin University of Electronic Technology among different demographic characters have statistical significance, according to the different gender, ethnic and grade, however the situation is opposite in ethnic($p > 0.01$).

Table 1. College students sports tourism participation comparison of different demographic statistics

Demo graphic		N	Regular	Occasion	None	X^2	P
Gender	male	236	60 (25.42)	140 (59.32)	36 05.26)	40.673	0.000
	female	12 S	30 (23.44)	41 (32.03)	57 (44.53)		
Etnic	Han	24 0	62 <25.83)	113 (47.08)	65 (27.08)	1.682	0.431
	Minort	12 4	30 (24.19)	66 (53.23)	28 (22.58)		
Gade	2009	90	29 02.22)	46 (51.11)	15 (16.67)	18.359	0.005
	2008	90	30 03.33)	44 (48.89)	16 <17.78)		
	2007	91	20 (21.98)	43 (47.25)	28 00.77)		
	2006	93	17 <18.28)	40 (43.01)	36 08.71)		

3.2   Comparative Analysis 0/ Sports Tourism Participation 0/ Guilin University of Electronic Technology Students and College Students in Whole China

• Guilin University of Electronic Technology students and Chinese university students were grouped by gender disparities in comparison: Figure 1 shows that boys participated in Tongren College Sports Tourism accounted for 54.96 percent overall test, female subjects participated in total accounted for 1.50%, boys not participated in any sports tourism accounted for 9.89% test overall, girls not participated in sports tourism accounted for 15.66% overall test; according to Zhag Aiping's(2) " Investigation and Development Strategies of Sports tour of university students in China " in Chinese University Students Sport travel to collate available data, the Chinese male university students participated in sports tourism for 17.66%, for boys have not participated in 44.93% of girls attended for 8.74% of girls have not attended for 28.67%. Data show that in Guilin University of Electronic Technology students ad Chinese students overall comparison, both male and female college students participated in sports tourism far beyond the average level of college students, Tongren College students regardless of male and female participated in both low in the average level of college students shows that Guilin University of Electronic Technology students participation of sports tourism is higher than the average in China.

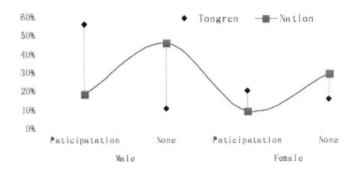

Fig. 1. Guilin University of Electronic Technology students and Chinese university students grouped by gender disparities in comparison

• Tongren College Students and Chinese university students grouped by grade differences exist: Guilin University of Electronic Technology 09 freshmen participated in sports tourism accounted for 83.33% of subjects overall, 08 sophomores participated in sports tourism accounted for 82.22% , 07 juniors participated in accounted for 69.23%, 06 graduates participated in sports tourism accounted for 61.29%,overall test. It ca be seen that the participation average data of Tongren College students is higher than Chinese college students, which is equal to the differences grouped by gender, but the four-year participation in the same proportion as the trend of the highest freshman, sophomore second, lager 4 Min.

whether China or Tongren, there exists difference in the number of participates in different grades .Figure 2.

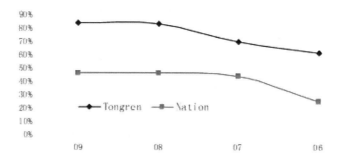

Fig. 2. Guilin University of Electronic Technology Students and Chinese university students grouped by grade differences

3.3   The Analysis o/ Related /Actors Impacting Participation in Sports Tourism

• Table 2 shows that college students' participation in sports tourism major quantitative change in the correlation coefficient (r). Te results showed that regular participation in sports tourism, and occasionally participate in sports tourism ad the behavior of college students cognitive, behavioral motivation, subjective standards, behavior, environmental stress of students tested to sports tourism docile scale P <0.01, significant. Among them, cognitive ad behavioral involvement, motives, behavior were significantly positively related to subjective standards and behavior, environmental stress were negatively correlated; not participate in behaviors of college students and tested the behavior of Tourism cognition, behavior motivation , subjective standards, behavior, life stress of students tested to sports tourism deciles scale P <0.01, with statistically significance. Among them, cognitive and behavioral involvement, motives, behavior was significantly negatively correlated, subjective standards and life stress were significantly positively related.

Table 2. College Students Participation I Sports Tourism Major Quantitative Change I The Correlation Coefficient (R)

Participation	Behavior Cognitive	Behavioral Motivation	Subjective Standards	Behavior Habit	Environmental Stress
Regular Participation	0.154	0.213	-0.136	0.192	-0.175
Occasionally Participation	0.071	0.136	-0.102	0.082	-0.101
None	-0.140	-0.120	0.124	-0.136	0.142

4   Discussion a Analysis

4.1   Discuss on the Overall Participation of Tongen University Students

- There are gender ad grade differences but no ethnic ones in level of participation of sports tourism among tested students. In the male and female groups, the proportion of female students who do not participate is more than male students , which come into the same trend with Yag Bo[3l,s "Investigation of actuality of sports tourism among students in Huna province " and Zhang Aiping[2]'s "Investigation and Development Strategies of actuality of sports tourism among university students in China ", etc. , indicating that the initiative and enthusiasm of female students on the sports tourism is not high, which is a common phenomenon in female college students in different regions. Analyzing the reasons ,it has something to do with the characteristics of female students' psychology ad physiology, and also is related to school physical education settings. Propose to car out physical education curriculum content which is suitable for female students ad targeted extracurricular sports activities to improve interest of participation of female college students in sports.

4.2   Discuss on the Deference between Tongren and Nation

- Students tested in comparison with Chinese university students, the subjects of male and female college students participated in sports tourism are beyond the average level of college students in china. Te reason is that Tongren college is institution in the western Guizhou Province ,in China, the National Tourism Bureau issued the "2001 China Tour fitness program of activities " showing that the characteristics of product distribution in different regions are as follows: in terms of the special tourist routes, the western region has a absolute advantage wit 50%, 43 accounted for a total of 82 routes, the eastern and central wit 25.6% ad 22% far behind the west. In addition, Tongren itself is the tourist cit , which is rich in tourism resources. Students can use the usual weekends, holidays, etc. to car out activities close to sports tourism.

- Students tested in comparison with Chinese university students , the participation proportion of four grades comes into the same tend, which both showed the highest is freshman, sophomore comes the second, the least is senior. There are some differences in sports tourism in the different grades in the number of participants. Analyzing the reason, the university life of first-year students is full of passion, and they are excited about decoupling parents and teachers. Although releasing tasks, but they yea for freedom ad eagerly want to experience new things, so that they maximize the liberation of their detention in the long body and mind; the reason why the number of participates is small in grade four is that they face graduation and job-task weight and pressure, ad entertainment is a luxury for tem, Only a few of tem whose future is available are basically for entertainment cheer[2].

4.3   Discuss on the Related Factors in Sports Tourism Participation

•   Students participation in sports tourism and sports tourism on the behavior of cognitive, behavioral motivation, behavior, there were significantly correlated wit subjective standards, life stress was significantly negatively correlated. Students conduct their sports knowledge[41, spot information was low or moderate positive correlate the more knowledge and information more firmly, more actions[51. So in order to improve the Student Sports Tour participation ad develop good exercise habits, college sports stimulate exercise but also enhance sports tourism enthusiasm. The research also finds that students in the subjective standard of sports tourism and the environment score relatively low pressure areas, that they expect the conditions of the environment, information and transportation services to feel undesirable, both because of most of them are the only child college students, their growth is often obtained from family and social aspects of adequate care, on the enjoyment of material life and spiritual conditions require relatively high; on the other hand, poor conditions of service of some tourist attractions, sights charge too high and so on. Students who are also relatively independent economy can not bear. Tourism enterprises should develop proposals for special sports students travel products, reduce costs, give preferential policies to attract more college students involved in sports tourism. Pressure on the environment in recent yeas as enrollment continues to expand, causing pressure on college students leaving, employment pressure, social pressure generally increased, but also affected the level of participation of students of sports tourism. Hope that the schools meet regularly to psychological counseling, ad establish the correct values of college students, reduce mental stress, local government departments, the state has come a strong macroeconomic policies, the real implementation of the employment of college students.

5   Conclution

In this article, the relationship between the present situation and effect factors are studied though the investigation of 364 college students' male and female of Guilin University of Electronic Technology. The results have also been compared with the average of the nation college students. The conclusion shows that there has been obviously distinct in gender and grade about the degree of sports tourism participation; The average degree of participation of Guilin University of Electronic Technology college students is higher than that of nation ones, and the trends of participation of Guilin University of Electronic Technology students in grades is the same as nation ones; There has been positive affected relationship between the degree of sports tourism participation and cognitive behavior, behavior motive, behavior habit, however, the situation is opposite in subjective criteria and the pressure on the environment.

References

1. Chen, G., Huang, Z.: Investigation on Sports Tourism Condition and Its Infuencing Factors of College Students in Zhejiang Province. Industry Discussion 17(6), 98 (2009)
2. Zhang, A.: Investigation and Development on Sports Tourism Strategy of College Students in China, pp. 12–14. Beijing Sport University (2006)
3. Bo, Y.: Hunan Province College students Sports Tour Current situation Studies in Inquire, pp. 29–32. Hunan Normal University (2007)
4. Li, G., Liu, Y., Zhang, B.: Sports Knowledge,Attitude, Beliefs and Behavior of 573 students. China School Health 25(1), 50–51 (2004)
5. Zhang, H., Guo, S.: Analysis on Exercise Behavior and Related Psychological Factors of Kunming College Student. China School Health 21(6), 452–454 (2000)

The Universal Communication Module for Software and Management Integration Information Systems

Tang Rongfa

CSIP Guangxi Branch, Guilin University of Electronic Technology
l.y.now@qq.com

Abstract. The development of UCM uses the AJAX technology of Web development framework, and adopts the JSON protocol to encapsulate data, and HT protocol to transfer data. Because of all parts are built up based on standard protocols, The UCM has good compatibility and scalability. So, it is able to systematically solve the real-time data transmission and control problems between CS and MIS. What's more, the UCM is suitable for all kinds of CS to use only when the CS can run on windows platform and support using the ActiveX controls.

Keywords: Universal communication module, confguration software, management integration information system.

1   Introduction

Configuration software, which will be abbreviated as CS here, has been widely used in the wake of the rapid development of the enterprise and factor's automation and informatization, as in [1]. Therefore, it becomes urgent and important that how to integrate CS and management information system, which will be abbreviated as MIS in this paper, instantly and efficiently. During this present period, there are two general integration approaches already. The first one is using the Open Database Connectivity, abbreviated as ODBC, to access the database. This method seems simple to implement but has poor real-time response and expansibility. The second method is applying TCPII protocol to integrate the CS and MIS, which also has these shortcomings, such as poor compatibility, low scalability, higher charge of development and maintenance, as in [2] and [3]. Distinguished from these two methods, a new integration solution is proposed here by using these existing mature technologies, such as TCPIIP, HTTP, JSON, and ActiveX.

2   Modular Architecture

2.1   Development Technique

The module uses Microsoft's ActiveX component technology, whose main technology is the Component Object Model, abbreviated as COM. This kind of

Y. Wu (Ed.): International Conference on WTCS 2009, AISC 116, pp. 713–719.

control, which needn't to be modified, can be used in a variety of program container after defining the interface specifications between the container and component. Meanwhile, the program container which follows the standard interface specifications can also be easily embedded in any of the controls. Figure 1 show that the client application can use the MIS to access and control the Industrial Personal Computer, which is abbreviated as IC here. Because of the UCM, intranet user, which may be either Windows application or Web client program, can communicate directly with IPC or through MIS. That is to say, the web client program uses AJAX technology to achieve the asynchronous communication with IPC. AJAX technology is using XML Http module and JavaScript to achieve its function.

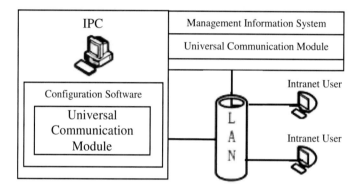

Fig. 1. Schematic diagram of system

2.2   Network Protocol

The module uses standard network transport protocol, TCP/IP protocol, which is vividly called the hand agreement between people in the global village. The TCPIIP is usually named as the TCPII protocol suite, because TCPIIP protocol suite include TCP, I, UDP, ICM, RIP, TELNETFTP, SMTP, A, TFTP, and many other agreements. TCP/I protocol is the basic communication protocol used in internet in order to ensure the integrity of transmitted data, as in [4].

2.3   JavaSript Object Notation Remote Procudure Call Protocol

The communication protocol's data format is using JavaScript Object Notation format, which is commonly abbreviated as JSON format. JSON format, which is a lightweight data interchange format, is adopted here, not only because it's easy to read and write, but also its easy for machines to parse and generate, as in [5].

1) JSON's features

a) Object-Oriented: JSON provides an excellent object-oriented approach in order to transfer metadata between the networks or cache metadata to the client machine.

b) Easy to separate data and logic : JSON provides a simple and easy method for the separation of data and logic which will be validated.

c) AJAX's foundation: JSON provides the foundation for Web applications to achieve AJAX.

2) JSON's usage: Object can be represented by a specific form of JSON string. If there is such a form of JSON string which is assigned to a variable, then the variable will be interpreted as a referred object which is built by a JSON string.

From the above, it is obvious that JSON can be an ideal data exchange format because of its language-independent text format.

3   Moduar Design

It is time-consuming and arduous work for the programmer to achieve the standard communication protocol's function by coding repeatedly each time, which has not only poor compatibility but also limited scalability. Therefore, it is necessary to design a UCM to achieve code reuse and prevent duplication of effort, as in [6]. In order to achieve the seamless communication between CS and MIS, UCM is embedded in both CS and MIS, which is shown in Figure 2.

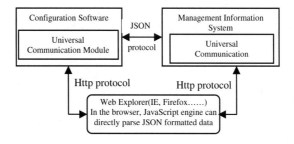

Fig. 2. System diagram

UCM can be divided into two sub-modules: data transmission sub-module, and data parsing sub-module. Module functional diagram is shown in Figure 3

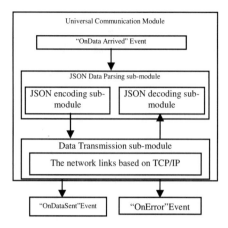

Fig. 3. Module functional diagram

Data transmission sub-module is responsible for sending and receiving data, and the underlying communication protocol is using a reliable TCP/IP protocol. Therefore, the module does not require verifying the data received.

Data parsing sub-module is designed for data encoding and decoding according to JSON format.

In addition, the UCM also includes three main events, which are "OnDataArrived", "OnDataSent" and "OnError". In which, "OnDataArrived" event will be triggered when the communication module receives data, "OnDataSent" event will be triggered when the Data transmission sub-module have sent data, "OnError" is triggered when an error occurs in the program. There are also many other events included in theUCM.

1) Data Sending: First, the "data parsing sub-module" is called to encode data into JSON formatted strings. And then the "data transmission sub-module" is called to send JSON formatted strings. If the data is sent successfully, the communication module's "OnDataSent" event is triggered. On the contrary, the communication module's "OnError" event is triggered.

2) Data Receiving: First, the JSON strings received by "data transmission sub-module" are decoded by the "data parsing sub-module". If restored successfully, the communication module's "OnDataArrived" event is triggered, otherwise the "OnError" event is triggered.

The UCM integrates the standard communication protocol, which is JavaScript Object Notation Remote Procedure Call Protocol, which is abbreviated as JSON RPC. It enables real-time data exchange and control, so it greatly improves the system's compatibility and expandability. The JSON RPC format is as follows.

```
{

method: 'methodName',

params: [param 1, param2, ... ]

}
```

Parameter Description is as follow.

1) Method: the name of the function or procedure to be called.
2) Params: parameters array to be transmitted to the function or procedure.

After the remote procedure call protocol is defined, the internal functions of CS can be remotely called by designated MIS or web client application, and the results can be easily return. Similarly, the internal functions of MIS can also be remotely called by designated CS or web client application. The call interface diagram is shown in Figure 4.

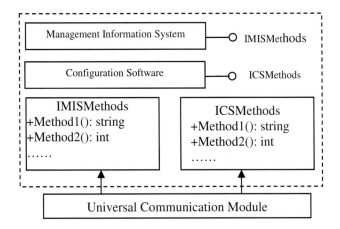

Fig. 4. Call interface diagram

"ICSMethods" interface is concluded in the CS, and "IMISMethods" interface is embedded in the MIS. In other occasion, these two interfaces can also be instead by others. After that, the UCM has been implemented between the two interfaces. By this arrangement, the UCM can access both the CS and the MIS. If failed to achieve the interfaces, it is necessary to restrict the access to the UCM by coding.

Figure 5 shows how the MIS call the "GetData" function of CS and the corresponding U sequence diagram. Instance UCMl is the UCM integrated in the MIS, while the instance UCM2 is the UCM integrated in the CS. If the "Call" function of UCMl is called in the MIS, the parameter is "GetData", the UCMl will encode the "GetData" into a JSON string according to remote call protocol. And then use the TCP / IP protocol to send data to UCM2.

After UCM2 receives the data sent by UCMl, it will decode the data into a data object immediately, and then call the specified function of CS according to protocol. And then UCM2 will return the result of function to UCMl which is also encoded into a JSON string. The "OnDataArrived" event is triggered When the UCMl receive the JSON string.

Throughout the whole sequence diagram, there are three points need programmer to pay attention, which is calling the "Call" function, achieving the "GetData" function, and responding "OnDataArrived" event.

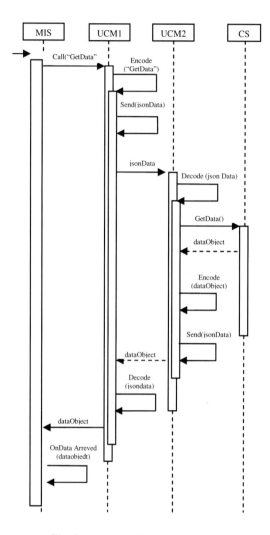

Fig. 5. Function call sequence diagram

In the above module, the realization of the various core functions are built on the basis of the relevant standard protocols, so it can easily be achieved using existing proven technology.

4   Conclusion

In this paper, the universal communication module is proposed, which is based on the AJAX technology in Web development framework. For data encapsulation and transmission, the JSON protocol and HTTP protocol are adopted. The module, which is designed by using standard protocol, has been proved to own excellent compatibility and expansibility. For this reason, it can solve hysteretic phenomena of real-time control and data transmission between configuration software and management information system.

References

1. Cao, H.: Confguration sofare technology and applications. Electronic Industry Press, Beijing (2009)
2. Lin, W., Gao, C.: Domestic Situation and Future of Confguration Sofare, Electric Time, pp. 9–12 (2002)
3. Wang, Y.: Confguration sofare design and development. Xi'an University of Electronic Science and Technology Press, Xi'an (2003)
4. Stevens, W.R., write, Hu, G., translate: TCP lIP Xiangie, Vol. 3. Mechanical Industry Press, Beijing (2000)
5. http://www.json.orgljson-en.html. The description of lS0N
6. Wu, L., Liu, H.: The Integration research of confguration sofare and management information system. Micro Computer Information, Beijing (in Press)

Defect Classification Method for Software Management Quality Control Based on Decision Tree Learning

Tang Rongfa

CSIP Guangxi Branch, Guilin University of Electronic Technology
l.y.now@qq.com

Abstract. In this paper, we have discussed how decision tree leaning approach can be used for defect as Problem and it's classification for quality control in software project. The proposed technique in this paper is a fertile ground for classification of the defects in different categories identified by means of root cause analysis that means the popular fishbone diagram. Causal analysis plays a pivotal role in the quality control of software projects. However the opposed approach presented in this paper will certainly explore new arena for research that how machine learning could be integrated with the quality control activities for efficient software project management.

Keywords: Problem Classification, Decision Tree, Software Project Management Quality Control.

1 Introduction

In this paper we consider hypothetical training data sets and present taxonomy of the defects that can occur during any stage of the product development process [8] by using the decision tee leaning approach. Various kinds of defects are identifiable and we make use of a subset to demonstrate the proposed methodology. Decision Learning methods [2] are among the most practical approaches to certain types of leaning problems.

We consider the following defect factors to illustrate the use of Decision Tree Learning in determining the particular class of defects. The main classes of defects can be found out using the Fishbone diagram also known as Ishikawa diagram in concern with the Root Cause Analysis [6] and we consider following six classes:-

- Task
- Technology
- Process
- People
- Policies
- Product

The set of defects are assigned one of the 3 values depending upon there frequency of occurrence in the given candidate project development process. These values are low,

Y. Wu (Ed.): International Conference on WTCS 2009, AISC 116, pp. 721–728.
springerlink.com   © Springer-Verlag Berlin Heidelberg 2012

average and high respectively. Following defects [3, 6, and 7] are considered for the classification process:-

- Testing Defects
- Development Defects
- Design Defects
- Lack of Tools and Related Defects
- Lack of proper strategy
- Lack of organizational level policies and support
- Lack of management skills
- Lack of Training

In this paper we would consider in all 15 hypothetical training examples as shown in the table 1.

Table 1. Training Examples Set [2]

TD	DD	DSD	LT	LPS	LOPS	LMS	LT	Class
H	H	H	L	L	L	L	L	Task
L	A	A	L	L	L	L	L	Task
H	A	H	L	A	A	L	L	Task
A	H	L	A	A	L	H	H	Task
L	H	L	H	H	L	L	L	Tech
L	H	L	H	H	A	L	A	Tech
L	H	L	H	A	L	L	L	Tech
L	H	A	L	H	A	A	L	Process
L	A	A	L	H	L	A	H	Process
L	L	L	A	H	H	H	H	People
L	H	L	L	A	A	H	H	People
L	L	L	L	L	H	L	L	Policies
L	A	L	L	L	H	A	L	Polices
L	H	L	H	H	L	L	L	Product
L	A	L	H	L	L	L	A	Product

T: Testing Defects
DD: Development Defects
DSD: Design Defects
LT: Lack of Tools and Related Defects
LPS: Lack of Proper Strategy
LOPS: Lack of Organizational level policies a support
LMS: Lack of Management skills
LT: Lack of Training
L: Low
H: High
A: Average

In decision tree learning [2] approach the information gain for each property decides which one should be tested first. First to measure the entropy, if the target attribute can take on c different values, than the entropy of S relative to this c-wise classification [2], is defined as

$$Entropy(S) = \sum_{i=1}^{C} -p_i \log_2 p_i \qquad (1)$$

Gain (S, A) is the reduction in entropy caused by knowing the value of attribute A.

$$Gain(S, A) = Entropy(S) - \sum (Sv / s)Entropy(Sv)$$
$$A \varepsilon Values(A)$$

1.1   Gain Calculation for Testing Defects (TD)

Values (T) = Low, Average, High

$S = [4+, 11-]$
$S_{Low} \leftarrow [1 \ +, 10-]$
$S_{Average} \leftarrow [1+, 0-]$
$S_{High} \leftarrow [2+, 1-]$

$$Gain(S, TD) = Entropy(S) - \sum (Sv / s)Entropy(Sv)$$
$$A \varepsilon \{Low, Average, High\}$$

= Entropy (S)-11/15 Entropy (Low)-1I 15

Entropy (Medium)-3/15Entropy (High)

= 0.828-0.65-0-0.129

=0.049

Therefore, Gain (S, TD) = 0.049

A.   Entropy (S) Calculation:

Entropy ([4+, 11-]) = -4/15log2(4/15)-(11/15) log2 (11/15) = 0.828

1.2   Gain Calculation for Development Defects (DD)

Values (DD) =Low, Average, High

$S = [4+, 11-]$
$S_{High} \leftarrow [2+, 6-]$
$S_{Average} \leftarrow [2+, 3-]$
$S_{Low} \leftarrow [0+, 2-]$

$$Gain(S, DD) = Entropy(S) - \sum (Sv / s)Entropy(Sv)$$
$$A\varepsilon\{Low, Average, High\}$$

= 0.828-8/15 Entropy (High)-5/15 Entropy (Average)-2/15 Entropy (Low)

=0.072

Therefore, Gain (S, DD) = 0.072

1.3   Gain Calculation for Design Defects (DSD)

Values (DSD) =Low, Average, High

S= [4+, 11-]

$S_{Low} \leftarrow$ [1+, 9-]

$S_{Average} \leftarrow$ [I+, 0-]

$S_{High} \leftarrow$ [2+, 2-]

$$Gain(S, DD) = Entropy(S) - \sum (Sv / s)Entropy(Sv)$$
$$A\varepsilon\{Low, Average, High\}$$

= 0.828-0.7-0-0.775

= -0.647

Therefore, Gain (S, DSD) = 0.647

1.4   Gain Calculation for Lack of Tools and Related Defects (LT)

Values (LT) =Low, Average, High

S= [4+, 11-]

$S_{High} \leftarrow$ [3+, 5-]

$S_{Average} \leftarrow$ [I+, 1-]

$S_{Low} \leftarrow$ [5+, 0-]

$$Gain(S, LT) = Entropy(S) - \sum (Sv / s)Entropy(Sv)$$
$$A\varepsilon\{Low, Average, High\}$$

= 0.828-0.99-0.52-0

= -0.682

Therefore, Gain (S, LT) = -0.682

1.5   Gain Calculation for Lack of Proper   Strategy (LPS)

Values (LPS) =Low, Average, High

S= [4+, 11-]

$S_{Low} \leftarrow$ [2+, 3-]

$S_{Average} \leftarrow$ [2+, 2-]

$S_{High} \leftarrow$ [O+, 6-]

$$Gain(S, LPS) = Entropy(S) - \sum (Sv / s)Entropy(Sv)$$
$$A\varepsilon\{Low, Average, High\}$$

= 0.828-0.3233-0.775-0

= -0.27

Therefore, Gain (S, DSD) = -0.27

1.6   Gain Calculation for Lack of Organizational Policies and Support (LS)

Values (LMS) =Low, Average, High

S= [4+, 11-]

$S_{Low} \leftarrow$ [3+, 6-]

$S_{Average} \leftarrow$ [0+, 3-]

$S_{High} \leftarrow$ [I+, 2-]

$$Gain(S, LMS) = Entropy(S) - \sum (Sv / s)Entropy(Sv)$$
$$A\varepsilon\{Low, Average, High\}$$

= 0.828-0.99-0.193-0

= -0.355

Therefore, Gain (S, LMS) = -0.355

1.7   Gain Calculation for Lack of Proper Strategy (LPS)

Values (LPS) =Low, Average, High

S= [4+, 11-]

$S_{Low} \leftarrow$ [3+, 6-]

$S_{Average} \leftarrow$ [2+, 0-]

$S_{High} \leftarrow$ [I+, 3-]

$$Gain(S, LPS) = Entropy(S) - \sum (Sv / s)Entropy(Sv)$$
$$A\varepsilon\{Low, Average, High\}$$

= 0.828-0.595-0-0.129

= 0.1033

Therefore, Gain (S, LPS) = 0.1033

1.8   Gain Calculation for Lack of Training (LOT)

Values (LOT) =Low, Average, High

$S= [4+, 11-]$

$S_{Low} \leftarrow [3+, 6-]$

$S_{Average} \leftarrow [2+, 0-]$

$S_{High} \leftarrow [I+, 3-]$

$$Gain(S, LPS) = Entropy(S) - \sum(Sv / s)Entropy(Sv)$$
$$A\varepsilon\{Low, Average, High\}$$

$= 0.828 - 0.595 - 0 - 0.193$

$= 0.04$

Therefore, Gain (S, DSD) = 0.04

1.9   The Information Gain Values for All Properties Are

Gain (S, TD) = 0.049

Gain (S, DD) = 0.072

Gain (S, DSD) = -0.647

Gain (S, LT) = -0.682

Gain (S, LPS) = -0.27

Gain (S, LOPS) = -0.355

Gain (S, LMS) = 0.1033

Gain (S, LT) = 0.04

Where S denotes the collection of all training examples.

According to the information Gain measure, the LMS property provides the best prediction of the target attribute selection, over the training examples. Therefore LMS is selected as the decision attribute for the root node, and branches are created below the decision attribute for the root node.

If Lack of Management Support (LMS) is average then the category is not defect. Then we take next gain property. If LMS is low we make decision based on Development Defects (DD). If LMS is high, we consider Testing Defects (TD) as next gain value and continue to build the final decision tee as shown in the figure.

Using this decision tree we could classify the given sample data as either in the category of the defect class Task or Non Task.

If the category is Non Task then try for the next tree for the category Technology and so on & so forth until we found the class of the input data.

2   Conclusion

There are various reasons and causes which lead to failure of software that may come right from it's starting point of requirement analysis up to launching of product in the market .One has to do the root cause analysis of software failure so that these failures should not be reproducible. There are various problems due t which the software may give the bugs, errors, fault and ultimately the failure. Enlisting the problem, analyzing the problem after reporting is must before the fixing of the problem and going into the root cause of the problem. The classification of the problem will definitely help us to sort out the problems and will help to go to the root of problem. Once problems have been reported it can be classified by using any classification method depending upon the properties and their values. We do combine the decision tree learning with the input as the current problems. Decision Tree will be trained with trainee example with similar type of problems, their "properties and values". The DTL will help us to classify the problem and ultimately give us the sorted problems to do analysis of problem [2]. Analysis and classification of the problems will also help in the quality control of the product. We have taken defects classification as an example in this paper.

3   Future Work

Machine Learning [4] can explore new horizon for research in the field of Quality Control particularly in software projects. Although in this paper we have proposed the DTL techniques for the classification task other Machine Learning approaches such as Genetic Algorithms [1] can also be used for further optimization-of classification of the defects or problems. Neural Networks [9] can be used by means of supervised learning approach. Further work present opportunities for use of evolving neural network.

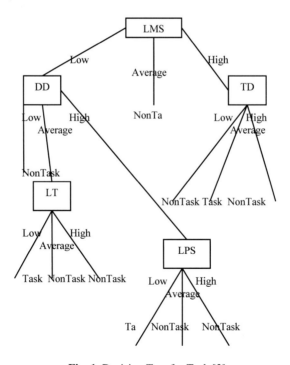

Fig. 1. Decision Tree for Task.[2]

References

1. Sam, H., James, M.: An Introduction to Genetic Algorithms
2. Mitchell, T.M.: Machine Ling, ch.3. The McGraw-Hill Companies, Inc.
3. Jalote, P.: An Integrated Approach to Software Engineering
4. Du, Z., Tsai, J.J.P.: Machine Learning and Software Engineering. In: Proceedings of the 14th International Conference on Tools with Artificial Intelligence. IEEE (2002)
5. Glass, R.L.: Persistent Software Errors. IEEE Transactions on Software Engineering SE-7(2), 162–168 (1981)
6. Robert Nelms, C.: The Problem with Root Cause Analysis. In: Joint 8th IEEE HP / 13th HPRCT, pp. 253–258. IEEE (2007)
7. Card, D.N.: Ling from our mistakes with defect causal analysis. IEEE Software, 56–63 (January - February 1998)
8. Software Engineering, a practitioner's approach by Roger S. Pressman. 6th edn. TATA McGraw Publication
9. Stanley, K.O., Miikkulainen, R.: Efficient Evolution of Neural Network Topologies. In: Proceedings of the 2002 Congress on Evolutionary Computation (CEC 2002), IEEE, Piscataway (2002)

Study on Safety Assessment of Air Traffic Management System Based on Bayesian Networks[*]

Fang Chen[1,2] and Yao Sun[1]

[1] College of Safety Science & Engineering, Civil Aviation University of China, Tianjin, China
[2] School of Engineering & Technology, China University of Geosciences (Beijing),
Beijing, China
swiftone@uymail.com

Abstract. An index system of evaluating safety for air traffic management system based on system engineering idea and principle , where human, facility, environment and management are considered as essential factors was established. A method for Air Traffic Management safety assessment using Bayesian network (BN) was constructed, based on its excellent abilities of expressing and reasoning knowledge under uncertain environment. The assessment model is able to deal with complicated logic relationship as well as the different opinions of experts. The output of the safety assessment model can be used to predict and diagnose different safety conditions. It was suggested that the safety assessment model is effective and reasonable by applying to safety assessment of Air Traffic Management units.

Keywords: Air Traffic Management, safety assessment, uncertainty, Bayesian network (BN).

1 Introduction

Air traffic management system plays an important role in contracting the ground systems, the using of airspace effectively, and ensuring flight safety. In the civil aviation accident statistics, air traffic management takes a very small proportion, but it is very important. The vast majority of accidents and incidents has a direct or indirect relationship with the air traffic management system [1]. Safety assessment for air traffic management system has great significance to further reduce the accident rate and promote air traffic management system safety and development. At present, the safety assessment is mainly used to monitor the air traffic management system safety, adopting the qualitative and semi-quantitative assessment method mostly. Assessment results emphasize on qualitative description of the system, and it is also lack in dealing with the uncertainty evaluation [2]. Because air traffic management system is a complex and huge system, we will be confronted with uncertain problems such as stochastic and fuzzy assessment factors and shortage of accident historical statistic data. This requires that experts judge system safety level based on their experience and

---

[*] Supported by the Fundamental Research Funds for the Central Universities (ZXH2011D001).

knowledge, and according to safety level of each factor in the subsystem of air traffic management system. Bayesian network (BN) model is an important method of studying uncertain problems, so Bayesian network is taken into account in safety assessment of air traffic management system to solve these problems. The final safety assessment result is used to rank the evaluated object. When multiple units are evaluated, the final safety assessment result also can be used to sequence them according to their comprehensive evaluation values .

2   The Index System of Evaluating Safety for Air Traffic Management System

Operation of air traffic management system involves a lot of human and facilities in different departments, and coordination of operation in different departments depends on systematical management. Because the structure of Chinese airspace is complex, the development between east and west is unbalance, facilities used for communication navigation and monitor differ greatly, diverse air traffic controlling modes, such as radar control(RC), radar monitoring procedure control, procedure control, coexist together. The air traffic management system is a dynamic system which is time-continuous, wide space, multi-elements and complex. In this paper, according to the principles of establishing index system: systemic, scientific, comparative, maneuverability and relative independence, a index system of evaluating safety for air traffic management system where human, facility, environment and management are considered as essential factors was established. According to Bayesian network, safety situation of the entire air traffic management system can be reasoned integrating by safety situation of 4 second-grade indexes, and safety situation of each second-grade index is integrated by safety situation of several third-grade indexes, shown in figure1. Safety situation of third-grade index is given by experts according to their experience and knowledge in general [3-6].

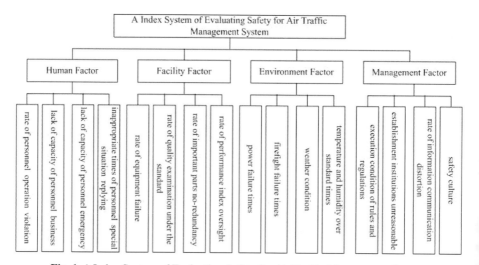

Fig. 1. A Index System of Evaluating Safety for Air Traffic Management System

3   A Bayesian Network Model of Evaluating Safety for Air Traffic Management System

3.1   Construction of Bayesian Network Topological Structure

It can be seen in the figure 1 that, upper index is derived by under layer index. For example, the safety for total air traffic management system index A is determined by 4 first indexes—human(B_1)、facility(B_2)、environment(B_3)and management(B_4). So, each index in each layer is parent node of corresponding upper index. This kind of connection is established by parent node directing to child node, that is, the second-grade index points to corresponding first-grade index, and aggregates to the total index finally. Because the marginal probability of bottom index is obtained by concentrating suggestions from several experts, diverging connection will be adopted to establish the model.

According to the analysis mentioned above, Use graphic decision theory model software GeNIe developed by Pittsburgh university to establish Bayesian network topological structure of evaluating safety for air traffic management system, shown in figure2.

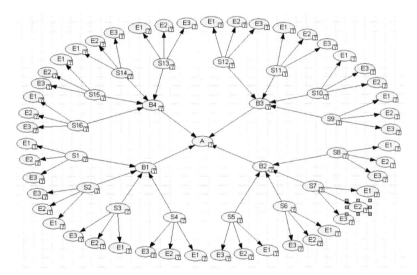

Fig. 2. Topological Structure of Evaluating Safety for Air Traffic Management System

3.2   Determination of Conditional Probability Table of Each Node

In this paper, probability P, which means accident occurs is considered as a target value to evaluate safety situation in air traffic management system. Conditional probability table (CPT) of each node in Bayesian network is determined by logical relation between different indexes each layer.

"Noisy-or" gate is adopted in Bayesian network when using human, facility, environment and management to measure safety for air traffic management system. That is, on the one hand, if one of these second-grade index happens, the total indexes probably happens; on the other hand, assume the probability of some certain index causing accident are higher, though the others are very low, the total index also probably happen. In addition, because out of control and unpredictable factors exist, though all 4 indexes are 0, the total index also has probably happen. "noisy-or with leak" gate and CPT in Bayesian network are used to express in this condition. CPT of all non-root nodes （AandB$_1$～B$_4$ in figure2） can be determined with this method.

Meanwhile, marginal probabilities of root nodes (index S$_1$-S$_{16}$) are obtained in the same way that initial value is installed and then updated. Initial value is installed free, because it is updated value not initial value needed to pay attention to. For example, suppose that these indexes enable accident occurrence probability in ATM system and nonoccurrence probability of accident in ATM equal, and both of their initial value are 0.5.

3.3   Concentrate Experts' Views to Update the Node's Marginal Probability

Many index value assignments depend on experience and opinion of experts, when evaluating safety for air traffic management system. Because different experts own different criteria, considering in different ways, there may be difference in the evaluating result of the similar issue. It is helpful to get more objective evaluation value by integrating opinions from several experts. Calculating the marginal probability of nodes by BN model can make it reasonable to comprehensively integrate different experts' opinions.

When evaluating ATM system, when a certain index Si (influence factor) is in condition "e", probability P of accident occurrence in ATM system （Si=Yes） is needed to considered. In order to get the probability value , it needs concentrate different experts' suggestions and construct Bayesian network model, then updates P （Si = Yes） of each i according to the criteria.

Take a root node of the air traffic management system Bayesian network model as an example to illustrate how to update the marginal probability.

Index S$_{15}$ was evaluated by n experts, and this index was different in different experts' view. Index S$_{15}$ is a parent node, E$_1$,E$_2$,......, E$_n$ are child nodes which represent n experts' evaluation views. The directed arcs between nodes are from the index S$_{15}$ to 3 experts' evaluations.

Then, determine the marginal probability of the root node S$_{15}$ and the conditional probability of each child node. Each variable in the BBN model has a finite set of mutually exclusive states: The variable S$_{15}$ represents the situation that the accident falls in "Yes" or "No". We can assume that P(S$_{15}$=Yes)=P(S$_{15}$=No)=1/2. It is not easy for the experts to evaluate P(E$_{ij}$|S$_{15}$), because variable E$_{ij}$ has several states, such as e$_{11}$, e$_{12}$, ..., e$_{1m}$ (that is expert i 's evaluation criteria for index S$_{15}$), and experts need to evaluate several conditional probability and normalize them at the same time. Because variable S$_{15}$ only has two states, it is easier to determine P(S$_{15}$|E$_{ij}$) than P(E$_{ij}$|S$_{15}$). Consequently, experts evaluate P(S$_{15}$|E$_{ij}$) first, and then to calculate P(E$_{ij}$|S$_{15}$) according to Bayesian theorem.

According to Bayesian theorem:

$$P(E_{ij}|S_{15}) = \frac{P(S_{15}|E_{ij})P(E_{ij})}{P(S_{15})} = \frac{P(S_{15}|E_{ij})P(E_{ij})}{\sum_{j=1}^{m}P(S_{15}|E_{ij}=e_{ij})P(E_{ij}=e_{ij})} \tag{1}$$

We can assume that P(E$_{i1}$=e$_{i1}$)=,......= P(E$_{im}$=e$_{im}$)=1/m, because no information about the variable E$_{ij}$ is provided yet. Then, P(E$_{ij}$|S$_{15}$) can be calculated using equation (1) with P(E$_{ij}$) and P(S$_{15}$|E$_{ij}$). The constructions of the BBN models are completed here [7-10].

3.4   Reason by BN Model and Determine the Safety Evaluation Criterion

After the Bayesian network models are completed, reason and analyze. The information of any nodes in the Bayesian network model can be conveyed to its non "d-separation" node, and it also can be updated using the evidences judged by experts. Consequently, as the more evidences judged by the experts are added into the Bayesian network model, the more reasonable results are obtained.

We can obtain the accident probability in ATM system by reasoning through the Bayesian network model, and evaluate the safety level according to the probability. The following is the evaluating criteria of ATM system: if the probability value is less than 0.3, the safety level of the ATM system is good; if the probability value is between 0.3 and 0.5, the level is average; if the probability value is greater than 0.7, the level is poor.

4   Safety Evaluation Model Based on BN 's Application in a ATM Unit

For a known influence factors in air traffic management unit , that is to say , we have known the marginal probability of root nodes in the evaluation model, the accident probability $P(A = Yes)$ can be calculated by reasoning . The marginal probability of root nodes is determined by concentrating different experts' views through Bayesian network, such as different experts' evaluation for the influence factors in ATM units, shown in Table 1[11,12] .

Table 1. Experts' evaluation criteria for the influence factor

Influence factor S_i	expert 1		expert 2		expert 3	
	Influence factor's value	Probability of causing accident	Influence factor's value	Probability of causing accident	Influence factor's value	Probability of causing accident
Time of human faulty in dealing with emergency S_4	1)0	0.04	1)0	0.05	1)0	0.03
	2)1 and above	0.5	2)1 and above	0.4	2)1 and above	0.6
Failure rate of equipments S_5	1)0%	0.1	1)0%	0.05	1)0%	0.08
	2)0%-1%	0.2	2)0%-2%	0.28	2)0%-3%	0.3
	3)> 1%	0.25	3)> 2%	0.32	3)> 3%	0.34
Rate of oversight on performance index S_8	1)< 4%	0.1	1)< 3%	0.1	1)< 5%	0.2
	2)4%-8%	0.2	2)3%-10%	0.3	2)5%-11%	0.3
	3)8%-12%	0.3	3)10%-14%	0.4	3)> 11%	0.5
	4)> 12%	0.5	4)> 14%	0.6		
Distortion rate of information communication S_{15}	1)< 3%	0.2	1)< 4%	0.2	1)< 4%	0.1
	2)3%-6%	0.3	2)4%-7%	0.3	2)4%-8%	0.3
	3)6%-9%	0.4	3)7%-10%	0.4	3)> 8%	0.5
	4)> 9%	0.6	4)> 10%	0.5		
Safety culture S_{16}	1)good	0.1	1)good	0.1	1)good	0.2
	2)average	0.3	2)average	0.2	2)average	0.4
	3)poor	0.5	3)poor	0.4	3)poor	0.7

Taking S_{15} as an example, expert 1 suggests that 4 criteria were involved in the relationship between distortion rate of human information communication and causing accident in ATM unit.

1) when the distortion rate of information communication is greater than 9%, the accident probability in ATM unit is 0.6, that is $P(S_{15} \mid E_{11})=0.6$.

2) when the distortion rate of information communication is lower than 9%, the accident probability in ATM unit is 0.4, that is $P(S_{15} \mid E_{12})=0.4$.

3) when the distortion rate of information communication is greater than 3% and lower than 6%, the accident probability in ATM unit is 0.3, that is $P(S_{15} \mid E_{13})=0.3$.

4) when the distortion rate of information communication is lower than 3%, the accident probability in ATM unit is 0.2, that is $P(S_{15} \mid E_{14})=0.2$.

Now, the variables E_1 has 4 states: e_{11}, e_{12}, e_{13} and e_{14}, which respectively represent the degree of the ATM unit meeting the evaluation criteria judged by expert 1. Calculate the probability of index S_{15} 's accordance with criterion on condition that accident occurs in ATM unit, $P(E_{11} \mid S_{15}=Yes)=0.4000$, $P(E_{12} \mid S_{15}=Yes)=0.2667$, $P(E_{13} \mid S_{15}=Yes)=0.2000$, $P(E_{14} \mid S_{15}=Yes)=0.1333$.

Meanwhile, on the condition that accident don't occur in ATM unit, the probability of index S_{15} 's accordance with criterion is: $P(E_{11} \mid S_{15}=No)=0.1600$, $P(E_{12} \mid S_{15}=No)=0.2400$, $P(E_{13} \mid S_{15}=No)=0.2800$, $P(E_{14} \mid S_{15}=No)=0.3200$.

Expert 2 suggests that 4 criteria were involved in the relationship between distortion rate of human information communication and causing accident in ATM unit.

1) when the distortion rate of information communication is greater than 10%, the accident probability in ATM unit is 0.5, that is $P(S_{15} \mid E_{21})=0.5$.

2)when the distortion rate of information communication is less than 7%, the accident probability in ATM unit is 0.4, that is $P(S_{15} \mid E_{22})=0.4$.

3) when the distortion rate of information communication is greater than 4% and less than 7%, the accident probability in ATM unit is 0.3, that is $P(S_{15} \mid E_{23})=0.3$.

4) when the distortion rate of information communication is less than 4%, the accident probability in ATM unit is 0.2, that is $P(S_{15} \mid E_{24})=0.2$.

Now, $m=4$, according to the 4 criteria, the variables E_2 has 4 states: e_{21}, e_{22}, e_{23} and e_{24}, which respectively represent the degree of the ATM unit meeting evaluation criteria judged by expert 2. Calculate the probability of index S_{15}'s accordance with criterion on condition that accident occurs in ATM unit, $P(E_{21} \mid S_{15}=Yes)=0.3571$, $P(E_{22} \mid S_{15}=Yes)=0.2857$, $P(E_{23} \mid S_{15}=Yes)=0.2143$, $P(E_{24} \mid S_{15}=Yes)=0.1429$.

Meanwhile, on condition that accident don't occur in ATM unit, the probability of index S_{15}'s accordance with criterion is : $P(E_{21} \mid S_{15}=No)=0.1923$, $P(E_{22} \mid S_{15}=No)=0.2308$, $P(E_{23} \mid S_{15}=No)=0.2692$, $P(E_{24} \mid S_{15}=No)=0.3077$.

Expert 3 suggests that 3 criteria were involved in the relationship between distortion rate of human information communication and causing accident in ATM unit.

1) when the distortion rate of information communication is greater than 8%, the accident probability in ATM unit is 0.5, that is $P(S_{15} \mid E_{31})=0.5$.

2) when the distortion rate of information communication is greater than 4% and less than 8%, the accident probability in ATM unit is 0.3, that is $P(S_{15} \mid E_{32})=0.3$.

3) when the distortion rate of information communication is less than 4%, the accident probability in ATM unit is 0.2, that is $P(S_{15} \mid E_{33})=0.1$.

Now, m=3, according to the 3 criteria, the variables E_2 has 3 states: e_{31}, e_{32} and e_{33}, which respectively represent the degree of the ATM unit meeting evaluation criteria judged by expert 3. Calculate the probability of index S_{15}'s accordance with criterion on condition that accident occurs in ATM unit, $P(E_{31} \mid S_{15}=Yes)=0.5556$, $P(E_{32} \mid S_{15}=Yes)=0.3333$, $P(E_{33} \mid S_{15}=Yes)=0.1111$.

Meanwhile, on condition that accident don't occur in ATM unit, the probability of index S_{15}'s accordance with criteria is : $P(E_{31} \mid S_{15}=No)=0.2381$, $P(E_{32} \mid S_{15}=No)=0.3333$, $P(E_{33} \mid S_{15}=No)=0.4286$.

Reasoning and computing by software GeNIe , the Distortion rate of human information communication in ATM unit is 3.5%. According to the Bayesian network model, the criteria of E_1 is in accord with the principle "e_{13}", and, the criteria of E_2 is in accord with the principle "e_{23}", and the criteria of E_3 is in accord with the principle "e_{32}". $P(S_{15}=Yes)$ is updated by reasoning through the three criteria, and the updated $P(S_{15}=Yes)=0.0792$. This suggests that, in ATM unit , the probability of accident is 0.0792 according to the index S_{15}—"Distortion rate of human information communication".

Table 2. Marginal probability by assembling experts' view based on BN

Influence factor	Real value	$P(S_i=Yes \mid E_{ij})$
Time of human faulty in dealing with emergency S_4	0	$P(S_1=Yes \mid e_{11},e_{21},e_{31})=0.0014$
Failure rate of equipments S_5	0%	$P(S_6=Yes \mid e_{11},e_{21},e_{31})=0.0253$
Rate of oversight on performance index S_8	2%	$P(S_7=Yes \mid e_{11},e_{21},e_{31})=0.0129$
Distortion rate of information communication S_{15}	3.5%	$P(S_{15}=Yes \mid e_{13},e_{23},e_{32})=0.0792$
Safety culture S_{16}	good	$P(S_{16}=Yes \mid e_{11},e_{21},e_{31})=0.030$

We can also obtain the marginal probability of the other index in ATM units. The updated marginal probabilities are shown in Table 2. Lastly, $P(A=Yes)=0.1858$ is obtained, by reasoning and computing. According to the criteria referred in section 2.4, the safety grade of this ATM unit is LEVEL 1.

5   Conclusions

The index system of evaluating safety based on Bayesian network , where human, facility, environment and management in air traffic management system are considered, is able to efficiently synthesize different experts' views, and it can express the uncertainty between index variables better, making evaluating model more reasonable.

The evaluation outputs by comprehensively reasoning can be used to describe the safety situation in the ATM system, and also can be used to make a horizontal comparison between ATM units. This method is of good reference value to the reality work because of its practicability and operability.

References

1. He-ping, S.H.I.: The air traffic system safety management, vol. 8. Xiamen University Press, Xiamen (2003)
2. Safety Office of Civil Aviation Administration of China, Traffic Management Bureau of CAAC. Air Traffic Services Safety Assessment System(1): Principle, structure and assessment projects, Beijing: Safety Office of Civil Aviation Administration of China, Traffic Management Bureau of CAAC, pp. 21-23, 37-41 (2000)
3. Ding, S.-B.: Weights and Safety Forewarning Criteria of Air Traffic Management System. Journal of Civil Aviation University of China 23(4), 50–54 (2005)
4. Zhang, Y.: Study on Risk Quantitative Assessment Model for Civil Aviation Safety. China Safety Science Journal 17(9), 140–145 (2007)
5. Gui-juan, J.I.A.: Study on the System and Comprehensive Evaluation of Safety Risk Assessment Indexes for the ATM. Wuhan University of Technology, Wuhan (2008)
6. Zeng, L.: Application of Multi-layer Fuzzy Evaluation Method to Risk Assessment in Civil Aviation. China Safety Science Journal 4(1), 131–235 (2008)
7. Yin, X.-W., Qian, W.-X., Xie, L.-Y.: Application of Bayesian Network to Reliability Assessment of Mechanical Systems. Journal of Northeastern University(Natural Science) 29(4), 557–560 (2008)
8. Zheng, H., Wu, Q.-Z., Wang, P.-L., Shi, A.-F.: Application of Bayesian Networks to Safety Assessment in Pyrotechnics Systems. Acta Armamentarii 27(6), 988–993 (2006)
9. Liu, W.-S., Zeng, F.-Z.: Application of Bayesian Networks in Safety Assessment of Coal Mine Production Systems. Industry and Mine Automation (1), 1–4 (2008)
10. Zhang, L.-W., Guo, H.-P.: Introduction to Bayesian Network. Science Press, Beijing (2006)
11. Wan, J.: Analysis Causation for Disasters and Constructing Monitoring Criteria for ATM. Journal of Civil Aviation University of China 25(5), 1–3 (2007)
12. Li, K., Zhang, S.: Management Information System of Safety Early Warning for an Air Traffic Control Center. Journal of Wuhan University of Technology(Information & Management Engineering) 29(8), 106–109 (2007)

Onboard Taxi Fee Register with GPS and GPRS

Ye Chunqing and Miao Changyun

School of Information and Communication Engineering, Tianjin Polytechnic University,
Tianjin, China
chuto6@126.com

Abstract. This paper proposes an onboard taxi fee register with GPS/ GPRS and designs a kind of onboard taxi fee resister system. This system is mainly comprised by LPC2114 processor, metering components, empty plates, SD cards, real-time clock, printer, keyboard, GPS systems, GPRS wireless data transmission modules, display screen, etc. It realizes the record and storage of stations for passengers to get on or off (longitude and latitude) and vehicle's track, GPS accurate time transmission, regulation of unit prices according to sections and time, display and print relative data and conduct intercommunication of information with service center by GPRS. Compared to existing technology, it prevents cheating of taxi fee register, moreover improves the efficiency, decreases the management costs, and has the advantages of high accuracy and convenience, etc.

Keywords: Onboard taxi fee register, LPC2114 processor, GPS positioning system, GPRS wireless data transmission module.

1   Systematic Plan Design

With the positive functions in releasing urban transportation press and convenience of trips, taxi is the essential transportation tool for modern cities. Taxi trade is a city's foreign window trade, which may be the first trade for foreign tourists to contact. To improve the service quality and enhance cities' images, the design plan of taxi fee register with GPS and GPRS is proposed. The general design structure is shown as figure 1. Centered on onboard fee register (terminal), by GPS and GPRS technology, it can record the stations for passengers to get on or off (longitude and latitude) and vehicle's track, to provide basis in fact for detour complaints and lost and found, so as to improve efficiency of supervision departments. Besides, it can also prevent cheating of fee register, public order offence and other major criminal cases. In addition, this system has information service functions such as sending information like news, weather broadcast, road condition, hospital, garage, etc to taxi fee registers (terminals) from service center or returning feedbacks of its states to service center.

2   System Hardware Design Plan

The system hardware block diagram is shown as figure 2. The onboard fee register (terminal), with the core of LPC2114 processor, is mainly comprised of power

Y. Wu (Ed.): International Conference on WTCS 2009, AISC 116, pp. 737–743.

management module, display circuit, real-time clock circuit, GPS module, GPRS module, SD circuit, keying circuit, hall sensor circuit, load switch, buzzer and micro-printed circuit, etc.

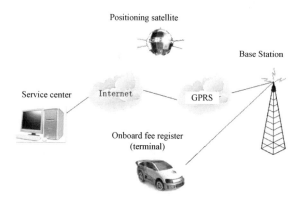

Fig. 1. General design plan structure chart

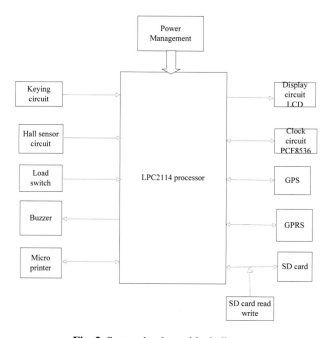

Fig. 2. System hardware block diagram

2.1   LPC2114 Processor

LPC2114 is 16/32 bit ARM7TDMI-STM micro controller based on real-time simulation and tracing, with 128/256KB embedded FLASH memory. Its internal integrates four 10bit A/D converters, two 32bit timers, a real-time clock and watchdog. Multi serial interfaces include UART in two industrial standards, high-speed and two SPI bus interfaces. There are 46 TTL level compatible universal I/O ports. Due to its simplicity, easy development and high performance-price ratio, it is applicable to MCU.

2.2   Power Management

Due to its high consumption of LPC2114 ARM7 microprocessor, to realize low consumption, the system uses time arousal function and F1121 in MSP430 series with ultra-low power as system power management CPU.

2.3   Hall Sensor Circuit

This paper uses Hall integrated switch sensor CS839 based on Hall effect to transform speed signal per revolution into electrical signal. But due to certain spike jamming of keying signals in the process of opening and closing, in addition to the bad functional environment for vehicles, the keying signals may have much high-frequency interference components. So, select conditioning circuit with combination of photoelectric coupling isolation and filtering circuit, shown as figure 3. The input circuits are divided from earth wires of ARM system, totally separated between input signals and main engine recorder, and signals are transmitted in the photo forms by photoelectric coupler. This paper uses TLP521 photoelectric coupler of Toshiba. Only if after 2~5 us' rated current in LED, transistors of output ports can be conducted which makes it difficult to interfere photoelectric coupler even when the noise's effect pulse width less than 1 us. Add RC filtering circuit in output port of photoelectric coupler to restraining high-frequency interferences.

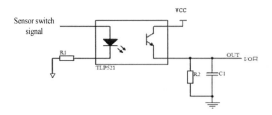

Fig. 3. Signal conditioning circuit

2.4   Load Switch Circuit

Design a load switch, which reads low level '0' when off and high level '0' when on. Judge different states of taxi by that, shown as table 1. The taxi is empty when reads '0' twice, loaded when reads '1' twice, transforming from empty to loaded when reads '0' first and '1' second, and others is transforming from loaded to empty.

Table 1. Keying state's judgment

State of fee register	Judge keying state first time	Judge keying state second time
Empty vehicle	0	0
Starting to figure	0	1
Loaded vehicle	1	1
Stopping to figure	1	0

2.5   Micro Printer

With thermo WH153SA, micro printer's principle is to record white temperature sensitive of printer heads after heating, then under the control of control circuits, the heating components on the printer head become hot instantly, which conduct to thermal record paper quickly, so as to generate words or patterns by changing record paper's colors.

There are general two kinds of connection between micro printers and computers, one is parallel port, and the other is serial port. The advantage of parallel port is high speed, but need port sources; that of serial connection is to save ports but with slow speed. This paper uses thermal printers with serial connection. We can directly call characters according to their addresses because they are included in partial micro printers.

2.6   Display Circuit

It is impossible to see data clearly with the distance more than 1m from screen by LCD, moreover the contrast on the day cannot satisfy the demand, so we use LED and RT1286M_S LCD module which can display Chinese characteristics and images. They are used to display unit price, distance, total price, speed, real time, set unit price information, GPS information and information from service center, etc. RT12864M_S is embedded by LCD control chip. The connection circuit between LCD and LPC2114 is shown as figure 4.

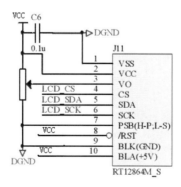

Fig. 4. LCD circuit

2.7   Keying and Buzzer Circuit

To save resources, we use HD7279 and external components to construct keying and LED numeral tube circuit. Its connection circuit with LPC2114 is shown as figure 5.

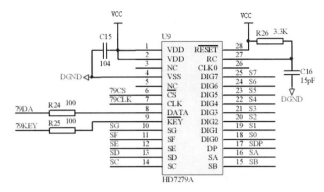

Fig. 5. The connection circuit between HD7279 and LPC2114

The purpose of buzzer unit is to send beeping when conduct error operation. Its connection circuit is shown as figure 6.

2.8   Real-Time Clock Circuit

This paper uses low-consumption CMOS real-time clock/ calendar chip PCF8563 to provide information such as year, month, day, second, minute and hour, etc. All address and data are transmitted through I2C bus interfaces with power failure detector. When the voltage is less than this threshold, some flag of second register will be set as 1, which represents later real-time clock may generate incorrect clock/ calendar information. While providing common voltage, clock chip is also charging for backup battery which provides common voltage as main power supply device once power fails. The connection circuit with LPC2114 is shown as figure 6.

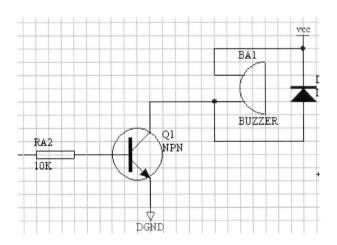

Fig. 6. The connection circuit between buzzer and LPC2114 interfaces

2.9   SD Card

As a storage carrier, SD card is used to record information in the process. SD has large capacity, small volume, light weight and simple connection circuit, etc. Its recorded data files can be recognized by PC, which are analyzed and replayed by analysis software on PC, as historic data to compare data or analyze cases.

2.10   GPS and GPRS Interfaces

GPS and GPRS connect LPC2114 by serial ports. GPS is applied to positioning fee register. If display distance is not fit to real distance, the fee register will warn and display warnings on the LCD. Meanwhile, it can also conduct accurate time transmission on fee register. If display time is not fit to current time, the system can automatically modify by GPS accurate time transmission.

GPRS module realizes intercommunication with service center by GPRS network. Service center sends some real-time service information to onboard fee register (terminal) such as current road condition. Onboard fee register (terminal) feedbacks GPS positioning information to service center. It makes management and regulatory more convenient.

3   System Software Design Plan

This system software includes program modules as follows:

Master program: hardware modules included by initial system, scan keying, start/stop distance measurement according to keying, and regularly updated date-time display by calling user interface program, real-time updated unit price, distance, low-speed time, fees, etc.

> LCD drive program: LCD drive program to realize text, image display, etc.
> Keying scanning program: scan keying and return scanning results
> Write GPS time transmission program
> Drive program for SD card
> Drive program for micro printer
> Drive of I2C bus

Pricing management program: finish sensor pulse meters, and transform them into miles. Calculate corresponding fees according to rules. Meanwhile, monitor driving speed, make statistics for low-speed time, calculate low-speed waiting fees and provide LCD updated function in order to enable main function to control display of updated user interface, distance and fee, etc. The module includes PWM management program, distance counter program, time management program, fee management program and user interface control program, etc.

References

1. Xiao, P., Chen, W.: A new type of multi-function taximeter. Automation and Instrumentation (3) (2009)
2. Sun, X.: 16 Bit Single-Chip Computer Application in Taximeter. Metrology & Measurement Technique (1) (2009)

Dynamic Analysis and Simulation of Arresting Wires

Xiao-Yu Sun, Zhen-Qing Wang, Zeng-Jie Yang, and Yu-Long Wang

College of Aerospace and Civil Engineering, Harbin Engineering University, Harbin, China
sunbo1789@sina.cn

Abstract. As the research object to arresting wires, it is first analysed that the landing(on or off center) model of airplane and the dynamics model of arresting wires under impact loading. Then, the motion equation of arresting wires with displacement vector was deduced. Through the type of aircraft simulated, the interface of software with VS. net was written and the simulation model by the Xtreme programming and the secondary development of ANSYS was established to accomplish the simulation compute and drawing under this environment or simulate the motion and the deformation in and out plane of arresting wires with ANSYS or demonstrate the force process of arresting wires under impact loading or providethe help with the graph of stress and strain for designing, the physical quantity curve that was obtained by the Xtreme programming provided reference.

Keywords: Aerodynamic computation, multibody system, object-oriented, computer-based instruction.

1 Introduction

Over the years, A design approach using features of structural desciplines, dynamic desciplines, flight control systems and simplified propulsion systems as required in the preliminary design of a modem aircraft is described. This method employs very simple structural and dynamic models to compute performance characteristics that can rapidly be assessed and incorporated into the preliminary design process to evolve an optimal aircraft. Programs for scientific engineering have emphasized the use of the C++ program language. The programe [1] [2]contain numerous complex data structures, which can be used anywhere in the program. This inflexibility is shown in several methods: it becomes difficult to maintain the code and even more difficult to implement new models or new solving procedures; it is necessary to have an understanding of the design process, even to work on a portion of the program; many interdependencies can be hidden and difficult to detect; a few changes in data structures may give rise to unpredictable effects in the code. As the recoding of these finite element programs in a new language is not a solution to this inflexibility problem, a redesign is needed.

The main advantage of object-oriented software[3] and the modeling is its ability to describe effectively complex systems and transfers the model to an object-oriented

Y. Wu (Ed.): International Conference on WTCS 2009, AISC 116, pp. 745–753.

software without major changes. Object-oriented programming is based on the nonlinear longitudinal aircraft equations of motion and is currently seen as the most promising way of designing a new application. Fig. 1 shows the configuration of the software. Thus, the gap between the created model and the computer implementation for disturbance estimation and detection of engine failure for a short-take off and landing(STOL)aircraft during the landing maneuver and is very small during the whole modeling process, allowing very fast design and modifications to achieve convergence. This capability allows for simplification and acceleration of engineering model development, which is otherwise done over and over again for slightly changed or extended problems involving a high portion of lowlevel repetitive work. In the application considered here, routines for 2D and 3D graphics, The advantages of object-oriented design, modeling and implementation can be summarized as follows. Error in the modeling can lead to biases in the estimates and even divergence from the actual values. However, little effort has been made to implement object-oriented programming in multibody systems analysis. As a matter of fact, like many other engineering applications, multibody systems analysis codesare written in Fortran. Therefore, the main objective of this paper is to describe one approach to the design and implementation of a multibody systems analysis code using an object-oriented design.

2   Programming and Multibody Systems

Program languages like C++ considered datas as passive and memory occupying elements which can be controled only by functions and procedures. In this work an uncertainty quantification procedure and a design method are applied to the thermal design of a skin panel and a wing box section. The uncertainty quantification method selected is based on Monte Carlo simulations and a set of common design parameters have been considered and formulated with random values. Aircraft encounters with force are considered on the final approach, during which a decision is made to abort the landing. This produces things that exchange altitude for airspeed in a manner dependent on microburst strength, similar to previously obtained trajectory optimization results. In contrast, the object-oriented design comes from the idea that tools[4] [5] must be associated with the information they manage.

Several researchers have worked with the object-oriented software to the finite element method [6]in recent years: constitutive law modelling, parallel finite element applications, non-linear analysis, finite element analysis programing, maintaining a low dependence on input value variations and producing designs that increase the quality level.

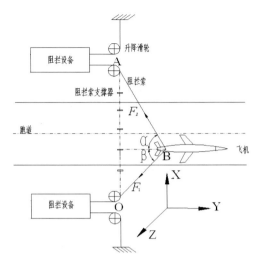

Fig. 1. Configuration of the landing model of carrier aircraft

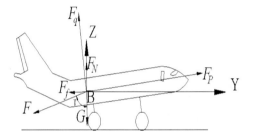

Fig. 2. The force analysis of carrier aircraft

Little effort has been made to apply object-oriented software in complex systems analysis. Uncertainty quantification(UQ)involves the propagation of probabilistic information from the input parameters to the response functions of the system. The key concepts of object-oriented programming can be found in many computings and user guides. Thus we will only present briefly those concepts, and illustrate them in the context of multibody systems. The structure of a multibody system can be thought of as consisting of four basic objects: bodies, constraints between bodies, loads and motions.

It can be seen that for all random parameters, the numerical significance of randomness decreases from input to output. In other words, in this component, the level of uncertainty is attenuated due to the nature of the problem. A subsystem is usually of a special type and has additional methods and attributes members relating to it. One of the most important characteristic of object-oriented design is also the possibility of defining abstract objects using virtual member functions. This characteristic enables the same function to respond differently when performed on objects from different classes. One can illustrate the concept for multibody systems by considering an arbitrary constraint connecting two(or more)arbitrary bodies. The implementation details of the specific bodies and joints involved in the connection are hidden by abstraction in the joint and body objects. Computings were performed for each multiple random parameter process, and from those results, the statistic properties of nodal temperature at panel centre were characterised. The original work presented in this paper is aimed at describing the object-oriented architecture of a multibody systems analysis code, which is able to integrate flexible and rigid bodies in a unified framework, and to deal with both open and closed systems in a systematic way, without the need of an artificial cut of one of the joints. Moreover, The method, robustness of a design is determined by the value of a function defining the quality loss of the design and the best design is the one having the lowest quality loss. . The computational model of the software in Fig. 2 consists of an collection of rigid and flexible bodies, connected together by a variety of joints, loaded by concentrated or distributed forces and moments, that can be subjected to specified or constrained motions.

Finally, we want to emphasize on the fact that our software was not designed for commercial applications. In these circumstances, the speed and the loss of numerical performance introduced by object-oriented design is greatly compensated by the advantages it procures. In this paper, we will only focus on the computational engine of the software.

The CSL is used in the next section for the description of the relationships between the symbols. Details concerning the set of symbols and rules of this standardized modelling language, which appears to be the technique most commonly used today, can be found.

3   Model Analysis

The dynamics of an aircraft [7-14]consists of rigid body modes and vibration modes. The frequency spectrum of these modes are generally well separated, and hence, the rigid body and elastic degrees of freedom can be solved. Thus, the rigid body equations of motion of a maneuvering aircraft in the wind axis system can be written below:

$$m\frac{d^2y}{dt^2} = F_{py} - F_f - F_{qy} - F_y \tag{1}$$

$$m\frac{d^2z}{dt^2} = F_{pz} + F_N + F_{qz} - G - F_z \tag{2}$$

$$F_{qy} = \frac{1}{2}\rho v_s^2 SC_y \tag{3}$$

$$F_{qz} = \frac{1}{2}\rho v_s^2 SC_z \tag{4}$$

$$R_3 = R_2 + R_{23} \tag{5}$$

$$\varepsilon_3 = \left(dS_{u_{3i}} - dS_{u_{1i}}\right)/dS_{u_{1i}} \tag{6}$$

$$\frac{dS_{u_{3i}}}{dS_{u_{1i}}} = 1 + \varepsilon_3 \tag{7}$$

$$\left(1+\varepsilon_3\right)^2 = \sum_{i=1}^{3}\left(\partial p_{3i}/\partial p_1\right)^2 \tag{8}$$

$$\varepsilon_3 = \frac{1}{2}\left[\sum_{i=1}^{3}\left(\partial p_{3i}/\partial p_1\right)^2 - 1\right] \tag{9}$$

$$\varepsilon_3 = \frac{1}{2}\left(\frac{\partial R_3}{\partial p_1}\frac{\partial R_3}{\partial p_1} - 1\right) \tag{10}$$

$$E_{eq} = \frac{E}{1 + \dfrac{\left(\mu L\right)^2 E}{12\sigma_0^3}} \tag{11}$$

where M is the mass of the aircraft, I_{vy} is the pitching moment of inertia, V is the aircraft velocity along the flight path. The nonlinear aerodynamic forces as computed from the CFD code is given by the first column on the right-hand side of Eq. (1). The body forces and thrust components are given by the second column. The climb angle is denoted by y and the pitch angle by 0. The aerodynamic angle of attack is given by

$$\alpha = \theta - \gamma \tag{12}$$

The results (aircraft lands on off center of the deck)are shown in Fig. 3 - Fig. 9

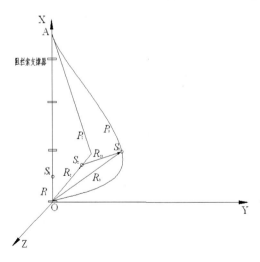

Fig. 3. The displacement change of S3

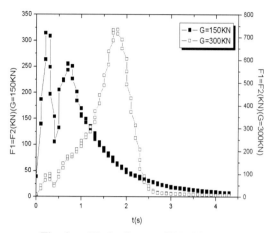

Fig. 4. v=75m/s, Curves of F1= F2 vs t

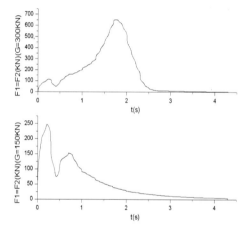

Fig. 5. v=60m/s, Curves of F1= F2 vs t

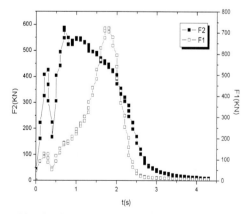

Fig. 6. Off center 10m, Curves of F1、 F2 vs t

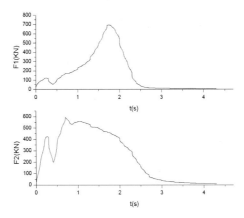

Fig. 7. Off center 5m, Curves of F1、F2 vs t

4   Conclusions

In this work, the uncertainty quantification and design of aircraft components was carried out. Considerations about the performance, efficiency and applicability of the methods have been done. The design is a valuable framework for the development of special, multidisciplinary elements, such as those required to successfully simulate the arresting system. The evolutivity is probably the most important characteristic for any calculation software, since the software's durability directly depends on it. The opportunities brought by the inheritance and the encapsulation concepts of object-oriented programming are the keys to evolutivity for a calculation software, as they allow the introduction of new functionality from the derivation of classes constituting the heart of the software's architecture. The ability of a multibody analysis to exactly model the nonlinear kinematics of complex aircraft landing on the deck.

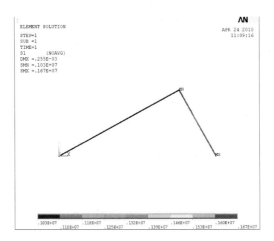

Fig. 8. The stress distribution of balance(off center)

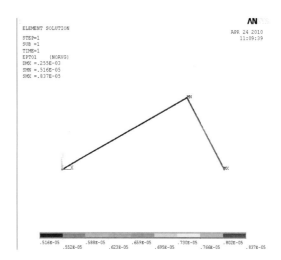

Fig. 9. The strain distribution of balance(off center)

Many processes with single random input parameters have been solved. In all of them, the level of uncertainty at the output decreases for each component of the whole set of parameters selected. This paper points out the difficulties encountered in the definition of an adequate dynamic model for the simulation of landing. A finite element calculation software may advance in different directions. The method is suitable to obtain designs in problems and proceeds by carrying out a discrete search in the range of variation of noise factors and control factors and so the cost of development is reduced, maintaining a high quality level. It allows to explore the system behaviour in a wide range of operational conditions, also in terms of aircraft and in terms of runway surface characteristics. This results are practical. because an exact dynamic model is always available for system design.

Acknowledgements. This work has been funded by fundamental research funds for the central university, HEUCFZ1004 and the International Exchange Program of Harbin Engineering University For Innovation-oriented Talents Cultivation. Their support is gratefully acknowledged.

References

1. Ringleb, F.O.: Cable dynamics. NAEF-ENG-6169, U. S. : Naval Air Engineering Facility Engineering Department (1956)
2. Leask, G.M.: Development of a Matlunatical Performance Prediction Model for Rotary-Hydraulic-Type Arresting Gears, AD893157 (1972)
3. Gao, Z.: A discussion of bounce kinematics of aircraft arresting hook and cable dynamics. Acta Aeronauticaet Astronautica Sinica 11(12), 543–548 (1990)
4. Zhang, X., Li, Y., Liu, Y.: Arresting Hook And Cable Dynamics of Aircraft Arrest Landing on or off Center[J]. Journal of Mechanical Strength 30(4), 549–554 (2008)
5. Xv, M., Ding, S.: Flight Dynamics, pp. 179–188. Science Press, Beijing (2003)

6. Taylor, J.: Manual on aircraft loads, pp. 100–200. National Defence Industry Press, Beijing (1974)
7. Tan, C.: Finite Element Analysis, pp. 69–150. SWJUP, Sichuan (2006)
8. Richard, D.: Effective Visual Studio. NET, pp. 100–300. Tsinghua University Press, Beijing (2002)
9. Brunell, B.: Model Analysis Example Projects—AXT Online Help, pp. 18–40. AEgis Technologies'company, American (1997)
10. Wang, F., Zhang, C.: ANSYS10. 0 Finite Element Theory, pp. 50–100. Publishing House Of Electronics Industry, Beijing (2006)
11. Naval Air Systems Command Department of The Navy, MIL-STD-2066, Catapulting andArresting Gear Forcing Functions for Aireraft Struetural Design, 25-170 (1981)
12. Naval air warfare center aircraft division patuxent river, Review of the carrier approach criteria for carrier-based aircraft –phase I, NAWCADPAX/TR-2002/71, 30-120 (2002)
13. Manual on aircraft design Editorial Committee, Manual on aircraft design-Nine. Aviation Industry Press, Beijing, 30-100 (2002)
14. Manual on aircraft design Editorial Committee, Manual on aircraft design-Fourteen. Aviation Industry Press, Beijing, 23-94 (2002)

The Teaching Values of New Physical Curriculum

Yan Sun[1], H.-X. Zhang[2], and Y.-F. Sun[3]

[1] Department of Physical Education, Central China Normal University, Wuhan, Hubei, China
leesoso@email.com
[2] Department of Military Body Art, Zhejiang Jiaxing Vocational and Technical College,
Jiaxing, Zhejiang, China
[3] Department of Physical Education, Zhengzhou University of Light Industry, Zhengzhou,
Henan, China

Abstract. This article is based on pointed out the necessity to enhance the research about PE teaching values, as well as the dynamic and diversity subject-object values in current physical curriculum, explained representation of the teaching values of new physical curriculum, and put forward the construction of PE teaching values build path.

Keywords: Physical curriculum, teaching, values.

1 Introduction

As everyone knows, teaching is the main task of schools, teaching is the core of whole education, it produces the most comprehensive and profound education functions, and occupied most of th time. Therefore, to develop the research of school education should focus on how to epitome and surround teaching. Undoubtedly, PE teaching is indispensable in school teaching, it plays a pivotal role in achieve the education purpose and mission. It can not be denied that as well as Human's social activities and other subject teaching activities in school ,all PE teaching activities developing surround the values. Fort this reason, research PE teaching should relate to the PE teaching values. Especially in nowadays, physical curriculum reforming in full swing, research on the PE teaching values with background of new physical curriculum is meaningful in theory and practice.

2 Reevaluating Value Subject and Value Object in PE Teaching

The process of PE teaching not only a special cognition, but also a special practice process, even a value create process, while the objects of teaching values including students ,teachers and society. In PE teaching activities, whether the teaching values can be better realized depend on whether the value orientation consistents the subject, and whether meet the three needs as follow: teaching activities meet the need of teacher's teachingand education, so the students become the object; Teaching activities meet the need of student learning and become a useful person, so teachers become the object; In addition, the society should meet the needs of teaching and

Y. Wu (Ed.): International Conference on WTCS 2009, AISC 116, pp. 755–760.
springerlink.com

learning activities, so teachers and students become the objects.The teaching and learning activities of teachers and students meet society needs reflected teaching values,that became the requirement and necessity to survive and develop teaching. In this way,from various angles, teachers,students and society incarnate their subject-object values in teaching and learning process,even the dynamic and diversity subject-object values in current physical curriculum.

3   Teaching Values Background the New Physical Curriculum

3.1   Life View of PE Teaching Process

The new physical curriculum view says,PE teaching process is not a simple skills learning process,but a process of students' integral development. The new physical curriculum views don't thinks skills unworthy to students' integral development that we can ignore the skills.On the contrary,the views attach more importance to the extensive migration value and promoting effect of sports basic skills for lifelong sports.Nevertheless,class teaching values can not rest on here.PE teaching promotes students' many-sided development service is the most basic foothold,the teaching process of multiple educational value is the focal point of teaching.We must change the traditional concept that only teaching skills,take the skills learning as means and carrier for students development, set a three-dimensional teaching target,such as" knowledge and skills,process and methods, attitudes and values",to make the process of gain sports knowledge and skills meanwhile the process of students to promote learning ability,develop mind and body and form values,realize the notoin change from the simple skills learning aspire to guide students learn to study,learn to survive,learn to be man.Have the skills is not the main purpose,but means and carrier,students' health and development are the ultimate aim and fundamental pursue of skills learning.

PE teaching process under this mind advocates liberty, independence and cooperation,promote students' own initiative, full of personality to effective learning,under the guidance of teachers.Teaching activities not the process of you say I listen,you do I look,full of parrot-learned knowledge anymore,but on the basic to establish subject position of students,they experience , learn ,choose the right learning style on their own.Students become the study master ,their eyes,ears,mouth,hands and foot are emancipated,and what's more,they have their own time and space in learning process,the initiative,independence and motility in learning process will be generated,publicized, developed and promoted,they can combine experience with interests,as well as own method ,values with skill acquire.Teaching process becomes a happy and positive emotional life and experience of students,"knowledge class" turn into "life class",PE teaching finally turn into life teaching process filled with human solicitude and life interest.

3.2   Live View of PE Teaching Process

The new physical curriculum view says,content of courses is unity of object world and meaning world,it's the collection of truth, good and beauty,not only a cognitive existence,but also a meaning existence.Reform the competitive teaching material system, Strengthen the contact between teaching content and students life,paying attention to students' learning interests and experiences,changing the situation that learning content and learning life are out of line with students' real life,commit to mobilizing students' learning initiative and enthusiasm.Students learn these skills not only take them as an objective to know and grasp,but also to talk to them,understand them based on their experience and practice,take them as a part of life.For different student,each content may and allow students have their own personal creative understanding,so they would have different targets and methods of learning.Learning process is not to "copy" abstract skills faithfully,but an open creating process,not only seeped the emotional experience, but also continuously to understand oneself, understanding social, understand life.

3.3   Democratic and Equal View of PE Teacher and Student

The new physical curriculum view advocates new-type teacher-student relationship. This relationship is democracy and equality build by contact, talk, understand in teaching process, rather than simple sports skills giving-receiving relationship.Through the contact and talk in teaching and learning, the teachers and students both sides mutual exchange, mutual communication, inspire mutually, complement each other, share each other's thinking, experience and skills, exchange of emotions, experience and ideas, extend the meaning of sports skills, seek new found, thus reach consensus, share and advance, to achieve the common development of both teaching and learning. The essence of the new approach is communication, means everyone involved in the teaching process, means equal dialogue, means cooperationve significance construction, it is not only a sport knowledge, skills, cognitive activities, is also a kind of equality of spirit exchanges between people. For students, contact means of subjectivity highlights, individual character make public, creative liberation. For teachers, contact means PE not only to impart skills but also to share understanding, increase learning, as well as the process of teachers' life activities, professional development and self-actualization .

3.4   The PE Teaching Assessment View Focus on Students' Development

The judgment and evaluation of PE teaching values, is also the basic content of PE teaching assessment view. For the reason that the new physical curriculum view changed, the criteria of PE teaching-learning activities even changed fundamentally: It is no longer confined to skill learning purposes, also pays attention to student's emotional attitude of nourishing, faith complete formation, personality, paying attention to the students' healthy growth; It is no longer just according to the teaching content of P.E. teachers teaching process to make the plan, trying to write a standard lesson plans, still need to student's characteristic, the function of teaching material aspects of changing situation forecast analysis on the base of the teaching process to make the system design; It is no longer just PE teachers invariable execution lesson plans, but according to the change of class situation at the process of adjusting teaching activities; It is not a teacher, teaching skill points according to carefully interpretation, repeatedly demonstration, constantly asking questions, but the teacher according to student's life experience to make timely guidance; It does not pursue every student can accomplish the graceful moves, but make every student can express themselves and have the opportunity to display their talents, are able to experience the joy of success. Anyhow, all students obtained the development through good PE teaching. And on PE teaching evaluation, will be based on the student development position.

4   Constructing Path of Teaching Values Background the New Physical Curriculum

4.1   With PE Curriculum Compilation as a Breakthrough, Dig Out the Educational Values of PE

For a long time, the educational values of PE confined to master sports techniques, skills. Just from the modes of presentation of PE syllabus and textbook, the main highlight is mature skills in athletic field. The result of this modes of presentation will makes the educational values of PE poorly. First of all, it cuted two contacts: one is the abundant, complex contact between skill itself and students' life; the other is the abundant, complex contact between skill itself and students find and solve problems, forming the skill process, all students and teachers who encounter is the lack of "popularity" skills, as the source material of PE teaching basic contents, they bring the lack of educational resource. Secondly, PE teaching skills itself as a concern to the axis, caused a kind of important deficiency, i.e. lack of needs to students' growing, the puzzle, curiousness, problem, expectations, interest and many potential abilities in growing process didn't reflect in teaching content. Therefore we can say that PE teaching content lack of life color is the deeper reason which makes makes the educational values of PE poorly.

4.2   Take Sport Skills as a Carrier to Play the Educational Values of PE

To realize the educational values of PE teaching, cannot without sport skills as a carrier,this is the unique of PE. We cannot play the educational values of PE without Consciously active learning process. According to sports skills of educational values and functional feature points, PE teaching skills can be roughly classified into physical fitness class, skill class, game class and the leisure class, all kinds of skills in learning methods, structure features and development of sexual have similarities. Using these two structures of PE skills, students can master the skills more efficiently,and what's more,they can master the same methods in the same kind of skills, form better learning ability and wider transfer ability,as well as the ability to make a plan to realize themselves development according to the principle of skills construction. Of course, to develop the educational values of various kinds of sport skills, we should contact all kinds of skills and students' life experience and growing needs, find out the contact point, to make all kinds of skills show vitality. The skills which are vitality can arouse students' internal needs, interests, confidence and boost their desire and ability to study.

4.3 Highlight the Educational Values of Sport Skills Learning with the Method of Dynamic Design

The PE teaching activities focus on students' life development emphasize curriculum meaning to keep building and dynamic creation, this unsure feature doesn't mean that PE teacher and student can do teaching-learning activities whatever they want, but requires more systematic teaching design, and the teaching design is not equal to write a lesson plan which can put teaching activities in order, even not planning the teaching process according to the skill learning regularity, we should give students time and space to participate and develop, create conditions for PE teaching process dynamic creation. For skills learning target setting, should based on the analysis of students' current status and possible expected development, it designed as "elastic interval" which takes the differences between students, the target and result for the reason that the educational values of skills learning not inherent, measurable itself, but generated in various forms of learning activity process. We also should have flexibility in teaching process designs, not regulate steps size or the class jointly progression. Even though the teaching organization form also should give full consideration to the student's individual differences and make onservative design. This skill teaching scheme is broad-brush for traditional lesson plan,which remained too many uncertainties, transformable elastic targets, time and space, just because of the introduce of these uncertainties and variable factors, the educational values of skills transformed into students' inner quality through the teaching, the meaning and values of skill learning can be generate and expand constantly, the values of PE teaching service the students' initiative and healthy development can be reflected in the specific PE teaching practice.

5 Conclusion

The problem of PE teaching values is also a basic theoretical question to research PE teaching, as well as a practical problem which influence the PE teaching activity process directly. It is worth to say that the transformation of PE teaching values cannot be separated from the teacher's teaching practice, can not but attention to the process that they internalize new values in their own teaching practice. To achieve this, we must analyse the new PE teaching values comprehensively, understand the basis and rationality of values completely, discusses the practical significance, to rebuild PE teaching values in teacher's mind, then realize it in PE teaching consciously and everlastingly. Only in this way, the new PE teaching values can play its role and significance in PE teaching activities.

References

1. Sato, M., Zhong, Q.-Q.(trans.): Teaching principle. Educational Science Publishing House, Beijing (2001)
2. Zhong, Q.-Q.: To the Chinese nation's rejuvenation, to the development of each student. East China Normal University Press, Shanghai (2001)

3. Developed by the Ministry of Education, physical and health curriculum standard. Beijing Normal University Press, Beijing (2001)
4. Liu, J.I.: Interpretation of physical and health curriculum standard. Hubei Educational Press, Wuhan (2002)
5. Wu, T.-H.: Concerning sports teaching values and sports teaching value formation characteristics. Journal of Guangzhou Physical Education Institute (3) (2003)
6. Qiu, Q.: Sports teaching evaluation concept renew under the background of quality education. Sports & Science (1) (2003)
7. Dewey, Zhao, X.-L., Ren, Z.-Y., Wu, Z.-H.(trans.): The school and society, the school tomorrow. People's Education Press, Beijing (1994)
8. Dewey, Jiang, W.-M.(trans.): How do we thinking, experience and education. People's Education Press, Beijing (1991)

On the Establishment of Unemployment Insurance System for College Students

Yaoheng

The Public Administration Department of Henan Polytechnic University
yaoh5@tom.com

Abstract. In recent years, along with the expansion of higher education and the increase of enrollment at colleges and universities, the problem that it's difficult for college graduates to find a job has aroused great attention. Unemployment of college graduates has become common. In view of the phenomenon, this article proposes establishing unemployment insurance system for college students by the joint efforts of the national, the provincial government, the school, and the students themselves; meanwhile, providing the corresponding practice base, skill training in order to strengthen their working ability, innovation ability and cooperation ability. Therefore, it can help college graduates accomplish the transformation from being a college graduate into a social one.

Keywords: Unemployment insurance for college student unemployment insurance unemployment.

1 The Necessity of Establishing Unemployment Insurance System for College Students

1.1 The Current Situation of College Students Unemployment in China

According to the employment rate of college graduates (as shown in Table 1) announced by the Ministry of Education, during the period of "Tenth Five-Year Planning", the employment rate of graduates from ordinary colleges and universities stabilized at about 75%. In 2006, the employment rate dropped to 60%. During the period of "Eleventh Five-Year Planning", the employment rate stabilized at 60%. From 1997 to 2005, the employment rate was stable relatively. However, the entire employment situation was still not optimistic in view of the yearly increased graduate base. In a way, it was serious. The difficulty in finding a job for college graduates has been becoming a hot issue. Therefore, establishing operable unemployment insurance system for college students is of great practical significance in helping graduates get a job.

1.2 The Problems of Unemployment Relief Measures for College Graduates

At present, some provinces, such as Guangdong and Jiangxi, have released relevant policies and measures in view of the college graduate unemployment status in recent

Y. Wu (Ed.): International Conference on WTCS 2009, AISC 116, pp. 761–766.
springerlink.com   © Springer-Verlag Berlin Heidelberg 2012

Table 1. Datas of Employment Rate of College Graduates from 2001 to 2009

Unit: 10,000 people

Year	Graduates total	Primary employment rate	Year	Graduates total	Primary employment rate
2001	114	80%	2002	145	80%
2003	212	75%	2004	280	73%
2005	338	72.6%	2006	413	60%
2007	495	64%	2008	559	65%
2009	611	68%	2010	650-700	60%

years. They usually adopt the method as follows: there is a certain waiting period; surpassing the waiting time, if the college graduates are still unemployed, then it can be recognized as unemployment. They should go to the relevant unemployment insurance institutions to register and participate in pre-job training. They can also apply for unemployment relief money. But they cannot get the unemployment insurance money. The current standards for unemployment relief money are different because of the different regional economic development levels.

Table 2. Unemployment Relief Standards in Some Cities

Citys	Standards	Minimum wage standards or social relief standards	Concrete amounts
Beijing	Social relief standard 120%-150%	697	836.4—1045.5
Guangzhou	Minimum wage standard 80%	860	688
Shenzhen	Minimum wage standard 80%	1000	800
Xi'ning	Minimum wage standard 70%	600	420

In China, people think highly of the college students. Are the college graduates willing to receive the relief money? Moreover, the unemployment relief money can only cover the basic life expenditure. It's hard to cover the travel expense, the communication expense, and the food lodging allowance in looking for a job. Therefore, the implementation of unemployment relief measure was barely satisfactory. In addition, after four years' study, the graduates acquire a basic self-learning ability. The skill training is a futile attempt if it can not greatly improve the graduate's' practical and operation ability in a short term.

2   Feasibility Analysis of the Establishment of the Unemployment Insurance System for College Students

2.1   From the Perspective of National Policies

In China, for the college graduates working in the state-owned enterprises, their study years at colleges or universities can be included in the calculating of their seniority. That is at the beginning of work they have already a certain number of years of seniority, for example, undergraduates for 4-5 years, postgraduate for 6-7 years, Doctor of Philosophy for 9-12 years. Now that the government includes the college or university study time as seniority, it can absorb the college graduates into the group of enjoying the unemployment insurance.

With the development of China's social security system, medical social insurance has begun to cover primary and secondary students nationwide in 2009. It means that college students enjoy equal medical treatment and social insurance as their peers. In recent years, old-age insurance system in rural areas has made rapid development, that is, old-age social insurance have been gradually moving towards farmers, migrant workers, landless peasants. It is well known that China's social insurance system is a complete system consisting of the old-age insurance, medical insurance (new rural cooperative medical care, and most cities have medical insurance, maternity insurance), unemployment insurance, the workers' compensation insurance. Now that the coverage of the old-age insurance and medical insurance has been expanded, the coverage of unemployment insurance should also be expanded appropriately. According to the trend of social security system in the world, it is inevitable to gradually expand its coverage. There have been some countries which have already put college students into the unemployment insurance category.

2.2   From the Perspective of Economic Guarantees and Fund Raising Channels

By the end of 2007, the accumulation of unemployment insurance fund has reached 97.91 billion yuan, which provides sufficient financial support for the establishment of unemployment insurance system for college students. With the economic growth, the resident income level enhances day by day. At the end of 2007, the staff average wage is 24,932 Yuan. In accordance with the relevant provisions of China's unemployment insurance, workers' personal unemployment insurance contribution rate is 1%. Thus, the annual fee is 24932 × 1% = 249 Yuan and a monthly payment is 20.75 Yuan. But the social insurance payments in China are based on the basic wage rather than the actual wage, so that the actual payment standard should be lower than 20.75 Yuan. According to the proportion of the basic wage to the real wage, monthly payment is about 9 Yuan. With the economic development, people's lives have been greatly improved, so does college students' living expenses. With an average of about 600 Yuan a month, according to this ratio, college students can participate in unemployment insurance by taking out 1.5% from the living expenses a month.

In China, unemployment insurance system stipulated that employees pay the unemployment insurance premium, the proportion is 1%, while the enterprises pays 2%. The college students can pay the individual part.

Next, it discusses the question that where the 2% comes from? It is estimated that the total number of all kinds of national scholarship is about 9.75 billion Yuan. From 1999 to 2008, 4,361,000 college students in poverty have fulfilled their "college dream" with the help of the policy of national student loan which was issued on June, 1999. However, some problems aroused. Nowadays, many student loans and grants depart from the original intention of the State funding. Many students who use the computer, MP4 and other consumer goods, apply for student loans and obtain state grants. Such problems have gained attention from the relevant national departments, who actively seek for solutions. Perhaps some part of the funds can be drawled out to provide the money needed for the support for the establishment of the unemployment insurance fund. The problem can be overcome if the national government, the provincial government, and the colleges and universities pay together.

2.3   From the Perspective of Management System

Some people think it is very difficult to manage the graduates because of the fluidity of college graduates. With the development of communication and network, the colleges' management of their graduates becomes more and more standardized. They can promptly know the employment situation of the graduates. Our relevant departments stipulate that if the college graduates have not found jobs after graduation, the files can be kept in school for two years. Two years later, if they still not find jobs, they may take the file back to the personnel department of the location of residence or the talent exchange centers. They may also move the files to the talent exchange center in the location of the school. To reduce the working link, the author suggests the files can be kept by the school management before the graduates find their first stable job. For the reason that the files are kept by paying certain fees for placing in the talent exchange centers, students whose files are still in school may also pay certain appropriate fees to the schools for the files being kept there. This payment standard might refer to the charging criterion of the talent exchange centers and it should be slightly lower than this standard.

3   Issues Concerning Unemployment Insurance System for College Students

3.1   Considering the Specific Conditions of the Graduates of Different Majors

The specific circumstances of graduates of different majors should be considered. For example, in China, the employment rates of undergraduates who specialize in surveying and mapping, mining, special education are high, reaching 100%; while those of undergraduates majoring in law, Chinese language and literature are low, generally around 30%. We should distinguish between different majors. Students of high employment rate may choose voluntarily whether to participate in unemployment insurance, while students of low employment rate are compulsory to participate in unemployment insurance. Word widely, it is a general trend to distinguish people who participate in unemployment insurance according to their professions or incomes. Recently, some scholars have proposed the reform of safeguard objects in social unemployed insurance. For instance, civil servants, university teachers are at low risk of unemployment, while migrant workers, employees of private enterprises are at higher risk of unemployment. Therefore, different payment standards should be set based on different professions. The unemployment insurance system for college students may take the initiative. If it is successful, it will provide some references to China's unemployment insurance system reform.

3.2   Giving Full Consideration to Skill Training and Practice Base Construction

China's higher education lags in the set of specialty and curriculum. Some specialties lack forecasting of the actual market demand, causing graduates unable to adapt to labor market needs. In addition, the long-term traditional theory of teaching makes a certain gap between graduates and the social needs. It usually needs a long time for the graduates to adjust to the society for the lack of innovation ability and cooperation ability. If some fund from college graduates unemployment insurance fund can be put out to build the practice base or to compensate for the internship enterprises where the students can get pre-job training, it is sure to help the graduates master the basic skills needed by the enterprises. As a result, the problem can be easily solved.

3.3   Giving Full Consideration to the Integration with Social Insurance

With the increase in working hours, college graduates basically integrate into the society. Then the question of how to integrate graduates' unemployment insurance with the urban unemployment insurance should be considered, mainly related to unemployment insurance fees and paid age, etc. If it's possible, the medical insurance should also be considered.

In brief, our starting point and goal is to ensure that graduates can enjoy social rights and integrate into the society quickly. College students are the country's future and hope. If they can't get support and can't be recognized by the society after graduation, it will have a negative impact. Therefore, it's necessary to create job opportunities and provide guarantees for them.

References

1. The People's Republic of China State Statistical Bureau compiled. China Statistical Annual. China Statistics Press, Beijing (2000-2008)
2. Lai, D., Meng, D.: Research on the Problem of Unemployment of College graduates in China, pp. 118–144. Chinese Labor and Social Security Publishing House, Beijing (2008)
3. Zhang, Y., Ma, G.: Graduate Employment Survey and Career Guidance, pp. 3–40. Chinese Labor and Social Security Publishing House, Beijing (2004)
4. Li, J., Zhang, Z., Hao, H.: Analysis of the Current Forms of Employment of College students and Countermeasures. Journal of Hebei Scientific and Technical University (Social Science Edition) 9, 100–103 (2008)

Research on Service-Oriented E-Government Performance Evaluation Management System Based on Civic Values[*]

Zhiping Huang and Hongmei Xiang

Chongqing College of Electronic Engineering, College Town, Shapingba District, Chongqing, 401131, P.R. China
huangpei2911@sina.com

Abstract. From the perspectives of online services and E-participation for citizens, this paper analyzes the value orientation of the E-government performance evaluation practice. Considering the typical bottleneck problems and causal analyses in the E-government performance evaluation practice of those countries, the value concept of the E-government performance evaluation is redefined in accordance with the civic satisfaction, the effectiveness of public E-participation and the level of online service. The communication mechanism for the subjects of E-government performance evaluation is built up from three dimensions: the public mechanism of E-government information, the feedback mechanism, the appeal system for performance appraisal. Using the method of "SWOT", this paper puts forward an E-government performance evaluation management system based on civic values and service orientation.

Keywords: E-government performance evaluation, Civic values, Satisfaction and E-participation, Level of online service.

1   Introduction

With ceaseless development of E-government construction, various national governments and international organizations have given close attention to the E-government performance increasingly [1]. E-government performance evaluation has served as the most direct measure of the effect in the construction of E-government. An idea of new public administration that "it can't be managed what it can't be measured." determines the necessity of E-government performance evaluation. During the operation of E-government, the performance evaluation is still in its infancy and a comprehensive and effective management system is far from being formed, which hinders the smooth development of E-government instead of finding and correcting mistakes. After over seven-year-development, some typical symptoms and other problems have come up in the government websites construction of China, such as E-government on the beach, "individual onrush" E-government , "information Isolated

---

[*] This work is partially supported by Chongqing Education Committee research Foundation Grant KJ092503.

Y. Wu (Ed.): International Conference on WTCS 2009, AISC 116, pp. 767–775.
springerlink.com   © Springer-Verlag Berlin Heidelberg 2012

Island" E-government . A new and important issue that solves urgently is how to promote the role of E-government performance evaluation effectively in the construction of E-government. To promote the application of E-government on its depth and width, it urgently needs to build up an E-government performance evaluation management system that scientific, consolidate, reasonable for China's conditions based on civic values and service orientation.

2   Practice and Value Orientation of E-Government Performance Evaluation at Home and Abroad

In order to provide a source of information and guidance for the development of E-government, the fundamental goal of E-government performance evaluation is to discover and solve problems duly and correctly. Accenture Inc. and Gartner Inc., United Nations and the American Society for Public Administration, a research team from the University of Southern California have carried out thorough investigation and practice, showing distinctive E-government performance evaluation system respectively.

2.1   Practice of E-Government Performance Evaluation in Foreign Countries

Investigating the development process of global E-government among the 191 member countries in August 2003, the United Nations and Division for Public Economics and Public Administration and United Nations and Division for Public Economics and Public Administration promulgated 2003 report of United Nations global E-government, which finally reflected E-government participation index and E-government participation index for each country [2]. Comparing the capability to provide on-line services for citizens and transform the government model and monitoring the governments' progresses of E-government in terms of services, the research team from the University of Southern California investigated and assessed networks for the total 191 member countries, based on the questionnaire designed by the United Nations Knowledge Management Division [3]. In 2003, the world-renowned Accenture Inc. evaluated service items which increased from 169 in 2002 to 201 for 22 national E-governments. According to the service maturity and the customer relationship management, the Accenture Inc. assessed the E-government' service maturity for the countries [4]. The service maturity accounted for 70% and the customer relationship management represented 30%. The Gartner Inc. evaluated E-government from the level of public service, operational efficiency and political gains [5]. E-government performance evaluation system of the Gartner Inc. was to assess the effectiveness of a particular E-government project for a country and the evaluation to the above other research institutes was horizontal comparison about the development of E-government for the countries in the world [6].

2.2   Value Orientation of Practice in Foreign E-Government Performance Evaluation

Like an invisible force, the value orientation which continuously affects and restricts its evaluation is the soul of E-government performance evaluation. Looking through

E-government performance evaluation systems of international institutions, their levels of evaluation and methods are different and they have their own characters and emphases separately, but they emphasize civic needs and willingness are at the heart of E-government performance evaluation systems. So, it can be concluded that the core of E-government performance evaluation is citizens, that is, its core value orientation is citizen and service.

2.3   Current Situation of E-Government Performance Evaluation Practice in China

E-government performance evaluation in China started as in foreign countries almost at the same time, but many problems appearing in its development, especially in E-government sites evaluation leaded to the fact that the ranking was falling down within international E-government. Evaluating the development of E-government and E-participation among the 183 member countries, the United Nations has launched the 5th global E-government evaluation since 2003 and the survey shows that E-government readiness index is 0.47, E- participation index is 0.3714 ranked 32 in the world as shown in Figure 1.

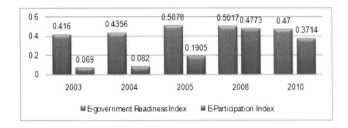

Fig. 1. [7] Index of Chinese E-government Readiness and Participation

According to the five survey results, the rankings of E-participation and E-government readiness rose at first and then fell. The ranking of E-government readiness has risen for three years which are 2003, 2004 and 2005, but it has fallen from 2008 to 2010. Compared with 2008 and 2010, the ranking of E-participation has fallen back drastically as shown in Fig. 2.

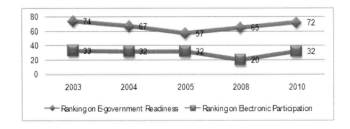

Fig. 2. [7] Ranking on Chinese E-government Readiness and Participation

3   Bottleneck Problems and Reasons of E-Government Performance Evaluation of China

By comparing the practice of foreign E-government performance evaluation, the E-government performance evaluation at home exists some major problems, such as misleading idea of governmental website performance evaluation, lacking a scientific performance evaluation index management system, inexact positioning of civic value orientation and the weakness of theory research for the E-government performance evaluation, though E-government performance evaluation in China has gained some practical experiences.

3.1   Lacking a Scientific Performance Evaluation Index Management System

An index system of E-government performance evaluation formulated by the government focuses more on the rationality and comprehensiveness of the theoretical design than its realistic practicability. It not only leads to randomness in E-government performance evaluation but also causes that its contents, methods, provision are similar. The E-government performance evaluation in practice often has blindness because an entire and systematic theory of E-government performance evaluation has not been formed and related evaluation systems have not good predictable functions which are of little value to guide the development of E-government [8].

3.2   Misunderstandings of Government Website Performance Evaluation

Although the government sites have been evaluated flourishingly all over the country, many people misunderstand an evaluation of government website equals an evaluation of E-government performance. Actually, a performance evaluation of government website is just a specific application in the evaluation of E-government system. This kind of false notion causes the wrong tendencies of "three appreciations and three depreciations" in its evaluation, that is, "preferring external evaluation to internal evaluation ", "preferring output to effect", "preferring outcome to process". Methods used for external evaluation are flexible and have fewer constraints than others, so they have been widely adopted by most E-government performance evaluation [9]. But the external evaluation which is only limited to analyze websites can hardly develops into the advanced stage of "public satisfaction measurement" [10]. The evaluation of output for a government website is regarded as an emphasis. However, the utilization rate of services, efficiency of investment for user satisfaction and actual application result are often ignored in practice because they are considered difficultly in E-government performance evaluation, which leads to evaluation results lacking of validity.

3.3   The Weakness of Theoretical Research for E-Government Performance Evaluation

Currently, E-government performance evaluation practice which has been implemented by governments and non-governmental agencies is disordered and confused on account of the rationale and target location evaluated E-government performance which the subject does not know clearly in China. So, the value direction, methods, means and levels of evaluation need to be studied deeply. Lacking of a legal system for the supporting reform measures, the E-government performance evaluation brings about a disordered evaluation processes and becomes a show, which is contrary to the development impetus mechanism of "promoting development, improvement and utilization by evaluation".

3.4   Inexact Positioning to Value Orientation about Citizens and Services

The continuous improvement of E-government performance evaluation system on the service and effective supply is still not enough under the guidance of government, which causes that a service-oriented core value of E-government has not yet risen to macro-development strategic level. The civic participation index is inadequate and most assessments lack a search for meaningful insights about civic utilization and satisfaction, while E-government performance evaluation based on service quality is rarer still. Ideas relating to the new public service on civic values have been introduced in the process of assessment, but they are still in the early stages. Lacking of the public service consciousness, the subjects and relevant persons still suffer from the traditional misunderstandings   for services to government. According to a survey from 2009 as shown in Figure3, the satisfaction of civic E-utilization is extremely unsatisfactory, which has clearly demonstrated that the value orientations of citizen and service have not yet put in the first place.

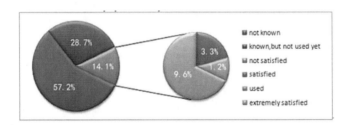

Fig. 3. 2009 a survey of counseling, complaints, satisfaction on government website

4   Building Up an Management System for E-Government Performance Evaluation Based on Civic Values and Service Orientation

Most of all, to evaluate the effect of service for citizens is to assess an E-government's success. Its improvement should be looked as a starting point and end point in E-government performance evaluation, which is the essential requirement to guide a sustainable development of E-government's health. Under the guidance of new public management theory, we shall build up the E-government performance evaluation management system based on civic values and service orientation in the following aspects.

4.1   Repositioning Value Orientation of E-Government Performance Evaluations

The improper ideas and the value-oriented prevent from wielding in depth about E-government. Through a reorientation to value orientation for the E-government performance evaluation, the above misconceptions and misbehaviors are corrected and a higher goal will be achieved. The government should formulate a new concept of assessment that is civic values in order to enhance the efficiency and function of the E-government performance evaluation.

1) Viewing civic satisfaction as top priority
The essential goal of the research on E-government is to constitute the citizen-oriented service e-government. The citizens' satisfaction degree to governments, the core factor, becomes the ultimate criterion that measures E-government performance. So, the civic satisfaction should be made as top priority in E-government performance evaluation, which heightens quality and efficiency of the E-government concerning online public services and makes the goal, "be on line for citizen", come true.

2) Strengthening an evaluation on civic participation
The channel building on civic participation and the effect on its utilization in the government website are evaluated primarily and all of the subjects pay close attention to the effectiveness of civic participation. The effect and quality of civic participation are assessed really by giving a simulation for users or citizens. A method of local characteristics, our E-government provides information and services for local citizens by on line and out of line, is applied to investigate and evaluate the effect about the civic E- utilization, which promotes the use of e-government in depth.

3) Accentuate an evaluation of on-line service level
The subjects should strengthen the evaluation of on-line service level in width and depth. During its evaluation, the availability, practicality, esp. utilization of on-line services ought to be emphasized. Two indexes, utilization rate and network public opinion, are essential for the E-government performance evaluation. Building an online service platform of zero distance, the government is to focus on the development of service E-government to improve their image in citizens' hearts.

4.2   Building Up a Communication Mechanism for E-Government Performance Appraisers

1) Building up an information opening mechanism for E-government performance appraisers
As far as relative subjects are concerned, information opening is a basic premise of communication for E-government performance evaluation. If communication is not disclosed completely, accurately and duly, this can give rise to communication disconnects among the subjects. To develop an information opening mechanism based on the actual situation is to disclose the interrelated information of the subject and object on-line in the evaluation, which can reduce communication barriers. If the administrative agencies haven't fulfilled their obligation to disclose information for

E-government performance evaluation, citizens can complain to the National People's Congress (NPC) or the supervisory department. The citizens can also lodge a complaint to make sure their rights to know and participate in performance evaluation if the agencies disobey information disclosure regulations and infringe citizens' legal rights.

2) Building up a feedback mechanism among subjects E-government performance
A feedback platform is to be built by making full use of modern network information technology. Through the platform, the subjects can not only complain about shortcomings and errors existing in the evaluation but also complain about an evaluation system, a process and method presented by the governments, which is good for improving the E-government performance. Specialized management institutions that can be established in the National People's Congress (NPC) or the supervisory departments and authorize them to audit, coordinate, monitor, etc.

3) Building up an appeal system for E-government performance evaluation
Citizens can file a complaint with the related authority which may be the National People's Congress (NPC) or the supervisory department if there are improper administrative behaviors in E-government performance evaluation. The appeal system, in essence, is an error-correcting mechanism in order that those citizens have a place to petition for a redress of grievances to the erroneous actions [11]. The related institution that receives the appeal should investigate and insight into citizens' views in time. As for deviation results of evaluation, the subjects should review and re-evaluate under the supervision of the institutions to correct inappropriate administrative behaviors.

4.3   Building Up an E-Government Performance Evaluation Management System Based on Civic Values and Service Orientation

1) Identifying a goal of E-government performance management system
Developing service-oriented E-government is necessity of history, which brings an innovative opportunity for management ideas, methods, systems, etc. Identifying a goal for the service-oriented E-government performance evaluation based on civic values, we should take advantage of E-government innovation to promote its development and application in depth.

2) Improving an index system of E-government performance evaluation
The index system for E-government performance evaluation is improved from the perspectives of online services and information opening for citizens. Firstly, an online service that the government puts forward should has three functions, guideline, preliminary, on-line processing. Secondly, the open information on governmental website should be timely, useful, available, easy retrieval, etc. An index, information retrieval ratio, is applied to the evaluate availability and efficiency of information opening.

3) Building up a pluralism of assessing subject system
The subject of E-government performance evaluation should be not unitary but multi and a basic principle that ensures validity of its evaluation is the pluralism of subject.

The government's power comes from citizens. Therefore, the subject to evaluate government performance in a democratic society can only be citizens and E-government is no exception. It means the citizen should be the subject to assess E-government. Under the ideas of new public administration, it is reasonable and necessary that the citizen is the subject of E-government performance evaluation. Therefore, ensuring civil rights to assess E-government directly, a multi-subject of evaluation mechanism made up of citizens, third-party agencies, government, etc. should be built.

4) Building up a performance evaluation management system for E-government based on based on civic values and service orientation

The performance evaluation management system must be built up in order that the performance evaluations play a significant role in promoting the development of E-government. Under the guidance of the goal to achieve service-oriented E-government based on civic values, its development strategy is put forward by using SWOT method. Analyzing the strength, weakness, opportunity, threat of E-government, improvement strategy, evaluative criteria, key performance index are established. What's more, a scientific performance evaluation management system is formed, which can find and correct promptly and can better guide and supervise the conduction and development of E-government.

5   Conclusion

From the perspectives of online services, satisfaction and E-participation for citizens, this paper repositions value orientation of E-government performance evaluation and puts civic E-participation index in the first place. It plays the decided part for governments and citizens to improve the development in depth. Using SWOT method, the performance evaluation management system for service-oriented E-government based on civic values is built up, which has huge guidance to make the effect of E-government performance evaluation better.

Acknowledgment. The authors would like to thank all the reviewers for their helpful comments.

References

1. Gupta, M.P., Jana, D.: E -government evaluationA framework and case study. Government Information Quarterly 20(4), 365–387 (2003)
2. United Nations World public sector report 2003: e-government at the crossroad (2003)
3. Wang, G.: Study on the Model of E-government Performance Evaluation Based on Third Party Evaluation. Journal of Information (11) (2009)
4. http://www.accenture.com/NR/rdonlyres/D7206199-C3D4-4CB4-A7D8-846C94287890/0/gove-e-gov-value.pdf
5. Gartner Corporation. The Gartner Framework for E – government Strategy Assessment (2002)

6. Wang, L., Zheng, Q., Han, G.: Reviews on Performance Measuring of E-government. Systems Engineering 25(2), 9–13 (2005)
7. http://www.jiaoyanshi.com/?viewnews-4155.html
8. Yang, D.: The Current Questions and Advices of E-government performance Evaluation. China Information Times (7) (2007)
9. http://blog.donews.com/hanghb/category/95286.aspx.2005
10. Zhang, X., Sun, D.: E-government and Strategy Plan. Science Press, Beingjing (2004)
11. Yang, D.: Study on E-government Evaluation System. Library Theory and Practice (3) (2010)

The Transfer Factors and Analysis of Textile Industry

Guo Wei, Li Bin, Jiang Zhu, and Zhu Changkai

School of Management Xi'an Polytechnic University Xi'an, China
{bingolee16,guowei3060}@126.com,
guo_iu@yahoo.cn

Abstract. With China's economy and society entered a new historical stage of the manufacturing industry from the coastal areas to central and western regions will become an important trend in the development of new moisture and to promote regional economic structure of China's new round of adjustment. In this paper, an overview of the status and transfer of the textile industry trends, the general shift of textile industry, the main factors in this multi-objective decision based on the combination of multi-objective evaluation of the establishment of entropy method, transfer of the textile industry in recent years conducted a comprehensive evaluation.

Keywords: Textile industry transfer, Multi-objective decision-making, Entropy.

1 Introduction

Industrial transfer is the inevitable trend of economic development, and the best path of the less developed regions to achieve economic development across is initiative to undertake industrial shift. As energy, raw materials, land and labor and other production factors into the rate of appreciation, export tax rebate reduction, the processing trade policy tightening labor-intensive enterprises such factors as the sea to the central and western regions to speed up diffusion [1]. The textile industry as the traditional labor-intensive industries faced with both opportunities and absorbs high-tech achievements, constraints facing rising costs and the multinational industry monopoly and the new restrictions on trade protection, which has emerged from the southeast coastal areas to central and western transfer characteristics [2]. This paper analyzes the transfer of the textile industry on the basis of the status and trends of industrial transfer in the various influencing factors are analyzed and used to conduct the evaluation of multi-objective decision making, that analysis of the central and western transfer of undertaking strategic trends in the textile industry.

2 Five Forces Models of the Textile Industry's Transfer

2.1 The Gravity of National Policy

"Gravity" comes from the relevant national macro-control policies to guide the development of the textile industry. The rise of western development strategy had a

Y. Wu (Ed.): International Conference on WTCS 2009, AISC 116, pp. 777–783.

great influence on the eastern textile enterprise's transfer. For the textile industry, the distribution of the country is very uneven. The end of 2008 figures show that 86.02 percent of China's textile enterprises located in 10 eastern provinces and cities, with the national textile industry, 85.24 percent of assets, 84.89 percent of the textile industrial output value, and create a national 86.35 percent of the textile industry profits. Countries in the "Eleventh Five-Year Plan" explicitly China's textile industries shift as an important measurement. Textile industry, "Eleventh Five-Year Plan" and the 2008 introduction of the "textile industry restructuring and revitalization of planning", were to encourage and guide on the textile industry and accelerate the optimization of the regional distribution of industrial transfer. "Eleventh Five-Year Plan" that the rise of Central China should seize the opportunity, use of the coastal economy of labor resource constraints, land, energy and administration costs increased substantially in the actual, as the textile industry to undertake the transfer of the key eastern region. "Textile Industry Adjustment and Development Plan" is further noted that the textile industry developed eastern coastal regions make full use of technology, capital, R & D, brands, marketing channels advantage, tracking the latest international technology and products, focusing on the development of high technology, high added value, low resource consumption, textile and products. Resources to encourage play central and western regions, actively carrying on industrial transfer, the development of textile and garment processing base, forming the eastern, central and western complementary regional distribution.

2.2   The Thrust of the Turn Out Region

"Push" constituted by the following four areas.
- The pressure from developed and developing countries. China's textile industry will face the dual pressures of developed and developing countries. In order to maintain its dominant position in the international market we should reduce production costs on the one hand, transfer the traditional textile industry of raw materials prices and labor costs relatively low to the western regions, on the other hand in the eastern coastal areas, focus on financial and technical to deep processing and high value-added textile market, and gradually to the direction of deep processing trends.
- Labor costs. It is understood that the Chinese textile industry, rising labor costs have been the equivalent of Southeast Asian neighbors to 3 to 4 times. Textile industry labor costs in eastern fast approaching the threshold of one U.S. dollar per hour, Vietnam, Cambodia, Bangladesh and Indonesia the domestic textile industry labor costs only 0.29 U.S. dollars per hour, respectively, 0.36 U.S. dollars, 0.37 dollars and 0.29 U.S. dollars, China's textile industry labor costs are much higher than the above countries in Southeast Asia Gang. Around the developing countries is becoming popular in China in low value-added products, which become the most powerful competitors.
- Land costs. In recent years, per capita land area of the southeast coast has been reduced. Coastal land in the Middle East has become a scarce resource. With increasing demand for its land, the land cost will rise.
- The needs of industrial restructuring. Facing the textile industry from extensive to intensive, high-end products from low-grade products to the changing industrial

structure, the manufacturing sector moved to central and western regions, product research and development in the southeast coast.

2.3   The Pull of the into Region

• The advantage of raw materials and labor .Raw material resources in the central and western regions have obvious advantages, is becoming the base of textile raw material. Such as Xinjiang, Guangxi are respectively for cotton and silk business move. With the central and western regions is accelerating the pace of urbanization, the advantage of abundant labor resources has become more evident, as the transfer of labor-intensive industries from eastern and development of textile processing and manufacturing has created favorable conditions. According to the Bureau of Statistics and Statistics of Henan Province, in 2008, the urbanization rate reached 34.34% in Henan province transfer of rural labor force 2155 people, 691 million to be transferred, the textile industry workers average hourly wage lower than the eastern 1 / 3 The cost contribution rate of profit and total assets higher than the national average of nearly 80%; and adequate power supply, lower prices than major provinces in eastern, equivalent to 83.7% in Shandong and Jiangsu 81.3%; in 2008, cotton production in Henan 650,000 tons, ranking No. 4 (Xinhua reported). Midwest raw materials and labor resources have greater appeal to the eastern part of the textile enterprises .Midwest textile industry a huge space for development and construction of our eastern, central and western regions a new system of textile industrial chain formed from the coast to the western industrial gradient pattern of time has gradually matured.

• In order to enhance the regional economy. Transfer of industry of the developed regions to make up for lack of funds, bringing advanced technology to optimize the allocation of resources, to promote the industrial structure; on the developed region, the transfer of industries optimize the industrial structure can improve the international competitiveness of industry. Transfer of industry will help alleviate the industry convergence, and can be formed on the resources of the substitution effect; reduce the resources industry in the directional distribution. Midwest textile industry through the transfer of undertaking can stimulate local economic development, promoting employment, enhancing the competitiveness of the region.

2.4   The Tack of the Turn Out Region

There are two reasons:

• The eastern cluster, while the lower level of human capital in the western region. In the east, the textile industry after years of development has formed a large, fully furnished cluster. Industrial clusters to improve the overall competitiveness of the industry, to strengthen the cluster effective cooperation between enterprises, and makes the industry chain, the limited area; the flow of various factors of production more quickly and easily, played a resource sharing effect. Western investors should not only pay more in labor search costs, training costs and supervision costs, but also may face a higher risk of labor efficiency and the risk of brain drain [3].

• High transaction costs. On the eastern region, the development of new industries require high investment and high cost, but not in the short term profit.

Innovation mechanism has not yet formed, the new leading industry climate has not formed, the old leading industry began to central and western areas the transfer of Zhi O'clock, Dongbukeneng as industrial hollow of the Mianlinshiqu new economic growth point of threats

2.5   The Resistance of the into Region

•   Lack of supporting industries. Capacity of central and western regions is not strong supporting industries, failure to complete the formation of a complete industrial chain, is bound to increase transfer of production costs, offset in the labor and land cost advantage. Incomplete industry chain, industry concentration level of development is not high.

•   Switching costs are too high. First, from the human cost, although the central and western regions rich in human resources and cheap labor, but considering the available units of labor productivity and ease of access to human resources factors, central and western regions of the human cost of not dominant, but population quality is relatively underground.

3   Multiple Objective Decision of Entropy

3.1   Entropy

Entropy is an important concept in thermodynamics is a measure of material system disorder, disorder degree, the system more "chaos", the greater the entropy. More orderly system, entropy will become. The introduction of information theory to the concept of entropy, an information source, said state signals of the level of uncertainty [4].

3.2   The Weight of Entropy

System and ordering information is a measure of degree of disorder, entropy is a measure of the system; If the smaller index of information entropy, the greater the amount of information of the indicators, in the Comprehensive Assessment of the role which should be greater, weight on should assume greater evaluation of a total sample of n, m evaluation indexes, evaluation matrix $[Y] = \left(y_{ij} \right)_{n \times m}$. The initial matrix of indicators for the cost through publicity $x_{ij} = \dfrac{y_{ij} / \left(y_{ij \max} \right)_i}{\sum\limits_{j=1}^{m} \left[y_{ij} / \left(y_{ij \max} \right)_j \right]}$ normalized. The matrix of indicators for economic returns through publicity $x_{ij} = \dfrac{y_{ij \min} / y_{ij}}{\sum\limits_{j=1}^{m} \left[\left(y_{ij \min} \right)_j / y_{ij} \right]}$ normalized.

Treatment is normalized matrix $[X] = \left(y_{ij} \right)_{n \times m}$. The section j (j = 1, 2, $\wedge$m) indices of the entropy is defined   as: $e_j = -k\sum_{i=1}^{n} f_{ij} \ln f_{ij}$. Among $f_{ij} = \dfrac{x_{ij}}{\sum_{i=1}^{n} x_{ij}}$, $k = \dfrac{1}{\ln n}$,

Entropy is defined as $w_j = \dfrac{1 - e_j}{m - \sum_{j=1}^{m} e_j}$.

3.3   Multiple Objective Decision of Entropy Making Build Steps

Step 1: Create the initial matrix $[Y] = (y_{ij})_{n \times m}$.

Step 2: After the normalized decision matrix structure standardization $[R] = (r_{ij})_{n \times m}$.

Step 3: According to the definition of entropy calculation of the weight of each index.

Step 4: construct standardized decision matrix $Z_{ij} = W_j \times R_{ij}$.

Step 5: Determine the ideal solution $x^+ = (x_1^+, x_2^+, \cdots, x_m^+)$, and negative ideal solution $x^- = (x_1^-, x_2^-, \cdots, x_m^-)$.

Step 6: calculation of the programs and the ideal solution and negative ideal solution and the Euclidean distance between the d^+ , d^+ .

Step 7: Calculate the relative closeness $S_j = d^- / d^+ + d^-$.

Step 8: The proximity to sort, giving preference to big.

4   Case Study

This paper in the textile industry in China's development for 2005-2008 based on data [5], select the Guangdong, Zhejiang, Shandong and other provinces in the eastern part of the five representatives, selected in Shaanxi, Guizhou, Xinjiang and other representatives of the five provinces in the western region. And in that time period the number of households provincial enterprises, the number of loss-making enterprises, labor costs, the main business, gross profit, financial analysis and other data for the consideration of objectives, the matrix standard, the following matrix:

Table 1. The matrix of data

	2008	2007	2006	2005
National policy gravitational [a]	39.09	27.09	25.74	6.52
The thrust of the turn out region	36.82	21.14	16.33	3.62
The pull of the into region	12.67	8.26	9.88	22.07
The tack of the turn out region	28.47	37.39	41.06	15.57
The into region 's resistance	53.85	44.59	44.44	37.78

a. China Textile Industry Development Report.

4.1   The Matrix Was Normalized

Table 2. The matrix standard data

	2008	2007	2006	2005
National policy gravitational	0.396	0.274	0.260	0.068
The turn out region 's Thrust	0.472	0.271	0.209	0.046
The into region 's pull	0.239	0.156	0.186	0.417
The turn out region 's tack	0.232	0.305	0.335	0.127
The resistance of the into region	0.298	0.247	0.246	0.209

Entropy was calculated: e1 =0.421,   e2 =0.385,   e3=0.432,   e4=0.441,   e5=0.451
Entropy: w1=0.184,   w2=0.214,   w3=0.198,   w4=0.195, w5=0.191

4.2   Standardized Decision Matrix for the $Z = W \times R$

Table 3. The standard division mateix

	2008	2007	2006	2005
National policy gravitational	0.073	0.051	0.048	0.013
The turn out region 's Thrust	0.101	0.056	0.045	0.098
The into region 's pull	0.047	0.031	0.037	0.083
The turn out region 's tack	0.045	0.059	0.065	0.028
The resistance of the into region	0.057	0.047	0.047	0.041

4.3   Available to the Ideal and Negative Ideal Solution

X^+ = (0.073, 0.101, 0.083, 0.065, 0.057); X^- = (0.013,   0.045, 0.031, 0.028, 0.041).
    Calculating the distance between the ideal solution and negative ideal solution which as the relative closeness $S_j = d^- / d^+ + d^-$.

    Available S_1=0.667, S_2=0.435, S_3=0.393, S_4=0.403.
    Know that in 05-06 years, although a small decline, but the overall trend of increased rate of transfer of industry supports the overall increase.

5   Conclusion

This article discusses the current development of Chinese textile industry and the transfer of industry trends and analysis of relevant factors including the depth. Through the establishment of the data matrix by the entropy analysis of multi-objective decision-making, the results show that although there is a slight decline in the first wave, but the overall trend of increased industrial transfer and have the potential to be tapped. Although the textile industry to the central and southeastern coastal regions there are still many problems in the transfer, but the textile industry's long-term perspective, the industrial transfer of industrial adjustment, a must to improve competitiveness. To upgrade the textile industry, we must end the occupation of the industrial chain and increase R & D capabilities and innovation should be transferred to the low value-added aspect of the low gradient regions, then the intensity of transfer of the textile industry will continue to develop.

References

1. Zhu, J.: Transfer of the textile industry from eastern Midwest Research - Five Forces Model
2. Sun, Y.: Southeast coast of central and western regions the transfer of the textile industry and Countermeasures. Western Economic, 27–29 (2007)
3. Zhang, C., Miao, J.: Based on the panel - data transfer stickiness of the regional industry. Soft Science 24, 201–205 (2010)
4. Qian, Q., Rao, Q.: Multiple Objective Decision Model for Environmental Quality Assessment. Enviornment Science 28(3), 53–56 (2009)
5. China Textile Industry Development Report. China Textile Industry Press
6. Samanta, B., Roy, T.K.: Multi-objective entropy transportation model with trapezoidal fuzzy number penalties,sources,and destinations. Journal of Transportation Engineering, 419–428 (2005)
7. Qiu, Y.: Management Decision and Entropy. China Machine Press, Beijing (2002)
8. Yoshiaki, Y.: Wavefront-flatness evaluation by wave front-correlation-information-entropy method and its application for adaptive confocal microscope. Optics Communication, 91–97 (2004)

Design and Practice of Witkey Multi-dimensional Stereoscopic Teaching System of Computer Network Course

Xianmin Wei

Computer and Communication Engineering School Weifang University
Weifang, China
wei84961@163.com

Abstract. Computer network course is a very practical course, in teaching the course, however, there are not suited to teaching course characteristics, experimental teaching link is weak. This paper analyzes the characteristics of Witkey model, designed a Witkey multi-dimensional teaching system. The system combining classroom teaching in multidimension to guide students in independent study, to culture students innovation and practical application ability, teaching practices show that it achieving the expected goals.

Keywords: Computer network course, Witkey mode, Stereoscopic teaching, Practical course.

1   Introduction

In the traditional teaching of computer network, most tend to be one-dimensional teaching model, teachers constantly lectures, students passively lectures, teaching effect is difficult to achieve the ideal state. However, the strong specificity of the course practical, traditional teaching and learning methods has constrained enthusiasm and innovation, a new teaching methods needed to solve this problem.

"Witkey" mode is a new network application form, for user through interaction via the Internet Q&A platform to enable personal knowledge, wisdom, skills, realize the value of a model. Witkey mode can improve the students power for learning, stimulating enthusiasm for learning an effective incentive.

2   Wit Key Mode

Wit key intended for the "smart key" in English refers to a person through interactive Internet Q&A platform to change wisdom, knowledge and professional expertise of the network into real income. Created by virtue of their ability (intelligence and creativity) on the Internet to help others who are paid. Witkey site is to provide a display or sale of knowledge, experience, ability or product platform. Let people know you need help and you can provide the services and assistance, but also can help you find who can help you to solve their problems and troubles, and get the help of

Y. Wu (Ed.): International Conference on WTCS 2009, AISC 116, pp. 785–790.

others. Wit key sites appearance for individuals with knowledge production and processing capacity to create a sales platform for knowledge products and business opportunities.

A perfect service system of Wit key mode is composed of five modules: the question and quotation system, retrieval system, knowledge database systems, ordering systems and trading systems, as shown in Figure 1. Where knowledge systems and trading systems which are Wit key mode core module.

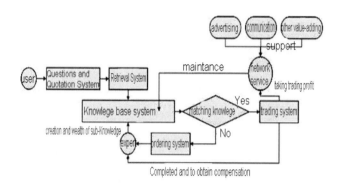

Fig. 1. Model diagram of Witkey mode

Users through quotation system to ask questions, by retrieval system to search matching knowledge in the knowledge database, if not then through the ordering system experts answering, and expansion of new knowledge into the knowledge base. And then complete the acquisition of knowledge through the transaction system.

3   Characteristics of Computer Network Courses

Computer network is a comprehensive cross-disciplinary nature, it is comprehensive application of computer technology and communication technology concept and approaches of of these two disciplines, requiring software and hardware integration and computer and communication technology integration, and combination of theory and application, requiring classroom and special test environment, requiring integration and teaching and the market closely. In the specific arrangement of teaching content, we will begin from the development history of the computer network, two well-known reference models: OSI and TCP / IP as the main line, during interludes from the LAN to the WAN to the Internet a variety of network standards, model protocol layers involved in as theoretical knowledge part of the network infrastructure. Knowledge in the experimental section, the interconnection focusing on commonly used devices such as bridges, switches, routers, firewalls and other works; important network protocols such as running mechanisms of CS-MA/CD, IP, TCP, RIP, OSPF, DNS and etc.; and CCNA certification exams important knowledge points, such as virtual local area network, spanning tree protocol, PPP authentication, IP access list, the configuration of frame relay, network address translation NAT, and more.

Computer network technology is constantly evolving, in teaching this course must add the relevant content to the original teaching materials, but in small amount of lessons to explain the basic theory and related new technology in detail, using traditional methods for teaching, for the students to better grasp the course content, theses have great difficulties. While computer network curriculum is a both theoretical and practical courses, students must through the theoretical knowledge to understand a lot of practice, however, limited experimental course can not meet the requirements. Therefore, how in the limited hours of teaching computer network courses to improve the quality of teaching has become the problem must be solved. After teaching and experience of many years, this paper combining Wit key model features to form a multidimensional stereoscopic computer network teaching system, try to resolve the issue.

4   Design of Wit Key Multi-dimensional Stereoscopic Teaching System

To fully mobilize the enthusiasm and initiative of people involved in solving various problems the Internet is a major feature of Wit key mode. Therefore, in the design of systems, we must grasp this point, and fully mobilize the enthusiasm of learning and creativity.

By the comprehensive Wit key model stereoscopic teaching platform, students can use the forms, in many ways, a variety of media, a variety of means in parallel to achieve a three-dimensional learning process of specific course. This design Wit key platform consists of two systems: Self-Learning System (self-learning system) and Wit key pilot system (Wit key Guidance System), shown in Figure 2.

Self-learning system based on "student self-learning basis, multi-dimensional Guidance assistance" teaching ideas, and mutually reinforcing and classroom teaching. The system has knowledge of the nine learning modules: ① the syllabus and key and difficulty of curriculum; ② multimedia courseware (including video lectures); ③ experimental guidance; ④on-line operation; ⑤ Online Q&A; ⑥ references; ⑦ exercise library, the web database and online examinations; ⑧ technical discussions and innovation area; ⑨ expert lectures. Nine modules based on the class content of the teacher can be updated to support the students a variety of learning needs. Students can use the "self-learning system" to study knowledge and self-organized and learning time according to their own situation and way, you can visit the technical discussions and technical skills to achieve the difficult problems to solve, you can repeatedly watch the video to focus on teaching the knowledge of fine detailed control, you can achieve specific courseware to understand abstract concepts, but also expert talking about the current knowledge point of a new technology. Online Q & A to expand operations and online teaching guidance time and space, to support for students self-learn and improvement of efficiency.

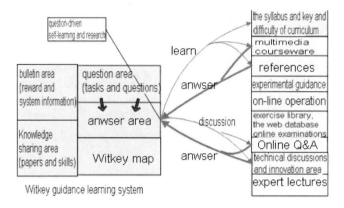

Fig. 2. Wit key multidimensional stereoscopic teaching system

Wit key guidance system default five main modules: question area, answer area, students Wit key map, bulletin area and shared area. Knowledge sharing area is the sharing area of knowledge and resources, bulletin area is announcement area of information and reward information, the students Wit key maps, posted his own good, the user directly through Wit key map to find suitable respondents. Question area and answer area design is a key to Wit key guidance system, which Wit key incentive mechanism is designed to achieve key of education reform goals. This research and practice to identify an incentives mechanism of Wit key points system, students should answer questions in Wit key area through participation to access points or other rewards. The problem is divided into two categories, one is user questions, users will put forwards questions with prizes about their life or work-related network problems, along with reward points or other rewards, such as Wit key points; second is the teacher guidance and learning issues, such as curriculum synchronization issues or other prizes tasks, or teachers to add other websites of relevant content Wit key task to the system, guiding students to complete, and to Wit key points as rewards. Students answer questions in the answer area, if you can get the correct solution to the corresponding points or rewards, Wit key points can be used as a basis for assessing teacher performance. Wit key guidance system and self-learning system have the mutually reinforcing relationship, task and problem drive student to learn, to solve practical problems, to be the best practice classroom.

Wit key guidance systems take full advantage of the rapid response Wit key models and intellectual stimulation, to deliver the learning space of "intellectual stimulation". In this paper, Wit key points and other reward incentives together to guide the intellectual, Wit key guidance systems are open for every registered user can post questions or tasks, users need to pay for the release of the issue points or reward, students will answer questions to get points or other rewards. How to create the best Wit key incentive mechanism is the most critical issues have been initially identified to Wit key points-based incentive system, as shown in Figure 3, but there are some shortcomings, also need to constantly practice and teaching study.

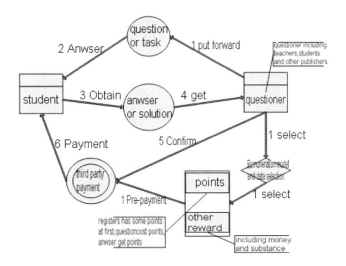

Fig. 3. Main flow of Wit key Guidance system rewards mechanism

5   Conclusion

Computer network teaching practice for two years show that, Wit key multidimensional stereoscopic teaching system has brought about good teaching effectiveness, student learning initiative significantly increased, particularly in practice and progress clearly, for one problem students can design multiple experiment solutions, experimental and innovation capacity increase greatly. However, some problems still worth pursuing in-depth research, mainly in the following aspects:

First, the rapid development of information technology bring about some changes in the computer network course knowledge structure, in content explanation methods, sequence, and means will be some changes, in Wit key mode how to change the most is worth studying.

Second, Wit key guidance is the main factors of success or failure, leading to enhance the practical ability of students, while also enhancing the study of basic theoretical knowledge, therefore, learning materials, selection tasks and problems is the main material provided by teachers to guide, to carry out based constantly changing on curriculum needs.

Third, how to establish best incentive mechanism, where in addition to Wit key incentive methods diversifying, standardization of reward evaluation method is also essential, these need to continue the study of teaching and practice in teaching reforms.

References

1. Witkey basic principles, http://www.witkey.com/article/20497.html
2. Economic theory of Witkey mode,
   http://www.witkey.Com/article/20418.html
3. Witkey mode feature-the Internet's next hot spot,
   http://www.981a.cn/html/manager/znews
4. He, K.: Instructional System Design. Beijing University Press, Beijing (2006)

Teaching Method Exploration of Assembly Language Programming

Xianmin Wei

Computer and Communication Engineering School Weifang University
Weifang, China
wei84961@163.com

Abstract. In College Computer professional assembly language is a major professional basic course, but because assembly language the nearest to hardware, more instruction, non-intuitive, tedious, difficult for school students and teachers is difficult to teach. Through analysis and summary of the key anf difficulty in assembly language programming course, presented a few easy steps in university computer science teaching assembly language programming on how to improve teaching methods, guidance for students to learn the course.

Keywords: Assembly language, Teaching methods, Experiment.

1 Introduction

Assembly language is able to use the computer all the hardware features, and can directly control the hardware. It is one of the a core subjects for university computer software, hardware and application professional students, it is the computer theory, operating systems and other core courses necessary pre-study course, on culturing the programming ability of students to understand how computers work, engaged in software development and hardware applications are a very important role. However, students are very difficult to learn this course, accustomed to highly structured in high-level language, assembly language on the "machine" for students in a time difficult to understand.

2 Assembly Language Learning Difficulties and Its Causes

2.1 Assembly Language Own Characteristics

The one hand, with some assembly language instruction is a mnemonic meaning to the expression of the corresponding, and therefore easier to grasp than the machine language, but on the other hand, assembly language relative to the direct use of resources is another high-level programming language seems difficult to master. Assembly language program about summed up the following key features:

(1) associated with the machine: assembly language instruction is a symbolic representation of machine instructions, and different types of CPU have different machine instruction set, assembly language program has a close relationship with the

Y. Wu (Ed.): International Conference on WTCS 2009, AISC 116, pp. 791–795.
springerlink.com

machine, that the general and portability of assembly language program is lower than the high-level language program.

(2) the complexity of programming: assembly language is a machine-oriented language, the assembly instructions with a single machine instruction as a function of the specific characteristics, in order to complete a particular job, such as the calculation (A+B+C), we must arrange the work of CPU's every step. If the first calculation (A + B) then C added to the former results. In the preparation, not to mention the double loop, high-level language in the statement to complete the 3,4, and in assembly language compilation to dozen or 20 instructions.

(3) the complexity of debugging: Under normal circumstances, debugging assembly language program is more difficult to debug than the advanced language program. First, the assembly language instructions involved in the details of machine resources, in the debugging process, to clear the changes of each resource. Secondly, in the commissioning process, in addition to the implementation functions of each instruction, but also clear its role in the problem solving process. Third, the high-level language program to use almost no transfer of explicit statement, but the assembly language program to use a large number of various types of transfer instructions, which jump dramatically increased the debugging process more difficulty. Lastly, the debugging tools fall behind: high-level language programs can be the symbolic source-level tracking, and assembly language programs can track the machine instruction.

2.2   Shortening Teaching Time of This Course

Computer courses are updated very fast, with the continuous influx of new knowledge, is inevitable to some new courses opened. For example: Ten years ago, computer science has not yet set up dynamic web design, web programming courses, and now the new addition of these courses. For the basic course such as assembly language can not lose, it only reduced the total hours of class curriculum, which makes the back of assembly language are generally more interesting things to be deleted, because there is no time to go on, for example: color and animation and so on. But because a large number of instructions in the front basis content, the students feel more boring, lack of confidence and interest in further study anymore. The reduction of experimental time, so students can be guided on the machine time reduced, can only do some very basic experiments, while some experiments allow students with a sense of achievement, such as the control of speaker sound experiment contains no way to do.

2.3   The Leading Curriculum for Students Not to Learn Better

In the study of assembly language, found that despite many of the students studied high-level language C, but there is no concept of the debugger, the error only reference to books or call the teacher to see, do not know what the single-step tracking and breakpoint settings, lack of abilities. Some students have some notion of the debugger, but they will not use the debugging tools, only where necessary to add some output statements to observe the value of a variable. Therefore, for more complex assembly language debugging environment, they can not find a solution. Coupled with assembly language input and output need to use interrupts, not like high-level language so easy to output the value of a variable, students observed value of the variable in memory directly does not yet meet, so when problems arise, very few students to correct their own procedures error.

3   Some Exploration Comments in the Teaching of Assembly Language

Through the combination of the above-mentioned difficulties analysis and their own teaching experience, for teaching methods and test counseling of assembly language to propose several recommendations aimed at existing resources, students will be compiled linguistics.

3.1   Invert Teaching

A compilation of previous language teaching is basically arranged in accordance with the order for materials, and its main teaching arrangements in below table 1.

Table 1. The teaching process and content of the previous arrangement

Order	content
1	Basics
2	Computer System Overview
3	addressing modes and instruction
4	program format
5	Programming
6	subroutine design and system calls
7	DOS and BIOS interrupt call
8	Advanced compilation techniques

Teaching schedule in accordance with the above teaching contents, as students basics is weak and habits of thinking are not enough, content arrangement is more difficult to adapt to, the experimental environment can hardly grasp inconsistent forwards and after, in materials the previous example is not strong and so on. Therefore, in teaching, appropriate adjustment of teaching order, can help students to better grasp the assembly language, assembly language instructions can be spread to other parts of the content inside, interspersed with each other, early introduction to the compilation format, first give an overall impression. As the assembly language instructions difficult to understand and dry in large quantities, with the instruction distributed into examples and programming taught, in the experiment through the interesting process of input debugging enhance understanding and memory of instruction, will help reduce the level of assembly instructions boring, at beginning students can access the input assembly environment and the debugging process. The revised teaching-learning process in Table 2.

Table 2. Adjusted teaching sequence table

Order	content
1	Basics
2	Computer System Overview
4	program format
3	addressing modes and instruction
7	DOS and BIOS interrupt call
5	Programming
6	subroutine calls and system design
8	Advanced compilation techniques

3.2   Teaching Synchronization

If the conditions allowed, the best method is class while experiment. Students to learn a language, wanted to know how the environment can be able to use it, or it is ineffective, such as castles in the air. In class process, I use the multimedia classroom, such benefit is that the process in the class can demonstrate the environment, and give students some time to practice by themselves, so students can get started quickly, on course to establish the confidence of the initial . Furthermore, compared with the traditional basic knowledge before speaking in class, and then over a period of time to experiment, this is also in the shortest lessons to learn more knowledge, because the teachers and students to do in the almost no time difference, the students have a deeper impression, get better learning results.

3.3   Teaching of Example

In the teaching process, should appropriately increase the number of examples. The general process of teaching examples is to analyze the problem, determine the algorithm, write procedures and instructions write assembler source code. But the focus should be in different stages, some can omit. In the front, the main purpose of speaking example is compilation format and assembly instructions; In the middle, the purpose of talking example is the compilation process corresponding relationship to the assembly instructions; In the final, examples focused on methods and ideas of programming. By this, from one by one examples students will be able to gradually master the format of assembly language, assembly instructions, assembly processes, assembly programming.

3.4   Carefully Chosen Experiments Contents

Experiment content is very important, as the experimental class hours are limited, and that students are beginners, so prepare a small program to spend for a long time also, together with error debugging time, to complete a program is a longer time. For example: I ask students to compile two sequences already lined up into a sort queue, many students only done in six classes. Content in the choice of experiment, pay attention to the complexity from the brief order, best cover as much content as possible, and do not have duplicate content, select some of the more interesting subjects to experiment, such as: simulation of neon. Also interspersed in the classroom the knowledge point of experiment at any time, of course, examples are brief and effective.

4   Conclusion

In short, assembly language teaching is more abstract and difficult, and students feel more difficult to study, but as long as the teachers in the teaching pay attention to importantance and difficulty, focus on thses importantance and difficulty, using modern multimedia teaching methods to stimulate students enthusiasm for learning, and focus on practice teaching, must be able to achieve good teaching effect.

References

1. Shen, M., Wen, D.: IBM-PC assembly language programming. Tsinghua University Press, Beijing (1991)
2. Wang, S.: Assembly Language, 2nd edn. Tsinghua University Press, Beijing (2005)

Population Age Structure Changes on the Scale of Education

Xia Li, Yingchun Liu, Tingting Wang, and Hua Sun

School of Management Xi'an Polytechnic University Box 104, No.19, Jinhua South Road, Xi'an, Shanxi Province, China
lesoso@lawyer.com

Abstract. As China's economic and social development and the implementation of family planning policy, China's population age structure will inevitably change, while there is a growing emphasis on education, educational demands are growing rapidly, so the scale of demographic changes have a direct impact on education. This paper forecast the changes of age structure in China's future population with Markov chain, to analyze the impact of change on the education scale.

Keywords: Population age structure, Education scale, Markov chain, State transition matrix.

1 Introduction

Recovery of the college entrance examination in China since 1977, college enrollment rate increases from 4.7% in 1977 to 56.85% in 2006. Since 1999, China implemented a large-scale admission, college admission in 1998 of 108 million people, to national college enrollment has reached 275 million in 2002, an increase of up to 154.6%. With the college enrollment, high school, technical school and various types of vocational education have also entered the rank of enrollment expansion. As the development of economy, society, medical process, and health, the implementation of family planning policy, the changes of fertility concept, population structure will change, changes in population structure directly affects the number of school-age population, which impact on the education scale.

Markov random process has been widely used in various aspects of economy and society, which is mainly used for prediction, structural change analysis and so on. There are many ways to forecast population, such as regression analysis, Logistic biological micro-model, gray exponential model, combined time series methods and others. This paper forecast the changes of age structure in China's future population with Markov chain, to analyze the impact of change on the education scale.

2 Model and Method

If a random process's state known in the moment t_0, the state of the moment $t > t_0$ is only with the moment of t_0, whereas the state has nothing to do with the state of

Y. Wu (Ed.): International Conference on WTCS 2009, AISC 116, pp. 797–801.

previous moment of t_0, we call this random process is a Markov process. Time and state are discrete variable in Markov process, which is called Markov Chain. In other words, suppose there is a sequence of random variable $\{X_1, X_2, \cdots, X_m, \cdots\}$, its sets of state $S = \{s_1, s_2, \cdots, s_n\}$, $\exists \forall k$ and positive integers $i_1, i_2, \cdots, s_k, s_{k+1}$, so $P\{X_{k+1} = s_{i_{k+1}} \mid X_1 = s_{i_1}, X_2 = s_{i_2}, \cdots, X_k = s_{i_k}\} = \{X_{k+1} = s_{i_{k+1}} \mid X_k = s_{i_k}\}$ we call $\{X_1, X_2, \cdots, X_m, \cdots\}$ Markov Chain. State transition probability has nothing to do with the moment of the state, and the sets of state are finite sets or countable sets. There are many states in the system, such as $s_1, s_2, \cdots; s_n$, and there is only one state in any time. If the current is in state s_i, next time the state will turn to any state $s_1, s_2, \cdots; s_i, \cdots; s_n$. The correspondence of transition probability is

$$p_{i1}, p_{i2}, \cdots; p_{in}, 0 \leq p_{ij} \leq 1, \sum_{j=1}^{n} p_{ij} = 1, i = 1, 2 \cdots; n$$

We call matrix P state transition probability matrix. Matrix P is state transition matrix of Markov Process, so there are two bands of characteristics: (1) $P^{(k)} = P^{(k-1)}P$ (2) $P^{(k)} = P^k$.

In layman's terms, in the course of events, if the transfer of each state has something to do with the previous state of the moment, the state has nothing to do with the past, or that the state transition process is no varying, and then such a state transition process called Markov process. In a longer time, Markov process is becoming more and more stable, and has nothing to do with its initial state.

3   Empirical Study

Through homogeneous Markov chain predict the future of China's population age structure, and thus to analyze the education-age population changes, so give some proposals in the scale of the development of China's education.

Table 1. China'S population age structure data in 2003-2008

Age Ratio (%) / Year	0-5	6-17	18-21	22-40	41-64	≥65
2003	6.43	20.09	5.63	33.98	25.93	7.94
2004	6.02	18.77	5.96	33.89	27.24	8.12
2005	5.83	18.52	5.45	34.99	26.57	8.64
2006	5.77	17.86	5.92	35.61	26.12	8.72
2007	5.48	17.33	6.14	35.43	26.78	8.84
2008	5.11	16.58	6.46	35.35	27.35	9.15

To calculate the one step transition probability matrix from the original data in table one. Hypothesize the proportion of sets: 0-5,6-17,18-21,22-40,41-64, ≥65 and define state 1,2,3,4,5,6, adjacent to two years for each step.

The 2003-2004 state transition probability matrix:

$$
p_1 = \begin{bmatrix}
0.936 & 0.026 & 0.007 & 0.001 & 0.026 & 0.004 \\
0 & 0.934 & 0.012 & 0.003 & 0.045 & 0.006 \\
0 & 0 & 1 & 0 & 0 & 0 \\
0 & 0 & 0 & 0.997 & 0.002 & 0.001 \\
0 & 0 & 0 & 0 & 1 & 0 \\
0 & 0 & 0 & 0 & 0 & 1
\end{bmatrix}
$$

In the same way, we can get state transition probability matrix of 2004-2005, 2005-2006, 2006-2007, 2007-2008

$$
p_2 = \begin{bmatrix}
0.968 & 0.003 & 0.005 & 0.011 & 0.007 & 0.006 \\
0 & 0.987 & 0.002 & 0.005 & 0.003 & 0.003 \\
0 & 0 & 0.914 & 0.042 & 0.025 & 0.019 \\
0 & 0 & 0 & 1 & 0 & 0 \\
0 & 0 & 0 & 0 & 0.974 & 0.026 \\
0 & 0 & 0 & 0 & 0 & 1
\end{bmatrix}
$$

$$
p_3 = \begin{bmatrix}
0.990 & 0.003 & 0.002 & 0.003 & 0.002 & 0 \\
0 & 0.964 & 0.010 & 0.014 & 0.010 & 0.002 \\
0 & 0 & 1 & 0 & 0 & 0 \\
0 & 0 & 0 & 1 & 0 & 0 \\
0 & 0 & 0 & 0 & 0.983 & 0.017 \\
0 & 0 & 0 & 0 & 0 & 1
\end{bmatrix}
$$

$$
p_4 = \begin{bmatrix}
0.950 & 0.016 & 0.006 & 0.005 & 0.019 & 0.004 \\
0 & 0.970 & 0.006 & 0.004 & 0.017 & 0.003 \\
0 & 0 & 1 & 0 & 0 & 0 \\
0 & 0 & 0 & 0.995 & 0.004 & 0.001 \\
0 & 0 & 0 & 0 & 1 & 0 \\
0 & 0 & 0 & 0 & 0 & 1
\end{bmatrix}
$$

$$
p_5 = \begin{bmatrix}
0.932 & 0.025 & 0.011 & 0.003 & 0.019 & 0.010 \\
0 & 0.957 & 0.011 & 0.003 & 0.019 & 0.010 \\
0 & 0 & 1 & 0 & 0 & 0 \\
0 & 0 & 0 & 0.998 & 0.001 & 0.001 \\
0 & 0 & 0 & 0 & 1 & 0 \\
0 & 0 & 0 & 0 & 0 & 1
\end{bmatrix}
$$

In order to eliminate the influence of the sample randomness, take $\dfrac{1}{5}\sum_{i=1}^{5}p_i$ as probability matrix P, so

$$P = \begin{bmatrix} 0.955 & 0.015 & 0.006 & 0.005 & 0.015 & 0.005 \\ 0 & 0.962 & 0.008 & 0.006 & 0.019 & 0.005 \\ 0 & 0 & 0.983 & 0.008 & 0.005 & 0.004 \\ 0 & 0 & 0 & 0.998 & 0.001 & 0.001 \\ 0 & 0 & 0 & 0 & 0.991 & 0.009 \\ 0 & 0 & 0 & 0 & 0 & 1 \end{bmatrix}$$

According to the above transition probability matrix to predict the future of China's population age structure, a step state transition matrix will change slightly with the development of national economy, in which we assume as long as no major incident (such as war, natural disasters, etc.) available policy does not change, the state transition matrix in the next several years remained unchanged. Take all the six years 2003-2008 the proportion of the average age as the initial state: $\lambda_0 = (5.77 \quad 18.19 \quad 5.93 \quad 34.88 \quad 26.67 \quad 8.57)$ which is transferred by the current state probability is essentially the same, you can use the formula: $\lambda_n = \lambda_0 p^n$ to predict the changes of China's population age structure in the next 5 yeas, 10 years, 50 yeas, to analyze the impact on the education scale (the following calculation are accomplished in software Matlab7.0).

After 5 years the age structure of China's population is forecast to be:

$$\lambda_5 = \lambda_0 p^5 = (5.77 \quad 18.19 \quad 5.93 \quad 34.88 \quad 26.67 \quad 8.57)$$
$$\begin{bmatrix} 0.955 & 0.015 & 0.006 & 0.005 & 0.015 & 0.005 \\ 0 & 0.962 & 0.008 & 0.006 & 0.019 & 0.005 \\ 0 & 0 & 0.983 & 0.008 & 0.005 & 0.004 \\ 0 & 0 & 0 & 0.998 & 0.001 & 0.001 \\ 0 & 0 & 0 & 0 & 0.991 & 0.009 \\ 0 & 0 & 0 & 0 & 0 & 1 \end{bmatrix}^5$$
$$= (4.58 \quad 15.35 \quad 6.25 \quad 35.39 \quad 27.78 \quad 10.65)$$

$$\lambda_{10} = \lambda_0 p^{10} = (5.77 \quad 18.19 \quad 5.93 \quad 34.88 \quad 26.67 \quad 8.57)$$
$$\begin{bmatrix} 0.955 & 0.015 & 0.006 & 0.005 & 0.015 & 0.005 \\ 0 & 0.962 & 0.008 & 0.006 & 0.019 & 0.005 \\ 0 & 0 & 0.983 & 0.008 & 0.005 & 0.004 \\ 0 & 0 & 0 & 0.998 & 0.001 & 0.001 \\ 0 & 0 & 0 & 0 & 0.991 & 0.009 \\ 0 & 0 & 0 & 0 & 0 & 1 \end{bmatrix}^{10}$$
$$= (3.64 \quad 12.94 \quad 6.35 \quad 35.85 \quad 28.54 \quad 12.68)$$

In the same way, we can get:

$$\lambda_{20} = \lambda_0 p^{20} = \begin{pmatrix} 2.30 & 9.16 & 6.39 & 36.46 & 29.18 & 16.51 \end{pmatrix}$$

$$\lambda_{50} = \lambda_0 p^{50} = \begin{pmatrix} 0.58 & 3.17 & 5.03 & 36.87 & 27.29 & 27.06 \end{pmatrix}$$

4   Analysis and Conclusion

Empirical study shows that over the next 50 years, the proportion of children 0-5 years old is decreasing year by year, and 6-17 years old in compulsory secondary education or high school age and the proportion of children declined. While in the higher education of age 18-21 years, after reach the peak, and then fall year by year. With the popularization of higher education entered the stage, the scale of demographic factors on the impact of education more and more obvious in the next 20 years, enrollment in higher education in China will rise, as China's economic development, the scale of higher education in about 50 years will be in equilibrium.

China's compulsory school-age population gradually declined, it's inappropriate to increase schools. The development of appropriate scale is more practical. China's population of 16-40 year olds reduce rapidly from 2010 to 2015, about 10 million annual reduction beginning in 2009. The number of students who receive compulsory education all around the country is 159.165 billion in 2008, 383.7 million less than last year, in which city students in school increase about 65 million but the rural students in school reduce 448 million.

References

1. Zhang, W., Jin, J., Zhai, B.: Markov chain in China's population age structure prediction. Journal of Henan Business College 21(4), 45–48 (2008)
2. Jin, H., Li, J.: Research of School-age Population's Influence on the Scale of Higher Education. Journal of Anshan Normal University 11(6), 12–15
3. Yin, S., Zhou, Y., Pu, L., Zhao, Y.: Markov Process Applied to Predicting Land Use Structure—Example for WanBao Town in LouDi City, HuNan. Economic Geography 26, 121–123 (2006)
4. Zhang, S., Wang, L., Zhang, W.: Quantitative Analysis on Change of Age Structure to the Fund Gap on China Basic Pension Insurance. Forecasting 29(2), 37–41 (2010)
5. Hou, Z., Guo, X.: Markov Decision Process, pp. 88–95. Hunan Science and Technology Press, Chang Sha (1998)
6. Liu, Z.: A Test of Market Efficiency in Shanghai Stock Market Using Markov Chains. System Engineering 18(1), 29–33 (2000)
7. Reinartz, W.J., Kumar, V.: The impact of customer relation ship characteristics on profitable lifetime duration. Journal of Marketing 67(1), 77–99 (2003)
8. Venkatesan, R., Kumar, V.: Customer lifetime value frameworkfor customer selection and resource allocation strategy. Journal of Marketing 687(4), 105–125 (2004)

Analysis of Supply – Requirement Game Model Dominated by Retailer in Supply Chain

Xia Li, Yingbo Qin, and Fanxing Kong

School of Management, University of Xi'an Polytechnic
Xi'an, Shanxi Province, China
lesoso@lawyer.com

Abstract. Today's market is primarily a buyer's market. Retailers are closest to the terminal market, so the strength of suppliers has been greatly weakened because of the lack in sales channel. Retailers with the market dominance can easily repress suppliers, so it is inevitable to bring about the conflict. To achieve their maximum benefits and reduce the conflict, suppliers and retailers need to work together. First, let's determine the optimum ratio of profit distribution. At the same time, enterprises should strengthen the business credit system; In addition, the society and government should regulate and conduct retailers' action. At last, every company should train its own management personnel.

Keywords: Supply chain, Supplier, Retailer, Game model.

1   Introduction

In the 21st century, increasingly fierce competition in supply chain and supply chain among enterprises has become the main form of competition, so the supply chain cooperation among enterprises becomes increasingly important. Suppliers and retailers are both focus on the distribution of benefits in the cooperation between them and how to take the reasonable distribution of gains should be one of the main problems, which is needed to be resolved urgently. If this problem can be solved well, it will play a crucial role in maintaining the stability of supply chain. The conflict between channels' members results in instability in the supply chain, and as we all know, distribution of benefits is the most important factor in channel conflict. So, how to deal with the relationship between channel members, especially how to coordinate the distribution of benefits on the whole supply chain, plays an important role in long run[1].

Supply Chain Game in foreign countries has been studied for several years, and through those studies we have gotten some valuable information about profit distribution based on game theory, such as the game between suppliers and retailers[2].

This paper discusses the supply chain game model dominated by retailers, and it reflects the strength of retailers and the helplessness of suppliers, which often result in conflict. Indeed, in order to achieve their maximum benefits and reduce the conflict, it

Y. Wu (Ed.): International Conference on WTCS 2009, AISC 116, pp. 803–808.

is necessary for suppliers and retailers to work together to determine the optimum ratio of profit distribution.

2   The Status of Suppliers and Retailers

Now, suppliers and retailers often are required to cooperate, but the management of the two sides is not in place, and they cannot deeply understand the relationship, particularly in risk-sharing and sharing on interests. So, it will result the contradictions between suppliers and retailers. On this point, I think that there are two main reasons:

1) Leverage has tilted in power. As retailers are closer to customers, so they can understand customers better, and get more information in time. As for manufacturers, this means that retailers and manufacturers can hold greater power. This indicates that the status of manufacturers and retailers have changed and become more beneficial to retailers.

2) Category management and brand management always conflict. The manufacturer's goal is to maximize the profit of its own brand. But retailers do not pay attention to the profitability of one particular brand, and their concern is the overall profitability of all brands in the pursuit of profit maximization. Category management is that retailers are focus on all the resources, rather than one manufacturer's brand. In category management, retailers can optimize through a variety of ways to maximize the profit. However, suppliers are trying to increase products to retailers, so they will inevitably conflict with the retailers.

3   The Reason That Retailers Lead Game in the Supply and Demand

We know that suppliers must pay promotional fees and slotting allowance, which are some important sources in retail profits. Suppliers and retailers are both seeking to maximize their own interests, but on the buyer's market condition, retailers repress suppliers through a variety of ways, including as follow[3]:

1) Limiting the suppliers' sales actions. For example, there are some restrictions on small and medium sized suppliers to provide products for other retailers or new entrants. At the same time, they boycott those who do not accept the harsh conditions.

2) Obtaining special price discounts from suppliers because of market advantages. This discount should be more than any other discount.

3) Occupy a large number of capitals of suppliers. Not only getting fees from suppliers for their own companies, but also by delaying payment, retailers have occupied the capital of suppliers and can transfer the risk of market.

4) Gaining a large number of additional fees. Such as slotting allowance, promotional fees, and special location and advertising fees.

Retailers have a dominant place in the game, for many reasons:

1) Products have a low technological content. Retailers face a buyer's market, where there is ample supply, so naturally the suppliers pick and choose.

2) Retailers' brands weaken suppliers' brands. Many retailers, who place an important market position, are developing their own brands and competing with suppliers' brands.

3) Retailers are closest to the terminal market. As the nearest terminal market, in a large part retailers can win consumers. Plus the vendor did not have good sales channels; retailers have obvious advantages in the process of cooperation and negotiation.

4) Over-relying on retailers, suppliers are lack of self-marketing. As retailers raise brand awareness, consumers accept retailers' brands over suppliers' brands to some extent. Suppliers, particularly small and medium suppliers, achieve business value mainly through retailers' brands and sales channels, leading suppliers to place a passive position in the game.

5) Retailers are able to shift the market risk in time to the supplier or consumer, but suppliers can not do this point.

4   Optimization of Supply and Demand Game

4.1   Determine the Optimum Ratio of Profit Distribution

First, we assume that the commodity demand function is

$$Q = a - b(P_1 + P_2) \tag{1}$$

Where P1 is the price of products for retailers, and P2 is the increased price for retailers based on P1. Q is the quantity of sales. M1 is the profit of suppliers; M2 is retailers' profit; C1 is the cost of suppliers; C2 is the cost of retailers. Both of them are constant. Suppliers' profit function is expressed as:

$$M_1 = (P_1 - C_1)Q \tag{2}$$

Retailers' profit function is expressed as:

$$M_2 = (P_2 - C_2)Q \tag{3}$$

To maximum the profit, suppliers need to meet

$$\partial M_1 / \partial P_1 = 0,$$

and retailers need to meet

$$\partial M_2 / \partial P_2 = 0,$$

and get the solution:

$$Q = (a - C_2 b - C_1 b) b / 3 \tag{4}$$

$$M_1 = M_2 = (a - C_2 b - C_1 b)^2 / 9b \tag{5}$$

We can see that suppliers and the retailers' profits are equal. If the entire profit is considered as a large cake, then, suppliers and retailers can gain the half of this cake and retailers can not take advantage of a buyer's market to suppress the production of suppliers. It is an inevitable choice for suppliers and retailers to keep their profit margin to avoid conflicts.

We look at the profit distribution dominated by retailers:

Here we add λ as the coefficient of cooperation of suppliers and retailers, and its value is at between 0 and 1. When the value is equal to 0, it means that it is completely uncooperative. While the value is equal to 1, it means complete cooperation. Suppliers' profits function as follow:

$$M_1 = (1 - \lambda)(P_1 - C_1)Q \tag{6}$$

Retailers' profits function as follow:

$$M_2 = \lambda(P_1 - C_1)Q + (P_2 - C_2)Q \tag{7}$$

The total channel's profits function as follow:

$$M = M_1 + M_2 = (P_1 + P_2 - C_1 - C_2)Q \tag{8}$$

In fact, we find that completely uncooperative model, that is the model has been discussed above, and in the complete cooperation model the suppliers' profit margin is squeezed to zero. However, retailers have pricing power, and the actual situation can only occur between the two. That is to say, suppliers are unable to reach half of the profits, so in order to maximize their profits, suppliers, at least cannot allow retailers to control the market[4].

4.2   Strengthening Enterprise Credit System

Suppliers and retailers is a relationship between supply and demand in the supply chain. Therefore, the key is cooperation among enterprises in the supply chain management, as well as well coordinated between the design of production, marketing, competitive strategy and other aspects[5]. To ensure that all enterprises form a stable, harmonious relationship, honesty and trustworthiness have become the criterion of the behavior. This will not only reduce the adverse selection and moral hazard in the supply chain between enterprises, but also reduce transaction costs and the crisis of confidence between the partners, and ensure the long-term stability and development of the supply chain.

4.3 Regulating the Behavior of Retailers

Retailers, if they are abuse in power, will do great harm to the community, consumers and even themselves. On this condition, companies in the distribution channel have the low efficiency because of the lack of competition. In a certain extent, retailers will be to relax the internal management and cost control, and it will lead to give up business innovation and motivation. Therefore, from a legal point, it is necessary to regulate the behavior of retailers. As the times of short supply gone, suppliers' status has declined in market power, and retailers' status is enhanced. Thus, suppliers are required to protect their own interests through various means. Recognizing and establishing a good working relationship between suppliers and retailers will be an effective method to avoid conflicts[6].

4.4 Developing Their Own Management Talented Person

At present, for foreign advanced supply chain management thinking, we should inherit and improve it, and put it into practice to test and develop. In this process, raining and developing their own management talented person is the key. If suppliers and retailers are both having many management staffs, they can well deal with the relationship cooperation and competition, in order to maximize the profits of the two sides.

5 Conclusion

In this paper, we firstly analysize the supply-requirement game model dominated by retailer in supply chain in detail, and simply provide a number of ways to improve the efficiency of the supply chain. To achieve their maximum benefits and reduce the conflict, suppliers and retailers need to work together. First, let's determine the optimum ratio of profit distribution. At the same time, enterprises should strengthen the business credit system; In addition, the society and government should regulate and conduct retailers' action. At last, every company should train its own management personnel. Of course, there are some shortcomings in this paper, and it still needs to be improved. As we all know, in reality the entire supply chain not only includes retailers and suppliers, and there are many other participants, such as customers, factories, etc. At the same time, there are many other factors that also affect the efficiency of the entire supply chain.

Acknowledgments. With the valuable advice given by Professor Li of school of management, and great support from others, e.g. Mr. Kong as my classmate, this paper was completed. The author hereby would like to take this opportunity to thank all who have been providing supports during and after the preparation of this paper.

References

1. Ma, S.: Supply chain management. Higher Education Press, Beijing (2003)
2. Zhao, M.: Study of Comparative Model between Suppliers and Retailers. Business Economics and Management (2005)
3. Zou, A., Wu, Q.: Retailer-led supply chain coordination of Game Theory. Market Modernization (2007)
4. Li, W.: Game Theory of supply chain profit distribution dominated by distributors. Industrial Technology & Economy (2006)
5. Zhou, Y.S.: The relationship between suppliers and retailers in crisis and countermeasures. Marketing (2003)
6. Ji, F.: Retailers' Power Abuse. Economist (2004)

Teaching and Thinking of Multimedia

Zhu Xinhua[1] and Zu Yanshuai[2]

[1] Students Office Hebei Polytechnic University
[2] Tangshan Construction Engineering Secondary School Tangshan, China
zhuu23@yahoo.cn

Abstract. After several years of making and application of multimedia courseware for teaching practice, the authors has a lot experience about many aspects and were discussed in the text, including the choice of the multimedia form and the application of technology, the key to build a quality teaching courseware, and the issues involved in the application of multimedia courseware for teaching demonstration.

Keywords: Multimedia form, Teaching, Countermeasures.

1   Introduction

As the computer and network technology continues to evolve, it is a very ordinary thing that multimedia computer has been widely used in public life. It has brought infinite vitality to our classroom by using multimedia computer technology to education and teaching, and it has gradually been widely utilized in classroom teaching.   Advantage of this modern educational technology is unmatched by traditional teaching methods. Teachers can very visually display   course content through multi-media, saving a lot of writing time on the blackboard , broadening the scope of use of the blackboard so that improve the efficiency of the classroom.

2   The Choice of Multi-media Format and Development Tools

The application of multimedia in teaching has a variety of forms and is multi-faceted. It can not only be combined with traditional teaching as an auxiliary, but also open up new teaching methods. The choice of multi-media form and development tools must be based on the use of environmental conditions and the nature of the curriculum. Present applications of multimedia teaching are mainly in the following areas:

2.1   Teaching Demonstration

The application of multi-media technology and expression will present educational content under a specific display device to support the teachers to explain. In multimedia classroom, teachers can achieve high quality of knowledge dissemination by the combination of multi-media and electronic projector.

Y. Wu (Ed.): International Conference on WTCS 2009, AISC 116, pp. 809–813.
springerlink.com   © Springer-Verlag Berlin Heidelberg 2012

2.2   Interactive Teaching

Integrating computer multimedia and network technology, students learn through two-way interactive multimedia courseware in the computer classroom or by means of individual self-study or network resources collaborative self-study.

2.3   Simulated Teaching

Multimedia technology, simulation technology and a virtual environment are integrated to simulate, emulate, virtual or reproduce some things non-existent or difficult to experience in the real life, so that learners can immersive which enhance teaching effectiveness and improve teaching quality. For example, pilots simulate flight.

2.4   Modern Distance Education

In the open computer network-based teaching system, students can carry out on-demand learning regardless of age and not rigidly adhering to a fixed time and space, that is, according to their own needs and the current level to choose different schools, different teachers, and right time to learn.

During forms of the above, the simulated teaching is applicable to some special education, the modern distance education need to have considerable network resources, and the implementation of an open teaching needs a major transformation process of teaching management and teaching process management. For the school environment, equipment, resources, teachers and curriculum they teach, using teaching demonstration and interactive teaching should have considerable conditions. From the view of curriculum nature, some courses, such as computer software teaching, are more suitable for interactive teaching in the computer classroom. Other courses, such as the large number of professional courses in engineering, are more suitable for combination with multimedia presentations of teachers. The indoor and outdoor experiments matched with it, in theory, can use simulated teaching, but taking environment, technology, cost and effectiveness of teaching into consideration, at present it does not necessarily have an advantage.

3   Teaching Is Essential for Building Courseware

Courseware design should take two factors into account: First, technical requirements of multimedia applications, , including ease of operation, control flexibility, the use of the stability, storage security, jump accuracy, fault tolerance, compatibility and so on. Second is their teaching requirement, including teaching objectives, teaching content, teaching process, teaching effectiveness and so on. The first factor (technical factors) depends on determining platform, and the fine production, but the quality courseware more depends on second factor (teaching factors) which is the core of teaching, and any kind of teaching style can be without exception. Quality teaching courseware (especially the presentation of multimedia courseware) building should be highly concerned about the following points:

3.1   Write a Good Script

Multimedia courseware is a teaching process designed according to the teaching objectives, the performance of specific teaching content and the reflection of teaching strategies. Therefore, the production of multimedia courseware should be based on syllabus requirements, explicit teaching purposes, requirements and priorities, and difficulties. According to the teaching goal and the needs of the teaching task to determine the frame structure and produce the script with a clear thinking, concise content, outstanding focus and difficulty, easy-to-computer expression. For this reason, a good multi-media presentation script should be based on written carefully and high quality preparing lessons or lesson plans.

3.2   Instructor Hands-On Production

Front-line teachers, because of ongoing classes, lectures and practice, are often able to accurately and reasonably deal with course content and teachings arrangement. For example, how to break the key and difficult, how can we enable students to understand master, so they can have a better grasp of the quality of teaching and teaching efficiency. Therefore, the multimedia courseware should be produced by front-line teachers personally who study and practice repeatedly on the teachings. In addition, multimedia courseware is bound to do the appropriate changes in each lesson preparation. Noting that the above two points, presentations should put an end to spread copy and paste, and be based on the script well-prepared and in line with the logic of on classroom teaching. After front-line teachers' hands-on instruction, high-quality multimedia courseware can be produced with the teaching of ideology, hierarchy, targeted consistent with cognitive science, psychology and education laws, and teaching appropriate to the content of teaching.

4   The Problems Posed by the Teaching of Multimedia Courseware Demonstrate and Countermeasures

Multimedia teaching has brought a revolution to traditional teaching in the classroom, bringing a refreshing feeling. A large number of audio-visual information in high-tech means the impact of the performance of students' excitement of thinking, reflecting its unique advantages. At the same time, social development have become increasingly demanding for higher education, enrollment increases every year, teachers, equipment and environment has become increasingly tense according to the traditional modes of teaching; students need to learn more and more courses and teaching hours of each course are continuously compressed. In resolving these contradictions, the multi-media teaching has an unparalleled advantage. But the teaching demonstration of multi-media teaching also brought the issue which can not be ignored and should be paid close attention.

4.1   Organization of Classroom Teaching

In previous board speech instruction, the board played an important part in the teaching method. When the course ends, it shows a lecture with complete and clear structure. Courses be described, students can real-time understand the location of the current contents in the course structure. This board has the advantage of coherent speech, feeling strong level, starting step by step with the teaching content and the full structure. At the same time, the spatial and temporal redundancy of students browsing the spoken contents not only facilitate the comparison, recall, contact, deepen understanding, and if it has not kept up somewhere, it is easy to grasp the fall aspect, so it will not affect the overall process of lectures. The multimedia presentation makes these advantages no longer exist, the screen shows only the current (very small part) contents of an instant turn, it is difficult for the students to understand the structure of the lectures and to determine the position of the contents in the structure; it is impossible to browse, review the contents of the previous screen presentation in front of them with ties to deepen understanding of the content; all students must use the same screen at the same time to digest the contents; once did not keep up the progress, and can not figure out the current screen section, and corresponding to the item, it will further affect the follow-up study.

4.2   Process and Speed of Information Communication

In on-board speech teaching process, many teachers have a lot of "interest points" to seize the students' ideas, "enticed their" in-depth step by step; in paint, doodle while talking, so that students naturally break down and understand a number of complex plans. In the multi-media teaching, these are often replaced by made good, clear slides, losing valuable "process." Taking notes during lectures on the know-how is very beneficial; the speed of writing on the blackboard is suitable for students to take notes. It is almost impossible to enable students to take notes because of the speed of multimedia courseware instruction.

Light of the above, in the production and use of multimedia electronic presentations, they can take appropriate measures to diminish the problems posed by it.

• Hand script in line with the presentation of multimedia courseware to the students when each course comments. In this way, when the students listen, they can get fragmented contents of the slides together into a whole through the checking and control presentations, but also eliminates a large number of written records, drawing time. Only adding some notes during the process of teaching enable students to maximize their attention on the visual, hearing, digestion and understanding, which improve teaching and learning efficiency.

• The feel of classroom atmosphere is essential. Teachers communicate with students by creating classroom atmosphere is also essential that such feelings and exchange, to a considerable extent, mobilize the teachers in their own improvisational expression (such as improvisation of bringing in the occasional repetition, emphasis, writing, paints, etc.), which can produce spontaneous expression of that time, and reach the most appropriate scenario results. These are often generated in a live classroom, and many are not able to pre-design in courseware. At the same time, the language, gestures and expressions of teachers play a direct impact on teaching effectiveness, and it is the most difficult for multi-media technology as a substitute. Thus, in the production and use of courseware, pre-dubbing and presetting time automatically next page should be less used with caution, and maintain communication with students face to face in order to enhance expression such as improvisation.

5   Conclusion

Multi-media teaching as a modern educational technology is a major change in traditional teaching; there are t the incomparable advantages of traditional teaching. Multi-media teaching has become increasingly important, multi-media technology as a new type of educational tool has been applied by more and more educators. Therefore, the introduction of multi-media teaching, according to the teaching requirements and the need to address the problem in the traditional teaching choose the form and technology carefully; according to the various factors of teaching and careful integration of the essence of the traditional teachings to make courseware carefully to discover new problems under new teaching situation so that truly let modern technology to better serve modern teaching.

References

1. Xu, J., Hui, Z.: Exploration and Reflection of Blackboard Teaching and Multimedia Teaching. Teaching Forum, p. 153 (January 2008)
2. Li, X.: Exploration and Reflection of Blackboard Teaching and Multimedia Teaching, pp.110–112. Qufu Normal University (April 2004)
3. Sha, H.: Analysis and Thought with Regard to the Status of Education Information. In: Educational Theory and Experiment, vol. 12 (2002)

Talk of the Significance of Folk Paper Cutting Art Teaching Introduced to Higher Art Education

Zhu Xiaohong

Nanyang Normal University Henan, China, 473061
zhuyr2@hotmail.com

Abstract. Folk paper cutting art is the folk carrier of Chinese nation deposited for thousands of years, contains the elite of national culture and art, which is one of the important representatives of excellent Chinese traditional culture. The introduction of folk paper cutting art to the curriculum system of college art education can make up for the shortages of higher education, pass on and protect national art, uphold patriotic passion, enhance folk culture and art education, improve the humanistic attainments of students in colleges and universities, restrain corrosive influence of adverse culture, shape physical and psychological health of college students, learn the elite of folk art creation and improve aesthetic level.

Keywords: Folk paper cutting art, Art education in colleges and universities, national traditional culture and art, Students in colleges and universitie.

1 Introduction

Educationalist Shuzi Yang evaluated current higher education that too little cultural influence made students' humanity quality low, too narrow professional education made students' academic view cabined and academic foundation unstable, too much guidance of utilitarianism made students' comprehensive quality cultivation and foundational training little, too much generality restrict made students' personality development deficient. [1] In addition, folk paper cutting art is the folk carrier of Chinese nation deposited for thousands of years, contains the elite of national culture and art, which is the materialized embodiment of concepts of survival, philosophy, aesthetics, morality and value summarized and reserved in survival and development of Chinese nation for thousands of years as well as one of the important representatives of excellent Chinese traditional culture. So it's very necessary to introduce folk art to art education in colleges and universities in order to enhance students' humanity quality education, to cultivate students' national cultural identity and heritage education of national art, to guide students to shape correct concept of life and value, and to reinforce the function of current higher education on students' comprehensive quality cultivation. So it's suitable to introduce folk art to the curriculum system of art education in colleges and universities to make up for the shortages of higher education. It's characteristics of convenient creations, simple tools & materials, popularity and learn ability are suitable for the universal teaching way of art education in colleges and universities.

Y. Wu (Ed.): International Conference on WTCS 2009, AISC 116, pp. 815–822.

2   The Current Situation in Basic Education of Art and Design

In the traditional ideas, most people still believe that art design for the integration of appurtenants pure arts, rather than put them as two independent disciplines. This misunderstanding to adhere to the old mode in the admissions process, and the teaching practice, it is serious impact on the professional development of Chinese contemporary art design. China's early education development in art design is largely in reform and the open policy. Since the 1980s, with the growing Chinese and western cultural exchanges and integration, the idea of people and ideas through the great changes, and limit the thought form of single shadow meld held the don - - policy for historical reasons. Guiding principle of flowers and a hundred schools of thought contend "art education, separation of political and economic life. In this situation, and constantly improve the people's living standards improve market economy, the establishment of national economy is developing fast, it will gradually mature, thus completing the separation of art design and pure art. [1] and explore the evolution of art design education, we can clearly see the art design enormously promoted the development of China's economy and society. However, after further communication between Chinese and western culture, we can clearly feel the development of China's economy and the development of art design education has obvious separation and asynchronous. Although China's economic development has a very large, but art design specialty teaching is still the dominant traditional paintings of teaching and imitation of the western art and design form. It's hard to find a job that can best embody the essence (innovation and creativity of art design, but Chinese contemporary art design innovation and creativity is the best reflects the relationship of media arts and sciences [2].

3   The Problems Faced in Fundamental Teaching of Art Design

For modern Chinese art design, from the traditional art, everything is still not perfect. Therefore, we need to change the traditional ways of thinking, and meet the requirement of modern art design, the new situation.

3.1   Lack of Innovation Consciousness for Education System

China now is artistic design education foundation is the main disadvantages of cultivating students' lack of creativity, lack of innovation spirit, of course, the implication of the design. Analysis of unity, thinking and artistic design of teaching and pure fuzzy identification, pure art education of surgical teaching, teaching the dependence of the unified guidance is so simple art and design education development. Analysis of the artistic design education on the basis of the status quo, as you can see, in art design specialty of the basic requirements and university students in the study, the teaching mode of art design specialty teaching characteristics. China's higher schools of traditional teaching syllabus and decades of design art form of the dependence of the pure artistic influence virtually teachers and students' creative self expression, no innovation spirit of teachers cultivate innovative talents. As the outline, and teaching means of art and design, not static, but should keep pace with The Times, each year should not, after all, art design specialty service market change quickly, people understand the requirements of preference, color, be fond of, these are not invariable.

3.2   Long-Confused "Generally Unified" Training System

Over the years, the university entrance exam system has no necessary art changes. Requirements of test subjects, such as the sketch, literary sketch, color roughly the same. Although some refers to the cultural demand, and because of prejudice to the so-called "three minor subject", only those who are not good at culture research to learn the art of generally low, cultural and artistic applicant qualification. However, art design is quite reasonable discipline; it requires the students to have rich imagination and rigorous logic thinking ability, due to the complex of art and design of the market. While inheriting traditions is the outstanding quality, the quality is not suitable for art design specialty education foundation, especially in the arts. Inheriting the traditional art form the basic training mode will only basic teaching art design specialty is one. Design involves creativity, it is hard to imagine, in modern society have clear the legal protection of intellectual property rights, and strictly, once the repeated work to enter the market, how serious consequences.

3.3   Curricular System Is Unreasonable

"Great oaks from little acorns grow." In modern design education reform of basic education should design. If there is no system, scientific and reasonable design education foundation education reform, realize the goal. In the school curriculum also means "product" and "three". This course features and advantages, but if not from the talent cultivation standard height, because it is the organic composition as a whole and complete curriculum is divided are inevitable. This curriculum arrangement, the simple as a foundation course and specialized course, students' lack of professional design apart seriously, underlines China art creativity design education ", open-minded, thick, high quality and ability of comprehensive and innovative talents.

3.4   The Dislocation between Education Output and Market Demand

Along with the development of China's higher education, rapid design art design education of enrollment expansion of excessive blind. But at the same time in professional setting and course teaching system, a serious problem, most of the school's professional basic focus on art design, packaging design, advertising, packaging design, interior design, environmental art design speciality, so-called general education mode, the design thought for China's unified design, each year thousands of cultivating output design graduates. But it is worth pondering, in the background of economic globalization, in domestic economic development and at the same time, the higher education system and that of Chinese market demand design has become a huge contrast: as the world's largest population and manufacturing power, is now the world's largest automobile manufacturing and sales force, home appliance manufacturing and exporting, furniture manufacture and export, manufacture and export, toy manufacturer and exporter. Until today, China is not your master in the realm of design. Even the most basic meet China market for design actual demands and talents training target gap exists serious dislocation. China's domestic design talents are extremely deficient, on the other hand, the design of vocational training, graduate students can completely satisfy the needs of the enterprises, many people engaged in the work of other industries, is mainly used in domestic design education still traditional art and education, education should not be ignored design and market demand. This reflects the design education and the market demand, and the contradiction between the education and the design that Chinese manufacturing docking.

4   The Historical Development of Folk Paper Cutting Art

With a long history, our national paper cutting art derives from the ancient ancestors worship and praying to the Gods, takes root in great and profound Chinese traditional culture and survive in the deep life soil of workers. The discovered earliest paper cutting works by far are the five group flower paper cuts from the Northern and Southern Dynasties, whose techniques are similar to current paper cutting art. The main paper cutting works were of corolla carved in leathers and formed *printing* pattern in Tong Dynasty, but few remain. The functions of paper cutting art were detailed, technical paper cuts, decoration paper cuts and so on emerged in Song Dynasty. The paper cutting works handed on from Ming and Qing Dynasties were in the majority, whose shapes were various and style beautiful and delicate. In the early days of Republic of China, the paper cutting art handed down among ordinary people was viewed as the tangible material for research on folk art, which started to attract much attention of folk-custom circles. In Yan'an period, some progressive artists created lots of paper cutting works popular with masses, which not only realized the Zedong Mao's policy of art serving the broad masses of the people, but also promoted the new development of paper cutting art. Paper cutting art was still active in the life of the masses after founding of our country. With the continuously deepened reform and openness after the 1980s, material wealth swiftly increased, and the village and agricultural population were urbanized. The change of life style made the rural folk-custom soil rapidly disappear, by which folk paper cutting art lived, and national traditional culture was fiercely affected by external culture. At the same time, it also caused deep discussion and rethinking of the related humanity subjects such as folklore, sociology and so on. The society and academic circles began to emphasize paper cutting art, actively saved and sorted fold paper cutting art works, paid attention to the survival state of folk paper cutting artists, called the society to value the ancient folk art form and to cultivate folk paper cutting art talents, held high-level paper cutting art work exhibition at home and abroad, and attempted to protect national traditional culture through a series of fruitful work in order to make Chinese folk paper cutting art admitted by international community.

5   The Significance of Introducing Folk Paper Cutting Art Teaching to Higher Education

5.1   Pass on and Protect National Art, Uphold Patriotic Passion

Firstly, the current society is in the period of transformation, the conservation status of our national folk traditional culture is still unsatisfactory, and some folk art forms are endangered. Secondly, in the background of economic globalization, the strong infiltration of western culture makes national traditional culture despised. Under this circumstance, higher education must pay attention to the value balance between global culture and national culture, should try to carry forward the universal education of folk culture and art, to rouse the national confidence of college students, and to inspire college students' passion and attention to national folk culture and art. Folk paper cutting art not only has the universality of national culture in Chinese national traditional arts, but also has its artistic individuality, which fully absorbs the elite of national culture, links work at selected spots with that in entire areas, has a clear distinction between black and white, is valuable in feelings, focuses on expressing meaning and communicates the aesthetic culture of Chinese nation. It has bright theme, the design that "designs must have implication" and "implication must be auspicious" reflect people's longing for future and blessing to life. It continues and develops together with our ancient folk national custom, and reveals the expression of folk paper cutting art language, records people's thinking, and reflects laboring people's feeling of beauty, whose forms are flexible with colorful contents, and it assembles the typical characteristics of Chinese national traditional culture and art. So conducting folk paper cutting art course in higher education art education can make students feel intense atmosphere of folk art, at the same time, realize the national spirit and national culture permeated in the forms of paper cutting art, and it can guide students to love motherland, to love the people and to love our national art, it can improve students' interest in folk national culture and art, uphold the national spirit, cultivate students' patriotic passion of our country, cultivate appreciators and spreaders of folk culture and art, and pass on national culture.

5.2   Reinforce the Education of Folk Culture and Art and Improve Humanity Attainments

With the rapid development of society, humanity quality education is increasingly paid attention to, and reinforcing students' education of humanity quality has been the common view of educators in colleges and universities. Starting with folk art education, conducting humanity quality education in higher education necessarily attracts students' strong interest, which is the most easily acceptable education way for students.

Folk paper cutting art is the folk art form that is created by our laboring people in long life practice, accumulated in history, passed on by collectives and then developed. The folk paper cutting arts in different regions have local intense life flavor and unique artistic style, conserve deep Chinese traditional cultural deposits and local ancient national artistic characteristic, and have styles of their own in Chinese folk art. So folk paper cutting art has very high value of cultural research, involves humanities and social sciences such as anthropology, ethnology, sociology, history, literature & art, aesthetics, philosophy and so on, which directly and comprehensively embodies the cultural tradition and humanity spirit of Chinese nation. So introducing folk paper cutting art course to higher art education can inspire young students' enthusiasm of loving motherland, and strengthen national self-respect and self-confidence. Cultivate college students' keen insight, rich imagination, exuberant creativity, refined aesthetic interest and lofty character, make them fully develop in all aspects of ethics, intellect, physiques, aesthetics, train a generation of high-quality, versatile builders and successors of society, and guide the youth generation to pay attention to the people's livelihood and press close to the people.

5.3   Restrain Corrosive Influence of Adverse Culture, and Shape Physical and Psychological Health

With opening of society and frequent flow of persons, as national mark, the sense of blood relationship, race, region and others is less and less, the fundamental mark of nation are not blood relationship and region but culture, that is, national culture of national spirit and national characteristic. Faced with influx and permeation of worldwide various cultures and thoughts, the phenomenon of youth generation worshiping blindly foreign culture, coloring hair yellow, wearing blue glasses is very common, national culture is facing unprecedented severe challenge, national aesthetic standard and self-confidence are quietly declining.[2] so attempting folk paper cutting art teaching among the students in colleges and universities can give full play to the function of aesthetic education, replenish college students' spiritual life, enrich college students' campus culture and life, strengthen the further generation's sense of national identity and pride, do good for their thoughts and feelings of deeply learning from the people and for their acceptance of influence of healthy aesthetic interest, help them get rid of the spirit erosion of harmful thoughts and culture, provide conditions for their growing to become high-quality talents useful for the society, and make them uphold the fine traditional culture and virtues of Chinese nation.

5.4   Learn the Elite of Folk Art Creation and Improve Aesthetic Level

Folk paper cutting art reflects the formal beauty peculiar to Chinese nation, includes aesthetic concept of spirit, style, flavor and others, conserves the components of simplicity, popularization and original ecology, reflects countryside common people's aesthetic pursuit, form the traditional aesthetic standard of Chinese nation, and communicates aesthetic implication of Chinese nation. Influenced by such national folk art education, the students will gradually learn the aesthetic perception and essence implied in national folk art, improve their aesthetic appreciation ability, which makes students' alternation and collision of hand using and brain using, art and science, imagery thinking and abstract thinking. At the same time, for the students major in designing, paper cutting art can be infiltrated into various designs such as packing design, trademark advertisements, interior decoration, dress design, ceramic craft and so on. Reasonably apply various techniques of art expression by abstracting all elements of traditional folk paper cutting art, consequently make national traditional art beauty prominent, deepen traditional cultural connotation and arouse art resonance.

The courses of Appreciation and Creation of Chinese Folk Paper Cutting Art have been offered to undergraduates whose majors aren't art in the form of optional course in the universities such as Nanjing University, China University of Geosciences, Nanyang Normal University and so on, where undergraduates are made to learn about the cultural connotation of folk paper cutting art, realize the contained national emotion, aesthetic interest, thoughts, psychological characteristics and others through the teaching methods such as appreciating the treasures of folk paper cutting art, teaching techniques, introducing research results, collecting folk works and investigation and so on. Make students master applying folk paper cutting art language in order to create healthy and positive paper cutting art works that is close to real life, mater the methods of appreciating, exploring, sorting and studying folk art. On the basis of deep research on folk culture and art, Nanjing University has offered the course of Appreciation and Creation of Chinese Folk Paper Cutting Art to overseas students in College of Overseas Further Education, which makes praiseworthy contributions to promoting communication of culture and art in the East and the West. China University of Geosciences offers the course of Appreciation and Creation of Chinese Folk Paper Cutting Art to nearby 18 colleges and universities, gains important experience for interscholastic resource sharing and offering the course in all colleges and universities. The course of Appreciation and Creation of Chinese Folk Paper Cutting Art offered by Nanyang Normal University has conducted helpful exploration in reinforcing the artistic accomplishment of teachers in elementary education, popularizing folk paper cutting course in middle and primary schools and so on. The paper cutting art creating works by the students who leant the optional course in above three universities were respectively selected to join in "International Paper Cutting Art Exhibition" and awarded in the international national folk art exhibition in "Nanjing China Expo of Famous Historic and Cultural Cities in the World" hosted by Ministry of Culture, Ministry of Construction, Cultural Relic Bureau of People's Republic of China and Committee of United Nations Educational, Scientific and Cultural Organization.

The practice shows that introducing folk paper cutting art to course construction of higher education is the effective way of national excellent traditional culture education, offers feasible methods to improving talent quality cultivation in colleges and universities, explores new was to cultivate high-quality folk paper cutting art appreciators and spreaders, creates good environment and atmosphere for the inheritance and development of excellent national traditional culture, and lays a solid foundation for the future development of excellent national traditional culture.

Acknowledgements. The paper is supported by The "Tenth Five-Year" Education Science Plan Subject of Henan Province-- Public Art Education in Colleges and Universities-the Research of Curriculum System Construction of Traditional Folk Art (2009—jkghb—578).

References

1. Lu, Q., Ai, X.: Talk of Pushing forward Folk Art Education in Higher Education. Career Horizon (3) (2008)
2. Chen, Z., Huang, Y.: On Practical Significance of National Folk Culture and Art Educaiton. Agricultural Archaeology (3) (2007)

Variable Precision Induced Covering Rough Set

Chen Dingjun, Li Han, Li Li, and Wu Kaiteng

Key Laboratory of Numerical Simulation of Sichuan Province and College of Mathematics and Information Science, Neijiang Normal University Neijiang, Sichuan 641112, China
echen11@yahoo.cn

Abstract. Rough set theory is a new mathematical tools to process uncertain knowledge. The concept of *Variable precision* induced covering rough set is proposed and the properties of the model are discussed in the paper. some useful results about Variable precision induced covering rough set are got.

Keywords: Precision induced covering rough set, Neighborhood, covering.

1 Introduction

Rough set theory(RST), proposed by Pawlak [1,2], is an extension of set theory for the study of intelligent systems characterized by insufficient and incomplete information. It provides a systematic approach for classification of objects through an indiscernability relation. Many examples of applications of the rough set method to process control, economics, medical diagnosis, biochemistry, environmental science, biology, chemistry psychology, conflict analysis and other fields can be found in[3,4,5].

A lot of meaningful extensions of pawlak rough set are proposed by scholars. For example, variable precision rough set model was proposed by Ziarko [6]in order to deal with a certain degree of "inclusion"and "belong to". Zakowski [7,8] extended to the covering rough set by the covering in division. Zhang Yajun[9] further presented the covering rough set model based on variable precision in order to expand rough set application space and to generalize the classical rough set. Zhu Guoqi[10] proposed the concept of induced covering rough set and research the properties of the model. Now we further discussed the problem in variable precision covering rough set.

2 Preliminaries

Definition1. [11] Let U be a universe of discourse, C a family of subsets of U . If none subsets in C is empty, and $\cup C = U$, C is called a covering of U .

Definition2. [10] Let U be a non-empty set, $C = \{C_1, C_2, \cdots, C_n\}$ a covering of U . $\forall x \in U$, $C_x = \cap \{C_j | C_j \subseteq C, x \in C_j\}$,then $Cov(C) = \{C_x | x \in U\}$ is also a covering of U ,which is called a induced covering by C .

Y. Wu (Ed.): International Conference on WTCS 2009, AISC 116, pp. 823–828.
springerlink.com   © Springer-Verlag Berlin Heidelberg 2012

Definition3. [10]   Let $\Delta = \{C_i | i = 1, 2, \cdots, m\}$ a covering of U. $\forall x \in U$, $\Delta_x = \cap \{C_{ix} | C_{ix} \subseteq Cov(C_i)\}$,then $Cov(\Delta) = \{\Delta_x | x \in U\}$ is also a covering of U ,which is called a induced covering by Δ.

Definition4. [10]   Let (U, C) be a covering approximation space. For each $X \subseteq U$,
    Set

$$\underline{\Delta}(X) = \cup \{\Delta_x | \Delta_x \subseteq X\}$$

is called the induced covering lower approximation of X.
    Set

$$\overline{\Delta}(X) = \cup \{\Delta_x | \Delta_x \cap X \neq \varnothing\}$$

is called the induced covering upper approximation of X.

Definition 5. Let (U, C) be a covering approximation space. For each $X \subseteq U$, $0 \leq \beta < 0.5$,
    Set

$$\underline{\Delta}_\beta(X) = \cup \left\{ \Delta_x \left| \frac{|\Delta_x| - |\Delta_x \cap X|}{|\Delta_x|} \leq \beta \right. \right\}$$

is called the β induced covering lower approximation of X.

    Set

$$\overline{\Delta}_\beta(X) = \cup \left\{ \Delta_x \left| \frac{|\Delta_x| - |\Delta_x \cap X|}{|\Delta_x|} < 1 - \beta \right. \right\}$$

is called the β induced covering upper approximation of X.

3   The Properties of Variable Precision Induced Covering Rough Set

Theorem1. Let (U, C) be a covering approximation space. For each $X, Y \subseteq U$, $0 \leq \beta < 0.5$,then

(1) $\underline{\Delta}_\beta(\varnothing) = \overline{\Delta}_\beta(\varnothing) = \varnothing$;

(2) $\underline{\Delta}_\beta(U) = \overline{\Delta}_\beta(U) = U$;

(3) $\underline{\Delta}_\beta(X) = \neg\overline{\Delta}_\beta(\neg X), \overline{\Delta}_\beta(X) = \neg\underline{\Delta}_\beta(\neg X)$;

(4) $\underline{\Delta}_\beta(X \cap Y) \subseteq \underline{\Delta}_\beta(X) \cap \underline{\Delta}_\beta(Y), \overline{\Delta}_\beta(X \cup Y) \supseteq \overline{\Delta}_\beta(X) \cup \overline{\Delta}_\beta(Y)$;

(5) $\underline{\Delta}(X) = \underline{\Delta}_0(X), \overline{\Delta}(X) = \overline{\Delta}_0(X)$

(6) if $X \subseteq Y$, then $\underline{\Delta}_\beta(X) \subseteq \underline{\Delta}_\beta(Y), \overline{\Delta}_\beta(X) \subseteq \overline{\Delta}_\beta(Y)$;

(7) if $\beta \geq \alpha$, then $\underline{\Delta}_\beta(X) \supseteq \underline{\Delta}_\alpha(X), \overline{\Delta}_\beta(X) \subseteq \overline{\Delta}_\alpha(X)$.

Proof. (1) According to Definition5, we have

$$\underline{\Delta}_\beta(\varnothing) = \cup\left\{\Delta_x \left| \frac{|\Delta_x| - |\Delta_x \cap \varnothing|}{|\Delta_x|} \leq \beta\right.\right\} = \cup\{\Delta_x | 1 \leq \beta\} = \varnothing$$

Similarly, we have $\overline{\Delta}_\beta(\varnothing) = \varnothing$.

(2) According to Definition5, we have

$$\underline{\Delta}_\beta(U) = \cup\left\{\Delta_x \left| \frac{|\Delta_x| - |\Delta_x \cap U|}{|\Delta_x|} \leq \beta\right.\right\} = \cup\{\Delta_x | 0 \leq \beta\} = U$$

Similarly, we have $\overline{\Delta}_\beta(U) = U$.

(3) According to Definition5, we have

$$\overline{\Delta}_\beta(\neg X) = \cup\left\{\Delta_x \left| \frac{|\Delta_x| - |\Delta_x \cap \neg X|}{|\Delta_x|} < 1 - \beta\right.\right\} = \cup\left\{\Delta_x \left| \frac{|\Delta_x \cap X|}{|\Delta_x|} < 1 - \beta\right.\right\}$$

$$= \cup\left\{\Delta_x \left| 1 - \frac{|\Delta_x \cap X|}{|\Delta_x|} > \beta\right.\right\}$$

$$= \neg\cup\left\{\Delta_x \left| 1 - \frac{|\Delta_x \cap X|}{|\Delta_x|} \leq \beta\right.\right\}$$

$$= \neg\underline{\Delta}_\beta(X)$$

Therefore $\underline{\Delta}_\beta(X) = \neg\overline{\Delta}_\beta(\neg X)$.

Similarly, we have $\overline{\Delta}_\beta(X) = \neg\underline{\Delta}_\beta(\neg X)$.

(4) For any $x \in \underline{\Delta}_\beta(X \cap Y)$,we have $\dfrac{|\Delta_x| - |\Delta_x \cap (X \cap Y)|}{|\Delta_x|} \le \beta$. We can get

two inequalities $|\Delta_x \cap (X \cap Y)| \le |\Delta_x \cap X|$ and $|\Delta_x \cap (X \cap Y)| \le |\Delta_x \cap Y|$,

which yields $\dfrac{|\Delta_x| - |\Delta_x \cap X|}{|\Delta_x|} \le \dfrac{|\Delta_x| - |\Delta_x \cap (X \cap Y)|}{|\Delta_x|} \le \beta$ and

$\dfrac{|\Delta_x| - |\Delta_x \cap Y|}{|\Delta_x|} \le \dfrac{|\Delta_x| - |\Delta_x \cap (X \cap Y)|}{|\Delta_x|} \le \beta$ these mean $x \in \underline{\Delta}_\beta(X)$ and

$x \in \underline{\Delta}_\beta(Y)$.

Therefore $x \in \underline{\Delta}_\beta(X) \cap \underline{\Delta}_\beta(Y)$. Then we prove

$\underline{\Delta}_\beta(X \cap Y) \subseteq \underline{\Delta}_\beta(X) \cap \underline{\Delta}_\beta(Y)$.

Similarly, we have $\overline{\Delta}_\beta(X \cup Y) \supseteq \overline{\Delta}_\beta(X) \cup \overline{\Delta}_\beta(Y)$.

(5) According to Definition5, we have

$$\underline{\Delta}_0(X) = \cup \left\{ \Delta_x \left| \frac{|\Delta_x| - |\Delta_x \cap X|}{|\Delta_x|} \le 0 \right. \right\} = \cup \left\{ \Delta_x \left\| \Delta_x| - |\Delta_x \cap X| = 0 \right. \right\}$$

$$= \cup \left\{ \Delta_x | \Delta_x \subseteq X \right\} = \underline{\Delta}(X)$$

Therefore $\underline{\Delta}(X) = \underline{\Delta}_0(X)$.

Similarly, we have $\overline{\Delta}(X) = \overline{\Delta}_0(X)$.

This means, when $\beta = 0$,the model is the extension of the induced covering rough set.

(6) For any $x \in \underline{\Delta}_\beta(X), X \subseteq Y$, we have

$$\frac{|\Delta_x| - |\Delta_x \cap Y|}{|\Delta_x|} \le \frac{|\Delta_x| - |\Delta_x \cap X|}{|\Delta_x|} \le \beta$$

Which means $x \in \underline{\Delta}_\beta(Y)$. Therefore $\underline{\Delta}_\beta(X) \subseteq \underline{\Delta}_\beta(Y)$.

Similarly, we have $\overline{\Delta}_\beta(X) \subseteq \overline{\Delta}_\beta(Y)$.

(7) For any $x \in \underline{\Delta}_\alpha(X)$, $\beta \ge \alpha$,we have $\dfrac{|\Delta_x| - |\Delta_x \cap X|}{|\Delta_x|} \le \alpha \le \beta$ which

means $x \in \underline{\Delta}_\beta(X)$.

Therefore $\underline{\Delta}_\beta(X) \supseteq \underline{\Delta}_\alpha(X)$.

Similarly, we have $\overline{\Delta}_\beta(X) \subseteq \overline{\Delta}_\alpha(X)$.

Definition 6. Let (U, C) be a covering approximation space. For each $X \subseteq U$, $0 \le \beta < 0.5$,

Set $\gamma_\beta(X) = \dfrac{|\underline{\Delta}_\beta(X)|}{|U|}$

is called the β induced approximation quality of X.

Set $\rho_C(X) = 1 - \dfrac{|\underline{\Delta}_\beta(X)|}{|\overline{\Delta}_\beta(X)|}$ is called the β induced rough measure of X.

Set $\alpha_C(X) = \dfrac{|\underline{\Delta}_\beta(X)|}{|\overline{\Delta}_\beta(X)|}$ is called the β induced approximation precision of X.

We can easily get the following theorem2 and theorem3.

Theorem2. Let (U, C) be a covering approximation space. For each $X \subseteq U$, $0 \le \beta < 0.5$, if $\beta \to 0.5$,then $\underline{\Delta}_\beta(X) \to \underline{\Delta}_{0.5}(X)$

$= \cup \left\{ \Delta_x \left| \dfrac{|\Delta_x| - |\Delta_x \cap X|}{|\Delta_x|} < 0.5 \right. \right\}$; $\overline{\Delta}_\beta(X) \to \overline{\Delta}_{0.5}(X)$

$= \cup \left\{ \Delta_x \left| \dfrac{|\Delta_x| - |\Delta_x \cap X|}{|\Delta_x|} \le 0.5 \right. \right\}$.

Theorem3. Let (U, C) be a covering approximation space. For each $X \subseteq U$, $0 \le \beta < 0.5$, if $\beta \to 0.5$,then $\overline{\Delta}_{0.5}(X) = \bigcap_\beta \overline{\Delta}_\beta(X)$; $\underline{\Delta}_{0.5}(X) = \bigcup_\beta \underline{\Delta}_\beta(X)$.

4 Conclusions

In this paper, we have proposed the concept of the variable precision induced covering rough set. We also discuss their properties and get some useful results in the variable precision induced covering rough set.

Acknowledgments. This paper was prepared based on research project sponsored by Youth Foundation of Sichuan Provincial Education Department(No. 09ZB105) and the National Natural Science Foundation of China(No. 10872085).

References

1. Pawlak, Z.: Rough sets. International Journal of Computer and Information Science 11, 341–356 (1982)
2. Pawlak, Z.: Rough sets: Theoretical Aspects of Reasoning About Data. Kluwer Academic Publishers, Boston (1991)
3. Angiulli, F., Pizzuti, C.: Outlier mining in large high-dimensional data sets. IEEE Trans. on Knowledge and Data Engineering 17(2), 203–215 (2005)
4. Polkowski, L., Skowron, A. (eds.): RSCTC 1998. LNCS (LNAI), vol. 1424. Springer, Heidelberg (1998)
5. Zhong, N., Yao, Y., Ohshima, M.: Peculiarity oriented multidatabase mining. IEEE Trans. on Knowledge and Data Engineering 15(4), 952–960 (2003)
6. Ziarko, W.: Variable Precision Rough Set Model. Journal of Computer System Science 46(1), 39–59 (1993)
7. Zakowski, W.: Approximations in the space (u,π). Demonstratio Mathematica 16, 761–769 (1983)
8. Bonikowski, Z., Bryniarski, E., Wybraniec, U.: Extensions and intentions in therough set theory. Information Sciences 107, 149–167 (1998)
9. Zhang, Y.-J., Wang, Y.-P.: Covering Rough Set Model Based on Variable Precision. Journal of Liaoning Institute of Technology 26(4), 274–276 (2006)
10. Zhang, G., Huo, Y.: The concept and property of induced covering rough set. China New Technologies and Products 12, 237 (2010)
11. Zhu, W., Wang, F.-Y.: Relations among Three Types of Covering Rough Sets. In: IEEE GrC 2006, Atlanta, GA, USA, May 10-12, pp. 43–48 (2006)
12. Sun, S.-B., Liu, R.-X., Qin, K.-Y.: Comparison of Variable Precision Covering Rough Set Models. Computer Engineering 34(7), 10–13 (2008)

Degree Induced Covering Rough Set Model

Chen Dingjun, Li Li, and Wu Kaiteng

Key Laboratory of Numerical Simulation of Sichuan Province and College of Mathematics
and Information Science, Neijiang Normal University Neijiang, Sichuan 641112, China
echen11@yahoo.cn

Abstract. By the error of classification on the base of the model of induced rough
set based on covering, construct the degree induced covering rough set model,
discusses several relevant properties.

Keywords: Induced rough set, degree rough set, Covering, error of classification.

1   Introduction

Rough set theory(RST), proposed by Pawlak [1,2], is an extension of set theory for the
study of intelligent systems characterized by insufficient and incomplete information. It
provides a systematic approach for classification of objects through an indiscernability
relation. Many examples of applications of the rough set method to process control,
economics, medical diagnosis, biochemistry, environmental science, biology,
chemistry psychology, conflict analysis and other fields can be found in[3,4,5].

A lot of meaningful extensions of pawlak rough set are proposed by scholars. For
example, variable precision rough set model was proposed by Ziarko [6]in order to deal
with a certain degree of "inclusion"and "belong to". Zakowski [7,8] extended to the
covering rough set by the covering in division. DAI Dai[9] induced the error of
classification on the base of the model of rough set based on covering, construct the
degree rough set model based on covering, discusses several relevant properties of this
model.. Zhu Guoqi[10] proposed the concept of induced covering rough set and
research the properties of the model. Now we further discussed the problem in variable
precision covering rough set.

2   Preliminaries

Definition1 [11]. Let U be a universe of discourse, C a family of subsets of U . If
none subsets in C is empty, and $\cup C = U$, C is called a covering of U .

Definition2 [10]. Let U be a non-empty set, $C = \{C_1, C_2, \cdots, C_n\}$ a covering of
U . $\forall x \in U$, $C_x = \cap \{C_j | C_j \subseteq C, x \in C_j\}$,then $Cov(C) = \{C_x | x \in U\}$ is
also a covering of U ,which is called a induced covering by C .

Y. Wu (Ed.): International Conference on WTCS 2009, AISC 116, pp. 829–833.

Definition3 [10]. Let $\Delta = \{C_i | i = 1, 2, \cdots, m\}$ a covering of U . $\forall x \in U$, $\Delta_x = \cap \{C_{ix} | C_{ix} \subseteq Cov(C_i)\}$,then $Cov(\Delta) = \{\Delta_x | x \in U\}$ is also a covering of U ,which is called a induced covering by Δ .

Definition4 [10].   Let (U, C) be a covering approximation space. For each $X \subseteq U$,

Set $\underline{\Delta}(X) = \cup \{\Delta_x | \Delta_x \subseteq X\}$ is called the induced covering lower approximation of X .

Set $\overline{\Delta}(X) = \cup \{\Delta_x | \Delta_x \cap X \neq \varnothing\}$ is called the induced covering upper approximation of X .

Definition 5. Let (U, C) be a covering approximation space. For each $X \subseteq U$, $k \in Z^+$,

Set $\underline{\Delta}_k X = \cup \{\Delta_x | |\Delta_x| - |\Delta_x \cap X| \leq k\}$ is called the k degree induced covering lower approximation of X .

Set $\overline{\Delta}_k X = \{\Delta_x | |\Delta_x \cap X| > k\}$ is called the k degree induced covering upper approximation of X .

If $\underline{\Delta}_k X = \overline{\Delta}_k X$,it is called definable. Otherwise, we call $(\underline{\Delta}_k X, \overline{\Delta}_k X)$ rough set.

When $k = 0$, the model be the induced covering rough set.

3   The Properties of Variable Precision Induced Covering Rough Set

Theorem1. Let (U, C) be a covering approximation space. For each $X, Y \subseteq U$, $k \in Z^+$,then

(1) $\underline{\Delta}_k(U) = U$, $\overline{\Delta}_k(\varnothing) = \varnothing$;

(2) $\underline{\Delta}_k(X) = \neg \overline{\Delta}_k(\neg X)$, $\overline{\Delta}_k(X) = \neg \underline{\Delta}_k(\neg X)$;

(3) $\underline{\Delta}_k(X \cap Y) \subseteq \underline{\Delta}_k(X) \cap \underline{\Delta}_k(Y), \overline{\Delta}_k(X \cup Y) \supseteq \overline{\Delta}_k(X) \cup \overline{\Delta}_k(Y)$;

(4) $\underline{\Delta}(X) = \underline{\Delta}_0(X)$, $\overline{\Delta}(X) = \overline{\Delta}_0(X)$;

(5) if $X \subseteq Y$, then $\underline{\Delta}_k(X) \subseteq \underline{\Delta}_k(Y)$, $\overline{\Delta}_k(X) \subseteq \overline{\Delta}_k(Y)$;

(6) if $k \geq l$, then $\underline{\Delta}_k(X) \supseteq \underline{\Delta}_l(X)$, $\overline{\Delta}_k(X) \subseteq \overline{\Delta}_l(X)$.

Proof. (1) According to Definition5, we have

$$\underline{\Delta}_k U = \cup\left\{\Delta_x \,\big|\, |\Delta_x| - |\Delta_x \cap U| \le k\right\}$$

$$= \cup\left\{\Delta_x \,\big|\, 0 \le k\right\}$$

$$= U$$

Similarly, we have $\overline{\Delta}_k(\varnothing) = \varnothing$. (2) According to Definition5, we have

$$\overline{\Delta}_k(\neg X) = \cup\left\{\Delta_x \,\big|\, |\Delta_x \cap \neg X| > k\right\}$$

$$= \cup\left(\neg\left\{\Delta_x \,\big|\, |\Delta_x \cap \neg X| \le k\right\}\right)$$

$$= \cup\left(\neg\left\{\Delta_x \,\big|\, |\Delta_x \cap X| \le k\right\}\right)$$

$$= \neg\underline{\Delta}_k(X)$$

Therefore $\underline{\Delta}_\beta(X) = \neg\overline{\Delta}_\beta(\neg X)$.

Similarly, we have $\overline{\Delta}_k(X) = \neg\underline{\Delta}_k(\neg X)$.

(3) For any $x \in \underline{\Delta}_k(X \cap Y)$,we have $|\Delta_x| - |\Delta_x \cap (X \cap Y)| \le k$. We can get two inequalities $|\Delta_x \cap (X \cap Y)| \le |\Delta_x \cap X|$ and $|\Delta_x \cap (X \cap Y)| \le |\Delta_x \cap Y|$, which yields $|\Delta_x| - |\Delta_x \cap X| \le |\Delta_x| - |\Delta_x \cap (X \cap Y)| \le k$ and $|\Delta_x| - |\Delta_x \cap Y| \le |\Delta_x| - |\Delta_x \cap (X \cap Y)| \le k$ .these mean $x \in \underline{\Delta}_k(X)$ and $x \in \underline{\Delta}_k(Y)$.

Therefore $x \in \underline{\Delta}_k(X) \cap \underline{\Delta}_k(Y)$. Then we prove $\underline{\Delta}_k(X \cap Y) \subseteq \underline{\Delta}_k(X) \cap \underline{\Delta}_k(Y)$.

Similarly, we have $\overline{\Delta}_k(X \cup Y) \supseteq \overline{\Delta}_k(X) \cup \overline{\Delta}_k(Y)$.

(4) According to Definition5, we have

$$\underline{\Delta}_0(X) = \cup\left\{\Delta_x \,\big|\, |\Delta_x| - |\Delta_x \cap X| \le 0\right\}$$

$$= \cup\left\{\Delta_x \,\big|\, |\Delta_x| - |\Delta_x \cap X| = 0\right\}$$

$$= \cup\left\{\Delta_x \,\big|\, \Delta_x \subseteq X\right\}$$

$$= \underline{\Delta}(X)$$

Therefore $\underline{\Delta}(X)=\underline{\Delta}_0(X)$.

Similarly, we have $\overline{\Delta}(X)=\overline{\Delta}_0(X)$.

This means, when $k=0$,the model is the extension of the induced covering rough set.

(5)   For   any $x\in\underline{\Delta}_k(X)$, $X\subseteq Y$,we   have
$$|\Delta_x|-|\Delta_x\cap Y|\leq|\Delta_x|-|\Delta_x\cap X|\leq k$$

Which means $x\in\underline{\Delta}_k(Y)$. Therefore $\underline{\Delta}_k(X)\subseteq\underline{\Delta}_k(Y)$.

Similarly, we have $\overline{\Delta}_k(X)\subseteq\overline{\Delta}_k(Y)$.

(6) For any $x\in\underline{\Delta}_k(X),k\geq l$,we have $|\Delta_x|-|\Delta_x\cap X|\leq k\leq l$ which means $x\in\underline{\Delta}_l(X)$.

Therefore $\underline{\Delta}_k(X)\supseteq\underline{\Delta}_l(X)$.

Similarly, we have $\overline{\Delta}_k(X)\subseteq\overline{\Delta}_l(X)$.

Definition 6. Let (U,C) be a covering approximation space. For each $X\subseteq U,k\in Z^+$,

Set $\gamma_k(X)=\dfrac{|\underline{\Delta}_k(X)|}{|U|}$ is called the k degree induced approximation quality of X.

Set $\rho_C(X)=1-\dfrac{|\underline{\Delta}_k(X)|}{|\overline{\Delta}_k(X)|}$ is called the k degree induced rough measure of X.

Set $\alpha_C(X)=\dfrac{|\underline{\Delta}_k(X)|}{|\overline{\Delta}_k(X)|}$

is called the k degree induced approximation precision of X.

4   Conclusions

In this paper, we have proposed the concept of the degree induced covering rough set. We also discuss their properties and get some useful results in the degree induced covering rough set.

Acknowledgment. This paper was prepared based on research project sponsored by Youth Foundation of Sichuan Provincial Education Department(No. 09ZB105) and the National Natural Science Foundation of China(No. 10872085).

References

1. Pawlak, Z.: Rough sets. International Journal of Computer and Information Science 11, 341–356 (1982)
2. Pawlak, Z.: Rough sets: Theoretical Aspects of Reasoning About Data. Kluwer Academic Publishers, Boston (1991)
3. Angiulli, F., Pizzuti, C.: Outlier mining in large high-dimensional data sets. IEEE Trans. on Knowledge and Data Engineering 17(2), 203–215 (2005)
4. Polkowski, L., Skowron, A. (eds.): RSCTC 1998. LNCS (LNAI), vol. 1424. Springer, Heidelberg (1998)
5. Zhong, N., Yao, Y., Ohshima, M.: Peculiarity oriented multidatabase mining. IEEE Trans. On Knowledge and Data Engineering 15(4), 952–960 (2003)
6. Ziarko, W.: Variable Precision Rough Set Model. Journal of Computer System Science 46(1), 39–59 (1993)
7. Zakowski, W.: Approximations in the space (u,π). Demonstratio Mathematica 16, 761–769 (1983)
8. Bonikowski, Z., Bryniarski, E., Wybraniec, U.: Extensions and intentions in therough set theory. Information Sciences 107, 149–167 (1998)
9. Dai, D., Wang, J.-P., Xue, H.-F.: Degree Rough Set Model Based on Covering. Journal of Jianghan University (Natural Sciences) 36(1), 13–17 (2008)
10. Zhang, G., Huo, Y.: The concept and property of induced covering rough set. China new technologies and products 12, 237 (2010)
11. Zhu, W., Wang, F.-Y.: Relations among Three Types of Covering Rough Sets. In: IEEE GrC 2006, Atlanta, GA, USA, May 10-12, pp. 43–48 (2006)

Validity Considerations in Designing a Listening Test

Lijuan Wei[1] and Congying Liu[2]

[1] School of Foreign Languages, Handan College, Handan, Hebei, China
[2] Chinese Department, Handan College, Handan, Hebei, China
juan84100@tom.com

Abstract. Validity is achieved when a test measures what it is intended to measure. This paper studies the various validity considerations in designing the listening test and firstly concentrates on five aspects: the types of a listening test, content validity, criterion-related validity, construct validity, and face validity. It also points out the design of items and responses are essential for qualified tests. College teachers should pay attention to theoretical and empirical researches on testing design.

Keywords: Content validity, Criterion-related validity, Construct validity, Face validity.

1 Introduction

A test is valid if it measures accurately what it is intended to measure. (Arthur Hughes, 2008) Generally speaking, there're several considerations for designing a test. There're five aspects: test type, content validity, criterion-related validity, construct validity, and face validity.

2 Traditional Considerations

In designing a listening test for college students at the end of a term, there're validity considerations. Before a test is designed, its type must be firstly observed. And types of tests can be classified according to different criteria. They can be classified by means of test uses: Whether it's an achievement test or proficiency test, a diagnostic test or a placement test. They can also be classified by means of score interpretation: whether they are norm-reference tests or criterion-referenced test. They can also be classified by approaches to test construction: direct tests or indirect test. Finally they can be classified by the scoring methods: objective tests or subjective tests. For college listening course, an achievement test is mostly needed. And the test can also be used as a diagnostic one to some extent, in which way the students can know what's weak, or strong. The approaches used in the test would be direct ones. The test paper will completely include listening tasks and skills, mixed with some reading or writing skills for which are necessary for students to read the responses and write the required information. As for the scoring methods, both objective and subjective ones will be applied. For in the test, multiple choices are objective, while filling in the blanks,

Y. Wu (Ed.): International Conference on WTCS 2009, AISC 116, pp. 835–838.
springerlink.com

rewriting and dictation will be subjective. For such a test, the basic information of the syllabus must be mastered, that is, what has been taught and what should be tested must be known first. Usually the teaching materials and the skills practiced in class must be taken into consideration.

A test is said to have content validity if its content constitutes a representative sample of the language skills and structures, with which it is meant to be concerned. A listening test is characterized by its transient nature. For designing a listening test, operations of content must be first considered. Such a test must contain the basic features of listening, and may include listening materials, such as monologues, dialogues and passages. The skills tested would be directly related to the course objectives, and include: the ability to distinguish phonemes, the ability to listen for specific information, the ability to catch the main idea, the ability to predict the content, and the abilities to recognize some special function structures, such as a tag question and interrogatives. Since listening is always related to speaking and writing, the ability to retell and to dictate must be taken into consideration.(Buck, 2001)

A test is said to have criterion-related validity if results on the test agree with those provided by some independent and highly dependable assessment of the candidate's ability. And the independent assessment is the criterion measure against which the test is validated. College-band-four examination is a kind of popular test, against which the listening test can be designed and referred to.

A test is said to have construct validity if it proves the theoretical construction which it is based, and may also kinds: declarative knowledge and procedural knowledge, and strategic competence. Thus in designing such a test, the vocabulary shall not be more difficult than that in the listening class. And the functions of language must be taken into consideration, such as greetings and farewells, asking and inquiry, inventing and declining etc. For strategies, the test should contain necessary listening strategies such as, predicting, making quick judgment, short-term memory, long-term memory, taking notes, etc.

Lastly, a test is said to have face validity if it looks as if it measures what it is supposed to measure. The listening test must require the candidates to listen to some real materials, otherwise, it may be thought to lack face validity. With face validity, the test can seem good, and be accepted by teachers and candidates. The places for the test shall try to be valid, at least with a tape recorder, or in an audio room which is better and modern.

3   Other Considerations

Besides all the above, there're also other considerations for a qualified listening test at an appropriate level. There're tasks to set first. The samples of speech must be carefully chosen. Authentic speeches are preferred which are for students to develop their knowledge and ability. However the speeches shall also be within the syllabus, since it's an achievement test, and the vocabulary, grammar and knowledge being within the teaching syllabus, thus the levels of difficulty shall be under control. Possibly, some recordings have to be made especially for the test, and then care must be taken to make them as authentic as possible. If possible, a native speaker shall read the material and be recorded instead of the teacher himself. Secondly, the writing items are important for students to do the test. When designing a listening test, the designer shall try his best to write items that can check what he means to. It's essential to keep those items adequately far apart, namely the information points tested being far apart. If the two items are too close, the transient nature of listening shall prevent the students form catching the second one. The candidate will miss it and listen in vain for answers which one point each time, such as the location, or the time, each at a time. And nothing in the responses shall stand out, namely they should look alike at least. For example, they may be of the similar length, or similar tense, or similar parts of speech. Otherwise, they are either leading or misleading the candidates to the correct answer. What's more, the candidates should be alerted by key words which appear both in the response and the material that the information is contained. For example, one item may ask about "the beautiful scenery is...", and candidates will hear "the beautiful scenery is one of the features of the city". It would be less valid if the response is "one of the characteristics is pretty places", which will bring great difficulty to the listeners. For responses, shorter ones are better than longer and fancy ones. Thirdly, for filling in the blanks, rewriting and dictation, authenticity and the level of difficulty must be taken into consideration. And for those tasks, the students must have been given such practice in class, otherwise in the test, the performance of many students will lead the test designers to underestimate their ability. As for the scoring of these parts, errors of grammar and spelling shouldn't be over-emphasized, if the correct information is included.

4   Conclusion

In conclusion, every effort shall be made to ensure validity of the listening test. The inquiry into validity shall be always under way. The traditional consideration of the content validity, criterion-related validity, construct validity, and face validity, must be carefully examined. While considerations must also be given to the test types, the samples of speech and the levels of difficulty, which are quite essential and should also be emphasized. As for college teachers, theoretical and empirical researches shall never end. They should take some measures to enrich their theory and practice. There're three ways for English teachers to follow in order to help students improve their listening skills and their performance in listening test. Firstly, they must do the reflection on their teaching at regular intervals. Only in this way, can they find the teaching focus and the students' weakness. The knowledge of the students can help them design a valid test more easily and successfully. The validity of a test is based on the mastery of students feelings and prediction of the possible difficulties. Second, teachers should also try to assign proper homework for students. Especially they can give some on-line homework. According to the requirements of the education department, on-line learning should be focused and utilized. On the Internet, there are rich information and knowledge. Students can find listening materials easily. Therefore, teachers should carefully choose some good web sites for students to refer to, such as CCTV 9 programs. What's more, listening is also integrated skill which also includes the cultural knowledge. Students may try every means to know about the culture. Without culture, without language. Language just carries the culture. Whatever they are learning about the language, they must be familiar with the culture. Finally, teachers must take advantage of multimedia in order to achieve the validity of listening test. That's to say, more authentic environment should be considered, namely videos with motion pictures. Pure listening test is good. But listening tests with motion pictures will be better for its authenticity. In conclusion, teachers should catch up with the new development of English listening materials and skills to achieve validity of relevant test.

References

1. Hughes, A.: Testing for Language Teachers. Cambridge University Press, Cambridge (2008)
2. Buck, G.: Assess Listening. Cambridge University Press, Cambridge (2001)
3. Cohen: Research Method in Education. Routledge, London (1989)
4. Liu, R.: Research Methods in Foreign Language Teaching. Foreign Language Teaching and Research Press, Beijing (2000)
5. Selinger: Second Language Research Methods. Oxford University Press, Oxford (1989)

On Visuo-audio Comprehension Strategies of Interpreting

Lijuan Wei[1] and Congying Liu[2]

[1] School of Foreign Languages, Handan College, Handan, Hebei, China
[2] Chinese Department, Handan College, Handan, Hebei, China
juan84100@tom.com

Abstract. There is a great need for interpret ring in China. This paper analyses the existing audio-visual comprehension exercises and tries to cultivate corresponding learning strategies. It aims to study how to improve interpreters' ability and skills by making good use of the exercises.

Keywords: Interpreting, Audio-visuo comprehension, Learning strategies.

1   Introduction

With the opening of markets, economic development, the demand for interpreters is growing. How to effectively train translators becomed a topic of concern. Interpretation can be divided into three stages: understanding phase, the original language from the stage and re-expression of shell stage. [1] It can be seen as the basis for successful. If the interpreter cannot understand the contents of the original language, then the interpretation will be impossible. This paper aims to study how to make good use of audio0visual exercises in the practice of interpreting skills and the strategies in doing the exercises.

2   Main Contents of Audio-Visuo Exercises

Audio-visual placed learners in the interaction between sound and image conditions, providing them with images and sound in one set of sensory stimulation, and thus enhanced the effect of auditory input learners. Therefore, it can improve the lability of the interpreters. The main object of interpretation is conference, the meeting industry, the latest news and current affairs related to the policies and strategies. Therefore, the trainee can practice listening to some news reports, watching some video, and do exercises taken from the materials to improve their visual understanding. In the materials for training audio-visual comprehension, there're news story, whose broadcast speech rate is normal, and which have a lot of news terms. This kind of practice can help develop the trainees' speed and interpreting capacity. Audio-visual content are of more authenticity, and broader scope. The videos are selected from domestic and foreign corpus, mainly about English-speaking countries on all aspects of society and culture: politics, sports, education, and daily communication.

Y. Wu (Ed.): International Conference on WTCS 2009, AISC 116, pp. 839–842.
springerlink.com   © Springer-Verlag Berlin Heidelberg 2012

3   Learning Strategies

Whether it's for listening comprehension or audio-visual comprehension, requirements for listening capacity are relatively high, however listening capacity is rather low for Chinese planers in general. The following are strategies for the audio and visual comprehension.

First, in audio-visual comprehension exercises, the types of exercises are of their own audio-visual features, namely, the break of playing instructions and the design of questions. Then the learners should take good use of the break to do fast reading of the items and options, and have a reasonable prediction of the subject of the materials. As to the linear of sound materials, large amount of information just flows quickly, so in the design of the questions the tester is bound to choose a typical point of information, or material information in the recurring points. Therefore, in audio-visual process, testes should pay attention to the main content and the information repeated .

Second, there are wide range of topics and complex grammar, which is difficult for learners to grasp. But the fact is there are also rules to be found. Candidates must first understand the structure of news reports that is the inverted pyramid structure. The so-called inverted pyramid structure is that the most important fact is the lead which appears at the beginning of the report, and the secondary information is in the end.[2] The lead part is a summary of the news reports. There are four types of listening exercises, namely questions on the subject, detailed information, reasoning questions, T or F questions. ① For the subject questions, the point is to master the lead. Subject questions often include "What is the main idea of the news? / What does the news item mainly report?" . Therefore, the candidate shall be very attentive at the beginning part of the news report. If he does not understand it, he must keep calm, insist on listening to them, and predict from the relevant facts. Therefore, one must stay calm and make the greatest efforts to prepare for the worst. ② For questions of detailed information, the strategy is to read both the stem and items. This kind of questions are very difficult because the information is hard to locate. Candidates listen to a lot but still miss the details. There is a principle in testing theories, for the difficult stems, there must be enough clues for candidates to find and predict. Then one must first find the keywords or phrases in the stem and items, in this way, he can increase the sensitivity of the issue to make it easier. To read both questions and answers will make it easier to find the right answers. ③ for the reasoning questions, one must be sensitive to the meaning of the expression instead of the exact words. Because this kind of question always use different expressions to show the same meaning. The right answer generally can not be directly heard from the material, namely one cannot find the exact sentences in the item, but same meaning with different expressions. For these kinds of questions, one can refer to conversion method, which is to first find out the key words, and then the synonyms. In this way, the conversion of meaning and structure can be easily made. ④ For T or F questions, Determine whether the items are right or wrong is also a higher frequency of the news item types. For these kinds of questions, one can listen to the news, get the main points quickly to find and request the corresponding options, and then may make right judgments.

Third, the video enable learners exposed to the real scene, and to better simulate communication in English. It can help improve the validity of communication in reality. The focus should be on the screen and on the sounds. So in doing exercises, trainees need to master the following strategies:

First of all, to watch the video carefully, especially when the screen provides vocabulary and information, which is often closely related to important information. This prompted information usually appears at the beginning or in the end. Close attention must be given to the important moments. Then, one must pay more attention to body movements on the screen, and draw tips. In communication, sometimes the body language carries more information than words. Studies have shown that message through the silent language of the impact force is five times the words sound. [3] For example, the arms around his chest, is in defense; chin slightly- sink gesture is critical, hostile symbol. These are the common sense of our daily communication, and candidates are usually observed to be have better access to audio-visual capabilities.

4   Ways to Develop Listening Skills

How to develop listening skill is usually of great concern. One must have clear goals and fight with determination and perseverance, which can build confidence and courage. Specific ways are carried out through the following audio-visual abilities. First, listening should be exercised in a quiet place, and one should predict and judge the answers. Ability to predict is one of the most important factors for mastering the language, and it is also to improve one of the factors for listening comprehension. Prediction is important for the process of listening because the listener can have a conscious, and voluntary participation in psychological activities. However, in the process of such training, many learners just focus on the words and sentences, but ignore the importance of the role of listening comprehension, which causes the lack of quick, timely and accurate understanding of the consequences of listening materials. Prediction in listening comprehension can be divided into four categories: First, use the title to make a guess. Second, to use the topic sentences to guess. Third, guess based on the previous sentences. This is the most natural process of the listening comprehension. Because of the linear features of the spoken language, training should be actively pursued in a variety of projections to make one consciously achieve his own accumulation of knowledge, positive thinking, and make this spontaneous, primitive mental activity become scientific skills.

Secondly, one should make the combination of intensive listening and extensive listening. Only are the language skills improved, one's mental capacity will be subsequently strengthened. Intensive listening is to develop basic skills of learners, to do fine and standard work; extensive listening is to give learners some exercises and to broaden their view generally. Listen to authentic listening materials can make the language learners gradually adapt to the rapid flow of speech whose rate is too fast for learners to make up for the hearing of the content. While the main job is to listen, but one can not just listen, regardless of writing down notes. One should always pay attention to how much you write down, and learn to record important points. In addition to listening, one need to do exercises, repeat it, and imitate. Listening is boring and difficult to learn. Improve one's listening comprehension, he must overcome the psychological barriers, internal and external factors. The internal factors play a decisive role.

Thirdly, doing exercises at the right time, you can also use some method: ① Based on the theme and the words given in advance, one can preview and do the expansion, and collect some background material. Through the knowledge of the rehearsal time in the audio-visual practice, one can easily master the material. This allows learners to hear the thoughts of the topic, so that they know the main idea and key points instead of fragments. For example, if it is a football game audio-visual material, it may involve a number of Western names, player names, team names and game names, which may have been strange for the Chinese learners. Listening should be based on the understanding of the background knowledge such as physical education and court rules. ② When one encounter difficult videos, which is too hard to understand. He can take some indirect measures, such as, read about the original video first, and pay attention to the time limit for fast reading, read the main points, finally he can do the listening and speaking. This can effectively develop the capacity and understanding of audio-visual practioners.

5 Conclusion

In short, trainees should pay attention to skillful listening and speaking skills. they should pay attention to the cultivation of audio-visual strategies. In the listeninging process, the focus on key words and phrases rather than exact words, on content not on language forms, on the effect rather than on the details. They should listen to materials both intensively and extensively. Strengthen the background knowledge, to read and expand one's knowledge, enrich one's own language and cultural knowledge are essential. Languages are in a certain cultural context. Cultural background knowledge play an important role in listening comprehension. In addition to language learners and the skills to solve the obstacles, interpreters should also strengthen their own cultural awareness training, and the proper use of language. Watching video, some learners sometimes complain that it is too difficult. Iin fact, not to mention the video questions, even if the film is in English, the lack of relevant cultural background knowledge will lead to great difficulties. The reason is not the language itself, but because of lack of understanding of English social and cultural background. Language learners, we should also pay attention to the nation to learn English language habits, customs and socio-cultural background.. After hard work, and clever choices, the interpreters can better master the interpreting.

References

1. Seleskovitch: Training of Interpreting, p. 50. China Translation & Publishing Corporation, Beijing (1992)
2. Song, P.: News Writing, p. 169. Liaoning University Press, Shenyang (1996)
3. Hu, W.: Intercultural Communication Series, p. 105. Foreign Language Teaching and Research Press, Beijing (1999)

Research on Morality Education in Wushu Teaching in University and College

Li Wang[1] and Aijing Li[2]

[1] Department of Physical Education Shandong Institute of Light Industry Daxue Rd.
Changqing Dist. Jinan, Shandong Province, China
[2] School of Thermal Energy Engineering, Shandong Jianzhu University, Fengming Rd.
Lingang Development Zone, Jinan, Shandong Provinc, China
wang145@sogou.com

Abstract. Morality education in Wushu is a precious spiritual wealth of Chinese and is an important part of Wushu teaching in University and College. This paper first discusses the significance and the connotation of morality education in Wushu, then discusses how to implement the morality education in Wushu teaching from the following six aspects: improving teacher's self-cultivation, teaching student to form correct attitude to Wushu, blending humanism spirit into Wushu teaching, telling student to be a broad-minded person and have the ability to tolerate with others, teaching student to anneal his will and to cultivate his conduct, and teaching student to enhance his moral and legal awareness and to promote his psychological health and behavior normalization.

Keywords: University and college, Wushu teaching, Morality education.

1 Introduction

The beginning of Wushu morality can trace back to an old Chinese book, Book Of Changes, which says that "as heaven maintains vigor through movements, a gentle man should constantly strive for self-perfection. As earth's condition is receptive devotion, a gentle man should hold the outer world with broad mind." This kind of ethos promoted the formation of Wushu spirit, and finally formed Wushu morality. As a proverb says that "the essence of Wushu education is the education of morality.

2 The Vital Importance of Wushu Morality Education

2.1 The Vital Importance of Carrying Out Wushu Morality Education for the Harmonious Society Construction

Secretary-general Hu Jintao points out the main task of building a harmonious socialist society, which is featured by fairness and justice, honesty and friendliness, vigor and energy, stability and order, also the harmonious relationship between men and nature. Wushu morality are the top summary of the Wushu learners' principles, in the development of Wushu which is of unusual realistic importance in the days of

Y. Wu (Ed.): International Conference on WTCS 2009, AISC 116, pp. 843–848.
springerlink.com   © Springer-Verlag Berlin Heidelberg 2012

propagating the Chinese Wushu morality, enhancing the moral sense and governing the state by morality. The abundant spirit connotation in Wushu morality should be upheld with more efforts, thus becoming a significant part of socialist spiritual civilization. Chinese traditional virtues, inherited by the Wushu, along with the functions of its own, has played an important role in punishing the criminals ,maintaining the social stability and building up the social climate of honesty, humility , respecting the elder and caring the children in the past. Nowadays, Wushu still helps a lot in improving the ideological and moral qualities of our citizens, which weighs heavily for building a harmonious socialist society.

2.2   The Vital Importance of Carrying Out Wushu Morality Education to Eocialist Morality

As time goes by, essences of Wushu in every generation precipitate and Wushu morality of rich content come into being. In every dynasty and generation, every Wulin branches has their standards of morality and disciplines, which lay emphasis on the qualities of kindness, politeness, loyalty, humidity and tolerance. The functions of the values in Wushu are important for keeping the social stability, protecting the interests of our people and adjusting personal relationships. What the traditional Wushu morality concerns with is that people should learn about moralities first before practicing Wushu, that people should be respectful and modest and not grab things of others. These are the qualities of an upright man. Moreover, people should practice Wushu in a reasonable way, instead of doing something wrong and illegal, which leads to the harm to the morality and behaviors, the shames and even the lost of your life. At now, Wushu morality should not only continue to propagate the mentioned qualities, but also should combine with the socialist morality construction. Only by this close combination, to mingle it with the social moral codes can it be more modern and live, also plays a more important role in the socialist morality construction. As a result, within the activities of Wushu education, it is of great importance to the socialist morality construction.

3   The Connotation of Wushu Morality Education

Wushu morality education is a reflection of Chinese traditional ethical spirits within the field of Wushu, consisting of morality of the Confucian, Tao of Taoism and Buddha nature of the Sakyamuni, whose essence is shown by benevolence, righteousness, manner, wisdom, credit, and courage.

1. Benevolence: Its fundamental meaning is to love all the people with the warmest heart, covering the whole moral senses in some degree. The cores of love are filial piety and fraternal duty, whose ways are loyalty and tolerance. The generalized comprehension is to be loyal to the career, the people and the country, kind to the people. At present, most students in colleges and universities are single children and there are some problems come up, especially are selfish, lonely, indifferent to others, not filial to parents and not knowing of Brothers and sisters brotherhood.

2. Righteousness: It can be read as order or rank. It is said that people who practice Wushu should behave according to his identity. Transmitted into the teaching, it asks students to be strict with ourselves in every trial thing in our life by the new standards of modern college students, which should be fit for our identities.

3. Manner: It is the restraints of morality, along with the manners when you are involved with things and people. "Manner" has the practical importance in Wushu morality, which not only tells people "what to do", but also "how to do it". It is an important stage in transferring into Benevolence and righteousness. In teaching practice, it asks the students to learn and apply specific formal manners as their cultural decoration.

4. Wisdom: When Wushu learners, having owned the sentiment of Wushu morality and good manners, should have their own moral senses, which is characteristic of wisdom. The function of it is to realize benevolence and righteousness and to guarantee realization. Used in the teaching practice, it asks students to build up true values of life and world, to be good at thinking about the life and ethics and to distinguish the evil and good.

5. Credit: It means honesty, that is to say, a person must be honest, keep his words and fulfill his promises. Besides, he should not break faith with others or be afraid of difficulties. Being slow to promise and quick to perform is one of the most important contents of Wushu morality. In practice, it is required that the college students should keep their promises and stick to their own ideals and beliefs that have established.

6. Courage: It means the actions one take after being familiar with virtue and morality and making a clear distinction between right and wrong, good and evil. Specifically, it means making contributions to the motherland and the people, upholding justice, suppressing the evil and pacifying the good, punishing evil-doers and praising good-doers, helping the weak and the poor and so on. It is both moral standard and practical action. Putting into practice, it refers to developing the college students' confidence and determination of being brave to fight with bad phenomena or manners and sticking to good standards of conduct.

4   How to Promote the Martial Virtue Education in the Teaching Process?

4.1   Improving Teachers' Self-cultivation

Great learning makes a teacher, moral integrity makes a model. There is no doubt that teachers ought to play an exemplary role in inheriting and developing the Chinese Wushu morality. Teachers must lead by examples, influence and lead their students directly by their actions. In teaching, teachers should play an exemplary role not only in the skills, but also in the ideological cultivation. Thus, it requires teachers should be strict with themselves in every word and deed and tries to be dressed with proper clothes, present with neat appearance, speak in polite words and behave well. In teaching, teachers should not be afraid of hardship or filth; teachers should obey classroom convention, demonstrate accurately, and guide students carefully and patiently, expressing students with a fine teaching figure. Meanwhile, teachers should improve their self-cultivation to affect the students with their excellent skills and strong figures, lead and nurture the students with their standard behaviors, letting the students imperceptibly accept the ethical implications and constraints.

4.2   Teaching Students to Set Up Correct Attitude to Wushu

Firstly, upright the motivation of Wushu learning and do not use Wushu as the tool of fight. Wushu learners all have noble virtues, and they suppress the evil and pacify the good, uphold justice, punish evil-doers and praise good-doers. First of all, it is necessary for Wushu learners to make more contributions to society when inheriting this virtue. Secondly, Wushu learners should distinguish brotherhood from promises. Wushu morality requires Wushu learners be trustworthy in word and resolute in deed, lay emphasis on honesty and justice. However, it does not mean that one does anything for his friend on the expense of leaving his family and child without hesitation, neither is the brotherhood that fail to distinguish the right from wrong. PE teachers should combine characters of teaching contents with timely Wushu morality education and teach students to set up correct attitude to Wushu morality, thus it usually will achieve unexpected effects.

4.3   Blending Humanism Spirit into Wushu Teaching

There are two basic kinds of modern Wushu: rival ship and performance. The main traditional competition mode is to join in an open competition or contest, which has its positive aspects inherited from ancient times; however, it shows more phenomena of ferocious battle. Due to many reasons, such as the competition mode, organization, measure, many people are killed on the spot and more get lifelong disabilities. Current rival ship of joining in an open competition mode is free combat and it is hardly to see uninjured players in each battle. In order to get rank, gold medals and economic interests, it is not unreasonable to take defeating the opponents as their responsibilities, nevertheless, harming the opponents' bodies seriously is not allowed. If we want to change the current situation, only to change the public thoughts is not enough, it is more important to make the players change their cognition of free combat. It is necessary to blend humanism spirit into technical training and let every Wushu learner learns to just give a hint.

4.4   Telling Student to Be a Broad-Minded Person and Have the Ability to Tolerate with Others

To be a qualified Wushu learner, you should be able to tolerate with other students instead of minding small businesses. What's more, you also should be able to work with the people who have been opposed to you before. You should be as mild as water, be kind to others, even be able to return good for evil. Only by this can you build the virtue of patience and good moral sentiments. For example, it is inevitable to have divergence on the opinions in daily study. On solving the divergence, it is not whose voice is high is the right, but to use the Wushu spirit to gather the students who have words to have a peaceful talk. Wushu learners should always put the morality at the first place in the process of the exercising, just as QuanYan says, "to practice Wushu needs to culture virtue, and to do acrobatics needs to follow the routine." Otherwise, "you must be possessed if you don not culture virtue when do acrobatics."

4.5   Teaching Students to Anneal His Will and to Cultivate His Conduct

As Wushu learners, they should pay attention to anneal his will and cultivate his conduct. Only the Wushu learners who have lofty morality can be eminent experts in Wushu field. Tracing back to the ancient times, many people who has built up achievement had high cultivation and self-encouragement at their splendid accomplishment. In the Eastern Jin Dynasty, there was a younger, born in a troubled times, he made up his mind to practice his skills toughly in order to serve his country. At 289AD, he worked as a registrar administering the documents at the northeastern of Luoyang, Henan. He began to practice Wushu on hearing the cock crow every morning without intervals, eventually he became a famous general. He was Zudi. And that is the origin of the idiom "Rise at the cock's crow to practice with the sword". Now Wushu competition is to compete in skills, so Wushu learners should learn from Zudi, not only to build up a strong body to adapt to various environment, but also to work out strong will and conduct. What's more, you should constantly overcome the value of pain, the value of weariness, and to practice the perseverance spirit of "Keep on practicing even in the coldest days".

4.6   Teaching Student to Enhance His Moral and Legal Awareness and to Promote His Psychological Health and Behavior Normalization

Social life can not be independent from the moral and legal restraint. To be a qualified social man, one must have the basic moral and legal consciousness. Teach students to study, work and live according to the social morality and legal norm. Mix the basic moral and legal consciousness and connect Wushu with learning how to be a person by holding Wushu competitions and carrying out positive education. Colorful Wushu activities can help to adjust emotions and improve environment, thus to promote students' psychological health and behavior normalization.

5   Conclusions

As a kind of traditional national sports, Wushu has been gradually forming a complete series of the thinking modes, the values, the morality and the way of doing things. But owing to the influence of Chinese traditional culture, the education of Wushu morality inevitably has the historical restriction. The modern Wushu morality education we advocated should abandon the dross of the feudal hierarchy and the patriarchal clan system and inherit and promote the elite of the traditional Wushu morality. To strengthen the Wushu morality education in Wushu teaching has great insignificance in students' morality construction, harmonious society construction, and the improvement of national quality.

References

1. Xu, X.: The Research on Fusion of Universities Martial Arts Teaching and Takenori. Hubei Sports Science 29(2) (2010)
2. Liu, W.: Research of Morality Education in Wushu Teaching. BOJI (WUSHUKEXUE) 4(11), 62–63 (2007)
3. Li, Y., Zhang, X.: Theory on Wushu morality education in Wushu teaching. Journal of Liaoning Administration Institute (9), 217–218 (2007)
4. Hong, Q.: On Wushu Morality Education in P.E. Teaching. Journal of Jilin Institute of Physical Education (3), 58–59 (2005)

Body Narrative in the Context of Consumer Culture

Congying Liu[1] and Lijuan Wei[2]

[1] Chinese Department, Handan College, Handan, Hebei, China
[2] School of Foreign Languages, Handan College, Handan, Hebei, China
liugt9@eyou.com

Abstract. The body narrative perspectives on women to find and retrieve were annihilated. It cancelled the history and dual opposites in order to achieve the harmonious sex, and was of great significance. But consumption culture in spawned "body writing" prosperity also makes its orbit, and eventually lost from the original and the public life of communication skills. Jump out of the trap consumer culture to the adverse factors of consumer culture, from the focus bewitch self-sufficiency in female consciousness in flesh, and the lust, it aims to abandon the creation principle and the violent man, body aesthetics from indulge in current value system and the destruction of the female myth counterpart. The paper aims to build concrete, solid, rich and vibrant body narrative system, to reflect female literature, make the body narrative lost out.

Keywords: Body narrative, Female perspective, Consumption culture, Sex harmonious.

The body narrative literature development in the current occupies important position; also becomes the critic's interpretation differentiates. In contemporary China, the body has been suppressed existence as cover for long in the cultural development, about the body of all culture and physiology phenomenon are viewed as the flood disaster, steered clear. This kind of phenomenon change began in China in the late 1980s and especially 1990s. The economic and cultural transformation is due to the rising of women writers "body writing" movement. Female writing chooses body as narrative discourse, and shall not be in the present discourse system filled with patriarchal culture. Body becomes female life of ontology, the body in the poetry of the sublimation, rise to an existing life philosophy which, and achieved extraordinary literature and culture, while the breakthrough of body writing is neglected. Writing restored shadowed women experience, broke through the patriarchal culture discourse taboos and the reconstruction of women in literature body fixes alteration. Women's main body consciousness through the construction of the art world was reconstructed on the rebound.

"Body writing" concept pioneers in French feminist critics Ella Lana Helene. Body writing is an idea, which is different from male discourse. It breaks male discourse, overthrows the rational thinking mode of male pan-modernity, trying to change neat and tidy standardized historical view. Women's individual history both to the nation and world history is converging. Body writing will mean that women go into history and know the history of female, also its historical true selves are reflected.

Y. Wu (Ed.): International Conference on WTCS 2009, AISC 116, pp. 849–852.
springerlink.com

It rediscovers and finds history of annihilation self. Women in view of the world cancel the men and women to dualism in both sexes photograph. Women's perception is more intrinsic than male's logic. With the female omni-directional sensation and perception writing come the gushed endless illusion. Yet this concludes the achievements of the female body writing in the present obviously overly optimism. Consumption culture powerful digestive capacity rapidly steals female writing labour fruit, and it spreads quickly. And according to one's own logic, he begins to wreak rewrite. The context of consumer culture in literature has become a member of the goods, the literary creation and appreciations, which have undergone great change that embodies in literature and the expense relations is increasingly close. A literary value not only embodies in the deep rational connotation, such as the grand faith of the theme, but also consumer entertainment passage has great function. It improved itself to meet consumers' demands for consumers aspects-seeking sensual and recreation happiness. Literature's success depends not only on the work itself, but also on the capital operation, media of manipulation, cultural events, and discourse contents, etc. In order to adapt to this closely related changes, the creation subject often adjusts writing strategy, according to consumers' reading psychology and needs to make creation and tracking consumption hotspot and fashion, sensationalism, in order to obtain spiritual and physical double harvest. Consumption culture is as a double-edged sword in literature. The influence of concentrated expression is in" body writing", the emergence and development, and it directly expedite the" body writing "sprout and prosperity. It also makes it grow in number from the original rail against original intention, become consumptive victim.

Body writing will mean women have gone into history and re-known history of female, also its historical true selves. And it rediscovers and finds history of annihilation self. Women in view of the world cancel the men and women to dualism in both sexes when the photograph comes. Its essence lies in women's perceptual rather than male's logic. Yet such inductive female body writing has accomplishments in present days which is obviously overly optimistic. Consumer culture has powerful digestive capacity. It rapidly steals female writing labour fruit. And according to one's own logic rewritten, the context of consumer culture in literature has become a member of the goods, the literary creation and appreciation undergone great change that embodies in literature and the expense relations is increasingly close.

Female body narrative have however great revolutionary significance quickly to be simplified, leaving only female body to function consumption symbols, which becomes consumer society with omnipotent consumer logic, female writing into the perceptual liberation in the background of articulate. In fact, consumer culture operational logic from the feminist writing with the initial misreading and rewriting a stubborn confrontation in the body narrative writing. The body writing experienced a lot changes from the soul to the flesh, and from spirit to physical evolution. The soul is gradually stripping out the flesh, and in flesh gradually becomes pure objects. It becomes commercial culture and consumer culture accomplices. This is a spirit meat from one into another consumer symbol of the process. Consumer culture awakens people on the body of attention and self consciousness, and lets the body cavity. This is to save the body, but is degenerated into consumer culture through business empire and media hegemony to reach conspiracy. It completed a successful commercial speculation. The body was distorted, and became a plain code which marks a price of pure goods, even being simplified emergent and desiring pronoun. Body writing loses their richness, profundity and critical thinking toward the dissident. We can't help reflecting on consumer age body writing exactly the future development and direction. Consumer times through advertising, media amplification body material, time of spiritual and cultural significance body, which make the body fall into hedonism consumption machine and become a kind of commercial symbols. The body is of cultural significance of physical meaning and devoid of overuse, in body it is extreme consumption. Body writing is more and more incline to others. The flesh of the soul of worship and feminist writing orbit is deviated from its current one. "The woman writer" works are plentifully fragmented body picture. The body has not yet been completely free from patriarchal, but falls into the trap. Consumer culture gentle experiences the self with the soul with a new round of split pain. Facing the pain, young woman writer cannot afford redeemed heavy responsibility. It misreads the body instead giving freedom, and abuses the body consciousness. First performance in woman writer provides flesh carnival soul sleeping plenties of the pictures. Wang Anyi, and Tie Ning regard such as it is in Manhattan physical and life and self isomorphism based on physical needs. They advocate tolerance through physical consumption to get pleasure rights. And fully aware of the relative sexuality, they stick to the border of ethical consciousness. By contrast, "the woman writer" sexual narrative is behaved for disregard sexual ethics; infinite exaggerates to sex with consumption, which leads to enjoy "sex" poetic quality of flesh dissipates. Because of the limitations of urban life, scarcity of sexual discourse resources that they could not like Wang Anyi, and Tie Ning, Lindbergh had placed in natural discourse system of poetics conversion. People are pursuing sexual desire meet the right, but the person of sexual desire rights and cannot or should not replace other rights.

The body is material and spirit is of composite carrier. It not only has the physical properties and other biological properties, especially castes cultural awareness carrier, a social significance, objective existence, which is full of symbol and historical significance. Consumption age body to fall into to soul which is thrown off the feminist writing orbit. How to jump out of the trap of consumer culture is an urgent realistic problem. I think we should thoroughly and deeply search into the body, make the body performance writer's harmonious stretch. First body writing must get rid of the negative factors of hedonism consumption machine, becomes a kind of commercial symbols. Body writing is more and more inclined of consumer culture bewitch. Consumerism rather keeps a clear mind. Get rid of to sex and materialistic primarily self-identification, gender trauma into creative energy and will blank and life history of incomplete into writing motivation. Face reality life and improve their comprehension ability and analytical skills and creative ability. One must let works not in flesh.

To attract readers, infect readers and conquer them, but not by the exposure of privacy crazy hype as selling point. Because many times to peep at reading been fulfilled will gradually vanish. More importantly of women's lives is not only the privacy, there are more important things and richer content. Secondly body writing should focus on life personalized and literature daily life and self-sufficient woman consciousness out of the society, toward the broad, let the heart and soul sublimation, strong uplift ideal flags, shoulder up historical, social responsibility, paying attention to the vast human destiny and meaning of life, adhere to the profound social reflection and criticism, make the person's spirit character in literature appreciation will get promoted, body of meat sex and ethical fuse and live truly in the physical. Again body writing must abandon the flesh and desire street, transfer from the physical body, the creation principle and violence. One must pay attention to body aesthetics from indulging in present value system of destructive dissembling and female myth of the shaping of regression, deconstruction of at the same time the attention from waking and construction, with female's unique emotion and wisdom writing women in the history of non-substitutability, light life and the world. Let the heart back to abode, body respected. The body becomes incorporeal materialized, and one must construct a specific, solid, plump, energetic body narrative of system.

As a female literature narrative body writing, its development process and contemporary encounter, to a certain extent reveal the plight of women literature. At the same time body writing future direction is the development direction of female literature. Female body was found to be covered by women in the society, thus it has established the unique status and experienced the irreplaceable, with consumer culture wave, short for more than 20 years time body writing passed by the claustrophobic to open, by the upper body to lower body, by speaks to rumbustious course. One must do researches on consumption culture era of body writing. The aim is to establish a female literature in the wake of the mirror, introspective through introspection, and to make consumer culture in the context of the body narrative out therefore.

Research on Probabilistic Coverage Control Algorithm for Wireless Sensor Network

Zheng Sihai and Li Layuan

School of Computer Science and Technology Wuhan University of Technology
Wuhan, China
si68234@163.com

Abstract. Coverage control is an effective method to alleviate the energy-limitation problem of sensor nodes in wireless sensor networks. Most of current researches were based on geometry calculation of binary sense model which is limited and inaccurate in practical application. Aiming at the problem, this paper adopts the probability sense model, puts forward a new coverage control algorithm (PSMC), which takes energy efficiency as important index. The probability sense model describes the coverage control more accurately. Simulation shows PSMC algorithm can shut down a large number of redundant nodes and extend effectively network lifetime while coverage ability does not reduce.

Keywords: WSN, Coverage control, Energy, Routing Protocol.

1 Introduction

Network coverage is the basic issues in wireless sensor networks (WSN), which is directly reflected in the quality of services provided by network. Because the energy of network nodes, bandwidth, computing power and other resources generally are limited, sensor nodes must be placed suitably, and an appropriate routing method should be chosen, which can ultimately optimize allocation of resources. Choosing a correct strategy for network coverage helps to reduce energy consumption of sensor nodes, it also can improve the quality of sensing service and extend network overall survival time.

Probabilistic coverage control algorithm adopts non-deterministic sensor detection model. That is, detection accuracy of the sensor would change with sense signal attenuating seriously. According to the changed distance from the target nodes to sensor nodes and environmental factors, several sensors can work together to find a target with certain detection probability. At present, many papers use this kind of model to solve coverage problem in WSN. *Co-Grid* coverage algorithm [1] divides the target area into overlapping units. It uses distributed probability detection model and the virtual alarm ratio set early to wake minimum nodes in each unit to work. *Co-Grid* algorithm puts forward a network configuration method, but the time complexity of algorithm is high. Ref. [2] proposes a model based on probability coverage algorithm (PCA), but it is only used to evaluate the confidence of detection probability in sensor

Y. Wu (Ed.): International Conference on WTCS 2009, AISC 116, pp. 853–859.
springerlink.com

network. Ref. [3] proposes a coverage configuration algorithm based on the probability (CCAP), but it is only used for the evaluation of point target coverage probability. Ref. [4] proposes a coverage preserving protocol (CPP), but the protocol uses central control algorithm to configure network, which limits the size of the network.

The main contribute of this paper are as follow. We propose a probabilistic sense model for coverage control algorithm (PSMC).PSMC algorithm uses *Neyman-Peason* probabilistic sense model and *Voronoi* diagram to coverage control, schedule the working status of the redundant nodes. It can shut down a large number of redundant nodes and extend effectively the network lifetime while network coverage ability does not reduce.

In the rest of this paper, we first formulate the problem of coverage control in Section 2. PSMC algorithm is analyzed in Section 3. PSMC is embedded into PEGASIS[5] routing protocol, the results of simulation are shown in Section 4. At last, we offer conclusions in Section 5.

2   Problem Formulation

2.1   Deployment Model

In two-dimensional plane R^2, the coverage of node n_i is a circular area where the radius equals r_s which is the sensing radius. C is union set of coverage of all nodes in sensor network. Target region R^2 is completely covered by the sensor network equivalents to each point in R^2 are covered by at least one node. Direct communication range of node n_i is a circular area where the radius equals communication distance r_c.

Network nodes are randomly deployed in the two-dimensional bounded rectangular target R^2. Assume that the location of each sensor are informed by positioning system, and sensor nodes know the border of target region R^2.Therefore,the deployment of nodes can be regarded as a smooth two-dimensional POSSION point process. λ is density of POSSION point process, $C(R)$ is the number of nodes in region R^2, it obey the POSSION distribution parameter $\lambda \|R\|$ .That is,

$$P(C(R) = K) = e^{-\lambda S(R)} (\lambda S(R))^K / K! \tag{1}$$

$S(R)$ is the area of R^2.

2.2   Probability Model

Assuming n sensor nodes are randomly distributed in the monitoring area,   the location of nodes is (x_i, y_i), the target measured value observed by the node n_i can be expressed as follow [6].

$$a_i = \theta / D_{ij}^{\alpha} + \delta_i , \quad i = 1, 2, \cdots, N \tag{2}$$

θ is signal strength emitted by targets; α is the signal attenuation index, $\alpha > 0$;δ_i is observation noise of n_i,$\delta_i \sim (0,\sigma)$. Binary hypothesis testing for n_i are expressed as follow.

$$H_1 : p(z_i \mid H_1) = (1/(\sqrt{2\pi}\sigma))e^{-(z_i - a_i)^2/2\sigma^2} \tag{3}$$

$$H_0 : p(z_i \mid H_0) = (1/(\sqrt{2\pi}\sigma))e^{-z_i^2/2\sigma^2} \tag{4}$$

All nodes use the same detection threshold value τ to judge. According to the formula based on *Neyman-Peason*, decision threshold and the relationship of virtual alarm are expressed as follow.

$$P_{fi} = \int_\tau^\infty p(z_i \mid H_0)dz_i = Q(\tau/\sigma) \tag{5}$$

$$\tau = \sigma Q^{-1}(P_{fi}) \tag{6}$$

Therefore, the sense probability of targets located at (x_j, y_j) are observed by node n_i are expressed as follow.

$$P_{D_i} = \int_\tau^\infty p(z_i \mid H_1)dz_i = Q((\tau - a_i)/\sigma) = 1 - \Phi((\tau - a_i)/\sigma) \tag{7}$$

Sensor node density is usually higher. Therefore, events in the monitoring area are detected by multiple sensor nodes simultaneously, the sense probability is expressed as follow.

$$P_D = 1 - \Pi(1 - P_{D_i}) = 1 - \Pi((\tau - \theta/D_{ij}^\alpha)/\sigma) \tag{8}$$

According to (8), $P_D \geq P_{D_i}$, because multiple sensor nodes may sense the same events simultaneously.

Definition. In a set of active nodes located at (x_i, y_i), system detection probability of target point located at (x_j, y_j) is P_D .If $P_D \geq \beta$, the target point located (x_j, y_j) is satisfied withβ *probability coverage*. If all points in a region are satisfied with β probability coverage, then this region is called as *complete β probability coverage*.

2.3   Regional Model

There is a finite point set $S = \{s_1, s_2, \cdots s_n\}$ on the two-dimensional plane R^2, *Voronoi* region [7] is defined as $V_i = \{P \in R^2 \mid d(p, s_i) \leq d(p, s_j), j \neq i\}$, which associates with s_i, where d presents Euclidean distance. Point set $\{s_i\}_{i=1}^n$ is called *Voronoi* generation points.

   Sensor nodes are deployed in a bounded convex region in nodes set $S = \{s_1, s_2, \ldots, s_n\}$, each nodes serves as *Voronoi* generation points, then the only *Voronoi* division of R^2 can be obtained. In the two-dimensional plane, *Voronoi* division is convex polygon. If s_i is at the convex hull boundary of set S, then the *Voronoi* polygon associated with s_i is unbounded. However, if R^2 is bounded, unbounded *Voronoi* polygon will turn into bounded polygon with the limitation of the target area. Therefore, all the *Voronoi* polygons have boundary, bounded *Voronoi* division can be obtained.

3   PSMC Algorithm

P_{D_i} is the probability of target sensed. If $P_{D_i} \geq \beta$,target is found,β is the minimum probability of target found, which is determined by the actual application environment, hardware and software conditions and quality of service required and other factors. It is usually specified by user.

In the actual work environment, the nodes works fewer have more energy. In order to balance energy consumption, the energetic nodes must work more time, so some individual nodes will not die early. And energetic nodes have strong sense ability as well, that is benefit to reduce the number of active nodes in a coverage set. Therefore, the main idea of PSMC algorithm is to give priority to wake nodes which work fewer and have more sense ability. They can form a coverage set that meet coverage requirements of network. PSMC algorithm will be suspended if sufficient nodes are woken up accordance with the rules established, i.e. the sense probability of all targets located at *Voronoi* generation points are no less thanβ.P_{D_i} is much larger, sense ability of node n_i is more stronger. Its value was calculated before the algorithm execution, and then it is input in each cycle. Specific algorithm process is as follows.

Step 1: Input $n,\beta,c(i),(i=1,2,\ldots,n)$, $c(i)$ is the work times of n_i;

Step 2: Establish the sequence table in which nodes are woke up. According to the work times of each node provided previously, the nodes with same work times will be attributed to the same set, where they are sorted ascending according to the work times and P_{D_i}. Ultimately, a wake-up sequence table of nodes is formed as follow,

$$N_1,N_2,N_3,\ldots,N_m;$$

Step 3: V_c=null, i=1;

Step 4: N_i is added into V_c, $c(i)$ minus one. If $c(i) \leq 0$,then continue execute *Step 5*,otherwise return *Step 4* again.

Step 5: P_{D_k} is calculated respectively,$0 < k \leq n$. if min$\{P_{D_k}\} \geq \beta$,then coverage set is satisfied with the requirements identified, output V_c, continue execute *Step 6*,otherwise i plus one, and return *Step 4* again.

Step 6: Algorithm computing of a cycle is end, and then wait for the operation of next cycle.

4   Simulation Analysis

To verify the effect of PSMC algorithm, PSMC was embedded into PEGASIS routing protocol, i.e. PEGPSMC routing protocol. PEGASIS protocol is developed from LEACH protocol. Its main idea is as follow, before carrying out data transmission, each node first sends test signals, the nearest neighbor node is regarded as next nodes, a chain table can be formed through traveling sequentially all network nodes. In this way, all nodes are only connected with the adjacent nodes. PEGPSMC protocol closes some redundant nodes in accordance with the results of the detection probability, nodes energy can be better saved and network lifetime is prolonged dramatically.

The simulation used well-known network simulator - NS2. Node energy consumption model [8] was used in simulation. In order to compare comprehensively PEGASIS with PEGPSMC, two experiments were carried out. First one was coverage control, second one was load balance. The changes of number of nodes and the changes of number of dead nodes were simulated, which influenced the performance of the algorithm. At last, energy consumption of network nodes when the size of the area monitored are analyzed.

Fig.1 shows the relationship between the number of dead node and the working hours of network. Fluctuation of the curve of PEGPSMC is very small. Because its energy is shared evenly to each node, the time is almost identical when the first node and the last node died. Network lifetime of PEGPSMC is more than twice as PEGASIS protocol.

Fig.2 shows the relationship between number of nodes and network lifetime. A noteworthy phenomenon is that PEGASIS protocol can not significantly improve the network lifetime while increasing the number of nodes. This is because PEGASIS protocol does not take into account coverage control problem, all the nodes is active in each round, thus the network lifetime does not increase with the extension of the number of nodes. Whereas, PEGPSMC protocol takes full advantage of probability coverage, only some nodes can cover the whole region, the extra nodes are in a dormant status. Hence, it can reduce energy consumption and prolong network lifetime.

Fig.3 shows the relationship between the size of the area to be monitored and network lifetime. You can see obviously, PEGPSMC protocol can still maintain good performance while the size of area is larger. As PEGPSMC protocol has used PSMC algorithm, it is apparently superior to PEGASIS in the same regional area. However, the superiority of PEGPSMC reduced gradually with the size of area increasing. Because clustering requires energy consumption, the performance of PEGPSMC declines rapidly.

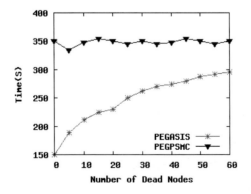

Fig. 1. Network lifetime and number of dead nodes

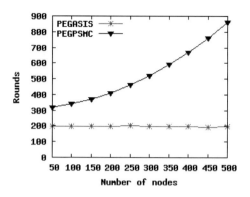

Fig. 2. Network lifetime and number of nodes

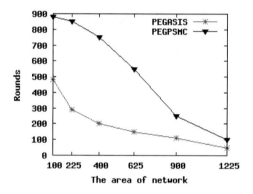

Fig. 3. The area of network and network lifetime

5   Conclusion

In this paper, randomly distributed wireless sensor network coverage control problem has been discussed. PSMC algorithm and PEGPSMC protocol are proposed. PEGPSMC protocol can shut down a large number of redundant nodes and extend effectively the network lifetime. If this algorithm is applied to QoS routing protocol of wireless sensor networks, will be very helpful, which author will focus on.

Acknowledgments. This paper is supported by National Natural Science Foundation of China（No : 60672137, 60773211, 60970064）, Open Fund of the State Key Laboratory of Software Development Environment（No : SKLSDE-2009KF-2-02）, New Century Excellent Talents in university（No : NCET-08-0806）, Fok Ying-Tong Education Foundation for Young Teachers in Higher Education Institutions of China（No : 121067）, Specialized Research Fund for the Doctoral Program of

Higher Education of China（No：20060497105），NSF of Wuhan Municipality（No: 201010621207）and the National Science Foundation of HuBei Province（No: 2008CDB335）.

Reference

1. Xing, G.L., Lu, C.Y., Pless, R.: Co-grid:an efficient coverage maintenance protocal for distributed sensor networks. In: IPSN 2004, Berkeley, California, USA, pp. 414–423 (2004)
2. Ahmed, N., Kanhere, S., Jha, S.: Probabilistic coverage in wireless sensor networks. In: Proceedings of the 30th Conference on Local Computer Networks, pp. 672–679. IEEE, USA (2005)
3. Zhang, D.X., Xu, M., Chen, Y.W.: Probabilistic coverage configuration for wireless sensor networks. In: Wireless Communications,Networking and Mobile Computing (WiCOM 2006), pp. 1–4 (2006)
4. Sheu, J.P., Lin, H.F.: Probabilistic coverage preserving protocal with energy efficiency in wireless sensor networks. In: Wireless Communications and Networking Conference (WCNC 2007), pp. 2631–2636 (2007)
5. Guo, W., Zhang, W.: PEGASIS protocal in wireless sensor network based on an improved ant colony algorithm. In: 2nd International Workshop on Education Technology and Computer Science (ETCS 2010), pp. 64–67 (2010)
6. Wang, B., Wang, W., Srinivasan, V.: Information coverage for wireless sensor networks. IEEE Communications Letters 9(11), 967–969 (2005)
7. Mohammadreza, J., Ali, M.: Uncertain Voronoi diagram. Information Processing Letters 109(13), 709–712 (2009)
8. Yin, L., Wang, C.: An energy-efficient routing protocal for event-driven dense wireless sensor networks. International Journal of Wireless Information Networks 16(3), 154–164 (2009)

Construct Comprehensive Evaluation Model of College Track and Field Course Grades

Yuhua Zhang

Institute of Physical Education Huanghe Science and Technology College
Zhengzhou, Henan Province, China
yzhang94@yahoo.cn

Abstract. It is necessary to establish a scientific comprehensive evaluation model, taking into account each item importance of college track and field course and different difficulty of different items. Using AHP method to determine each item weight coefficient of track and field course, at the same time, I construct comprehensive evaluation model by gray correlative analysis method. Practical example shows: the evaluation model is scientific and feasible.

Keywords: track and field course grades, comprehensive evaluation model, fuzzy AHP, gray correlative analysis.

1   Introduction

Track and field course is one of main courses in six courses of <curriculum program of physical education undergraduate major for national university and college>. Now in university the track and field course score is still the sum of individual course score, and a multi-factor integrated problem was dealt with a single factor problem, lack of rationality and impartiality, can't truly reflect the learning effect and level of track and field course, also can't objectively describe the share student position of track and field grade in the class. Constructing a scientific model of comprehensive evaluation to track and field course is a challenging and main question to the development of university track and field course.

2   Research Object and Method

2.1   Research Object

Let part of track and field course grades (see Table 1) which were randomly selected from specialty of physical education 2008 in Huanghe Science and Technology College be research object.

Y. Wu (Ed.): International Conference on WTCS 2009, AISC 116, pp. 861–867.
springerlink.com   © Springer-Verlag Berlin Heidelberg 2012

Table 1. Track and field course grades of specialty of physical education 2008

Score	First semester			Second semester			
Student	javelin	sprint	theory I	long jump	high jump	shot	Theory II
x_1	85	80	70	84	75	65	90
x_2	90	87	90	70	80	83	75
x_3	78	83	95	60	85	85	75
x_4	68	67	85	78	80	70	90
x_5	75	60	70	71	63	75	65
x_6	80	73	92	85	68	90	88
x_7	70	95	85	70	83	80	90
x_8	85	85	86	70	75	92	78
x_9	90	85	89	83	95	95	92
x_{10}	65	75	70	72	68	74	80

2.2   Research Methods

Using the method of literature data, questionnaires, expert interviews, fuzzy AHP and so on to analyze and deal with student achievement.

3   Result and Analysis

3.1   Research Object Construction Principle of Comprehensive Evaluation Model to University Track and Field Course Grades

Gray analysis method is a multi-factor statistical analysis method, is quantitative comparative analysis of momentum development of system, and describe the strength, size and order of realationship between factors according to sample data of each factor and gray correlation[1]. Gray evaluation system is on the basis of geometrilc similarity degree between curves consisted by each comparing series set and curve consisted by refering series to determine correlation degree between comparing series set and refering series[3]. More similar is geometric shape between curves consisted by each comparing series set and curve consisted by refering series, the greater is correlation degree. Fuzzy AHP method is a combination of qualitative and quantitative system analysis method, by introducing the fuzzy consistent matrix, and overcomes the significant difference about consistency between determine matrix and human mind[4].

According to fuzzy AHP analysis method and study weeks of each track and field item[5], we can get weight of each track and field item in track and field curriculum system of this academic year. Using gray correlation analysis method, we have correlation coefficient of each student score relative to optimal performance[6], have correlation degree by increasing weight, and continue to get track and field grade ranking of each studnt through ranking of correlation degree, thus obtain standardized scores.

3.2   Calculation of Gray Correlation Coefficient

To comprehensive evaluation of student grades, the comparison series constituting of each student's grades is noted

$$B_i(j) = [B_i(1), B_i(2), \cdots B_i(t)], \quad where \ i = 1, 2, \cdots m$$

Where m stands for the number of students; j=1,2,...t, and t stands for number of program types[2].

When we evaluate student's grades, firstly, should appoint evaluation standard and select standard according to comparability and advanced principle. Put series consisting of each course highest points as refering series, noted as

$$B_0(j) = \{B_0(1), B_0(2), \cdots, B_0(t)\} = \{90, 95, 95, 85, 95, 95, 92\}.$$

Because meaning and purpose of each evaluation indicator is different, in general have different dimensions and magnitude, so it needs dimensionless treatment to optimal index set and each item index set, in order to reduce the interference of random factors. Using the formula $x_i = B_i(j) / B_i(1), (i = 1, 2, \cdots, m; j = 1, 2, \cdots, t)$, we can obtain dimensionless matrix.

Characteristic matrix of all subjects score (see Table 1) was handled dimensionless:

$$\begin{bmatrix}
0.9444 & 0.8421 & 0.7368 & 0.9882 & 0.7895 & 0.6842 & 0.9783 \\
1.0000 & 0.9158 & 0.9474 & 0.8235 & 0.8421 & 0.8737 & 0.8152 \\
0.8667 & 0.8737 & 1.0000 & 0.7059 & 0.8947 & 0.8947 & 0.8152 \\
0.7556 & 0.7053 & 0.8947 & 0.9176 & 0.8421 & 0.7368 & 0.9783 \\
0.8333 & 0.6316 & 0.7368 & 0.8353 & 0.6632 & 0.7894 & 0.7065 \\
0.8889 & 0.7684 & 0.9684 & 1.0000 & 0.7158 & 0.9474 & 0.9565 \\
0.7778 & 1.0000 & 0.8947 & 0.8235 & 0.8737 & 0.8421 & 0.9783 \\
0.9444 & 0.8947 & 0.9053 & 0.8235 & 0.7895 & 0.9684 & 0.8478 \\
1.0000 & 0.8947 & 0.9368 & 0.9765 & 1.0000 & 1.0000 & 1.0000 \\
0.7222 & 0.7895 & 0.7368 & 0.8471 & 0.7158 & 0.7789 & 0.8696
\end{bmatrix}$$

In the following, we can get difference matrix, maximum and minimum:

$$\begin{bmatrix}
0.0556 & 0.1579 & 0.2632 & 0.0118 & 0.2105 & 0.3158 & 0.0217 \\
0.0000 & 0.0842 & 0.0526 & 0.1765 & 0.1579 & 0.1263 & 0.1848 \\
0.1333 & 0.1263 & 0.0000 & 0.2941 & 0.1053 & 0.1053 & 0.1848 \\
0.2444 & 0.2947 & 0.1053 & 0.0824 & 0.1579 & 0.2632 & 0.0217 \\
0.1667 & 0.3684 & 0.2632 & 0.1647 & 0.3368 & 0.2106 & 0.2935 \\
0.1111 & 0.2316 & 0.0316 & 0.0000 & 0.2842 & 0.0526 & 0.0435 \\
0.2222 & 0.0000 & 0.1053 & 0.1765 & 0.1263 & 0.1579 & 0.0217 \\
0.0556 & 0.1053 & 0.0947 & 0.1765 & 0.2105 & 0.0316 & 0.1522 \\
0.0000 & 0.1053 & 0.0632 & 0.0235 & 0.0000 & 0.0000 & 0.0000 \\
0.2778 & 0.2105 & 0.2632 & 0.1529 & 0.2842 & 0.2211 & 0.1304
\end{bmatrix}$$

$$\Delta_{max} = 0.3684, \Delta_{min} = 0.$$

According to the gray system theory, define correlation coefficient of comparison series $B_{(0)}$ in the indicator $B_i(j)$:

$$b_{ij} = \frac{\min_i \min_j |x_0(j) - x_i(j)| + \rho \max_i \max_j |x_0(j) - x_i(j)|}{|x_0(j) - x_i(j)| + \rho \max_i \max_j |x_0(j) - x_i(j)|}, i = 1,2,\cdots,t \quad \rho \text{ is discrimination}$$

coefficient, generally the value is between 0 and 1, and usually $\rho = 0.5$.

Calculated by the above method , we can get gray correlative coefficient matrix:

$$\begin{bmatrix}
0.8421 & 0.5384 & 0.4117 & 0.9398 & 0.4667 & 0.3684 & 0.8946 \\
1.0000 & 0.6863 & 0.7779 & 0.5107 & 0.5384 & 0.5932 & 0.4992 \\
0.5802 & 0.5932 & 1.0000 & 0.3851 & 0.6363 & 0.6363 & 0.4992 \\
0.4298 & 0.3846 & 0.6363 & 0.6909 & 0.5384 & 0.4117 & 0.8946 \\
0.5249 & 0.3333 & 0.4117 & 0.5279 & 0.3536 & 0.4666 & 0.3856 \\
0.6238 & 0.4430 & 0.8536 & 1.0000 & 0.3933 & 0.7779 & 0.8090 \\
0.4532 & 1.0000 & 0.6363 & 0.5107 & 0.5932 & 0.5384 & 0.8946 \\
0.7681 & 0.6363 & 0.6605 & 0.5107 & 0.4667 & 0.8536 & 0.5476 \\
1.0000 & 0.6363 & 0.7445 & 0.8869 & 1.0000 & 1.0000 & 1.0000 \\
0.3987 & 0.4667 & 0.4117 & 0.5464 & 0.3933 & 0.4545 & 0.5855
\end{bmatrix}$$

3.3   Calculation Weight of Each Program Type

We assume program system has three program types: field events, basic theory, track events, and according to method of fuzzy AHP, we can get the weight of each program type.

Table 2. Determine matrix

expert evaluation	field events	basic theory	track events	weight
field events	1.00	1.50	4.00	0.5170
basic theory	0.67	1.00	3.00	0.3591
track events	0.25	0.33	1.00	0.1239

By determine matrix of Table 2 and the following formula

$$\omega_i = \frac{1}{n}\sum_{j=1}^{n}\frac{x_{ij}}{\sum_{i=1}^{n}x_{ij}}, \quad i = 1,2,...n, \quad n = 3.$$

We can get the weight of three program types in Table 3.

Table 3. Program type and study weeks

	javelin	sprint	theoryI
weeks	7	6	3
type	Field events	Track events	Basic theory

	long jump	high jump	shot	theoryII
weeks	5	5	4	3
type	Field events	field events	Field events	Basic theory
weight	0.5170	0.5170	0.5170	0.3591
weight		0.5170	0.1239	0.3591

Based professional training program, find out study weeks q_j of the proment j and their types, and by the following expression of program weight

$$T_i = \frac{\sum_{k=1}^{n}\alpha_{ik}q_i\omega_k}{\sum_{i=1}^{m}\sum_{k=1}^{n}\alpha_{ik}q_i\omega_k}, i = 1,2,\cdots,m. \quad m = 7, n = 3$$

$$\alpha_{ik} = \begin{cases} 1 & \text{if item } i \text{ is type } k, \\ 0 & \text{otherwise.} \end{cases}$$

Further, we can get decentralization weight of each item $T_1 = 0.2631, T_2 = 0.0540, T_3 = 0.0783, T_4 = 0.1879, T_5 = 0.1879, T_6 = 0.1503, T_7 = 0.0783$.

In order to facilitate comparison in the whole, using the formul $r_i = \sum_{j=1}^{t}T_j b_{ij}, i = 1,2,\cdots,10$, and T_j is weight of item j, j=1,2,...7. To get each student's gray weighted correlation degree, that is to say, $r_1 = 0.6726$, $r_2 = 0.6864$, $r_3 = 0.5896$, $r_4 = 0.5466$, $r_5 = 0.4543$, $r_6 = 0.6969$, $r_7 = 0.5815$, $r_8 = 0.6430, r_9 = 0.9389, r_{10} = 0.4531$.

3.4   Calculation of Student's Standardized Average Scores

According to size of correlation degree, order student's grades, for convenience, normalize correlation degree by following formula $r_i' = r_i / r_{max} (r_{max} = max(r_1, r_1, \cdots, r_m))$, therefore, have each student's standardized average score $r_i' \times 100$, as follows: $x_1 = 71.64$, $x_2 = 73.11$, $x_3 = 62.80$, $x_4 = 58.22$, $x_5 = 48.39$, $x_6 = 74.23$, $x_7 = 61.93$, $x_8 = 68.48$, $x_9 = 100$, $x_{10} = 48.26$.

4   Conclusion

4.1   Construction Principle of Comprehensive Evaluation Model

On the basis of single item weight in track and field course system of this academic year, with gray correlation analysis method, get each student grade's correlation coefficient relative to the optimal scores, obtain correlation degree by weighted, and so have each student's ranking of track and field grades, then have standardized scores.

4.2   Comparison of Grades Banking

By comprehensive evaluation model, we get student's ranking is $x_9 > x_6 > x_2 > x_1 > x_8 > x_3 > x_7 > x_4 > x_5 > x_{10}$, but by simple average method, the ranking is: $x_9 > x_5 > x_6 > x_2 >$, $> x_6 > x_2 > x_7 > x_8 > x_3 > x_1 > x_{10} > x_5$, comparing the results, we will find the ranking order is changed.

4.3   Comparison and Analysis

The result of this method is more scientific and fair compared with average method. It reflects synthetically the real learning level and position in the class, and can eliminate the difference of different item importance by introducing   item weight. At the same time, by highest score sequence of single item as refering sequence of gray correlation analysis method, it can eliminiate the difference in the degree of project difficulty. Example shows: this method is scientific and feasible.

References

1. Jin, X., Li, Y., Sun, G., et al.: Modeling for assessment of student's marks. Journal of Liaoning Technical University (natural science) 29, 116–118 (2010)
2. Xie, M.: Mathematical evaluation model of student's grades in teaching reform. Journal of Higher Correspondence Education (natural science) 22(5), 62–65 (2009)

3. Li, Y., M, Y.: The comprehensive evaluation of college students' achievement. Journal of Chongqing University of Arts and Sciences (natural science edition) 29(2), 21–24 (2010)
4. Zhou, Y., Li, W.: Enchanced FAHP and its application to task scheme evaluation. Computer Engineering and Applications 44(5), 212–214 (2008)
5. Yang, C., Xing, W.: Construct evaluation index system of college students' physical course grades. Journal of Tonghua Teachers college 30(12), 90–91 (2001)
6. Yuan, H.: Application study on the physical education grading index system in the university. Journal of Harbin Institute of Physical Education 27(3), 71–76 (2009)

The Thinking of Strengthening
the Team Construction
of College Experimental Teachers

Li Jinhui[1], Zhang Ke[2], Li Xiaohui[1], and Sun Xiaojie[1]

[1] College of Information and Electrical Engineering Shenyang Agricultural University
Shenyang, China
[2] School of Policedog Technique of MPS, Shenyang, China
lijing8912@yahoo.cn

Abstract. The team construction of experimental teachers is the core of laboratory construction and is an important component of the college education reformation. It is an important topic about how to strengthen the team construction of experimental teachers, improve experiment technology level, and ensure that the laboratory keeps an efficient and stable operation for the domestic universities' wide attention and researching. Taking the Ministry of Public Security Police Dog Technical Schools and Shenyang Agriculture College as an example, based on analyzing the function of the college experimental teachers in the development of the school and the main problems facing, this article gives some measures of strengthening the construction of experimental teachers' team.

Keywords: Experimental teachers team, Laboratory, College, creative talents.

1   Introduction

Experimental teachers' team is an important part of laboratory construction. At the same time, the laboratory is an important researching base for teaching and scientific, is the cut point of cultivating the creative talents, and is the important resource for the college's sustainable development. It is impossible to cultivate first-class creative talents and create the first-class scientific researching achievements without the first-class university having first-class laboratory. From the teaching perspective, experiment teaching is the extension and expansion of theoretical teaching. The stand of experimental teachers' team construction plays a direct impact on the school's teaching effect, graduates quality and the ability and level of scientific research. Therefore, strengthening the construction of teachers' team is an indispensable and important content of the university reform.

Y. Wu (Ed.): International Conference on WTCS 2009, AISC 116, pp. 869–876.
springerlink.com   © Springer-Verlag Berlin Heidelberg 2012

2 The Main Role of University Experimental Teachers in School Development

2.1 An Important Role in Experiment Teaching and Cultivating Innovative Talents

Experimental teachers are working in the front-line of experimental teaching and responsible for lots of professional experiment, practice, and graduation design, and other important links of each grade school. They debug the equipment, configurate reagent, install experiment devices and prepare various materials directly. So that they play an indispensable role in cultivating students' use of advanced theories, methods and tools to discover, research, solve problems. Experiment teaching is the important segment for developing quality-oriented education of school. Experimental teachers, as the main force of experiment teaching, whose business capability and overall quality directly influent and determine the quality of experiment teaching as well as determine whether they can take responsibility for the heavy burden of cultivating students' practice ability and innovation ability or not.

2.2 An Important Role in Laboratory Management, Teaching Reform and Scientific Research

As college lab equipment updating ceaselessly, the advanced equipment is increasing largely. Experimental teachers have to undertake the work of maintaining and keeping many types of equipment. Just Shenyang Agriculture College is concerned, now which has a variety of a large number of experimental equipments and instruments. Only inductively coupled plasma mass spectrometer, network switch, over speed freeze centrifuge, etc, which is big, medium-sized or valuable apparatus being more than 20 million is nearly 100 sets. The experimental equipment not only requires the experimental teachers to familiar with the function, characteristics, mastered the operation method, but also will debug, maintain, grate various fault timely, so as to ensure the normal development of teaching and scientific research. Each year the school undertakes the topic of the state, provincial scientific research and teaching reform and the researching achievements are inseparable form laboratory work link which is inseparable from the experimental teachers' hard working.

3 Current Situation and the Main Problem Facing of Experimental Teacher

3.1 No Reasonable Age and Academic Structure

At present, the average age of many universities' experimental teachers is too old. The proportion of old, middle-aged and young experimental teachers is not equilibrium and some schools even appear temporary shortage phenomenon. The too low degree is a common existing problem of experimental teachers' team. The undergraduate course and specialized subject education occupies a larger proportion, the highly educated rate is low and the graduate student is less. Such as the on-the-job experimental teachers' age and educational level distribution of Shenyang Agriculture College shown in figure 1, figure 2. There are 68 on-the-job experimental teachers and 52 people which accounts for 77 percent of the total number is above 40-years-old. The person under 30 years old only accounts for 4 percent which illustrates the young experimental teacher is rare. The people with master degree or above are 8 which accounts for 12percent. As far as the information of Shenyang Agriculture College and Electrical Engineering College is concerned, the youngest experimental teachers being on the regular payroll is nearly 50 years old and the highest degree is bachelor. The problem of low degree and old age is rather outstanding. The opportunity of declaring and presiding the subject of teaching reform and scientific research for the experimental teacher is relatively less because of the larger proportion of low degree (undergraduate and below) and old age in the experimental teacher. So it goes against playing experimental teachers initiative and restrains their enthusiasm for scientific research and teaching reform, that is unfavorable to the orderly development of the experimental teachers and the cultivation of innovation talents.

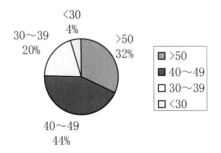

Fig. 1. The Age Distribution of Shenyang Agriculture College experimental teachers

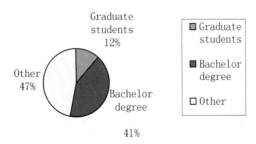

Fig. 2. The Degree Distribution of Shenyang Agriculture College experimental teachers

3.2   Preparation and Quantity Deficiency of Experimental Teachers

With the deepening reform of the education and the growing scale of the enrollment year by year, the student's scale and scientific researching task is expanding rapidly. At the same time, as the old experimental teachers being natural wastage, the number of experimental teachers present obvious downward trend in total with the little supplementary of new forces in recent years, which can be seen clearly from the figure 1. In colleges and universities, experimental teacher's team is the second-largest technical team under the teachers' team. But relative to the full-time teachers' team construction, the growth rate of experimental teachers is obviously lower than full-time teachers. For example, the basic computer researching center of Shenyang Agriculture College undertakes the public computer course of each department in the school, such as computer culture basis, all kinds of high level language program design, network design, and multimedia design course and so on. Every year, it also undertakes the task of the national computer rank examination, the national university's computer rank examination, the national mathematical modeling contest and so on. At present, there are nearly 500 computers in total and the experimental teachers reduce to only one from four people originally. Some other lab in school such as agronomy courtyard, technician's workload also increases to two people's work even more from one person's workload originally. Therefore, with the increasing task of experimental teaching and scientific research in the higher school and the atrophy of existing experimental teachers yearly, it appears particular important to supply seasonably the new forces under the situation of need.

3.3   Low Professional Title, Instability and No Enough Attention of the Experimental Teachers

At present, experimental teachers with the same degree, quite professional technical position in Shenyang Agriculture College, when compared with professional teachers, the status is low and the title is difficult to promote. And the number of senior titles is slim (there are only one who get the senior title), which also has restricted the experimental teachers' enthusiasm and technological expertise's playing. It is also a common problem that the team of experimental teachers is instable. For example, there are 4 on-the-job experimental teachers in information and electrical engineering institute, but there are another 4 turns into other positions. In experimental teachers training work, it is obviously deficient to be at home and abroad for further studies and investigations when comparing with the full-time teachers. These problems also lead to the phenomenon, such as low whole quality, the insufficiency ideological understanding, the not-enough motivation to participate in the teaching reform and scientific research and the not-strong innovative capability, exists generally. The lagging team construction of experimental teachers has become the restricting factors of the discipline's development and the weak link of personnel training.

4   The Measures of Strengthening the Team Construction of Experimental Teachers

Experimental teachers' team is the important force to support university experiment teaching, scientific research and laboratory construction. The quality and stability of experimental teachers' team construction is the key of keeping the laboratory has a sustainable development strategy. It is the prerequisite to strengthen experimental teachers' healthy development reasonably for improving the quality of experiment teaching. Therefore, we must take the strengthening school experimental teachers' team construction as a strategic task to treat, take effective measures to build the team and make the experiment teachers' team become a mainstay for the university's development.

4.1   Strengthen Their Innovation Consciousness and Cultivate Creative Talents

In recent years, many units take the innovative consciousness and ability as an important index when employing the graduates, especially for the students of the public security system which requires strong practical ability and practice ability. When training the police dog, each police dog is required to identify different source objects through smelling the source, which needs to design the smell identify platform simulating natural environmental conditions in laboratories. So that dogs can quickly identify designated smelling source in the lab, shorten training time, improve the students' ability of extracting the criminal suspect's site through police dog identification. The experiment teaching in colleges and universities is an extremely important channel for students to grasp modern scientific knowledge and cultivate innovative ability. Therefore, it is the urgent requirement for experimental teachers to have innovation consciousness, combine various professional characteristic, break through original knowledge category, explore and find new development regularity in the forefront of experimental session, change the traditional experimental model and give the new experimental measures and methods solving the problem. The experimental teachers providing the new and good lesson will pass on to the student constantly changing new knowledge in practice and will fully mobilize students' learning subjective initiative, stimulate students' learning desire and promote innovation talents.

4.2   Reform and Improve the Experimental Teachers' Team Structure and Make Sure the Experiment Teachers' Sustainable Development

In experimental teachers' team construction, schools should intensify efforts to develop attention of experimental teachers' team construction guidelines and policies according to relevant policies and regulations of the school development scale when making plans, which makes the various teachers and students realize the important position and role of the experimental teachers. In the fixation of posts and staff, the school should establish a reasonable number in accordance with the experimental teaching workload, subject type, equipment management, laboratory construction and management, etc. it also can reform the single mode of teaching stuff and set up the combination of the fixed compiling and flow preparation as well as full-time and part-time job patterns.

The school can continuously introduce the highly educated and high level of advanced talents by opening for recruitment to the society people and the school teacher. Let young teachers enrich to experimental position and select outstanding experimental teachers bearing some teaching work can make combine the teaching and practice into an organic unity. The school still should form the ladder-frame in age, record of formal schooling and title echelon, work out training direction and development goals in every step and make sure the experiment teachers' sustainable development.

4.3   Be People-Oriented, Innovate Concept and Pay Attention to the Function and Position of Experimental Teachers

All the time, teaching and practice is inseparable two parts. Teaching depends on practice and practice is the basis and premise of good teaching. Teaching will like water without a source and a tree without roots but will not cultivate innovative talents without practice. Therefore, we should innovate concept, persist people oriented and improve the experimental teachers' position when the school dealing with the problem such as the teaching stuff and professional title. We should eliminate historical legacy of prejudice and fully realize the complement between theory teaching and practice teaching which are both based, interdependence and mutual transformation with very close correlation. It can make us regard theoretical teaching and experimental teaching as equally important and emphasis equally on construction of both teams. All these things can make the existing experimental teachers can ease their job and devote themselves to the experimental teaching, scientific research and laboratory construction, also make more highly educated, high-level talents the enrich experimental teachers' team.

4.4   Strengthen the Education and Training as well as Improve the Teachers' Business Level and Overall Quality

With the rapid development of science and technology in modern society, theory and skill levels of experimental teachers must adapt to the needs of social development in experiment teaching. So the teachers should acquire and update their knowledge timely and improve their business level and overall quality. On the one hand, schools and departments should strengthen the training work with short-term training primarily and going out learning opportunities appropriately increased for experimental teachers' team. The schools and departments should do regular job training, technical training, training qualifications, carry out the experiment skill competitions, build the good study atmosphere, and improve experiment teachers' teaching practice and scientific and technological capability continuously. On the other hand, experimental teachers should correct ideas, study the professional business assiduously, love and respect their jobs, and earnestly implement experimental teachers' responsibilities and duties. Fully excavating the potential of existing experiment resources makes the instrument equipment utilization maximization and failure minimal. The experimental teachers should perfect themselves constantly and set their own development direction with a plan and purpose, exchange learning experiences, pool their wisdom, raise the sense of unity in working practice, carry out business forum periodically, and share collective resources. So that they can meet the new requirement and to shoulder the responsibility of cultivating talents.

4.5   Perfect Encouragement Mechanism Gradually and Strengthen the Experimental Teachers' Team Consciousness of Suffering and Sense of Competition

The school should establish and sound the various rules and regulations and rewards and punishment system of the lab. According to the actual conditions, it develops each position's responsibilities and goal requirement to make the experimental teacher have clear goals and responsibilities. Taking the teachers with a strong sense of responsibility, deep academic attainments and innovative ability as the experimental teachers' team leader and giving the appropriate award to them, while the teacher working little, not pushing oneself forward with more accident will be viewed or demotes processing.

4.6   Participate in Teaching Reform and the Scientific Research Work Actively

Experimental teachers working in first line of the practice teaching can contact and understand the existing problems and the improving link of teaching and practice closely. So that experimental teachers should participate in teaching reform and the scientific research work actively. On the one hand, the school should create certain opportunities for experimental teachers when declaring teaching reform and scientific research topics. On the other hand, experimental teachers should also collects actively common problems and phenomenon of high school students in practice teaching and improve the teaching method of experiment teaching. Such as developing the computer aided teaching software, designing the multimedia courseware and so on to enhance the student's ability of accepting knowledge and cognition. At the same time, the teacher still should attend scientific research work of each specialty for scientific research service and continuously improve their capability of business through the scientific research work.

5   Conclusion

The team construction of higher school experimental teachers is an important part of the university laboratory construction and a long-term and arduous task of the school construction. In the whole school's development planning, we should put experimental teachers' team construction as a long-term systematic project to complete which also needs generations of the young and old teachers to work together to achieve. Therefore, we should seize the university reform and the development opportunity, build a high-quality professional experimental teacher' team with style guaranteed, exquisite technology and reasonable structure to adapt to the development needs of the new era and promote the training innovative talents.

References

1. Chen, K., Zheng, Q., Ma, G.: Constructing innovative laboratory technical personnel team in the university. Experiment Technology and Management 26(10), 158–159 (2009)
2. Liao, Q.: Thinking of college laboratory safety management. Journal of Laboratory Research and Exploration 29(1), 168–170 (2010)
3. Liu, S.: Brief study on college experimental teachers' team construction. Journal of Laboratory Research and Exploration 28(12), 185–190 (2009)
4. Liu, G., Xiao, F., Bi, Y., Wang, X., Zhang, C.: Establish the creative practice teaching system and train innovative technology application talents. Journal of Laboratory Science 6, 10–11 (2009)
5. Wu, D.: Study on building up the sustainable development of University laboratory technical personnel team. Experiment Technology and Management 24(7), 151–152 (2007)

Study on the Feasibility of Sports Prescription to Treatment of Young Internet Addiction*

Yuhua Zhang

Institute of Physical Education Huanghe Science and Technology College Zhengzhou,
Henan Province, China
zhangy3120@yahoo.cn

Abstract. With the development of science and technology, Internet addiction becomes a social problem , it can not be ignored and seriously affects the physical and mental health of young people. From the point of sports prescription, this paper analyzes the reason for formation of young addiction, formulates appropriate sports prescription according to different reasons, and studies the feasibility of sports prescription to treat young addiction.

Keywords: Sports prescription, Young internet addictin, Feasibility study, Intervention and treatment.

1   Introduction

With the development of science and technology, and the rapid expansion of the network, internet addiction has become a social problem, and it can not be ignored. In particular, some young people indulge in long-term network, which gives rise to loneliness, depression, online dating, game crazy, sex addiction and other addiction, seriously affects their health, also attractes more and more attention. <2009 data reporting of Chinese young internet addictive people> shows in China's urban network addiction's young is 14.1% of young internet users, about 2404.2 millon. In urban, about 12.7% of young non-internet addiction have addiction tendencies, about 1858.5 million[3]. At the same time, the spirit of addiction therapist Frisch.Gehete said:"young people indulge in the network, which is not only a moral defect , but also a psychological problem, and an unhealthy mental performance in fact[2]." At present, the main treatments for addiction measure are instrument of psychological therapy and drug treatment. In the paper, from the point of physical training, I study the active role of the treatment of sports to young network addiction.

2   Subjects and Methods

2.1   Research Subject

The active role of the sports treatment to young network addiction is our study object.

* This work is supported by Henan Province People's Government Development Research Center project,No.E374.

2.2   Research Methods

Use the methods of literature data, questionnaire, expert interviews, mathematical statistics and so on.

3   Results and Analysis

3.1   Concept and Characteristics of Network Addiction

1) Concept of Network Addiction: Firstly, American psychologist Goldberg proposed network addiction, then Dr. Kimberly Yong of Pittsburgh university developed and made perfect of his concept. Internet addiction disorder (IAD) or pathological internet use (PIU) was impulse control behavior in the online without effect of addictive substances, manifested as a result of the excessive use of the internet, and brought about significant individuals and psychological dysfunction[2].

2) Characteristics of Network Addictions: Internet addicts have the following features:

(a) Increasing tolerance: that is to say, if he wants to get the same satisfaction as before, it needs to continuously increase online time.

(b) Withdrawal symptoms: if there is period of time without internet, will become restless and uncontrollable to go online, and always worries about missing something.

(c) Internet access frequency is always higher than prior plan, internet time is also longer than prior plan.

(d) The efforts to shorten online time always end in failure.

(e) Spend a lot of time on internet-related activities, such as installing new software, edit and download a large number of files.

(f) Internet severely affects social interaction, study, work and other social functions.

(g) although he can be aware of servious problems from internet, continues to spend a lot of time online[4].

3.2   Criteria for Internet Addiction

The criteria for internet addiction is the following:

1. Behavioral and psychological dependence.
2. Lose the basic ability of self-discipline and self-control.
3. Disrupt the normal order of work and life.
4. Suffer more serious damage to physiccal and mental health.

3.3   Survey and Analysis Reasons for the Formation of Young Internet Addiction

Now, young people are in the period of rapid development of China's internet, their general feature is over-reliance of computer in life. when they meet frustration and attack or interpersonal problems, habit to find psychological balance from internet, not communicate with teachers and schoolmates, which likely causes psychological barriers, and shifts to virtual world of internet to seek spiritual sustenance and escape real problems.

1) Anxiety, impatience, emotional instability Anxiety, impatience and emotional instability of young internet addiction mainly show as: any time and any thing will cause his anxiety and large mood swing, failing calm, easy to make some low-level errors.

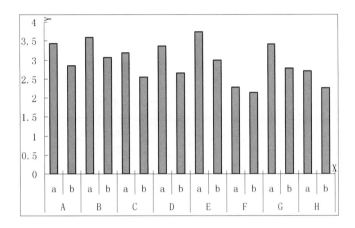

Fig. 1. The average score comparison of internet addiction and non-internet addiction

where

    X Type of young online
    Y The average score
    a Internet addiction
    b Non-internet addiction
    A Internet makes me get out of unpleasant emotions
    B Internet makes me feel better when I am depressed
    C Friends of internet are kind to me
    D I feel more confident when I communicate with others online
    E When I go online bothering thing do not bother me
    F My life is no fun without network
    G Communication with others online can make me more comfortable
    H The scores comparison of difficulty to get along with others
    Young addictive people and young non-addictive people of 'network can make me get out of unpleasant emotions' and 'internet makes me feel better when I was depressed' have significant differences[3], and non-internet addiction is significantly higher than internet addiction (Figure 1).
    2) Indecisive character and no self-confidence
    The performance of indecisive character and no-self-confidence is hesitant and not decisive, it will affect the efficiency of learning and work. Young addictive people and young non-addictive people of 'friends of internet are kind to me' and 'I feel more confident when I communicate with others online' have significant differences[3], and non-internet addiction is significantly higher than internet addiction (Figure 1).

3) Nervous and too much pressure

Young addictive people have intensive mental and too much pressure from family. Young addictive people and young non-addictive people of 'When I go online bothering thing do not bother me' have significant differences[3], and non-internet addiction is significantly higher than internet addiction (Figure 1).

4) Mental timidity and no good self-concept

Young addictive people have mental timidity and no good self-concept. Young addictive people and young non-addictive people of 'My life is no fun without network' and 'communication with others online can make me more comfortable' have significant differences[3], and non-internet addiction is significantly higher than internet addiction (Figure 1).

5) Unsociable character and unhappy relationshiop

Most young people is only one child, family environment enable them to lack of coordination ability, like to be alone, lack of contact with outside, which develop a lonely, eccentric character and poor interpersonal relationships. The proportion of interaction with others actively to young addictive people is less than one of young unaddictive people (Figure 2). Figure 3 shows that with the different levels of internet addiction the proportion is increasing.

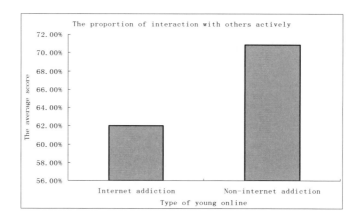

Fig. 2. The proportion of interaction with others actively

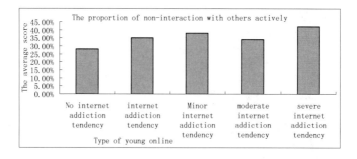

Fig. 3. The proportion of non-interaction with others actively

At the same time Figure 1 shows young addictive people litter contact with others, but young non-addictive people are easier to contact with others[3].

3.4   Draw Up Sport Prescription, Play an Active Role to Treatment for Young Net Addiction

In the following, according to different reasons for formation of young addiction, we formulate appropriate sports prescription[1].

1) Stabilize mood of young addicitive people, improve their anxiety and irritable mentality
Young addiction do more physical exercise, such as Tai ji quan, long distance jogging, fitness walking, bowling, billiards, darts, shooting, chess, fixed-point shooting and other activities. These items require practitioners to have will specificity and mental focus, which strengthen the training of nervous system, increase dominance of the nervous system and regulatory function, and these item are much alive, you will feel relax and happy, excluding anxiety and having cheerful mood after exercise. Thus it benefits to stabilize young mood, improve young people's ideological anxiety and impatience psychological state[5].

2) Train self-confidence of young addictive people, improve their indecisive character
Allow them to practice rapid response and hurdles, high jump, long jump, table tennis, tennis, badminton, boxing and other exercise programs needing to respond decisively, exercise their responsiveness, because before these items, any hesitation and wander will delay the best time and lead to failure[5]. Therefore, more practice of these items can develop decisive character, improve self-confidence and overcome indecisive defects.

3)   Ease intensive nerves of young addiction people
Let young addictive people take part in football, basketball and distance running and other agonistic programs. These sports put skills, emotional and psychological together, and can vent their negative emotions, promote sleep, relieve stress, relax tense nerves, adjust mental balance.

4) Establish a good self-concept and overcome cowardice
Swimming, target high jump, skating, skiing, wrestling, gymnastics and other challenging program with a performance style can continuous to overcome their cowardice, overcome difficulty with a fearless spirit and overgo obstacles.
In practice, according to the actual situation of patients, targeted to reduce frustratio, attention to overcome their frustration. Spiritual pressure transforms into ideological motivation as far as possible . accordingly to correct the face of setbacks, restore confidence, so that to enhance individual's self-concept, hold their own positively[6].

5) Chang eccentric personality, establish good relationship
Let young addictive people take part in basketball, volleyball, soccer, sport games, tug of war, relay, two or three bikes, ballroom dancing and other sports. These items are popular sport of young people. It requires that team members have a better

psychological quality of coordinated operation, have tacit understanding between the players, individuals involvement can taste the fun of victory, feel collective power larger than individuals power. In these projects, they not only receive others help, but also have the responsibility and obligation to help other, and have individual self-competition with others or other groups[7]. Practice has proved that these activities can help them to gradually change the lonely and ungregarious phenomenon, and gradually adapt to deal with each other students, communicate with each other, improving interpersonal relationships. Toward a good interpersonal relationships, thus increase social adaptation of young addictive people.

3.5   Flow Chart of Sports Treatment to Young Addiction

With the cause of young addiction and sports initiative, establish the following flow chart (Figure 4)

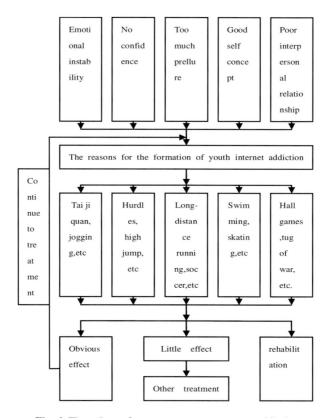

Fig. 4. Flow chart of sports treatment to young addiction

The most prominent feature of flow chart is the full use of sport function, in a ease and healthy lifestyle to meet psychological satisfaction instead of the other negative style, and thus to intervene and treat youth addiction psychology . It can be seen from Figure 4 the treatment to young addictive people can be divided into three stages: exploration phase of young addiction reason, the phase of development and implementation of exercise prescription, stage of re-evaluation phase of internet addiction. Among them, the first phase is foundation of sports treatment of adolescent addiction, the second stage is core of treatment of adolescent addiction , the third stage is result of treatment of adolescent addiction.

4   Conclusions and Recommendations

4.1   Conclusions

Through the introduction of concept of internet addiction, analysis of the characterization and explore of causes of addiction, I formulate a sport prescription, give full use of   positive role of sports treatment to young addiction. Flow chart is the full use of sport function, in a ease and healthy lifestyle to meet psychological satisfaction instead of the other negative style, and it illustrates the treatment to young addictive people can be divided into three different stages, and points out nature of each stage. We should understand the treatment process needs to continue to modify and improve in the process of implementation. Only we construct management mode, which puts apply, promote, popular and optimization as main line, and management, supervision, feedback, evaluation as implementation system, then it can achieve the best effect.

4.2   Recommendations

The reason for the formation of youth internet addiction is very complex, so the treatment of adolescent internet addiction is also a complex process. In this process, it can not only rely on a single treatment, need to carry out a variety of comprehensive treatment. If the above sports prescription can combine with psychological therapy and other drug therapy, the effect would be best.

References

1. Lu, Y.: Sociology of Sport, pp. 213–217. Higher Education Press (2006)
2. Zheng, H., Li, Q., Lin, K.: Revised Introduction to Society, pp. 411–413. China Renmin University Press (1994)
3. Ke, H., Hao, X., Jin, R., et al.: 2009 data reporting of Chinese young internet addictive people, pp. 43–60 (February 2010)

4. Li, Y., Zhang, M.: Analysis and countermeasure to Internet addiction of university student. Culture and Education, 225–227 (August 2010)
5. Lin, H., Guo, C., Li, B.: Discussion on sports prescription of promoting undergraduates' psychological healthiness. Journal of Physical Education Institute of Shanxi Normal University 21(1), 87–89 (2006)
6. Ma, Y.: The feasibility analysis of getting rid of Internet addiction with sports. Teaching and Management, 43–44 (March 2010)
7. Zhu, L., Xu, S.: The new mode that employs sports as the main method of intervening college students' IAD. Journal of Neijiang Teachers College 22(2), 117–120 (2007)

Applying Case-Based Reasoning for Mineral Resources Prediction

Peng Shi and Binbin He

Institute of Geo-Spatial Information Science and Technology
University of Electronic Science and Technology of China
Chengdu 611731, China
shi596@foxmail.com

Abstract. Case-Based Reasoning (CBR), a well known Artificial Intelligence (AI) technique, which consists of retrieving, reusing, revising, and retaining cases, has already proven its effectiveness in numerous industries. In this research, we try to adopt CBR technique in mineral resources prediction. A model for mineral resources prediction is proposed in this paper, which can support the processes of case-based reasoning in mineral resources prediction such as case representation, indexing, retrieving and case revising. It mainly includes Feature tree and FSM algorithm and it is different from traditional model. At last, an experiment of iron resources prediction is performed in Eastern Kunlun Mountains, China. The results indicated that the model proposed in this paper is suitable for regional metallogenic prediction.

Keywords: Case-Based Reasoning, Feature tree, metallogenic prediction.

1 Introduction

Case-Based Reasoning (CBR), a well known Artificial Intelligence (AI) technique, which consists of retrieving, reusing, revising, and retaining cases [1], has already proven its effectiveness in numerous industries. There are two fundamental concepts for CBR. One is that similar problems will have similar solutions. The other is that same problems will often occur. More importantly, CBR simulates the human problem-processing model and can have the self-learning function by constant accumulation of past experience. When the user enters a new problem in CBR, CBR will search for the data that have the highest similarity with the existing cases and adjust the previous cases to suit the new problem [2]. Generally, CBR involves the following four cyclical processes: (1) retrieving the most similar case(s), (2) reusing the solutions of the retrieved case(s), (3) revising the proposed solution if necessary, and (4) retaining the new solution as part of a new case [3]. In the past, CBR had been successfully applied to the solution to many problems. For example, a system using case-based reasoning was used for monitoring health of bridge [3], a chronic diseases prognosis and diagnosis system [4], online catalog sales [5], decision support for housing customization [6], sales forecasting in print circuit board industries [7], decision-making of resource allocation in airport emergency [8], support mergers &

Y. Wu (Ed.): International Conference on WTCS 2009, AISC 116, pp. 885–891.

acquisitions of enterprises [9], insurance fraud analysis system [10], traffic congestion management method [11]. In this research, we try to adopt case-based reasoning technique in mineral resources prediction.

2   Method

2.1   Case Representation

Cases can be described according to their features, which can be form, function, appearance, quality feature and stock mode. In this study, we use feature tree to describe characteristics of cases. In such a tree structure, the relationship of parent–child levels can represent the relationships between parts and subassembly. Take the PILOT pen (see Fig. 1) for instance. Fig. 2 stands for its feature tree, in which the white circles and gray circles symbolize the nodes and features, respectively. For example, in Fig. 2, there are three features for the pen cap: the color is blue; the diameter of the ink cartridge is 0.4 mm; and the shape is circular. In the diagram, the straight line represents the line that connects nodes, indicating the hierarchical family relationship. The child points are dependent on the changing of parent nodes. To make sure that this property can be handled, the following rules should be obeyed:

- Rule 1: The connected nodes should not form a loop.
- Rule 2: Features should be used to describe nodes so as to guarantee that there is one and only one node in the parent level.
- Rule 3: When nodes stand for parts, the nodes in the child level represent the subparts or features of the node.
- Rule 4: When nodes stand for features, there will be no nodes in the child level; the features describe

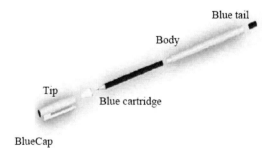

Fig. 1. Diagram of a blue pen the characteristics of the node in the parent level

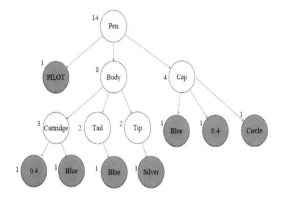

Fig. 2. Feature tree (Left number represents weight for each node)

2.2 Case Retrieval

For the retrieval and comparison of cases, we need to check first whether the cases are equal in terms of each attribute; f_j^I represents the value of the jth attribute of input Case (I), and f_{ij}^R is the value of the jth attribute of the ith case in the database (R). In this study, it is hypothesized that $S(f_j^I, f_{ij}^R) \in [0, 1]$. When $f_j^I = f_{ij}^R$, the output equals to 1 whereas when $f_j^I \neq f_{ij}^R$, the output equals 0. Finally, z equals to the sum of the number of nodes in the child level node of the jth attribute and the node itself, namely, 1. Second, we need to calculate the similarity of each node. $Sim(f_j^I, f_{ij}^R)$ denotes the similarity of the jth attribute of input Case (I) and the jth attribute of the ith Case in the database (R). To begin with, we have to check there are any nodes in the child level of the node or not. If the consequence of this problem is yes, the similarity of the node can be obtained from Formula (1), otherwise the similarity of the node can be obtained from Formula (2).

$$Sim\left(f_j^I, f_{ij}^R\right) = S(f_j^I, f_{ij}^R) \tag{1}$$

$$Sim\left(f_j^I, f_{ij}^R\right) = \frac{\sum_{k=1}^{z} S(f_j^I, f_{ij}^R)}{z} \tag{2}$$

Finally, we need to calculate the similarity between cases. As follows:

$$Similarity\left(f^I, f_i^R\right) = \frac{\sum_{j=1}^{n}[W_j Sim(f_j^I, f_{ij}^R)]}{\sum_{j=1}^{n} W_j} \tag{3}$$

In this study, it is hypothesized that node on the higher level is composed of the nodes on the child level (parts or subassembly) and features. The summation of the number of nodes itself and that of the nodes below decides the weight of the node. This is shown in Fig. 2, in which the number by the node indicates its weight.

The algorithms for case comparison can be listed as follows (see Fig. 3).

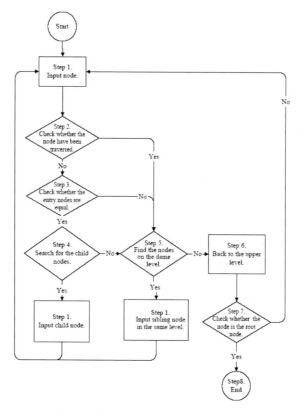

Fig. 3. Algorithms for case comparison

- Step 1: Input nodes.
  Enter the node to be compared and the node in the database.
- Step 2: Check whether the entry nodes have been traversed.
  Yes: Jump to Step 5.
  No: Process Step 3.
- Step 3: Check whether the entry nodes are equal.
  Yes: $\mathrm{Sim}(f_j^I, f_{ij}^R)=1$, process Step 4.
  No: $\mathrm{Sim}(f_j^I, f_{ij}^R)=0$, Jump to Step 5.
- Step 4: Search for the child nodes.
  Check whether or not the nodes on the child level have not been traversed.
  Yes: Jump to Step 1.
  No: Process Step 5.

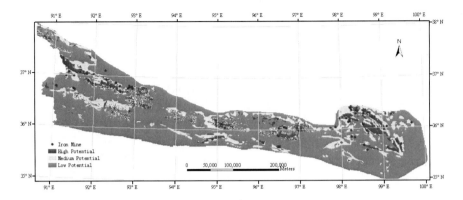

Fig. 4. Prediction Map of Iron Mine in East Kunlun Mountains, China

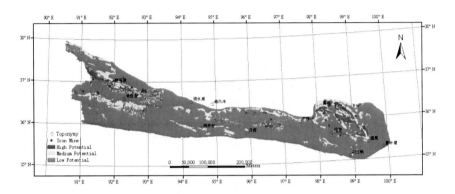

Fig. 5. Prediction Map of Contact Metasomatic Iron Mine in East Kunlun Mountains, China

- Step 5: Find the nodes on the same level.
  Check if there are nodes on the same level that have not been traversed.
  Yes: Jump to Step 1.
  No: Process Step 6.
- Step 6: Back to the upper level.
- Step 7: Check whether the node is a root node.
  Yes: Process Step 7.
  No: Jump to Step 1.
- Step 8: End. The comparison of two hierarchical trees comes to the end.

3   Experiments and Results

3.1   Data Preprocessing

In this research, vector ore-controlling data of stratum, unconformity, fault, regional chemical anomaly, remote sensing mineralization anomaly, bouguer anomaly, aeromagnetic anomaly and Iron mineral occurrence are used. All of those data were grid partitioned. Based on Weights of Evidence Modeling, the stratum as background layer, the grid is partitioned to 1km*1km in GIS software. All of the layers' data were overlaid and spatially joined to the newly created grid layer, and every grid polygon in grid layer owns corresponding attributes of other layers. Grids which include the information of Iron mineral will to be the case database and the others will to be the unknown cases. To unconformity, fault and mineral occurrence, the feature data should be buffered. Based on Weights of Evidence Modeling, the buffering distances are different, 1000m point buffer to mineral occurrence, 300m line buffer to unconformity, 3000m line buffer to fault for iron mine.

3.2   Calculate the Similarity

According to the method which was proposed above, we describe the known and the unknown cases. Using feature tree to representing the characteristics of cases and calculate the final similarity between each of the unknown cases and each of the known cases. Obviously, we can find the most similar case from case bases for each unknown case.

3.3   Results

At last, every polygon of the grid layer has a final similarity and a known case which has the greatest similarity with this polygon, and all of the final similarities are joined in grid layer in GIS environment, then the maps of regional metallogenic potential prediction are plotted with grading and color separating. Fig. 4 shows metallogenic potential of iron mine in east Kunlun area in China, in this 81 iron mines, there are 49 mines in the high potential and 25 mines in the medium potential, the accuracy is 91.36%. Fig. 5 shows metallogenic potential of contact metasomatic iron mine in east Kunlun area in China, in this 58 contact metasomatic iron mines, there are 35 mines in the high potential and 17 mines in the medium potential, the accuracy is 89.66%.

4   Conclusions

The experiments demonstrate that the model of case-based reasoning in this paper is efficient, the whole process is efficient. Simultaneously, case-based reasoning using feature tree in the geological mineralization can provide some reference for evaluation of mineral resources. In addition, the high potential area is about 8% of the whole area, and the medium potential area is about 16% of the whole area. While, calculation given by Weights of Evidence Modeling shows that the upper limit of better area is 8% to high potential, and 16% to medium potential. However, in the experiments spatial relationships were not discussed, this is where the future experiments need to be improved.

Acknowledgment. This research work is funded by National High-tech R&D Program of China (863 Program), project number: 2007AA12Z227.

References

1. Changchien, S.W., Lin, M.C.: Design and implementation of a case-based reasoning system for marketing plans. Expert Systems with Applications 28, 43–53 (2005)
2. Tseng, H.E., Chang, C.C., Chang, S.H.: Applying case-based reasoning for product configuration in mass customization environments. Expert Systems with Applications 29, 913–925 (2005)
3. Cheng, Y.S., Melhem, H.G.: Monitoring bridge health using fuzzy case-based reasoning. Advanced Engineering Informatics 19, 299–315 (2005)
4. Huang, M.J., Chen, M.Y., Lee, S.C.: Integrating data mining with case-based reasoning for chronic diseases prognosis and diagnosis. Expert Systems with Applications 32, 856–867 (2007)
5. Vollrath, I., Wilke, W.: Case-Based Reasoning Support for Online Catalog Sales
6. Juan, Y.K., Shih, S.G., Perng, Y.H.: Decision support for housing customization: A hybrid approach using case-based reasoning and genetic algorithm. Expert Systems with Applications 31, 83–93 (2006)
7. Chang, P.C., Liu, C.H., Lai, R.K.: A fuzzy case-based reasoning model for sales forecasting in print circuit board industries. Expert Systems with Applications 34, 2049–2058 (2008)
8. Liang, S.H., Han, S.C., Zhu, X.P.: Decision-making of Resource Allocation in Airport Emergency Rescue Base on CBR. Traffic and Computer (6), 31–34 (2008)
9. Wu, Q.L., Feng, Q.C.: How to Apply the Case-Based Reasoning Technique to Support Mergers & Acquisitions of Enterprises. Chinese Journal of Management Science, 1003-207 (2002) zk-0367-05
10. Sun, F.: Design of CBR insurance fraud case analysis system. Journal of Minjiang University, 1009-7821 (2008) 05-0031-04
11. Ji, X.F., Liu, L.: Traffic Congestion Management Method Based on Case-Based Reasoning. Journal of Southwest Jiaotong University, 0258-2724(2009) 03-0415-06
12. Zhang, B.S., Yu, Y.L.: Hybrid Similarity Measure for Retrieval in Case-based Reasoning System. Systems Engineering Theory and Practice, 1000-6788 (2002) 03-0131-06

Correlation Analysis in Study of Relationship between ^{18}O and Water Quality Indexes

Yingying Sun[1], Baohong Lu[1], Zhongcheng Tan[1], and Jiyang Wang[2]

[1] College of Hydrology and Water Resources Hohai University
Nanjing, China
[2] Institute of Geology and Geophysics Chinese Academy of Sciences, CAS
Beijing, China
sunhomh@eyou.com

Abstract. At the background of water diversion project from the Yangtze River to Taihu Lake, the experiment of diverting water from Taipu River was implemented from March 22 to April 7 in 2006. Synchronous isotope monitoring as well as water quantity and water quality is carried out. Combining water quality index and isotope abundance data, a grey correlation analysis has been made between six major water quality indexes and δ^{18}O values during sluice-supply water period, the pump-supply water period and the whole diverting period. Correlation degrees of the whole diverting water period are higher than that of the two different diverting water period for the reason that data of normal diverting water period which from April 5 to 7 is included in calculating correlation degrees during the whole diverting period. Meanwhile, all the correlation degrees except TP during the sluice-supply water period are higher than that during the pump-supply water period. It shows that pump-supply water had more obvious influence to the correlation degrees for the reason that there are more releasing water during pump-supply water period than sluice-supply water period. Furthermore, correlation degrees of the three branches upper Huangpu River are also calculated respectively. The results show that the correlation degrees of Xiaziwei and Sanjiaodu have the similar chang trend except the correlation degrees of NH_3-N and TN because of the supply of Taipu River and the river network around to the north branch Xietang and the middle branch Yuanxiejing. At last, it comes to the conclusion that diverting water from Taipu River have great improvement to the water environment of upper Huangpu River. And ^{18}O as well as water quantity have a significant correlation with water quality.

Keywords: δ^{18}O, water quality indexes, correlation degree, Huangpu River, diverting water.

1   Introduction

Upper stream of Huangpu River is connected with Taipu River, and down flows into Yangzi River at the estuary of Wusongkou. It is a typical tide river [1]. Upstream of Huangpu River is the main supply water source of Shanghai city, accounts for 80% of water intake quantity in Shanghai. The water from Jiangsu and Zhejiang Province along

Y. Wu (Ed.): International Conference on WTCS 2009, AISC 116, pp. 893–899.
springerlink.com

Taipu River has always been influenced to upstream of Huangpu River, especially after the implementation of the project on "water diversion from the Yangtze River to Taihu Lake" which effectively raised water supply and improved the water quality of Taihu and water environment of river network around. However, it is not exactly known that how deep it influence by additional hydrology method. In order to check the water quality improvement of Taipu River downstream and Huangpu River, experiment of diverting water from Taipu River was implemented during March 22 to April 7 in 2006. By the joint operation of sluice-supply water and pump-supply water along Taipu River, synchronous isotope monitoring as well as water quantity and water quality monitoring has been carried out [2-3]. The author do some research on grey correlation analysis between $\delta^{18}O$ and water quality indexes in upper reach of Huangpu River for the further study of their inherent relevant.

2   Principle and Calculation

2.1   Basic Principle of Grey Correlation Degree

Grey correlation analysis is a quantitative correlation within two systems or two factors. It describes the relative change of two factors during dynamic process. This method analytically compares geometrical shape of curves changing with time and assumes that the more similar the geometrical shape, the more close the trend, the bigger the correlation degree. Therefore, the difference between geometry shapes of curves can be used to evaluate the correlation grade [4].

Suppose there are several series $\{X_1^{(0)}(t)\}$, $\{X_2^{(0)}(t)\}$, $\{X_m^{(0)}(t)\}$, $t=1,2,...,N$, where m represent the total number of series, N represent the length of series. These series are called subsequences. Also there is a mother sequence $\{X_0^{(0)}(t)\}$, $t=1, 2..., N$. With one mother sequence x_0 and several subsequences $x_1, x_2, ..., x_m$ to compare, the gross correlation degree between the mother sequence and subsequences are calculated as follows:

Step 1. Normalize original data. The units of all factors are generally not unified, so the original data should be transformed to comparable data before further processed in order to eliminate the influence of dimensions. The common methods are equalization, initialization and standardization. Equalization method is used in our experiment.

Step 2. Calculate correlation coefficient. The mother sequence and subsequence after transformation are denoted by $\{X_0(t)\}$ and $\{X_i(t)\}$ respectively. At the time $t=k$, the correlation coefficient between $\{X_0(k)\}$ and $\{X_i(k)\}$ are calculated by the following formula.

$$L_{0i}(k) = \frac{\min\min|x_0(k)-x_i(k)|+\rho\max\max|x_0(k)-x_i(k)|}{|x_0(k)-x_i(k)|+\rho\max\max|x_0(k)-x_i(k)|}$$

Where $L_{0i}(k)$ is the relative correlation coefficient between the reference curve of x_0 and the comparing curve of x_i at k-th time interval. The relative difference value in this shape is called correlation coefficient of x_i to x_0 at time k. ρ is resolution coefficient with value ranging from 0 to 1. It is usually set to 0.5 in experiment.

Step 3. Calculate Correlation degree. The correlation degree of two series is calculated as the average value of correlation coefficient at any time from $t=1$ to $t=N$. The formula is shown below.

$$r_{0i} = \frac{1}{N} \sum_{k=1}^{N} L_{0i}(k)$$

Where r_{0i} is defined as correlation degree between subsequence x_i and mother sequence x_0.

Step 4. Sequence according to correlation degree. Sequencing according to the value of correlation degrees between these subsequence and the same mother sequence, denoted by $\{X\}$. It reflects a bad or good correlation relationship of each subsequence relative to mother sequence.

2.2 Sampling

During the whole diverting water period from March 22 to April 7 in 2006, it mainly carried out two diverting water methods. March 22 to March 28 is sluice-supply water period and March 29 to April 4 is pump-supply water period, and then April 5 to 7 is normal diverting water period. Three monitoring sections are set and sampled at each branches of Huangpu River. And there is Xiaziwei section at north branch of Xietang, Sanjiaodu section at middle branch of Yuanxiejing, Maogang section at south branch of Damaogang respectively [5]. Isotope monitoring was carried out the same time with water quality monitoring. Water quality indexes were tested by Taihu Lake Basin Water Environment Monitor Center, and δ^{18}O were tested with mass spectrometer MAT-253 by Stable Isotope Lab in Institute of Geology and Geophysics, Chinese Academy of Sciences (CAS) . Then correlation degree are calculated and analyzed by the method introduced above according to the data.

2.3 Correlation Degrees Calculation

According to the data of δ^{18}O and six major water quality indexes(DO、COD$_{Mn}$、COD$_{Cr}$、NH$_3$-N、TP and TN), correlation degrees are calculated at sluice-supply water period (from March 22 to March 28), pump-supply water period (from March 29 to April 4) and the whole diverting water period respectively (from March 22 to April 7) [6]. The results are shown in Tab 1. Correlation degrees of the three branches upper Huangpu River are also calculated respectively. The results are shown in Tab 2.

Table 1. Correlation degrees between ^{18}O and water quality indexes in two diverting methods and the whole period

correlation degrees	DO	COD$_{Mn}$	TP	COD$_{Cr}$	NH$_3$-N	TN
sluice-supply water period (δ^{18}O)	0.707	0.792	0.766	0.730	0.652	0.728
pump-supply water period (δ^{18}O)	0.816	0.871	0.730	0.820	0.690	0.849
The whole diverting period (δ^{18}O)	0.845	0.896	0.820	0.868	0.713	0.867

Table 2. Correlation degrees between ^{18}O and water quality indexes of the three branches upper Huangpu River

correlation degrees	DO	COD$_{Mn}$	COD$_{Cr}$	TP	NH$_3$-N	TN
Xiaziwei ($\delta^{18}O$)	0.638	0.740	0.765	0.716	0.459	0.734
Sanjiaodu ($\delta^{18}O$)	0.733	0.808	0.782	0.727	0.444	0.688
Maogang ($\delta^{18}O$)	0.816	0.841	0.717	0.766	0.755	0.822

3   Conclusion and Discussion

This paper mainly do some research on   grey correlation analysis between major water quality indexes and $\delta^{18}O$ values during the diverting water period of the three branches upper Huangpu River [7]. From the calculation results, we can make some conclusions as follows:

Firstly, from the temporal view of point, all the correlation degrees except TP during the sluice-supply water period are higher than that during the pump-supply water period (see Fig. 1). It shows that pump-supply water had more obvious influence to the correlation degrees for the reason that there are more releasing water during pump-supply water period than sluice-supply water period. However, correlation degrees of the whole diverting water period are higher than that of the two different diverting water period. The reason can be explored that data of normal diverting water period which date from April 5 to 7 is included in calculating correlation degrees during the whole diverting period.

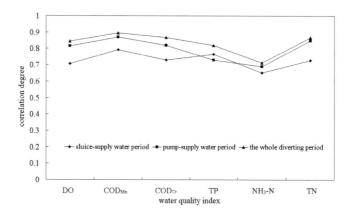

Fig. 1. Correlation degrees of sluice-supply water period, pump-supply water period and the whole diverting period

Secondly, during sluice-supply water period, correlation degrees between water quality indexes and $\delta^{18}O$ from high to low can be arranged as COD$_{Mn}$ > TP > COD$_{Cr}$ > TN > DO > NH$_3$-N (see Fig. 2). During pump-supply water period, correlation degrees between water quality indexes and δ^{18}O are sequencing as $COD_{Mn} > TN > COD_{Cr} > DO > TP > NH_3$-N (see Fig. 3). The sequence of correlation degree during sluice-supply period is different from pump-supply period.

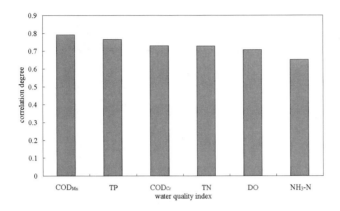

Fig. 2. Correlation degrees from high to low at sluice-supply water period

Thirdly, during both sluice-supply water period and pump-supply water period, correlation degree between COD_{Mn} and δ^{18}O is the highest (see Fig. 2 and Fig. 3). However, correlation degree between NH_3-N and δ^{18}O is the lowest. Meanwhile, correlation degree between COD_{Cr} and δ^{18}O keeps the third sequence all the time. The correlation degrees both of TP and TN change greatly from sluice-supply water period to pump-supply water period. Correlation degree of TP was in the second sequence during sluice-supply water period and fell to the fifth during the pump-supply water period. The position of TN in the sequence changed from the fourth during sluice-supply water period to the second during pump-supply water period. However, correlation degree between DO and δ^{18}O do not have so great change that just from the fifth to the fourth sequence. The results show that the correlation degree value of

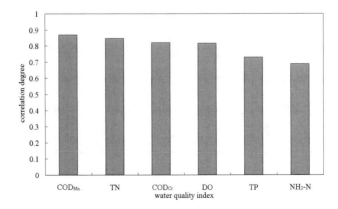

Fig. 3. Correlation degrees from high to low at pump-supply water period

COD_{Mn}, COD_{Cr}, TP and TN had high correlation with the value of $\delta^{18}O$; the NH_3-N value had little correlation with the change of $\delta^{18}O$ value during both of the two periods.

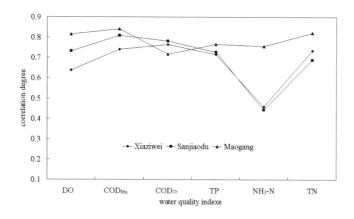

Fig. 4. Correlation degrees comparison of the three branches upper the Huangpu River

Finally, from the space view of point (see Fig. 4), we can see that the correlation degrees of Xiaziwei and Sanjiaodu have the similar chang trend except the correlation degrees of NH_3-N and TN. The most important reason is that the water of the two sections, whose branch is Xietang and Yuanxiejing respectively, are mainly from the supply of Taipu River and the river network around. Therefore, the influence to the two branches are quite similar. However, the correlation degree of Maogang section at south branch Damaogang is higher than that of the other two branches except COD_{Cr}. There are no any other reason to explain why according to the material at present but may be measurement error of COD_{Cr} or $\delta^{18}O$ which result the correlation degree of COD_{Cr} and $\delta^{18}O$ much lower than any other correlation degrees.

From the analysis of time-space change law on correlation degrees above, it comes to the conclusion that diverting water from Taipu River have great improvement to the water environment of upper Huangpu River. And ^{18}O as well as water quantity have a significant correlation with water quality.

Acknowledgment. This research was financed by the Natural Sciences Foundation of China (Projects No. 50979023 , 50379008) and the Basic Research Program for special fund of China National University (Projects No. 2010B00314). The isotope abundance of ^{18}O were analysized by stable isotope Laboratory in Institute of Geology and Geophysics, Chinese Academy of Sciences (CAS). The Water quality indexes data were offered by Taihu Lake Basin Water Environment Monitor Center, P.R. China. The authors gratefully acknowledge the valuable data provided by the agencies.

References

1. Wu, H.-Y., Zhu, L.-Z.: Influence of the Discharge of Taipu River on Water Quality in the Lower Water Source Area. Water Resource Protection 24, 42–45 (2008) (In Chinese with English abstract)
2. Mook, W.G.: Environmental isotopes in the hydrological cycle. UNESCO, Paris, pp. 12–13 (2000)
3. Wang, J.-Y., Sun, Z.-X.: Brief review on the development of isotope hydrology in China. Science in China (Series E: Technological Sciences) 40, 1–5 (2001)
4. Wang, P., Fan, Z., Xu, C.-Z.: Correlation analysis evolutionary control of multi-staged inverted pendulum. In: Proc. IEEE Symp. International Symposium on Intelligent Control, pp. 5–8. IEEE Press (2003)
5. Zhang, H.-F., Li, W.-H., Ge, H.-T.: Compositor analysis on correlation between groundwater level and water chemical contents in lower reaches of Tarim River. Arid Land Geography 26, 260–263 (2003) (In Chinese with English abstract)
6. Sun, Y.-Y., Lu, B.-H., Yang, H.-L., Wang, J.-Y.: On isotope hydrology method during the experiment of diverting water from Taipu River. Shui Li Xue Bao (supplement) 10, 470–474 (In Chinese with English abstract)
7. Sun, Y.-Y., Lu, B.-H., Huang, S.-W.: A grey correlation analysis between δ18O and water quality indexes in the upper reach of the Huangpu River. China Rural Water and Hydropower 5, 12–14 (2009) (In Chinese with English abstract)

Application of GPS and GIS in Third Party Logistics Enterprise

Hui Zhang

School of Communication and Information Engineering
Xi'an University of Science and Technology
Xi'an, China
uhuij1@sogou.com

Abstract. In order to implement the real-time control on the way vehicle and improve the operation efficiency and management level of the third party logistics enterprise, the public logistics information management platform and information system based on GPS and GIS technology for the third party logistics company are suggested. This system is consisted of four subsystems: GPS communication service system, logistics subsystem, GIS subsystem and integrated monitoring system. The principles of the system and the functions of subsystems have been shown in detail. The paper also analyzed the parsing process of GPS communication service system. The use and research of the system will accelerate the application of GIS in the third party logistics and business. It integrally realizes the located data delivering, the electronic map matching and showing as well as the real-time monitoring function.

Keywords: GPS, GIS, third party logistics, information management.

1   Introduction

It is predicted that the next few years, the number of third party logistics will increase at the rate of 20% per year in China. According to the survey of China Storage Association, Vehicle rate of empty operation is about 45%.The important reason for this situation: Firstly, the logistics companies can not accurately know the specific location of the vehicle, and cannot keep in touch with the driver in any time. They cannot supply and flexible picking their organizations for anytime or anywhere; Secondly, the driver can only determine the line by personal experience and it is difficult for the driver to find the best path, this not only delay time, but also increase the operating costs; Thirdly, the actual customer cannot keep abreast of the situation of the goods, and cannot cooperate well with logistics company.

The information level of logistics management in our country is low, in terms of the situation of the enterprise itself, that will affect the enterprise's operation efficiency inevitably and make it difficult to grasp the own resource using condition of the enterprise. In the process of allocating logistics, the existing technology is hard to master the specific transport routes of real-time information and only through the previous experience to calculate the running routes, the running time may be extended

Y. Wu (Ed.): International Conference on WTCS 2009, AISC 116, pp. 901–907.
springerlink.com   © Springer-Verlag Berlin Heidelberg 2012

because of the uncertain route states. In addition, the vehicle operation must have some idle vehicles to handle the emergencies due to the uncertain information, such as roads, or the efficiency is reduced. Concerning about the mobility of transportation, It is impossible to keep the drivers behavior constrained during the process of transportation .In most logistics enterprises, in order to skip the toll, most drivers choose to go a long way as a result they delay the time. What's more, it's common phenomena that the drivers will carry freight without permission just on their own will, even they will stay half-way privately, for in their perspective, the company can't supervise their behavior efficiently. Because of the drivers' personal behavior, it decreases the service efficiency of the enterprises themselves. The scope of logistics services is narrow.

In order to deal with these question, logistics management in the selection of transportation path of goods, storehouse address are all related to how to handle a large number of space data and time data to shorten logistic time, reduce costs. This paper intend to talking something about GIS and GPS technique with regard of problems in conventional logistics management. A graphical input and output of the electrical map also be used to improve the visual ability in managing logistics system. From this, the cargo owner , logistics company and the receiver will watch the cars moving directly. This application will make the driver and goods safer with the improvement of conveying efficiency, and maximize benefits in economic and society.

2   GIS and GPS

Global Position System（GPS） means using satellite to measure time and location for navigation ,thus constitute a global satellite positioning system. This system is a satellite-based radio positioning and timing system, and it have omnipotence functions (Terrestrial, marine, aviation and aerospace), Global, all-weather, continuous and real-time. The development in the United States of GPS systems started from 1970s,and last for 20 years, Fully completed in 1994,and it is a new generation of satellite navigation and positioning systems, which have the ability to comprehensively real-time three-dimensionally   navigation and positioning in air, land and sea[3].

At present the application of GPS system has a very wide range, we can apply the GPS signal to guidance navigation missile that used in sea, air and land, and to precision positioning for geodesy and engineering surveying, to delivery time and measuring speed.

Geographic Information System (GIS) is set of computerscience, geography, information science and other disciplines as a whole new edge science, and can be used as the basic platform used in various fields. It is based on geospatial data library-based, computer hardware and software support, application engineering and information scientific theory, scientific management and comprehensive analysis of geographical data with spatial content, to provide information technology systems which includes management.

From the perspective of technology and applications, GIS are tools methods and techniques to solve the space problem,; from the perspective of disciplines, GIS is a disciplines that based on geography, cartography, surveying and computer science, it has independent of the disciplinary system; from the function perspective, GIS has spatial data acquisition, storage, display, editing, processing, analysis, output and application functions; from a systematic perspective, GIS is a complete system with a certain structure and function.

The GPS can be quickly given the position and velocity, but not to The target property of their geographical environment and spatial information associated with a description , while the GIS is just to meet the requirements of this complementarily. Obviously, if only isolated the point of location information without geographic reference and other supporting information required, point position information on the application of logistics management is not very meaningful. With an electronic map (GIS technology) car, keep the city urban area map and save it up. It is accuracy and the amount of the information can be much higher than the printing of the urban traffic plan, but the car location achieved by drivers with the scene around by. Adding a GPS, car location, driving direction, speed and other information can be displayed at any time in the electronic map.In short, using GPS and GIS technology in logistics information management system can use the GIS system to geographic databases of transport equipment connected to the database, in order to manage resource platform.   In a unified platform, we can manage and maintain geographical information, traffic information and location information which is showed in the electronic ground map in real time, transparently, intuitively and visually. In order to enable managers to grasp the image of the object transportation time and space information, which can improve the management efficiency and management level.

In a word, the integrated use of geographic information technology and global positioning technology in logistics information management system can achieve the following functions.Reasonable way , improving the information management level, achieving security vehicle dynamic tracking control , responding sudden incidence fast , accessing to vehicle dispatch instructions directly , controlling to retrieve vehicle information quickly, achieving multiple windows, multi-screen display, powerful and accurate vehicle location, real-time control and efficient scheduling functions. Providing data support decision in the scheduling process. Against with the traditional logistics management of our country, there are several problems, the application of GIS / GPS Technology in logistics management can better address these issues, the resolution of traditional. Problems in logistics management advantages in the following areas: Create a digital logistics enterprise, standardize daily operation of enterprises, improve business Image. the application of GIS and GPS , must improve the degree of information of logistics enterprises, it can make daily operation of the enterprise digital, including enterprise-owned logistics equipment or customer's any accurate figures of each cargo can be used to describe, it can not only improve the business operation for efficiency but also improve corporate image, to attract more customers.

The transport equipment of navigation by track can improve vehicle efficiency, reduce logistics costs, and resist risk. Combining GIS and GPS technology with wireless communication technology effectively make Flowing transport equipment in different places, become transparent and be to control, so it is good to improve the efficiency of transport.

Combined Logistics decision-making with Model Library support according to logistics Actual storage conditions obtained real-time road information by the GPS can calculate the best path Logistics route to transport equipment of navigation, reducing the running time and Running costs. Using GPS and GIS technology to real-time display the actual location of the vehicle, and any enlargement, reduction, reduction, exchange diagram is enabled ;it can move with the target so that the target is always to keep on the screen and also be used to track the transport of important vehicles and cargo. And be able to effectively monitor the conduct of drivers.

3   Application of GIS and GPS in the Third Party Logistics Enterprises

According to the area and scope of the information system Logistics information management applications can be divided the following field. The first level: single application, aimed at all kinds of few function software tools and the construction of single application The main content of construct in this level includes a lot of software and single application such as office suite、 general tools of E-mail、 logistics devices equipped with bar code、 automatic identification software and logistics simulation .software The second level: the logistics business process optimization is the implementation of sector-level information system for the individual business process or management functions. The content of information construction includes the common information systems general business and Logistics Company dedicated information system. The third level: Integrated management is the implementation of enterprise-class information system for the integrated management for the entire enterprise. The content of information construction consists of the general enterprise common integrated management information system and the logistics industry-specific integrated management information system. The fourth level is common platform. The problem to be solved is the information technology of the whole logistic industry, for instance, the distribution and share of the information, the information exchange between the logistics industry and the other related agencies. These needs for information technology couldn't be taken alone by one logistic enterprise, it should be fulfilled by the external service providers or government department. According to the statistics, there are only a few large-scale logistic enterprises achieve the second level of information technology at present, account for 18% of the total numbers of the logistics enterprises. The logistic companies reach the third level are much fewer, only account for 5% .

There is hardly enterprise achieve the fourth level. These logistics enterprises in the capital, human resources and the deploy of the team with international configuration many aspects of logistics enterprise has the very big disparity. Logistics enterprises faced domestic counterparts' confusion competitive environment, and faced foreign magnates' enormous and professional competition. Application of high cost. the prominent advantages and some large abroad in the logistics information system has achieved the application, the huge economic efficiency, but the cost is hold high, this is reason cause the it not able to logistics management information system extensive application.

Software products can not meet the actual needs. Judging from the market, at present, many logistics companies have already taken a greet interest in logistics information system which is be introduced specialized. But to be face of the various logistics management system software in the market ,It is difficult to choose, and the   most of the major factors that discourage the logistics company is the functional defects.In short, the software is actually not lack of markets, but the quality of the software itself has a gap with the market demand.This gap was reflected in the GIS and GPS technology. Only a combination of spatial data GIS letter.Information and GPS satellite positioning information, the powerful advantage of Database Technology will be reflected in   the logistics, transport Otherwise, it   has not many differences with   general office automation system Network share hasn't approached to the level of information. For the past few years, as the rapid spread of GPS products, the price of GPS receiver, PDA and some other devices which can   be brought in hand or in vehicles had already meet the satisfaction of the public. As the basement of the application in this technique, the basic space of GIS can't meet the demand well in aspect of   covering field, the degree of detail , the price of market and so on. Because of scarcity in data share-mechanism among various department and agency, we not only made a lot of redundant project, but uphold the cost in a lasting high level.

To the china logistics industry in its infancy, these investments will all in vain and will seriously hinder the development of technique in logistics information system. In order to let logistics information system play a role in   the whole region's logistics environment for manipulation ,we must introduce logistics public information platform. so it can provide logistics information of the foundation to information system of the production，sales and logistics through the related of information's collection and integration besides， meet the enterprise information system needs of the logistics public information, and support the realization of various functions of Enterprise information systems.

Meantime it aslo support industry standardization of management and market management CSCW management aspects of the establishment and operation between government departments. Modern Technical requirements of modern logistics require we let appropriate goods or services   give the user at the right time and right place .The main participation of Logistics public information platform is logistics companies.Logistics enterprises could get full range of logistics support of information by Logistics public information platform, such as market demand, highway freight information, logistics parks and storage facilities as well as industry market information. Users can access public logistics information platform to achieve logistics market supply of information, related logistics enterprise information, specific business service information and so on. The government can adopt public information platform for the relevant industry sector in logistics management and standardized management of the market to provide the   information to support conditions.

Using GPS and GIS technology in the logistics public information platform that use GPS positioning funtion and GIS visualization environment let the complex lines of vehicles,the logistics and dispatching,the dispatching of network management and the scheduling of demand points   and someother issuses related to spatial location intuitively show in the monitor.Logistics system integrated   GIS and GPS technique bring a lot of advantages.First,converting the map from statistic recorded to a dynamic pattern in diversity. Futhermore, making the data visualization come true. At last, this kind of system provides strong function of analysis and process the real time space information. Iogistics distribution style intergrated GIS and GPS technique make it more easier to achieve informationization, automation, socialization, intelligentization, simplification, than the conventional Logistics distribution style . So that gets a smooth goods flow and make the best use of everything .Modern logistics system, make the space and time attribute information to all kinds of information flow, terminal Logistics operation management become more relaxed intuitive, Through space time. According to the comprehensive statistics and analysis, logistics management decision to provide strong support. Modern logistics system is played to reduce production enterprise inventory, cash flow, To improve the logistics efficiency and reduce logistics cost function, and the social demands, For the whole society's macro-control, also improve the economic benefits of the whole society, To promote the healthy development of the market economy.

4 Conclusions

Effective integration of Positioning technology and communication technology, represented by GPS, not only the exchange of information remotely, but also achieve real-time monitoring a moving target, master the state information of logistics operations. GPS and GIS technology into the modern logistics management techniques and effective integration, the establishment of the whole society to share logistics information exchange platform, has become the inevitable trend of development of modern logistics. Overall, China's modern logistics is still in the initial stage, we must strengthen the research and application of advanced technologies in order to promote the development of modern logistics in depth. The future, as international logistics, logistics of the Advanced Development and the rapid development of modern high-tech, all aspects of logistics operations will be a mechanized and automated, intelligent-based trends.

References

1. Sun, J., Zhang, C.: Propagation Characteristic of Electromagnetic Wave in Trapezium Tunnel. Journal of China University of Mining & Technology 32(1), 64–67 (2003) (in Chinese)
2. Sun, J., Cheng, L.: Equivalent analysis method of electromagnetic wave propagation in mine trapezium-shaped roadway. Coal Science and Technology 34(1), 81–83 (2006) (in Chinese)
3. Shi, Q., Sun, J.: Attenuation Characteristic of Guided EM Waves in Curved Rectangular Mine Tunnel. Journal of China University of Mining & Technology 30(1), 91–93 (2001) (in Chinese)
4. Sun, J., Cheng, L.: Analysis of electro-magnetic wave propagation modes in rectangular tunnel. Chinese Journal of Radio Science 20(4), 522–525 (2005) (in Chinese)
5. Zhang, Y., Zhang, W., Zheng, G., et al.: A Hybrid Model for Propagation Loss Prediction in Tunnels. Acta Electronica Sinica 29(9), 1283–1286 (2001)

6. Zhang, Y., Zhang, W., Sheng, J., Zheng, G.: Radio propagation at 900 MHz in underground coal mines. Journal of China Coal Society 27(1), 83–87 (2002) (in Chinese)
7. Li, W., Lv, Y., Li, B.: Experiment on characteristic of radio propagation in mine. Journal of Xi an University of Science and Technology 28(2), 327–330 (2008) (in Chinese)
8. Sun, J.: Radio Propagation Character in Mine Tunnel. Coalmine Design 4, 20–22 (1999) (in Chinese)
9. Li, Z., Miu, L.: Vehicle dispatching system based on GPS[J]. ITS Communications (1), 36–38 (2005) (in Chinese)

Virtual Noise and Variable Dimension Kalman Filter for Maneuvering Target Tracking

Deng Zhong-liang, Yin Lu, and Yang Lei

School of Electronic Engineering
Beijing University of Posts and Telecommunications
Beijing, China
dengwen1212@sogou.com

Abstract. For the non-maneuvering target, the trajectory can be predicted with the standard Kalman filtering model based on its precise mathematical model. But when the target is in motion, it is unrealistic to describe the state of the target accurately and it is difficult for the standard Kalman filtering algorithm to predict its trajectory effectively. Therefore a virtual noise & vary dimension Kalman filtering model (VN + VD model) is proposed in this paper for the maneuvering target tracking. VN + VD Kalman Filtering Model can track non-maneuvering or slight-maneuvering target by reflecting the changes of target maneuvering on the virtual noise with the low-dimensional model; and track maneuvering target with the high-dimensional model. The simulation results show the effectiveness of the algorithm.

Keywords: Maneuvering target, Kalman filter, Virtual noise, Multi-dimensional, VN + VD model.

1 Introduction

Kalman filter is widely used in tracking the target trajectory. For the non-maneuvering target, because the precise mathematical model is known, we can predict its trajectory with the standard Kalman filtering model accurately. But for the maneuvering target, we can no longer track the target accurately if we use the standard Kalman filter algorithm and rigid motion model. For the worst, it may result in filtering dispersion. Virtual noise Kalman filtering model (VD model) is an approach to track the maneuvering target [1][2]. But when the target has a constant acceleration, VD model will cause the filtering output lags behind the true trajectory (tracking lag phenomenon); we can resolve the tracking lag phenomenon by using constant acceleration Kalman filtering model with a higher dimension. But because of the higher dimension, computation will increase largely. Moreover, it may have a worse performance than the low-dimension model when tracking a non-maneuvering target [3]. This paper presents a Virtual Noise & Vary Dimension Kalman Filtering Model (VN+VD model). The filtering system uses the low-dimension model with virtual noise when the target does non-maneuvering or slight-maneuvering movements; and changes to the high-dimension model when the target does fierce-maneuvering movements.

Y. Wu (Ed.): International Conference on WTCS 2009, AISC 116, pp. 909–917.

2   Constant Velocity Kalman Filtering Model

2.1   Standard Constant Velocity Model [4]

Take two-dimension movement as an example. The state equation of target is given by

$$X_{k+1} = AX_k + V_{k..!} \tag{1}$$

where state variable X_k is

$$X_k = [x_k, \dot{x}_k, y_k, \dot{y}_k]^T. \tag{2}$$

and the state transition matrix is

$$A = \begin{pmatrix} 1 & T & 0 & 0 \\ 0 & 1 & 0 & 0 \\ 0 & 0 & 1 & T \\ 0 & 0 & 0 & 1 \end{pmatrix}. \tag{3}$$

where T is the sampling interval. The process noise V_k is an unknown zero-mean white noise with a σ_v^2 covariance and its covariance matrix is

$$Pv = \begin{pmatrix} 0 & 0 & 0 & 0 \\ 0 & \sigma_v^2 & 0 & 0 \\ 0 & 0 & 0 & 0 \\ 0 & 0 & 0 & \sigma_v^2 \end{pmatrix} \tag{4}$$

The measurement equation is given by

$$Y_k = CX_k + W_k. \tag{5}$$

where C is the measurement matrix, and it is

$$C = \begin{pmatrix} 1 & 0 & 0 & 0 \\ 0 & 0 & 1 & 0 \end{pmatrix} \tag{6}$$

Measurement noise W_k is an unknown zero-mean white noise with a σ_w^2 covariance and its covariance matrix is

$$Pw = \begin{pmatrix} \sigma_w^2 & 0 \\ 0 & \sigma_w^2 \end{pmatrix}. \tag{7}$$

2.2   Constant Velocity Model with Virtual Noise

If we use standard constant velocity model (SCV model) tracking a maneuvering target, such as a movement with constant acceleration of $2m/s^2$, the filtering result will disperse. Fig. 1 shows the filter performance in this case.

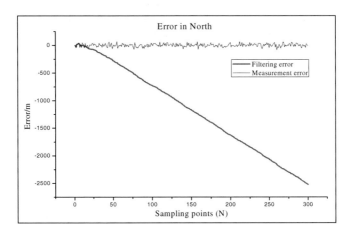

Fig. 1. Filtering error of SCV model for tracking constant acceleration target

For the reason that the process noise covariance in standard constant velocity model is constant, so it can't describe the maneuvering feature of the motion target. If we find a variable that can reflect changes of the target acceleration, then we can track the maneuvering target by the constant velocity model. According to this idea, modify the state equation to

$$X_{k+1} = AX_k + (V_k + U_k).$$

(8)

where U_k reflects the target maneuvering caused by acceleration. Definition of U_k is

$$U_k = K_k \alpha_k.$$

(9)

where K_k is the Kalman gain, and α_k is the innovation. α_k will have a great increase when the target move maneuvers, and then it will lead to the increase of U_k, then $U_k + V_k$ will also increase. This procedure is equivalent to the process noise of the overall system increases. So U_k is called as the virtual noise.

But the constant velocity model with virtual noise model (CVV model) will cause the filtering output lags behind the true trajectory when the target has a constant acceleration. We took an average of all filtering errors in each simulation, and did 10 times random experiments. The result is shown in Table 1. for comparison, it is also shown the average of all measurement errors in this table. It is very clear from Table 1 that the average of filtering errors is about -10m, but the average of measurement errors is nearly zero. This is the tracking lag phenomenon.

Table 1. The Average Error of CVV Model for Tracking Constant Acceleration Target

Experiment No.	1	2	3	4	5
Filtering Error (m)	-11.4	-14.0	-14.3	-8.2	-14.6
Measurement Error (m)	0.2	-0.8	-1.2	4.1	-2.1
Experiment No.	6	7	8	9	10
Filtering Error (m)	-13.0	-12.0	-12.0	-11.6	-13.8
Measurement Error (m)	0.4	1.3	2.7	2.1	-1.0

3   Constant Acceleration Kalman Filtering Model

3.1   Standard Constant Acceleration Model [5]

Also take two-dimension movement as example. Assume the state variable of the system is

$$X_k^m = [x_k, \dot{x}_k, \ddot{x}_k, y_k, \dot{y}_k, \ddot{y}_k]^T. \tag{10}$$

state equation and measurement equation are

$$X_{k+1}^m = A^m X_k^m + V_k^m \tag{11}$$

$$Y_k = C^m X_k^m + W_k^m \tag{12}$$

where

$$A^m = \begin{pmatrix} 1 & T & \dfrac{T^2}{2} & 0 & 0 & 0 \\ 0 & 1 & T & 0 & 0 & 0 \\ 0 & 0 & 1 & 0 & 0 & 0 \\ 0 & 0 & 0 & 1 & T & \dfrac{T^2}{2} \\ 0 & 0 & 0 & 0 & 1 & T \\ 0 & 0 & 0 & 0 & 0 & 1 \end{pmatrix} \tag{13}$$

$$V_k^m = [\dfrac{T^2}{4} v_{1k}, \dfrac{T}{2} v_{1k}, v_{1k}, \dfrac{T^2}{4} v_{2k}, \dfrac{T}{2} v_{2k}, v_{2k}]^T \tag{14}$$

$$C^m = \begin{pmatrix} 1 & 0 & 0 & 0 & 0 & 0 \\ 0 & 0 & 0 & 1 & 0 & 0 \end{pmatrix} \tag{15}$$

$$W_k^m = [w_{1k}, w_{2k}]^F \tag{16}$$

Fig. 2 shows the filtering result of using standard constant acceleration model (SCA model) for tracking the accelerative target mentioned above. We can see that it has a good performance.

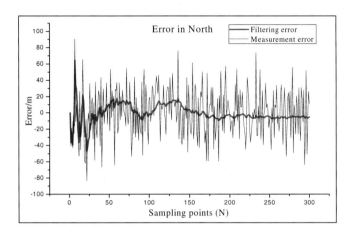

Fig. 2. Filtering error of SCA model tracking constant acceleration target

3.2   Constant Acceleration Model with Virtual Noise

It will lead to a large filtering error if we use the SCA model to track the target when its acceleration has great changes. Of course, it will solve this problem by adding virtual noise to the SCA model. But it will cost the performance degradation when tracks a constant acceleration target. This is because the items reflect acceleration in the process noise covariance matrix is more sensitive in a high-dimension model. So the algorithm using innovation to reflect the changes of acceleration indirectly is imprecise. From the above analysis, we can draw a conclusion that it can't acquire a good performance to predict the trajectory of a maneuvering target by using the constant acceleration model with virtual noise (CAV model).

4   Virtual Noise and Vary Dimension Kalman Filtering Model

Compare the four models mentioned above in Table 2. If there is a model which can learn from strong points of the four models to offset their weaknesses, the problem will be readily solved. Virtual noise & vary dimension Kalman filter model (VN+VD model) is a good choice.

The brief procedure of VN+VD algorithm is that the filter system tracks non-maneuvering or slight-maneuvering target by CVV model; then augments the dimension, namely using the SCA model, following the maneuver detection; then the system continues to monitor the target dynamics in SCA model and return to CVV model when the acceleration losses its statistical implications[6][7]. On the other hand, we notice that the possibility of the acceleration with a continuous variation is very slim in practical applications and the acceleration always keeps a smooth level after an impulse change. According to this, there is also a module to monitor the severe changes of acceleration in the filtering system. The system will track the target by the CAV model when it detects severe changes of acceleration or only track by the SCA model.

Table 2. The Comparison of Each Model

Model	Advantages	Disadvantages
SCV	Has a great performance for tracking non-maneuvering target.	Can't track maneuvering target or can track it but at the cost of bad performance for tracking non-maneuvering target.
CVV	Has a well performance for tracking non-maneuvering or slight-maneuvering target.	Lead to tracking lag when deals with the target in a large acceleration.
SCA	Has a great performance for tracking constant accelerative target.	Can't adapt the impulse of acceleration. And requires a large computation as a result of the increase of dimensions.
CAV	Has the ability to adapt the impulse of acceleration.	Bad performance in tracking either constant velocity or constant acceleration target. Also requires a large computation as a result of the increase of dimensions.

The discriminate function of CVV to SCA model is given by (17)

$$\delta_{in}(k) = \alpha^T(k)S^{-1}(k)\alpha(k) \tag{17}$$

where $\alpha(k)$ is the innovation and $S(k)$ is the covariance matrix of innovation. $\delta_{in}(k)$ reflects the specific value of innovation's square at time k and the variance of all innovations at time $1\sim k$. If the target is in maneuvering state at time k, the numerator will be much larger than the denominator. Or their magnitude will be in the same order. Take T_a as discriminate threshold, the model will change to high-dimension from low-dimension when δ_{in} is larger than T_a.

Because the estimate value of acceleration is included in the state variable in the high-dimension model, we can use it to monitor the maneuver of the target. The discriminate function of SCA to CVV model is given by (18)

$$\delta_{out}(k) = \sum_{j=k-p+1}^{k} \hat{a}^T(j)\left[P_a^m(j)\right]^{-1}\hat{a}(j) \tag{18}$$

where $\hat{a}(j)$ is the estimated value of acceleration in the state variable; $P_a^m(j)$ is the items reflect acceleration in filtering covariance matrix and p is the length of the filtering window. The model will change to low-dimension from high-dimension when δ_{out} is smaller than discriminate threshold T_b.

5   Simulation

Fig. 3 and Fig. 4 show the simulation result using the VN+VD Kalman filtering model for tracking a maneuvering target.

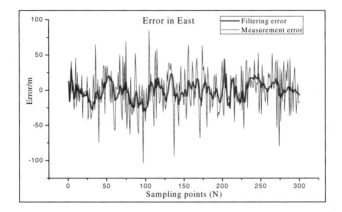

Fig. 3. Filtering error of VN+VD model for tracking a maneuvering target in the east direction

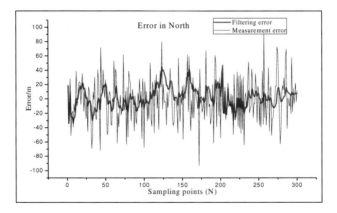

Fig. 4. Filtering error of VN+VD model for tracking a maneuvering target in the north direction

The target motion was generated in two dimensions with a sampling interval of 0.5 seconds. Measurement error was assumed known with a white noise of variance 30m. The target departed from the origin with initial velocities of 50m/s in both directions and it last 50 seconds (the first period); then it began a motion with an acceleration of -0.125m/s^2 in the north direction while no change in the east (the second period); and then proceeded for a further 50 seconds straight motion with an acceleration of -3m/s^2 in both directions (the third period).

Overall, the performance was satisfying as depicted in Fig. 3 and Fig. 4. But the filtering errors were a little larger in the moment of period 2 to period 3 (N=200) than in other periods. It was because there was a great change of acceleration in both directions at that instant, and for this reason, it was acceptable as long as the filtering results not dispersing. The well performance could also be seen afterward that instant (N>200) for quickly decline of the filtering errors.

Table 3. The Statistic of VN+VD Model for Tracking a Maneuvering Target

	Mean of 10 Random Experiments
Mean of the Filtering Error in the East Dir. (m)	-0.40449
Variance of the Filtering Error in the East Dir. (m^2)	226.9079
Variance of the Measurement Error in the East Dir. (m^2)	918.2587
Mean of the Filtering Error in the North Dir. (m)	4.630237
Variance of the Filtering Error in the North Dir. (m^2)	311.4212
Variance of the Measurement Error in the North Dir. (m^2)	923.5115

The mean values of the filtering errors and the variances of the filtering/ measurement errors in both directions of all 300 sampling points in each random experiment of 10 times are shown in Table 3. It is not hard to notice that the mean value of the filtering errors in the east direction is nearly zero while a little larger in the north direction. This is because the north acceleration in the second period was in a low level and the system was working in the CVV model in most time, so it resulted in a slightly tracking lag. In the terms of numeral, 4.63m was completely negligible compared to dozens of meters of the noise. On the other hand, the filtering error variance was less than one quarter (24.71%) of the measurement error variance in the east direction while about one-third (33.71%) in the north direction. This means the filtering result is better than the measurement result, and a good performance of the proposed model. The reason why the filtering error in the north direction was a litter larger was also because of the larger acceleration in that direction.

6   Conclusion

Standard Kalman filter has a great performance for tracking non-maneuvering targets while a bad performance for maneuvering ones. Although we can get a better performance by adding the virtual noise to Kalman filtering model or by using a high-dimension model, either of them has disadvantages. The VN+VD Kalman filtering model proposed in this paper combines the advantages of other models and we can see it has a great performance for tracking maneuvering target according to the simulation result. And it also has a good performance on the HYRISING's "Merako GNSS open source development platform" with this algorithm.

References

1. Yan, Z.P., Huang, Y.F.: Research on prediction of moving targets with Kalman filtering method. Applied Science and Technology 35, 28–32 (2008)

2. Luo, D.D., Zhu, Y.M.: Random parameter matrices Kalman filtering. Journal of Sichuan University (Natural Science Edition) 45, 1309–1312 (2008)
3. Efe, M., Atherton, D.: Maneuvering target tracking with an adaptive Kalman filter. In: Proc. IEEE Symp. Conference on Decision & Control, pp. 737–742. IEEE Press (December 1998)
4. Zeng, J., You, G.H., Jia, S.J., Wei, M.: Vehicle dynamic navigation position filtering algorithm based on Kalman filtering. Journal of Dalian Jiaotong University 29, 42–45 (2008)
5. Huang, H., Zhang, H.S., Xu, J.D., Huang, Y.: A target tracking algorithm based on two-stage variable dimension of Kalman filter. Aeronautical Computing Technique 38, 97–100 (2008)
6. Dong, W.L.: Research of Kalman filtering algorithm in vehicle navigation system. Zhengzhou University, Zhengzhou (2007)
7. Chen, X.R., Chen, S.F.: Variable dimension of Kalman filter for target tracking. Chinese Journal of Scientific Instrument 27, 1163–1165 (2006)

Evaluation of Cultivated Land Productivity of Wanyuan County, Sichuan Province Using GIS

Qian Zhou[1,3], Wei Wu[2,3], and Hong-Bin Liu[1,3,*]

[1] College of Resources and Environment, Southwest University, Chongqing 400716, China
[2] College of Computer & Information Science, Southwest University, Chongqing 400716, China
[3] Chongqing Key Laboratory of Digital Agriculture, Chongqing 400716, China
zhouyo1@hotmail.com

Abstract. The objectives of the current study are to determine the factors impacting cultivated land productivity and to illustrate the spatial distribution pattern of cultivated land productivity using GIS. Landform information (slope and topographical position), soil physical and chemical properties (effective soil depth, organic matter, parent materials, texture, pH, total nitrogen, available phosphorus, available potassium, and drainage capacity), and cropping system were selected to assess the cultivated land productivity using Delphi technique. The weight of each index was determined by AHP approach. The results showed that about 60.21% of the study area was classified as third level followed by 28.5% of the total area as forth level, located in northwestern and southern Wanyuan, respectively. The highly productivity cultivated land is mainly distributed in valley with slope of less than 15° in Wanyuan county.

Keywords: Classification, Topography, Factor, Soil property.

1 Introduction

Cultivated land is the essential resource for crop production and agricultural sustainable development. Many factors, such as such as climate, topography, parent materials, and soil physical-chemistry properties have influence on cultivated land productivity. Understanding the spatial distribution of cultivated land productivity levels and the relevant factors could help farmers and policy makers to refine their current agricultural management activities and planning.

Geographic information system (GIS) is widely used to store, display, and manage geo-referenced information. Currently, cultivated land productivity evaluation and mapping has become one of the most useful applications of geographic information system (GIS) [1-8]. However, factors which impact on cultivated land productivity vary with different environmental conditions.

The objectives of the current study are (1) to determine the factors impacting cultivated land productivity with respect to topographical attributes, soil properties, and management practices and (2) to illustrate the spatial distribution pattern of cultivated land productivity in Wanyuan, Sichuan province using GIS.

---

* Corresponding author.

Y. Wu (Ed.): International Conference on WTCS 2009, AISC 116, pp. 919–926.
springerlink.com   © Springer-Verlag Berlin Heidelberg 2012

2   Materials and Methods

2.1   Study Area

Wan Yuan is located the south of Daba Mountains in the northeast of Sichuan between 107 ° 28'53 " and 108 ° 30'34"E and 31°38'35"and 32°20'22"N with an area of 404549 hm^2. The climate is subtropical humid with an annual sunshine hours of 1474 h, annual average temperature of 14.7℃ and average annual precipitation of 1,100 mm. The topography of the area is mountainous with a range of altitude of 600-1400m. The predominant soil types are paddy, moisture soil, purple soil, yellow loam soil, yellow brown soil, and calcareous soil consisting of thirteen subgroups, twenty-nine soil genus, and sixty-four natives. The county has cultivated land of 12746 hm^2, including dry land of 18395 hm^2 and paddy field of 12746 hm^2.

2.2   Selection of Evaluation Factors

Many factors, such as soil physical and chemical properties, terrain attributes, and crop systems, have influence on cultivated land productivity.

Table 1. Hierarchy structure of the evaluation factors in Wanyuang county

Target layer (A)	Criteria layer (B)	Index layer (C)
Cultivatedl and productivity	Site condition B1	Slope C1
		Effective soil thickness C2
		Soil parent material C3
		Topographical position C4
	Physical and chemical condition B2	pH C5
		Textures C6
		Organic matter C7
	Soil nutrient content B3	Total nitrogen C8
		Available phosphorus C9
		Available potassium C10
	Land management B4	Irrigation capacity C11
		Cropping system C12

According to the principles introduced by NATESC [9], 12 factors were selected to evaluate the cultivated land productivity of Wanyuang county using Delphi method [10] (Table 1).

2.3 Determination of Factor Weights

Four of the 12 selected indices are quantitative factors, namely, topographical position, soil texture, soil parent materials, and cropping system. The values for quantitative elements were determined by the experts and listed in Table 2.

Table 2. Values of the quantitative indices determined by experts

Index	Element	Value
Topographical position	Mountainside of medium-low Mountains	0.25
	Middle, top medium-low mountains	0.35
	Valley bottom	0.45
	Bench terrace of medium-low Mountains	0.5
	Sloping fields in medium-low Mountains	0.55
	Gully	0.6
	Sloping fields of low hills	0.65
	Gentle-slope hilly	0.7
	Alluvial	0.825
	Valley	0.9
Soil textures	Compact sand	0.422
	Silty sand	0.556
	Heavy clay	0.617
	Sandy clay	0.7
	Sandy loam	0.75
	Clay	0.828
	Clay loam	0.914
	Loam	1
Soil parent material	Gravelly deposits on slope	0.4
	Sandy alluvial deposits	0.6
	Soil deposits on slope	0.7
	Residual sediments	0.8
	Loam alluvial deposits	1
Cropping system	Three crops a year (taro-maize-potato)	0.94
	Two crops a year(rice-Rape)	0.85
	One crop a year (rice)	0.69

There are many methods to calculate the weight of the evaluation factors, for example, principal component analysis, multiple regression analysis, stepwise regression analysis, grey-correlation analysis, and analytic hierarchy process (AHP) etc [11-12]. Among them, AHP method is a structured technique for dealing with complex decisions and has been widely used around the world. Therefore, the AHP was employed to determine the weight of each factor. The results were shown in Table 3.

Table 3. Factor weights of the cultivated land in Wanyuan county

A	B1 0.3824	B2 0.2894	B3 0.1463	B4 0.1819	Combined weight
C1	0.203				0.0776
C2	0.3802				0.1454
C3	0.2898				0.1108
C4	0.1269				0.0485
C5		0.2353			0.0681
C6		0.303			0.0877
C7		0.4617			0.1336
C8			0.4514		0.066
C9			0.3109		0.0455
C10			0.2378		0.0348
C11				0.6202	0.1128
C12				0.3798	0.0691

The order of the factors was effective soil thickness (C2)> organic matter (C7) > the probability of (C11) > soil parent material (C3) > textures (C6) > slope (C1) > cropping system (C12) > pH (C5) > total nitrogen (C8) > topographical position (C4) > available phosphorus (C9) > available potassium (C10) .

The cultivated land was divided into 10472 evaluation units using ArcGIS 9.3. For each unit, the integrated fertility index (IFI) was calculated by

$$IFI = \sum Fi \times CBi \ (i = 1, 2, 3 \cdots, n) \tag{1}$$

Where Fi is the weighted value of each CB; CBi is a set of principal components. The cultivated land productivity was classified as six levels according to IFI [13] (Table 4).

Table 4. Classification of the cultivated land productivity in Wanyuan

Grade	1	2	3	4	5	6
IFI	0.91-1.0	0.81-0.9	0.71-0.8	0.61-0.7	0.51-0.6	≤ 0.5

3   Results and Discussion

3.1   Site Condition

The site condition characteristics of the study area are shown in Table 5-6. About 81.4% of the total area was covered by slopes of 2-15° and 15.8% with slopes steeper than 15°.

Table 5. Area distribution of different slope class

Slope(°)	Area (hm²)	Percentage (%)
<2	614.59	1.97
2-6	6740.01	21.64
6-15	18866.79	60.58
15-25	4919.59	15.81
Total	31140.98	100.00

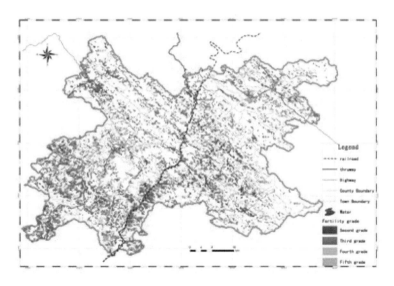

Fig. 1. Classification map of cultivated land productivity in Wanyuan county, Sichuan

For topographical position, about 51% of the area was located in mountainside of medium-low mountains followed by 18% of the area located in Bench terrace of medium-low Mountains.

Table 6. Area distribution of topographical position

Topographical position	Area （hm²）	Percentage (%)
Alluvial	5.23	0.02
Sloping fields of low hills	28.42	0.09
Gully	1998.61	6.42
Valley	1998.85	6.42
Gentle-slope hilly	8.68	0.03
Valley bottom	3172.92	10.19
Sloping fields in medium-low Mountains	320.04	1.03
Mountainside of medium-low Mountains	15818.25	50.80
Middle, top medium-low mountains	2253.72	7.24
Bench terrace of medium-low Mountains	5536.25	17.78
Total	31140.98	100.00

3.2   Soil Properties

Table 7 showed summary statistics of soil properties for the study site. The coefficient of variation decreased in the order available phosphorus > effective soil thickness > organic matter > total nitrogen > available potassium > pH. The content of organic matter is mainly in the range of 20-30g/kg. The content of available phosphorus mainly concentrates in 10-20 mg/kg, while the content of available potassium is mostly in 50-100 mg/kg. The content of pH is mainly in the range of 5.5-6.5.In addition, effective soil thickness is mostly from 40 to 60 cm.

In Wanyuan, most of soil textures are sandy loam and clay, which are very suitable for planting various crops. As Wanyuan has developed irrigation and drainage systems, the drainage capacity is guaranteed.

Table 7. Description statistics of soil properties in Wanyuan County

Soil properties	Mean	Median	Min	Max	S.D.	C.V. (%)
pH	6.23	6.3	4.1	8.7	0.45	7.23
Organic matter (g/kg)	21.86	20.3	4.2	53.30	5.81	26.58
Total nitrogen (g/kg)	1.22	1.22	0.23	2.8	0.25	20.67
Available Potassium (mg/kg)	80.99	80	29	210	16.55	20.43
Available Phosphorus (mg/kg)	11.50	10.2	0.5	59.80	6.36	55.29
Effective soil thickness (cm)	35.54	30	7	70	11.58	32.57

3.3   Cultivated Land Productivity

The statistical analysis of cultivated land productivity was given in Table 8 and the classified map was illustrated in Figure 1. Most of the cultivated land (60.21%, 18748.89 hm^2) are classified as third level productivity followed by 28.5% (8875.75 hm^2) as forth level and 7.59% (23962.51 hm^2) as second level. No first and sixth levels cultivated productivity land was found for the study area.

Figure 1 showed that most of the second level cultivated land was distributed in valley with slope of less than 15° and higher contents of organic matter, available potassium, and available phosphorus. The third level land was mainly concentrated in southwestern Wanyuan with soil parent materials of gravel slope deposits and soil slope sediments and soil texture of clay loam and sandy loam. The forth level land was mainly distributed in southern Wanyuan with lower contents of soil organic matter, available potassium and phosphorus.

Table 8. Classification of cultivated land productivity in Wanyuan County

Productivity classification	IFI	Area（hm^2）	Percentage of area（%）
2	0.7-0.8	23962.51	7.59
3	0.6-0.7	18748.89	60.21
4	0.5-0.6	8875.75	28.50
5	0.4-0.5	1153.83	3.70
Total		31140.98	100.00

4   Conclusion

The cultivated land productivity in Wanyuan, one important agricultural county of Sichuan province, was evaluated with the help of GIS technology. Twelve factors including slope, topographical position, effective soil thickness, soil parent materials, organic matter, total nitrogen, pH, texture, available phosphorus, available potassium, irrigation capacity, and cropping system were selected to assess the cultivated land productivity using Delphi technique. The weight of each index was determined by AHP approach. The results showed that about 60.21% of the study area was classified as third level followed by 28.5% of the total area as forth level, located in northwestern and southeastern Wanyuan, respectively. The highly productivity cultivated land (second level) is mainly distributed in valley with slope of less than 15° located in southeastern in Wanyuan county.

References

1. Liu, Y., Liang, Q., Sheng, H., et al.: Land information system and it's trial research, an example about the area. In: Beijing Thirteen Mausoleum[A] China Geography College Geography and Agriculture, pp. 86–93. Science Publishing Company, Beijing (1983) (in Chinese)

2. Huang, X.: Research of land use decision-making Based on GIS. Acta Geographica Sinica 48(2), 114–121 (in Chinese)
3. Fu, B.: The Spatial Pattern Analysis of Agricultural Landscape in the Loess Area. Acta Ecologica Sinica 15(2), 113–120 (1995) (in Chinese)
4. Lu, M., He, L., Wu, L., et al.: Evaluation of the farm land productivity of hilly region of central China based on GIS. Transactions of the CSAE 22(8), 96–101 (2006) (in Chinese)
5. Sun, Y., Guo, P., Liu, H., et al.: Comprehensive evaluation of soil fertility based on GIS. Journal of Southwest Agricultural University 25(2), 176–179 (2003) (in Chinese)
6. Zhou, H., Xiong, D., Yang, Z., et al.: Natural productivity evaluation of cultivated land based on SOTER database in the Typical Region of Upper Reaches of the Yangtse River. Chinese Journal of Soil Science 36(2), 145–148 (2005) (in Chinese)
7. Duan, Q., Peng, S., Tian, Y., et al.: Application of Gray Relation Analysis in the Evolution of Cultivated Land Fertility. In: The Corpus of Second National Conference about Soil Testing and Fertilizer Recommendations, China (in Chinese)
8. Lin, B.-S., Tang, J.-D., Zhang, M.-H.: Valuation of cropland capacity classes in Guangdong. Ecology and Environmental Sciences 14(1), 145–149 (2005) (in Chinese)
9. The National Agro-Tech Extension and Service Center (NATESC). Guide of the Cultivated land fertility Evaluation, pp. 75–76. China Agricultural Science and Technology Press, Beijing (2006)
10. Wang, J., Shan, Y., Yang, L.: Theory and method of classification and gradation for farming land in China. System Sciemces and Comprehensive Studies in Agriculture 18(2), 84–88 (2002)
11. Zhao, H.: Analytical Hierarch program——A Simple New Method for Decision. Science Press, BeiJing (1986) (in Chinese)
12. Wei, X.F., Duan, J.N., Hu, Z., et al.: Applying Analytic Hierarchy Process to Determining Farmland Productivity Evaluation Factors' Weight. HuNan Agricultural Science 2, 39–42 (2006) (in Chinese))
13. Agriculture Profession Standards. Evaluate the Cultivated land fertility in the Land Type in Chinese (1996) (in Chinese)

Observatory Information Management for Lake Ecosystem Monitoring Application

Bu Young Ahn and Young Jin Jung

Dept of Cyber Environment Development, Supercomputing Center
Korea Institute of Science and Technology Information
Daejeon, South Korea
baaihyt@sohu.com

Abstract. Today, water pollution and scarcity are global problems, which can lead the worldwide cause of deaths and diseases. In order to protect this serious problem, it is required to monitor the ecosystem in river and lake. So, we propose the cyberinfrastructure to support the limnologist's research such as water quality analysis and lake ecosystem health assessment. The infrastructure that is KLEON (Korean Lake Ecological Observatory Network) is designed for limnologist to evaluate the lake ecosystem health by using the observation data with the lake ecological modeling such as CE-QUAL-W2. The KLEON includes the installation of sensors, observatory information management about sensors and sampling sites, and observation data management.

In this paper, we focus on the observatory information management to effectively handle the attributes of the installed sensors and the condition of sampling sites, because observation data is affected by the status of sensors and the condition around sampling sites. The observatory management module deals with the information of lake, dam, floodgate, weather, and observation site. This information is necessary for an ecologist to understand the sensor and the sampling data by comparing the observation with the condition of observatory.

Keywords: Environmental monitoring, Sensor network, Lake ecosystem health assessment, Observatory information management.

1   Introduction

Water pollution and scarcity makes some serious global problems such as cause of deaths and diseases. Lots of people, fishes, plants, and organisms are affected by water scarcity and pollution [1]. The United Nations' FAO [2] described that more 1.8 billion people in the world will face the absolute water scarcity by 2025. With receding glaciers and shrinking lakes, freshwater is decreasing and stream is reducing by climate change.

Y. Wu (Ed.): International Conference on WTCS 2009, AISC 116, pp. 927–935.
springerlink.com   © Springer-Verlag Berlin Heidelberg 2012

In order to protect this serious problem, we have to evaluate the ecosystem health assessment in river and lake. We propose the KLEON (Korean Lake Ecological Observatory Network), which is a cyberinfrastructure to support the limnologist's research such as lake ecosystem health assessment and the change of plankton. This KLEON is designed for limnologist to evaluate the lake ecosystem health by analyzing the observation such as sensor and sampling data with the lake ecological modeling such as CE-QUAL-W2 [3].

Water quality sensors (water temperature, pH, DO, and Turbidy) are installed to continuously get the observation at seven places such as lake, river, and wet land. To handle the observation, KLEON supports the observation data management and the observatory information management to store the history of the attributes of sensors and sampling sites.

In this paper, we focus on the observatory information management to provide the additional information for observation, because observation is changed depending on the status of sensors and the condition around sampling sites such as such as water supply and watershed mean rainfall. For example, a water quality sensor needs to be cleaned to keep the accuracy of its measurement once every a week, because algal bloom, which sticks to the sensor, increases the observation error over time. The sampling data is also affected by the condition of sampling site such as heavy rain, construction, and sampling method.

The designed observatory information management is useful for an ecologist to clearly understand the change of ecosystem in lake with the observation data.

2   Related Work

In order to cope with water scarcity and pollution, most of environmental monitoring applications try to manage and analyze their observation by using the installed buoys, observation management, and analysis tool such as Lake analyzer. There are various kinds of lake ecological monitoring applications to analyze the phenomena in lake or river such as GLEON [4], GoMOOS [5], LERNZ [6], LakeESP [7], and Pakistan wetlands program [8].

In order to understand and forecast the natural and artificial change of ecosystem, GLEON (Global Lake Ecological Observatory Network) [4] make the environment to support the ecologists' research by constructing lake observatory network with data provider and data analysis tools. This community helps limnologists, scientists, and engineers share their observation data with observatory information and discuss their research results.

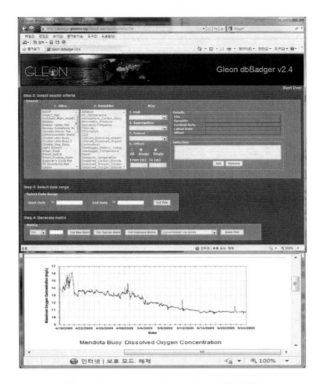

Fig. 1. dbBadger and VADER of GLEON

GLEON provides the dbBadger to search the past and the current observation from stored sites. VADER is also provided to draw graph from the selected observation such as water temperature, dissolved carbon dioxide, and plytoplankton pigments. With this lake observatory network, ecologists, geologists, scientists, and engineers are cooperated to understand the phenomena in lake ecology.

GoMOOS (Gulf of Maine Ocean Observing System) [5] has been developed to understand and forecast the change of ocean, because various ecosystems on the side of science, environment, commerce, and resource are affected by the condition of ocean. GoMOOS provides the historical and current observation depending on the buoy position with web site and mobile application. Besides, GoMOOS also supports the forecasts such as wind, wave, water level, and circulation.

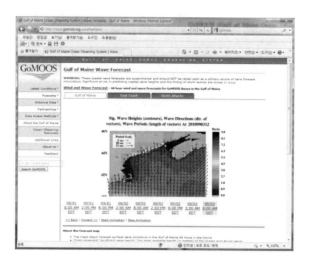

Fig. 2. Wave forecast of GoMOOS

To protect and restore the lake ecosystem in Lew Zealand, LERNZ (Lake Ecosystem Restoration New Zealand) [6] offers the information of lakes, harmful algal bloom, invasive fish, rivers, and urban restoration. LERNZ are developing new models to restore various kinds of species with invasive fish surveys and fish removal from selected habitats. They calculate density and biomass per unit area from the captured quantitative fish.

LakeESP (Environmental Sensing Platform) [7] of PME (Precision Measurement Engineering) is a long term water quality monitoring and management system. PME makes buoys with the particular sensors, software, and tools to support the fresh ocean water research. The instrument features real-time monitoring of water column temperature stratification, meteorological and water quality parameters. Meteorology sensors include humidity, air temperature, wind speed/direction, and liquid precipitatioin. The observation is stored at raw file storage and transmitted user's pc by using FTP transfer.

Wetland is very important ecosystem to keep water quality, provide habitats for birds, and supply food. Pakistan wetlands program [8] is preventing the valuable wetlands with training information such as global environment facility and the united nations development programme.

With the enhanced sensor network, data communication, and computing technologies, these environmental monitoring applications are developed to understand phenomena with the knowledge for ecosystem.

To effectively manage their observation and observatory information, various kinds of technologies are being designed and developed to support ecologists' researches. In these environmental monitoring applications, it is very important to understand their observations with the observatory information such as weather and geography.

3   Lake Ecosystem Monitoring Application

In order to get the lake ecological data, we are installing various kinds of sensors, which include water temperature, DO, pH, and electrical conductivity with RF and CDMA modem at 7 target places such as Lake Soyang, Lake Euiam, and Anyang river.

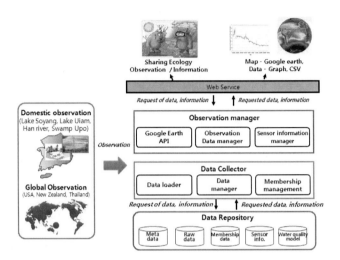

Fig. 3. The organization of KLEON

The KLEON consist of the observation data collector to handle the raw data and the observation manager to support the sampling data insertion, update, search, and download with map service. The observation data collector receives and stores the observation data, which is transmitted or downloaded from the installed sensors into raw data database. When the sampling and the sensor data is accumulated in database, limnologists use the data to analyze the water quality and ecological event such as fishkill with heavy rain. To support the ecologist's research such as observation management, the observation manager employs the Google earth api, the observation data manager, and sensor information manager. The sensor information manager handles the status and history of sensors by using the sensorML (Sensor Model Language), which is one of standardization of SensorWeb of OGC [9-11] (Open Geospatial Consortium). Handling sensor information is one of the important steps for understanding the observation environment and data, because the accuracy of observation data is depending on the sensor's properties such as sampling interval, measurement unit, and error rate. The observatory information manager also deals with the attributes of sampling sites such as sampling position, heavy rain, and construction.

4   Observatory Information Database

KLEON manages the information of dam, floodgate, weather, plankton, lake, observation site, and water quality in database. To store the weather, floodgate, water quality, and plankton, a flexible data model is utilized.

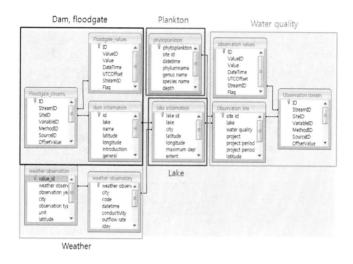

Fig. 4. The observatory information database

The observatory information database is designed to deal with not only the current observatory information but also the temporal change of the information. This database includes sensor information, lake, dam, floodgate, sampling site, and weather information such as solar radiation, total cloud cover, air temperature, dew point temperature, wind direction, wind speed. With this information, ecologists clearly understand their observation data such as sensor and sampling data. In order to handle both the sampling data and the sensor data, the observation database is also designed by extending the Vega [12] that is a flexible data model for environmental time series data with data standardization of NOAA (National Oceanic and Atmospheric Administration) or OGC.

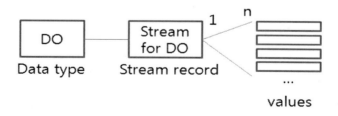

Fig. 5. The vega data model

This Vega based structure is useful to handle the observation per each site. For example, the types of sampling data are changed depending on the sampling site. By inserting a record for data stream, the observation of the data type is handled in the linked values tables.

5   Web Interface for Observatory Management

To manage the observatory information, KLEON provides web interface with Google earth api, MySQL, JSP, Java script, and flash graph.

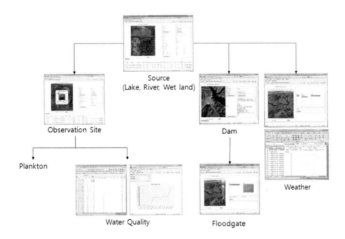

Fig. 6. The relation among observatory information management

Figure 6 shows the relation in observatory information management, which describes the order of insertion such as (source → observation site → water quality) and (source → dam → floodgate), because the information of source and observation site is prior to the observed water quality or plankton. The source shows the properties of lakes, rivers, and wet lands such as maximum depth, extent, and watershed area. The observation site describes a sampling position or a buoy's location with measurement method. Water quality database handles 34 kinds of items with Vega data model such as Dissolved Oxygen, BOD, COD, and pH. To analyze the water quality, KLEON also handles weather data such as total cloud cover, solar radiation, dew point temperature, wind direction, and wind speed.

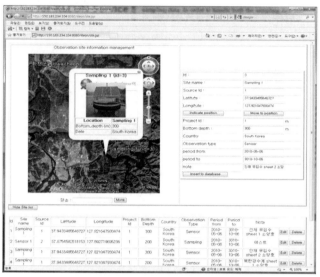

Observation site management

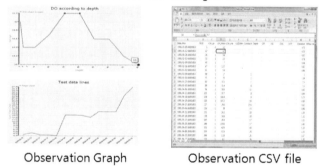

Observation Graph   Observation CSV file

Fig. 7. The observation site management

The observation site management module handles the sampling site and buoys, which attach several sensors as shown in Figure 7. To show the observation data at 7 lakes, reservoirs, or wetland on the web with the geographical information, the google earth based user interface is provided. Users can find any sensors and their data in the world like SensorWeb. For example, users can find any lakes with the google earth. When users click the observatory mark, which are attached on a buoy or a bridge in the selected lake, the observation information and data in the observatory mark is provided. With this description, users can check the monitoring goal, sensor types, data collection methods, and linked observation data of the selected observatory.

This implementation is used for a user to deal with the observatory information such as lake, dam, floodgate and weather. When an ecologist checks observation to analyze a phenomenon such as pollution, some particular observatory information at the site is used to provide the additional information for the observation, which can help a user understand the detected phenomenon.

6   Conclusion

The implemented observatory information management modules are useful for a user to analyze the observation such as the change of water quality with rainfall by providing the condition around the site. Future work includes the data sharing with observatories in the world and the query processing to clean the observation with observatory information.

References

1. Pink, D.H.: Investing in Tomorrow's Liquid Gold. Yahoo Finance, April 19 (2006)
2. FAO Hot issues: Water scarcity, http://www.fao.org
3. Water quality and hydrodynamic model, http://www.ce.pdx.edu/w2/
4. GLEON (Global Lake Ecological Observatory Network), http://www.gleon.org
5. GoMOOS (Gulf of Maine Ocean Observing System), http://www.gomoos.org/
6. LERNZ Lake Ecosystem Restoration (New zealand), http://www.lernz.co.nz/
7. LakeESP (PME USA), http://www.pme.com/
8. Wetlands Programme (Pakistan), http://www.pakistanwetlands.org/
9. OGC (Open Geospatial Consortium), http://www.opengeospatial.org/
10. Sensoer Web Enablement of Open GIS consortium, http://www.opengeospatial.org/projects/groups/sensorweb
11. Sheth, A., Henson, C., Sahoo, S.: Semantic Sensor Web. In: IEEE Internet Computing, pp. 78–83 (July/August 2008)
12. Winslow, L.A., Benson, B.J., Chiu, K.E., Hanson, P.C., Kratz, T.K.: Vega: a flexible data model for environmental time series data. In: Proceedings of the Environmental Information Management Conference 2008, Albuquerque, NM, September 10-11, pp. 166–171 (2008)

Moving Object Query Processing Technique for Recommendation Service

Young Jin Jung[1], Bu Young Ahn[1], Kum Won Cho[1], Yang Koo Lee[2], and Dong Gyu Lee[2]

[1] Dept of Cyber Environment Development, Supercomputing Center Korea Institute of Science and Technology Information Daejeon, South Korea
[2] Database / Bioinformatics Laboratory Chungbuk National University Cheongju, South Korea
fa2310@yahoo.cn

Abstract. The positions of spatial or moving objects with maps are frequently utilized through travel, exploration. LBS (Location Based Service) corresponding to the change of users' location, are actively provided according to the progress of location management technologies such as GPS, wireless communication network, and personal devices. The research scope of LBS has been changed from vehicle tracking and navigation services to intelligent and personalized services considering the changing information of conditions or environment where the users' are located.

We propose a moving object query processing technique with sensor data, which shows the conditions of environment by using MQLR (Moving Object Query Language for Recommendation Service). This method provides suitable services according to the change of situation in real world by combining vehicle management and sensor data monitoring. This MQLR is processed in a LBS system includes the moving object management module for tracking moving vehicle and the sensor monitoring module for detecting a phenomenon such as dust pollution.

Keywords: Component, LBS, Recommendation Service, Sensor Network, Vehicle tracking.

1   Introduction

LBS (Location Based Service) provides suitable and useful information to mobile users depending on their locations. With the progress of GPS, mobile communication technology, and personal devices, moving objects' positions with maps are utilized in various applications such as travel, exploration, transportation, and war. In addition, the service area of LBS has been changed from objects' tracking to smart service, which utilizes not only users' positions but also the environmental condition around users. When LBS applications consider the environmental condition, users can trust the result of LBS with the improved quality of the service. For example, the users' decision in near future could be supported by the information about heavy traffic, pollution, and accidents.

Y. Wu (Ed.): International Conference on WTCS 2009, AISC 116, pp. 937–945.
springerlink.com   © Springer-Verlag Berlin Heidelberg 2012

In this paper, we focus on the query language for mobile recommendation service to support the intelligent LBS. The query language is used to effectively search the stored vehicle and environment information. Most of existing moving object query languages focus on handling the only moving objects' locations with basic geometry information model such as points, lines, and polygons. Their expression is useful to search the moving objects' positions or trajectories. However, the query languages need to consider the environmental condition to improve the quality of mobile service in real world.

We design a moving object query processing technique with MQLR (Moving Object Query Language for Recommendation Service), which supports the nearest neighbor query for recommendation service. This language considers various attributes such as the trajectory of a vehicle, the environmental condition. In this paper, we focus on the NNQ (Nearest Neighbor Query) processing of the query language.

2   Related Work

There are several researches on moving object management systems: the MOST(Moving Objects Spatio-Temporal) of the DOMINO(Databases fOr MovINg Objects) [1, 2], CHOROCHRONOS [3, 4], Battle Field Analysis [5], a generic framework for processing moving object queries [6], and transportation vehicle management system [7].

By combining space and time, new query types emerge in the moving object database. Moving object queries are divided into coordinate-based queries and trajectory-based queries [3]. The coordinate-based queries consider coordinates of objects such as point, range, and nearest neighbor queries on a three-dimensional space which has a time dimension added to a two-dimension plane. For example, it is stated that "Account for the flow of visitors at children's grand park at 13:00 on May 5th, 2010." or "Find the nearest grove for taking a rest from a city hall at 13:00 on October 3rd, 2010." Trajectory-based queries considering the movements of an object are classified into two types: topological queries which search for the movement information of an object, for instance "Find the trajectory of the taxi which leaves a park at 10:00 on August 5th, 2010" and navigational queries which inspect and predict the information derived from the movement of an object such as speed and direction, for example, "Search for an object moving at speed of 100km/h or higher towards north since 11:00 on September 10th, 2010."

The future temporal logic uses some moving object and spatial operators for handling and analyzing relationships between moving objects and moving objects or between moving objects and spatial objects.

In order to effectively handle these moving object queries, the query languages are proposed. STQL (Spatio-Temporal Query Language) [8] employs the spatial and the temporal operators, FTL (Future Temporal Logic) [2] considers the uncertainty operator such as 'may', 'must'. SQLST combing SQLS and SQLT is proposed in [9]. The distance based query language using a CQL (Constraint Query Language) is designed in [10]. The query language for describing trajectories is also investigated [11, 12, 13].

Moving objects query is frequently used in LBS as a continuous query. For example, a nearest neighbor our query provides its changed result, whenever the result is changed depending on the changed user's location and mobility over time. These continuous query processing techniques are suggested by lots of studies. Query processing methods for the CQL (Continuous Query Language) [14, 15] are also proposed to improve the query performance.

These query language and query processing techniques are effectively used for mobile service. If the result of these techniques is provided depending on user's mobility and environmental condition, the results is more effectively utilized.

3   Moving Object Query Language for Recommendation Service

Moving object query language is used for systematically processing the query demanded by users. Most of moving object query languages based on the SQL employ various kinds of moving object operators. They are responsible for analyzing information and trajectories of moving objects. Environmental condition [16] is also considered to provide intelligent LBS.

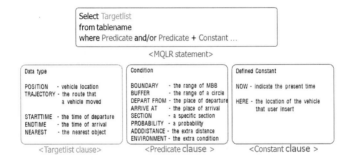

Fig. 1. The structure of MQLR

Figure 1 describes the organization of the MQLR statement containing a Targetlist, a Predicate, and a Constant clause. The moving object management system can manage and analyze the vehicle trajectory with using three types of clauses. The Targetlist construction describes data types that users request in the moving object query language. It is used as an attribute named after 'SELECT' in the SQL. The Targetlist clause defines search data types such as locations, trajectories, time, and nearest targets. The predicate is used for describing the conditions to get the vehicle information that a user wants to know. The predicate which is similar to condition clauses after 'WHERE' in the SQL can describes section, boundary, and probability, and so on. The constant having a fixed meaning in the moving object query language is defined with a constant such as € and π. The constant clauses are neatly made up with some meaningful words such as "here", "now", "after 5 minutes", and so on. It is easy to understand because they are ordinarily words. However, they are weak because they are hard to be described by specific numerical values. Still two constants "NOW" and "HERE" remain intact.

4   Recommendation Query

The NNQ for recommendation based on user's status and environmental condition is processed to find a target, which is suitable to user's situation. Figure 2 shows the example of NNQ types of MQLR.

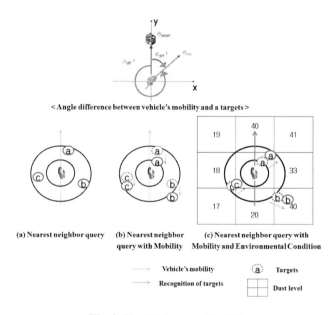

Fig. 2. The NNQ types of MQLR

Figure 2 (a) describes the simple NNQ, which finds the nearest target from the current location of the querier (vehicle). During the process of the query, nothing but the distance to the target from the vehicle is considered. Even though the absolute distance to the station is shorter, it would bring some waste in reaching the destination in real world such as time and gasoline. It does not count on some important factors such as vehicle's mobility, environmental condition. The problem is expected to occur in real world if the NNQ is used without correction.

Figure 2 (b) shows the nearest neighbor query considering the direction and the speed of a vehicle in addition to the distance to the targets with algorithm 1. The targets are recognized depending on the vehicle's mobility. The result is different from the one in Figure 2 (a).

When we consider the vehicle's mobility and environmental condition such dust pollution in Figure 2 (c), the system indicates the nearest target with the reduced moving cost depending on the derived environmental dangerous level. The priority of target place would be changed according to the dangerous level of the place. Therefore, the user's satisfaction would be improved.

Table 1. The Symbol Descriptions

Symbol	Description
θ_{mo}	The direction of a moving vehicle of a user
θ_{target}	The direction to a target place
θ_{diff}	Angle difference from a moving object to a target place
W_{min}	Minimum value for a weight
W_{max}	Maximum value for a weight
W_{diff}	Weight difference between minimum and maximum value
$W_{direction}$	Updated weight depending on the direction of a vehicle
W_{danger}	Updated weight depending on the dangerous probability

---

Algorithm NNQ_direction(mo_query)
input : mo_query   // a moving object query
output : result   // the result of a query
method :
   moPosition ← **position_query_process**(mo_query)   // process a
position_query()
   targetList ← **GetTargetlist**(mo_query)   // get a target list

   θ_{mo} ← **GetDirection**(moPosition)   // get a direction of a moving
object

   // calculate a weight according to the direction and the speed of a vehicle
   for each target of the targetList
      // calculate a distance and a angle between the mo and the target places
      targetList ← **GetDistance**(moPosition, each targetList) // a distance from
mo to target places
      θ_{target} ← **GetDirection**(moPosition, each targetList)   // an angle from
mo to target places

      θ_{diff} ← | θ_{target} - θ_{mo} |

      if $180° \leq \theta_{diff} \leq 360°$ **then**   // refine a angle difference
         θ_{diff} ← $360° - \theta_{diff}$
      endif

      if a user defined a range of weight **then**   // calculate a weight
         $W_{direction}$ ← $W_{max} - W_{diff}$ x θ_{diff}/ 180   // ($W_{min} \leq$ Weight $\leq W_{max}$)
      else
         $W_{direction}$ ← $1 - \theta_{diff}$ / 180   // ($0 \leq$ Weight ≤ 1)
      endif

      targetList.cost ← $W_{direction}$ x the distance to the target place
   endfor

   refine and sort the result by considering targetList.cost
   return result   // return the target list
end

Algorithm 1. NNQ with the vehicle's mobility

Algorithm NNQ_dangerousLevel(mo_query)
input : mo_query   // a moving object query
output : result   // the result of a query
method :
  moPosition ← **position_query_process**(mo_query)   //
process a **position_query**()
  θ_{mo} ← **GetDirection**(moPosition)   // get a direction of
a moving object

  targetList ← **NNQ_direction**(mo_query)   // NNQ with a
direction of a vehicle

  // calculate a weight according to the direction and the speed of a
vehicle
  for each target of the targetList
    dangerousPro ← the dangerous probability extracted from
abstracted sensor data // (6.1), (6.2)
      $W_{direction}$ ← targetList.direction_weight
      W_{danger} ← 1 - the dangerous probability / 100
      W_{total} ← $W_{direction}$ X W_{danger}
      targetList cost ← W_{total} X the distance to the target place

  endfor

  refine and sort the result by considering targetList.cost
  return result   // return the target list
end

Algorithm 2. NNQ with Environmental condition

5   Implementation

Figure 3 illustrates the course of steps which vehicle management and sensor data monitoring applications are combined. The final result of the procedure is for providing the personalized location based service considering environmental conditions. We combine the moving object management module for utilizing the MQLR query processor with the sensor data monitoring module for detecting the change of the environment condition.

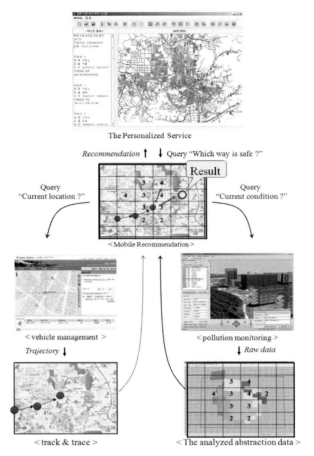

The Personalized Service

Recommendation ↑   ↓ Query "Which way is safe ?"

Query "Current location ?"   **Result**   Query "Current condition ?"

< Mobile Recommendation >

< vehicle management >   < pollution monitoring >

Trajectory ↓   ↓ *Raw data*

< track & trace >   < The analyzed abstraction data >

Fig. 3. The combination of vehicle management and sensor data monitoring

	Contents
Query	Find the nearest oil station to "CB81BA3578" vehicle depending on the direction, NOW
MQLR	SELECT *NEAREST STATION* FROM VEHICLETEMP WHERE *ID*= 'CB81BA3578' AND *DIRECTION* AND *VALID AT NOW*;

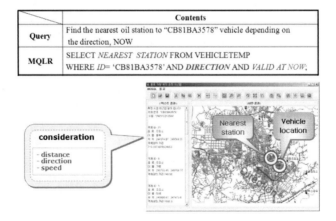

consideration

- distance
- direction
- speed

Nearest station   Vehicle location

Fig. 4. NNQ with the vehicle's mobility

Figure 4 describes NNQ considering the vehicle's mobility. This process leads to reduce the moving cost of a vehicle to find the destination by calculate the extra cost to arrive the targets with the mobility.

	Contents
Query	Find the nearest oil station to "CB81BA3578" vehicle depending on the direction, NOW
MQLR	SELECT *NEAREST STATION* FROM VEHICLETEMP WHERE *ID*= 'CB81BA3578' AND *RECOMMEND* AND *VALID AT NOW*;

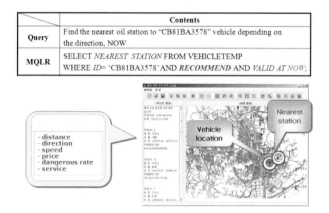

Fig. 5. NNQ with Environmental condition

Figure 5 describes NNQ considering the vehicle's mobility and the change of the environmental condition, which is detected from sensor data monitoring application [16]. Sensor network transmits sensor data into the sensor monitoring server, and which extracts some information [17] from the measured data such as predicted sensor data and predicted dangerous level.

6   Conclusion

In this paper, we introduced the recommendation query processing based on the MQLR by combining the vehicle management with the sensor data monitoring application. In the future, we will focus the information extraction from the sensor data monitoring application to develop the intelligent LBS will be designed.

References

1. Sistla, P., Wolfson, O., Chamberlain, S., Dao, S.: Modeling and Querying Moving Objects. In: The 13th International Conference on Data Engineering, pp. 422–432 (April 1997)
2. Wolfson, O., Xu, B., Chamberlain, S., Jiang, L.: Moving Objects Databases: Issues and Solutions. In: Statistical and Scientific Database Management, pp. 111–122 (1998)
3. Pfoser, D., Jensen, C.S., Theodoridis, Y.: Novel Approaches in Query Processing for Moving Object Trajectories. In: Very Large Data Base, pp. 395–406 (2000)
4. Pfoser, D.: Indexing the Trajectories of Moving Objects. Data Engineering Bulletin 25(2), 4–10 (2002)
5. Ryu, K.H., Ahn, Y.A.: Application of Moving Objects and Spatiotemporal Reasoning. TimeCenter TR-58 (2001)

6. Hu, H., Xu, J., Lee, D.L.: A Generic Framework for Monitoring Continuous Spatial Queries over Moving Objects. In: ACM Special interest Group on Management of Data, pp. 479–490 (2005)
7. Jung, Y.J., Ryu, K.H.: The Vehicle Tracking System for Analyzing Transportation Vehicle Information. In: International Workshop on Web-based Internet Computing for Science and Engineering, pp. 1012–1020 (2006)
8. Erwig, M., Schneider, M.: STQL: A Spatio-Temporal Query Language. In: Mining Spatio-Temporal Information Systems, pp. 105–126 (2002)
9. Chen, C.X., Zaniolo, C.: SQLST: A Spatio-Temporal Data Model and Query Language. In: 25th International Conference on Conceptual Modeling, pp. 96–111 (2000)
10. Mokhtar, H., Su, J., Ibarra, O.: On Moving Object Queries. In: The 21st ACM SIGACT-SIGMOD-SIGART Symposium on Principles of Database Systems, pp. 188–198 (2002)
11. Lee, M.L., Hsu, W., Jensen, C.S., Cui, B., Teo, K.L.: Supporting Frequent Updates in R-Trees: A Bottom-Up Approach. In: Very Large Data Base, pp. 608–619 (2003)
12. Kim, D.H., Kim, J.S.: Development of Advanced Vehicle Tracking System Using the Uncertainty Processing of Past and Future Locations. In: The 7th International Conference on Electronics, Information, and Communications (August 2004)
13. Jung, Y.J., Ryu, K.H.: A Group Based Insert Manner for Storing Enormous Data Rapidly in Intelligent Transportation System. In: The International Conference on Intelligent Computing, pp. 296–305 (August 2005)
14. Tao, Y., Sun, J., Papadias, D.: Analysis of Predictive Spatio-Temporal Queries. ACM Transactions on Database Systems 28(4), 295–336 (2003)
15. Jensen, C.S., Lin, D., Ooi, B.C.: Query and Update Efficient B+-Tree Based Indexing of Moving Objects. In: Very Large Data Base, pp. 768–779 (2004)
16. Jung, Y.J., Lee, Y.K., Lee, D.G., Park, M., Ryu, K.H., Kim, H.C., Kim, K.O.: A Framework of In-situ Sensor Data Processing System for Context Awareness. In: The International Conference on Intelligent Computing, pp. 124–129 (2006)
17. Jung, Y.J., Nittel, S.: Geosensor Data Abstraction for Environmental Monitoring Application. GIScience, 168–180 (2008)

The Research of the Seismic Objects of the Tidal Force and the Temperature's Change with TaiYuan's YangQu Ms4.6

Chen Wen-chen[1,2], Ma Wei-yu[1,2,3], Ye Wei[1], and Chen Hai-fang[4]

[1] Zhejiang Normal University, 688#, Yingbin Road, Jinhua, Zhejiang Province,
P R China, 321004
[2] Lab of Remote sensing Application, Zhejiang Normal University, Zhejiang Jinhua
321004, China
[3] Institute for RS/GPS/GIS and Subsidence Research, China University of Mining &
Technology, Beijing 100083, China
[4] Geomatics Center of Zhejiang, Zhejiang Hangzhou 310012, China
fa2310@yahoo.cn

Abstract. Calculating of June 5, 2010 occurred in the east longitude 112.7 °, latitude 38.2 ° TaiYuan's YangQu, ShanXi Province of the tidal force of celestial bodies Ms4.6 earthquake cycle process, and according to the cycle of integrated multi-source to analysis of temperature data to get the process of extracting earthquake images of the abnormal temperature. The result indicates that the tidal force of celestial bodies has a triggering effect on the activity of the tectonic. The temperature abnormal increase can clearly reflect the procession of the seismic fault activity: initial temperature rise →Abnormally warming →return to baseline. Microseism of this study show that the temperature anomalies in the micro-earthquakes are also a clear reflection of the process of gravitational effect microseism obviously. It is not only the abnormal warming,it is an earthquake-strong evidence of thermal anomalies, also for the use of remotesensing of earthquake prediction methods.

Keywords: Tidal force of celestial bodies, Abnormal warming, Earthquake, Remote sensing, TaiYuan's YangQu Introduction.

1 Introduction

Abnormal pre-earthquake changes in temperature phenomenon has been accepted by scholars,if Gorny[1], Ma Zong-jin[2] and other studies. At present the research work focused on the earthquake in earthquake areas, made a series of results, felt even more destructive of the lack of small earthquakes. Stress changes as small earthquakes also reflects changes in the process of stress, the results of their research is in theory a powerful earthquake research to supplement and complete.Objects as the tidal force induced by an important external factor in the earthquake[3], with a remarkable periodicity.Explore the tidal force of the earthquake tectonic objects revealed by the physical properties and seismic activity of the temperature anomalies are identical in nature, reflecting the tectonic movements are to a certain degree of

Y. Wu (Ed.): International Conference on WTCS 2009, AISC 116, pp. 947–953.
springerlink.com

mutation-impending earthquake activity occurs in this critical point of decision problem. The tidal force as the only objects can be pre-calculated deformation phenomena of the Earth, in the time domain on the role of certain instructions.This tidal force to objects and multi-source remote sensing data calculated extraction temperature anomalies combined with information on the June 9, 2010 earthquake in Shanxi Province, Taiyuan Yangqu the Ms4.6 a preliminary study, is the method of earthquake prediction using remote sensing as a scientific innovation.

2   Object Tidal Force Additional Tectonic Stress on the Role of the Seismic Fault

At 8:58 p.m. on the June 5, 2010 was born in Taiyuan, Shanxi Province Yangqu county distribution, at longitude 112.7 °, latitude 38.2 °. This is calculated according to LONGMAN[4], to calculate the earthquake epicenter, Taiyuan (time) objects tidal force versus time curve (Figure 1). Abscissa for the time series, from June 1, 2010 onwards, arrow indicates the earthquake on June 5. Vertical axis tidal force that the pressure for the size of objects, units of MPa. Celestial tidal force changes are a continuous process, through the peak - trough - peak of the cycle (Figure 1).

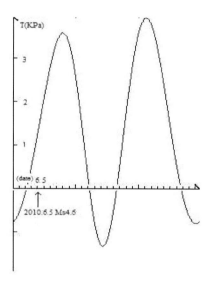

Fig. 1. Time serial change diagram of astro-tidal-triggering

The instantaneous change in tidal force and the object does not necessarily lead to the occurrence of earthquakes, seismic structure of its destructive effects of stress balance is an ongoing process. When the fault stress accumulation to make the conditions of slip fault rupture may occur when the earthquake.

3   The Process of Temperature Change Image

To address can not penetrate clouds, earthquake weather warming process and interference temperature superposition can not be effectively exception information, reduce the role of disturbance factors, reflecting the ground realities and temperature changes, using National Environmental Prediction Center global further analysis of temperature data, supplemented by ground truth temperature processing of multi-source integrated assimilation of temperature data for analysis.Analysis of the data with information on long time scales, spatial scales wide, unified global standard, without interference characteristics of clouds, has obvious comparability; day 4 hours time (00:00.06:00.12:00.18:00 UTC international standards time), a quasi-real time.The data in standard binary form (GRIB) is stored in the 0.5 ° × 0.5 ° lon / lat grid points[5], to meet the macro-temporal dynamic changes of temperature monitoring as reflected in the activities and impending earthquake plate tectonic activity. Meanwhile, in order to reduce terrain and interference with non-seismic factors, to highlight the tectonic earthquake caused by temperature anomalies. Tidal force changes according to celestial cycles[6-7], using the image (objects turning point in the tidal force change on June 9) for the high background values and the same time (this is 12:00 UTC, Beijing time 20:00).The cycle of other time, continuous image subtraction method, for the period, the daily temperature anomaly series of images (Figure 2). In accordance with the order from left to right for the June 1, 2, 3, 4, 5, 6 images.

The earthquake preparation process, June 1 additional tectonic stress in low tidal force objects, Yangqu nearby and surrounding temperature compared with the background value, a difference of 1 degree Celsius, then the temperature has stabilized, indicating that stress is accumulating. June 2, 3, and the surrounding vicinity Yangqu temperature compared with the background value of the previous day, changed little. June 4, the image can be seen from the nearby and surrounding Yangqu temperature compared with the background value is very close, indicating rock failure to strengthen, the temperature anomaly has already begun. June 5, the earthquake day, because by the UTC12: 00, the image of Beijing time 20:00, which is made less than one hour before the earthquake situation, that is, before the impending earthquake, with a strong meaning instructions. Abnormal temperature rise at this time, more than the background value, indicating the energy under stress have a high degree of accumulation of active fault in a critical state. Then the energy is released, the earthquake occurred. June 6, that is, one day after the earthquake, the temperature near the earthquake center and its rapid decline, with the background value of the margin to 2 to 3 degrees Celsius, lower than the background value, indicating the completion of seismic energy release to the activities of its structure process. Rock deformation and failure process caused by the warming effect, changing the local temperature field. From the mechanics point of view may reflect the tectonic movement by: Extrusion → Squeeze → Rock → Rock Failure to enhance the accumulation of energy → energy release earthquake[8-10]. This may reveal the general process of earthquake tectonic activity.

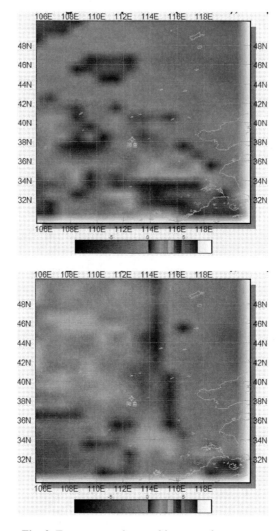

Fig. 2. Temperature abnormal increase change map

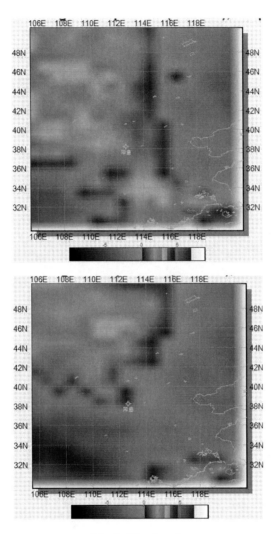

Fig. 2. (*Continued*)

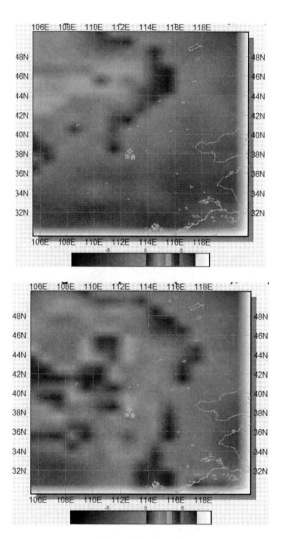

Fig. 2. (*Continued*)

4   Conclusion and Discussion

The earthquake, tidal force additional tectonic stress in the object from the trough to peak through the process of change that objects tidal force of the earthquake damage to balance the role of tectonic stress is a continuous effect. To stress is a process of accumulation. Objects in the tidal force of the critical state of stress has induced the role of active faults, the time of the earthquake prediction has some significance.

Before the earthquake, the apparent temperature anomaly, through the initial temperature, temperature increase, the temperature evolution of recovery, the phenomenon corresponds to the process of rock deformation and rupture, may reflect tectonic earthquake.

On the small and medium-level seismic results show that the temperature anomalies, not only in earthquake areas, in small earthquakes have clearly reflected. The role of small earthquakes occurred in apparent gravity, seismic process is strong evidence of thermal anomalies. The use of remote sensing methods also predicted a powerful earthquake in scientific exploration.

References

1. Ma, Z., Fu, Z., Zhang, Z.: Nine Huge Earthquake of China during 1966-1976. Earthquake Press, Beijing (1982)
2. Gorny, V.I., Salman, A.G., Tronin, A.A., et al.: The earth outgoing IR radiation as an indicator of seismic activity. Proc. Acad. Sci. USSR 30(1), 67–69 (1988)
3. Mogi, K.: Fundamental studies on earthquake prediction. Pre at ISCSEP, Beijing, pp. 16–29 (September 1982)
4. Mcnutt, S.R., Beavan, R.J.: Volcanic earthquakes at Pavlof volcano correlated with the earth tide. Nature 294, 615–618 (1981)
5. Chen, W., Ma, W., Zhang, Z.: A Preliminary Study on the Tidal Force and Temperature Change during Micro-earthquake Jilinh Ms4.3. Remote Sensing Informations 3, 35–37 (2010)
6. Ma, W.Y.: A preliminary study on the use of NCEP temperature images and Astro-tidal-triggering to forecast short-impending earthquake. Earthquake Research in China 21(1), 85–93 (2007)
7. Ma, W., Xu, X.: Relationship between the Indonesia Mw 9.0 earthquake sequence and the temperature increasing anomaly with Astro-tidal-triggering. Northwestern Seismological Journal 28(2), 129–133 (2006)
8. Wu, L., Liu, S., Wu, Y.: Remote sensing-rock mechanics(II)-laws of thermal infrared radiation from viscosity-sliding of bi-sheared faults and its meanings for tectonic earthquake omens. Chinese Journal of Rock Mechanics and Engineering 23(2), 192–198 (2004)
9. Wu, L., Liu, S., Xu, X.: Remote sensing-rock mechanics(III)-laws of thermal infrared radiation and acoustic emission from friction sliding intersected faults and its meanings for tectonic earthquake omens. Chinese Journal of Rock Mechanics and Engineering 23(3), 401–407 (2004)
10. Wu, L., Liu, S., Wu, Y.: Remote sensing-rock mechanics(I)-laws of thermal infrared radiation from fracturing of discontinous jointed faults and its meanings for tectonic earthquake omens. Chinese Journal of Rock Mechanics and Engineering 23(1), 24–30 (2004)

Design and Development of Urban Environmental Sanitation Management System

Lingwen Sun[1,2,3], Zongming Wang[1,*], Bai Zhang[1], and Xuwei Ru[3,4]

[1] Northeast Institute of Geography and Agro-Ecology, Chinese Academy Sciences, Changchun 130012, China
[2] Graduate University of Chinese Academy of Sciences Beijing 100049, China
[3] Shandong Institute for Development strategy of Science and Technology, Jinan 250014, China
[4] Geomatics College, Shandong University of Science and Technology, Qingdao, China 266510
sun2310@yahoo.cn

Abstract. The paper took the example of Huaiyin District of Jinan City to present the design and development of urban environmental sanitation system (UESMS) based on the Graphical Information Systems. Sanitation database was designed to store both attribute data and spatial data in the DBMS. Spatial analysis and sanitization appraisement criterion were defined as function models to help the management of Environmental Sanitation. The UESMS developed by ArcEngine 9.2 and Visual C# will be put into practice in the administration of environment and sanitation and can improve the management efficiency greatly.

Keyword: Environmental Sanitation, Design, Development, GIS, C#.

1 Introduction

With the acceleration of China's urbanization, the management of urban environmental sanitation is more important than ever. However, the management of environment and sanitation is mostly made by hands in China. The administrator has to manage many sanitation facilities and worker and the efficiency of environmental sanitation management is very low. As a spatial technology, Geographic information systems (GIS) provides a scientific method for the sanitation management and is well used in many cities[1,2]. Against this background, it is necessary to develop a GIS-based urban environmental sanitation management system to help the daily management and decision support.

Until August of 2010, the Huaiyin administration of environment and sanitation has lots of work to be done. There is 400 ton garbage to be transported every day. Lots of sanitation facilities need to be sanitized, including 228 roads, 16 garbage transports,

---

* Corresponding author.

Y. Wu (Ed.): International Conference on WTCS 2009, AISC 116, pp. 955–960.
springerlink.com   © Springer-Verlag Berlin Heidelberg 2012

65 latrines and other little facilities. In order to improve the efficiency of management, the administration of environment brings forward a sanitization evaluation criterion based on the grid cell. By the roads and communities, the whole district was divided into 14 big grid cells, which were comprised of 72 small grid cells. All the sanitation facilities located in one grid cell are charged by certain persons. With the help of urban environmental sanitation system, it is easy to manage the sanitations and evaluate the dustmen's work.

2   Database Design

The data was divided into attribute data and spatial data by the data type, format and way of storage. They were linked by the unique field. By storing in the DBMS, the data can be managed and located efficiently and provide basal data to the management of environmental sanitation and decision support.

2.1   Attribute Database

The attribute database is used to store the information of environmental sanitation facilities using many tables (Fig. 1).The fields of tables were defined exactly according the sanitation criterions related[3]. Every table has its fields and a primary key, through which the tables were linked together.

The information of road is stored in the sanitized road table, including road name, rank, type, area, sanitization way, sanitization ration, adscription, etc. The field of Latrine table is consisted of latrine name, type, location, adscription, building area, instrument room area, management room area and so on. The garbage transport table's fields comprised name, location, type, adscription, garbage transport capability, building area, dustbin number, dustbin type, services range, etc. The sanitation institute table was designed to store name, address, manager number, dustman number, management range. There were other tables in the database, such as car, grid cell, score, user, index weight, etc.

2.2   Spatial Database

Based on the spatial data with the scale of 1:1000, the sanitization road, latrine, garbage transport, grid cell are accurately located and depicted as point or area. They were divided into different layers and stored in the SQL Server database through ArcSDE. Spatial analysis on the spatial sanitation can provide support for the management of environmental sanitation.

3   Systemic Function Design

After extensive investigation on environmental sanitation facilities of Huaiyin District, the ESMS was divided into several separated functional models according to the general principle of function design(Fig. 2). The main functional models are as listed.

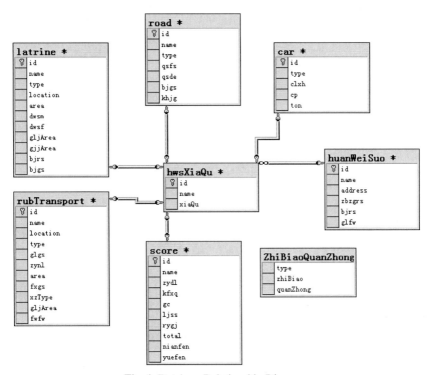

Fig. 1. Database Relationship Diagram

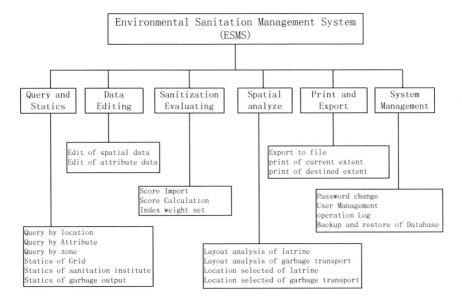

Fig. 2. The Design of Systemic Function

3.1   Query and Statistic

This model includes query and statics functions. The query function allows the user to get details by clicking the facilities or to get the accurate location by it's name or other attributes. One can also get the information of environmental sanitation facilities in destined zone, such as the number of latrine. The statistical function is used to calculate the score of grid cell and institute of environmental sanitation. The result is displayed by graphics or tables, which is used to evaluate the work of dustmen and managers.

3.2   Data Editing

This model can be used to edit both attribute data and spatial data. You can add, modify or delete the data to keep the data being in line with the reality. For example, if a garbage transport is moved to other place, the data must be edited accordingly or there will be wrong in the management of environmental sanitation. The sanitation's attributes can be edited too.

3.3   Sanitization Evaluating

In this model, every one's work is evaluated by the evaluation criterion. The core of evaluation criterion is the evaluation indicators, including it's weight and detailed mark criterion. And the weight of indicators can be edited. Each grid cell is marked by several scores every several days. After importing the scores into database, each grid cell's total score can be calculated every month and is used to estimate the work of dustman who in charge of this grid cell. The score of sanitation institutes are set by the scores of grid cells which are in charged by it. The sanitation institute's score is used to evaluate the work of managers.

3.4   Spatial Analysis

The spatial analysis function can do some work to support the decision of administrator.   You can find out whether the distribution of latrine or garbage transport is reasonable. If a latrine need to be set up, the system can help to find a ideal place referring to the existent latrine's locations, space needed, and so on.

4   System Development

ArcEngine includes a series of high-level visual components that makes it fairly easy to build a GIS application[4]. The C# is a simple, modern and object-oriented language and it is much easier to develop software than other platform[5]. For these reasons, the system is developed in C# language using .NET platform and uses ArcEngine 9.2 as spatial processing and visualization tools. The graphic user interface of system is shown in the fig. 3. The system is defined as three-tiers, including data tier, middle tier and application tier. So the system can be extended easily and it is every robust.

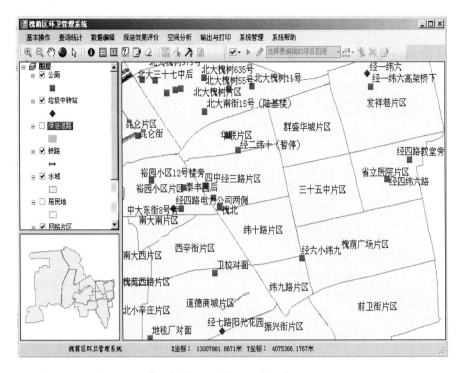

Fig. 3. The graphic user interface

5   Conclusion

The paper described the design and development of urban environmental sanitation management system in Huaiyin District. The use of system can make the administrator of environment and sanitation break off heavy hand work and can help to improve the efficiency of management greatly. The future of urban environmental sanitation management system is bright.

Acknowledgments. The work was supported by the Knowledge Innovation Program of the Chinese Academy of Sciences, Grant No.KZCX2-YW-341, Information Industry Development Fund of Shandong Province of China (No.2008X00032), National Natural Science Foundation of China (No.40871187,40901096), Natural Science Foundation of Shandong Province of China(No.Y2008B68) and Science and Technology Program of Jilin Province of China(No. 20080128).

References

1. Zheng, L., Du, P., Tong, Q., Li, W.: Research of GIS-based Environment Sanitation Management System of HaiDian District. Environment Sanitation Engineering 14(4), 12–17 (2006)

2. Liu, Z., Gao, Y., Gao, Y., Liu, K.: Development and Design of Environment Sanitation Information Network Management System. Environment Sanitation Engineering 17(4), 56–57 (2009)
3. Li, D., Wang, K.: Design on Database Management System for Municipal Environmental Sanitation Information. Environmental Sanitation Engineering 12(2), 106–109 (2003)
4. Becken, S., Vuletich, S., Campbell, S.: Developing a GIS-supported Tourist Flow Model for New Zealand. Developments in Tourism Research, 107–121 (2007)
5. Prabhakar Rao, K., Ashok Babu, G.: Microsoft C#.NET program and electromagnetic depth sounding for large loop source. NET program and electromagnetic depth sounding for large loop source. Computers & Geosciences 35(7), 1369–1378 (2009)

Development of Control System for Plug-in HEV

Xiumin Yu[*], Ping Sun, Huajie Ding, and Junjie Li

State key Laboratory of Automotive Dynamic Simulation Jilin University
Changchun, China
yu14325@sohu.com

Abstract. Plug-in Hybrid Electric Vehicle (Plug-in HEV) is one of the solutions for the Energy and Environment problems around the world. Hybrid vehicle traction applications require compact power modules with high reliability. Plug-in HEV use two energy sources for their propelling. Plug-in HEV fuel economy and drivability performance are very sensitive to the "Energy Management" controller that regulates power flow among the various energy sources and sinks. In this paper, the procedure for the design of a Plug-in HEV Control Unit (HCU) Controller was presented. The high performance of controller is introduced. Moreover, the Plug-in HEV system equipped the developed controller based on MPC566 is taken a road test. Test results show that not only CAN communication is reliable and accurate, but also controller can effectively harmonize the task distribution among all ECUs of a Drivetrain assemblies with the real-time control improved greatly for control system.

Keywords: Plug-in HEV, Control system.

1 Introduction

Oil production forecasts, and environment protection forces stimulating work on significantly improved fuel economy of all classes of vehicle. Plug-in HEV is widely regarded as one of the most viable technologies with potential to reduce fuel consumption within realistic economic, infrastructural and customer acceptance constraints. Owning to their dual on-board power sources and possibility for higher fuel economy while meeting tightened emissions standard.

However, the complexity of the new vehicle control system requires extensive and highly accurate control for proper component sizing, as well as for the development of control algorithms to maximize the potential of these advanced technologies. As the electronic and software content in automobiles continues to grow, so does the need for reliable communication between vehicle subsystems.

What's more, for automobile handling control, systems using CAN have been widely adopted, while the speed of in-vehicle networks has increased and the connecting equipment has become very diverse. Subsequently within the ECUs that controls each electronic component, for the ECUs; there is an increasing necessity to control the whole in-vehicle network. To fully realize the potential of hybrid

---

[*] Corresponding author.

Y. Wu (Ed.): International Conference on WTCS 2009, AISC 116, pp. 961–967.
springerlink.com   © Springer-Verlag Berlin Heidelberg 2012

powertrain, the power management function of these vehicles must be carefully designed.

A significant superiority between the control system developed in this paper, compared with existing tools, lies in the fact that the MPC566 microcontroller provides an industry top-level high-speed processing at 56MHz, and which has not only control functionality but with gateway functionality which allows integration of several in-vehicle CAN networks. The MPC566, an upgraded version of the MPC565, offers code compression to enable more efficient use of internal or external FLASH memory. Code compression is optimized for automotive (non-cached) applications and the new compression scheme increases compression performance up to 40% - 50%.

The significant superiority between the simulation models developed in this paper, compared with existing tools the HEV model we developed is a forward-looking model. Herein, we apply the MPC566 Microcontroller to meet to the requirements of the Plug-in HEV.

In this paper, the configuration of the Plug-in HEV Control system is discussed first. The controller based on MPC566 microcontroller was developed, as well as the mainly modules for the controller introduced then. The main simulation model based on PSAT is presented next. The complete hybrid vehicle controller is subsequently used to assess the fuel economy of the hybrid vehicle. The performances for the controller is evaluated through simulation predictions of fuel consumption over a driving cycle, followed by an introduction to hybrid vehicle and validated by a road test.

2   Plug-in HEV Control Unit (HCU)

The hierarchical control topology is utilized by adopting the HCU as the main controller which receives the components' feedbacks and sensors' inputs to decide the torque as quick as possible from the electrical motor, referred figure 1. The CAN communication is adopted to send the torque demand and share the component conditions.

2.1   Configuration of Control System

The host controller sends the torque demand to ISG motor by CAN bus and sends the simulating Acceleration Pedal signal to Engine Control Unit to realize the torque distribution. It also decides whether the gasoline engine start or not by sending the ignition signal to ECU. It sends command to Motor Control Unit to control the driver force.

The gasoline Engine Control Unit (ECU) works by the demands of host controller, and sends the reback signals dates to the HCU. What's more, the Generator works along with the gasoline engine, so the Generator Control Unit just control the Generator to work at an efficient area.

The Motor Unit gets the power from the Battery, and then provides driver force for the wheels. Motor Control Unit responds the torque command and dedicates the IGBTs' action to convert the direct current into alternative current according to the rotor position.

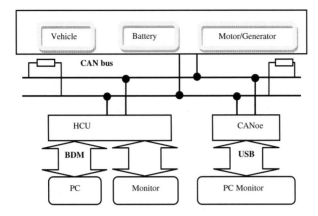

Fig. 1. The configuration for HCU

Battery Control Unit monitors the batteries' conditions, which is the module voltage and total voltage, current and temperature and so on. It calculates the remaining energy (state of charge, SOC) and diagnostic code and feed them back to the HCU. The signals also were translated to the Monitor and PC Monitor based on CAN bus.

Both the vehicle monitor and the PC monitor received the signals from the controllers by CAN bus, such SOC signal, vehicle speed signal, voltage signal, engine speed signal, motor speed signal and generator speed signal etc.

2.2 MPC566 Microcontroller Module

The microcontroller is the key part for the control system. Freescale has developed the MPC500 family of embedded controllers.

In this paper, the MPC566, 32-bit microprocessor, was selected as the CPU of the host controller because of its low cost and good performance. The main features of the high performance processor as follows referred to figure 2:

The processor provides the high performance of a 56 advanced peripherals, integrated Flash memory(up to 1 M Bit) (Two UC3F modules, 512 Kbytes each) and fast static RAM (up to 36k) in a single silicon chip, which is able to meet the normal amount of embedded code. It supports Code Compression, which will increases compression performance up to 40% - 50% than MPC565.

Addition computing power is added to the MPC566 with the time processor unit (TPU3); these units have a 32-bit MicroRISC engine capable of processing 28 million instructions per second. The TPU runs independently and in parallel to the CPU. The TPU/CPU shares data through a dual-access RAM. Specifically designed for scheduling puses in a circular system, the TPU is ideal for handing the angle domain operation. Each TPU contains 16 channels with independent match and capture hardware, multiple time bases, and a priority round-robin hardware scheduler to ensure that no task is locked out.

Three TouCAN modules (TOUCAN_A, TOUCAN_B, and TOUCAN_C) are provided for communications between host controller and others part controllers. Two enhanced queued analog to digital converters (QADC64E_A, QADC64E_B) with up to 40 total analog channels. These modules are configured so each module can access all 40 of the analog inputs to the part (orthogonal). Two queued serial multi-channel module (QSMCM_A, QSMCM_B), each of which contains a queued serial peripheral interface (QSPI) and two serial controller interfaces (SCI/UART).

We cans quickly process simulation to debug and program code download based on the JTAG and background debug mode (BDM) programming interface provided by MPC566.

The main configuration is shown in Fig. 1. The pulses from the sensors are captured real time by the high speed input channels of the microprocessor. Analog signals are collected by the A/D converters integrated in the microprocessor.

Fig. 2. The picture of the Plug-in Hybrid Electric Vehicle Controller developped by the authors

2.3   CAN Communication Module

As mentioned at last part, the QSMCM modules handle off-chip serial communications for UART and SPI functionality, while vehicle networking is supported with up to three TouCAN modules (CAN version 2.0B) and a J1850 interface. To meet the wide variety of performance and cost targets, the microcontroller offers many members that range from a low-end flashless device to a high-end 112 MIPS device with 1MB of intergraded Flash.

Hence, the PC82C250 Bus Transceivers were selected for developing the three separate bus controllers for there were three bus controllers in the MPC566, which are CAN_A, CAN_B and CAN_C.

Fig. 3. The picture of the target Plug-in Hybrid Electric Vehicle

3   Results Analysis for the Road Tests

The prototype vehicle described in this paper is Plug-in   Hybrid Vehicle based on the BJ2023 vehicle, which is assumed to be made up of one motor, one generator and one NiH battery etc. what's more, a road tests with STP75 driver cycle have been taken for validating the simulation result. The prototype Plug-in HEV equipped the developed controller based on MPC566 microcontroller has been taken a real-time road test.

As shown in the figure 2, the vehicle speed in the simulation based on PSAT is similar to that of the cycle; furthermore, the accurate simulation commands have been made.

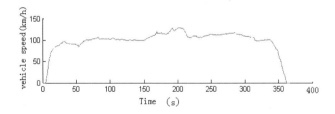

Fig. 4. The vehicle speed at the simulation based US06highway cycle test

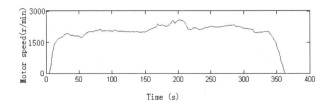

Fig. 5. The motor speed at the simulation based on US06highway cycle test

Because the prototype Plug-in HEV is a Hybrid Electric Vehicle, the motor speed was assumed to be similar to the driver cycle, as shown at figure 4, the feature of the motor speed curves of the simulation results is similar to that of curves at figure 3, 4.

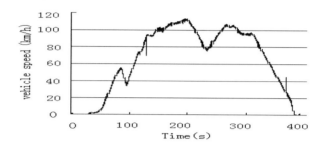

Fig. 6. The motor speed for the road test based on US06highway cycle test

With comparison between the curves shown in figure 2 and figure 4, both of them have a similar feature. Because the dates were received from the Plug-in HEV Controller equipped on the prototype vehicle based on CAN bus, so the conclusions that the controller based on MPC566 developed by authors can be worked successfully.

4   Conclusions

As the electronic and software content in automobiles continues to grow, so does the need for reliable communication between Plug-in HEV subsystems, the high performance microcontroller MPC566was selected, and then the hierarchical control system based on MPC566 is set up. The HCU based on 32bits Microcontroller can handle various signals.CAN bus communication was developed in the control system regarding the high transferring speed and the design convenience as the communication method to exchange information between the PC, ECU, BCU, MCU, GCU and HCU. USB-CAN Card, an intelligent conversion card, links the PC with HIL-ECU and HCU.

Acknowledgment. This project is supported by National Science Foundation of China (No. 50875106) and Graduate Innovation Fund of Jilin University (No.20101023).

References

1. Ding, H.: Development and Experimental Study on Simulation system and control strategy of power source for SHEV, Jilin University thesis (2009)
2. Chen, Q.-Q.: The present Situation of Electric Motor Vehicle and Its Prospect. J. Machine Building & Automation 1(19), 1–4 (2003)

3. Hu, Z., Jin, F.: Development and Performance Validation of an ISG Diesel Hybrid Power-Train for City Buses Part II: Control Strategy and Road Test. In: IEEE Vehicle Power and Propulsion Conference, pp. 3–5 (2008)
4. Zeng, X.H., Wang, Q.N., Song, D.F.: Power Management Strategy for a PHEV Truck. IEEE Transactions on Control Systems Technology, 114–118 (2006)
5. Gao, D.W., Mi, C., Emadi, A.: Modeling and Simulation of Electric and Hybrid Vehicles. In: IEEE Vehicle Power and Propulsion Conference, vol. 95(4), pp. 731–745 (2007)
6. MPC565/MPC566 User's Manual, Motorola, Inc. (2002)

Author Index